SUPERCONDUCTIVITY AND APPLICATIONS

SUPERCONDUCTIVITY AND APPLICATIONS

Edited by
Hoi S. Kwok
Yi-Han Kao
David T. Shaw

New York State Institute on Superconductivity
State University of New York at Buffalo
Buffalo, New York

SPRINGER SCIENCE+BUSINESS MEDIA, LLC

Library of Congress Cataloging in Publication Data

Conference on Superconductivity and Applications (3rd: 1989: Buffalo, N.Y.)
  Superconductivity and applications / edited by Hoi S. Kwok, Yi-Han Kao, and David T. Shaw.
    p. cm.
  "Proceedings of the Third Annual Conference on Superconductivity and Applications, held September 19–21, 1989, in Buffalo, New York"—T.p. verso.
  Includes bibliographical references.
  ISBN 978-1-4684-7567-8   ISBN 978-1-4684-7565-4 (eBook)
  DOI 10.1007/978-1-4684-7565-4
  1. Superconductivity—Congresses. 2. High temperature superconductivity—Congresses. 3. High temperature superconductors—Congresses. 4. Thin film devices—Congresses. I. Kwok, Hoi S. II. Kao, Yi-Han. III. Shaw, David T. IV. Title.
  QC611.9.C66 1989   90-32937
  621.3—dc20   CIP

Proceedings of the Third Annual Conference on Superconductivity
and Applications, held September 19–21, 1989, in Buffalo, New York

© 1989 Springer Science+Business Media New York
Originally published by Plenum Press, New York in 1989

Softcover reprint of the hardcover 1st edition 1989

All rights reserved

No part of this book may be reproduced, stored in a retrieval system, or transmitted in any form or by any means, electronic, mechanical, photocopying, microfilming, recording, or otherwise, without written permission from the Publisher

PREFACE

This Proceedings is a collection of papers presented at the Third Annual Conference on Superconductivity and Applications organized by the New York State Institute on Superconductivity. This year the Conference was held at the Buffalo Hilton Hotel on September 19-21, 1989, with previous meetings on September 28-29, 1987, and April 18-20, 1988. As in previous years, this meeting was highly successful, with an attendance of over three hundred researchers participating in lively scientific exchanges and discussions. The high quality of the talks is evident in this Proceedings.

The field of high temperature superconductivity has matured considerably since its early days of media frenzy and rapid new discoveries. However, the enthusiasm and pace of research have not slowed down. A much better picture of the nature of high temperature superconductivity, the properties of these new materials and where they may find their eventual use has emerged. Processing techniques, especially thin film deposition, have been perfected nearly to the point of allowing commercial applications. We expect continued phenomenal growth of the field of high temperature superconductivity, both in terms of research and applications for many years to come.

At the third Buffalo Conference, eighty invited and contributed talks and fifty-eight poster papers were given. The poster session, held simultaneously with the Conference reception, was highly successful. It allowed Principal Investigators of NYSIS and other Conference participants to communicate their ideas and results very effectively. There were also two workshops held at the end of the Conference for NYSIS researchers, one on Thin Film Processing and one on Wire and Cables.

H.S. Kwok
Y.H. Kao
D.T. Shaw

ACKNOWLEDGEMENTS

We wish to thank the plenary speakers for excellent reviews of the status of high temperature superconductivity. They were:

R.B. Laibowitz, IBM Corporation
J.C. Phillips, AT&T Bell Laboratories
M. Schluter, AT&T Bell Laboratories
M. Suenaga, Brookhaven National Laboratory

The program committee assisted in finalizing the program. Their excellent work is appreciated. They are:

R.D. Blaugher, Intermagnetics General Corporation
A. DasGupta, Department of Energy
H.R. Hart, General Electric Company
R. McConnell, Solar Energy Research Institute
R.L. Snyder, Alfred University
T. Venkatesan, Bell Communications Research
S.A. Wolf, Naval Research Laboratory
R. Wright, Rensselaer Polytechnic Institute
M.K. Wu, Columbia University

Thanks are due to the invited speakers who delivered excellent talks. We would also like to acknowledge the session presiders who did their part in keeping the Conference running smoothly. They are:

R.D. Blaugher, Intermagnetics General Corporation
R. Buhrman, Cornell University
D.D.L. Chung, University at Buffalo
J.R. Clem, Iowa State University
P. Coppens, University at Buffalo
A. DasGupta, Department of Energy
N. Dhere, Solar Energy Research Institute
T.F. George, University at Buffalo
Y.S. Gou, National Chiao-Tung University (China)
M. Gurvitch, AT&T Bell Laboratories
H.R. Hart, General Electric Company
S.H. Liu, Oak Ridge National Laboratory

M.J. Naughton, University at Buffalo
R.L. Snyder, Alfred University
C. Uher, University of Michigan
R.N. Wright, Rensselaer Polytechnic Institute

Finally, we would like to dedicate this Proceedings to the staff of NYSIS who are instrumental in putting the Conference together.

CONTENTS

PLENARY REVIEWS

The Electronic Structure of High T_c Copper Oxides ... 1
  M. Schluter

Flux Pinning and Microstructure in $YBa_2Cu_3O_7$ ... 27
  M. Suenaga, D.O. Welch, Y. Xu, Y. Zhu, A.K. Ghosh, and A.R. Moodenbaugh

Quantum Percolation Theory of High-T_c Superconductors ... 41
  J.C. Phillips

The Fabrication and Application of High T_c Superconducting Thin Films ... 45
  R.B. Laibowitz

THIN FILMS

Physics of In-Situ Oxide Superconducting Thin Films Deposition ... 47
  H.S. Kwok, D.T. Shaw, Q.Y. Ying, J.P. Zheng, H.S. Kim, and N.H. Cheung

Tailored Thin Films of Superconducting Bi-Sr-Ca-Cu Oxide Prepared by
    the Incorporation of Exotic Atoms ... 61
  H. Tabata and T. Kawai

In-Situ Processing and Theoretical Model for Deposition of Laser Ablated
    High-T_c $YBa_2Cu_3O_7$ Superconducting Thin Films ... 71
  R.K. Singh and J. Narayan

RF Plasma Deposition of $YBa_2Cu_3O_{7-x}$ Films ... 99
  S. Patel, A. Shah, and D.T. Shaw

Superconducting Films of Bi-Sr-Ca-Cu-O by Laser Deposition ... 109
  E. Narumi, S. Patel, and D.T. Shaw

In-Situ Growth of $Y_1Ba_2Cu_3O_{7-x}$ Thin Films on Three-Inch Wafers
    Using Laser-Ablation and an Atomic Oxygen Source ... 117
  J.A. Greer

Low-Pressure Metalorganic Chemical Vapor Deposition and Characterization
    of $YBa_2Cu_3O_{7-x}$ Thin Films ... 127
  P. Zawadzki, G.S. Tompa, P. Norris, D.W. Noh, B. Gallois, C. Chern,
  R. Caracciolo, and B. Kear

Electrical Transport Measurements on Polycrystalline Superconducting Y-Ba-Cu-O Films ... 139
  M.A. Stan, S.A. Alterovitz, and D. Ignjatovic

Studies of Neutral and Ion Transport During Laser Ablation of 1:2:3
Superconductors by Optical Absorption Spectroscopy 153
D.B. Geohegan and D.N. Mashburn

Study of Deposition of $YBa_2Cu_3O_{7-x}$ on Cubic Zirconia 163
J.D. Warner, J.E. Meola, and K.A. Jenkins

Interdiffusion Between Sputtered MgO Films and Sapphire Substrates 169
J. Van Hook

Processing and Substrate Effects on YBaCuO Thin Films Formed by Rapid
Thermal Annealing of $Cu/BaO/Y_2O_3$ Layered Structures 175
Q.Y. Ma, M.T. Schmidt, T.J. Licata, D.V. Rossi, E.S. Yang, C.-A. Chang,
and C.E. Farrell

Low-Temperature Metal-Organic Chemical Vapor Deposition (LTMOCVD)
Route to the Fabrication of Thin Films of High Temperature Oxide
Superconductors 185
A.E. Kaloyeros, A. Feng, J. Garhart, M. Holma, K.C. Brooks, and
W.S. Williams

PHYSICAL PROPERTIES

Josephson and Quasiparticle Tunneling in High-T_c Superconductors 193
M. Gurvitch

Electrodynamics of Superconducting $Y_1Ba_2Cu_3O_y$ 207
S. Sridhar

Mechanisms of Heat Conduction in High-T_c Superconductors 217
C. Uher

Acoustic Response of $YBa_2Cu_3O_x$ Films 241
S.-G. Lee, C.C. Chi, G. Koren, A. Gupta, and A. Segmüller

Magnetization Study of High T_c Superconductors 249
W.-Y. Guan, Y.-H. Xu, and K. Zeibig

Microstructure and Electrical Properties of Bulk High-T_c Superconductors 265
U. Balachandran, M.J. McGuire, K.C. Goretta, C.A. Youngdahl, D. Shi,
R.B. Poeppel, and S. Danyluk

Electronic Structure Studies of Cuprate Superconductors 273
N. Nücker, H. Romberg, M. Alexander, S. Nakai, P. Adelmann, and J. Fink

Studies of Microstructures in High-T_c Superconductors by X-Ray Absorption
Techniques 281
A. Krol, C.J. Sher, Z.H. Ming, C.S. Lin, L.W. Song, Y.H. Kao, G.C. Smith,
Y.Z. Zhu, and D.T. Shaw

Structural Complications in Superconducting Solids: Chemical Disorder,
Structural Modulation, and Low Temperature Phase Transitions in the
Bi-Sr-Ca Cuprates 285
P. Coppens, Y. Gao, P. Lee, H. Graafsma, J. Ye, and P. Bush

Local Structure and Distortions in Pure and Doped $Y_1Ba_2Cu_3O_{7-\delta}$:
X-Ray Absorption Studies
J.B. Boyce, F. Bridges, and T. Claeson 303

Probing Charge on Cu in Oxide Superconductors ... 313
E.E. Alp, S.M. Mini, M. Ramanathan, G.L. Goodman, and O.B. Hyun

Andreev Reflection and Tunneling Results on $YBa_2Cu_3O_7$ and
$Nd_{2-x}Ce_xCuO_{4-y}$ Single Crystals ... 323
T.W. Jing, Z.Z. Wang, N.P. Ong, J.M. Tarascon, and E. Wang

Positron Annihilation Studies of High Temperature Superconductors ... 335
C.S. Sundar, A. Bharathi, L. Hao, Y.C. Jean, P.H. Hor, R.L. Meng,
Z.J. Huang, and C.W. Chu

Weak Links and Planar Defects in Superconducting Cuprates ... 351
J. Halbritter

Magnetic Field Dependence of Critical Current Density in Silver-Doped
Y-Ba-Cu-O Superconductors ... 363
L.W. Song and Y.H. Kao

Description of the Tl, Pb and Bi Cuprate High-T_c Superconductors (HTSC's)
as Intercalation Compounds and Classification of all HTSC's According to
Doping Mechanism ... 371
B.C. Giessen and R.S. Markiewicz

$La_2MO_{4+\delta}$ (M=Cu, Ni, Co): Phase Separation and Superconductivity
Resulting from Excess Oxygen Defects ... 379
B. Dabrowski, J.D. Jorgensen, D.G. Hinks, S. Pei, D.R. Richards,
K.G. Vandervoort, G.W. Crabtree, H.B. Vanfleet, and D.L. Decker

Compositional Dependence and Characteristics of the Superconductive
Bi-Sr-Ca-Cu-O System ... 389
H.L. Luo, S.M. Green, Y. Mei, and A.E. Manzi

Of Grain Boundary Diffusion and Superconductivity in
Bi-Pb-Sr-Ca-Cu-O Compound ... 399
T.K. Chaki and S.C. Tseng

Dissipative Structures in Thin Film Superconducting Tl-Ca-Ba-Cu-O ... 405
M.L. Chu, H.L. Chang, C. Wang, T.C. Wang, T.M. Uen, and Y.S. Gou

Grain Orientation in High T_c Ceramic Superconductors ... 411
S. Gopalakrishnan and W.A. Schulze

Mössbauer Effect Studies in $YBa_2Cu_{3-y}Sn_yO_{7-x}$ ... 419
M. DeMarco, G. Trbovich, X.W. Wang, P.G. Mattocks, and M. Naughton

Energetics of Superstructures in the Bi-Series of Superconducting Compounds ... 425
A. Dwivedi and A.N. Cormack

Microstructural Effects in Oxide Superconductors ... 441
J.G. Darab, R. Garcia, R.K. MacCrone, and K. Rajan

THEORY

Theory of Strongly Fluctuating Superconductivity ... 455
S.H. Liu

Correlation in Metallic Copper Oxide Superconductors: How Large Is It? ... 463
W.E. Pickett, H. Krakauer, R.E. Cohen, D. Singh, and
D.A. Papaconstantopoulos

Squeezed Polaron Model of High T_c Superconductivity 475
   D.L. Lin and H. Zheng

Local Pairing and Antiferromagnetism in High-T_c Superconductors 485
   H.R. Lee and T.F. George

Superconductivity Due to the Softening of the Breathing Mode in $Ba_{1-x}K_xBiO_3$
   Compounds 495
   C.S. Ting and Z.Y. Weng

Evidence for Pairing of Semions in Finite Systems 507
   W. Wu, C. Kallin, and A. Brass

PROCESSING

Oxygen Stabilized Zero-Resistance States with Transition Temperatures
   Above 200 K in YBaCuO 517
   J.T. Chen, L.-X. Qian, L.-Q. Wang, L.E. Wenger, and E.M. Logothetis

Study of High T_c Superconductor/Metal-Oxide Composites 531
   M.K. Wu, C.Y. Huang, and F.Z. Chien

Magnetic Alignment of Bi and Tl Cuprates Containing Rare Earth Elements 541
   F. Chen, R.S. Markiewicz and B.C. Giessen

The Removal of Organics from Tape Cast $YBa_2Cu_3O_{7-x}$ 547
   B.A. Whiteley and J.A.T. Taylor

Anisotropic Critical Currents in Aligned Sintered Compacts of $YBa_2Cu_3O_{7-\delta}$ . . . 557
   J.E. Tkaczyk, K.W. Lay, and H.R. Hart

$YBa_2Cu_3O_{7-\delta}$: Enhancing J_c by Field Orientation 567
   R.S. Markiewicz, K. Chen, B. Maheswaran, Y.P. Liu, B.C. Giessen,
   and C. Chan

Surface Passivation of Y-Ba-Cu-O Oxide Using Chemical Treatment 573
   Q.X. Jia and W.A. Anderson

Carbon Fiber Reinforced Tin-Superconductor Composites 581
   C.T. Ho and D.D.L. Chung

Single Crystal Growth of 123 YBCO and 2122-BCSCO Superconductors 591
   K.W. Goeking, R.K. Pandey, G.R. Gilbert, S. Nigli, P.J. Squattrito, and
   A. Clearfield

The Effect of Ag/Ag_2O Doping on the Low Temperature Sintering
   of Superconducting Composites 599
   A. Goyal, S.J. Burns, and P.D. Funkenbusch

Effect of Doping on Defect Structure and Phase Stability in Y-Based
   Cuprates: *In-Situ* TEM Studies 611
   K. Rajan, R. Garcia, L.C. Gupta, and R. Vijayaraghavan

Stability of Yttrium Barium Cuprate in Molten Salts 621
   D.B. Knorr and C.H. Raeder

Differential Thermal Analysis of the $Y_1Ba_2Cu_3O_x$ Complex 629
   G.A. Kitzmann, R.J. Tofte, W.E. Woo, and L.E. Rowe

Experimental Research of the Construction and Structural Relaxation
of Bi-Sr-Ca-Cu-O Superconductivity System ... 635
J. Enyong, Q. Yuanfang, Y. Zhiying, Y. Zhian, W. Yubin, and L. Wenxi

Anodic Oxidation of Metallic Superconducting Precursor ... 641
J.F. Chiang

A New Phase in the Na/Cu/O System Prepared by High Pressure Synthesis ... 647
X. Liu, R.C. Liebermann, L. Mihaly, and P.B. Allen

Effect of Magnetic Fields on Silver Contacts to YBCO and YBCO:Ag
Composite Superconductors ... 653
Y. Tzeng and M. Belser

Fracture Behavior of Superconducting Films Prepared by Plasma-Assisted
Laser Deposition ... 661
D.D.L. Chung

APPLICATIONS

Multilayer Flexible Oxide Superconducting Tapes ... 665
S. Witanachchi, D.T. Shaw, H.S. Kwok, E. Narumi, Y.Z. Zhu,
and S. Patel

Josephson AC Effect in Tl-Based High-T_c Superconducting Bridge ... 677
C. Wang, H.L. Chang, M.L. Chu, J.Y. Juang, T.M. Uen, and Y.S. Gou

Sputtered High-T_c Superconducting Films as Fast Optically Triggered Switches ... 685
P.H. Ballentine, A.M. Kadin, and W.R. Donaldson

Radiation Detection Mechanisms in High Temperature Superconductors ... 695
Y. Jeong and K. Rose

RF Surface Resistance of a Magnetically Aligned Sintered Sample of $YBa_2Cu_3O_7$ ... 705
H. Padamsee, J. Kirchgessner, D. Moffat, D. Rubin and Q.S. Shu,
H.R. Hart, Jr., and A.R. Gaddipati

Fabrication of a Technologically Useful Conductor Using Ceramic Superconductor ... 709
D.W. Hazelton, R.D. Blaugher, and M.S. Walker

Rapidly Solidified Superconducting Film on Metallic Wire ... 719
S.A. Akbar, M.L. Chretien, and J. Huang

Superconducting Projectile Accelerator ... 727
X.W. Wang and J.D. Royston

Millimeter Wave Transmission Studies of $YBa_2Cu_3O_{7-\partial}$ Thin Films in the
26.5 to 40.0 GHz Frequency Range ... 735
F.A. Miranda, W.L. Gordon, K.B. Bhasin, V.O. Heinen, J.D. Warner, and G.J. Valco

Fabrication of $YBa_2Cu_3O_{7-x}$ Superconducting Wires by Spray Deposition ... 749
J.G. Wang and R.T. Yang

Microwave Loss Measurements on YBCO Films Using a Stripline Resonator ... 757
A.M. Kadin, D.S. Mallory, P.H. Ballentine, M. Rottersman, and R.C. Rath

Response of High Temperature Superconductors to a Step
in Magnetic Field ... 767
C.P. Bean

Development of a Composite Tape Conductor of Y-Ba-Cu-O ... 773
R.D. Blaugher, D.W. Hazelton, and J.A. Rice

Fabrication of High-T_c Superconducting Coatings by Electrostatic
Fluidized-Bed Deposition ... 785
D.W. Kraft, K.S. Gottschalck, and B. Hajek

Superconducting Power Transmission Lines Cost Assessment ... 795
J. Wegrzyn, R. Thomas, M. Kroon, and T. Lee

Contributers ... 809

Index ... 813

THE ELECTRONIC STRUCTURE OF HIGH T_c COPPER OXIDES

M. Schluter

AT&T Bell Laboratories
Murray Hill, New Jersey 07974

ABSTRACT

A consistent picture of the electronic structure of copper oxide materials is developed in this review. The approach starts with a first principles Density Functional description of the materials properties. Then, in two successive mapping procedures simplified models and their parameters are extracted. These models focus on the description of the electron dynamics on a small energy scale appropriate for the study of magnetism and superconductivity. At each stage of the renormalization process, observables are calculated and found in agreement with experiments. The final models are of the one-band, either Heisenberg (insulator) or "t-t'-J" (metal) type. These findings have strong implications for the possible nature of high temperature superconductivity; they also need to be further tested against critical experiments.

INTRODUCTION

The discovery of high temperature superconductivity in the Cu-oxide materials has stimulated much theoretical work on correlated electron systems.[1,2] Many different simplified models have been proposed for the electronic structure of the Cu-O planes with the goal to describe or at least rationalize the occurrence of high temperature superconductivity as well as number of other, rather anomalous properties in the normal state of these materials.[3-28] The trend has been to explore the phenomenology of a particular model as its parameters are varied with less emphasis on material-specific properties. In this review we follow the opposite approach and describe the selection of a generic electronic structure model and its parameters from explicit considerations of the Cu oxide material properties. This approach, which has been developed by us in series of papers,[29-31] begins with an accurate quantum chemical description of a Cu-O material (i.e. La_2CuO_4) and then proceeds to renormalize down in energy considering a selected set of simplified models. In this fashion one arrives at a final "fixed-point" Hamiltonian and its parameters which should explicitly reflect the properties of the materials. The validity of the final model Hamiltonian has to be tested by considering a variety of low energy scale physical properties, such as normal state resistivity $\rho(T)$, optical conductivity $\sigma(\omega)$, Raman scattering intensity $S(\omega)$ and nuclear magnetic relaxation rate $T_1^{-1}(T)$. Some of these tests are currently being carried out. The model should also be searched for the possible occurrence of high temperature superconductivity. It should be kept in mind, however, that while the final model is the one among those considered at the outset, that represents the Cu-O materials best, it does not, of course, guarantee the occurrence of high T_c superconductivity.

The remainder of this paper is organized as follows. In section II the results of local density functional (LDA) calculations are analyzed, stressing both their successes and failures. On the basis of these results and using general quantum chemical considerations we select the electronic degrees of freedom which should be important for the relevant low energy scale electron dynamics. In a first mapping process, described in section III, a generalized multi-band Hubbard Hamiltonian is selected and its parameters are derived from selected, trustworthy LDA results. This Hamiltonian is then studied in the limit of small clusters and several experiments on the 1-10 eV energy scale are analyzed testing the validity of the derived parameters. A well defined picture of the chemical nature of La_2CuO_4 emerges. In section IV, in a second step the low energy spectrum of this generalized Hubbard Hamiltonian and its associated dynamics are mapped onto several simplified model Hamiltonians. Good mappings are found for the insulating parent material as well as for the electron or hole doped cases. Various response functions of these low energy Hamiltonians on a 0-1 eV energy scale are finally tested against experimental measurements. Some of this work is still in progress. The general picture of the electronic structure of the Cu-O materials that emerges from this approach will be contrasted at each step with other proposed generic models and its implications for superconductivity will be discussed. The review ends in section IV with a caveat by recalling the various approximations made in the renormalization process and by assessing several alternative proposed scenarios.

QUANTUM CHEMISTRY AND DENSITY FUNCTIONAL RESULTS

A rough first impression of the electronic structure of e.g. La_2CuO_4 may be obtained by considering its crystalline structure together with the ionization potentials of the constituent elements.[32] Figure 1 reproduces a compilation of the relevant ionic 3d transition metal (TM) ionization potentials in comparison to that of oxygen. Going along the 3d series from Ti to Cu the TM ionization potentials increase and approach that of oxygen due to the increase of nuclear charge. This finding together with the short Cu-O bond length of <2 Å, observed in all high T_c oxides, signals the presence of strong covalency. This is in contrast to early TM oxides, e.g. V_2O_3, where the oxygen derived states are found significantly below the TM 3d-states. The strong covalency in the Cu-O materials, i.e. the strong mixing of TM d-states and oxygen p-states is clearly born out by Local Density Approximation (LDA)-type Density Functional (DF) results. Figure 2 shows a typical LDA band structure[33] of La_2CuO_4. A complex, about 10 eV wide, of 17 bands corresponding to mainly Cu d (5 states) and O p (4×3 states) is seen well separated from higher lying, mainly La d and f states. The large width of this complex derives from the aforementioned strong hybridization. Distributing the 33 valence electrons per unit cell of La_2CuO_4 among these bands leaves one hole in the complex. This makes La_2CuO_4 metallic, in contrast to the observed AF insulating behavior. Various attempts[34-36] have been made to remedy this situation by using spin-polarized versions of the LDA, but generally no spin-density wave (SDW) state could be stabilized. Moreover, the moment extracted from data on the AF state is nearly the full Cu moment, which is too large to be accounted for by a SDW distortion. This failure signals the existence of strong electronic correlations among the electrons in the Cu-O planar complex which can not be described adequately within the LDA picture. As a consequence, electron dynamics on a low energy scale, such as magnetism and superconductivity and certain excitations on a high energy scale, such as localized photoemission shake-up processes are not accounted for in this approximation.

However, it has been found that the LDA gives a rather accurate description of the ground state electronic charge distribution, with similar quality as found for less correlated systems. For instance, cohesive properties, such as the equilibrium lattice structure and rather delicate

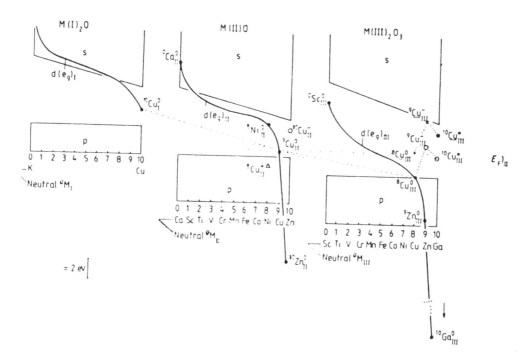

Fig. 1 Variation of the ionization potentials of 3d transition elements with respect to that of oxygen. (from ref. 32).

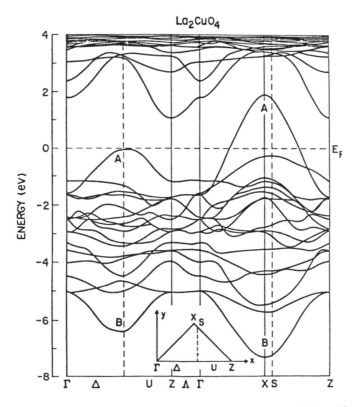

Fig. 2 Local Density Functional (LDA) band structure of La_2CuO_4. (from ref. 33).

quantities like phonon frequencies[37] and electric field gradients[38] have been well reproduced. This is the motivation and justification to use selected LDA results further, e.g. to calculate the effective screening of the various bare parameters as they appear in the different model Hamiltonians we shall discuss. The LDA bands are also used to select among many all those electronic degrees of freedom which seem important for small energy excitations. This selection is based on quantitative crystal-chemical LDA-derived data, such as strength of hybridization and crystal field electron energies.

Before we proceed to describe the mapping procedures onto model Hamiltonians we shall briefly comment on the communalities and differences of the electronic structures of different Cu-O materials. LDA calculations have been carried out for a variety of Cu-O materials, including the T-structure (La_2CuO_4), the T'-structure (Nd_2CuO_4), the 123-structure ($YBa_2Cu_3O_7$), the Aurivillius derived structures containing Bi ($CaBi_2Sr_2Cu_2O_8$) or Tl ($Ba_2Tl_2CuO_6$) and containing different numbers of Cu-O planes per unit cell (see Fig. 3).[39] Due to the similarities in structure, the planar Cu-O derived electronic bands are very similar in all cases. What differs is the exact hole count in the anti-bonding portion of these bands. In La_2CuO_4 (Fig. 2)[33] or Nd_2CuO_4 (Fig. 4)[40] no additional bands appear near the top of the Cu-O complex, leading to the AF state with exactly one hole per complex i.e. one half-filled hole band. Additional carriers are introduced by chemical doping on the cation (La, Nd) sites with e.g. Sr or Ce ions. In $YBa_2Cu_3O_7$ (Fig. 5)[41] a one-dimensional band due to the Cu-O chains, existing in this structure also crosses the Fermi level. This allows for some charge transfer and changes the hole count in the planar Cu-O bands away from half-filling and metallic behavior does exist even without chemical doping. In the various Aurivillius phases studied by LDA, small additional electron pockets are found near E_F arising from dispersive Bi-O or Tl-O antibonding bands above E_F (Figs. 6,7).[42,43] The possibility of "self-doping" is thus suggested by a small charge transfer from these pockets to the Cu-O sheets. It should be noted, however, that experimentally many of these phases show complicated superlattice and/or defect structures not considered in the present LDA calculations. Conclusions based on the theoretical findings are therefore not firm and the possibility of carrier doping primarily arising from non-stoichiometry is also given in these structures. Nonetheless, the LDA calculations suggest that the Aurivillius phases can be viewed as intercalated structures of Cu-O sheets and slightly metallic Bi(Tl)-O layers with their Fermi levels lined up electrostatically. The observed increase of T_c with increasing number of Cu-O sheets can be due to a variety of reasons: a simple increase in the total density of states near E_F, a higher density of carriers per Cu-O sheet or an increased metallicity in the Bi(Tl)-O layers giving better 3-D coupling. The answer to these questions depends on many subtleties and the LDA results can only suggest possibilities. For further details we refer to several rather comprehensive reviews on LDA results of Cu-O materials.[44,45] On a more general scale, the LDA results clearly support the point of view that in all these Cu-O materials superconductivity occurs in the strongly correlated Cu-O sheets when these are doped away from the half-filled hole-band situation.

MAPPING ONTO A GENERALIZED HUBBARD TYPE HAMILTONIAN

Most correlated electron systems can be successfully studied by using some form of localized orbital representation and by retaining the appropriate interaction terms. In its most general form the interacting electron Hamiltonian reads:

$$H = \sum_{ij\sigma} \varepsilon_{ij} \left\{ c^+_{i\sigma} c_{j\sigma} + h.c. \right\} + \frac{1}{2} \sum_{\substack{ijkl \\ \sigma\sigma'}} <ij | \frac{1}{r} | kl> c^+_{i\sigma} c^+_{j\sigma'} c_{l\sigma'} c_{k\sigma} \qquad (1)$$

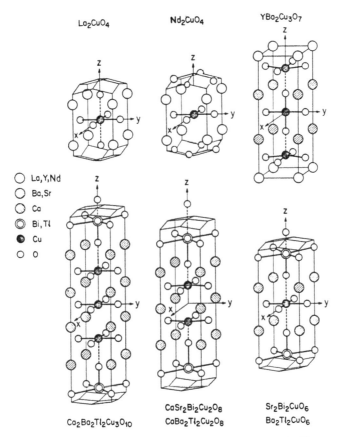

Fig. 3   Structural diagrams of various Cu-O compounds (from ref. 39).

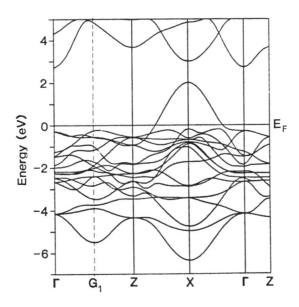

Fig. 4   LDA band structure of Nd_2CuO_4 (from ref. 40).

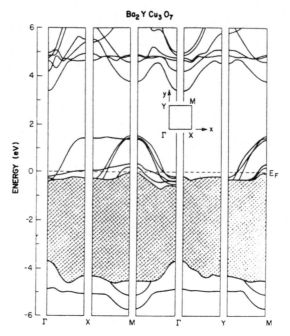

Fig. 5   LDA band structure of YBa$_2$Cu$_3$O$_7$ (from ref. 41).

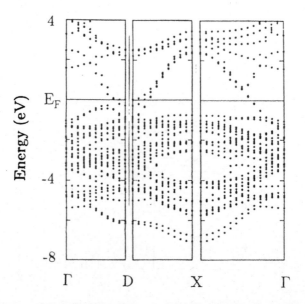

Fig. 6   LDA band structure of CaBi$_2$Sr$_2$Cu$_2$O$_8$ (from ref. 42).

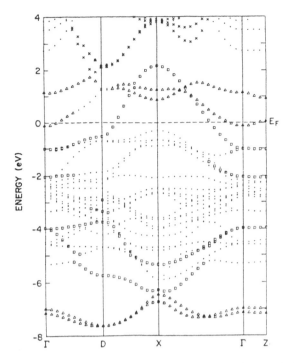

Fig. 7   LDA band structure of $Ba_2Tl_2CuO_6$ (from ref. 43).

where the indices i, j, k. l run over all suitably defined orthogonal electron orbitals centered on different sites in the lattice and where σ, σ′ denote the spin indices. The "one-electron" terms contain on-site energies $\varepsilon_{ii} \equiv \varepsilon_i$ and hopping terms $\varepsilon_{ij} \equiv t_{ij}$. The interaction terms contain all direct Coulomb and exchange contributions arising from up to four different orbitals. In a first approximation we retain only nearest (nn) and second nearest (nnn) neighbor terms. This is motivated by e.g. a successful analysis of the LDA band structure in terms of nn and nnn one-electron terms and by the observation that longer range interaction terms should be more strongly screened. Next we select, based on the LDA results, the Cu $d_{x^2-y^2}$ and O p_σ planar orbitals as essential degrees of freedom. However, nearby in energy are the crystal field split Cu d-states, the planar and apical O p_π-states and the apical O p_σ-states. These will be eliminated at this stage, but we will investigate their role separately later on.

By reducing the Hamiltonian (1) this way we implicitly assume the remaining interaction terms $\langle ij \mid \frac{1}{r} \mid kl \rangle$ to be screened by the removed "inactive" degrees of freedom in the solid and to become effective parameters U_{ijkl} (Coulomb) and K_{ijkl} (exchange). In a further approximation we retain only those terms involving at most two different orbitals. Other terms are higher order as can easily be seen by assuming non-orthogonal orbitals and expanding in powers of the overlap S_{ij} of orbitals. This leaves the following set of interaction parameters to be retained: i) the screened on-site Coulomb energies arising from $\langle ii \mid \frac{1}{r} \mid ii \rangle$, namely U_d and U_p for Cu and O respectively, ii) the analogous intersite parameters U_{pd} and $U_{pp'}$ and the intersite exchange terms K_{pd}, $K_{pp'}$ arising from $\langle ij \mid \frac{1}{r} \mid ji \rangle$ and $\langle ii \mid \frac{1}{r} \mid jj \rangle$. There is one further interaction term, Δt arising from $\langle ii \mid \frac{1}{r} \mid ij \rangle$ which has been invoked by Hirsch[16] to be essential for the high T_c phenomenon. This term can be approximated by $\Delta t \approx \frac{1}{2} S_{ij} (U_{ii} + U_{ij})$ illustrating its nature as an overlap modulated Coulomb term. Transforming to orthogonal Wannier-like orbitals for which the Hamiltonian (equ. 1) is defined, leaves Δt 3rd order in S_{ij} while all other retained terms are either 0th (direct Coulomb and hopping) or 2nd (exchange) order in S_{ij}. With S_{ij} to be of order 10% or less, as estimated from atomic/ionic cluster calculations,[7] we feel we can safely neglect Δt at this stage. This finally leaves us with the following effective, multi-band Hubbard Hamiltonian:

$$H = \sum_\sigma \left\{ \varepsilon_p n_{p\sigma} + \varepsilon_d n_{d\sigma} \right.$$

$$+ t_{pd} c^+_{p\sigma} c_{d\sigma} + t_{pp} c^+_{p\sigma} c_{p'\sigma} + h.c.$$

$$+ \frac{1}{2} U_d n_{d\sigma} n_{d-\sigma} + \frac{1}{2} U_p n_{p\sigma} n_{p-\sigma}$$

$$+ \sum_{\sigma'} \left[U_{pd} n_{p\sigma} n_{d\sigma'} + U_{pp'} n_{p\sigma} n_{p'\sigma'} \right] \quad (2)$$

$$+ K_{pd} c^+_{p\sigma} c_{p\sigma'} c^+_{d\sigma'} c_{d\sigma}$$

$$\left. + K_{pp'} c^+_{p\sigma} c_{p\sigma'} c^+_{p'\sigma'} c_{p'\sigma} \right\}$$

where $c^+_{d\sigma}$ creates a *hole* in the Cu $d_{x^2-y^2}$ orbital of spin σ and $c^+_{p\sigma}$ creates an O p_x or p_y hole of spin σ as shown in Fig. 8. A vacuum state $d^{10}p^6$ has been assumed.

The approach for calculating the parameters in eqn. 2 is described in detail in ref. 30. The complexity of the problem arises from i) the effective screening of the interaction parameters and ii) the strong hybridization between the Cu and O orbitals. The Cu and O orbitals must be treated on an equal footing which makes this case qualitatively different from previously studied systems of a localized orbital weakly coupled to a metallic continuum.[46] To solve these problems we adopt the following point of view: the LDA calculations represent a particular mean-field solution to the problem. To treat the Hubbard Hamiltonian (eqn. 2) as analogously as possible for comparison we apply the usual mean-field (MF) (spin-less Hartree or Hartree-Fock) approximation to its interaction terms. This generates an effective one-electron tight-binding Hamiltonian with diagonal energies and hopping terms which depend self-consistently on the average orbital occupation numbers and the underlying bare parameters of eqn. 1. One has e.g. in Hartree-Fock $\varepsilon^{MF}_{d\sigma} = \varepsilon_d + <n_{d-\sigma}> U_d + 4 U_{pd} \sum_{\sigma'} <n_{d\sigma'}>$.

There are two classes of parameters: i) the MF-screened one-electron on-site and hopping energies and ii) the interaction parameters. They are determined by two different, but coupled procedures.

The one-electron parameters are derived by assuming that the energy bands of the MF solution of eqn. 2 correspond to the pdσ-derived bands in the LDA bandstructure. In this procedure one has to realize that the LDA pdσ bands are also coupled to degrees of freedom not contained in eqn 2 and therefore one has to appropriately correct (down-fold) the parameters obtained from a direct mapping. Moreover, the LDA approach contains interactions in mean-field somewhere between a pure Hartree and Hartree-Fock treatment. This results in upper and lower bounds for the bare energies after the un-screening procedure. Details are discussed in ref. 30.

The interaction parameters are determined in a different way, the so-called Constrained Density Functional (CDF) approach.[47] In the standard DF approach the total energy of the system is minimized constraining the total number of electrons only to be preserved. This yields the electronic ground state charge density and the band structure. For the purpose of calculating local interaction parameters like U_d one has to consider local d-charge fluctuations

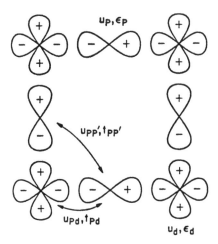

Fig. 8   Schematic representation of the orbitals and interactions included in the 3-band Hubbard model, Eqn. 2. (from ref. 30).

(δn_d) away from this ground state density and calculate the associated energy change. This can be done variationally in DFT by constraining n_d to a chosen value N_{di} at an isolated site i:

$$E(N_{di}) = \min \left\{ E[n] + \lambda [\int dr \, n_{di}(r) - N_{di}] \right\} \quad (3)$$

where $n_{di}(r)$ is appropriately projected out of the DF solutions that produce n(r). This variational procedure allows calculation of the energy associated with local charge (static) fluctuations which are optimally screened by the crystalline environment. Because of the strong hybridization, local Cu and O charge fluctuations are not independent of each other and one obtains an energy surface as a functional of charge profiles in these channels. The LDA calculations of this CDF energy surface are implemented using the linear muffin-tin orbital (LMTO) technique and up to (2×2) supercells to spatially isolate the "frozen" charge fluctuations.

Completely analogous calculations are also carried out for the MF-Hubbard model. The two energy surfaces $E^{LDA}[\delta n_i]$ and $E^{MF-H}[\delta n_i]$ are now mapped onto each other by expanding them around the ground state. This mapping yields the interaction parameters U_{ij} in the Hubbard model. As mentioned above, these depend implicitly on the chosen one-electron parameters ε_{ij}. The crucial one-electron parameter of the problem is the difference in on-site energies $\varepsilon = \varepsilon_p - \varepsilon_d$; the dependence of the U_{ij} on this "bare" ε is shown in Fig. 9, illustrating the interdependence of the parameters. Numerical noise in the mapping procedure does not allow the direct extraction of the small intersite exchange (K_{pd} and $K_{pp'}$) parameters. Therefore, we have separately estimated these parameters from ionic cluster wavefunctions, appropriately correcting for overlap.[7] The final self-consistent set of parameters is listed in Table I. Notable are large on-site Coulomb energies, strong p-d and p-p hybridization terms and relatively small intersite Coulomb energies. The Coulomb energies are consistent with estimated atomic values, suitably reduced by screening. For instance, the free ion Cu^{++} value of 16.5 eV is reduced to 10.5 eV. The value of $U_p \approx 4$ eV arises mainly from the compression of O^{--}-like wavefunctions in the Madelung potential of the crystal. The value is

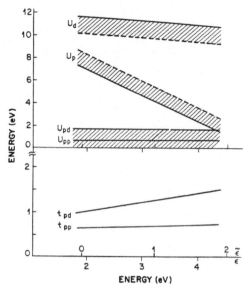

Fig. 9 Dependence of the Coulomb interaction parameters on the one electron on-site energy differences (bare ε and screened ε^{MF}). (from ref. 30).

consistent with accepted values for other oxides, but there is no simple free ion equivalent. The small intersite Coulomb energy is consistent with the interaction of screened (by a background $\varepsilon \approx 4$) ionic charges. Estimates based on simple point charges cannot be used as bounds because of wavefunction spread. The bare on-site energy difference ε is larger than sometimes anticipated, though still smaller than U_d. It should be kept in mind, however, that this bare value of 3.6 eV derives from a MF-screened value of only 1.3 eV which is obtained from careful analysis of the LDA band structure.[30] It is obtained assuming a Hartree-Fock-like screening and is therefore a lower bound.

The values of Table I allows us to draw a simple zeroth order picture of the electronic structure of the Cu-O sheets (see Fig. 10). For its explanation we begin by neglecting all hopping energies, switching them on later. Starting with the "theoretical" vacuum of $d^{10}p^6$ the first ionization occurs out of Cu from $d^{10} \rightarrow d^9$ at $E = \varepsilon_d \equiv 0$. This would be the ground state of La_2CuO_4 with one hole per Cu atom. Extra holes as e.g. created in a photoemission experiment have the following energies: d^9p^5 at $E = \varepsilon_p + 2U_{pd} = \varepsilon_d + 6$ eV, d^8p^6 at $E = 2\varepsilon_d + U_d = 2\varepsilon_d + 10.5$ eV etc. This leads to the general picture shown in Fig. 10. An upper and a lower Hubbard band mainly Cu-derived are interleaved with the oxygen band complex. The Fermi level falls between the upper Cu Hubbard band and the oxygen band; La_2CuO_4 is thus, according to the classification scheme proposed by Zaanen et al.,[48] a charge transfer insulator. This picture for La_2CuO_4 was first proposed by Emery.[4] Note, however, that this situation is different from that of an ordinary ionic insulator, since its insulating behavior depends on a large finite U_d. The $t_{pd} = t_{pp} = 0$ energies are also shown in Fig. 11 where they are compared to the peak-positions (band centers) of direct and inverse photoemission experiments.[49,50] The overall agreement is satisfactory supporting the derived values of the parameters. For a more direct comparison the (strong!) hybridization cannot be neglected. In this case eqn. 2 can still be solved exactly but only on small clusters. Moreover, for detailed spectroscopic signatures on the high (1-10 eV) energy scale, the remaining Cu and O orbitals, we discarded at the beginning of this discussion, have to be reconsidered. This has been done by Eskes and Sawatzky[51] for a Hubbard parameter set quite similar to ours. It was found that the detailed spectroscopic features derive from complex multiplet structures as a consequence of both large Coulomb and hybridization energies. In fact Eskes and Sawatzky used various

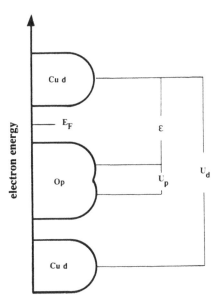

Fig. 10   Schematic diagram of the electronic structure of La_2CuO_4 as derived from the 3-band Hubbard model.

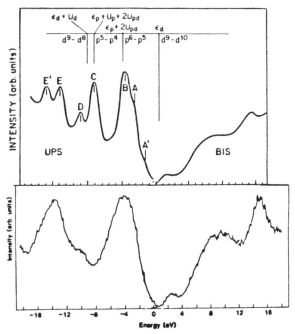

Fig. 11   Comparison of the calculated 3-band Hubbard parameters with direct and inverse photoemission data (refs. 49,50). The "band center" energies (t=0) are indicated.

spectroscopic data and their experience with other Cu and O containing materials to derive a parameter set appropriate for the Cu-O layers in the high T_c materials. We are currently in the process of computing photoemission spectra for Cu_nO_m clusters based on our derived parameter set (table I) and we expect results qualitatively similar to those of ref. 51. Photoemission spectra derived from band structures of course fail to reproduce the rich satellite structures but they generally have success in reproducing structures due to Van Hove singularities in those parts of the spectrum that are dominated by one-electron type excitations.[52]

In addition to photoemission, optical absorption or reflectivity can also be used to probe the validity of the derived parameter set. In particular the size of the optical gap, if non zero, is a delicate test for quantities like ε, t_{pd}, t_{pp} and U_{pd}. We have studied the total energy of clusters (with non-periodic boundary conditions) with N, N−1 and N+1 holes to determine the gap as:

$$\Delta = E(N+1) + E(N-1) - 2E(N) \quad (4)$$

Finite size effects are found to be important and scaling to the thermodynamic limit is necessary. This was done using simple confinement scaling appropriate for the single particle kinetic energy (i.e. $\Delta \sim (size)^{-2}$). The results for Cu_1 to Cu_5 clusters are shown in Fig. 12,

Table I Calculated set of parameters of the 3-band Hubbard model describing La_2CuO_4. Hole notation is used in connection with the phase conventions shown in Fig. 8 (from ref. 30)

U_d	U_p	U_{pd}	$U_{pp'}$	K_{pd}	$K_{pp'}$	$\varepsilon_p - \varepsilon_d$	t_{pd}	t_{pp}
10.5	4	1.2	0	−0.18	−0.04	3.6	1.3	0.65

suggesting a charge transfer gap of $E_g = 2.4 \pm 0.3$ eV which is in excellent agreement with ellipsometry data on $YBa_2Cu_3O_6$ (Fig. 13).[53] Here, on the basis of the measured temperature dependence, the structure near 2.6 eV is interpreted as the charge transfer absorption edge while the lower energy structure at 1.8 eV is assigned to the corresponding excitonic transition. This too, corresponds well with our parameters, since the excitonic binding energy should be on the scale of U_{pd}.

In concluding this section we note that our set of Hubbard parameters, given in Table I and derived from a mapping onto LDA results is consistent with available data on a large (1-10 eV) energy scale. As mentioned above, independent efforts to extract parameters from cluster fits to several types of spectroscopic data yield very similar values.[51,54] Also, an independent calculation by McMahan et al,[55] similar to ours in spirit but different in detail comes to the same overall conclusions.

In the next section we shall then use our results and try to inspect in more detail the dynamics of electron/holes on a smaller energy scale (0.1-1 eV) which is the scale of interest for superconducting behavior. This is done with the goal in mind to identify a simple "fixed point" to which the present 3-band Hamiltonian with its parameters would be renormalized.

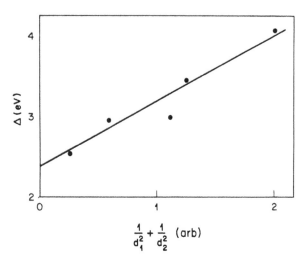

Fig. 12   Scaling of the charge transfer gap Δ as calculated from clusters of different size (from ref. 31).

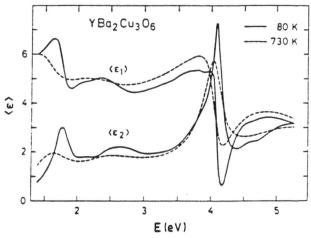

Fig. 13   Real and imaginary parts of the dielectric function of $YBa_2Cu_3O_6$ measured at 80 and 730K. (from ref. 53).

MAPPING OF FINITE CLUSTER RESULTS

We begin by solving the generalized Hubbard Hamiltonian (eqn. 2) exactly for a sequence of clusters:[31] CuO_4, Cu_2O_7, Cu_3O_{10}, Cu_4O_{12} and Cu_5O_{16}. The clusters are embedded in an array of Cu d^9 sites which shift the effective on-site energy of the outer O orbitals due to the intersite Coulomb energy U_{pd}. The results are insensitive to details of this procedure. We do not use periodic boundary conditions which gives more stable trends as a function of cluster size. The resulting sparse Hamiltonian matrices are symmetry block-diagonalized according to total spin S_z and then numerically diagonalized using a block Lanczos technique. The largest matrix considered is of rank 80000 and up to 32 low lying states are calculated. For each cluster, three cases are considered: N−1, N and N+1 holes, where N is the number of Cu sites. As expected from table 1 and Fig. 10, the chemical character of the holes in the insulating ground state is copper dominated, Cu (80%) and O (20%). This is quite consistent with the observed magnetic moment in the AF Néel state of $\mu = 0.48 \pm 0.15 \mu_B$[56] which is reduced from the full moment of $1.14 \mu_B$ for Cu^{2+} both by hybridization (~ factor of 0.8)[30] and by quantum fluctuations (~ factor of 0.6).[57] Drastic variations in the parameter set, such as a reduction in ε would clearly destroy this agreement. As mentioned above in section II, the experimental result is quite inconsistent with the small SDW moment claimed in some LSD-type calculations.

The Cu orbital occupation (electrons) as found from the mean field Hubbard (and LDA) calculations is somewhat larger, about $d^{9.5}$ instead of the correlated density of $d^{9.2}$ obtained for these clusters. This difference can be understood as the strong correlations tending to localize the hole wavefunction. Nevertheless, the differences are small on the scale of the total electronic charge which is quite well described by the LDA (see Section II). The chemical character of an extra hole introduced is 80% O and 20% Cu while that of an extra electron is 80% Cu and 20% O again reflecting the chemical asymmetry in the Cu-O materials. Since these numbers are derived from exact diagonalization of the many-body Hamiltonian (for clusters) they contain all many-body polarization and hybridization effects. There is thought to be some experimental evidence for the prevalent oxygen character for added holes obtained from several different core level spectroscopics, apparently confirming our findings.[58] We caution, however, that these types of spectroscopies do not probe the ground state but an "impurity"-like excited state with a localized core-hole and, therefore strong excitonic effects associated with this core-hole may modify the results. Quantum chemical cluster calculations with core-holes present could help in interpreting these spectra.

We now turn to the question of the nature of the low lying excitations. For the insulating phase, the calculated low energy cluster spectra with N holes show the correct number of lines for a spin ½ Heisenberg model separated by a large (~2.5 eV) gap from the charge transfer excitations, as already found in ref. 7. Mapping these spectra to the Hamiltonian

$$H = \sum_{<ij>} J_{ij} \vec{S}_i \cdot \vec{S}_j \qquad (5)$$

for all clusters yields a nearest neighbor $J = 128 \pm 5$ meV and a next nearest neighbor $J' = 3 \pm 1$ meV, error bars indicating the variations with cluster size. The mapping is illustrated in Fig. 14 where the RMS errors are 1 meV. Note that $J' \ll J$, and smaller than other estimates.[59] However, since the parameters in table 1 have a precision of only about 10%, the calculated values of J, J' are uncertain within ≈ 50%. It should be noted, however, that drastic variations of the original 3-band Hubbard parameters would not only modify the numerical value of J but also distort the spectrum away from the simple Heisenberg case. Experimentally, spin excitations in the AF have been measured by Raman scattering (see Fig. 15).[60,61] These spectra have been shown to arise from two-magnon type excitations of a simple 2D spin ½ Heisenberg AF on a square lattice, strongly influenced by quantum

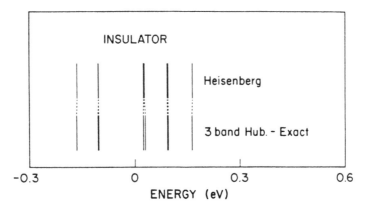

Fig. 14   The low energy spectrum of an insulating Cu_5 cluster calculated in the 3-band Hubbard model and compared to an effective one-band 2D spin ½ Heisenberg Hamiltonian. (from ref. 31).

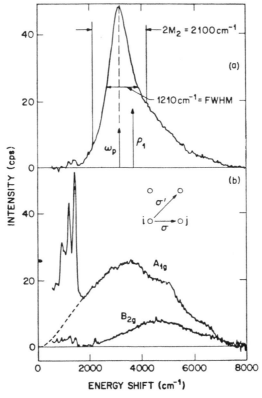

Fig. 15   Experimental Raman scattering data on magnetic fluctuations observed in insulating La_2CuO_4 (from ref. 61).

fluctuations. An excellent fit of the first three spectral moments derived from an approximate ground state wavefunction was obtained for a value of J=128±6 meV (YBa$_2$Cu$_3$O$_6$).[61] The spectral moments derived from our cluster calculations reproduce these findings within 10%. The Raman measurements are resonant enhanced ($\hbar\omega \approx 2$ eV). A detailed analysis involving symmetry selection rules also indicates strong extra scattering intensity, at higher energies, most likely due to second nearest neighbor pair spin-flips. These can occur due to quantum fluctuations and should be resonantly enhanced for excitations frequencies near the charge transfer continuum.[61] We are presently using our cluster results to compute these resonantly enhanced Raman cross sections. It is furthermore experimentally found that the Raman spectra differ little (~10%) in shape and resonant behavior for the various Cu-O insulators (i.e. La$_2$CuO$_4$, YBa$_2$Cu$_3$O$_6$, Nd$_2$CuO$_4$, ...).[62] This is significant, since these materials differ in their Cu-O coordination *outside* the Cu-O planes (2-fold, 1-fold, 0-fold respectively). We interpret this result as indicating that the low lying magnetic states derive from planar Cu-O orbitals and that excitations into other out-of-plane degrees of freedom (Cu d_{z^2}, apical O_{P_z}, etc.) are in the insulator enough removed in energy so as not to destroy the observed simple Heisenberg spectrum.

We now turn to the examination of the low energy spectra of the doped phases. For the clusters we investigated, adding one extra hole would formally correspond to doping levels of 20-25%. However, as will be discussed below we have retained a constant spin-spin interaction value of J which is the appropriate limit for small doping concentrations. We again find for all clusters a spectral gap of ≥ 2 eV between the lowest manifold and higher lying states. This separation suggest the existence of a simplified Hamiltonian whose Hilbert space comprises only the lowest manifold. We find good mapping onto a generalized "t–J" type Hamiltonian, first proposed for these materials by Anderson[3] and by Zhang and Rice,[6] which has the form:

$$H = \sum_{ij\sigma} t_{ij} c_{i\sigma}^+ c_{j\sigma} + \sum_{<ij>} J_{ij} \vec{S}_i \cdot \vec{S}_j \qquad (6)$$

where $c_{j\sigma}$ creates an effective carrier at site j centered about a Cu in the plane. No double occupancy of these sites is admitted. Sites without carriers contain spins that interact in analogy to the insulating Heisenberg case. The effective carriers move in this spin background. The mapping consists of three steps. First the spectrum is mapped (see Fig. 16) constraining J, J' to the values found for the insulator. Nearest neighbor (t) and second nearest neighbor (t') hopping is needed to obtain mappings with RMS error of ~60 meV (about 3% of the bandwidth). The values are given in table II for both electron and hole doping. The error bars again reflect cluster to cluster consistency. The signs of t, t' refer to hole notation. This convention means that for the electron doped case, the effective carrier is a vacancy, while for hole doping it is a singlet bound state of two holes. The values in table II show that the effective low energy carrier dynamics exhibits remarkable electron-hole symmetry, despite the underlying strong chemical asymmetry. We cannot identify any underlying symmetry for this, but believe it to be a consequence of the materials parameters.

Although the spectral mapping is well defined for the clusters we studied, the large value of t suggests that in the limit of many sites the top of the "tt'-J" band will merge with the higher energy continuum. Therefore, the "t-t'-J" model strictly applies only to the region near the bottom of the band. The mappings to these lowest states may be further improved by weighing the low energy portion of the spectra increasing t' to ≈ -100 meV.

To test the validity of the mapping process, in a second step the subspace for the "t-t'-J" model has been identified. We used products of one-hole and two-hole wavefunction, individually derived from CuO$_4$ cluster calculations. The two-hole case for this cluster corresponds formally to the singlet proposed by Zhang and Rice,[6] formed between two holes

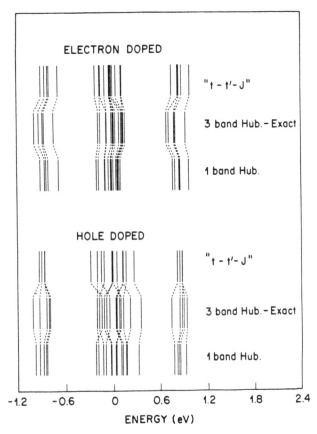

Fig. 16   The low energy spectrum of a doped Cu_5 cluster (electrons or holes) as calculated in the 3-band Hubbard model and compared to an effective one-band "t-t'-J" model or a one-band Hubbard model. (from ref. 31).

Table II   Calculated set of parameters of the one-band "t-t'-J" model describing electron and hole doped La_2CuO_4. Also shown are the parameters for a one-band Hubbard model. All energies in meV. (from ref. 31).

	t-t'-J electrons	t-t'-J holes	Hubbard
t	410±5	440±5	430
t'	−70±10	−60±20	−70
J	128±5	128±5	−
J'	3±1	3±1	−
U_{eff}	−	−	~5400

in the antibonding linear combination of the Cu $d_{x^2-y^2}$ and the $O_p\sigma$ orbitals. The first excited state, the triplet is about 2.8 eV higher in energy. Since this two-hole ground state is derived from exactly diagonalizing a CuO_4 cluster is contains full local correlations beyond the simple original singlet model. Using these product wavefuntions as basis set we find that the projection onto the space spanned by the low energy wavefunctions of the original three-band Hubbard model is at least 93% across the entire manifold. In a third step we checked whether the internal dynamics in this subspace is correctly described by the "t-t'-J" wavefunctions. For this we expressed the "t-t'-J" eigenfunctions in the simple CuO_4 product basis. Again the mapping onto the low-lying Hubbard-model eigenfunctions was in the 80%-90% range. These are rather strong tests of the validity of the "t-t'-J" model for the low-energy electronic structure of the (doped) three-band Hubbard model with parameters appropriate for La_2CuO_4. Our numerical results thus confirm the original proposals by Anderson[3] and by Zhang and Rice[6] and are also in agreement with a recent derivation of the model by Shastry.[63]

One of the restrictions in the "t-t'-J" model is that double occupancy is not allowed. We therefore also considered a simple one-band Hubbard model with a single effective site per Cu-atom. As for the "t-t'-J" model the internal oxygen spin degrees are effectively frozen out. All three low-energy spectra (N−1, N, N+1 holes) can be accurately reproduced with a single parameter set of $U_{eff} = 5.4$ eV, $t = 0.43$ eV and $t' = -0.07$ eV (see table II). Because of the enlarged configurational subspace the mapping is somewhat better than for the "t-t'-J" model. The similarity of the t,t' values between the two models underlines the consistency of the mappings.

Our results indicate several important points. There is a remarkable symmetry in the dynamics between electron and hole doped cases, in spite of the pronounced chemical asymmetry of these carriers as we discussed earlier. This is a somewhat unexpected finding and, as mentioned, most likely a result of materials parameters. It is of relevance for the description of superconductivity in view of the experimental confirmation of both hole- and electron-type high T_c superconductors. The effective carriers have very different internal structures. In the electron doped case, they are ~80% Cu $d_{x^2-y^2}$ electrons, i.e. simple vacancies in the Heisenberg lattice. In the hole doped case, they are locally correlated, strongly bound singlets between the original spins (80% Cu) and the doping holes (80% oxygen). These carriers are found to propagate with comparable amplitudes. The second main finding is that one needs a sizeable second nearest neighbor hopping amplitude t' to obtain good mapping. This t' will make a substantial contribution to the carrier motion, since the intra-sublattice hopping does not, in contrast to t, require spin flips if one naively assumes a Heisenberg-like AF background. The hopping amplitudes t' arise from two interfering channels, involving t_{pp} on the one hand and a higher order process via a third Cu site. No attempts have been made to extract hopping amplitudes for further distant neighbors.

All these findings should be tested against critical experiments. On an energy scale of t (~0.5 eV) these are the Raman scattering results (see Fig. 17)[64] analogous to those obtained for the insulator, and the frequency dependent conductivity results (see Fig. 18)[65] which are dominated by a low frequency Drude-like feature and a mid-infrared structure. We are presently studying spectral functions obtained from "t-t'-J" clusters to interpret these experiments in a fashion similar to what was done for the insulators. On the smaller energy scale of kT there are the linear-in-T resistivity data (Fig. 19),[66] the linear-in-V tunneling data (Fig. 20)[66] and the anomalous NMR data for Cu and O sites (Fig. 21).[67] The kT energy scale is too small for the "graininess" of cluster studies and analytic studies of the model are needed. There are many attempts in this direction which shall not be discussed here. While many of the questions are still open, there is no strong indication at this point that a "t-t'-J" model *cannot* at least qualitatively explain all these experimental findings. An interesting case are e.g. the NMR data which indicate that i) Cu and O relaxation times have a very

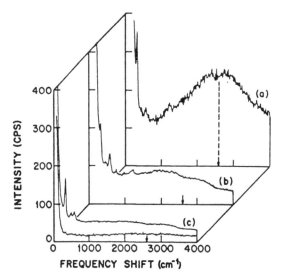

Fig. 17 Experimental Raman scattering data on magnetic fluctuations in $YBa_2Cu_3O_{6+x}$ and their modifications as dopant carriers are introduced (from ref. 64). Curve a) $x=0$; b) $x \approx 0.6$; c) $x \approx 0.9$. Samples b and c are superconducting with $T_c = 60K$ and 88K respectively.

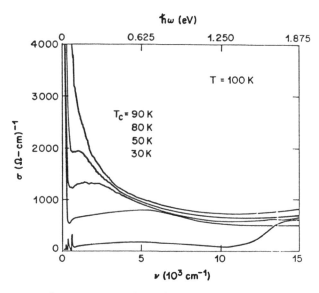

Fig. 18 Frequency and temperature dependent conductivity for $YBa_2Cu_3O_{6+x}$ samples with different superconducting transition temperatures. (from ref. 65).

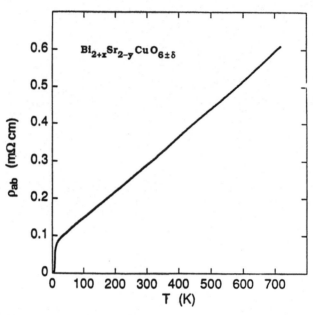

Fig. 19 In-plane resistivity of $Bi_2Sr_2CuO_6$ as a function of temperature. (from ref. 66).

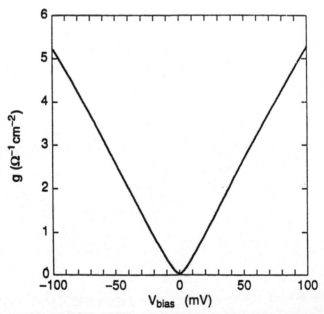

Fig. 20 Differential specific conductance vs. bias voltages of a $Bi_2Sr_2CuO_6$/Pb tunneling junction (from ref. 66).

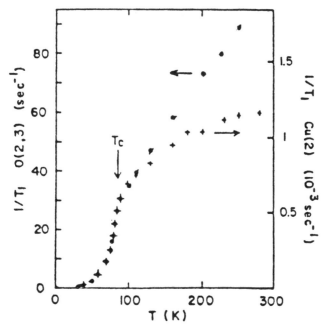

Fig. 21 Temperature dependence of the NMR relaxation rates for in-plane Cu (right scale) and O (left scale) in $YBa_2Cu_3O_7$. The Cu rate has been scaled down to coincide with the O rate at T_c. (from ref. 67).

different temperature behavior above T_c, ii) the absolute rate for Cu is much larger than for O and iii) below T_c both rates decay in the same fashion without coherence enhancement (see Fig. 21). As pointed out by Mila and Rice,[68] by Shastry[63] and by Bulut et al.[69] there is enough flexibility in a "t-J" type one-band model for drastic differences between Cu and O due to geometrically determined structure factors. For these factors to be effective one has, however, to assume that there remain spin fluctuations with reasonably long coherence length to contribute to the wavevector-dependent susceptibility even in the doped metallic regime. While there is some evidence for this from recent neutron scattering[70] and NMR[71] experiments, the issue is still widely debated. On the other hand, direct support for an effective one-band Hamiltonian comes from the scaled decay of the Cu and O relaxation rates below T_c in the superconducting state. The data there seem to indicate the condensation of *one* quantum-spin system and not a system in which Cu and O spins have separate dynamics.

To address these questions in more detail and also the question of a sufficiently strong pairing interaction that could lead to high T_c superconductivity, solutions to one-band Hubbard or " t-t'-J "-type models, analytic, variational, numerical, etc. are currently being studied, with emphasis on conventional states as well as new, exotic flux-phases.[72-88]

CONCLUSIONS AND CAVEATS

The consecutive mapping procedures described in this paper are neither exact nor unique. There are several sources for systematic errors. In the first step, the selection of the relevant, active degrees of freedom is, of course, crucial. We selected the obvious three-band model, but there are competing channels for the accommodation of the dopant holes. Earlier studies of Guo et. al.[10] suggested that the planar oxygen π-(rather than σ-) orbitals should accommodate the dopant holes. The local crystal field favours these orbitals, but hybridization is weaker for them and disfavours them. Based on our analysis of the LDA band structure in combination with finite cluster Hubbard-studies[7] we would place the lowest hole states involving π-orbitals at least 0.5-1.0 eV above the ground state. The results are strongly dependent on the crystal field energies, which can be estimated by fits to band structures, but which are not known too precisely.

Another competing channel involves the Cu d_{z^2} orbitals which couple to the apical oxygen p_σ orbitals. Again the local crystal field may favour this strongly coupled molecular complex. Direct exchange would stabilize a triplet state between one hole of x^2-y^2 symmetry and one hole of z^2 symmetry.[12] However, the intraplanar hopping is considerably smaller for this triplet than for the singlet we discussed above. An accurate estimate of this difference is difficult to obtain from cluster studies, since the energy difference converges slowly with cluster size. However, using various approximations we would place the triplet again about 0.5-1 eV above the ground state. We like to note in this context, that in $YBa_2Cu_3O_7$ only one and in $Nd_{2-x}Ce_xCuO_4$ no apical oxygen is present per Cu atom, which should increase the energy of the triplet state significantly. Finally there is the lower tail of the Cu-O excitonic charge transfer spectrum[5] of the original three-band Hubbard model which also overlaps into this general energy range. Another source of possible systematic errors is associated with the lack of uniqueness of simple model Hamiltonians onto which the original Hubbard results can be mapped. There may be small interactions which do not trivially renormalize the effective parameters and therefore remain important for the low energy dynamics.

Inclusion of these extra degrees of freedom or new interactions would likely invalidate the final one-band models and would result in more complicated irreducible "fixed-point" models. Many of those have been suggested in the literature.[1-28] Finally, there is the question whether the metallic superconducting state can at all meaningfully be derived from the insulating state. When pushed too far into the metallic regime the doped Mott insulator description, given here could lose its validity. Attempts have been made to approach the superconductivity regime from the metallic or Fermi liquid side using weak coupling expansions.[24,25] The materials parameters considered in these cases are generally rather far from those we calculated for the Cu-O materials. This general set of questions, however, remains open and it may well be that high T_c superconductivity has its home in the crossover regime between a doped Mott insulator and a normal Fermi liquid.

On the basis of our analysis, however, we come to the conclusion that the low energy dynamics of moderately electron or hole doped Cu-O planes should be describable by a "t-t'-J" like *one-band* model, with additional degrees of freedom occurring at energies above about 0.5 eV. This point of view is explicitly derived from materials properties and is supported by several experimental findings. While there is no obvious experimental contradiction the model has still to be tested against many other experiments.

ACKNOWLEDGEMENTS

I would like to acknowledge close collaboration with M. S. Hybertsen on all aspects of the work described in this review. We have also benefited from collaborations with N. C. Christensen on the LDA part of the problem and with E. B. Stechel and D. W. Jennison on the cluster part of the work. Finally I would like to thank the many experimental colleagues at Bell Laboratories for numerous valuable discussions.

REFERENCES

1) Proc. Intl. Conf. "High Temp. Supercond. Mat. Mechan. Supercond." Eds. J. Muller and J. L. Olsen (North Holland, Amsterdam, 1988), (Physica C, 153-155, 1988).

2) Proc. Intl. Conf. "High Temp. Supercond. Mat. Mechan. Supercond." Eds., N. E. Phillips, R. N. Shelton and W. A. Harrison (North Holland, Amsterdam) to be published.

3) P. W. Anderson, Science, *235*, 1196 (1987).

4) V. J. Emery, Phys. Rev. Lett. *58*, 3759 (1987).

5) C. M. Varma, S. Schmitt-Rink and E. Abrahams, Solid. Stat. Comm. *62*, 681 (1987).

6) F. C. Zhang and T. M. Rice, Phys. Rev. *B37*, 3759 (1988).

7) E. B. Stechel and D. R. Jennison, Phys. Rev. *B38*, 4632 (1988) and Phys. Rev. *B38*, 8873 (1988).

8) H. B. Schuttler and A. J. Fedro, J. Less Comm. Met. *149*, 385 (1989).

9) P. A. Lee, Phys. Rev. Lett. *63*, 680 (1989).

10) Y. Guo, J. M. Langlois and W. A. Goddard III, Science *239*, 896 (1988).

11) W. Weber, Z. Phys. *B70*, 323 (1988).

12) H. Kamimura, S. Matsuno and R. Saito, "Mech. High Temp. Supercond." Eds. H. Kamimura and A. Oshiyama (Springer, 1989) (Series Mat. Sci. 11, p. 8).

13) D. L. Cox, M. Jarrell, C. Jayaprakash, H. R. Krishnamurthy and T. Deisz, Phys. Rev. Lett. *62*, 2188 (1989).

14) J. R. Schrieffer, X. G. Wen and S. C. Zhang, Phys. Rev. Lett. *60*, 944 (1988).

15) J. E. Hirsch, Phys. Rev. Lett. *59*, 228 (1987).

16) J. E. Hirsch, Physica *C158*, 326 (1989).

17) C. A. Balseiro, A. G. Rojo, E. R. Gagliano and B. Alascio, Phys. Rev. *B38*, 9315 (1988).

18) L. P. Gorkov and G. M. Eliashberg, JETP Lett. *46*, Suppl. 584 (1987).

19) I. E. Dzyaloshinskii, JETP *66*, 848 (1987).

20) B. K. Chakraverty, D. Feinberg, Z. Hang and M. Avignon, Sol. Stat. Comm. *64*, 1147 (1987).

21) J. Ruvalds, Phys. Rev. *B35*, 8869 (1987).

22) V. Z. Kresin, Phys. Rev. *B35*, 8716 (1987).

23) R. B. Laughlin, Science *242*, 525 (1988).

24) D. M. Newns, P. Pattnaik, M. Rasolt and D. A. Papaconstantopoulos, ref. 1, p 1287.

25) P. B. Littlewood, to be published.

26) D. Emin, Phys. Rev. Lett. *62*, 1544 (1989).

27) J. C. Phillips, "Physics of High T_c Superconductors", Academic Press, New York, 1989.

28) J. R. Hardy and J. W. Flocken, Phys. Rev. Lett. *60*, 2191 (1988).

29) M. Schlüter, M. S. Hybertsen and N. E. Christensen, ref. 1, p. 1217.

30) M. S. Hybertsen, M. Schlüter and N. E. Christensen, Phys. Rev. *B39*, 9028 (1989).

31) M. S. Hybertsen, E. B. Stechel, M. Schlüter and D. R. Jennison, to be published.

32) J. A. Wilson, J. Phys. C *21*, 2067 (1988).

33) L. F. Mattheiss, Phys. Rev. Lett. *58*, 1028 (1987).

34) T. C. Leung, X. W. Wang and B. N. Harmon, Phys. Rev. *B37*, 384 (1988).

35) G. Y. Guo, W. M. Temmerman and G. M. Stocks, J. Phys. *C21*, 1103 (1988).

36) T. Kasuya and K. Takegahava, Jpn. J. Appl. Phys. Ser. *1*, 251 (1988).

37) R. E. Cohen, W. E. Pickett and H. Krakauer, Phys. Rev. Lett. *62*, 831 (1989).

38) C. Ambrosch-Draxl, P. Blaha and V. Schwarz, J. Phys. C *1*, 4491 (1989).

39) L. F. Mattheiss, unpublished.

40) S. Massida, N. Hamada, J. Yu and A. J. Freeman, Physica *C157*, 571 (1989).

41) L. F. Mattheiss and D. R. Hamann, Sol. Stat. Comm. *63*, 395 (1987).

42) M. S. Hybertsen and L. F. Mattheiss, Phys. Rev. Lett. *60*, 1661 (1988).

43) D. R. Hamann and L. F. Mattheiss, Phys. Rev. *B38*, 5138 (1988).

44) K. C. Hass, "Sol. Stat. Phys." Eds. H. Ehrenreich and D. Turnbull, (Academic Press, Orlando) Vol. 42, (1989).

45) W. E. Pickett, Rev. Mod. Phys., *61*, 433 (1989).

46) O. Gunnarsson, O. V. Andersen, O. Jepsen and J. Zaanen, Phys. Rev. *B39*, 1708 (1989).

47) P. H. Dederichs, S. Blugel, R. Zeller and H. Akai, Phys. Rev. Lett. 253, 2512 (1984).

48) J. Zaanen, G. A. Sawatzky and J. W. Allen, Phys. Rev. Lett. *55*, 418 (1985).

49) P. Thiry, G. Rossi, Y. Petroff, A. Revcolevschi and J. Jegoudez, Europhys. Lett. *5*, 55 (1988).

50) D. van der Marel, J. van Elp, G. A. Sawatzky and D. Heitmann, Phys. Rev. *B37*, 5136 (1988).

51) H. Eskes and G. A. Sawatzky, Phys. Rev. Lett. *61*, 1415 (1988), H. Eskes, L. H. Tjeng and G. A. Sawatzky, to be published.

52) R. S. List, A. J. Arko, Z. Fisk, S. W. Cheong, J. D. Thompson, J. A. O'Rourke, C. G. Olson, A. B. Young, T. W. Pi, J. E. Schirber and N. D. Shinn, to be published.

53) J. Humlicek, M. Garriga and M. Cardona, Sol. Stat. Comm. *7*, 589 (1988).

54) F. Mila, Phys. Rev. *B38*, 11358 (1988).

55) A. K. McMahan, R. M. Martin and S. Satpathy, Phys. Rev. *B38*, 6650 (1988).

56) D. Vankin, S. K. Sinha, D. E. Moncton, D. C. Johnston, J. M. Newsam, C. R. Safinya and H. E. King Jr., Phys. Rev. Lett. *58*, 2802 (1987).

57) J. D. Reger and A. P. Young, Phys. Rev. *B37*, 5978 (1988).

58) N. Nücker, J. Fink, J. C. Fuggle, P. J. Durham and W. M. Temmerman, Phys. Rev. *B37*, 5158 (1988).

59) J. F. Annett, R. Martin, A. K. McMahan and S. Satpathy, Phys. Rev. *B40*, 2620 (1989).

60) K. B. Lyons, P. A. Fleury, J. P. Remeika, A. S. Cooper and T. J. Negran, Phys. Rev. *B37*, 2353 (1988).

61) R. R. P. Singh, P. A. Fleury, V. B. Lyons and P. E. Sulewsky, Phys. Rev. Lett. *62*, 2736 (1989).

62) P. E. Sulewsky, P. A. Fleury, U. B. Lyons, S. W. Cheong and Z. Fisk, to be published.

63) B. S. Shastry, Phys. Rev. Lett. *63*, 1288 (1989).

64) K. B. Lyons, P. A. Fleury, L. F. Schneemeyer and J. V. Waszczak, Phys. Rev. Lett. *60*, 732 (1988).

65) J. Orenstein, G. A. Thomas, A. J. Millis, S. L. Cooper, D. H. Rapkine and T. Timusk, to be published.

66) A. T. Fiory, S. Martin, R. M. Fleming, L. F. Schneemeyer, J. V. Waszczak, A. F. Hebard and S. A. Sunshine, ref. 2.

67) P. C. Hammel, M. Takigawa, R. H. Heffner, Z. Fisk and K. C. Ott, ref. 2.

68) F. Mila and T. M. Rice, Physica *157C*, 561 (1989).

69) N. Bulut, D. Hone, D. J. Scalapino and N. E. Bickers, to be published.

70) G. Shirane, R. J. Birgenau, Y. Endoh, P. Gehring, M. A. Kastner, K. Kitazawa, H. Kojima, I. Tanaka, T. R. Thurston and K. Yamada, Phys. Rev. Lett. *63*, 330 (1989).

71) P. C. Hammel, M. Takigawa, R. H. Heffner, Z. Fisk and K. C. Ott, Phys. Rev. Lett. *63*, 1992 (1989).

72) J. Wagner, R. Putz, G. Dopf, B. Ehlers, L. Lilly, A. Muramatsu and W. Hanke, to be published.

73) M. Ogata and H. Shiba, J. Phys. Soc. Jpn. *57*, 3074 (1988).

74) S. N. Coppersmith, to be published.

75) S. R. White, D. J. Scalapino, R. L. Sugar, E. Y. Loh, J. E. Gubernatis and R. T. Scaletar, to be published.

76) S. Liang and N. Trivedi, to be published.

77) P. Lederer, D. Poilblanc and T. M. Rice, to be published.

78) P. B. Wiegmann, Phys. Rev. Lett. *60*, 821 (1988).

79) G. Kotliar, Phys. Rev. *B37*, 3664 (1988).

80) I. Affleck, Z. Zou, T. Hsu and P. W. Anderson, Phys. Rev. *B38*, 745 (1988).

81) S. A. Trugman, Phys. Rev. *B37*, 1597 (1988).

82) B. I. Shraiman and E. D. Siggia, Phys. Rev. Lett. *62*, 1564 (1989).

83) J. Bonca, P. Prevlovsek and I. Sega, Phys. Rev. *B39*, 7074 (1989).

84) E. Dagotto, A. Moreo and T. Barnes, to be published.

85) S. Sachdev, Phys. Rev. *B39*, 12232 (1989).

86) Z. B. Su, Y. M. Li, W. Y. Lai, L. Yu, Phys. Rev. Lett. *63*, 1318 (1989).

87) C. X. Chen, H. B. Schüttler and A. J. Fedro, to be published.

88) C. Jayaprakash, H. R. Krishnamurthy and S. Sarker, Phys. Rev. *B40*, 2610 (1989).

Flux Pinning and Microstructure in $YBa_2Cu_3O_7$

M. Suenaga, D. O. Welch, Youwen Xu, Y. Zhu,
A. K. Ghosh, and A. R. Moodenbaugh

Materials Science Division
Department of Applied Science
Brookhaven National Laboratory
Upton, New York 11973

INTRODUCTION

Critical current densities J_c of superconductors are one of the most important properties of superconductors in the majority of applications. Unfortunately, in spite of the fact that tremendous efforts have been mounted internationally in studies of the high temperature oxide superconductors, the current values of J_c are disappointingly low except in single crystal bulk specimens and films with the current flowing in the basal planes.[1,2] The majority of the studies of J_c in bulk superconductors have emphasized that the difficulties may have been due to the poor intergranular superconductivity and/or the large anisotropy in the value of J_c, depending on the orientation of the crystallographic axis with applied magnetic fields.[3-5] In order to minimize this problem, Jin et al.[6] and more recently Murakami et al.[7] have synthesized grain-aligned bulk specimens which show values J_c (~10 A/mm^2 in applied magnetic fields of a few tesla and at 77 K) significantly larger than those in the sintered materials.

As point out by Jin et al.[6] however, this value of J_c is still <u>at least</u> an order of magnitude lower than that which is required for use in applications to produce useful magnetic fields. This indicates that critical current densities within the grains of $YBa_2Cu_3O_7$ require significant improvements, in addition to further improvements in the alignment. Thus, it is also very important to understand the factors which influence intragrain critical current densities or the flux pinning in the oxide. As shown below, these are, for example, superconducting parameters such as the thermodynamic critical magnetic field H_c, the superconducting coherence lengths ξ, and the size and the density of the crystallographic defects which act as the pinning centers. For the purpose of illustrating how these factors may influence the value of critical current density J_c in the oxide, the effects of small levels of substitutions for Cu in $YBa_2Cu_3O_7$ on its critical current are considered here.

As shown in Fig. 1,[8] when 2% of the Cu in $YBa_2Cu_3O_7$ is replaced by either Al or Fe, it is found that the intragranular critical current densities $[J_c(H) \propto d\Delta M(H)$ where d is the size of the specimen and ΔM is the width of the magnetic hysteresis] of these alloyed oxides were reduced. On the other hand, it was also found that substitution for Cu by 2% Ni did not show any effect on J_c at 10 K (although its J_c was lowered below that of the pure specimen when the temperature was higher than ~30 K). Similar decreases in J_c were also reported by others.[9] In order to illustrate what are the primary controlling parameters in reducing the values of J_c when small amounts of Cu were replaced by Fe or Al, we will first review some elementary aspects of flux pinning mechanisms. Then it

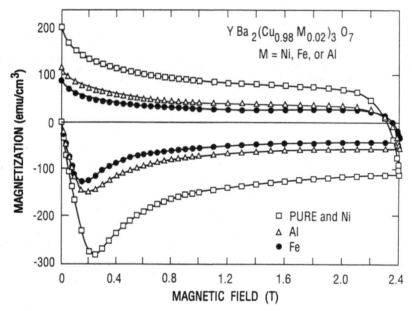

Fig. 1. Magnetic hysteresis curves for a pure and alloyed $YBa_2Cu_3O_7$ powders at 10 K which are aligned to the c axis and applied field H was oriented such that C is parallel to H. Note that the width of the hysteresis $\Delta M(H)$ is proportional to $J_c(H)$ and the powder size are approximately equal for all of the cases and ~30μm.

will be shown how those parameters which are pertinent to flux pinning may be estimated for these specimens. Finally, we will deduce what is probably the most prominent cause for the reduction in J_c for the specimens discussed above.

FLUX PINNING

Critical current densities J_c in superconductors are primarily determined by the effectiveness of crystal structural defects to pin magnetic flux lines to avoid their motion under the Lorentz force J x B. In this context, the interaction energy between the line and a defect is the difference in the energy of a local superconducting region depending on whether the line is situated at and outside of a defect. For simplicity, in this discussion we will only consider the interaction of a single flux line against a planar defect. Two of the strongest types of the pinning interactions are core pinning and $\Delta\kappa$ pinning, where κ is the Ginzburg-

Landau constant. The first is due to the variation in the thermodynamic critical field H_c at or near the defect, and in the case of a planar normal precipitate of a thickness $d \geq \xi$ parallel to the flux line, the elementary pinning energy per unit length of a flux line is

$$U_p/L \simeq \frac{1}{2} \pi \mu_o H_c^2 \xi^2 \qquad (1)$$

where ξ is the Ginzburg-Landau coherence length. The second type is the pinning force at a planar defect such as a grain boundary due to a variation in the Ginzburg-Landau constant ($\kappa \simeq \lambda/\xi$), where λ is the magnetic field penetration depth. Recently, a number of studies were carried out to assess its effectiveness in pinning a flux line by a planar boundary[10-13,38] and the elementary pinning energy from this interaction is found to be of the form

$$U_p/L = \mu_o H_c^2 f(p, \xi/d) \qquad (2)$$

where the factor $\mu_o H_c^2$, which also appears in the core pinning strength above, is multiplied by a function f which depends on the electron-scattering strength p of the boundary and the range d of its effectiveness. The electron scattering alters near the boundary, and hence, this is called the $\Delta\kappa/\kappa$ pinning.

From Eqs. (1) and (2), it is clear that the pinning force in both cases is strongly dependent on the coherence length and the thermodynamic critical field. Here we will only consider core pinning in this discussion. Furthermore, this discussion is limited to consideration of the interaction of a single flux line with a defect. In reality, one has to consider the interaction of the flux lattice with three-dimensionally distributed pinning centers as was considered by, for example, Yamafuji et al.[14] However, the essential idea of flux pinning is contained in this simplified form.

From the discussion above, it is clear that the two primary superconducting parameters of importance to the flux pinning are the thermodynamic critical field H_c and the coherence length ξ. Here, we will briefly discuss how these parameters are estimated using the alloyed $YBa_2Cu_3O_7$ specimens as examples. This will also indicate what type of substitutions are more effective in varying these values.

If we assume the standard Ginzburg-Landau description of Type II superconductivity is applicable to these oxides, the above parameters can be calculated using:

$$H_{c2}(0) \simeq 0.7 (dH_{c2}/dT)_{T_c} \qquad (3)$$

$$\xi(0) \simeq [\phi_o/4\sqrt{3}\, H_{c2}(0)]^{1/2} \qquad (4)$$

$$d(\Delta M_{n,s})/dH|_{H_{c2}} \simeq [4\pi(2\kappa_2 - 1)]^{-1} \qquad (5)$$

and $\quad H_c(0) = H_{c2}(0)/\sqrt{2}\, \kappa_1(0) \qquad (6)$

where 0 indicates the value at $T = 0$. For calculations of H_c and ξ, two independently accessible and measurable parameters are $(dH_{c2}/dT)_{T_c}$ and $d(\Delta M_{n,s})/dH|_{H_{c2}}$. These values are determined by measuring the magnetic moments (or susceptibility) of the specimens as a function of temperature in the <u>reversible</u> temperature region for a series of values of H. (An example for $H = 1T$ is shown in Fig. 2.[16]) From these data, dH_{c2}/dT and dM/dH are determined to calculate pertinent superconducting parameters using the above relationships and these are listed in Table I.[16]

One notable change due to the substitutions is the substantial decrease in the values of $H_c(0)$ while T_c and $H_{c2}(0)$ stayed relatively

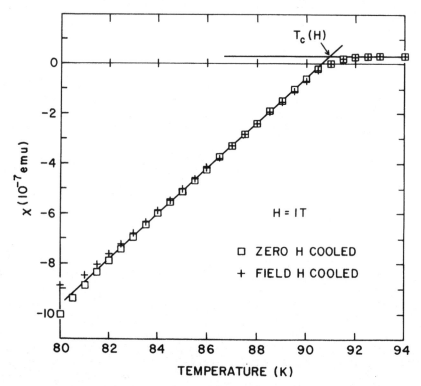

Fig. 2. Magnetic susceptibility vs temperature for an oriented powder of $YBa_2Cu_3O_7$ (c∥H).

unchanged. The cause of the decrease in H_c is related to the drop in the size of the specific heat jump at T_c, $\Delta C/T_c$ [$= (dH_{c2}/dT)^2 (d\Delta M_{n,s}/dT)$], which is proportional to the carrier density of states, $N(0)$. [Although the values of the specific heat jumps which are directly measured by specific heat techniques are somewhat smaller (~67 mJ/mole K^2 for a pure $YBa_2Cu_3O_7$[17]) than the present values, the general trend of its decrease by alloying was shown to be consistent by specific heat measurements for similar specimens with Co,[18] Cr,[19] and Zn[19] substitutions.]

Table I. Measured and calculated parameters for $YBa_2(Cu_{0.98}M_{0.02})_3O_7$.

M	$T_c(K)$	dH_{c2}/dT*	dM/dH#	$H_{c2}(0)$**	κ	$H_c(0)$**	$\Delta C/T_c$+	$\xi(0)$++
Cu	91.5	-2.2	0.198	141	42	1.6	99	1.4
Ni	83.5	-2.4	0.109	140	56	1.2	65	1.4
Al	87.0	-2.8	0.015	170	150	0.6	12	1.2

*(T/K). #(emu/T). **(tesla). +(mJ/K²mole). ++(nm).

It is also noteworthy that the coherence length in the a-b plane for the substituted oxides did not change in spite of a larger increase in the values of κ by the substitution. Thus, the effect of the substitution on the pinning potential is primarily due to the decreased H_c and one can assume that this is a likely source of decreased values of J_c (or the width of magnetic hysteresis) as shown in Fig. 1.

In the above discussion, the changes in the pinning energy due to the substitutions were inferred from the measurements of $dH_{c2}/dT|_{T_c}$ and $d(\Delta M_{n,s})/dH|_{H_{c2}}$. A more direct way of studying variations in the maximum pinning energy by the substitutions is to determine the Gibbs free energy per unit length $\delta G/L$ of a flux line for the oxides. As it was shown shown earlier,[20] $\delta G/L$ can also be measured from the set of M vs H curves in the reversible region since $G_H - G_N = \int MdH$ and $\delta G/L = (G_H - G_N)\phi_0/H$ where G_H and G_N are the Gibbs free energy of the material in the superconducting and the normal state under applied field H, respectively. For an ideal pinning site which is considered here $\delta G/L = -U_p/L$. Athreya et al.[21] have measured $\delta G/L$ for a series of Fe substituted $YBa_2Cu_3O_7$ and as shown in Fig. 3, they have found that not only the free energy decreased vary rapidly with the increasing Fe substitution, but also the value of J_c (as measured by ΔM) decreased nearly proportional to $\delta G/L$. Thus, this also confirms that the thermodynamic properties have a strong influence on the intragrain critical current density of $YBa_2Cu_3O_7$.

Although the above discussion provides an indication of a physical basis for understanding the general trends/changes in critical current densities with processing variables in terms of Eq. (1), this U_p is only the maximum pinning energy achievable under ideal conditions. Thus, it is of great interest to measure directly the pinning energy for these oxides. Unfortunately, measured values of the pinning energies in these oxides are, at present, a subject of active debate. One approach which was taken by a number of groups is to measure the resistivity of a specimen as a function of temperature (or the broadening of the resistance transition under magnetic field) in great detail near the transition.[22-24] This type of measurement leads to values of U_p of the order of 1-4 eV (for H = 1-4 T). However, others have employed measurements of the decay in the magnetic moment with time (flux creep) to estimate the pinning energy.[25-27] They found that U_p is of the order of a few tens of meV. Hence, there exists a large discrepancy in the values of U_p depending on how it is measured. However, recent experiments[28] have shown that the magnetic field effect on resistivity is independent of the orientation of H with respect to the current, this bringing into question the basis of the analysis used to obtain U_p. On the other hand, the previous analysis of the creep data neglected the fact that the pinning potential $U(\nabla B, B)$ [$= (U_p - BJ_cVX)$ where V is the activation volume and X is the jump distance] is a nonlinear function of J_c. Additionally, it was also shown that even if this is included in the analysis, the applicability of the existing flux creep theory to the oxide is questionable.[27] This is due to the fact that even in the case for a single crystal it is not clear whether the induced circulating current in the specimen is around the entire specimen or the subdivided islands making up a specimen at all T and H conditions. Unfortunately, any of these difficulties in determining U_p is not easily overcome. Therefore, at this time, the values of the true pinning potential are elusive and cannot unambiguously guide us to understand the variations in J_c due to changes in conditions of the oxides by measuring these values.

FLUX CREEP

Since the above mentioned difficulties in the use of the flux creep phenomena to determine U_p appear to be related to fundamental

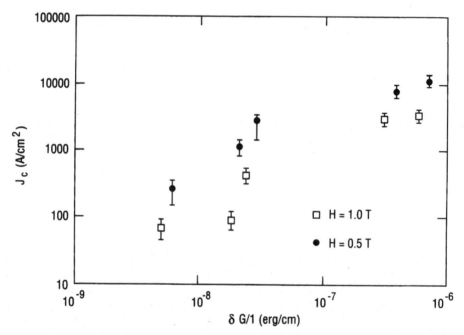

Fig. 3. Dependence of J_c on the free energy of the flux lines $\delta G/\ell$ at 60 K.

properties of the oxides (at least $YBa_2Cu_3O_7$), a more detailed discussion on flux creep in $YBa_2Cu_3O_7$ is given below.

Magnetization of most of Type II superconductors is hysteretic when cycled in magnetic field, and the spatial distribution of magnetic flux density in a superconductor in an applied field ($H>H_{c1}$) is nonuniform, and a nonequilibrium magnetization ΔM (the width of the hysteresis curve) is related to the magnetic flux gradient ∇B in the specimen. The gradient is established due to the fact that the flux lines acted on by the Lorentz force $F(= -\gamma B \nabla B/4\pi$ where $\gamma = \partial B/\partial H$) are pinned against potential barriers in the specimen. At $T > 0$, the flux lines can creep along the gradient by thermally activated processes, and this is manifested in the often observed time-dependent decrease in magnetic moments of superconductors.[30] As often observed,[25-27] the time dependent decrease follows a relationship,

$$dM/Md\ln t \simeq -kT/U_0^*. \tag{7}$$

This or a similar relationship has also been derived theoretically[31-33] and the value of U_0^* was taken as the pinning potential U_p. If we do

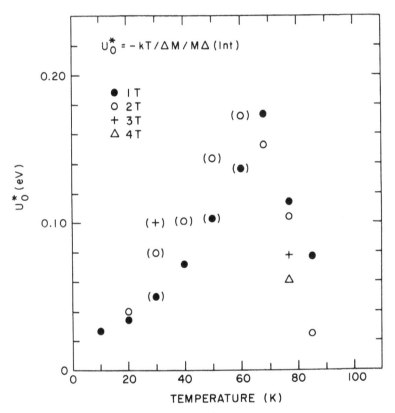

Fig. 4. The temperature dependence of the <u>apparent</u> pinning potential U_0^* for a melt textured $YBa_2Cu_3O_7$.

this, we find a temperature dependence of the pinning potential for a $YBa_2Cu_3O_7$ which is shown in Fig. 4.[34]

A generally accepted temperature dependence of the pinning potential is that it will slowly decrease with increasing temperature until $T \simeq T_c$ where expected to rapid by decrease to zero. The observed temperature dependence of U_0^* is markedly different from the generally accepted dependence. As discussed in detail,[29] this discrepancy is due to the fact that $U_0^* = U_p$ only if the activation potential $U(B,\nabla B)[= U_p - FVX]$ is a linear function of ∇B or the force on flux lines. As schematically illustrated in Fig. 5, the linear relationship exists only for a V notch-shaped potential which is nonphysical. As originally pointed out by

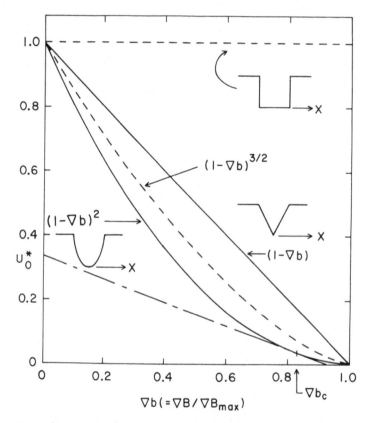

Fig. 5. The dependence of the pinning potential U on ∇B for different potential shapes.

Beasley et al.[31] U is a nonlinear function of ∇B (or J). In fact, explicit calculations[35] were made for a wide variety of assumed shapes and spacings for potential wells, ranging from closely-spaced double wells to widely-spaced isolated wells, with shapes ranging from very sharp to very flat, and the results show that to a good approximation:

$$U(B,\nabla B) \simeq U_p(B)(1-\frac{\nabla B}{\nabla B_{max}})^n \tag{8}$$

with $3/2 < n < 2$, where ∇B_{max} corresponds to the maximum force which the pinning potential can sustain in the absence of thermal activation. Thus, the apparent pinning potential U_o^* which appears in Eq. (7) is given by:

$$U_o^* \simeq nU_p \left| \frac{\nabla B}{\nabla B_{max}} \right| (1 - \frac{\nabla B}{\nabla B_{max}})^{(n-1)} \qquad (9)$$

Note that only if the potential-versus-driving-force curve [Eq. (8)] is linear (n=1), and then only when $J_c(T) \simeq J_c(o)$ so that $\nabla B/\nabla B_{max} \simeq 1$, does U_o^* equal the true pinning potential U_p. Our model calculations show that n>1 is more realistic than n=1 so that U_o^* is expected to be smaller (perhaps by a large factor) than U_p. The difference between U_o^* and U_p is graphically illustrated in Fig. 5. Thus, in determining the true pinning potential U_p, Eq. (7) should be used in conjunction with Eq. (9). Furthermore, the same argument also leads to the fact that M(T) in Eq. (7) should be M(0).

Unfortunately, the above modification for the expression for the relationship between the creep rate and the pinning potential did not result in a meaningful temperature dependence of U_p. In order to determine whether this difficulty is due to the inadequacy of the theories which are applied to type II superconductors in general or to the magnetic properties of $YBa_2Cu_3O_7$, we have performed the creep measurements for Nb_3Sn wires,[36] and it was found that the pinning potential U_p varied with temperature as expected from the temperature dependence of the coherence length indicating that the theory for the creep appears to be generally sound. It appears that the difficulty in attempting to obtain U_p from the creep measurements for $YBa_2Cu_3O_7$ is related to the temperature dependence of the critical current density J_c. In conventional superconductors, $J_c(H)$ is linearly dependent on temperature for most of the temperature range while that for $YBa_2Cu_3O_7$ is very strongly dependent on temperature. In fact, in many instances, the temperature dependence of $J_c(H)$ is described as an exponentially decreasing function of T. It is also possible that this rapid decrease is related to the subdivision of the single grain crystals at microstructural defects. However, this rapid decrease in $J_c(H)$ with temperature may also be related to the fundamental nature of the flux line behavior in the oxide such as, flux lattice melting and entanglement as discussed, for example, by Nelson.[37]

MICROSTRUCTURES

The other important factor in the flux pinning is the size and the density of the pinning sites. From Eq. (1) it is clear that the most effective size of the defects is the value which is equal to the coherence length ξ. As shown in Table I, the value of $\xi(0)$ is approximately 1.5 nm and the substitutions do not affect the value. Also, the optimum distribution of the pinning sites is to have the sites distributed such that each flux line is pinned by the defects for a given magnetic field. This implies that the ideal separation between the defects are, for example, d~17 and ~5 nm at 1 and 10 T, respectively. Thus, rather small and closely spaced crystallographic defects are desirable for a strongly pinned superconductor. However, transmission electron

Table II. The Twin Boundary Layer Thickness t, Twin Spacing, and Superconducting Critical Temperature T_c.

	n^a	$t(nm)^a$	Twin Spacing $(nm)^b$	$T_c(k)$ (midpoint)
$YBa_2Cu_3O_7{}^c$	3.6-5.4	1.0-1.5	89	90.5
$YBa_2Cu_3O_7{}^d$	5.3-8.0	1.4-2.1	23	90
$YBa_2(Cu_{0.98}Zn_{0.02})_3O_7$	3.3-5.0	0.9-1.35	31	67
$YBa_2(Cu_{0.98}Ni_{0.02})_3O_7$	2.6-3.9	0.7-1.0	40	82.
$YBa_2(Cu_{0.98}Fe_{0.02})_3O_7$	5.9-8.9	1.6-2.4	20	86.
$YBa_2(Cu_{0.98}Al_{0.02})_3O_7$	9.5-14	2.6-3.9	40	88

The width t of the boundary layer can be calculated using the following relationship: $t \simeq A(R_{110}/\ell_{st}) d(110)$, where R_{110} and ℓ_{st} are the distance between the center and the <110> diffraction spot and the length of the streak at <000> in mm, respectively, and are measured off the films. d(110) is the (110) lattice spacing and A is a constant.
(a) The range in n and t are given for A=1 and 1.5 in the above equation. $n = t/d_{110}$; $d_{110} \simeq 0.27$ nm.
(b) These twin boundary spacings are from the areas where SAD were taken to make the measurements of the streaks.
(c) The oxygen content in a specimen, which was identically prepared to this, was determined to be 7 by neutron powder diffraction.
(d) A fine grained $YBa_2Cu_3O_7$ heated at 950°C for 48 h in O_2. The exact oxygen content is unknown.

microscopy of pure $YBa_2Cu_3O_7$ has shown that the only microstructural defects observable in it are twin boundaries, as shown in Fig. 6. Partly due to the fact that this is the only type of defect which is commonly seen in a stoichiometric $YBa_2Cu_3O_7$, these are thought to be the pinning centers responsible for high current densities in single crystals.[38,39] Thus, we will examine the nature of the twin boundary at atomic scales and discuss why these may act as the pinning centers. Then, we will illustrate how the density of boundaries change and what effect this may have on the critical current density.

Most discussions concerning the twin boundaries assume an ideal boundary i.e., the boundary consists of a mirror plane along the (110) plane without any atomic distortion. However, it was recently shown that the twin boundary in $YBa_2Cu_3O_7$ contains a distorted thin boundary layer of 1-1.5 nm thick, as shown in Fig. 7.[40] It was also shown that the substitutions for Cu by such elements as Al and Fe widen the layer thickness to ~3 nm, while Ni and Zn additions (at least at the level of 2% of Cu) caused very little change in the thickness as shown in Table II.[41] Since superconductivity in $YBa_2Cu_3O_7$ is strongly related to the oxygen stoichiometry and its crystallographic order, the distorted region at the twin boundary is expected to be superconductively weak under some conditions of temperature and magnetic field. Also, the width of the distorted layer is approximately equal to or greater than the size of ξ. Thus, it is not unreasonable to expect that a twin boundary in $YBa_2Cu_3O_7$ would be a strong pinning center, since there will likely be contributions toward pinning from both the core and the Δκ pinning. Unfortunately, the density of boundaries is very low for pure and high quality specimens (see Fig. 6). However, it was also found that the substitutions of Al, Fe, Co,... for Cu significantly reduced the boundary spacing as shown in Fig. 8 as an example.[42] On the other hand, the substitutions of Ni and Zn did not change the spacing. One should note

Fig. 6. A transmission electron micrograph of $YBa_2Cu_3O_7$ showing the (110) twin boundaries.

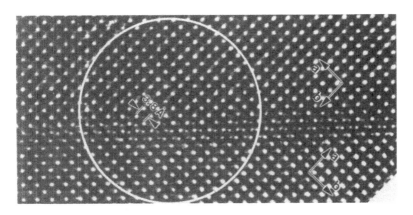

Fig. 7. A high resolution electron micrograph of a twin boundary in a pure $YBa_2Cu_3O_7$ showing a distorted boundary layer.

here that the first set of elements primarily substitutes for Cu in the chain site while the second set is in the plane site.

From TEM images such as that shown in Fig. 8, the large reduction in the twin spacing found in Al- and Fe-substituted $YBa_2Cu_3O_7$, should be expected to increase intragrain-critical current densities in these specimens. On the contrary, as shown in Fig. 1, the additions of these elements resulted in large <u>reductions</u> in the intragrain current densities. What these results appear to imply is that the reduction in the pinning energy due to the decreased H_c overwhelms the increased pinning site density and this results in lowered J_c in the alloyed $YBa_2Cu_3O_7$.

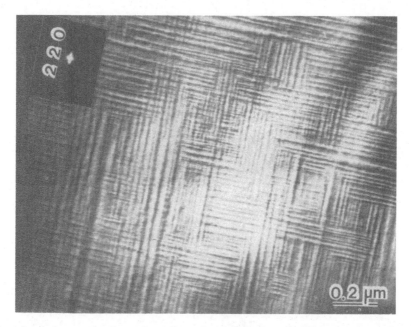

Fig. 8. A transmission electron micrograph of $YBa_2(Cu_{0.98}Fe_{0.02})_3O_{7+\delta}$ showing a very fine set of the (110) twin boundaries.

CONCLUSION

From the simplified discussion above and the examples of alloying-induced variations in the intragrain critical current density in $YBa_2Cu_3O_7$ which are oriented such that $c\|H$, it is suggested that the strength of the pinning in the oxide is strongly influenced by the thermodynamic properties of $YBa_2Cu_3O_7$. Thus, any attempts to introduce additional pinning sites must be done without lowering the value of H_c. Small precipitates of Y_2BaCuO_5 in $YBa_2Cu_3O_7$ may be such a candidate, as pointed out by Murakami.[43] It is also interesting to note that processing procedures applied to produce fine pinning centers did not result in a large decrease in H_c in the conventional superconductors suchas Nb_3Sn when their values of T_c were reduced by small amounts in the process.

ACKNOWLEDGMENTS

This work was performed under the auspices of the U.S Department of Energy, Division of Materials Sciences, Office of Basic Energy Sciences under Contract No. DE-AC02-76CH00016.

REFERENCES

1. For example, see T. R. Dinger, T. K. Warthington, W. J. Gallagher, and R. L. Sandstrom, Phys. Rev. Lett. 58, 2687 (1987).
2. For example, see Y. Enomoto, T. Murakami, M. Suzuki, and K. Moriwaki, Jpn. J. Appl. Phys. 26, L1248 (1987).
3. M. Suenaga, A. Ghosh, T. Asano, R. L. Sabatini, and A. R. Moodenbaugh, in Proc. Mater. Res. Soc. Meeting, Anaheim, April, 1987.
4. D. C. Larbalestier, M. Daeumling, X. Cai, J. Seuntjens, J. McKinnell, D. Hampshire, P. Lee, T. Willis, H. Muller, R. D. Ray, R. D. Dillenburg, E. E. Hellstrom, and R. Jount, J. Appl. Phys. 62, 3308 (1987).
5. J. W. Ekin, A. I. Braginski, A. J. Panson, M. A. Janocko, D. W. Capone, II, N. J. Zaluzec, B. Flandermeyer, O. F. de Lima, M. Hong, J. Kwo, and S. H. Liou, J. Appl. Phys. 62, 4821 (1987).
6. S. Jin, T. H. Tiefel, R. C. Sherwood, M. E. Davis, R. B. Van Dover, G. W. Kammloff, R. A. Fastnacht, and H. D. Keith, Appl. Phys. Lett. 52, 2075 (1988).
7. M. Murakami, M. Morita, K. Miyamoto, and S. Matsuda, in Proc. Intern. Symp. on High T_c Superconductors, Osaka, Oct. 1988, to be published.
8. Youwen Xu, unpublished. These data were taken from "c" axis oriented powder specimens with applied fields parallel to the c axis. Thus, the induced current circulates in the a-b plane.
9. R. Wordenweber, G. V. S. Sastry, K. Heinemenn, and H. C. Freyhardt, J. Appl. Phys. 65, 1649 (1989).
10. G. Zerweck, J. Low Temp. Phys. 42, 1 (1981).
11. D. O. Welch, IEEE Trans. Magn. MAG-21, 827 (1985).
12. E. V. Thuneberg, J. Low Temp. Phys. 57, 415 (1984).
13. A. Pruymboom and P. H. Kes, Jpn. J. Appl. Phys. 26, Suppl. 26-3, 1533 (1987).
14. K. Yamafuji, T. Fujiyoshi, K. Toko, and T. Matsushita, Physical C 159, 743 (1989).
15. B. Serin, in Superconductivity, R. D. Parks, editor, p. 925, Marcel Dekker, New York, 1969.
16. Youwen Xu, M. Suenaga, A. R. Moodenbaugh, J. L. Peng, and R. N. Shelton, Mater. and Mech. of High T_c Superconductivity, Stanford, CA 1989, to be published.
17. A. Junod, D. Ekert, T. Graf, G. Triscone, and J. Muller, Mater. and Mech. of High T_c Superconductivity, Stanford, CA 1989, to be published.
18. F. J. Blunt, A. M. Campbell, P. P. Edwards, J. E. Evetts, P. Freeman, J. Johnson, J. Loram, K. Mirza, A. Putnig, E. Salije, and W. Schmall, Mater. and Mech. of High T_c Superconductivity, Stanford, CA 1989, to be published.
19. S. Kim, R. A. Fisher, N. E. Phillips, and J. E. Gordon, Mater. and Mech. of High T_c Superconductivity, Stanford, CA 1989, to be published.
20. K. S. Athreya, O. B. Hynn, J. E. Ostenson, J. R. Clem, and D. K. Finnemore, Phys. Rev. B 38, 1846 (1988).
21. K. S. Athreya, S. C. Sanders, D. Hofreiter, D. K. Finnemore, and Youwen Xu, M. Suenaga, Phys. Rev. B, to be published.
22. T. T. M. Palstra, B. Batlogg, R. B. van Dover, L. F. Schneemeyer, and J. V. Waszczak, Appl. Phys. Lett. 54, 763 (1989).

23. J. D. Hettinger, A. G. Swanson, W. J. Skocpol, J. S. Brooks, J. M. Graybeal, P. M. Mankiewich, R. E. Howard, B. L. Straughn, and E. G. Burkhardt, Phys. Rev. Lett. <u>62</u>, 2044 (1989).
24. E. Zeldov, N. M. Amer, G. Koren, A. Gupta, R. J. Gambino, and M. W. McElfresh, Phys. Rev. Lett. <u>62</u>, 3093 (1989).
25. Y. Yeshurun and A. P. Malozemoff, Phys. Rev. Lett. <u>60</u>, 2202 (1988).
26. A. P. Malozemoff, L. Krusin-Elbaum, D. C. Cronemeyer, Y. Yeshurun, and F. Holtzberg, Phys. Rev. B <u>38</u>, 6490 (1988).
27. R. Griessen, C. F. J. Flipse, C. W. Hagen, J. Lensink, B. Dam, and G. M. Stollman, J. Less-Comm. Metals <u>151</u>, 39 (1989).
28. Y. Iye, S. Nakamura, and T. Tamegai, Physica C <u>159</u>, 433 (1989).
29. Youwen Xu, M. Suenaga, A. R. Moodenbaugh, and D. O. Welch, Phys. Rev. B (1989), to be published.
30. Y. B. Kim, C. F. Hempstead, and A. R. Strnad, Phys. Rev. <u>129</u>, 528 (1963).
31. M. R. Beasley, R. Labush, and W. W. Webb, Phys. Rev. <u>181</u>, 682 (1969).
32. M. Tinkham and C. J. Lobb, Solid State Phys. Vol. 42.
33. C. W. Hagen, R. Friessen, and E. Salomons, Phys. Rev. B, to be published.
34. M. Suenaga and M. Murakami, unpublished.
35. D. O. Welch, unpublished.
36. M. Suenaga, A. K. Ghosh, and D. O. Welch, unpublished.
37. D. R. Nelson, Phys. Rev. Lett. <u>60</u>, 1973 (1988).
38. P. H. Kes, Physica C <u>153-5</u>, 1121 (1988).
39. T. Matsushita, K. Fumaki, M. Takeo, and K. Yamafuji, Jpn. J. Appl. Phys. <u>26</u>, L1524 (1987).
40. Yimei Zhu, M. Suenaga, and Youwen Xu, Philos. Mag. Lett. <u>60</u>, 51 (1989).
41. Y. Zhu, M. Suenaga, Youwen Xu, R. L. Sabatini, and A. R. Moodenbaugh, Appl. Phys. Lett. <u>54</u>, 374 (1989).
42. Youwen Xu, M. Suenaga, J. Tafto, R. L. Sabatini, A. R. Moodenbaugh, and P. Zolliker, Phys. Rev. B <u>39</u>, 6667 (1989).
43. M. Murakami, private communication.

QUANTUM PERCOLATION THEORY OF HIGH-T_c SUPERCONDUCTORS

J. C. Phillips
AT&T Bell Laboratories
Murray Hill, NJ 07974

INTRODUCTION

I have developed an entirely new theory of the electronic properties of layered cuprates, like LSCO, YBCO, BCSCO, which explains their anomalous properties in both the normal and superconductive states. It is not possible to describe this theory in a few pages here. However, to lead the reader to a better understanding of its salient features, I present below selected material from my papers and book.

Why Localized and Extended Impurity Band States Can Coexist and Be Separated
(Solid State Communications, Vol. 47, No. 3, pp. 191, 1983.)

It has been assumed that in the impurity band metallic regime there exists an energy E_c below which all states are localized and above which all states are extended. This discontinuous model contradicts the results of recent experiments. Using the quantum theory of measurement I resolve these difficulties as my title indicates.

Separable Spectral Model of the Metal–Insulator Transition in Si:P (Phil. Mag. B 47, No. 4, 407-418, 1983).

The two-fluid spectral model based on separable localized and extended state predicts that near but above threshold the impurity-band conductivity, σ, of a randomly doped semiconductor is proportional to $(\rho - \rho_c)^s$ where ρ_c is the critical density and $s(d) = (d-2)/2$ where d is the dimensionality. Localization, quantum and scaling theories have predicted $s(3) \sim 1$, $=0$ and $\geq 2/3$, respectively, whereas the present model gives $s(3) = \frac{1}{2}$, in agreement with recent experiments which give $s(3) = 0.48 \pm 0.07$. The theory is extended to predict $s^K(d)$, the conductivity index for compensated samples. The result is $s^K(d) = 1(\frac{1}{2})$ for d>2 and d odd (even). This again is in good agreement with experiment, which gives $s^K(3) = 1 \cdot 15 \pm 0.15$.

Giant Defect–Enhanced Electron–Phonon Interactions in Ternary Copper Oxide Superconductors (Phys. Rev. Lett. 59, 1856 (1987))

A microscopic but schematic model is discussed for defects (such as oxygen vacancies $O^{\square}$) in compounds such as $YBa_2Cu_3O_{7+\delta}O^{\square}_{2\frac{1}{2}-\delta}$ and

$La_{3-x}Ba_{3+x}Cu_6O_{14+\delta}O^7_{4-\delta}$ in relation to the superconductive properties of these materials. The discussion shows in detail how T_c can reach its maximum value at or near the metal-semiconductor transition T_{ms}. It introduces a new mechanism for enhancement of electron-phonon interactions that is separate from, and empirically superior to, Fermi-surface nesting.

Physics of High–T_c Superconductors (Academic Press, Boston 1989, p. 150)

One of the most striking aspects of high-T_c oxide structures is that increases in T_c occur as the layered character of the structure becomes more pronounced. The study of electronic conductivity and the nature of electronic states in the presence of substantial disorder has led, since the pioneering early work of Mott and Anderson, to the conclusion that the ability of electrons to carry metallic current ballistically (rather than by diffusive hopping) in strongly disordered systems (such as impurity bands) is essentially dependent on dimensionality. Because the present materials are often so close to the metal-semiconductor transition in terms of their normal-state conductivity $\sigma(T)$ and $d\sigma/dT_m = 0$ for T_m close to T_c in many cases (III.7-III.9), these questions of electron localization are of critical importance. Here we are far from good metals ("weak" localization), and are instead concerned with "strong" localization.

The essential state of the theory at present is that there is metallic conductivity for dimensionality d=3, but that all states are localized for d=1. Dimensionality d=2 is marginal, and relatively small structural changes can have drastic effects on the existence of extended electronic states which are capable of carrying ballistic metallic currents. This viewpoint is highly suggestive for layered superconductive oxides, as we shall now see.

Direct Evidence for the Quantum Interlayer Defect Assisted Percolation Model of Cuprate High–T_c Superconductivity (Phys. Rev. *B39*, 7456 (1989))

Recent Raman and infrared data are used to argue that in $YBa_2Cu_3O_7$ metallic paths are not confined to Cu planes but must cross Ba planes at some points, possibly because of the presence of defects such as oxygen vacancies near those points.

Quantum Percolation and Lattice Instabilities in high–T_c Cuprate Superconductors (Phys. Rev. B 1 Nov. 1989)

Lattice instabilities are the mechanisms that place upper bounds on T_c in theoretical models of superconductivity based on dynamical electron-phonon interactions. In percolative metals this upper bound can be much larger than in ballistic metals. For parameters appropriate to high-T_c cuprates, I show that the maximum value λ_{max} of the percolative electron-phonon coupling constant is about four times larger than the value $\lambda_{max} \sim 1$ appropriate to ballistic metals. I also review values of λ in "anomalous" ballistic metals, such as Nb_3Ge homoepitaxial films, and show that these may be defect enhanced.

Reconciliation of Normal-State and Superconductive Specific Heat, Optical, Tunneling and Transport Data on YBCO (Phys. Rev. *B40*, 7348 (1989)).

Quantum percolation theory reconciles a variety of data on $YBa_2Cu_3O_{7-\delta}$, and in particular explains the linear temperature dependence of planar resistivity in the normal state.

THE FABRICATION AND APPLICATION OF HIGH T_C SUPERCONDUCTING THIN FILMS

R.B. Laibowitz

IBM Research Division
P.O. Box 218
Yorktown Heights, NY 10598

Many techniques have been used to produce high quality, high T_c superconducting films[1]. Some of the more popular techniques include laser ablation, a variety of sputtering techniques and electron beam vapor deposition. Resistance heating of the source material is also used in conjunction with these techniques, e.g. electron beam heating. These methods of film preparation have been applied to the three main classes of high T_c materials, i.e. $YBa_2Cu_3O_{7-y}$ (YBCO), Tl-based and Bi-based films. While there are several ways of evaluating these films, a reasonable measure of the quality of the material is the value of the critical current density (J_c) in zero magnetic field. J_c (measured magnetically and by transport) appears to depend strongly on fabrication parameters but in general the highest values are obtained for epitaxial films of YBCO on polished, single crystal, (100) oriented substrates of $SrTiO_3$. At 4 K values as high as about $5 \times 10^7 A/cm^2$ have been obtained and at 77 K, J_c drops by about a factor of 10. Polycrystalline films on other substrates generally exhibit much lower currents. Thus, choice and quality of the substrate is still quite important in determining film quality.

In order to consider using these films in a variety of thin film applications, the reproducibility of the J_c and ease of fabrication (or reliability) are also important considerations. Laser ablation[2-4] has recently demonstrated these properties although such films can be very rough due to particulates formed during the deposition[5]. Thus, another important criterion for the high T_c films is surface smoothness. In fact, much work continues to be needed in order to control and understand the surface properties of the high T_c films. There is also a need for compositional and thickness uniformity over reasonably large areas (several cm^2).

Thin film applications involving circuits and devices will require lithographic patterning and superconducting materials that do not deteriorate as a result of the processing. We have been mostly interested in developing thin films SQUIDs (Superconducting QUantum Interference Device) from the high T_c materials. These devices consist of a loop of superconductor containing one or two weak links which exhibit Josephson coupling[6,7]. The single level SQUIDs that we have produced generally have line widths as small as 2 um. A substractive technique is used wherein the selected film areas are protected by resist and the unwanted regions are ion milled using Ar at a pressure of 8×10^{-5} torr. In this way, single level SQUIDs have been fabricated from both YBCO and Tl-based films. While these SQUIDs have very good sensitivity to magnetic flux, the weak link in these devices is not as yet well understood as it appears across naturally occurring grain boundaries in the films themselves.

Recently we have attempted to fabricate tunnel junction-type devices, both as single junctions and as SQUIDs[8]. As expected such efforts required extending the technology to multilevel processing, adding many steps to the fabrication. These all high T_c (YBCO) junctions have been formed in cross stripe, in-line and edge junction geometries. We have used plasma oxidation and oxy-flouridation to attempt to form a barrier between the two superconducting electrodes. Deposited barriers such as MgO have also been attempted. Supercurrents have been observed across the barrier regions in both the single junctions and the SQUIDs which modulate in a magnetic field. The SQUIDs show good sensitivity to magnetic field but noise measurements have not yet been made on these devices. As yet it has not been possible to identify higher voltage structure with such phenomena as a superconducting energy gap. I-V characteristics that we have measured on the all high T_c junctions do not show the sharp quasiparticle rise as observed in low T_c materials. Much more effort is needed on determining the quality of the superconductor-barrier interface both as deposited and after high temperature processing. Of course, it is important to develop techniques which lower the maximum temperature to which the barrier is exposed. Thus it would be very desirable in future work to have more information about the surfaces (internal and external) of the high T_c oxide superconductors in both single level and multilevel structures.

REFERENCES

1. R.B. Laibowitz, High T_c Superconducting Thin Films, Materials Research Society MRS Bull., XIV:58 (1989), and references therein.
2. X.D. Wu, A. Inam, T. Venkatesan, C.C. Chang, P. Barboux, J.M. Tarascon and B. Wilkens, Low Temperature Preparation of High T_c Superconducting Thin Films, Appl. Phys. Lett., 52:754 (1988).
3. G. Koren, A. Gupta, E.A. Giess, A. Segmuller and R.B. Laibowitz, Epitaxial Films of $YBa_2Cu_3O_{7-y}$ on $NdGaO_3$, $LaGaO_3$ and $SrTiO_3$ Substrates deposited by Laser Ablation, Appl. Phys. Lett., 54:1054 (1989).
4. T.R. McGuire, A. Gupta, G. Koren, R.B. Laibowitz and D. Dimos, Magnetic Study of Superconductivity in YBCO Thin Films, Proc. Int. Conf. Mat. and Mech. of Superconductivity High T_c Supercond., Stanford, July, 1989, to be published in Physica C.
5. G. Koren, A. Gupta, R.J. Baseman, M.I. Lutwyche and R.B. Laibowitz, Laser Wavelength Properties of YBCO Thin Film Deposited By Laser Ablation, to be published Appl. Phys. Lett., Dec. 1989.
6. R.H. Koch, C.P. Umbach, G.J. Clark, P. Chaudhari and R.B. Laibowitz, Quantum Interference Devices Made From Superconducting Oxide Thin Films, Appl. Phys. Lett., 51:200 (1987).
7. R.H. Koch, W.J. Gallagher, B. Bumble and W. Lee, Low Noise Thin Film TlBaCaCuO dc SQUIDs Operated at 77 K, Appl. Phys. Lett., 54:951 (1989).
8. R.B. Laibowitz, R.H. Koch, W.J. Gallagher, V. Foglietti, J.M. Viggiano, G. Koren, A. Gupta, M.I. Lutwyche and B. Oh, to be published.

PHYSICS OF IN-SITU OXIDE SUPERCONDUCTING THIN FILMS DEPOSITION

H.S. Kwok, D.T. Shaw, Q.Y. Ying, J.P. Zheng, H.S. Kim and N.H. Cheung[*]
Institute on Superconductivity
State University of New York at Buffalo
Bonner Hall, Amherst, NY 14260

ABSTRACT

The physics of the in-situ laser deposition process is reviewed. Emphasis will be placed on the use of excimer lasers, and the deposition of Y-Ba-Cu-O thin films. It is shown that the laser target interaction conditions, and the properties of the laser generated plume are inducive to the formation of high quality films. In addition to the formation of energetic atomic beams, laser deposition is also highly compatible with reactive deposition which make it suitable for oxide and nitride films. An in-situ diagnostic technique is introduced which is capable of detecting the interfacial boundary layer between the film and the substrate. It is also shown that post-deposition in-situ oxidation is necessary for the formation of superconducting films..

1. INTRODUCTION

Laser deposition is by now a well-established technique for the growth of thin films. Materials ranging from insulators to semi- and superconductors have been deposited with high quality [1-9]. Even though commercial laser deposition systems are not yet available, home-made varieties have proliferated in various research laboratories recently because of the many attractiveness of this novel deposition scheme. Laser deposition has come a long way since the early days of giant pulse lasers, in terms producing useful films [1,2].

Fig. 1 shows the schematic of a typical laser deposition system. A pulsed laser is focused onto a target to evaporate materials out of it. Both long wavelength lasers such as CO_2 and Nd:YAG, and short wavelength ones such as excimers and harmonics of the Nd:YAG laser have been used. In this paper we shall concentrate on the latter since it has been proven to produce better results. Typically, an energy fluence of 1-5 J/cm^2 is used. The substrate is usually held at an elevated temperature for improved film quality. Typical values are 450-650°C for the superconducting films.

---

[*] Permanent address: Department of Physics, Hong Kong Baptist College.

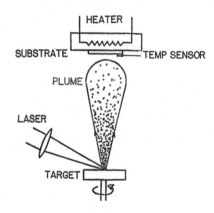

Fig. 1 Schematic diagram of the basic laser evaporation deposition (LEDE) system.

For the deposition of oxide or nitride films, an ambient of oxygen or nitrogen is usually provided to enhance oxidation or nitridation of the thin film. Various schemes, such as plasma-assist [9] or ion-assist can be incorporated as well. Even though laser deposition has been quite successful in providing good films, the physics of this process has not yet been fully understood. It is the intention of this paper to review some recent results aimed at understanding this process, and perhaps shed some lights for further optimization. Several excellent review papers have also appeared recently [10].

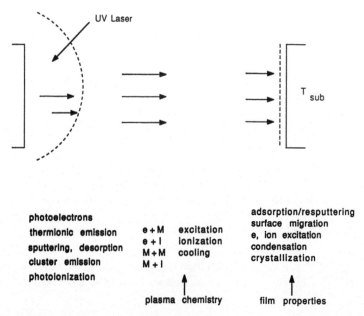

Fig. 2 Schematic diagram of the LEDE processes. There are 3 separate regions of interactions.

It should be emphasized that the term pulsed laser deposition (PLD) here means physical deposition, as opposed to laser chemical vapor deposition (LCVD). LCVD is an important thin film deposition technique which is very different from PLD. Another term for PLD is perhaps the more suggestive laser evaporation deposition (LEDE) [8,11].

Fig. 2 schematically represents some of the processes that can occur during the laser deposition process. Roughly speaking, the thin film deposition process can be divided into three regions of interaction: (1) laser-target interaction at the target surface, (2) freely expanding plasma plume region where ion-atom chemistry occurs, and (3) thin film formation at the substrate surface. We shall discuss the processes occurring in these three different regions separately. Finally, we shall discuss a novel method for the in-situ diagnostics of the laser deposition process. This method can be extended easily to other deposition techniques where a high pressure ambient gas is present.

2. LASER-TARGET INTERACTION ZONE-PLASMA DYNAMICS

In the laser-target interaction zone, which extends to perhaps 1 mm above the target surface, the usual laser-target interaction physics occurs. The situation is common to all branches of laser-materials interactions studies, such as laser damage, laser fusion, laser drilling and welding. Essentially, a dense plasma is formed during the initial part of the laser pulse. The plasma can be due to thermionic emission, desorption, evaporation, photoelectric effects etc. Normally, for Nd:YAG and longer wavelength lasers, this plasma shields the material from the laser by plasma screening. Therefore the remaining major portion of the laser pulse interacts only with this plasma. Interaction with the target will only occur via this plasma as the intermediate. However, for the excimer laser, the situation is considerably different. For the plasma to become overdense, the plasma density has to exceed the critical density N_c, given by

$$N_c = m\omega_L^2/4\pi e^2 \qquad (1)$$

where m,e are the electronic mass and charge, and ω_L is the laser frequency. At 248 nm, N_c is $1.8 \times 10^{22} cm^{-3}$, which is not too different from the material atomic density. For the ArF laser, N_c is $3.0 \times 10^{22} cm^{-3}$. Hence, it can be said that in excimer laser deposition, the laser interacts with the target during the entire laser pulse, with little plasma attenuation.

Besides continuous heating and evaporation of the target for the case of short wavelength lasers, the other major difference is that the plasma tends to be "colder", and less explosive. Fig. 3 shows the emission spectrum of the plasma with a $YBa_2Cu_3O_7$ target using the excimer laser (upper trace) and the Nd:YAG laser (lower trace). The two spectra are normalized at the peak of the Ba II ion. It can be seen that the excimer laser induced spectrum consists of much more neutral species than the case of the 1.06 μm laser [11]. A simple explanation can be offered for this observation. The laser interacts mostly with electrons, and the energy absorbed by the electrons in the plasma is given by [12]

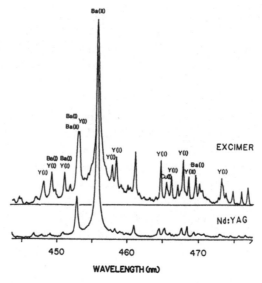

Fig. 3 Emission spectra of YBCO laser plume for Nd:YAG and excimer (ArF) laser excitations.

$$\frac{d\varepsilon}{dt} = \frac{eE^2 \omega_c}{m(\omega_L^2 + \omega_c^2)} \qquad (2)$$

where E is the rms laser electric field and ω_c is the collision frequency in the plasma. It can be seen that the energy absorbed by the electrons is much smaller for the case of short wavelength lasers. Hence electron impact ionization, or cascade ionization is much smaller as well, giving rise to less ions in the plasma. The same argument can be presented in terms of the equivalent quiver energy ε_Q of the photon field:

$$\varepsilon_Q = eE^2/m \omega_L^2 \qquad (3)$$

ε_Q can be estimated to be 1.21 meV for the KrF laser at 5 J/cm^2, the typical laser deposition energy. This ε_Q is much smaller than the typical ionization energy of the atoms in the plasma. Hence ionization in the excimer laser plasma is achieved mainly by photoionization. For comparison, ε_Q for the Nd:YAG laser is 22 meV. This is the main reason why the ions are more prominent for the Nd:YAG laser in Fig. 3.

Actually, the plasma dynamics occurring in the deposition system is very similar to the case of laser fusion, which has been studied intensively for obvious reasons. It was found that in laser implosion studies that long wavelength lasers tended to transfer too much energy to the plasma in front of the target, producing very high energy electrons (keV). Short wavelength lasers, on the other hand, can deliver more energy to the solid pellet, producing better implosion conditions. This is the main reason why excimer lasers and the fourth harmonic of the Nd:YAG laser are used exclusively for the laser fusion projects instead of the more energy efficient CO_2 laser.

In summary, in excimer laser deposition, the laser continuously heat up the target, with negligible plasma shielding. Materials are ejected or evaporated from the target throughout the laser pulse. The dense plasma formed in the space immediately above the target interacts with the laser only weakly, and therefore maintains its outward momentum and expands into free space. Because of this continual heating, evaporation is expected to be highly congruent, with stoichiometric composition in the plasma as in the target.

We should also make a comment here about particulates on the film. It is found generally that the films produced by excimer lasers are very smooth, with occasional ~ 1 μm large particles randomly dispersed on the surface. On the other hand, the long wavelength laser produced films are very rough and not shiny. The reason is related to the explosive nature of long wavelength laser-plasma interaction. Because of the strong plasma coupling, shock waves and other forms of detonation waves are produced inadvertently immediately above the target. This tends to crack the target surface and produce particles in the plasma. The excimer laser, for reasons explained above, is much more gentle in the evaporation process, thus produces less particles. The particles observed for the excimer laser case is actually due to liquid droplet ejected from the target [10]. The situation is akin to laser damage studies, where it was found that excimer lasers produce "soft" accumulative damages, while other longer wavelength lasers tend to create craters.

3. FREE EXPANSION ZONE

The materials ejects from the target expand into free space to form the main laser-plume. There are two considerations for this plume: the longitudinal velocity distribution and the angular spread. Near the target, the density is high, implying that collisions among different species can occur frequently. In the free space, beyond the Knudsten layer collisions become less frequent. The expansion of this plasma has been studied extensively in the past. Various hydrodynamic descriptions such as shock barrels, Knudtsen layers, Mach disks have been used to describe the boundary between the dense plasma and the free expansion zone [13].

Within the dense plasma, various collisions can occur

$e + h\nu + A$	$\rightarrow e^* + A$	inverse Bremstrahlung
$e^* + A$	$\rightarrow e + A^*$	excitation
$e^* + A$	$\rightarrow e + e + A^+$	ionization
$A^* + B$	$\rightarrow A + B^*$	energy transfer
$A + B$	$\rightarrow A + B$	elastic collisions

Here A, B stand for atomic species in the plasma. The last process is the most important one in determining the direction of travel for the various species. Since fast atoms will collide with slow atoms, there is always a tendency in the plasma for the velocities to equalize. Also, because electrons have little momentum, their function is mostly in providing internal excitation energy and ionization, rather than determining the velocity distributions. Notice that the unit of energy exchange between the photon field and the electron is the quiver energy ε_Q, rather than $h\nu$.

For reactive deposition, such as the formation of Y-Ba-Cu-O films, an ambient oxygen environment is present. These oxygen molecules are affected by the expanding plasma via many different kinds of collisions:

$e^* + O_2$	$\rightarrow e + O + O$	dissociation
$e^* + O_2 + A$	$\rightarrow O_2^- + A$	attachment
$e^* + O_2$	$\rightarrow e + e + O_2^+$	ionization
$A^* + O_2$	$\rightarrow A + O + O$	dissociation
$A^* + O_2$	$\rightarrow A + O_2^+ + e$	ionization
$A + O$	$\rightarrow A + O$	elastic collision

Effectively, the originally stagnant oxygen molecules become part of the expanding plasma plume. The last type of collision, involving only momentum exchange, slows down the Ba, Y and Cu atoms and speeds up the oxygen atoms in the Y-Ba-Cu-O case.

A time-of-flight study of the various species reveals the plasma expansion dynamics [14,15]. A detailed study shows that the velocities of the various atoms are quite fast, at ~ 10^6 cm/s. This translates to kinetic energies of 10-40 eV for the various atoms. The velocity distribution of these atomic beams can be modeled quite well using a free expansion model. This is the same principle as in the formation of supersonic molecular beams in chemical dynamic studies [16]. It can be shown that the velocity distribution can be described very well by

$$f(v) = A\, v^3 \exp(m(v - v_0)^2/2kT_s) \qquad (4)$$

where A is a normalization constant, v_0 and T_s are parameters describing the peak velocity and the effective temperature of the atomic beam. Fig. 4 shows a TOF spectrum of Cu and the theoretical fit.

Detailed studies of the velocity distributions of Ba, Y, Cu and O and their dependence on distance have been performed [15,16]. Basically the spatial dependence comes from constant longitudinal collisions of the various atoms along the beam direction, which tend to equalize their velocities. Of special interest is the velocity of the ions. Fig. 5 shows the change of the velocity of BaII and other species as a function of distance from the target. The behavior of the ion is intimately related to that of the electrons. Detail plasma dynamic calculations in principle can be carried out to model such changes. It should be pointed out that the electrons and ions can be manipulated by external electric potentials, and therefore can be used to change the characteristics of the other atomic beams. This happens, for example, in the case of plasma-assisted deposition [9].

While the longitudinal collisions tend to equalize the velocities of the species. The transverse collisions will tend to scatter the plume away from the main direction of travel. The angular distribution of the laser plume has also been studied by several authors [17,18]. It is generally agreed that the $\cos\theta$ fit suitable for the description of thermal evaporation, is not adequate here. Because of the intense transverse collisions, the angular distribution is generally narrowed and shows considerable forward peaking. A $\cos^n\theta$ fit is more appropriate, with n ranging from 4 to 48 [19].

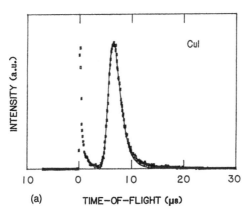

Fig. 4 Time-of-flight spectrum of Cu. The solid line is a theoretical fit using the supersonic expansion model.

For a multi-component system such as Y,Ba,Cu,O, the value of n is expected to be different for each component. The difference will depend on the collision cross section of the various species. This variation in angular distribution is detrimental to stoichiometric deposition, and also would affect film uniformity. Venkatesan et al [18] studied this problem for Y- Ba-Cu-O films and found that the stoichiometry is generally maintained for $15° < \theta < 15°$. Beyond that angle, Y tends to be less than Cu or Ba. The implication of this result to the scaling up of laser deposition to large areas is obvious.

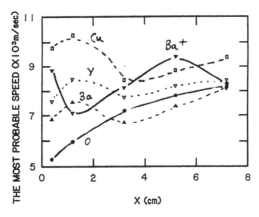

Fig. 5 Change of speed as a function of distance from the target. Note the distinct behavior of the ions.

In addition to emission spectroscopy, absorption spectroscopy can also be performed on the laser plume. It was revealed by TOF absorption spectroscopy that there exist two components in the velocity distribution - a fast one and a slow one. The fast component agrees with the emission TOF, while the slow component is not present in the emission spectrum. This is understandable because emission depends on the continuous excitation of the atoms via the reaction:

$$e^* + A \rightarrow e + A^*$$
$$A^* \rightarrow A + h\nu$$

This is because of the short fluorescence lifetimes (~ tens of ns) of most atomic lines. For the slow component, there is no energetic electrons around to provide the excitation. Hence no emission can be observed. This points to one major weakness of emission spectroscopy in characterizing the plasma plume.

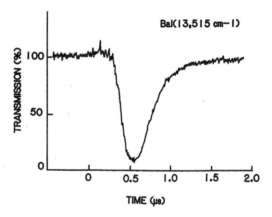

Fig. 6   Transmission time-of-flight of Ba atoms at 5 mm above the target. Note the deep absorption minimum.

Fig. 6 shows the absorption TOF spectrum of the plasma plume using a dye laser tuned to the 13,515 cm^{-1} absorption band of BaI. The slow tail is evident in that data. Geohegan et al used an incoherent source and observed similar TOF absorption spectra [20]. Another advantage of absorption spectroscopy of course is the ability to measure the absorption linewidth, which can be correlated to the plasma temperature.

4. THIN FILM FORMATION

The laser induced plasma plume consists of atomic beams of the various species, electrons and some amount of ions. Because of the congruent evaporation, it can be assumed that the atomic and ion beams arrive at the substrate surface with the correct stoichiometric proportions. The question of stoichiometric film formation, therefore, is whether the sticking coefficients are the same for all the species.

For Y,Ba,Cu,O it is obvious that O, having a much higher vapor pressure, will not stick as well as Y,Ba and Cu. As a matter of fact, the O atoms on the surface will combine with other O atom and form stable O_2 molecules and evaporate from the surface. Therefore, increasing the strength of the O- atomic beam disproportionately is important in maintaining a high O concentration on the film surface. Experimentally, it was found that filling the chamber with 5 mtorr of O_2 increases the O-atomic beam by 50 times, as compared to the O from the target. In-situ films could be formed under this situation. For O_2 pressures below 1 mtorr, in-situ films could no longer be produced.

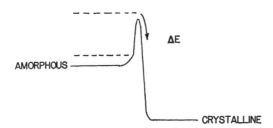

Fig. 7 Schematic diagram of the free energy of a layer of atoms the substrate.

A few generalities about thin film deposition can be mentioned here to help us appreciate the merits of laser deposition. In any method of deposition, atoms are laid on the surface of the substrate, and it is desired that somehow they will form the proper crystal structure. This problem is actually akin to chemical reactions where chemicals are put together and it is then wished that the proper reactants will result. Borrowing the terminology from chemical kinetics, we can plot an energy diagram for the film formation process. Fig. 7 shows the free energy of the new layer of atoms on top of a crystalline substrate. As the atoms arrive at the substrate, they would lie randomly on the surface. It is obvious that the free energy of the new layer of atoms is higher if they are amorphous rather than crystalline. This is because of the additional entropy in the amorphous structure. An activation energy ΔE is needed to bring the amorphous state to the crystalline state. This activation can be accomplished by heating the surface, since the reaction rate K is given by the Arhenius expression

$$K = K_0 \exp(-\Delta E/kT) \qquad (5)$$

There are other ways of nonthermally activating the reaction by increasing the mobility of the atoms on the surface. This can be accomplished, for example, by ion or electron irradiation, or by photo-excitation similar to photochemistry.

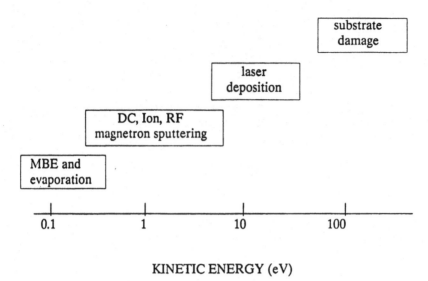

Fig. 8   Kinetic energy of emitted particles for various deposition techniques

This chemical picture of thin film formation should be used together with Fig. 8 in order to appreciate the advantage of laser deposition. In Fig. 8, we plot the kinetic energy of the particles (atoms) impinging on the substrate under several different deposition schemes. For large particle momentum higher than 100 eV, substrate damage will result and is therefore undesirable. It can be seen that the atoms from laser deposition are generally more energetic than those from other methods, but not energetic enough to cause substrate damage.

The combination of high substrate temperature and energetic particles enables the atoms to move around the surface and eventually form a crystalline lattice compatible with the substrate. For single phase or single component films, the crystallinity is unique. However, for Y-Ba-Cu-O and other multi-component films, there could be several stable crystal structures. Using the chemical picture in Fig. 7 again, it can be assumed that the phase with the lowest free energy would prevail. For the case of Y-Ba-Cu-O, this implies the tetragonal phase will be formed under the usual deposition temperature of 600°C. An in-situ diagnostic measurement of the thin film formation process which supports this conclusion will be discussed next.

5. IN-SITU DIAGNOSTICS OF THIN FILMS

In-situ diagnostics is paramount in any deposition system. For molecular beam epitaxy (MBE), reflective high energy electron diffraction (RHEED) is commonly used. However, for laser deposition, and reactive evaporation or sputtering, where there is a high pressure (mtorr) background, RHEED is difficult, if not impossible. Various diagnostic technique, such as using optical beams in reflection and transmission have been proposed [21]. In here, we shall discuss an in-situ resistivity measurement which is capable of measuring minute changes on the substrate surface.

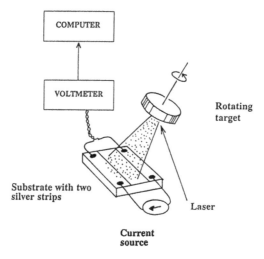

Fig. 9 Schematic set up of the in-situ resistivity measurement.

Fig. 9 shows the experimental setup. Basically the resistance of the film is measured as the film grows [22]. The use of a fast voltage measurement system allows the sampling of electrical resistance in less than 1 second. Fig. 10 shows the resistivity of a Y-Ba-Cu-O film grown on ZrO_2 starting from the time of deposition. It can be seen that the resistivity drops from a high value to a minimum of 0.1 mΩ-cm in less than 10 seconds, and then increases steadily as the film grows thicker. The most interesting feature is the rapid drop of the resistivity immediately after deposition, when the chamber is immediately filled with 100 torr of O_2. It demonstrates that the film deposited is not the superconducting phase. The in-situ post annealing is responsible for forming the final proper orthorhombic phase.

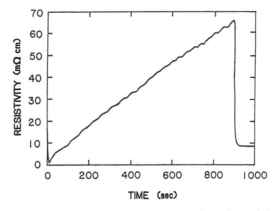

Fig. 10 Resistivity of a YBCO film on YSZ as a function of deposition time. The film is about 1000 Å thick.

The behavior of the resistivity curve is quite different for other substrates. Even though the final drop due to the 100 torr O_2 in-situ anneal is the same in all cases, the change of ρ during the deposition is different. This difference can be used to study the interfacial compound for different substrates. Assuming that the interfacial compound has a resistivity ρ_1 and thickness t_1, and the deposited YBCO film has a resistivity ρ_2 and a thickness t_2, then the measured resistivity is given by

$$\rho = \rho_1 \rho_2 (t_1 + t_2) / (\rho_1 t_2 + \rho_2 t_1) \tag{6}$$

The time dependence of ρ can be obtained by letting $t_2 = \alpha t$ where α is the deposition rate.

From eq.(6) depending on the relative values of ρ_1 and ρ_2, the temporal behavior of ρ measured can be quite different. There are 4 cases:

(A) No interface layer, $t_1 = 0$
$$\rho = \rho_2 \tag{7}$$

(B) $\rho_1 << \rho_2$ for all t_2
$$\rho = \rho_1 (t_1 + t_2)/t_1 \tag{8}$$

(C) $\rho_1 >> \rho_2$
$$\rho = \rho_2 (t_1 + t_2)/t_2 \tag{9}$$

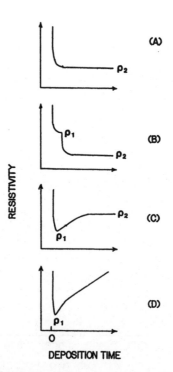

Fig. 11 Expected behavior of ρ vs deposition time for
(A) $t_1 = 0$, (B) $\rho_1 > \rho_2$, (C) $\rho_1 < \rho_2$ and (D) $\rho_1 << \rho_2$.

(D) $\rho_1 \geq \rho_2$

ρ = case (C) for small t_2
= case (B) for large t_2

Fig. 11 shows the expected behavior of ρ for these 4 cases. It can be seen that Fig. 10 corresponds to the case of (B). Other substrates can also be described by the models in Fig. 11. For example, it was found that SrTiO$_3$ is described by (D), and MgO can be described by (C).

It can be seen that in-situ resistivity measurement can be a very sensitive measure of the interface compound, and the quality of the deposited YBCO film. Other studies can also be performed using this in-situ technique, such as measurement of the tetragonal to orthorhombic transformation rate.

6. CONCLUSION

In this paper, we have discussed some of the detailed physics that can occur during the pulsed laser deposition process. In general, the plasma dynamics is highly dependent on the laser source and operating conditions. In turn, the quality of the thin film is affected by the atomic beam source. For such a complicated system, many parameters have to be optimized simultaneously. We have touched upon surface activation by e-beam and ion beams. Clearly, much remains to be done in terms of completely characterizing laser deposition to improve its reproducibility. It is expected that laser deposition will rank among other thin film techniques as a viable growth method for most insulators and superconductors. For semiconductors, MBE and MOCVD may still be the method of choice, for a while.

This research was supported by the New York State Institute on Superconductivity.

REFERENCES

1. H.W. Smith and A.F. Turner, Appl. Opt. 4, 147(1965).
2. V.S. Ban, D.A. Kramer, J. Mat. Sci. 5, 978(1970).
3. V.G. Dneprovskii and B.A. Osadin, Sov. Phys. Tech. Phys. 19, 275(1974).
4. J.T. Cheung and D.T. Cheung, J. Vac. Sci. Tech. 21, 182(1982).
5. J.J. Dubowski and D.F. Williams, Thin Solid Films, 117, 289(1984).
6. M.I. Baleva, M.H. Maksimov, S.M. Metev and M.S. Sendova, J. Mat. Res. Lett. 5, 533(1986).
7. X.D. Wu, D. Dijkkamp, S.B. Ogale, A. Inam, E.W. Chase, P.F. Miceli, C.C. Chang, J.M. Tarascon and T. Venkatesan, Appl. Phys. Lett. 51, 861(1987).
8. H.S. Kwok, J.P. Zheng, S. Witanachchi, L. Shi and D.T. Shaw, Appl. Phys. Lett. 52, 1815(1988).
9. S. Witanachchi, H.S. Kwok, X.W. Wang and D.T. Shaw, Appl. Phys. Lett. 53, 234(1988).
10. J.T. Cheung and H. Sankur, CRC Critical Reviews in Solid State and Materials Sciences, 15, 63(1988).

11. H.S. Kwok, P. Mattocks, D.T. Shaw, L. Shi, X.W. Wang, S. Witanachchi and J.P. Zheng, Appl. Phys. Lett. 52, 1825(1988).
12. I.P. Shkarofsky, RCA Review, 35, 49(1974).
13. R. Kelley and R.W. Dreyfus, Surface Science, 198, 263(1988).
14. J.P. Zheng, Z.Q. Huang, Q.Y. Ying, D.T. Shaw and H.S. Kwok, Appl. Phys. Lett. 54, 280(1989).
15. J.P. Zheng, Z.Q. Huang, Q.Y. Ying, S.Witanachchi and D.T. Shaw, Appl. Phys. Lett. 54, 954(1989).
16. J.B. Anderson, R.P. Andres and J.B. Fenn, Adv. Chem. Phys. 10, 275(1966).
17. A. Namiki, T. Kawai and K. Ichige, Surface Science, 166, 129(1986).
18. T. Venkatesan, X.D. Wu, A. Inam and J.B. Wachtman, Appl. Phys. Lett. 52, 1193(1988).
19. R. Kelly in **Photochemistry in Thin Films**, p.258, ed. by T.F. George, SPIE Proc. vol.1056, 1989.
20. D.B. Geohegan and D.N. Mashburn, presented at Third Annual Buffalo Conference on Superconductivity and Applications, Buffalo, NY 1989.
21. M. Fraeastoro-Decker, J.S. Ferreira, N. Gomes and F. Decker, Thin Solid Films, 147, 291(1989).
22. Q.Y. Ying, H.S. Kim, D.T. Shaw and H.S. Kwok, Appl. Phys. Lett. 55, 1041(1989).

TAILORED THIN FILMS OF SUPERCONDUCTING Bi-Sr-Ca-Cu OXIDE PREPARED BY THE INCORPORATION OF EXOTIC ATOMS
Superconductivity and the distance between CuO_2 layers

Hitoshi TABATA[*] and Tomoji KAWAI

The Institute of Scientific and Industrial Research
Osaka University, Mihogaoka, Ibaraki, Osaka 567
[*] Technical Institute, Kawasaki Heavy Industries Ltd.
Kawasakicho, Akashi, Hyogo 673

(ABSTRACT)
Layer-by-layer successive deposition utilizing laser ablation has been applied to the site-selective substitution of +1, +2 and +3 ions having different ionic radii at the Ca and Sr site of $Bi_2Sr_2Ca_1Cu_2O_y$ superconducting films. The substitutions at Sr and Ca site exhibit quite different behaviors in the superconductivity and the structural parameters. For the substitution of larger ions at the Ca site, lattice parameter c increases showing the higher Tc value, while the substitution of at the Sr site does not elongate the c axis and does not increase the Tc. These changes have been explained by the structure models, and the correlation between Tc and the lattice parameters is explained by the changes of the distance between CuO_2 layers, indicating the importance of the interaction between CuO_2 layers.

I. INTRODUCTION

In the copper-oxide based superconductor, the essential structural feature is CuO_2 sheets. The high Tc superconductivity appears by doping of charge carriers into the CuO_2 sheets. Positive ions, such as Y, Ba, Bi, Sr, Ca and Tl, are playing roles to construct structural framework and to control the amount of charge carriers in the superconductor. Thus, the control of the structures based on the CuO_2 sheets together with the control of charge carriers are necessary for the elucidation and improvement of the superconductivity.

Fig. 1 shows important structural factors of the CuO_2 based structures for the example of $Bi_2Sr_2Ca_1Cu_2O_8$;
(1) the numbers of the CuO_2 layers, (2) the distance between CuO_2 layers, (3) the bond length of Cu-O and (4) the distance between Cu and O(apex). These factors can be controlled either by accumulation of CuO_2 layers or the insertion of exotic atoms with different ionic radii into Ca layer or SrO layer.

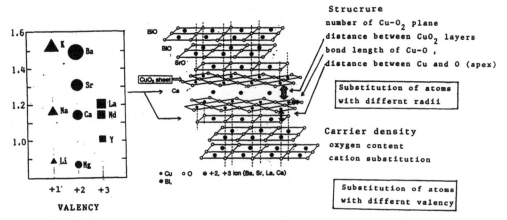

Fig. 1. Schematic structure of layered $Bi_2Sr_2Ca_1Cu_2O_8$ superconductor. Important structural factors based on CuO_2 sheets and carrier concentration can be changed by an incorporation of ions with different radii and valence state.

The numbers of carriers can be changed by the control of; (1) the amount of oxygen and (2) the substitution of the atoms with different valence state. This is performed by the incorporation of ions with different valence state into Ca, SrO or Bi_2O_2 layer.

For these controls of the composition at the atomic layer scale, we have demonstrated that layer-by-layer successive deposition is an excellent method in which each constituent layer is piled up on a substrate layer-by-layer to form the artificial crystal structure and compositions.[1,2]

In this paper, we will report on the site selective substitution of ions with different ionic radii and valence state taking advantage of layer-by-layer successive deposition. As shown in Fig.1, the ions we have chosen for the substitutions are , Mg, Ca, Sr and Ba, the 2+ ions, Li, Na and K, the 1+ ions, and Y, Nd and La, the 3+ ions. These ions are systematically substituted at either Ca site or Sr site, and the changes of the properties such as lattice parameter and the superconductivity are thoroughly studied. These "tailored superconducting films" have been proved to be useful to elucidate the mechanism of the superconductivity and be useful as a sophisticated technique for the device application of the superconductor.

II. STRUCTURAL MODELS FOR THE SITE SELECTIVE SUBSTITUTION AT Ca AND Sr SITES IN $Bi_2Sr_2Ca_1Cu_2O_8$ CRYSTAL

Fig.2(a) shows estimated changes of the lattice parameter c with the incorporation of Ba, the larger ion(r_{ion}=1.49Å), and Mg, the smaller ion(r_{ion}=0.86Å) at the Ca site. The Ca ion is osculating with the neighboring oxygen ions in the standard $Bi_2Sr_2Ca_1Cu_2O_8$ structure[3,4]. Consequently, for the incorporation of small Mg ion, the c axis is expected to shrink down, while for the incorporation of large Ba ion, expansion of the c axis is expected. (see Fig.2(a))

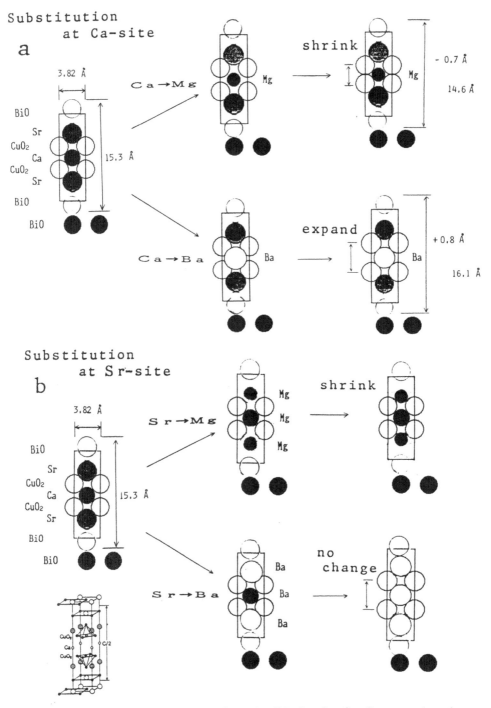

Fig. 2. Structure model of $Bi_2Sr_2Ca_1Cu_2O_8$, and the changes of the lattice parameter c by the incorporation of Mg and Ba. (a) substitution at the Ca site, (b) substitution at the Sr site. For Mg incorporation, the c axis is shrunk. For Ba incorporation at Ca site c axis becomes longer, while it does not change much for the Sr site substitution.

As shown in Fig. 2(b), on the other hand, it is reported from the crystallographic analysis of $Bi_2Sr_2Ca_1Cu_2O_8$ single crystal that there is a room between Sr and oxygen above, since close contact is achieved by Bi and oxygen beside them. Accordingly, it is expected that incorporation of small Mg ion may lead to the shrinking of the c axis, while the incorporation of Ba may give a small changes for the c axis length. (see Fig. 2(b))

The most important difference between the Ca site substitution and Sr site substitution is that large expansion of the distance between CuO_2 layers is expected for the Ba incorporation at the Ca site, while it is not expected for the incorporation at the Sr site. The present experiment is aimed to perform these site-selective substitution to see whether these expected crystallographic changes are achieved or not, and to see the correlations between these structural changes and the superconductivities.

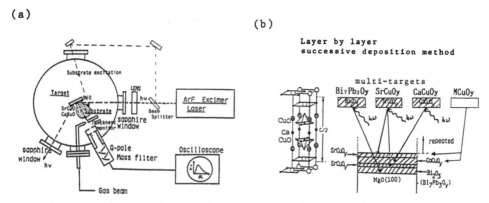

Fig. 3. Schematic representation of the apparatus for the laser deposition. The procedure for the layer-by-layer deposition using multi-targets to form layered Bi-Sr-Ca-Cu-O structure are also shown.

III. EXPERIMENTAL

The films are prepared by a layer-by-layer successive deposition method using a pulsed ArF excimer laser (193nm). (Fig. 3) The laser beam is sequentially focused on the multi-targets to form a film on a substrate placed at the opposite side in the presence of N_2O and/or O_2 gas atmosphere.[1] Typical experimental conditions are : energy density for ablation after focusing = 0.5 J/cm^2 , and repetition rate = 10 Hz. The substrate used was a MgO(100) single crystal heated at 580 - 600 °C.

The thickness of the thin film was measured by an oscillating film thickness monitor in situ, and the total thickness was measured by optical thickness monitor.

For the deposition, sintered disks of $Bi_7Pb_3O_y$, $Sr_1Cu_1O_y$, $Ca_1Cu_1O_y$, $Ba_1Cu_1O_y$ and $Y_1Cu_1O_y$ (nominal compositions) are used as targets, and the layers are successively deposited as we desire. These targets are prepared by calcining the mixture of Bi_2O_3, PbO, $SrCO_3$, $CaCO_3$, Li_2CO_3 Na_2CO_3, K_2CO_3, $MgCO_3$, $BaCO_3$, Y_2O_3, La_2O_3 and/or $NdCO_3$ with appropriate ratios. The $Bi_7Pb_3O_y$, NaCuO and KCuO target are prepared by heating in air at 630 °C for three hours, and the rest by heating at 800 °C for 10 hours.

III-1 Substitution at the Ca-site

Partial substitution of mono-(Li, Na and K), di-(Mg, Sr and Ba), and tri-valent(Y, La and Nd) ions for Ca has been examined. In this case, the deposition was carried out successively from the targets, BiPbO-SrCuO-CaCuO-MCuO-CaCuO-SrCuO-BiPbO (M=Li, Na, K, Mg, Sr, Ba, Y, La and Nd). This consisted of one cycle which was repeated. Typical irradiation periods for successive deposition are 25 s for $BiPbO_y$, 15 s for $SrCuO_y$, 10 s for $MCuO_y$ and 28 s for $CaCuO_y$.

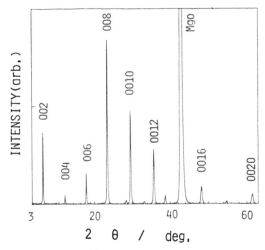

Fig. 4. Typical x-ray diffraction pattern for the Bi-Sr-Ca-Cu-O films after incorporation of Na^+ and annealed at 800C.

III-2 Substitution at the Sr-site

Partial substitution of the ions for Sr has been performed as for the Ca site: The sequence is similar to the Ca substitution, except that the deposit from the $MCuO_y$ (M=Li, Na, K, Mg, Ca, Ba, Y, Nd and La) is sandwiched between the deposits from $SrCuO_y$. Typical irradiation periods for the successive deposition are 25 s for BiPbO(A), 5 s for SrCuO(B) or MCuO(M) and 65 s for CaCuO(C).

IV. RESULTS AND DISCUSSION

IV-1. Partial substitution of Ca-site and Sr-site by the divalent ions: (Mg, Ca, Sr and Ba)

The partial substitution of Mg, Sr and Ba ions for Ca by about 30% shows broad X-ray diffraction patterns of double $Cu-O_2$ layered structure for all substituted atoms. After annealing at 800°C, the films showed x-ray pattern of the single phase of double CuO_2 layers structure with their c axis perpendicular to the substrate surface. (Fig. 4) No apparent impurity peaks have been observed, indicating that these atoms are incorporated into the BSCCO crystal structure.

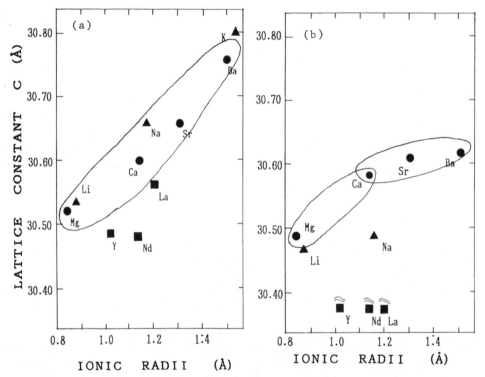

Fig. 5. (a) Lattice constant c versus ionic radii of the ions incorporated at the Ca site in the BSCCO films, and (b) lattice constant c versus ionic radii at the Sr site. The c axis is expanded most evidently for the Ba substitution at Ca site.

The lattice parameter c is found to be dependent on the ionic radii of the incorporated ions. (Fig. 5) When small Mg ion is incorporated at the Ca site, the c axis shrunk, as expected in the section II. On the other hand, with the incorporation of larger ions such as Sr and Ba, the lattice constant c increases.

That is, the c axis is expanded with the increase the ionic radii of the substituted ions. It would be reasonably understood that the distance between CuO_2 planes increases with inserting the large ions and decreases with small ions.

On the other hand, when the Sr site is substituted by the Mg, Ca and Ba ions, the lattice constant c decreases only in case of the small ion (Mg). The lattice constant c does not change much even by the substitution by the large ion like Ba. These behaviors have been expected as discussed in section II, and the differences are actually observed in the present site-selective substitution experiment.

IV-2. Substitution by mono- and tri-valent ions at Sr or Ca-site: (Li, Na, K , Y, Nd and La)

The changes of the lattice parameter c by the incorporation of +1 ions, Li, Na and K and +3 ions, Y, Nd and La have also been examined. (Fig. 5) The mono-valent ions show similar behavior to the divalent alkaline earth ions, that is , the larger ions leads to the larger lattice constant. The trivalent ions, on the other hand, show distinct shrinking of the c axis. This is presumably because highly positive valency works to attract negative oxygen ions.

IV-3. Superconductive properties of the tailored films.

These tailored films in which Sr or Ca ions are substituted by +1, +2 or +3 ions show systematic changes in the resistance-temperature curves. For example, the R-T curves for the incorporation of +2 ions, Mg, Ca, Sr and Ba, at the Ca site are shown in Fig. 6. It is clearly observed that the larger the substituted ion is, the higher the Tc_{mid} is. The highest Tc is observed for the Ba substitution, Tc_{onset} to be 90K and Tc_{mid} to be 83K. Accordingly the correlations are observed between the lattice constant c (Fig. 5) and the Tc, as shown in Fig. 7. For the substitution of Ba at the Sr site, we have shown in the previous paper that Tc becomes lower.[2] This is consistent with the smaller expansion of lattice constant c for the Sr site substitution as shown in Fig. 2 (b).

In this manner, the site-selective partial substitution of exotic atoms at the Sr and the Ca site shows drastic difference in the superconductivity. The substitution by larger ions at the Ca site make longer c axis, while that at the Sr site does not elongate the c axis. Accordingly, the spacing between double CuO_2 planes can be changed artificially by this technique, and we found that this spacing between CuO_2 planes has an important effect on the Tc value. It has been reported that there is a hopping of Cooper pairs between CuO_2 layers and this kind of interaction between layers is important for the Tc value.[5] The present experimental results indicate that the distance between the layers is really important as the determining factor of the Tc value.

(V) Conclusion
1. With the layer-by-layer successive deposition method utilizing laser ablation, site selective partial substitution of +1, +2 and +3 ions at the Ca and Sr site has been performed.

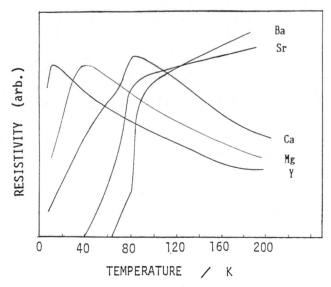

Fig. 6 Temperature-resistivity curves for the BSCCO films which contain various +2 ions at the Ca site.

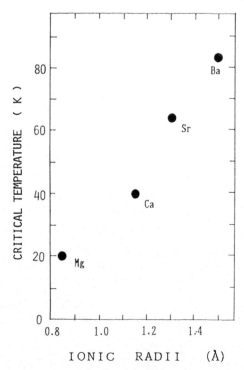

Fig. 7. Critical temperature versus ionic radii of the ions incorporated at the Ca site.

2. The incorporation of smaller +2 ion, **Mg**, at the Ca site makes the c axis shrink, while that of larger Sr and Ba ions makes the c axis longer. On the other hand, the incorporation of even Ba at the Sr site does not elongate the c axis so much.
3. This phenomenon is explained by the structure model in which there is a capacity to incorporate the large Ba ion at the Sr site.
4. The Tc value has correlation with the expansion of c axis due to the incorporation of exotic atoms. The larger lattice parameter c produces higher Tc.
5. It is suggested from these results that the spacing between CuO_2 planes is important structural factor for the Tc, suggesting the strong interaction between CuO_2 layers.

References

1) M. Kanai, T. Kawai, S. Kawai and H. Tabata: Appl. Phys. Lett., 54, 1802 (1989).
2) H. Tabata, T. Kawai, M. Kanai, O. Murata and S. Kawai, Jpn. J. Appl. Phys., 28, L823 (1989).
3) M. A. Subramanian, C. C. Torardi, J. C. Calabrese, J. Gopalakrishnan, K. Morrissey, T. R. Askew, R. B. Flippen, U. Chowdhry and A. W. Sleight, Science, 26, 1015 (1988).
4) K. Imai, I. Nkai, T. Kawashima, S. Sueno and A. Ono, Jpn. J. Appl. Phys., 27, L1661 (1988).

5) J. M. Wheatley, T. C. Hsu and P. W. Anderson, Nature, 333, 121 (1988).

8) M. Kanai, T. Kawai, M. Kawai and S. Kawai: Jpn. J. Appl. Phys., 27 (1988) L1293.

IN-SITU PROCESSING AND THEORETICAL MODEL FOR DEPOSITION OF LASER ABLATED HIGH-T_c $YBa_2Cu_3O_7$ SUPERCONDUCTING THIN FILMS

Rajiv K. Singh and J. Narayan

Department of Materials Science and Engineering
North Carolina State University
Raleigh, NC 27695-7916

ABSTRACT

We have theoretically analyzed and developed a model for the interaction of pulsed laser beams with bulk $YBa_2Cu_3O_7$ targets resulting in evaporation, plasma formation and deposition of thin films. In this model, the laser generated plasma is treated as an ideal gas initially at high temperature and pressure, and is then allowed to expand in vacuum. The three-dimensional isoentropic expansion of this plasma gives rise to spatial thickness and composition variations observed in $YBa_2Cu_3O_7$ thin films. The solution of the hydrodynamic gas equations show that the deposition characteristics are dependent on the dimensions and temperature of the plasma, and the mass of the plasma species. The effect of pulse energy density and other parameters on the deposition characteristics is also analyzed. Some of the theoretical results are compared with the experimental values obtained by pulsed excimer laser evaporation of bulk $YBa_2Cu_3O_7$ targets. We also discuss some of the superconducting and the microstructural characteristics of in-situ processed epitaxial films fabricated at substrate temperatures between 500-650°C by XeCl laser irradiation on bulk $YBa_2Cu_3O_7$ targets in an oxygen ambient of 200 millitorr. An application of a positively biased ring between the substrate and target was found to reduce the deposition temperature to 500°C, although the quality of epitaxial growth decreased significantly below 500°C. Excellent quality epitaxial single crystal superconducting thin films were fabricated on (100) SrTiO$_3$ and (100) LaAlO$_3$ substrates at temperatures between 550-650°C. The films were nearly defect free with minimum ion channeling yields of 3% and 3.5% for YBa$_2$Cu$_3$O$_7$ films on (100) LaAlO$_3$ and (100) SrTiO$_3$, respectively. Superconducting YBa$_2$Cu$_3$O$_7$ films on (100) YSZ substrates, although epitaxial, possessed much larger defect concentrations as a result of lattice mismatch. Very high critical current densities (at 77K and zero magnetic field) exceeding 6.0×10^6 Amps/cm^2 were obtained on YBa$_2$Cu$_3$O$_7$ films on (100) LaAlO$_3$ substrates which represents the best values quoted in the literature. The effect of silver doping on critical current densities in these films is also discussed.

I. INTRODUCTION

The pulsed laser evaporation (PLE) technique for deposition of high T$_c$ superconducting thin films is distinguished by a number of unique properties[1-5]: excellent control over stoichiometry; nonequilibrium evaporation with plasma formation; high deposition rates; relatively inexpensive and versatile process, thus making it commercially viable. The nonequilibrium nature of this technique has facilitated "in-situ" processing of superconducting thin films at substrate temperature below 650°C[6-9]. A modification of this technique known as the biased PLE method can be used to deposit superconducting thin films at temperature as low as 500°C[9]. Presently, in-situ PLE films exhibit the best superconducting properties quoted in the lierature[9,10].

Although the PLE is an evaporation technique, the deposition characteristics of thin films are significantly different from films produced by other vaporization methods. Higher surface temperatures, formation of a high temperature plasma of partially ionized species and forward directed nature of the the deposited material are some of the characteristics of the PLE technique[11-17]. In the first section of this paper, we will analyze the complex nature of the laser-solid-plasma interactions resulting in the unique deposition characteristics. The effect of various beam and substrate parameters including pulse energy density, substrate-target distance, irradiated spot size, etc. on the deposition characteristics is quantitatively analyzed. In the second section, we will discuss the in-situ processing, microstructure and superconducting properties of epitaxial $YBa_2Cu_3O_7$ superconducting thin films on $SrTiO_3$, $LaAlO_3$ and yttria stabilized zirconia (YSZ) substrates. Films with excellent crystalline quality (minimum ion channeling yields of 3%) and critical current densities over 6.0×10^6 Amps/cm^2 (at 77K and zero magnetic field) have been produced on (100) $LaAlO_3$ substrates. A modified (biased) laser deposition method which involves placement of a positively biased ring between the substrate and the target can be used to deposit superconducting thin films as low as 500°C, however, the crystalline quality of the film decreases sharply below 550°C. The microstructural properties of these films are correlated with the superconducting properties.

II. THEORITICAL MODEL FOR PULSED LASER DEPOSITION TECHNIQUE

To develop a model for the PLE technique for the thin film deposition, one must first analyze the salient experimental features of the deposition process. Fig. 1 shows a time integrated picture of the partially ionized plasma formed after excimer laser irradiation of the 1-2-3 targets with an energy density of 1.8 J/cm^2. This figure suggests that the expansion velocity is much greater in the direction perpendicular to the target compared to the velocity in direction parallel to the target surface. It has been observed that the preferential expansion of the laser plasma is always directed perpendicular to the target surface, regardless of the angle of incidence of the incoming laser beam. Near the target surface, a more intense glow is observed indicating a much higher temperature. Experimental studies have shown that the maximum temperatures are in the region of 1-2 eV for excimer laser irradiated 123 targets[12-13]. This plasma temperature has been found to be dependent on pulse energy density (more precisely, the power density), wavelength, and material parameters as well as dimensions of the irradiated spot. The velocity distributions of the laser ablated species measured optically and by ion time of

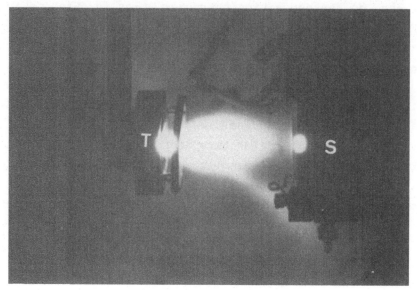

Fig.1 Time integrated photograph of the laser induced plume generated by excimer laser irradiation of 1-2-3 target with pulse energy density of 1.8 J/cm^2. In this figure, S is the substrate and T is the target.

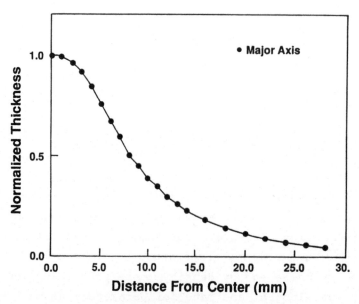

Fig.2 Experimental spatial thickness variations in film deposited by XeCl laser irradiation of 1-2-3 target with 250 pulses of energy density of 1.8 J/cm^2

flight experiments show[13,18] that the velocities of the atomic and ionic species are about 10^6 cm/sec, which is much higher than the equilibrium thermal values. Also, the velocity distribution is much broader than the Maxwell-Boltzmann distribution expected from a conventional thermal evaporation process. Thus, it is extremely important to understand the effect of the beam and materials parameters on the PLE films.

The spatial thickness variations have been found to be significantly different from a conventional thermal evaporation process. Unlike a $\cos \theta$ variation expected from a thermal evaporation process, the laser ablation process is characterized by a forward directed deposit with spatial thickness variations[19] of the order of $(\cos \theta)^{8-12}$. Fig. 2 shows the spatial thickness variations of the $YBa_2Cu_3O_7$ film deposited on silicon substrate at an energy density of 1.8 J/cm^2 and substrate-target distance of 3 cm. The shape of the deposit is approximately Gaussian with full width at half maximum (FWHM) of 14 mm with most the deposit occuring near the center of the film. The FWHM increases slightly with decreasing pulse energy density, but varies strongly with other two parameters namely the dimensions of the irradiated spot, and the substrate-target distance[20].

The above characteristics of the PLE technique can be explained by a hydrodynamical model based on the iso-entropic expansion of the plasma in vacuum[21]. The partially ionized plasma is mathematically treated as an high temperature and high pressure (HT-HP) gas which is initially confined to small dimensions, and then is allowed to expand in vacuum. The schematic diagram of the laser-solid plasma interaction is shown schematically in Fig. 3. Four separate areas can be identified: A is the region unaffected by the laser beam; B is the evaporated target material; C is the region in the plasma which absorbs the incoming laser beam and D is the rapidly expanding outer edge which initiates after the termination of the laser pulse. Based on the nature of the laser interaction, three regimes are postulated which are, (i) evaporation regime in which the laser beam interacts with the target resulting in evaporation, (ii) isothermal regime in which the laser energy is absorbed by the evaporating material, leading to formation and initial isothermal expansion of the partially ionized plasma, and (iii) adiabatic regime in which the iso-entropic expansion of the plasma takes place resulting in film deposition. In the evaporation regime, the thermal effects of pulsed nanosecond laser irradiation can be analyzed by solving the one dimensional heat flow partial differential equation with non-linear boundary conditions which take into account the presence of moving liquid/evaporation interface. For this problem, however, detailed solution of the heat flow equation is not necessary, and approximate dependence of the evaporation rate on the beam and material parameters is required. Simple energy considerations can be applied to compute the evaporation rate[22]. By applying energy balance considerations, the evaporated thickness, Δx_t and the target evaporation rate, F_t, for highly absorbing laser beams is given by

$$F_t = \Delta x_t/\tau = (1-R)(E-E_{th})/\tau(\Delta H + C_v \Delta T) \qquad (1)$$

where, τ, ΔH, C_v and ΔT, R are the pulse duration, latent heat, volume heat capacity, rise in target temperature and reflectivity of the target, respectively. In this equation, E_{th} correponds to the minimum energy above which appreciable evaporation is observed. The energy threshold term takes into account the heat conduction losses in the target and the plasma losses due to absorption of the laser beam by the evaporating material. Although, this equation shows a linear increase in the evaporated thickness as a function of energy density, this necessarily may not be true as a result of increased plasma losses and change in reflectivity of the target. Thus, at higher energy densities, a deviation from the linear regime may be observed[11]. From the above equation, we see that the target evaporation rate is dependent on the beam (energy density, wavelength, pulse duration) and substrate parameters (thermal conductivity, latent heat, evaporation temperature, specific heat capacity, etc.). The evaporation characteristics of the target in turn control the interaction of the laser beam with the evaporating material.

In the isothermal regime, the evaporated material is further heated by the the incoming laser radiation resulting in the formation of a partially ionized high temperature plasma. In thermal equilibrium, the density of ions in the plasma can be

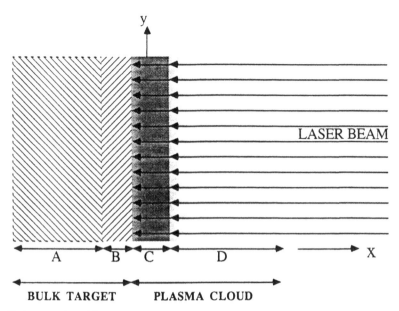

Fig. 3 Schematic diagram showing the different phases present during laser irradiation of a target: (A) unaffected target, (B) evaporated target material, (C) dense plasma absorbing laser radiation, and (D) expanding plasma outer edge transparent to the laser beam.

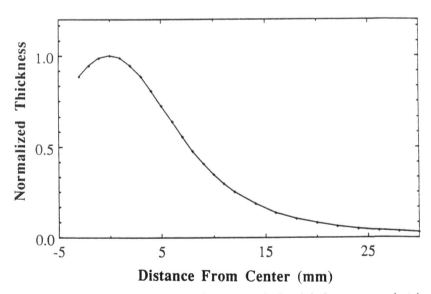

Fig. 4 Simulated curve showing spatial thickness variation as a function of distance from the center of the film deposited at a plasma temperature of 10000K and a substrate-target distance of 3 cm. An atomic weight of 89.0 was assumed throughout the calculations.

predicted by the Saha Equation[24], which shows an exponential dependence of ion (or electron) concentration with temperature. However, other mechanisms, for example, direct photoionization where photon energies are greater than binding energies, may be responsible for higher density of ionized species than expected from a purely thermal process. Even for photon enegies less than the binding energies, the neutral atoms can be excited to higher energy levels where they can subsequently be thermally ionized[25]. The main absorption mechanism of the evaporating material by the laser beam is the inverse Bremsstrahlung which involves the absorption of a photon by an electron near an ion[26]. However, during the initial stages of the laser pulse when very low electron and ion densities and a large number of neutral atoms are present in the plasma, free-free transitions involving neutral atoms provide a primary laser absorption mechanism. With slight increase in the ion densities, the free-free transitions involving ions takes over and becomes the dominant mechanism for laser absorption. The absorption coefficient for the plasma, α_p, for free-free transitions involving ions is given by

$$\alpha_p = 3.69 \times 10^8 \left(Z^3 n_i^2 / T^{1.5} \nu^2 \right) \quad h\nu/kT \ll 1 \qquad (2)$$
$$\alpha_p = 3.69 \times 10^8 \left(Z^3 n_i^2 / T^{0.5} \nu^3 \right) \quad h\nu/kT \geq 1$$

where n_i is the ion concentration, ν the laser wavelength, T the plasma temperature, Z the average charge, and k and h the Boltzmann and Planck constants, respectively. In the isothermal regime, the number of particles in the plasma is constantly augmented by the evaporated species from the target. From the above equation, we see that higher photon energy beams should have a smaller absorption coefficient. But experimentally higher ionization has been observed for high photon energies, primarily due to the effect of bound-free transitions which scales up with increasing photon energies[25]. Another factor affecting the degree of ionization is electron impact process.

As the absorption coefficient is dependent on the density of ionized species, most of the the laser energy is absorbed near the inner surface where the plasma density is the highest. The density and the pressure gradients control the expansion velocities of the plasma in the three orthogonal directions. If we assume an expansion velocity of 10^5 cm/sec, the dimension of the plasma in the direction perpendicular to the substrate is about 20-100 μm depending on pulse length. The solution of detailed hydrodynamical calculations[21] shows that in the isothermal regime the expansion velocity of the plasma edge in the x-direction (perpendicular to the target surface) reaches the isothermal velocity of sound which is given by $(\gamma RT/M)^{0.5}$, where, γ is the ratio of the specific heats, R is the gas constant, and M is the molecular weight of the species. In the other orthogonal directions (y and z), the expansion velocities are much lower, primarily due to the large transverse dimensions of the plasma in relation to the evaporated thickness.

After the termination of the laser pulse there is no more heat input into the plasma which then expands iso-entropically into vacuum. In this regime, the thermal energy is converted into kinetic energy with the plasma expanding rapidly outwards. To determine the deposition characteristics, we solve the differential equation equations governing the motion of the expanding plasma[27]. The expansion of the plasma in the adiabatic regime is controlled by the equation of continuity (conservation of mass) which is given by

$$\frac{d\rho}{dt} + \nabla \cdot (\rho \bar{v}) = 0 \qquad (3)$$

and the equation of motion (conservation of linear momentum) is

$$\frac{\partial \bar{v}}{\partial t} + (\bar{v} \cdot \nabla)\bar{v} + \frac{1}{\rho}\nabla P = 0 \qquad (4)$$

In these equations $\bar{v}$, ρ and P denote the spatially and temporal varying velocity, density and pressure in the plasma, respectively. The adiabatic equation of state is given by

$$\frac{1}{P}\left(\frac{\partial P}{\partial t} + \bar{v} \cdot \nabla P\right) - \frac{\gamma}{\rho}\left(\frac{\partial \rho}{\partial t} + \bar{v} \cdot \nabla \rho\right) = 0 \qquad (5)$$

and the equation of temperature is

$$\frac{\partial T}{\partial t} + \bar{v} \cdot \nabla T = (1-\gamma) T \nabla \cdot \bar{v} \qquad (6)$$

We assume that there are no spatial variations in the plasma temperature, and the velocity and density profiles to have the same shape throughout the course of the expansion. The concept of self similar motion has been extensively used in compressible hydrodynamic motion[25]. The plasma expansion in vacuum is driven by the density gradients in it. In this model, we adopt a linear increase in the plasma velocities and exponentially decreasing plasma density with increasing distance from the inner edge of the plasma. If X, Y, Z correspond to the outer edge of the plasma in three orthogonal directions x, y, z respectively, the plasma velocity is represented by

$$\bar{v}(x,y,z,t) = \frac{x}{X(t)} \frac{dX(t)}{dt} i + \frac{y}{Y(t)} \frac{dY(t)}{dt} j + \frac{z}{Z(t)} \frac{dZ(t)}{dt} k \qquad (7)$$

and the particle density profile of the plasma is given by

$$n(x,y,z,t) = \frac{N_T}{2^{0.5} \pi^{1.5} X(t) Y(t) Z(t)} \exp\left(-\frac{x^2}{2 X(t)^2} - \frac{y^2}{2 Y(t)^2} - \frac{z^2}{2 Z(t)^2}\right) \quad t > \tau \qquad (8)$$

where, the particle density is n, and N_T is the total number of particles in the plasma. The particle density is related to mass density by the equation $\rho = nM/N_a$, where N_a is the Avogadro's number and M the molecular weight of the species. For an ideal gas, the pressure can be related to the particle density by the equation, $P = nkT$. It should be noted that for iso-entropic expansion, the pressure and the density distributions[25] should be dependent on γ. However, it turns out that for different particle density profiles, the final results are similar and can explain the salient features of the PLE process. For adiabatic expansion, the average temperature of the plasma is related to its dimensions by the expression $T[XYZ]^{\gamma-1}$ = constant. If the substitute the density, pressure and velocity profiles in the equation of motion, we obtain

$$X(t)\left[\frac{d^2X}{dt^2}\right] = Y(t)\left[\frac{d^2Y}{dt^2}\right] = Z(t)\left[\frac{d^2Z}{dt^2}\right] = \frac{RT_0}{M}\left[\frac{X_o Y_o Z_o}{X(t) Y(t) Z(t)}\right]^{\gamma-1} \quad t > t \quad (9)$$

where, X_o, Y_o, Z_o correpond to the initial dimensions of the plasma at the end of the isothermal regime. The above equation shows that the acceleration of the plasma species depends on the temperature and the orthogonal dimensions of the plasma, and the mass of the plasma species. It should be noted that this is a hydrodynamic model and applies to all atomic and molecular species in the plasma. However, the expansion velocities will not be the same for different species and will depend on their masses.

(B) RESULTS AND DISCUSSION

From eqn. 9 we see that the maximum acceleration occurs in the direction of the smallest dimension. If x is the direction perpendicular to the target, the initial dimension in the x direction corresponds to the evaporated thickness (approximately 10-40 nm) while the transverse dimensions are of the order of mm. Thus, the maximum acceleration takes place in the x direction resulting in the forward directed nature of the deposition process. Similar final solutions with different constant prefactors values[21] will be obtained if we assume that the pressure and density distribution obey a linear decreasing profile instead of an exponentially decreasing profile.

(i) Deposition Characteristics. Detailed solutions of the plasma expansion equations were adopted to study the deposition characteristics and compare it with the experimental results. Runge Kutta method with small time steps (0.1-10 nanoseconds) was employed to solve the differential equation in the adiabatic expansion regimes. Fig. 4 shows a simulated normalized thickness obtained for an initial plasma temperature of 10^4 K, surface evaporation temperature of 2200 K, laser irradiated spot size of 2 mm, an average atomic weight of 88.9 (corresponding to Y in 123), γ equal to 5/3, and substrate target distance of 3 cm. The trends of this graph are similar to the experimental results obtained in Fig. 2, and shows the characteristic forward directed nature of the deposition process.

(ii) Deposition Dynamics. Fig. 5 shows the theoretical obtained velocity distribution of the of Y atomic species for plasma temperatures of 4000, 8000 and 15000 K. The other parameters used in he solution of the equation (9) are the same as in Fig. 4. In this figure, the most probabale velocity is much higher than expected from a conventional Maxwell-Boltzmann distribution, and also the velocity distribution is much broader. This type of velocity distribution has been earlier described as a Maxwellian distribution superimposed on a drift velocity[26]. The time of flight curves obtained from these theoretical calculations show a broad tail corresponding to low kinetic energies. Similar curve shapes for excimer laser irradiated superconductors have been obtained from time resolved optical measurements[13,18]. The average kinetic energy of the species are almost an order of magnitude greater than the thermal energies. For one dimensional iso-entropic expansion in vacuum, the maximum velocity attained by the plasma edge is given by

$$v_{max} = 2a/\gamma-1 \qquad (10)$$

where a is the isothermal velocity of sound given by $(\gamma RT_s/M)^{0.5}$, where T_s is the stagnation temperature, and is related to the plasma temperature T_o by the equation, $T_o/T_s = 2/\gamma+1$. Optical and ion time of flight measurements[13,18] have shown that the most probable velocities are of the order of 1.0×10^6 cm/sec, while in these theoretical calculations the most probable velocities are much lower. This may be explained by the effect of three factors. Firstly in the calculations, we have assumed a value of $\gamma = 5/3$ which corresponds to specific heat ratios of unexcited monatomic species. Because of the high plasma temperature, we expect a high degree of excitation and ionization which decreases the value of γ. It has been estimated that the value of γ is in the range of 1.2 to 1.3 for excited monatomic species[25] which has a strong effect on the maximum velocites of the species as shown by equation (10). The second factor which affects the maximum velocity and the velocity distributions is the assumption that the exponentially varying density distribution is independent of γ. Detailed calculations have shown that this assumption would lead to a $(\gamma-1)^{-0.5}$ dependence

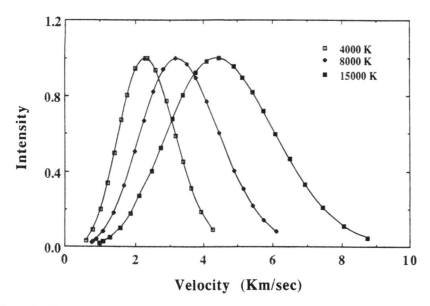

Fig. 5 The intensity distribution for Y atomic velocities in the plasma generated by excimer laser irradiation of 1-2-3 targets for plasma temperatures of 4000K, 8000K and 15000K.

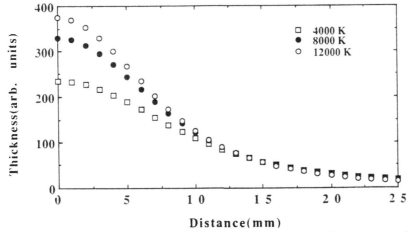

Fig. 6 Simulated thickness variations as a function of distance from the center of the film, for plasma temperatures corresponding to 4000, 8000, and 12000 K respectively.

of expansion velocity instead of the expected $(\gamma-1)^{-1.0}$. Thirdly, equation 10 is strictly valid for one dimensional isoentropic expansion. The dimensionality of the laser generated plasma although close to one, may deviate from one depending on the plasma temperature, irradiated spot size, and atomic mass of the species.

(iii) Effect of Pulse Energy Density. Due to the complicated nature of the laser interaction process, it becomes extremely difficult to exactly quantify the dependence of energy density on the deposition characteristics. The effect of pulse energy density on the deposition characteristics can be understood by determining the relation between plasma temperature and pulse energy density in the isothermal regime. This is a very complicated problem because the plasma absorption coefficient related to different factors including, concentration of ionized species, plasma temperature, and laser wavelength. The density of ionized species in turn is dependent on evaporation rate, rate of initial plasma expansion, degree of ionization, etc. Moreover, with varying plasma temperatures and laser wavelengths, the absorption coefficient shows a different functional dependence (eqn. 2). If we assume that the degree of ionization, χ, and particle density N_T are proportional to $(E-E_{th})^\varepsilon$, and $(E-E_{th})^\beta$, respectively, where E is the energy density and E_{th} is the vaporization threshold, the relation between plasma temperature and ion velocity with pulse energy density can be derived from energy balance in the plasma and is given by [21]

$$T \alpha \left[\frac{E(E-E_{th})^{2\varepsilon+\beta}}{\tau^2}\right]^{\frac{2}{2\eta+3}} \quad (11)$$

and

$$v \alpha \left[\frac{E(E-E_{th})^{2\varepsilon+\beta}}{\tau^2}\right]^{\frac{1}{2\eta+3}} \quad (12)$$

In this equation, η corresponds to either 0.5 (low temperature plasma) or 1.5 (high temperature plasma). This equation shows that the plasma temperature and expansion velocity as a function of energy density is controlled by the target evaporation rate and the degree of ionization. At low plasma temperatures ($h\nu/kT \sim 1$), $\eta = 0.5$, and ionization and evaporation rates scale linearly with energy density. This gives rise to a linear increase in velocity as a function of energy density. However, in the high plasma temperature regime, $\eta = 1.5$, and assuming $\varepsilon = 0.5$ [13] and evaporation rate independent of energy density, the above equations reduce to

$$T \propto E^{2/3} \text{ and } v \propto E^{1/3} \qquad (13)$$

Similar dependence of ion velocity with pulse energy density has been observed for CO_2 laser irradiated YSZ targets[28]. Because of this complex dependence, detailed experiments are necessary to quantify the plasma temperature as a function of pulse energy density. We have calculated spatial thickness variations corresponding to plasma temperatures of 4000, 8000, and 12000 K (shown in the figure 6) corresponding to different energy densities. The maximum thickness of the deposit increases with increasing plasma temperature. The maximum spatial thickness variations which occur at a distance of about 5-10 mm away from the center, also increase with increasing plasma temperatures. These results have been verified experimentally by us and other groups[16,20]. It should be noted that in these calculations we have assumed that a fixed number of particles are evaporating at all energy densities.

(iv) Spatial Compositional Variations. Spatial compositional variations as a function of energy density in the thin films can also be predicted by determining the expansion velocities as a function of plasma temperature. According to the eqn. 9, the asymptotic velocity of the plasma species should be inversely proportional to the square root of its atomic weight. However, this dependence should be followed for non interacting species. In the case of multicomponent targets (for example 1-2-3 materials), the measured velocities show a general trend of lower asymptotic velocities for higher masses, however, the

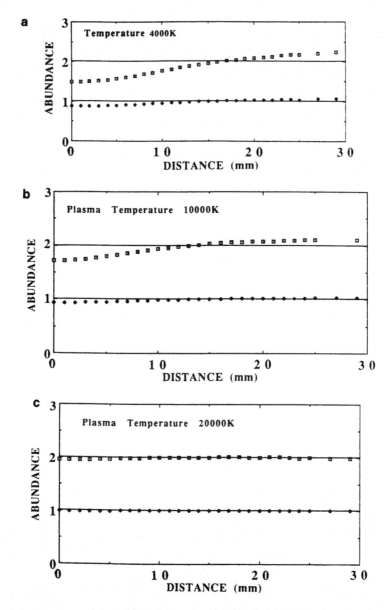

Fig. 7 Spatial compositional deviations from stoichiometric values of Y, Ba, and Cu in interacting plasmas for plasma temperatures of (a) 4000K, (b) 10000K, and (c) 20000K. A stoichiometric evaporation of the 1-2-3 target is assumed in the calculations.

dependence is much weaker than the inverse square root relation[28]. This is probably due to the interaction of different species in the plasma. During the initial stages of expansion all the plasma species of varying masses initially expand with uniform velocity. However, in the later stages of expansion, the acceleration is dependent on the respective atomic mass of the species, thus resulting in a less than the predicted inverse square root dependence. Fig. 7(a), (b), and (c) show the simulated spatial compositional variation curves of 123 films for plasma temperatures of 4000, 10000, and 20000K., respectively. These calculations asssume stoichiometric evaporation from the laser irradiated target. The results show that for low energy densities, the composition of the lightest species is larger than the stoichiometric value near the center of the deposit. As the plasma temperature(or energy density) is increased the compositional variation decrease. The shape and trend of these curves are in good agreement with experimetally obtained spatial compositional variation profiles[20].

III. IN-SITU PROCESSING OF SUPERCONDUCTING THIN FILMS

We have earlier seen that nanosecond pulsed laser irradiation of 123 targets results in the formation of high temperature partially ionized plasma with kinetic energies of the order of 10-100 ev. The non-equilibrium nature of deposition can be advantageously applied for in-situ processing of $YBa_2Cu_3O_7$ superconducting thin films at substrate temperature ranging from 600-650°C in an oxygen ambient. The incorporation of a positively biased ring between the substrate and the target can be used to further lower the deposition temperature to 500°C, however, the film quality deteriorates considerably below 550°C. Presently, the films exhibiting the highest critical current densities in presence or absence of magnetic field can be repoducibly produced by this method[10]. In this section, we will discusss some of the effects of the processing parameters on the superconducting properties and microstructure. As the coherence length in these material are extremely small, it is expected that small defects will play a major role in controlling the superocnducting properties especially the critical current densities.

(A) EXPERIMENTAL

In the experimental setup for deposition of thin films, a pulsed excimer laser (λ=308 nm, τ=45 x10^{-9} sec, energy density, 2-3 J/cm^2) is used to evaporate a bulk 1-2-3 target, and the deposition occurs on the substrate placed at a distance of about 5 cm from the target. The ring is biased positively (+300-400 V) with respect to the target, while the substrate is grounded or kept at a floating potential. A schematic diagram of the deposition setup is shown in Fig. 8. The voltage has been so chosen that no plasma discharge occurs in the chamber. During the deposition process the oxygen partial pressure is maintained between 180-200 mTorr with the oxygen flow rate of 20-25 sccm. In the above pressure and substrate-target-ring geometry, the bias voltage is not high enough to introduce a gaseous discharge in the plasma, and current of only a few microamperes flows across the ring during the time duration of the laser pulse. This is in contrast to a discharge assisted deposition scheme where high current may flow across the ring. The oxygen nozzle is directed toward the substrate. We have have observed this is the optimized geometry for in-situ laser deposition of high T_C superconducting thin films[29].

(B) RESULTS AND DISCUSSION

Fig. 9 shows the normalized resistance as a function of temperature for $YBa_2Cu_3O_7$ films deposited on (100) YSZ substrates for different nozzle positions. A substrate temperature of 550°C and oxygen partial pressure of 200 mtorr was maintained during the biased laser deposition process. The films deposited with the nozzle directed on the substrate exhibited metallic T_c curves with sharp resistive transitions. In contrast, the films deposited with the nozzle directed on the target showed weaker metallic behavior with much larger superconducting transition widths suggesting lack of incorporation of oxygen and poorer crystallinity in the films.

Fig. 10 shows resistance plots for films deposited on (100) $SrTiO_3$ and (100) YSZ substrates at different temperatures. The samples exhibit normal metallic behavior above T_c with sharp

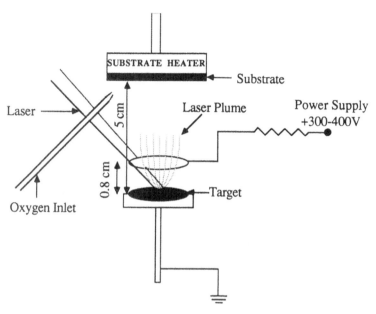

Fig.8 Schematic diagram of the biased laser deposition method showing the layout of the substrate, target and the positively biased ring.

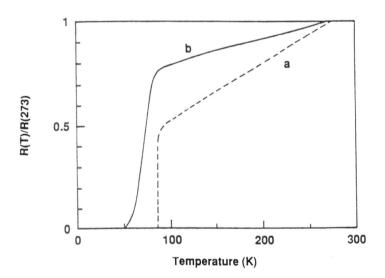

Fig.9 Resistance plots as a function of temperature for superconducting thin films deposited at 550°C on (100) YSZ with different oxygen nozzle geometries (a) nozzle directed towards the substrate and (b) nozzle directed towards the target

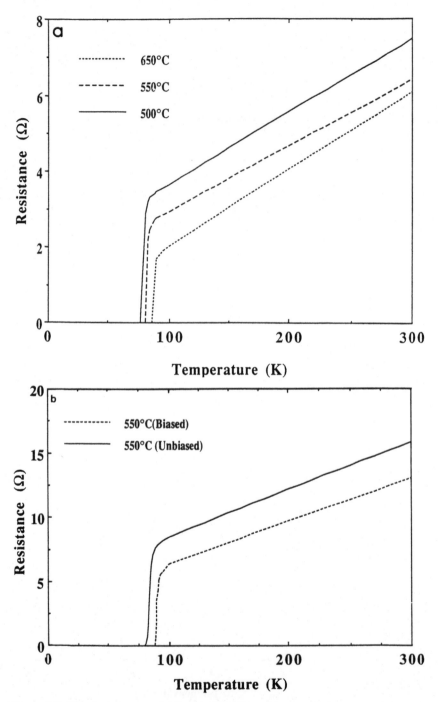

Fig. 10 Resistance plots as a function of temperature for YBa$_2$Cu$_3$O$_7$ thin films deposited on (a) (100) Strontium titanate substrates at 650 (unbiased), 550 (biased) and 500°C (biased). (b) Yttria stabilized zirconia at 550°C under biased and unbiased conditions.

superconducting transitions. We have observed that with decreasing substrate temperature, the transition width, ΔT_c, increased with rapid deterioration of superconducting properties at temperatures below 550°C. There is a clear improvement in the microstructural properties particularly below 600°C as a result of the interposing ring[9]. At 650°C, the improvement as a result of biased deposition was found to be small. The epitaxial nature of growth of as-deposited thin films was studied using x-ray diffraction, Rutherford backscattering spectrometry, ion channeling, electron channeling and transmission electron microscopy.

The nature of epitaxial growth was studied using RBS ion channeling and SEM electron channeling patterns (ECP), each probing a relatively large area ($\sim 2mm^2$ in RBS). The ion channeling results provide information on the perfection of the lattice, alignment of the planes and orientation of the film with respect to the substrate. The ratio of the channeling yields in the aligned to the random direction (χ_{min}) provides a direct measure on the degree of alignment of different planes of the film with the substrate. The random and aligned RBS spectra for $YBa_2Cu_3O_7$ films deposited on (100) $SrTiO_3$ and (100) $LaAlO_3$ substrates at 650°C is shown in Fig. 11(a) and 11(b), respectively. The values of χ_{min} of 3.5% for a 1300 Å thick film on $SrTiO_3$ and about 3% for (100) $LaAlO_3$ substrate is indicative of the perfect crystallinity of the films. The χ_{min} values corresponds to the ideal values expected from defect free $YBa_2Cu_3O_7$ single crystals.

We have also observed that incorporation of silver into the films result in improvement in both the crystallinity and the superconducting properties especially the critical current densities. Fig. 12(a) and 12(b) show random and aligned RBS spectra for thin undoped and silver doped $YBa_2Cu_3O_7$ films deposited on (100) YSZ substrates at 650°C. The minimum ion channeling yields for undoped 123 films was in the range of 15-25% due to the larger number of defects as a result of lattice mismatch between the film and the substrate. The minimum channeling directions for the film and the substrate was displaced by about 1.2°. In contrast, films which where doped with silver, the minimum channeling yields were in the range of 10-15% suggesting an improvement in the crystalline quality. We

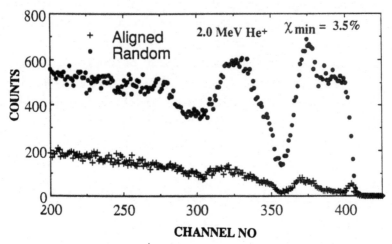

Fig. 11(a) 2.0 MeV He$^+$ channeled and random spectra from superconducting YBa$_2$Cu$_3$O$_7$ thin film deposited on (100) SrTiO$_3$ substrate at 650°C. The minimum channeling yield is roughly 3.5%.

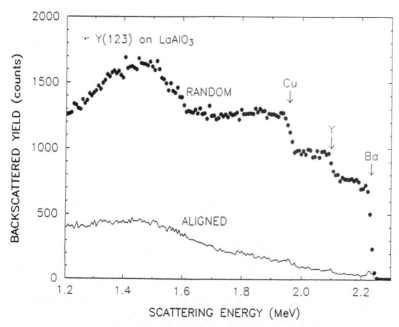

Fig. 11(b) 2.5 MeV He$^+$ channeled and random spectra from a superconducting YBa$_2$Cu$_3$O$_7$ thin film deposited on (100) LaAlO$_3$ substrate at 650°C. The minimum channeling yield is roughly 3.0%.

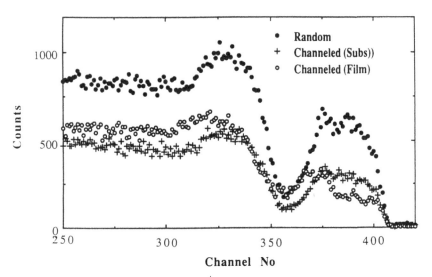

Fig. 12(a) Typical 2.0 MeV He$^+$ channeled and random spectra from a superconducting YBa$_2$Cu$_3$O$_7$ thin film deposited on (100) YSZ substrate at 650°C.

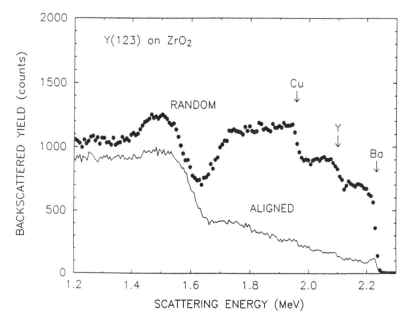

Fig. 12(b) Typical 2.5 MeV He$^+$ channeled and random spectra from a superconducting YBa$_2$Cu$_3$O$_7$ thin film deposited on (100) YSZ substrate at 650°C.

speculate that some of the silver is incorporated into the film in the form of small precipitates, while the remainder is segregated and helps in the improvement of crystalline quality of the films. Experiments are in progress to determine the exact role of silver in the films.

Fig. 13 is a cross-section TEM micrograph showing directly the alignment of (001) or c planes parallel to the (001) $SrTiO_3$ substrate. The high resolution micrograph also showed the perfect alignment of (010) planes of the film with those of the substrate. The selected area diffraction pattern contains diffraction spots from both the substrate and the film, from which the following epitaxial relationships are directly established: $[001]_{123}$ // $[[001]_{SrTiO3}$, and $[100]_{123}$ // $[100]_{SrTiO3}$. However, more accurate three dimensional X-ray measurements has shown that instead of the [100] directions, the [110] directions of the of the substrate and the film are aligned to each other. The interface is sharp within 10Å, indicating interdiffusion to be minimum at these temperatures. The epitaxial films contained (001) stacking faults but no subgrain boundaries or twins. The defect content increases sharply below 550°C.

The critical transport current densities were measured in 1000Å thick film patterns (0.5 mm wide x 3 mm long). In (100) $SrTiO_3$ films J_c values exceeded 5.0×10^6 at 77 K and the corresponding values for (100) YSZ films were 1.0×10^6 A/cm^2 in zero magnetic field. The films deposited with the nominal target composition of $YBa_2Cu_3O_7$-Ag_3 on (100) $LaAlO_3$ substrates exhibited critical current densities greater than 6.0×10^6 $Amps/cm^2$ at 77K. We speculate that the presence of very small silver precipitates act as pinning sites for the current generated fluxiod lattice.

IV CONCLUSIONS

We have developed a hydrodynamic model for PLE thin film deposition. In this model, the laser generated partially ionized plasma is treated as a HT-HP gas, which is allowed to expand adibatically after the termination of the laser pulse. The anisotropic expansion velocites in the three dimensions dictate the nature of the deposition process. The expansion velocities

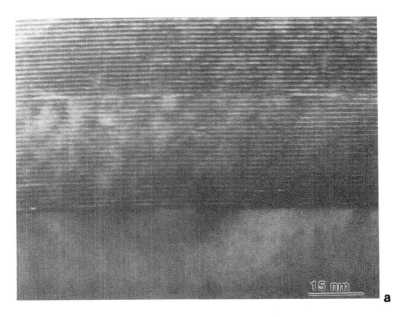

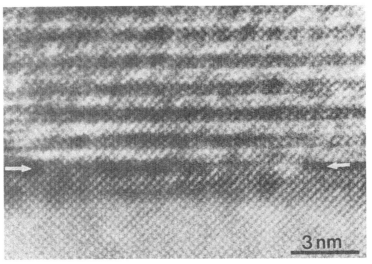

Fig.13 Cross-section (a) [100] and (b) [110] high resolution TEM micrograph of the c-axis oriented epitaxial superconducting $YBa_2Cu_3O_7$ film deposited on (100) $SrTiO_3$ at 650°C.

were found to be dependent on the plasma dimensions, plasma temperature and the atomic weight of the plasma species. The effect of energy density on the deposition characteristics has been quantitatively analyzed. Detailed solutions of the equations governing the plasma expansion characteristics were performed and compared with experimental results. The theoretical results were in good agreement with the experimental values obtained by us and other authors.

In-situ processing of $YBa_2Cu_3O_7$ superconducting thin films on various substrates has been studies in detail. Using a biased laser method the deposition temperatures could be reduced to 500°C, however the epitaxial quality of thin films was found to deteriorate below 550°C. The films are shown to be epitaxial on (100) $SrTiO_3$ (100) $LaAlO_3$ and (100) YSZ with minimum interdiffusion between the substrate and the film. The films deposited at 650°C showed J_c values (at 77 K and zero magnetic field) of 5.0×10^6 A cm^{-2} and 6.0×10^6 Acm^{-2} on (100) $SrTiO_3$ and (100) $LaAlO_3$ substrates, respectively. As the temperature of "in-situ" processing was lowered, the quality of the films decreased but the films still remained epitaxial. The use of low temperature "in-situ" processing techniques should be an important advancement in fabrication of high-T_c superconducting thin films for device applications.

ACKNOWLEDGEMENTS

Part of this project is sponsored by NSF # MSM 8818994, DARPA and ECUT program of DOE, Martin Marietta Energy Systems ORNL. The authors would like to acknowledge P. Tiwari, L. Ganapathi, C. B. Lee and D. Bhattacharya for their technical help.

REFERENCES

1. D. Dijjkamp, T. Venkatesan, X. D. Wu, S. A. Shaheen, N. Jisrawi, Y. H. Min-lee, W. L. McLean, and M. Croft, Appl. Phys. Lett. **51**, 619 (1987)
2. J. Narayan, N. Biunno, R. Singh, O. W. Holland and O. Auchiello, Appl. Phys. Lett. **51**, 1845 (1987)
3. A. M. Desantolo, M. L. Mandich, S. Sunshine, B. A. Davidson, R. M. Fleming, P. Marsh, and T. Y. Kometani, Appl. Phys. Lett. **52**, 1995 (1988)

4. K. Moorjani, J. Bohandy, F. J. Adrian, B. F. Kim, R.D.Shull, C. K. Chiang, L. J. Swartzebdruber, and L. H. Benett, Phys. Rev. B **36**,4036 (1988)
5. X. D. Wu, A Inam, T. Venkatesan, C. C. Chang, E. W. Chase, P. Barboux, J. M. Tarascon, and B. Wilkens, Appl. Phys. Lett., **52**, 754 (1988)
6. B. Roas, L. Schultz and G. Endres Appl. Phys. Lett.**53**, 1557 (1988)
7. G. Koren, E. Polturak, B. Fisher, D. Cohen and G. Kimel, Appl. Phys. Lett., **53**, 2330 (1988)
8. S. Witanachchi, H. S. Kwok, X. W. Wang, and D. T. Shaw, Appl. Phys. Lett., **53**, 234 (1988)
9. R.K Singh, J. Narayan, A. K. Singh, and J. Krishnaswamy, **54**, 2271 (1989)
10. J. Narayan, R.K. Singh, P. Tiwari and C.B. Lee , Appl. Phys. Lett. (in press)
11. A. Inam, X.D. Wu, T. Venkatesan, S. B. Ogale, C. C. Chang, D. Dijjkamp, Appl. Phys. Lett., **51**, 1112 (1987)
12. Q.Y. Ying, D. T. Shaw, H. S. Kwok, Appl. Phys. Lett., **50**, 359 (1987)
13. P. E. Dyer, S. D. Jenkins, and S. Sidhu, Appl. Phys. Lett. **49**, 453, (1988)
14. W. A. Weiner, Appl. Phys. Lett. **52**, 2171 (1988)
15. C.H. Becker and J. B. Pallix, J. Appl. Phys. **64**, 5152 (1988)
16. T. Venkatesan, X. D. Wu, A. Inam, and J. B. Watchman, Appl. Phys. Lett. **52**, 1193 (1987)
17. R.K. Singh, N. Biunno, and J. Narayan, Appl. Phys. Lett., **53**, 1013, (1988)
18. J. P. Zheng, Z. Q. Huang, D. T. Shaw, and H. S. Kwok, Appl. Phys. Lett. **54**, 280,1989
19. R. A. Neifeld, S. Gunapala, C. Liang, S. A. Shaheen , M. Croft, J. Price, D. Simons, and W. T. Hill III, Appl. Phys. lett. **53**, 703, 1988
20. R. K. Singh and J. Narayan, Physical Review B (in press)
21. R. K. Singh and J. Narayan, Phys. Rev B (in press)
22. R.K. Singh and J. Narayan, Mat. Sci. and Engr. B **3**, 217 (1989)
23. S.G. Hansen and T.E. Robitaille, Appl. Phys. Lett, **50**, 359(1987)
24. J. F. Ready, Effects of High Power Laser Radiation, Academic Press, New York, 1971

25. Zel'dovich and Raizer, Physic of Shock Waves and High Temperature Phenomena, Academic Press, NY (1966)
26. T.P Hughes, Plasma and Laser Light, John Wiley, NY 1975
27. A.J. Chapman and W.F. Walker, Introductory Gas Dynamics, Holt, Rinehart and Winston, NY 1971
28. H. Sankur, J. Denatale, W. Gunning, and J. G. Nelson, J. Vac.Sci. Tech. **A 5**, 2869 (1987)
29. R. K. Singh, J. Narayan, L. Ganapathi, and P. Tiwari Appl. Phys. Lett, Dec 5, 1989

RF PLASMA DEPOSITION OF $YBa_2Cu_3O_{7-x}$ FILMS

S. Patel, A. Shah and D. T. Shaw

New York State Institute on Superconductivity
State University of New York at Buffalo
Bonner Hall, Amherst, New York 14260

Abstract

Superconducting $YBa_2Cu_3O_{7-x}$ films have been grown on single crystal yttria stabilized zirconia by rf plasma deposition from an aqueous solution of the nitrates of Y, Ba and Cu. The as-grown films are highly c-axis oriented perpendicular to the substrate with critical temperature and current density of 85K and ~ 1 x 10^4 amp/cm^2 at 77K, respectively. The influence of the process parameters on the microstructural and transport properties will be discussed.

Introduction

Numerous techniques such as e-beam[1,2], sputtering[3], MOCVD[4], laser deposition[5,6] and rf plasma deposition[7,8], have successfully been employed to fabricate thin films of the oxide superconductor. In this report, we describe results of $YBa_2Cu_3O_{7-x}$ (YBCO) films grown by the novel rf plasma deposition process. The process has some inherent advantages that the other deposition processes do not possess. These advantages include the electrodeless configuration of the inductively coupled plasma, which eliminates the possibility of electrode contamination, operation at atmospheric pressures and the availability of high rf power generators. The first report using rf deposition of as-grown superconducting YBCO thick films made from powders had a T_c (99.9% drop of resistivity) of 57K with a deposition rate of 10 μm/min[7]. A reduced deposition rate of 0.2 μm/min has recently been successful in growing YBCO films from a vapor phase with a T_c of 87K and a current density of ~1 x 10^6 amp/cm^2 at 77K[10]. With high temperature post annealing, YBCO films from an aqueous Y, Ba and Cu nitrate precursor have been grown both by spray pyrolysis[9] and by rf plasma deposition [10]. Using this technique, we report some of the effects of the processing parameters on the crystalline structure and transport properties of the YBCO films grown from the nitrates of Y, Ba and Cu nitrates in solution. The effect of substrate location relative to the torch, substrate temperature and solution concentration of film transport and structural properties will be discussed.

Experimental

Figure 1 shows the experimental set-up of the rf plasma deposition system. A two turn Tafa 56 torch (Tafa, Inc.) is powered by the 0.10kW variable 13.56 MHz rf generator via a matching network. The yttria-stabilized zirconia substrate is mounted on a grounded heater block located above the movable mask, 10.5 cm from the plasma torch nozzle. A

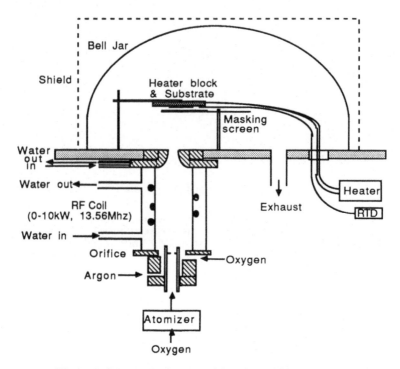

Figure 1. Schematic diagram of the rf deposition system.

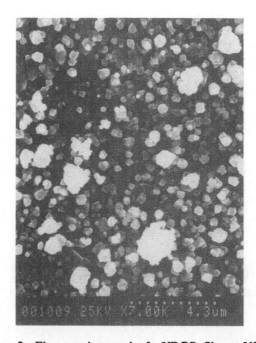

Figure 2a. Electronmicrograph of a YBCO film on YSZ.

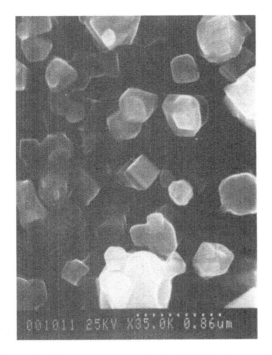

Figure 2b. Electronmicrograph of a YBCO film on YSZ.

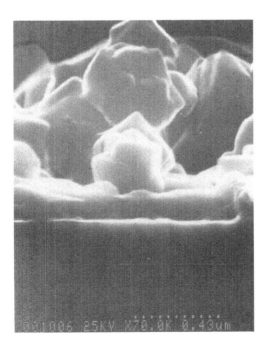

Figure 2c. Cross-sectional view of a YBCO film on YSZ.

solution containing the nitrates of Y, Ba, and Cu is atomized and injected into the Ar-O_2 plasma. The deposition conditions are listed in Table 1. The substrate temperature listed in Table 1 corresponds to that maintained by resistive heating without the plasma. The plasma further heats the substrate of 150°C to 200°C as measured by a pyrometer. This resistive heating was essential for the slow cooling process (5°C/min) necessary to grow quality films. Plasma heating must be coupled with resistive heating, to avoid a rapid drop in the substrate temperature when the plasma is turned off following deposition.

The following procedure was followed to grow as-deposited superconducting films. The substrate, after mounting on the heater block, was preheated to 550°C in an argon atmosphere. The plasma is ignited in argon gas, stabilized, and slowly switched to a 50:50 Ar-O_2 atmosphere. The atomized nitrate solution is injected into the plasma and deposition of the aerosol on the substrate is allowed for 30 minutes. This allows the formation of a 0.3 µm to 0.5 µm thick film. The deposition rate by this process is comparable to that by laser deposition[5].

Table 1. Operating conditions.

RF power	4 kW
Total plasma gas (50% O_2:50% Ar)	10-12 lpm
Pressure	760 torr
Aerosol flowrate (O_2 carrier gas)	2 lpm
Y, Ba, Cu nitrate solution concentration	9.7 gm/100 ml
$\quad$ Y(NO$_3$)$_3$.7H$_2$O	2.2 gm
$\quad$ Ba(NO$_3$)$_2$	3.5 gm
$\quad$ Cu(NO$_3$)$_2$.3H$_2$O	4.0 gm
Substrate	YSZ[100]
Substrate to torch distance	10.5 cm
Substrate temperature	550°C
Deposition time	30 min.

Results and Discussion

Electronmicrographs of a typical YBCO film on YSZ is shown in Figure 2a. The film is composed of YBCO single crystals which are fused together as shown in Figure 2b. The size of these crystals are typically 0.25 µm. In addition, there is 0.05 µm to 0.1 µm thick underlying continuous film seen in the cross sectional view in Figure 2c. These films are extremely rough. In comparison, 0.1 µm to 0.15 µm thick films, formed after a 15 minute of deposition, are continuous and smooth and do not show the single crystal particle formation. A study of these thin films is in progress. Figure 3 shows the size distribution of the nitrate solution aerosol at the entrance and the exit of the plasma torch. There is transformation from the spherical nitrate particles to single crystal cubic structured YBCO particles. The particle sizing was done by a Laser Aerosol Spectrometer Model LAS-X (Particle Measuring System, Inc.). The average droplet diameter at the entrance of the torch and the particle diameter exiting the torch were 1.5 µm and 0.15 µm, respectively.

Temperature dependence versus resistance in these films is measured by the four point method as shown in Figure 4. The substrate-to-torch distance is 10.5 cm. The T_c (R=0) and current density are 80K and ~ 6.0×10^3 amp/cm^2 at 77K, respectively. Cu deficient YBCO film

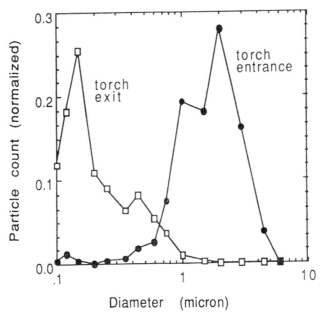

Figure 3. Size distribution of (Y, Ba and Cu) nitrate solution aerosol at entrance of torch and YBCO particles exiting the torch.

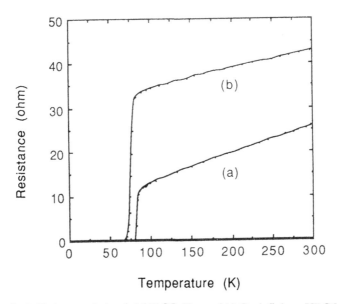

Figure 4. R-T characteristic of a) YBCO film and b) Cu deficient YBCO film.

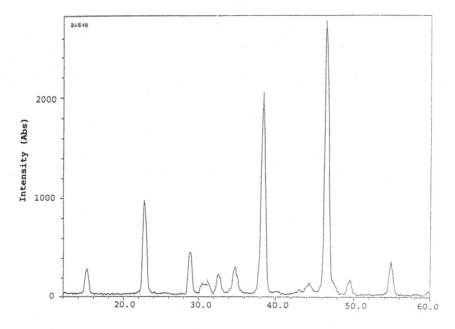

Figure 5. X-ray diffraction of a YBCO film

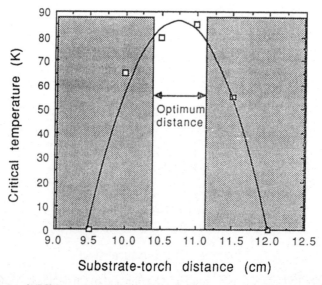

Figure 6. Effect on substrate to torch distance on critical temperature.

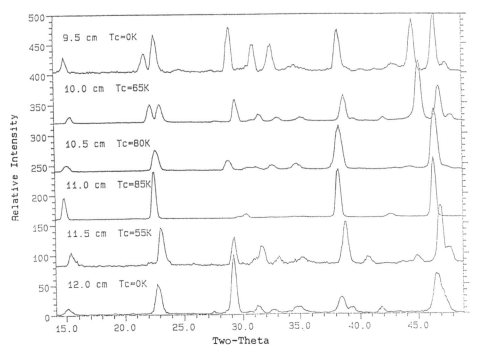

Figure 7. X-ray diffraction of YBCO films grown at differnt substrate to torch distance.

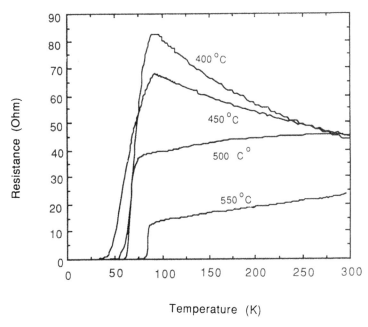

Figure 8. Effect of substrate temperature on R-T characteristic of YBCO films.

with a T_c (R=0) of 70K is made with a $Cu(NO_3)_2 \cdot 3H_2O$ content in solution of 3.8 gm. The x-ray diffraction (XRD) pattern of a typical film with all the [00L] peaks is shown in Figure 5. The preferential orientation of the c-axis perpendicular to the substrate is attributed to the underlying film.

Figures 6 and 7 show the dependence of the location of substrate in the plasma tail on the critical temperature and the corresponding XRD patterns. The optimum distance for growing good quality films is 10.5 to 11.0 cm. At 11.0 cm, a T_c (R=0) of 85 K was obtained with a J_c of ~ 1 x 10^4 amp/cm^2 at 77K. Small substrate-torch distance produced low T_c films. These films are rough with numerous particulates in comparison to films grown at 10.5 cm and 11.0 cm substrate-torch distance. XRD shows the presence of other phases, such as BaO, $BaCuO_2$ and Y_2O_3 peaks. For substrate-torch distances greater than 11.5 cm, the deposition rate is small, resulting in films of 0.05 μm to 0.15 μm thick for 30 min. deposition. The T_c is low although XRD shows all the OOL peaks indicating a possible insufficiency of oxygen in the film.

Figure 8 shows the effect of substrate temperature on the R-T characteristics. The substrate temperature is the predeposition temperature in the absence of the plasma. At low substrate temperature, 400°C and 450°C, T_c (R=0) is below 60K with a semiconductor behavior prior to T_c onset. At 500°C and 550°C substrate temperature, T_c (R=0) is above 60K with a metallic behavior prior to T_c onset.

In conclusion, we have grown by rf plasma deposition, as-deposited c-axis oriented YBCO films on YSZ from a nitrate precursor. T_c (R=0) of 85K and current density of ~ 1 x 10^4 amp/cm^2 at 77K was obtained at a substrate-torch distance of 11 cm and substrate temperature of 500°C. The morphology of the film is not as good as the morphology of films made by laser ablation and sputtering. Optimization of process variables such as the Y, Ba and Cu nitrate solution composition, nitrate aerosol size distribution to a more monodisperse size, aerosol injection into the plasma and rf power will aid in the fabrication of high quality superconducting oxide films.

Acknowledgement

This research is supported by an award from the New York State Institute on Superconductivity in conjunction with the New York State Energy Research and Development Authority.

References

1. R. B. Laibowitz, R. H. Koch, P. Chaudari and R. J. Gambino, Phys. Rev. B 35, 1822 (1987).

2. A. Mogro-Campero, B. D. Hunt, L. G. Turner, M. C. Burrell, and W. E. Balz, Appl. Phys. Lett., 52 (7) (1988).

3. H. Adachi, K. Setsune, T. Mitsuyu, K. Hirochi, Y. Ichikawa, T. Kamada and K. Wasa, Jpn. J. Appl. Phys., 26, L709 (1987).

4. T. Tsuruka, H. Takahashi, R. Kawasaki and T. Kanamori, Appl. Phys. Lett., 54 (18), 1808 (1989).

5. S. Witanachchi, H. S. Kwok, X. W. Wang and D. T. Shaw, Appl. Phys. Lett., 53, 234 (1988).

6. X. D. Wu, D. Dijkkamp, S. B. Ogale, A. Inam, E. W. Chase, P. F. Miceli, C. C. Chang, J. M. Tarascon and T. Venkjatesan, Appl. Phys. Lett., 51, 861 (1987).

7. K. Terashima, K. Eguchi, T. Yoshida and K. Akashi, Appl. Phys. Lett., 52, 1274 (1988).

8. H. Komaki and T. Yoshida, Proc. of 9th Int'l Sym. on Plasma Chemistry, Pugnochiuso, Italy, Volume 3, 1491, (1989).

9. M. Kawai, T. Kawai, H. Masuhira and M. Takahasi, Jpn. J. Appl. Phys., 26, (10) L1740 (1987).

10. H. Zhu, Y. C. Lau and E. Pfender, Proc. of 9th Int'l Sym. on Plasma Chemistry, Pugnochiuso, Italy, Vol. 3, 1497, 1989.

SUPERCONDUCTING FILMS OF Bi-Sr-Ca-Cu-O
BY LASER DEPOSITION

E. Narumi, S. Patel and D. T. Shaw
New York State Institute on Superconductivity
State University of New York at Buffalo
Bonner Hall, Amherst, New York 14260

Abstract

Superconducting Bi-Sr-Ca-Cu-O (BSCCO) films have been grown on single crystal [100] MgO by laser deposition. The films are highly c-axis oriented perpendicular to the substrate. The 2212, 2223 and 2234 phases of the BSCCO system have been grown independently in film formed from a sintered 2223 BSCCO target. By post annealing at 870°C for five minutes, the critical temperature of as-grown predominantly single phased films was increased from below 77K with semiconductor behavior to above 80K with metallic behavior. Post-annealed BSCCO films were transformed from a single phase to a multiphase film. The current density of these films ranged from 10^3 to 10^4 amp/cm^2 at 77K.

Introduction

The high critical temperature 2223 phase of the Bi-Sr-Ca-Cu-O (BSCCO) system has prompted numerous studies in the processing of these compounds in both bulk and film forms. This paper will emphasize thin film fabrication. Films of BSCCO have been grown by sputtering[1-4], e-beam[5,6], MOCVD[7,8], pyrolyis[9] and laser deposition[10-12]. Laser deposition is a well established and highly successful process for growing as-deposited highly oriented $YBa_2Cu_3O_{x-7}$ films[13,14]. Using this technique, we report some of the effects of the processing parameters on the crystalline structure and transport properties of the BSCCO films.

Experimental

Figure 1 shows the experimental set-up for laser deposition. A 193 nm ArF excimer laser is focused on a sintered 1" diameter 2223 BSCCO target. The angle of incidence is 45° normal to the target. The 1 cm x 1 cm [100] MgO substrate is mounted on a heater block located 3" away and normal to the target. BSCCO films of 0.4 micron thickness are grown at a rate of ~1 Å/sec with a laser fluence and repetition rate of 1 J/cm^2 and 5Hz, respectively.

Results and Discussion

The 2212, 2223, and 2234 phase of the BSCCO system can be independently grown by laser deposition. Figure 2 shows the oxygen pressure and substrate temperature dependence for different phases of the BSCCO films. Lower pressures and higher substrate temperatures are required to make larger number structured films. The 2234 phase has a smaller window than the high T_c 2223 phase. Figure 3 shows the x-ray diffraction (XRD) results of the as-grown 2212, 2223 and 2234 BSCCO films with different lattice parameters of the c-axis. The lattice parameters for the 2212, 2223, 2234 phases are 31Å, 37Å, and 43Å, respectively. The films are highly c-axis oriented normal to the substrate. However, these as-grown films always show T_c <77K with a semiconducting behavior. The resistivity drop around 110K cannot be observed during the T_c measurement, even though the XRD results show that the film is a 2223 phase (110K phase of BSCCO). This is due to lack of oxygen in the as-grown film. Figure 4 shows the XRD pattern and the R-T characteristics of an annealed 2212 phase film. This film

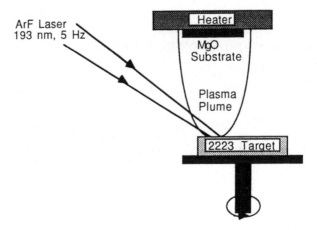

Figure 1. Schematic diagram of the laser deposition system.

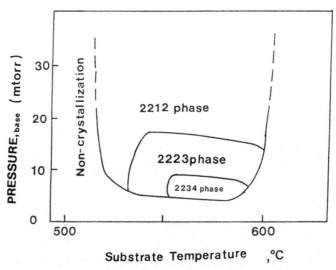

Figure 2. Effect of pressure and substrate temperature on phase formation

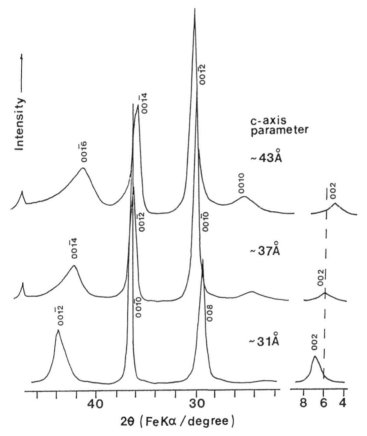

Figure 3. XRD patterns af as-grown Bi-Sr-Ca-Cu-O films. Lattice parameters: 2212 (~31Å), 2223 (~37Å) and 2234 (~43Å).

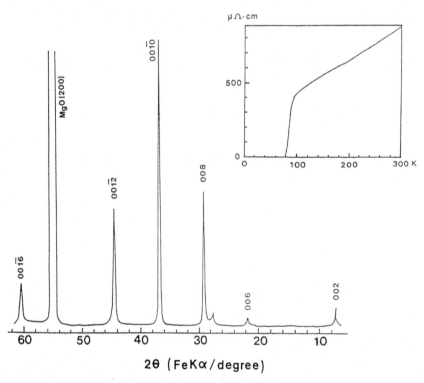

Figure 4. XRD pattern and R-T behavior of an annealed 2212 BSCCO film on MgO.

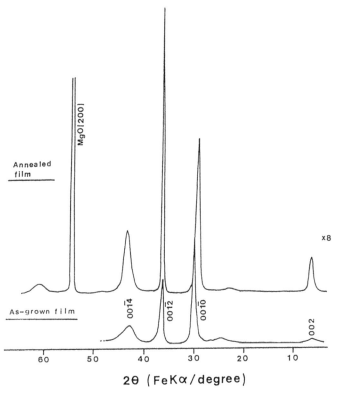

Figure 5. XRD pattern of an as-grown and annealed 2223 BSCCO film on MgO.

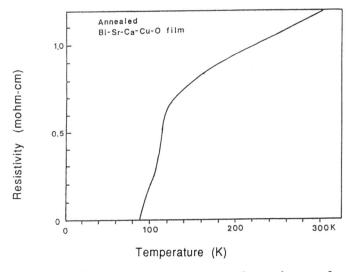

Figure 6. Resistivity-temperature dependence of an annealed 2223 BSCCO film on MgO.

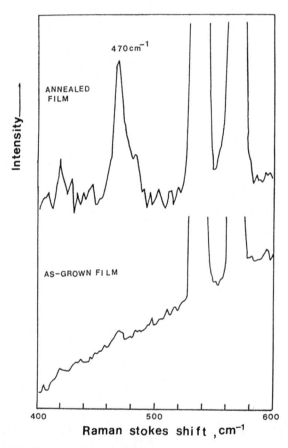

Figure 7. Raman backscattering measurement of an as-grown and annealed 2223 BSCCO film on MgO.

was annealed at 870°C in air for five minutes. T_c was improved from 30K to 80K by annealing. Figure 5 shows the XRD results of a 2223 phase film before and after annealing. The peaks of the c-axis became sharper and stronger due to annealing. In comparison, the XRD peaks of the as-grown films are broad due to the possible presence of other phases and to internal stresses in the film. The R-T characteristic of the film was improved from a semiconducting behavior to a metallic behavior as shown in Fig. 6. Resistivity of the film dropped initially at 110K indicating the presence of the 2223 phase, and finally to a zero resistivity at 90K. We studied the difference between as-grown and annealed films using Raman backscattering technique. As shown in Figure 7, no oxygen shift peak could be found in the measurement of the as-grown film. The first indication of an oxygen shift occurred after 30 minutes of annealing at 870°C, and after 60 minutes annealing the 470 cm^{-1} oxygen shift peak was easily observed.

In conclusion, we have grown predominantly single phase 2212, 2223 and 2234 BSCCO films by proper control of oxygen pressure and substrate temperature. The T_c of the films was increased to above 80K by annealing and consequently the single phase films were transformed into a multiphase film.

Acknowledgement

This research is supported by an award from the New York State Institute on Superconductivity in conjunction with the New York State Energy Research and Development Authority. We also thank Dr. A. Petrou for the Raman backscattering measurements.

References

1. M. Fukutomi, J. Machida, Y. Tanaka, T. Asano, T. Yamamoto, and H. Maeda, Jpn. J. Appl. Phys., 27 (8), L1484 (1988).
2. B. T. Sullivan, N. R. Osborne, W. N. Hardy, J. F. Carolan and B. X. Wang, Appl. Phys. Lett. 52 (23), 1992 (1988).
3. H. Koinuma, H. Nagata, A. Takano, M. Kawasaki and M. Yoshimoto, Jpn. J. Appl. Phys., 27 (10), L1887 (1988).
4. Y. Hakuraku, Y. Arodome and T. Agushi, Jpn. J. Appl. Phys., 27 (10), L1892 (1988).
5. T. Yoshitake, T. Satoh, Y. Kuboand, H. Igarashi, Jpn. J. Appl. Phys., 27 (7), L1262 (1988).
6. J. Steinbeck, B-Y Tsaur, A. C. Anderson and A.J. Strauss, Appl. Phys. Lett. 54, (5), 466 (1989).
7. J. A. Agostinelli, G. R. Paz-Pujalt and A. K. Mehrotra, Physica C 156, 208 (1988).
8. H. Takeya and H. Takei, Jpn. J. Appl. Phys., 28 (2), L229 (1989).
9. H. Shimojima, K. Tsukamato and C. Yamagishi, Jpn. J. Appl. Phys., 28 (2), L226 (1989).
10. C. R. Guarnieri, R. A. Roy, K. L. Saenger, S. A. Shivashankar, D. S. Yee and J. J. Cuomo, Appl. Phys. Lett. 53 (6), 532 (1988).
11. N. K. Jaggi, M. Meskoob, S. F. Wahid and C. J. Rollins, Appl. Phys. Lett. 53 (16) 17 October 1988.
12. H. Tabata, T. Kawai, M. Kanai, O. Murata and S. Kawai, Jpn. J. Appl. Phys., 28 (5), L823 (1989).
13. S. Witanachchi, H. S. Kwok, X. W. Wang and D. T. Shaw, Appl. Phys. Lett., 53, 234 (1988).
14. X. D. Wu, D. Dijkkamp, S. B. Ogale, A. Inam, E. W. Chase, P. F. Miceli, C. C. Chang, J. M. Tarascon and T. Venkatesan, Appl. Phys. Lett., 51, 861 (1987).

IN-SITU GROWTH OF $Y_1Ba_2Cu_3O_{7-x}$ THIN FILMS ON THREE-INCH WAFERS USING LASER-ABLATION AND AN ATOMIC OXYGEN SOURCE

James A. Greer

Research Division Raytheon Company
131 Spring Street
Lexington, MA 02173

INTRODUCTION

Pulsed laser-ablation is emerging as the deposition technique of choice for producing high quality, high temperature superconducting (HTS) thin films. Since the first report of $Y_1Ba_2Cu_3O_{7-x}$ (YBCO) thin film deposition using pulsed laser-ablation[1], a number of HTS compounds and novel in-situ laser-ablation approaches have been described in the literature[2,3,4,5,6,7,8,9,10]. The laser-ablation process, when used in conjunction with a variety of activated oxygen sources, has produced HTS films with excellent superconducting properties, principally on single crystal substrates of $SrTiO_3$, MgO, ZrO_2, and more recently $LaAlO_3$, $NdGaO_3$, and $LaGaO_3$. However, the laser-ablation technique has, until now, only been used to coat relatively small substrates (up to about one square inch) due to the rather narrow angular distribution for the ablated material, which is a fundamental property of the ablation process. This narrow angular distribution is also responsible for the noticeably poor thickness uniformity obtained over even smaller substrate sizes. Practical electronic applications of HTS materials will require larger areas to be uniformly coated in order to increase the device size and/or the number of devices obtainable per substrate. Film thickness uniformity is extremely important when defining HTS structures with micron size features using photolithography and dry etching techniques. Also, by utilizing larger area substrates, more practical use can be made of existing semiconductor processing equipment to fabricate HTS devices in an efficient and cost effective manner.

This paper deals with the first demonstration of coating three-inch substrates with $Y_1Ba_2Cu_3O_{7-x}$ (YBCO) using an in-situ, laser-ablation process. The physical and electrical uniformity of a variety of the film properties are addressed. The results demonstrate that the laser-ablation process can readily be scaled up to coat three inch substrates while maintaining excellent compositional and thickness uniformity.

EXPERIMENTAL APPARATUS

In order obtain uniform laser-ablated films of YBCO over three-inch substrates, a special vacuum system was designed and built, as shown in Fig. 1. An excimer laser and focusing lens are not shown in the figure. The programmable mirror is used to reflect the incident laser beam through a quartz window and down onto a rotating (6 RPM), high density stoichiometric YBCO ablation target[11]. The angular position of the mirror is controlled using a computerized linear actuator to scroll the laser beam across a predetermined section of the 89 mm diameter target. The angular velocity of the mirror is adjusted to allow the beam dwell at larger radii for longer periods of time in order to obtain good thickness uniformity across three-inch substrates. The incident laser radiation makes an angle of 60° with respect to the target normal when the beam is directed at the target center. The laser can provide a maximum of 250 mJ/pulse at 248 nm (KrF) with a pulse width of 15 ns. Pulse repetition rates up to 200 Hz are possible. The focused spot size is about 1 mm by 3 mm, resulting in a fluence of up to 2 J/cm^2 at the target. The ablation plume is directed at the rotating (10 RPM) substrate which is held 12.7 cm above and concentric with the YBCO target. The substrate can be heated to temperatures in excess of 700°C. A quartz crystal microbalance is used to monitor both the growth rate and film thickness.

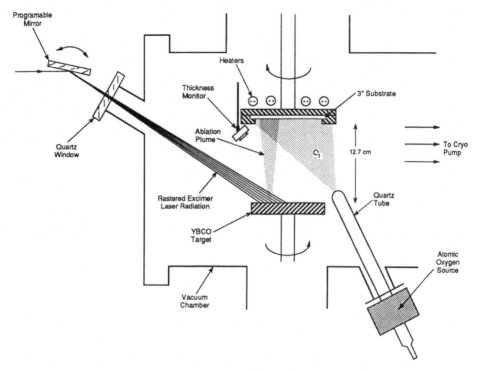

Fig. 1. Schematic of the laser-ablation deposition system. An excimer laser and focusing lens sit off to the left.

In order to facilitate in-situ growth a beam of atomic oxygen is directed at the substrate during the laser-ablation deposition process. The atomic oxygen is produced in an one-inch diameter quartz tube by a microwave discharge source[12] which is just external to the chamber. The dissociated oxygen diffuses down from the discharge region and exits a 3 mm hole in the end of the quartz tube which sits about 14 cm from the center of the substrate. This source can provide a flux of up to 5×10^{16} atoms/cm^2/sec at the substrate surface with a background oxygen over pressure of 5×10^{-4} Torr. This atomic flux was calculated by making a rough calibration of the amount of atomic oxygen exiting the quartz tube using a residual gas analyzer. The system is pumped with a 1500 L/s cryo pump and a base pressure below 1×10^{-8} Torr is readily achievable.

An important parameter during in-situ HTS film growth is the substrate temperature. Unfortunately, most of the useful substrates for YBCO growth are transparent in the infrared making them difficult to heat by radiation. Thus, the substrate was clamped to a stainless steel holder with molybdenum springs and a layer of indium was used between the substrate and holder to insure excellent thermal contact. An alternative substrate holder can also be used to hold several smaller substrates such as $SrTiO_3$ and MgO. A calibration of the substrate holders temperature versus heater power was made by inserting a small thermocouple into a hole drilled into the body of the holder. This calibration was checked periodically and the temperature was not found to vary by more than $\pm 5°C$ with input power. Due to radiative losses, the actual substrate temperature will be slightly less than that of the holder and the temperatures reported below should be taken as an upper limit of the growth temperature.

Typical operating conditions for YBCO film growth are as follows: laser fluence of 1 J/cm^2 at 248 nm (KrF), 20 PPS, chamber pressure of 5×10^{-4} Torr O_2 with 500 watts of RF power in the microwave discharge. Typical growth rates obtained with these conditions are 0.5 μm/hour. After growth, the heaters are turned off and the chamber is back filled with O_2 to a pressure of about 100 Torr. The sample then cools to room temperature in a period of about two hours.

RESULTS AND DISCUSSION

The issues of uniformity across three-inch wafers is first presented. Then, the electrical characteristics of the YBCO thin films on sapphire and $SrTiO_3$ substrates are discussed.

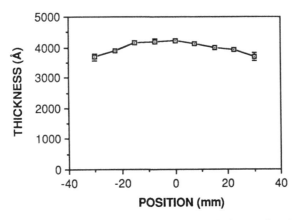

Fig. 2. Thickness versus position of a YBCO film laser-deposited onto a three-inch wafer. The thickness variation across the wafer is ± 5%.

Fig. 2 displays the film thickness of YBCO deposited onto a three-inch Si wafer. The thickness measurement was obtained by first patterning the film using photolithography and then wet chemical etched to remove narrow lines of the YBCO. Each data point displayed in Fig. 2 represents the average of four thickness measurements taken over a small segment centered around the indicated position point using a profilomiter. Error bars representing the standard deviation are included. As shown in Fig. 2, the film thickness is found to vary by no more than ± 5% over the three-inch diameter wafer. It should be pointed out that larger substrates could readily be coated uniformly with the laser-ablation process, if required.

The YBCO films produced, using the rastered laser beam, also show excellent compositional uniformity as well. Rutherford backscattering spectroscopy (RBS) was used to determine the chemical composition of an in-situ YBCO film deposited onto a (1102) sapphire wafer and the results yielded a

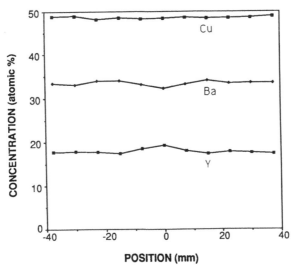

Fig. 3. The metal ion composition versus position across a three-inch wafer. The nominal film composition is $Y_{1.2}Ba_{2.0}Cu_{3.0}O_{6.6}$ as determined by RBS.

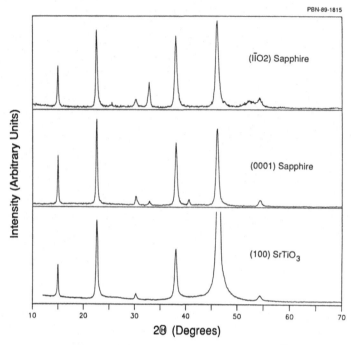

Fig. 4. X-ray diffraction pattern of YBCO films laser ablated onto ($1\bar{1}02$) sapphire, (0001) sapphire, and (100) SrTiO$_3$ substrates.

chemical composition of $Y_{1.2}Ba_{2.0}Cu_{3.0}O_{6.6}$. The slightly increased Y content of this film is consistent with other RBS data reported in the literature for both as-deposited and annealed YBCO laser-ablated films[1,8,13]. The composition of this film was then used as a standard for energy dispersive X-ray analysis (EDX) to determine the metal ion composition of YBCO films deposited onto a three inch Si wafer and the results are plotted in Fig. 3. The film composition does not vary by more than ± 0.6% over the diameter of the three-inch substrate. Depth profiling using SIMS analysis indicates that the metal ion composition remains constant through the bulk of the film down to the film-substrate interface. The uniformity of the metal ion composition obtained with this technique is considered to be state-of-the-art for laser ablation, as well as for any other physical vapor deposition technique.

X-ray powder diffraction of YBCO films deposited on both ($1\bar{1}02$) and (0001) sapphire as well as (100) SrTiO$_3$ indicate that the films are highly oriented with the $\hat{c}$-axis perpendicular to the substrate as seen in Fig. 4. The spacing of the (00l) peaks obtained from these scans indicate that the films are oxygen deficient ($\hat{c} \simeq 11.8$ Å)[14]. A rocking curve obtained from the (005) peak of a YBCO film grown on SrTiO$_3$ at 575°C has a full-width half-maximum of 1.5°C.

Fig. 5 shows a photograph of a 3500 Å in-situ laser-ablated YBCO film deposited on a three-inch sapphire wafer at 575°C with a fluence of 1 J/cm^2. The reflection of the ruler seen in this figure illustrates the excellent film morphology obtained using this in-situ deposition technique. However, this film, as well as most other laser-ablated YBCO films reported to date, show a number of particles which are ejected from the target during the ablation process and become incorporated into the film[8,15]. Figs. 6a and 6b show scanning electron micrographs (SEMs) of a YBCO film deposited onto a three inch sapphire wafer at magnifications of 2000 X and 20,000 X, respectively. The film is 5500 Å thick and the deposition conditions for this film were identical to those for the film shown in Fig. 5. As seen in Fig. 6a, there are 9 particles larger than 0.5 μm in size which are within an area of 1.2×10^{-5} cm^2, yielding a number density of 7.5×10^5 particles/cm^2. This particle number density was measured as a function of position across a three-inch wafer and the data is shown plotted in Fig. 7. The number densities of the particulate material plotted in Fig. 7 are representative of several wafers, and no effort was made by the SEM operator to find areas with higher or lower numbers of particles. It is interesting to note that the outer edges of the wafer appear to have a higher density of this particulate material. This may be due to the fact that the laser beam dwells at the outer edge of the ablation target for longer periods of time (in order to obtain good thickness uniformity) and the angular distribution of these ejected particles is even more forward scattered than the atoms and molecules

Fig. 5. Photograph of an in-situ laser-ablated YBCO film deposited onto a three-inch ($1\bar{1}02$) sapphire wafer.

making up the bulk of the ablation plume. The maximum particle size observed in these films was about 2 µm. It should be pointed out that these large particles will make patterning a large number of small structures (less than a few microns) using photo-lithography and dry etching difficult. However, for microwave applications typical device features are several hundreds of microns across, and these particles do not pose a significant patterning problem. Fig. 6b shows that aside from the small particles the films deposited with this technique are essentially featureless, displaying excellent surface morphology.

A variety of the electrical properties of these films have been measured as well. First, the room temperature resistivity was measured using a standard four point resistance probe as function of position across a three-inch ($1\bar{1}02$) sapphire wafer and the results are shown in Fig. 8. The film was grown at 600°C and is 5500 Å thick. The nominal room temperature resistivity of this film is about 1 mΩ-cm and is considered to be typical for films grown on sapphire with this technique. Similar films grown on $SrTiO_3$ have room temperature resistivities of about 0.7 mΩ-cm. The normal state resistivity of these films are the lowest

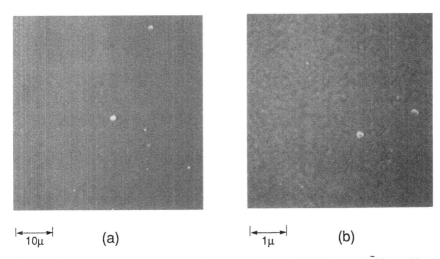

Fig. 6. Scanning electron micrograph of in-situ laser-ablated YBCO films on ($1\bar{1}02$) sapphire.

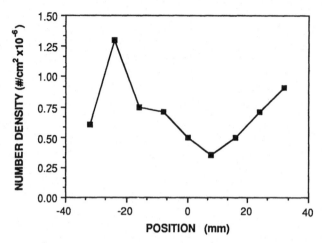

Fig. 7. The number density of laser-ejected particles larger than 0.5 μm as a function of position across a three-inch wafer.

reported to date for in-situ YBCO films grown using an RF source to produce the activated oxygen species[16,17]. The critical temperature (defined as the temperature that the film obtains zero resistance) measured as a function of position across half of a three-inch wafer is shown in Fig. 9. This data was obtained from small specimens cut from the sapphire wafer shown in Fig. 5. T_c was measured using a standard four point arrangement and is seen to increase at the outer edge of the wafer. This non-uniformity in T_c is not well understood at this time, but may be due to radial temperature gradients across the three-inch wafer. The best T_c's obtained with the atomic oxygen source have been 65 K on both ($1\bar{1}02$) sapphire and (100) $SrTiO_3$ as shown in Figure 10 for films grown at 600°C. As noted, the resistance of these films show metallic behavior as a function of temperature.

The reason for the low T_c's obtained with this technique is not clear at this time, although both RBS and X-ray powder diffraction indicate that the films are oxygen deficient. The flux of atomic oxygen provided by the microwave source is two orders of magnitude greater than the average metal ion flux produced by the ablation plume at a deposition rate of 1 Å/s. However, the metal ions arrive over a time

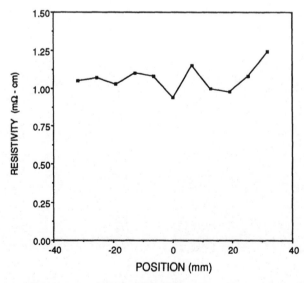

Fig. 8. Room temperature resistivity versus position of an in-situ laser-ablated YBCO film deposited onto a three-inch ($1\bar{1}02$) sapphire wafer.

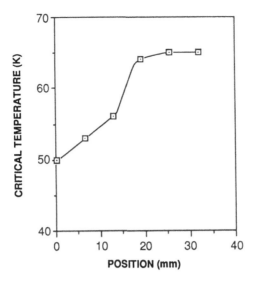

Fig. 9. Critical temperature versus position of an in-situ laser-ablated film on a three-inch ($1\bar{1}02$) sapphire wafer.

period of about 10 μs in a pulsed fashion (every 50 ms) assuming that the spread in the velocity of the atomic and molecular species ablated from the target is of the same order of their ejection velocity (1 X 10^6 cm/sec)[18,19]. Thus, the instantaneous growth rate is calculated to be 5000 Å/s and the atomic oxygen flux from the microwave source is then two orders of magnitude too low to fully oxidize the instantaneous metal ion flux. Other workers have shown that laser-ablated films only incorporate atomic oxygen during the arrival of the metal ions[5]. On the other hand, at a deposition rate of 1 Å/s, it takes about 12 seconds (240 pulses) to grow one monolayer of ĉ-axis oriented YBCO and it would appear that the atomic oxygen flux

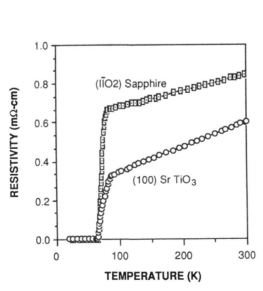

Fig. 10. Resistance versus temperature for in-situ laser-ablated YBCO films on (100) $SrTiO_3$ and ($1\bar{1}02$) sapphire grown at 600°C.

provided by the source would be sufficient to oxidize the growing film as each laser pulse provides only a small fraction of the growing YBCO lattice. Another possibility for the oxygen deficiency is the diffusion of oxygen out of the film during growth due to the elevated growth temperatures.

Attempts to increase the oxygen content of the films have been conducted without realizing any measurable improvement in film quality. First, simply increasing the flow of oxygen through the quartz tube by increasing the background pressure to 1 mTorr doubles the flux (assuming the recombination rate of O_1 in the tube is negligible). Second, by adding a second cryo pump (doubling the pumping speed of the system) the atomic oxygen flux was doubled while operating at the same background pressure. In either case, the increase in oxygen flow was found to be deleterious to the film's electrical properties. Also, several attempts at slow cooling[17] and post annealing did not improve the electrical properties of our YBCO films.

Microwave surface resistance measurements as a function of frequency were obtained for an in-situ laser-ablated YBCO film grown on a three-inch ($1\bar{1}02$) sapphire substrate, and the results are shown in Fig. 11. The film was grown at 575°C and is 5000 Å thick. The measurement was obtained by cutting a 1.00" by 0.5" piece from the coated three-inch wafer. This film was then incorporated as one ground plane in a three piece, superconducting Nb stripline resonator[20]. This RF measurement indicates that the film has a surface resistance which is 8 times better than that of copper at 10 GHz and a temperature of 4 K. The surface resistance of the film improves as the frequency is decreased, but deviates from the expected f^2 slope due to connector problems at low frequency. This is the lowest surface resistance for YBCO films grown on sapphire reported in the literature to date. This result is important for several reasons. Sapphire is inexpensive, has both excellent microwave and thermal properties, and is readily available in three-inch wafers. The advantages of sapphire clearly contrasts with those of newer substrates suggested for microwave applications such as $LaAlO_3$, $LaGaO_3$. These new substrates provide a good lattice match for YBCO, but are expensive, fragile, highly twinned, and are currently available in sizes only to 55 mm and 25 mm for $LaAlO_3$ and $LaGaO_3$, respectively.

CONCLUSIONS

In-situ YBCO HTS thin films have been deposited over three-inch wafers using the pulsed laser-ablation technique and excellent thickness (± 5%) and composition (± 0.6%) uniformity has been demonstrated. This is the first demonstration that the laser-ablation process can be scaled up to substrate sizes compatible with equipment used in the semiconductor industry.

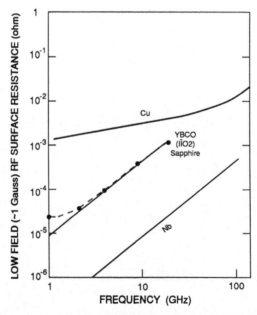

Fig. 11. Microwave surface resistance of an in-situ laser-ablated YBCO film on ($1\bar{1}02$) sapphire wafer measured at 4 K. The film used for this measurement has a critical temperature of 65 K.

The laser-ablated YBCO films grown with the atomic oxygen source appear to be deficient in oxygen which accounts for the relatively low critical temperatures of these films. The reason for the oxygen deficiency is unclear, but may be due to the pulsed nature of the deposition process. However, the films grown with this technique are highly ordered, display excellent surface morphology, and relatively low, normal state resistivity. This later fact may account for the excellent microwave properties measured at 4 K, as demonstrated by films grown on sapphire substrates. If the critical temperature of these films can be improved significantly above that of 77 K while maintaining excellent microwave properties, then sapphire may once again become a substrate of significant interest for high temperature superconducting thin films.

ACKNOWLEDGEMENTS

Thanks to Drs. D. Oates and R. Withers from M.I.T. Lincoln Labs for providing the microwave surface resistivity measurements. Also, thanks to D. Howe and B. Majewski for their patience and assistance with the material analysis work. Special thanks also to Huy Ngyuen for assisting in all aspects of this effort.

REFERENCES

1. D. Dijkkamp, T. Venkatesan, X. D. Wu, S. A. Shaheen, N. Jisrani, Y. M. Min-Lee, W. L. McLean, and M. Croft, "Preparation of Y-Ba-Cu Oxide Superconductor Thin Films Using Pulsed Laser Evaporation from High T_c Bulk Material," Appl. Phys. Lett. 51:619 (1987).

2. S. Witanachchi, H. S. Kwok, X. W. Wang, and D. T. Shaw "Deposition of Superconducting Y-Ba-Cu-O Films a 400]C Without Post-Annealing," Appl. Phys. Lett. 53:234 (1988).

3. G. Koren, A. Gupta, and R. J. Baseman, "Role of Atomic Oxygen in the Low-Temperature Growth of $YBa_2Cu_3O_{7-x}$ Thin Films by Laser Ablation Deposition," Appl. Phys. Lett. 54:1920 (1989).

4. B. Roas, L. Schultz, and G. Endres, "Epitaxial Growth of $YBa_2Cu_3O_{7-x}$ Thin Films by a Laser Evaporation Process," Appl. Phys. Lett. 53:1557 (1988).

5. G. Koren, A. Gupta, E. A. Geiss, A. Segmuller, and R. B. Laibowitz, "Epitaxial Films of $YBa_2Cu_3O_{7-x}$ on $NdGaO_3$, $LaGaO_3$, and $SrTiO_3$ Substrates Deposited by Laser Ablation," Appl. Phys. Lett. 54:1054 (1989).

6. A. Inam, M. S. Hegde, X. D. Wu, T. Venkatesan, P. England, P. F. Miceli, E. W. Chase, C. C. Chang, J. M. Tarascon, and J. B. Wachtman, "As-deposited High T and J_c Superconducting Thin Films Made at Low Temperatures," Appl. Phys. Lett. 53:908, (1988).

7. N. K. Jaggi, M. Meskoob, S. F. Wahid, and C. J. Rollins, "Superconductivity in Thin Films of Bi-Sr-Ca-Cu Oxide Deposited Via Laser Ablation of Oxide Pellets," Appl. Phys. Lett. 53:1551, (1988).

8. D. B. Geohegon, D. N. Mashburn, R. J. Culbertson, S. J. Pennycook, J. D. Budai, R. E. Valiga, B. C. Sales, D. H. Lowndes, L. A. Batner, E. Sunder, D. Eres, D. K. Christen, and W. H. Christe, "Pulsed Laser Deposition of Thin Superconducting Films of $Ho_1Ba_2Cu_3O_{7-x}$ and $Y_1Ba_2Cu_3O_{7-x}$," J. Mater. Res. 3(6):1169 (1988).

9. S. H. Liou, K. D. Aylesworth, N. J. Ianno, B. Johs, D. Thompson, D. Meyer, J. A. Woollam, and C. Barry, "Highly Oriented $Tl_2Ba_2Ca_2Cu_3O_{10}$ Thin Films by Pulsed Laser Evaporation," Appl. Phys. Lett. 54:760 (1989).

10. R. W. Simon, A. E. Lee, C. E. Platt, K. P. Daly, J. A. Luine, C. B. Eom, P. A. Rosenthal, X. D. Wu, and T. Venkatesan, "Growth of High-Temperature Superconductor Thin Films on Lanthanum Aluminate Substrates," to be published in the Proceedings of the Conference on Science and Technology of Thin Film Superconductors, Colorado Springs, Nov. 1988.

11. W. R. Grace and Company, Columbia, MD.

12. Applied Science and Technology, Inc., Woburn, MA.

13. X. D. Wu, D. Dijkkamp, S. B. Ogale, A. Inam, E. W. Chase, P. F. Micelli, C. C. Chang, J. M. Tarascon, and T. Venkatesan, "Epitaxial Ordering of Oxide Superconductor Thin Films on (100) SrTiO$_3$ Prepared by Pulsed Laser Evaporation," Appl. Phys. Lett. 51:861 (1987).

14. R. J. Cava, B. Batlog, C. H. Chen, E. A. Rietman, S. M. Zahurak, and D. Werder, "Single-Phase 60-K Bulk Superconductor in Annealed Ba$_2$YCu$_3$O$_{7-x}$ ($0.3<x<0.4$) with Correlated Oxygen Vacancies in the Cu-O Chains," Phys. Rev. B 36:5719 (1987).

15. N. Klein, G. Muller, H. Piel, B. Roas, L. Schultz, U. Klein, and M. Peiniger, "Millimeter Wave Surface Resistance of Epitaxially Grown YBa$_2$Cu$_3$O$_{7-x}$ Thin Films," Appl. Phys. Lett. 54:757 (1989).

16. J. N. Eckstein, D. G. Schlom, E. S. Hellman, K. E. vonDessonneck, Z. J. Chen, C. Webb, F. Turner, J. S. Harris, Jr., M. R. Beasley, and T. H. Geballe, "Epitaxial Growth of High-Temperature Superconducting Thin Films," J. Vac. Sci. Technol. B 7:319 (1989).

17. R. J. Spah, H. F. Hess, H. L. Stormer, A. E. White, and K. T. Short, "Parameters for In Situ Growth of High T$_c$ Superconducting Thin Films Using an Oxygen Plasma Source," Appl. Phys. Lett. 53:441 (1988).

18. J. P. Zheng, Q. Y. Ying, S. Witanachchi, A. Q. Huang, D. T. Shaw, and H. S. Kwok, "Role of the Oxygen Atomic Beam in Low-Temperature Growth of Superconducting Films by Laser Deposition," Appl. Phys. Lett. 54:954 (1989).

19. C. Girault, D. Damiani, J. Aubreton, and A. Catherinot, "Influence of Oxygen Pressure on the Characteristics of the KrF-Laser-Induced Plasma Plume Created Above an YBaCuO Superconducting Target," Appl. Phys. Lett. 54: 2035 (1989).

20. D. E. Oates and A. C. Anderson, "Superconducting Stripline Resonators and High T$_c$ Materials," Proceedings of MTT Long Beach, CA 267 (1989).

Low-pressure Metalorganic Chemical Vapor Deposition and Characterization of YBa$_2$Cu$_3$O$_{7-x}$ Thin Films

P. Zawadzki*, G. S. Tompa*, P. Norris*, D. W. Noh**, B. Gallois**, C. Chern***, R. Caracciolo*** and B. Kear***

* EMCORE Corporation, 35 Elizabeth Ave., Somerset, N.J. 08873; ** Stevens Institute of Technology, Hoboken, N.J. 07030; *** Rutgers, The State University, New Brunswick, N.J. 08854

ABSTRACT

Low-pressure metalorganic chemical vapor deposition (LP-MOCVD) is a technique which has been used with great success for the growth of compound semiconductors. In this study, we have successfully applied this technique to the deposition of yttrium barium cuprate (YBCO) superconducting thin films grown on a variety of substrate materials. These superconducting films have been deposited in a commercial-scale MOCVD reactor with a capacity of over 100 cm^2 per growth run. The EMCORE System 5000 reactor is based on a high-speed rotating disk susceptor (0-2000 rpm) design.

Thin films of the binary oxides were initially deposited in order to determine the respective growth rates. This data was then used to calculate trial values of reactant flows for YBCO growth. YBCO layers were grown at 76 Torr on substrates held at 500-550 °C. The ß-diketonates of Y, Ba and Cu, heated to 130, 240 and 120 °C, respectively, were used as sources. A 4.5 slpm flow of pure oxygen, representing a partial pressure of 38 Torr, was introduced uniformly at the top of the reaction chamber.

Post-growth annealing was performed in pure oxygen for 30-60 minutes at 900-950 °C. After annealing, initial YBCO layers grown on (100) YSZ substrates exhibited semiconducting behavior above the 90 K onset of the superconducting transition. The zero resistance temperature, $T_c(R=0)$, was 40 K. Under slightly different growth conditions, YBCO deposited on (100) SrTiO$_3$ exhibited a metallic characteristic, with onset and $T_C(R=0)$ values of 95 and 85 K, respectively. These layers also showed a rather smooth morphology compared to previously reported results using LP-MOCVD. Results of characterization by x-ray diffraction, SEM, EDAX and XPS are reviewed. Evidence of low carbon/carbonate contamination was observed, however a significant substrate/layer interaction was detected in the case of YSZ.

1. INTRODUCTION

Since the discovery of high critical temperature oxide based superconductors (HTSC)[1,2] many techniques have been investigated to prepare bulk and thin film samples of these materials. Development of high quality epitaxial thin films on suitable substrates is necessary for the application of HTSCs in electronic applications. Potential applications are SQUIDs, Josephson junctions and passive microwave devices such as ring resonators, filters and delay lines. In the future ultrastructured devices incorporating alternating layers of insulators, semi-conductors and superconductors will be outgrowth of present developments.

Presently, several techniques have been used to produce thin film HTSCs. These techniques include sol-gel[3], sputtering[4,5], evaporation[6,7], laser ablation[8,9], Molecular Beam Epitaxy (MBE)[10] and Metal Organic Chemical Vapor Deposition (MOCVD). Among these, the last three have shown the best results. Laser ablation, however, is limited by the control of large molecule (or droplet) formation and has not been shown to be suitable for large area epitaxy. MBE is limited by low process thruput and the inability to incorporate significant levels of oxygen during growth.

MOCVD is a fast growing, mature, and very promising technology in the compound semiconductor industry for epitaxial growth of III-V and II-VI materials and extremely versitile for the future applications. An important advantage of the MOCVD technique is the ability to perform growth under a high oxygen overpressure. The MOCVD growth mechanism involves the pyrolysis of organometallic compounds which have appropriate vapor pressure. The vaporized organometallic source materials (precursors) are transported by a carrier gas and injected into the reaction chamber. After the film deposition on the heated substrates, the by-products are carried away in the vapor phase and safely exhausted via a scrubbing system.

Several groups have demonstrated MOCVD-grown films of superconducting $YBa_2Cu_3O_{7-x}$[11,12] (YBCO) as well as for the Bi[13]-and Tl-[14] based superconducting oxide films. We present here our initial results for the films grown in a commercially available production scale growth system.

2. EXPERIMENTAL PROCEDURES

All depositions were done in a prototype EMCORE series 5000 MOCVD system. The growth chamber was a vertical, high speed rotating disk reactor. The wafer carrier was radiatively heated from below by an electrical resistance filament. A schematic drawing of the reactor is shown in Figure 1. The high rate of rotation produces a pumping action because of viscous drag at the surface of the rotating disk. This results in a lateral flow of gas near the substrate surface, promoting uniform growth. A uniform downward flow of carrier gas and/or oxygen (between 4.5 and 6 slm.) was introduced uniformly from the top of the reactor. The chamber lid contains two zones a small quartz, perforated appeture disk and a concentric, annular, perforated, stainless steel plate. The gas flow rate could be controlled to each zone with the flows being kept proportional to the suface area of the zone. Reactant distribution was further controlled through the use of dividing flow injection manifolds(injector tubes)[15]. Both the injector tubes and the uniform flow of gas from the top of the reactor result in a reduction of convection cells (recirculation) giving sharp interfaces in multilayer structures. The internal components of the growth chamber were chosen for their ablility to withstand the high temperature oxidizing environment. The 5 inch diameter carrier which held the substrates was machined so that depositions could be carried out on a variety of substrate materials (i.e., (100)YSZ, (100)MgO, (100)SrTiO$_3$, (100)Si, and (1$\bar{1}$02) Saphire) within a single growth run.

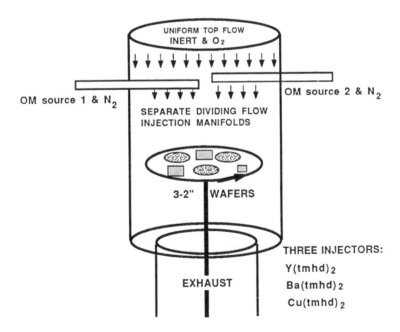

Fig. 1 A schematic drawing of a vertical, high-speed rotating disk MOCVD reactor chamber

The ß-diketonates (2,2,6,6-tetramethyl-3,5-heptanedionates) of yttrium, barium, and copper were chosen as precursors because of their previously successful use.[16] Each of the precursors was held in a uniformly heated, stainless steel cylinder (vapor source). The vapor source temperatures were maintained at 130, 240, and 120°C for Y-, Ba-, and Cu- precursors respectively. Nitrogen, preheated to the source temperature, was used as the carrier gas to transport precursors. Typical nitrogen flows for the sources varied from 100-500, 100-1200, and 500-800sccm for the Y, Ba and Cu precursors, respectively. All lines in contact with the precursors were heated at approximately 10°C above the source temperature to avoid condensation. Also additional preheated nitrogen was added to the vapor carrier mixture downstream from the source (bringing the total flow to 1.5 slm) to decrease the dew point temperature.

The precursor vapor and nitrogen gas mixture entered the chamber through three separate injector tubes placed 120° apart to minimize the possibility of prereactions between the source materials. The multiple tube arrangement was also necessary because of the wide range of temperatures used to volatize the precursors.

The vapor sources are configured to allow three modes of operation; idle, vent, and run. In the idle state, carrier gas by-passes the precursor source, flowing directly to the reaction chamber. Vent mode introduces the carrier gas into the vapor source, but the carrier/precursor mixture is routed through the vent line to the system vacuum pump. This configuration allows the mass transport flux of the precursor to stablize before being switched to the growth chamber. The run mode allows the precursor mixture to enter the growth chamber. The run-vent mode was operated under pressure balanced operation which is important for sharp interfaces on multilayer structures. Figure 2 shows a schematic of the series 5000 MOCVD system.

Initial growth runs were single component metallic or binary oxide films deposited on (100) Si wafers at 500 °C. These films were used to investigate precursor transport and growth rates. This information was then used to estimate the proper values of the three precursor flows in order to obtain stoichiometric YBCO films. The actual depositions of the YBCO films were performed at 500 °C, with a growth chamber pressure of 76 Torr, and rotation speed of 1100 rpm. Pure oxygen was introduced through the top of the reactor at 4.5 slm. corresponding to a partial pressure of 38 Torr. At the completion of the growth run, the film was cooled to 100 °C under the mixture of nitrogen and oxygen (1:1 flow ratio). The samples were then annealed from 900 to 950 °C for 30 to 60 min. in pure atmospheric pressure oxygen and then slowly cooled ($\approx 3°C/min$) to room temperature.

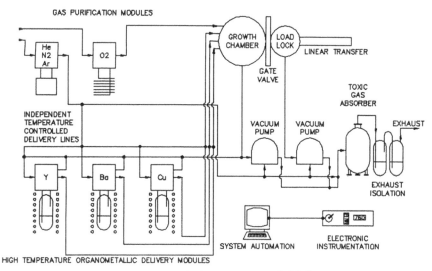

Fig. 2 A schematic of the system of high temperature superconducting thin film MOCVD

3. RESULTS

Fig. 3 (a), (b) and (c) show the Scanning Electron Microscopy (SEM) micrographs of the Y, Ba, and Cu oxide films on (100) Si respectively. In general, the growth rate was proportional to the flow rate. Each film was grown for 2 hours. The yttrium oxide film appeared bright and shiny metallic yellow. The copper oxide appeared dull reddish-orange and the barium oxide had a dull black color. The cross section of the micrographs apparently shows columnar growth. The top surfaces of the the oxide films show clusters of small domes with a diameter ~0.1 - 0.2 μm in agreement with the cross sectional views.

X-ray photoelectron spectroscopy (XPS) was also used to investigate the single oxide films. Full XPS spectra for the Y, Ba, and Cu oxide films are shown in Fig. 4 (a), (b), and (c) respectively. In Fig. 4 (a), we observed the Y $3d_{5/2}$ located at 156.4 eV with a 2 eV gap between Y $3d_{3/2}$ and Y $3d_{5/2}$. This is commonly observed for a standard Y_2O_3 spectrum[17]. The O 1s peak is located at about 531.5 eV which confirms the existence of the Y_2O_3. Fig. 4 (b) shows the Ba $3d_{5/2}$ located at ~ 779.6 eV correspond to the standard BaO spectrum[17]. Fig. 4 (c) shows the Cu $2p_{3/2}$ located at ~933.4 eV which indicates a standard CuO spectrum. All three spectra show that a trace of carbon impurity was incorporated in the film. The carbon impurity may be due to fragments of the metal organic precursor molecules. The effect of varing the O_2/N_2 pressure on carbon incorporation was not investigated at this stage.

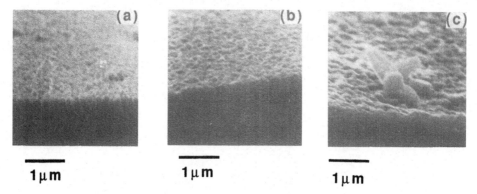

Fig. 3 Scanning electron micographs of a) yttrium oxide, b) barium oxide, and c) copper oxide deposited on (100) Si substrate.

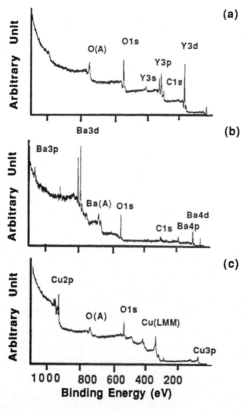

Fig. 4 XPS spectrum of a) yttrium oxide, b) barium oxide, and c) copper oxide deposited on (100) Si substrate

Eneregy Dispersive Analysis of X-ray (EDAX) was used to optimize the composition of the films after the growth and annealing. Table 1 lists the atomic ratios in percent and compositional ratios with respect to Y, Ba and Cu using a semi-quantitative software program. For comparison purposes, an $YBa_2Cu_3O_{7-x}$ superconducting thin film on MgO, deposited by pulsed laser ablation,[18] was used as a reference standard. The low T_c of the standard film reflects the substrate and not the composition. Sample I is an MOCVD grown YBCO film deposited on (100) Y-stabilized zirconia (YSZ), and sample II is on (100) $SrTiO_3$. The proportions of Y, Ba, and Cu were adjusted to match the measured composition of the standard thin film sample between sample I and sample II growth run. The areas under the peaks of Y Lα, Ba Lα and Cu Kα were compared for the relative compositions. The standard film (T_c = 82K) and the sample II (T_c = 85K) agree both in atomic and compositional ratios. However, sample I was barium-rich compared to the other two thin films and the T_c was lower, e. g., 40K.

After the film growth, the sample I was annealed at 900 °C in 1 atmospheric pressure of oxygen for 1 hour. The resistance versus temperature characteristics are shown in Fig. 5 (b). A semiconductor-like resistivity behavior was observed between 300 K and T_{onset}, 90 K. The T_c (R = 0) of this film was 40 K. X-ray diffraction (XRD) data of the sample I showed the existance of $BaZrO_3$ in addition to the (00l) peaks, indicating a substrate/film interaction. Also relatively weak peaks of impurities were observed, suggesting the existence of small volumes of mixed oxides like Y_2CuO_5 and $BaCuO_5$. The impurities might cause semiconducting behavior as well as the long tail in the superconducting transition.

TABLE 1. Energy dispersive analysis of X-ray data comparison of the YBCO superconducting thin films

	ATOMIC RATIO (%)			COMPOSITIONAL RATIO: [C]/Y			T_c
	Y	Ba	Cu	Y	Ba	Cu	
LASER ABLATED FILM ON MgO (STANDARD)	17.0	23.6	59.0	1	1.36	3.38	82*
SAMPLE I ON YSZ	17.0	33.4	48.3	1	1.96	2.38	40
SAMPLE II ON $SrTiO_3$	18.6	25.0	56.4	1	1.34	3.03	85

* equivalent film on $SrTiO_3$ had T_c = 92K

SrTiO$_3$ (100) substrates were also investigated because of the close lattice match to the YBCO superconducting phase. Using optimized growth condition, sample II was annealed at 950 °C for 30 min. Fig. 5 (a) shows the resultant resistance versus temperature characteristics. The film showed a metallic resistivity behavior with the $T_{onset} \approx 95$ K and T_c (R=0) = 85 K. The SEM micrograph of the annealed film, Fig. 6 (a), shows a plate-like morphology which is different from the columnar morphology of the as-deposited films. Some sharp boundaries, however, between the plate-like grains were seen.

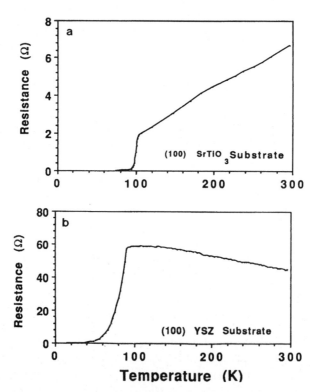

Fig. 5 Resistance vs. temperature curves of oxygen post-annealed a) YBCO films on (100) SrTiO$_3$ (Tc = 85K) and b) YBCO film on YSZ (Tc = 40k)

XRD pattern of the sample II is in Fig. 6(b), which indicates the c-axis perpendicular to the substrate surface orientation. The relative intensity of the (00l) reflections is consistent with those previously reported for MOCVD grown YBCO superconducting films on SrTiO$_3$[11].

The discontinuities at 23~24° and 46~50° of 2 theta are caused by intentionally skipping over the strong substrate peaks to protect the X-ray detector. The peak at ~41.7° indicated the existence of an unknown impurity phase[19,20], possibly resulting from a substrate/film interaction. The c-axis lattice parameter of the orthorhombic unit cell based on the (002), (005), (007) calculations varied from 11.65 to 11.70 Å.

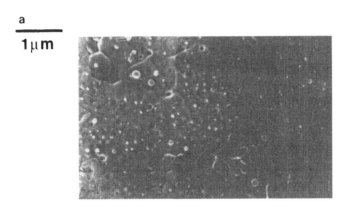

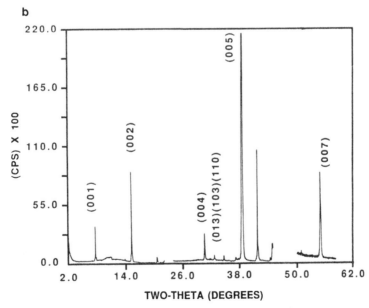

Fig. 6 a) Scanning electron micrographs of YBCO films on (100) SrTiO$_3$ (Tc = 85K) and b) X-ray dif-fraction pattern of (a) showing (00l) preferred orientation.

4. CONCLUSION

Superconducting thin films of $YBa_2CU_3O_{7-x}$ were deposited successfully by MOCVD at 76 torr in a vertical, high-speed rotating-disk reactor. The initial results on superconducting thin films on (100) YSZ and (100) $SrTiO_3$, after annealing, showed the onset temperature above 90K and zero resistace occured at 40K and 85K for (100) YSZ and (100) $SrTiO_3$ respectively. YBCO superconducting thin films deposited on (100) $SrTiO_3$ exhibited c-axis perpendicular to the substrate orien-tation and a metallic behavior in normal state. Both films showed evidence of an interaction between the substrate and the film.

ACKNOWLEDGEMENT

The authors would like to acknowledge Dennis Stucky and Glen DeRose for their technical assistances during the system construction. The authors also wish to acknowledge Dr. X. D. Wu of Rutgers the State University of NJ for his contribution in resistance measurement. This work was partly supported by the New Jersey Commision on Science and Technology and the Department of the Air Force, Aeronautical Systems Division, Wright Paterson AFB under contract No. F33615-C-5464

REFERENCES

1. J. G. Bednorz and K. A. Muller, Possible high T_c superconductivity in the Ba-La-Cu-O, Z. Phys. B64, 189 (1986)
2. C. W. Chu P. H. Hor, R. L. Meng, L. Gao, Z. J. Huang, and Y. Q. Wang, Evidence for superconductivity above 40K in the La-Ba-Cu-O compound system, Phys. Rev. Lett. 58, 405 (1987).
3. S. Shibata, T. Kitagawa, H. Okagaki, and T. Kimura, C-axis oriented superconducting oxide film by the sol-gel method, Jpn. J. Appl. Phys. 27, L646 (1988)
4. M. Hong, S. H. Liou, J. Kwo, and T. R. McGuire, Superconducting Y-Ba-Cu-O oxide by sputtering, Appl. Phys. Lett. 51, 694 (1987)
5. G. C. Xiong and S. Z. Wang, Epitaxial growth of superconducting $YBa_2Cu_3O_{7-x}$ thin films by reactive magnetron sputtering, Appl. Phys. Lett. 55, 902 (1989)
6. F. H. Garzon, J. G. Beery, D. R. Brown, R. J. Sherman, and I. D. Ruistrick, Characterization of Y-Ba-Cu-O superconducting thin films prepared by coevaporation of Y, Cu, and BaF_2, Appl. Phys. Lett. 54, 1365 (1989)
7. C. B. Rice, A. F. J. Levi, R. M. Fleming, P. Marsh, K. W. Baldwin, M. A. Anzlowar, A. E. White, K. T. Short, S. Nakahara, and H. L. Stormer, Preparation of superconducting thin films of

calcium strontium bismuth copper oxides by co-evaporation, <u>Appl. Phys. Lett.</u> 52, 1828 (1988)

8. A. Inam, M. S. Hegde, X. D. Wu, T. Venkatesan, P. England, P. F. Miceli, E. W. Chase, C. C. Chang, J. M. Tarascon, and J. B. Wachtman, As-deposited high T_c and J_c superconducting thin films made at low temperature, <u>Appl. Phys. Lett.</u> 53, 908 (1988)

9. G. Koren, A. Gupta, E. A. Giess, A. Segmuller, and R. B. Laibowitz, Epitaxial growth of 9. G. Koren, A. Gupta, E. A. Giess, A. Segmuller, and R. B. Laibowitz, Epitaxial growth of $YBa_2Cu_3O_{7-x}$ on NdGaO, LaGaO, and SrTiO substrate deposited by laser ablation, <u>Appl. Phys. Lett.</u> 54, 1054 (1989)

10. J. Kwo, J. C. Hsieh, R. M. Felmeng, M. Hong, S. H. Liou, B. A. Davidson, and L. C. Feldman, Structural and superconducting properties of orientation-ordered $YBa_2Cu_3O_{7-x}$ films prepared by molecular beam epitaxy, <u>Phys. Rev.</u> B36, 4039 (1987)

11. H. Yamane, H. Masumoto, T. Hirai, H. Iwasaki, K. Watanabe, N. Kobayashi, and Y. Muto, Y-Ba-Cu-O superconducting films prepared on $SrTiO_3$ substrates by chemical vapor deposition, <u>Appl. Phys. Lett.</u> 53, 1548 (1988)

12. K. Zhang, B. S. Kwak, E. P. Boyd, A. C. Wright, and A. Erbill, C-axis oriented $YBa_2Cu_3O_{7-x}$ superconducting films by metalorganic chemical vapor deposition, <u>Appl. Phys. Lett.</u> 54, 380 (1989)

13. A. D. Berry, R. T. Holm, E. J. Cukauskas, M. Fatami, D. Gaskill, R. Kaplan, and W. Fox, Formation of high T_c superconducting films by organometallic chemical vapor deposition, <u>J. Cryst. Growth</u> 92, 344 (1988)

14. J. Zhang, J. Zhao, H. Marcy, and L. Tonge, Organometallic chemical vapor deposition routes to high T_c superconducting Tl-Ba-Ca-Cu-O films, <u>Appl. Phys. Lett.</u> 54, 167 (1989)

15. G. S. Tompa, M. McKee, C. Beckham, P. A. Zawadzki, J. M. Colabella, P.D. Reinert, K. Capuder, R. A. Stall, and P. E. Norris, A parametic investigation of GaAs epitaxial growth uniformity in a high-speed, rotating disk MOCVD reactor, <u>J. Cryst. Growth</u>, 93, 220 (1988)

16. H. Yamane, H. Kurosawa, and T. Hirai, Preparation of $YBa_2Cu_3O_{7-x}$ films by chemical vapor deposition, <u>Chem. Lett.</u>, 939 (1989)

17. C. Wangner, W. M. Riggs, L. E. Davis, J. F. Moulder, and G. E. Muilenberg, "Handbook of X-ray photoelectron spectroscopy", Perkin-Elver Corp., Eden Prairie, MN (1979)

18. Courtesy of X. D. Wu, Rutgers, The State University, NJ

19. M. Naito, R. H. Hammond, B. Oh, M. R. Hahn, J. W. P. Hsu, P. Rosenthal, A. F. Marshal, M. R. Beasley, T. H. Geballe, and A.

Kapitulink, Thin film systhesis of the high Tc oxide superconductor $YBa_2Cu_3O_{7-x}$ by electron-beam codeposition, J. Mater. Res., 2, 713 (1987)

20 C. T. Cheung and E. Ruckstein, Superconductor-substrate interaction of the Y-Ba-Cu-O oxide, J. Mater. Res., 4, 1 (1989)

ELECTRICAL TRANSPORT MEASUREMENTS ON POLYCRYSTALLINE

SUPERCONDUCTING Y-Ba-Cu-O FILMS

M. A. Stan, S. A. Alterovitz, and D. Ignjatovic

NASA - Lewis Research Center
Cleveland, Ohio

ABSTRACT

The current-voltage, I-V, characteristics of polycrystalline Y-Ba-Cu-O films have been measured as a function of temperature. The I-V characteristics are interpreted using a model based upon an array of weak links with a statistical distribution of critical currents. In addition, we find evidence that the supercurrents flow in nearly independent filaments near T_c. Various criteria are discussed with respect to the definition of the transport critical current, I_c, in these films. A temperature dependence for I_c has also been deduced from the I-V data by appealing to an empirical scaling law. We propose that this temperature dependence, $I_c \propto (1-T/T_c)^{2.2}$, is representative of the weaker links within the critical current distribution.

INTRODUCTION

The T_c measurement is the most commonly used method for the characterization of superconducting materials. However, I_c is a more effective parameter in determining technological usefulness. In particular it is desirable to measure the temperature dependence of I_c in order to illuminate the mechanism limiting I_c, such as depairing in a single crystal sample, or perhaps tunnelling through grain boundaries in a polycrystalline sample. In the presence of a magnetic field, single crystal high temperature superconducting (HTS) samples have been shown to exhibit flux creep[1], while, in polycrystalline samples a variety of effects ranging from weak link limited I_c behavior at low magnetic fields to flux flow limited I_c behavior in high magnetic fields[2] have been observed. Implicit in any of these I_c measurements is some criterion for the definition of I_c. The selection of a criterion for I_c has been particularly difficult in the HTS materials because, in many instances, the resistive transition is not sharp. Several attempts to standardize the criteria for I_c in the presence of a magnetic field have been proposed[3,4]. However, for thin films in zero magnetic field, a constant field criterion ranging from 0.1 µV/mm to 1 µV/mm is widely used. While this criterion may be satisfactory for epitaxial thin films, where the critical supercurrent density, J_c, is of the order of 10^6 A/cm^2 at 77 K and the transition is sharp, the meaning of an I_c obtained in this way for polycrystalline films is not clear, owing to the broadness of the resistive transition in these films. The lack of a consensus on the temperature dependence of I_c, and in particular, the exponent in the expression $(1-T/T_c)^n$ ($1.5 \leq n \leq 2.0$), could be symptomatic of the arbitrariness of an I_c obtained from a constant voltage criterion[5,6,7,8]. While the various exponents reported for the temperature dependence of I_c may relate in some way to the microstructure in the films, we believe that one must first understand the mechanism responsible for the transition in order for the measured I_c to have meaning.

To our knowledge, only one attempt has been made to understand the resistive transition in HTS polycrystalline films. This was the work of England et al.[9], where they suggested that the films underwent a phase locking transition similar to that found in compacted polycrystalline Ta samples[10]. Within this model, the phase, θ_i, of the wave function of each of the grains is uncorrelated from grain to grain when $T > T_c$ (T_c being defined at R = 0). At T_c, the phase difference, $\theta_i - \theta_j$, between the neighboring grains becomes fixed because the thermal fluctuations ($k_B T_c$) are exceeded by the Josephson coupling energy ($hi_c/2e$). Here i_c is the intergrain critical current. The strongest evidence for this mechanism in polycrystalline HTS films came from the I-V data at T_c, where it was shown that $V(T_c) \propto I^2$. The quadratic dependence of V on I is a prediction of the phase locking model[10].

Recently the resistive transition of sintered Y-Ba-Cu-O wires in a magnetic field has been modeled by Evetts et al.[11] using extensions to the conventional model given for superconducting multifilamentary composites[12]. The extended model, hereafter referred to as the Weak Link Filament Array (WLFA) model, treats the sample as an array of weak links with a normal distribution of I_c's. Essential features of our I-V data are discussed in terms of this model. The implications of this model with respect to defining I_c are then discussed.

Finally, by appealing to an empirical scaling law for the I-V data, we can deduce a temperature dependence for an I_c which is representative of the weakest links in the I_c distribution.

EXPERIMENTAL

Sample Preparation

The films were prepared by sequential evaporation of Cu, Y, and BaF_2 onto (100) $SrTiO_3$. The films had a superstructure period of three, for a total of twelve layers, and a nominal thickness of 1 μm prior to annealing. Details of the deposition parameters and annealing have been published elsewhere[13]. Scanning electron micrographs of the films showed rectangular shaped grains with dimensions of 0.3 μm X 1.0 μm having no epitaxy with the substrate. X-ray diffraction analysis of the films indicated a polycrystalline nature with some a-axis texturing, as well as the presence of $BaCuO_2$ and Y_2O_3 phases. Electrical contact to the films was made by depositing a 1 μm layer of Ag through a shadow mask, resulting in four equally spaced 0.1 cm wide strips traversing the short dimension ($\simeq$ 0.3 cm) of the sample. The contacts were subsequently annealed in dry O_2 at 500° C for approximately two hours resulting in contact resistances of 10 μΩ or less. Wires were then attached to the Ag strips via In-soldering.

Electrical Transport Measurement

Transport measurements in low magnetic fields (0 - 70 Oe) were made in a ^{4}He closed cycle refrigerator with an externally mounted Cu wire Hemholtz coil around the sample chamber. The samples were attached to an OFHC copper holder and housed in an OFHC copper radiation shield. Cooling of the assembly was done by convection through 1 atm (STP) of ^{4}He gas to the refrigerator cold head ($\simeq$ 16 K). The sample temperature was determined by a Si diode which is epoxied into the sample holder, and has a rated accuracy of ± 0.25 K over the temperature interval of the measurements.

All the transport measurements discussed below were made using the conventional four-probe method. The T_c measurement of the films was made with a typical measuring current density of 0.2 A/cm^2 and a voltage sensitivity of approximately 10 nV. This voltage sensitivity was obtained using a nanovoltmeter in conjunction with signal averaging.

TRANSPORT DATA AND DISCUSSION

Nine samples were measured, but we will present detailed results on one typical film. A common feature of the films is that they have a zero resistance temperature, T_c, in the range 60 K - 70 K. The films exhibit a metallic nature above the onset temperature of 90K but the resistance is not linear in temperature, presumably because of the various grain orientations and presence of other phases.

I-V data, obtained at six different temperatures, is displayed in Fig. 1. The T_c of this film was 62.3 K. The upper and lower horizontal dashed lines

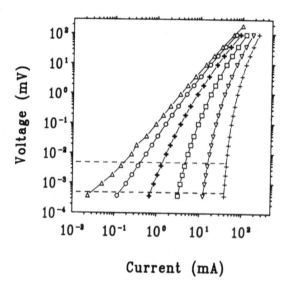

Fig. 1 Voltage against current obtained for a sequence of temperatures. From right to left: 20 K, 40 K, 51 K, 58 K, 62 K, and 64 K.

in Fig. 1 represent I_c voltage criteria of 2.5 μV/mm and 0.25 μV/mm, respectively. A vertical line drawn from the intersection of one these lines with the I-V data yields a value for I_c. At 20 K the two criteria yield essentially the same I_c while at 58 K they yield numbers differing by a factor of two. Obviously, an important parameter in defining I_c is the logarithmic derivative dLog(V)/dLog(I) commonly referred to as the n value of the resistive transition. The smaller the n value at the chosen voltage criterion the more sensitive I_c is to the voltage criterion. Because the n value is temperature dependent and approaches one at T_c, it is inevitable that this type of criterion will give I_c a weaker temperature dependence the nearer one gets to T_c. This effect is demonstrated in the I_c vs. $1-T/T_c$ characteristic shown in Fig. 2. The upper four sets of I_c data in Fig. 2 were obtained using different voltage criteria. The solid lines are guides to the eye. The lowest line in Fig. 2 was obtained from a data fitting procedure and will be described later. Notice that the upper four data sets are straight lines at the low temperature end of the graph with a slope of $\simeq 2$. The data show a departure from linear behavior at diff-

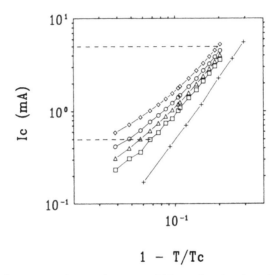

Fig. 2 I_c against $(1-T/T_c)$ obtained using different criteria. From top to bottom: 2.5 $\mu V/mm$, 1 $\mu V/mm$, 0.5 $\mu V/mm$, 0.25 $\mu V/mm$, and data from a least-squares fit of the I-V data to $V = A(I-I_c)^m$.

erent temperatures depending upon the voltage criterion. One final note with regard to the constant voltage criterion concerns the lowest meaningful I_c measurement. To be self consistent with the T_c measurement the I_c data must be greater than or equal to V_c/R_c, where R_c is the smallest detectable resistance in the T_c measurement and V_c is the voltage criterion used in the I_c measurement. The upper and lower horizontal dashed lines in Fig. 2 represent the lowest self consistent values of I_c for the I_c voltage criteria of 2.5 $\mu V/mm$ and 0.25 $\mu V/mm$, respectively. Failure to make the T_c and I_c measurements consistent with one another may result in assignment of I_c values to very resistive samples[14]. Regardless of the I_c criterion, one must first understand the nature of the resistive transition before one can attach any physical significance, such as tunneling, to I_c measurements on polycrystalline HTS films.

It has long been recognized that the broad resistive transitions into the flux flow state in type-II superconductors can be explained in terms of the variation in the local I_c along the sample length, where the flux-flow voltage is determined by that fraction of material whose I_c is lower than the applied current[15,16]. The I-V characteristic is then determined by

$$V = \left(\frac{\rho}{A}\right) \int_0^I \int_{i_{cm}}^i f(i_c) \, di_c \, di \qquad (1)$$

where $f(i_c)$, ρ, and A are the critical current distribution function, the flux-flow resistivity, and the cross-sectional area, respectively. The lower limit, i_{cm}, on the second integral in Eq. 1 is the smallest i_c in the distribution.

In type-II multifilamentary composites the transition into the flux-flow state is broad as a result of a distribution of localized constrictions along the individual filaments. For such structures the empirical relation $V \propto I^n$ seems to characterize the resistive transition quite well. Recently a theoretical understanding of this relation was obtained by Plummer and Evetts[12] by assuming the constrictions were normally distributed. The calculations were made using Eq. 1 and were based on either independent filaments or nearest neighbor coupling. There are three important results which apply for either limit: the scaling law $V \propto I^n$ is a natural consequence of a normal distribution of inhomogeneities; the n value is proportional to $(<I_c>/\sigma)^{5/3}$ where $<I_c>$ and σ are the mean critical current and the width of the distribution, respectively; and the measured I_c, obtained by a constant voltage criterion, will increasingly underestimate $<I_c>$ as the n value decreases.

Recently, Evetts[11] has been successful in qualitatively explaining the resistive transition of sintered Y-Ba-Cu-O wires in a magnetic field using the WLFA model. The WLFA model is an extension of the independent filament model to include long range coupling and the tunnelling nature of the weak links. In the WLFA model, the conductor geometry is essentially a multiply connected weak link network. However, in low density samples near the percolation threshold, there are many junctions in series between parallel interconnections and one has, effectively, an array of independent filaments, in complete analogy to the multifilamentary composite system. If a normal distribution of junction I_c's is assumed, n bears the same significance to $<I_c>$ and σ that it does in the multifilamentary composite system.

In the multiply connected limit of the WLFA model it is argued[11] that there is a length scale, Λ, in the direction perpendicular to the direction of applied current, which is used to describe the degree to which junctions within Λ switch to the normal state at the same current. For low values of magnetic field, temperature, and current the network is isotropic and Λ extends across the entire sample. If M represents the number of connected junctions within Λ, then it is argued that the n value is increased by a factor $\sqrt{M}$, as compared to the single filament result, and the effective width of the normal distribution becomes $\sigma/\sqrt{M}$. In the WLFA model Λ falls with increasing current since larger applied currents require larger transverse balancing currents which in turn

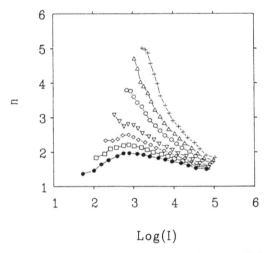

Fig. 3 Logarithmic derivative of the I-V data for a sequence of temperatures near T_c. From top to bottom: 54 K, 56 K, 58 K, 60 K, 61 K, 62 K, and 63 K.

increase the probability that the transverse junctions will be broken. The reduction in Λ manifests itself as a current dependent reduction of the n value, a result which is qualitatively different from multifilamentary composite systems where n is independent of the current. At this point we caution the reader that when the junctions are multiply connected Eq. 1 no longer applies, and one must resort to numerical techniques[11] in order to obtain an I-V relation.

Referring once again to Fig. 1, we believe that the I-V data qualitatively exhibits the main features of the WLFA model. These features are: the larger the n value, the larger the current required to drive the film normal, and at a given temperature, the n value falls monotonically with increasing current. In addition, the I-V data shows evidence near T_c that the films are nearly in the independent filament limit, i.e. that the films are made up of independent chains of weak links. If one looks closely at the low voltage portion of the I-V characteristic for the two left most curves, a tail in the data is evident. The data designated by the open circles was obtained at T_c and the left most data about 1 K higher. To make the tail more prevalent we plot the n value against Log(I) in Fig. 3 for seven different temperatures spaced 1 K apart, with the lowest temperature being at the top of the graph. It is evident that the tail appears very abruptly within approximately 1 K of the T_c value determined in the resistivity measurement. In the experiments of Evetts et al.[11], where the magnetic field was varied and the temperature held constant, the same feature was observed to appear suddenly at a field of 2 mT for sintered Y-Ba-Cu-O wires immersed in LN_2. An obvious explanation for the tail is that the measured voltage contains an ohmic component and a nonlinear component. The ohmic component could be a result of tunnel junctions, which have been driven normal, in series with the nonlinear component from the remaining portion of the weak link connected filaments. The fact that an ohmic component of the voltage exists near T_c implies that the weak links are not multiply connected. If the junction array was multiply connected there would always be a superconducting path across the sample and the ohmic contribution from the normal junctions would not be visible.

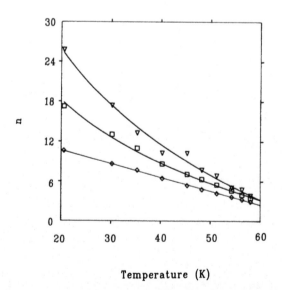

Fig. 4 Resistive transition n value against temperature obtained using different criteria. From top to bottom: 0.4 μV, 4 μV, and 40 μV.

An implication of the WLFA model on the measurement of the intergranular critical current is that the measured I_c is nearly equal to $<I_c>$ at any given voltage criterion provided that the n value at the voltage criterion is large. However, when n becomes small one tends to underestimate $<I_c>$ at any given voltage criterion. The latter point indicates that the I-V curves are being controlled by the weakest links, a result also deduced from magnetic field measurements[2]. In WLFA model, or more specifically, if the I-V characteristic results from switching of the weakest links which have a statistical distribution, we expect the temperature dependence of I_c, as determined using a constant field criterion, to be a convolution of the junction and statistical distribution temperature dependencies, casting doubt on interpretations of the temperature dependence based solely on the tunnelling model.

Continuing under the assumption that n is determined by $<I_c>$ and σ, we plot in Fig. 4 the n value vs. temperature at three different voltage criteria spanning two orders of magnitude. The solid lines are guides to the eye. For large n values the dependence of n with temperature is of a nonlinear nature, while the small n values show a more or less linear temperature dependence. When n is large the I_c distribution is sharp and we expect that many weak links will be broken simultaneously resulting in a rapid reduction of n with increasing temperature. Small n values indicate a broad I_c distribution therefore few weak links switch co-operatively and n has a weaker temperature dependence.

Although the WLFA model provides no analytical expression for the I-V characteristic, and the temperature dependence of I_c obtained using a constant voltage criterion is open to question in light of a statistical interpretation of the I-V data, we have been very successful in obtaining the temperature dependance of I_c by appealing to the empirical relation $V = A(I-I_c)^m$, first applied to such films by England et al.[9] In this relation I_c and m are temperature dependent, and m was found to vary continuously from approximately m = 3 at low temperature to $m(T_c) = 2.2 \pm 0.1$, and finally to m = 1 at 90 K. This empirical relation has been applied with similar success on both Tl and Bi films[7,8] to describe the I-V data. We also find that the I-V characteristic of our films is reasonably described by this relation. In Fig. 5 we show the results of a least-squares fit of the data from Fig. 1 to the equation $V = A(I-I_c)^m$. The solid lines represent the least-squares fit at each temperature. In Fig. 6 we show the temperature dependence of the exponent, m, determined from the least-squares fit for the entire range of temperatures in which the I-V data were obtained. After performing least-squares fits on five such films we find that the value of m in the low temperature limit, $m(20\ K) = 3.13 \pm 0.05$ is determined with far more confidence than $m(T_c) = 1.90 \pm 0.28$. It would seem that this is a natural consequence of the fact that $(dm/dT)_{20K} \ll (dm/dT)_{T_c}$. The I_c data of Fig. 5 are plotted in Fig. 2 and are denoted with a + symbol. This fitted data shows a simple power law behavior, with an exponent of 2.22, for all temperatures. The power law temperature dependence is seen at low temperatures in the upper four I_c data sets where the constant voltage criterion is expected to provide a more accurate

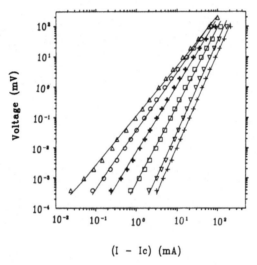

Fig. 5 Voltage against excess current obtained for a sequence of temperatures. From right to left: 20 K, 40 K, 51 K, 58 K, 62 K, and 64 K.

value for I_c. The temperature dependence of I_c, deduced from the least-squares fitting procedure, on three separate films is shown in Fig. 7. The slope of the lines is 2.20 ± 0.05. The fit in Fig. 7 illustrates two points. The first point is that the temperature dependence is the same even for films whose J_c's differ by more than an order of magnitude, and second, that the power law fits the data from near T_c down to the lowest temperature measured (20 K). If one accepts that the I-V characteristic is controlled by a distribution of weak links, as described by the WLFA model, then the I_c obtained from fitting the I-V data to $V \propto (I-I_c)^m$ is characteristic of the weakest links in the distribution. It is not clear whether the $(1-T/T_c)^2$ dependence of I_c, deduced from $V \propto (I-I_c)^m$, is indicative of S-N-S tunnelling[17]. In other words, the temperature dependence of the I_c distribution may be admixed with the weak link

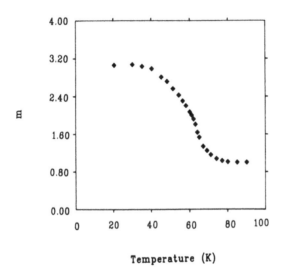

Fig. 6 Temperature dependence of the exponent, m, obtained from a least-squares fit of the I-V data to $V = A(I-I_c)^m$.

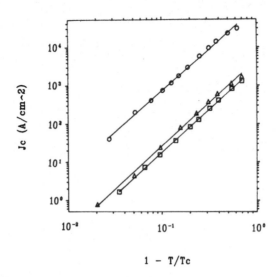

Fig. 7 J_c vs. $(1-T/T_c)$ for three films. In each case J_c was obtained from a least-squares fit of the I-V data to $V = A(I-I_c)^m$. The best fit straight lines through the data have a slope of 2.2.

temperature dependence. In fact, what is puzzling about the $(1-T/T_c)^2$ dependence of I_c is the wide range of temperatures over which it fits the data, an observation which has been made in Tl- and Bi- based HTS films as well[7,8]. In any type of tunnelling model, whether it is S-N-S, S-I-N-S, or S-I-S, one expects to see a saturation at low temperatures because the energy gap and the decay length both become independent of temperature for $T \leq 0.5T_c$. Our measurements are made down to $T = 0.2T_c$ and still show no evidence of saturation in the I_c data. One other possibility is that there is a distribution of T_c's in the material, but again, it is not clear what type of a temperature dependence in I_c one is expected to observe from such an effect.

CONCLUSIONS

We have shown that the I-V characteristics of the polycrystalline HTS films are qualitatively explained within the Weak Link Filament Array model (WLFA). In addition, our films show evidence of being in the filamentary limit of the WLFA model near T_c. The I-V data is well described by the empirical relation $V \propto (I-I_c)^m$ for the temperature range $20\ K \leq T \leq 90\ K$. From this empirical I-V relation we find that $I_c \propto (1-T/T_c)^{2.2}$, and that this temperature dependence is characteristic of the weaker links in the I_c distribution.

ACKNOWLEDGEMENTS

The authors wish to thank Drs. G. J. Valco, Kul B. Bhasin, and Mr. N. J. Rohrer for providing the thin film samples used in these experiments.

REFERENCES

1. T. T. M. Palstra, B. Batlogg, L. F. Schneemeyer, and J. V. Wazczak, Thermally Activated Dissipation in Bi-Sr-Ca-Cu-O, Phys. Rev. Lett., 60:1662 (1988).

2. J. W. Ekin, T. M. Larson, A. M. Hermann, Z. Z. Sheng, K. Togano, and H. Kumakura, Double-Step Behavior of Critical Current vs. Magnetic Field in Y-, Bi-, and Tl- Based Bulk High-Tc Superconductors, (preprint submitted to Physica C).

3. T. T. M. Palstra, B. Batlogg, R. B. van Dover, L. F. Schneemeyer, and J. V. Wazczak, Critical Currents and Thermally Activated Flux Motion in High-Temperature Superconductors, Appl. Phys. Lett., 54:763 (1989).

4. J. W. Ekin, Offset Criterion for Determining Superconductor Critical Current, Appl. Phys. Lett., 55:905 (1989).

5. S. B. Ogale, D. Dijkkamp, and T. Venkatesan, Current Transport in High-T_c Polycrystalline Films of Y-Ba-Cu-O, Phys. Rev. B, 36:7210 (1987).

6. J. W. C. deVries, M. A. M. Gijs, G. M. Stollman, T. S. Baller, and G. N. A. Van Veen, Critical Current as a Function of Temperature in Thin Y-Ba-Cu-O Films, J. Appl. Phys., 64:426 (1988).

7. J. F. Kwak, E. L. Venturini, R. J. Baughman, B. Morosin, and D. S. Ginley, High Critical Currents in Polycrystalline Tl-Ca-Ba-Cu-O Films, Cryogenics, 29:291 (1989).

8. H. E. Horng, J. C. Jao, H. C. Chen, H. C. Yang, H. H. Sung, and F. C. Chen, Critical Current in Polycrystalline Bi-Ca-Sr-Cu-O Films, Phys. Rev. B, 39:9624 (1989).

9. P. England, T. Venkatesan, X. D. Wu, and A. Inam, Granular Superconductivity in R-Ba-Cu-O Thin Films, Phys. Rev. B, 38:7125 (1988).

10. C. Lebeau, J. Rosenblatt, A. Raboutou, and P. Peyral, Current-Voltage Hyperscaling in Arrays of Josephson Junctions, Europhys. Lett., 1:313 (1986).

11. J. E. Evetts, B. A. Glowacki, P. L. Sampson, M. G. Blamire, N. McN. Alford, and M. A. Harmer, Relation of the N-Value of the Resistive Transition to Microstructure and Inhomogeniety for Y-Ba-A$_2$-Cu$_3$-O$_7$ Wires, IEEE Trans. Magn., 25:2041 (1989).

12. C. J. G. Plummer and J. E. Evetts, Dependence of the Shape of the Resistive Transition on Composite Inhomogeneity in Multifilamentary Wires, IEEE Trans. Magn., 23:1179 (1987).

13. G. J. Valco, N. J. Rohrer, J. D. Warner, and K. B. Bhasin, Composition and Processing Effects in Sequentially Evaporated Y-Ba-Cu-O Thin films, A.I.P. Proceedings, 182:147 (1989).

14. S. S. Yom, T. S. Hahn, Y. H. Kim, H. Chu, and S. S. Choi, Exponential Temperature Dependence of the Critical Current in Y-Ba-Cu-O Films, Appl. Phys. Lett., 54:2370 (1989).

15. R. G. Jones, E. H. Rhoderick, and A. C. Rose-Innes, Non-Linearity in the Voltage-Current Characteristic of a Type-2 Superconductor, Phys. Lett. A, 24:318 (1967).

16. J. Baixeras and G. Fournet, Pertes par Deplacement de Vortex dans un Supraconducteur de Type II Non Ideal, J. Phys. Chem. Sol., 28:1541 (1967).

17. J. Clarke, Supercurrents in Lead-Copper-Lead Sandwiches, Proc. Roy. Soc. A, 308:447 (1969).

STUDIES OF NEUTRAL AND ION TRANSPORT DURING LASER ABLATION OF

1:2:3 SUPERCONDUCTORS BY OPTICAL ABSORPTION SPECTROSCOPY

D. B. Geohegan and D. N. Mashburn

Solid State Division, Oak Ridge National Laboratory
Oak Ridge, Tennessee 37831-6056

ABSTRACT

Transient optical absorption spectroscopy, a new diagnostic technique for examining the density of *ground state* neutrals and ions following laser ablation, has been utilized for the first time to study the transport of ground state Y, Ba, Cu, Y+, and Ba+ following laser ablation of $Y_1Ba_2Cu_3O_x$ pellets under film deposition conditions. The technique reveals significantly broadened velocity distributions with a low velocity component which is not observed in velocity distributions inferred from monitoring excited state plume fluorescence. Ion probe measurements confirm this low velocity component. High resolution emission spectroscopy is also utilized to examine the spectral broadening arising from collision processes in the laser plasma as well as obtain estimates of plasma densities.

INTRODUCTION

The successful deposition of epitaxial, high-T_c superconductor thin films by pulsed laser deposition has ignited a renewed interest in the gas phase processes occurring in the plume of excited material ejected from the solid target. However, advances in the properties of the films deposited by this technique have far outpaced an understanding of the deposition process. The kinetic energy of the transported vapor, the composition of the expanding plasma and gas-phase molecular formation are several unique properties of the laser plume which are suspected as keys to successful film deposition.

Efforts to monitor the transport of the ablated vapor have concentrated upon spectroscopy of the fluorescence from the electronically excited atoms, ions and molecules in the expanding laser plasma.[1-12] The appearance of the laser plume is quite sensitive to variations of the laser energy density, the ambient pressure in the chamber and the distance from the target. Several spectroscopic studies have catalogued the fluorescence under various irradiation conditions[1-5] and spatially resolved measurements have been used to obtain estimates of expansion velocities.[6-8] Venkatesan, et al.,[9] measured velocity distributions of Y and BaO by laser ionization mass spectroscopy (following excimer laser ablation of $Y_1Ba_2Cu_3O_{7-x}$ at 7×10^9 W cm^{-2}) which indicated velocities slower ($v < 6 \times 10^5$ cm s^{-1}) than those measured by fluorescence.[6,7] Recently, Zheng, et al.[6,7] measured the time dependence of the fluorescence from excited Y, Ba, Cu, and O atoms in the plume and found that the velocity distribution of each species could be modeled by a supersonic expansion mechanism characteristic of a free expansion jet.

However, the monitoring of excited states by emission spectroscopy relies upon production processes in the expanding plasma to populate the observed states, which may be more representative of the production processes of the plasma than the density of neutrals in the plume.[13-15] The bulk of the plume is thought to be comprised of ground state neutrals and ions, which should not be visible directly by fluorescence spectroscopy.

In this paper, the spatially-resolved, temporal profiles of ground state Y, Ba, Cu, Y^+, and Ba^+ are investigated by transient optical absorption spectroscopy. The data reveal a previously unobserved long-lived component to the laser plume, with absorption lasting much longer than fluorescence at all distances studied. The absorption technique is non-intrusive and can be used as an in situ monitor during deposition of thin films by laser ablation. Ion probe studies are also described which confirm the new, long-lived component.

EXPERIMENTAL

The laser ablation/optical absorption apparatus is shown schematically in Fig. 1 and described in detail elsewhere.[16] Bulk superconducting $Y_1Ba_2Cu_3O_x$ pellets were rotated inside a turbopumped stainless steel bell jar at base pressures of $< 2 \times 10^{-6}$ Torr and irradiated at an incident angle of 60° by focused 193 nm or 248 nm pulses from an excimer laser. Energy densities were determined by measuring the energy transmitted through tilted apertures placed at the pellet position.

A high pressure, pulsed Xe arc lamp provided 500 ns full width at half maximum (FWHM) pulses of structured continuum emission. Two apertures formed a restricted optical axis through the ablation region and a quartz lens focused the light to probe a $\sim 1 \times 1$ mm region through the plume in a plane perpendicular to the irradiated pellet spot. Another quartz lens collected and focused the probe beam onto the entrance slit of a 1 meter spectrometer (spectral resolution of 0.04 Å). A boxcar averager sampled the photo-multiplier (2.2 ns rise time) output during the peak of the lamp pulse. Translation stages on the lenses and apertures were used to probe the plume at different distances parallel to the pellet face. For fluorescence collection, a lens was substituted for the second aperture and the plume fluorescence was focused on the entrance slit of the spectrometer. The slit defined a vertical slab of fluorescence at a distance, d, along the normal to the pellet face.

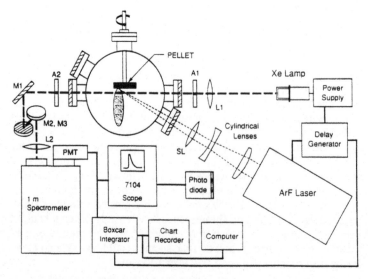

Fig. 1. Schematic diagram of the transient optical absorption / laser ablation experimental apparatus.

A digital delay generator allowed the laser-lamp delay to be continuously varied at a fixed wavelength for temporal scans of the absorption or emission. The temporal resolution was determined by the boxcar gate widths (typically set to 50 ns) necessary to reduce the statistical noise. Absorption and emission spectra of a selected atomic line were obtained by scanning the spectrometer at a fixed time delay.

OPTICAL ABSORPTION TECHNIQUE

The absorption of a uniform vapor of density N, over a path length L is determined by

$$I(\omega) = I_0(\omega) \, e^{-N\sigma(\omega)L}$$

where $I_0(\omega)$ and $I(\omega)$ are the intensity of the light before and after the absorption medium, ω is the frequency of the light and $\sigma(\omega)$ is the frequency dependent absorption cross section. The absorbance of the medium, A, is defined as

$$A(\omega) = \ln[I_0(\omega)/I(\omega)] = N\sigma(\omega)L$$

and varies, at a fixed wavelength, directly as NL. A spectral scan of $A(\omega)$ reveals the lineshape of $\sigma(\omega)$.

For investigations of the laser ablation plume, the spectrometer is tuned to the strongest atomic transitions from the ground state of the selected atom or ion. As the laser plume expands from the pellet, it intersects the optical axis with increasing path length and variations in density. The measured absorbance therefore represents an averaged number density over a given path length. The same considerations also apply to the emission from the traveling plume.

The extremely narrow line widths of atomic lines would normally prohibit optical absorption spectroscopy. For example, the natural line width of the Ba $6s^2\ ^1S_0 - 6p\ ^1P_1^0$ transition at 5535.48 Å is only 0.00032 Å. However, several features of the laser ablation process broaden the spectral widths to well above the resolution of available spectrometers.

As shown in Fig. 2, both absorption and emission spectral profiles are broadened in the plume. In the figure, ground state yttrium $^2F_{7/2} \leftarrow\ ^2D_{5/2}$ absorption and excited state Y $^2F_{7/2} \rightarrow\ ^2D_{5/2}$ emission were measured at a distance of 2 mm from the pellet face and a delay time of 400 ns following 1 J cm^{-2} ArF laser irradiation. The delay time was chosen such that both absorption and fluorescence existed simultaneously. The lamp spectrum over the same spectral region is essentially flat, while the absorption profile displays a broadened (0.74 Å FWHM) and somewhat asymmetric line shape. The spectral emission displays a spectral broadening nearly identical to the absorption profile.

The region of greatest broadening in the plume corresponded spatially to the hottest region of the laser plasma, where the fluorescence was a bright white continuum near to the pellet. At an energy density of 1 J cm^{-2} and distances < 1 mm from the pellet, the absorption profiles were typically broadened to from 0.5 to 3 Å (FWHM). The absorption widths decreased quickly to a nearly constant 0.15 to 0.3 Å for distances > 5 mm from the pellet.

A combination of several mechanisms is likely responsible for the observed line broadening. The dominant effect in the dense plasma near the pellet is probably Stark broadening from electron impact with neutral and singly ionized atoms.[13,17-19] Electron densities of $10^{16}-10^{18}$ cm^{-3} are typical for laser plasmas at energy densities similar to those used in these experiments.[14] Figure 3 is an emission scan of the oxygen 4368.25 Å line measured at a distance of 2 mm from the pellet surface following 1 J cm^{-2} ArF laser irradiation. By comparison of the measured spectral width of 0.39 Å with the tabulated electron impact half widths[18] for Stark broadening of this line, an estimate of the local electron density of 5×10^{16} cm^{-3} is obtained.

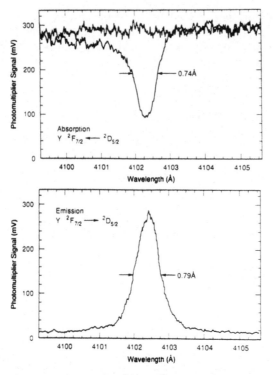

Fig. 2. Spectral profiles of the Y $^2F_{7/2}$–$^2D_{5/2}$ transition at 4102.38 Å in absorption (upper) and emission (lower) at a distance of 2 mm from the surface of the YBaCuO pellet, 400 ns following irradiation by 1 J cm^{-2} ArF (193 nm) laser pulses. The lamp spectrum is included for comparison with the absorption profile (upper).

In addition, since the transitions monitored in absorption were the strongest dipole allowed transitions from the ground state, resonance broadening from collisions between ground state species should contribute strongly to the broadened resonance lines. Finally, the component of the plume velocity parallel to the probe beam, resulting from either the initial expansion or turbulent effects of subsequent collisions, results in Doppler broadening of the observed lines.

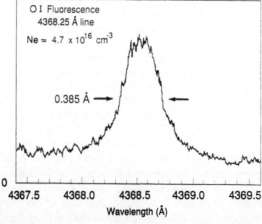

Fig. 3. Spectral profile of atomic oxygen fluorescence at 4368.25 Å measured at a distance of 2 mm from the pellet surface following 1 J cm^{-2} ArF laser irradiation.

The broadening of the spectral lines both close to and several centimeters away from the superconductor target indicate the importance of collisions in the laser ablation process at all distances. The pronounced broadening for distances < 3 mm from the pellet indicate a dense collisional boundary layer close to the pellet accompanied by high electron densities.

TRANSPORT MEASUREMENTS

By tuning to the center of the absorption line and scanning the laser-lamp delay, the temporal profile of the absorbing species can be obtained as it moves past the probed region. The absorption of ground state Ba (5 mm from the $Y_1Ba_2Cu_3O_{7-x}$ pellet face following 1 J cm^{-2} ArF laser irradiation at 2×10^{-6} Torr) is given in Fig. 4. Emission on the same Ba 6p $^1P_1 \leftrightarrow$ 6s^2 $^1S_0$ transition at 5535.48 Å was separately recorded and the fluorescence has been scaled and plotted for comparison with the measured absorbance profiles. Each point represents the average of 200 measurements.

The onset of the absorption precedes that of the fluorescence by approximately 100 ns, however the leading edges of both density profiles are in good agreement considering the possible experimental errors. Both profiles peak between 1 and 2 µs. At 4 µs, however, the fluorescence is gone while the absorption measurements indicate a ground state Ba density still 43% of the peak population. Measurable absorption lasts out to 150 µs.

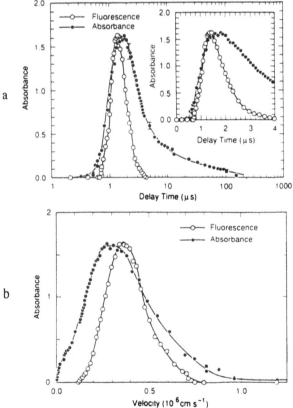

Fig. 4. (a) Comparison between the time dependences of atomic Ba, 5 mm from the irradiated pellet, as monitored by Ba* emission and ground state Ba absorption on the 6p $^1P_1 \leftrightarrow$ 6s^2 $^1S_0$ transition at 5535.48 Å. The peak of the fluorescence signal was normalized to the peak of the measured absorbance, [ln(I_0/I]. The initiation of the 1 J cm^{-2} ArF laser pulse at the pellet is defined as t = 0. (b) The calculated absorbance TOF velocity distribution shows a pronounced low velocity component not revealed by the plume fluorescence.

If the delay time is regarded as a time-of-flight (TOF), then the velocity ($v \equiv d/\tau$, where τ is the laser-probe delay time) distribution from the absorbance measurements is broader, peaked at lower velocities, and possesses a low-velocity component that is absent in the velocity distribution deduced from the fluorescence (Fig. 4, lower). The fluorescence wave forms were in near agreement with the studies of Zheng, et al.,[7] and could be fit by a Maxwellian free expansion distribution model. However, the absorbance velocity profiles were not fit by a simple free expansion model but appeared to require at least two components: a fast component, matching the fluorescence, and a slow component.

Each of the atomic and ionic species investigated by the absorption technique displayed a similar, additional low velocity component to the plume transport. In each case, the high velocity component of the fluorescence from the associated excited state was in reasonable agreement with that of the absorbance. However, while the fluorescence waveforms shared the same temporal shapes, the absorbance profiles displayed some marked differences.

Figure 5 gives a comparison of the absorbance profiles and the corresponding TOF velocity distributions for ground state Y, Ba, Cu, and Ba$^+$ at d = 5mm following 1 J cm^{-2} ArF laser irradiation at 2×10^{-6} Torr. The absorbance measurements indicate that the Y disappears first, followed by Ba$^+$, Cu, and Ba. The most pronounced low-velocity component is observed for Cu, however, Ba absorption persists for approximately twice as long. The Ba absorbance profile is shifted noticeably to lower velocities from that of Ba$^+$, Cu, and Y. A more detailed analysis of the evolution of the velocity distributions as a function of distance, laser energy density and background pressure will be presented elsewhere.

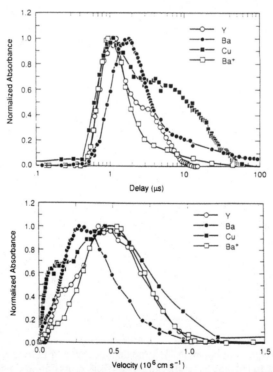

Fig. 5. The velocity distributions of ground state Y, Ba, Cu, and Ba$^+$ at d = 5 mm as measured by optical absorption. The monitored transitions and multiplicative factors to arrive at the measured absorbances are, respectively: Y(4102.38 Å, 1.00), Ba (5535.48 Å, 1.62), Cu (3247.54 Å, 0.428) Ba$^+$ (4554.55 Å, 1.45).

Close to the pellet (d < 2 mm), within the white region of the plasma, significant densities of YO, BaO, and CuO were also observed in optical absorption. Additionally, Y^+ was observed. Cu^+ and O were not investigated since the transitions required vacuum ultraviolet wavelengths. Significantly, no absorption from excited states was observed.

The explanation for the second, slow component to the ablation process may be evaporation. Venkatesan, et al., found stoichiometry and thickness variations in as-deposited YBaCuO films which displayed two distinct components to the ablation process: a forward-directed, on-stoichiometry ablation component and a cos θ, off-stoichiometry evaporative component.[11] In the data reported here, the atomic densities at long times are extinguished according to descending vapor pressures of the individual elements or their oxides, in agreement with cooling of the pellet following irradiation.

However, an alternative explanation for the long lifetimes may be the formation of an extended Knudsen layer as suggested by Kelly and Dreyfus[20]. High collision rates in the dense vapor following ablation are predicted to lead to a forward peaking of the angular distribution and also negative velocities, resulting in recondensation of the vapor on the pellet surface. The presence of vapor at times long compared to the fluorescence might then be interpreted as vapor which expanded away from the pellet, underwent collisions and essentially stopped or was traveling back toward the pellet.

Several observations fit this hypothesis. (1) The spectral broadening data shows a region < 5 mm from the pellet (at 1 J cm^{-2}) where collisions are very frequent. (2) Visible recondensation is observed (elemental or compounds containing Y, Ba, Cu as observed by EDX) on the Cu ring used to mount the pellet. (3) The spatial analysis of film deposits in the forward direction are sharply peaked (e.g., $\cos^{10}\theta$).[11]

However, other observations support the observed density vs time data as representative of actual TOF velocities. For example, the absorption at late times persists for all distances measured (d < 5 cm). This indicates sufficient collisions for spectral line broadening even at late times, however, it seems to contradict the existence of a well defined Knudsen layer followed by a collisionless TOF region.

The confusion regarding the time-of-flight interpretation of the absorption density data results from the fact that the optical absorption technique detects density rather than flux at a given distance and time. Similarly, the collected fluorescence also reflects the density of excited states. By way of contrast, flux sensitive detectors measure only densities moving at velocities. A comparison of the two methods can resolve the TOF interpretation.

ION PROBE MEASUREMENTS

Several ion probes were constructed to investigate the temporal behavior of the plume transport with a flux sensitive detector. All gave nearly identical temporal profiles. A detailed description of the probes, various plasma measurements and temporal characteristics of the charged particles in the plume will be published elsewhere.

In this experiment, the temporal behavior of Y^+ ions was measured by both optical absorption and a simple ion probe for direct investigation of the long time behavior. A yttrium pellet (99.99%) was used to provide ions of yttrium for comparative detection. The Y^+ ions were detected optically by absorption (and emission) at 3633 Å. The same temporal relationship between the absorption and emission was observed; the fluorescence concluding quickly following the peak absorption signal with a long absorption tail.

The ion probe was simply constructed of two parallel 4 cm × 6 cm copper plates, held at 2 cm separation by nylon spacers. The plates were positioned lengthwise along the normal to the pellet face starting at a distance of 5.2 cm. A 1 cm × 2 mm slit on a grounded copper plane was positioned 3.6 cm from the pellet to define a plasma slab which would pass through the center of the plates. A grounded metal box with slots to permit the entrance and exit of the plume was used to mount the plates and maintain shielded connections to the plates. A bias of -30 V was applied to one of the plates using a battery. The time response of the probe was measured in situ at less than 10 ns by time domain reflectometry.

The flashlamp beam was aligned behind the slit at d = 3.7 cm for simultaneous absorption measurements. The pressure was maintained at 2×10^{-5} Torr and KrF irradiation at 2 J cm^{-2} was used.

The electric field between the plates collects singly charged ions arriving between the plates after a time $t = d(2m/qV)^{1/2}$ regardless of the incoming velocity. For Y$^+$ ions, the collection time is 2.48 μs which limits the maximum velocity of the ions to $< 2.4 \times 10^6$ cm/s for the 6 cm long plates.

The time dependences of both the ground state Y$^+$ optical absorption at 3.7 cm and the ion probe signal at 5.2 cm are given in Fig. 6. The ion probe signal at early times displays an additional peak which is attributed to Y^{++} and thus is not observed at the Y$^+$ probe wavelength by optical absorption. This additional peak was a strong function of energy density and could be eliminated by lowering the energy density. The ion probe curve was scaled such that the peak (130 mV into 50 Ω) was normalized to the peak absorbance value. The curves are in qualitative agreement, with both the absorbance and ion flux lasting to long times.

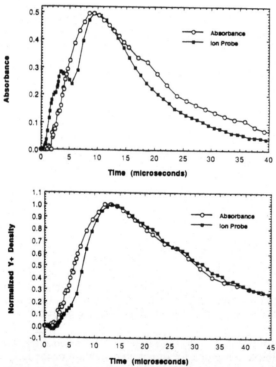

Fig. 6. A comparison between Y$^+$ ion time dependences as measured by the optical absorption technique and an ion probe following laser ablation of yttrium metal by KrF laser irradiation at 2 J cm^{-2} and 2×10^{-5} Torr. The digitized signals of the density-sensitive absorption measurement at d = 3.7 cm and the flux-sensitive ion probe at d = 5.2 cm are given in the upper plot while the TOF-corrected Y$^+$ density comparison at d = 5.2 cm is given in the lower plot. The disparity at early times is attributed to Y^{++}.

To compare Y^+ densities at d = 5.2 cm, time of flight velocities $v_x(t) = x/t$ are assumed. When the absorbance is corrected for distance and the ion probe flux signal is converted to density, the time dependences of the Y^+ densities at 5.2 cm in Fig. 6 (lower) are obtained. Aside from the difference at early times (attributed to Y^{++}), the densities derived from the two different detection methods agree very well.

In summary, the ion probe measurements verify the additional, slow component to the plume transport revealed by optical absorption. Since the ion probe responds only to an actual flux of particles, and not a stopped or negative-going vapor, the time-of-flight analysis is justified for these conditions.

CONCLUSIONS

Optical absorption spectroscopy has been used for the first time to examine the transport of *ground state* neutrals and ions following laser ablation of 123 superconductors. The spectral line broadening responsible for the atomic absorption has been investigated by high resolution absorption and emission spectroscopy. Close to the pellet, Stark broadening in the dense laser plasma is likely responsible for the broadened emission and absorption lines. Electron densities of 5×10^{16} cm^{-3} were measured 2 mm from the pellet utilizing Stark-broadened oxygen emission. Much farther from the pellet, several mechanisms including resonance broadening and Doppler broadening may contribute to the maintained line widths.

The transport of the atomic species was studied by the absorption technique and the data reveals a low velocity component to the plume transport which is not observed by monitoring fluorescence from the excited states. Ion probe measurements were performed to verify this low velocity component. The velocity distributions that are obtained assuming time-of-flight analysis are not fit by a simple free expansion velocity distribution but appear to contain at least two components.

The addition of optical absorption spectroscopy, high resolution emission spectroscopy and ion probe measurements to the fluorescence techniques used so far in the literature allow several new insights into the laser ablation process at the experimental conditions used for growth of superconductor films. Within the white plasma near the pellet, the ablation plume is a dense laser plasma with electron densities ranging from 10^{16} to 10^{18} cm^{-3}. Several centimeters from the pellet, the densities are estimated at 10^{13} cm^{-3} in vacuum. The rapid expansion is accompanied by a dense collisional environment which extends up to 5 mm from the pellet at just 1 J cm^{-2}. The leading edge of the expanding plume contains fast electrons and ions, expanding under the constraints of space charge fields. The fluorescence from the neutrals and ions on this leading edge reflects the overlap between the time dependences of both the excitation mechanism (such as electron impact excitation, or recombination) and the density of neutrals or ions. The bulk of the neutrals and ions are in the ground state following the cooling of the plasma after the initial expansion. Although the plume is not emitting fluorescence at late times, the transport of the weakly ionized plasma continues at low velocity. This secondary component to the plume transport likely reflects both an evaporative component and the results of formation of slower velocity components formed in the Knudsen layer.

The optical absorption technique presents a new view of the transport of monatomic neutrals and ions during film deposition. The data suggest several consequences for film growth. First, the deposition lasts typically an order of magnitude longer than the fluorescence from the plume. This results in a larger low-energy tail to the kinetic energy distribution than previously thought. Second, in addition to the differences in the arrival times of the bulk of the deposition fluxes, the significant differences between the slow tails to the velocity distributions results in an additional layering after each shot.

ACKNOWLEDGEMENTS

The authors gratefully acknowledge G. E. Jellison, Jr. for assistance with the computer modeling. The authors also acknowledge the technical assistance of D. H. Lowndes, J. W. McCamy and D. P. Norton. Research sponsored by the Division of Materials Sciences, U.S. Department of Energy under contract DE-AC05-84OR21400 with Martin Marietta Energy Systems, Inc.

REFERENCES

1. X. D. Wu, B. Dutta, M. S. Hedge, A. Inam, T. Venkatesan, E. W. Chase, C. C. Chang, and R. Howard, *Appl. Phys. Lett.* **54**, 179 (1989).
2. Q. Y. Ying, D. T. Shaw, and H. S. Kwok, *Appl. Phys. Lett.* **53**, 1762 (1988).
3. O. Auciello, S. Athavale, O. E. Hankins, M. Sito, A. F. Schreiner, and N. Biunno, *Appl. Phys. Lett.* **53**, 72 (1988).
4. Wayne A. Weimer, *Appl. Phys Lett.* **52**, 2171 (1988).
5. T. J. Geyer and W. A. Weimer, *Appl. Phys. Lett.* **54**, 469 (1989).
6. J. P. Zheng, Z. Q. Huang, D. T. Shaw, and H. S. Kwok, *Appl. Phys. Lett.* **54**, 280 (1989).
7. J. P. Zheng, Q. Y. Ying, S. Witanachchi, Z. Q. Huang, D. T. Shaw, and H. S. Kwok, *Appl. Phys. Lett.* **54**, 954 (1989).
8. P. E. Dyer, R. D. Greenough, A. Issa, and P. H. Key, *Appl. Phys. Lett.* **53**, 534 (1988).
9. T. Venkatesan, X. D. Wu, A. Inam, Y. Jeon, M. Croft, E. W. Chase, C. C. Chang, J. B. Wachtman, R. W. Odom, F. Radicati di Brozolo, and C. A. Magee, *Appl. Phys. Lett.* **53**, 1431 (1988).
10. C. H. Becker and J. B. Pallix, *J. Appl. Phys.* **64**, 5152 (1988).
11. T. Venkatesan, X. D. Wu, A. Inam, and J. B. Wachtman, *Appl. Phys. Lett.* **52**, 1193 (1988).
12. C. H. Chen, M. P. McCann and R. C. Phillips, *Appl. Phys. Lett.* **53**, 2701 (1988).
13. *Principles of Laser Plasmas*, edited by George Bekefi, John Wiley and Sons, p.549 (1976).
14. T. P. Hughes, *Plasmas and Laser Light*, John Wiley and Sons, New York, (1975).
15. R. E. Walkup, J. M. Jasinski, and R. W. Dreyfus *Appl. Phys. Lett.* **48**, 1690 (1986).
16. D. B. Geohegan and D. N. Mashburn, *Appl. Phys. Lett.*, in press.
17. Hans R. Griem, *Spectral Line Broadening by Plasmas*, Academic Press, London (1974).
18. Hans R. Griem, *Plasma Spectroscopy*, McGraw-Hill, New York (1964).
19. J. F. Ready, *Effects of High Power Laser Radiation*, Academic Press, London (1971).
20. Roger Kelly and R. W. Dreyfus, *Nucl. Instr. and Meth. B* **32**, 341 (1988)

STUDY OF DEPOSITION OF YBa$_2$Cu$_3$O$_{7-x}$ ON CUBIC ZIRCONIA

Joseph D. Warner and Joseph E. Meola
National Aeronautics and Space Administration
Lewis Research Center
Cleveland, Ohio 44135

Kimberly A. Jenkins*
Ohio University
Athens, Ohio 45701

ABSTRACT

Films of YBa$_2$Cu$_3$O$_{7-x}$ have been grown on (100) cubic zirconia with 8 percent yttria by laser ablation from sintered targets of YBa$_2$Cu$_3$O$_{7-x}$. The temperature of the zirconia substrate during growth was varied between 700 and 780 °C. The atmosphere during growth was 170 mtorr of oxygen. The films were subsequently slowly cooled in-situ in 1 atm of oxygen. The best films were c-axis aligned and had a transition temperature of 87.7 K. The superconducting transition temperature and the x-ray diffraction analysis will be reported as a function of the substrate temperature and of the angle between the laser beam and the target's normal.

INTRODUCTION

Laser ablation of high temperature superconducting (HTS) thin films has been shown to give high-quality films on SrTiO$_3$.[1-6] For microwave applications, however, a high-quality film is not enough. The substrate must also have low losses. Unfortunately, SrTiO$_3$ does not have low microwave losses. Therefore, in investigating HTS films for microwave applications, we have chosen as a substrate cubic zirconia stabilized with 8 percent yttria. It has a real dielectric constant of 25.4 and an imaginary dielectric constant of 1.74 at 33 GHz.[7] This is suitable for many microwave applications for space communications.

EXPERIMENTAL PROCEDURE

The laser ablation technique is similar to that which other researchers have used.[1-6] The details of the geometry are shown in Fig. 1. The substrates used were polished (100) cubic zirconia obtained from Atomergic Corporation. The samples were degreased in acetone and methanol prior to being glued with silver paint to the stainless steel sample holder. The paint was cured at 200 °C for 15 min in air and

---

*Summer Intern at NASA Lewis.

Superconductivity and Applications
Edited by H. S. Kwok *et al.*
Plenum Press, New York, 1990

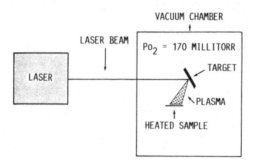

Figure 1. – Schematic of laser ablation experiment.

allowed to cool before being placed in the vacuum chamber. The chamber was evacuated to 2×10^{-7} torr or less using a liquid nitrogen cold trapped diffusion pump before the sample was warmed to 500 °C within 30 min. A continuous flow of oxygen (120 sccm) was then introduced into the chamber, and the sample was heated to its final temperature using resistive heating. The temperature was measured by a type K thermocouple that was welded to the stainless plate. The samples were within 5 mm of the thermocouple. During deposition the pressure was 170 mtorr, the laser wavelength was 248 nm, the energy density was 1.5 J/cm^2/pulse, the pulse rate was 4 pps, the distance between the target and the sample was 7.5 cm, and the laser beam was scanned up and down 1 cm over the target using an external lens on a translator. The angle between the normal of the target and the laser beam α was either 15°, 20°, or 45°.

After deposition the oxygen pressure was raised to 1 atm and the temperature was lowered to 450 °C at a rate of 2 °C/min. The temperature was held at 450 °C for 2 hr before it was lowered to 250 °C at a rate of 2 °C/min. The heater power was then turned off and the sample was allowed to cool to 40 °C or less before it was removed from the chamber.

Resistance was measured using a standard 4-point probe technique and with a current density between 4 to 10 A/cm^2. The contacts to the sample were made by wire bonding directly to the samples with 1 by 2 mil gold ribbon. The spacing between the voltage leads was approximately 3 mm.

RESULTS

The transition temperatures T_c of several samples for different α's and different deposition temperatures T_d are given in Table 1. As can be seen for the samples deposited, when $\alpha = 20°$, a change of 5 °C in T_d can result in a change in T_c of 3 to 10 K. The best films were for $\alpha = 45°$ and $T_d = 772$ °C. When the films made at $\alpha = 15°$ and $\alpha = 20°$ are compared, it can be seen that the lower angle gave a higher T_c for a lower T_d.

We observe that the intercept on the normal resistance axis is correlated with T_c. The closer the intercept is to 0 the higher T_c is. This is illustrated in Fig. 2 by samples 1 and 5. Film 1 had a T_d of 772 °C and a T_c of 87.7 K, while sample 5 had a T_d of 751 °C and a T_c of 75.8 K.

The difference between high T_c HTS films and lower T_c HTS films can also be seen in the morphology. Figure 3(a) is a scanning tunneling micrograph (SEM) of a film grown at $\alpha = 20°$ and $T_d = 751$ °C with $T_c = 80.1$ K. One can see the extreme roughness in the surface caused by 1 µ-size crystals growing on the surface. In contrast, in Fig. 3(b) the

TABLE 1. – TRANSITION TEMPERATURE T_c OF $YBu_2Cu_3O_x$ THIN FILMS ON CUBIC ZIRCONIA FOR VARIOUS DEPOSITION TEMPERATURES AND ANGLES BETWEEN THE TARGET'S NORMAL AND THE LASER BEAM

Sample number	Angle between laser beam and target's normal, deg	Deposition temperature, T_d, °C	Transition temperature, T_c, K
1	45	772	87.7
2	45	772	87.7
3	20	764	83.1
4	20	755	86.0
5	20	751	75.8
6	15	719	83.5
7	15	699	71.3
8	15	696	74.0

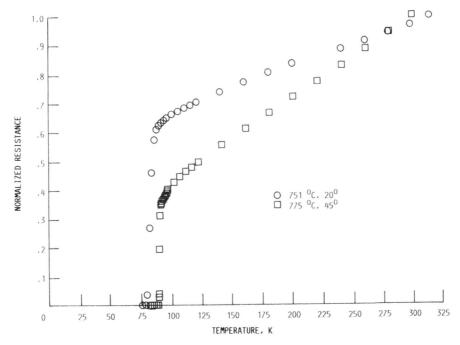

Figure 2. – Normalized resistance of $YBa_2Cu_3O_x$ films on cubic zirconia made by laser ablation different deposition temperature and different angle between the laser beam and the targets normal. +(751 °C and 20°), (775 °C and 45°)

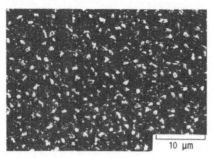
(a) FILM DEPOSITED AT 751 °C WITH A T_c OF 81 K.

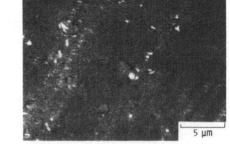
(b) FILM DEPOSITED AT 775 °C WITH A T_c AT 86.4 K.

Figure 3. - Scanning electron micrograph of laser ablated $YBa_2Cu_3O_x$ thin film on cubic zirconia. (a) Film was deposited at 751 °C and had a T_c of 81 K. (b) Film was deposited at 775 °C and had a T_c of 86.4 K.

film was grown at $\alpha = 45°$ and $T_d = 772$ °C with $T_c = 87.7$ K. This film was smooth with a very low density of particles on the surface.

The x-ray diffraction pattern of the film in Fig. 3(a) is shown in Fig. 4. All the major peaks are either from $YBa_2Cu_3O_x$ or from cubic zirconia. The 2-θ angle full width of half maximum of the $YBa_2Cu_3O_x$ (005) line is 1.3°. This condition indicates that the film was oriented.

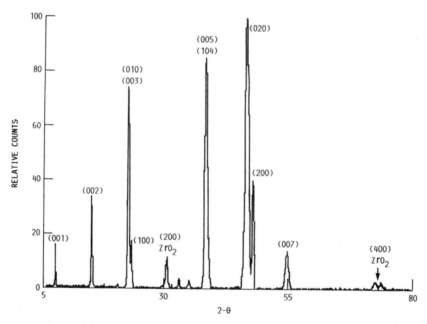

Figure 4. - X-ray diffraction of $YBa_2Cu_3O_x$ film on cubic zirconia. The film was deposited at 751 °C with a T_c of 81 K.

CONCLUSIONS

To conclude, high-quality $YBa_2Cu_3O_x$ thin films have been grown on cubic zirconia $((ZrO_2)_{0.92} (Y_2O_3)_{0.08})$ with a T_c as high as 87.7 K. The best films were smooth and had few particles on the surface. The transition temperature of the film was found to be highly dependent on both the deposition temperature and the angle between the normal of the target and the laser beam. To obtain the highest T_c and have it be reproducible, the temperature of the substrate must be held within 3 °C. To grow films with high T_c at lower temperatures in a partial oxygen atmosphere, the angle α must be as small as possible.

REFERENCES

1. D. Dijkkamp, T. Venkatesan, X.D. Wu, S.A. Shaheen, N. Jisrawi, Y.H. Min-Lee, W.L. McLean, M. Croft, Appl. Phys. Lett. 51:619 (1987).

2. X.D. Wu, D. Dijkkamp, S.B. Ogale, A. Inam, E.W. Chase, P.F. Miceli, C.C. Chang, J.M. Tarascon, and T. Venkatesan, Appl. Phys. Lett. 51:861 (1987).

3. J. Narayan, N. Biunno, R. Singh, O.W. Holland, and O. Auciello, Appl. Phys. Lett. 51:1845 (1987).

4. L. Lynds, B.R. Weinberger, G.G. Peterson, and H.A. Krasinski, Appl. Phys. Lett. 52:320 (1988).

5. T. Venkatesan, C.C. Chang, D. Dijkkamp, S.B. Ogale, E.W. Chase, L.A. Farrow, D.M. Hwang, P.F. Miceli, S.A. Schwarz, J.M. Tarascon, X.D. Wu, and A. Inam, J. Appl. Phys. 63:4591 (1988).

6. A.M. Desantolo, M.L. Mandich, S. Sunshine, B.A. Davidson, R.M. Fleming, P. Marsh, and T.Y. Kometani, Appl. Phys. Lett. 52:1995 (1988).

7. F.A. Mirandi, W.L. Gordon, V.O. Heinen, B.T. Ebihara, and K.B. Bhasin, Measurements of Complex Permittivity of Microwave Substrates in the 20-300 K Temperature Range from 26.5 to 40.0 GHz, NASA TM-102123 (1989).

INTERDIFFUSION BETWEEN SPUTTERED MgO FILMS AND SAPPHIRE SUBSTRATES

Jerrold Van Hook

Raytheon Company
Lexington, MA 02173

INTRODUCTION

Single crystalline αAl_2O_3 (sapphire) is a useful substrate material for a variety of thin film applications because of its inertness to processing chemicals and most film materials and the availability of high purity single crystalline substrates at reasonable size and cost. However, for films of the high temperature superconducting (HTS) oxides such as $YBa_2Cu_3O_{7-x}$ (YBCO) grown directly on sapphire, there is a degradation in HTS properties occurring when deposition or subsequent annealing is carried out at high temperatures. In the early studies[1] of YBCO on sapphire, the films were annealed at temperatures in the range of $850°-1000°C$ where interaction with sapphire was appreciable. More recent work[2] has emphasized processing at temperatures as low as $400°C$ which significantly reduces but does not eliminate substrate-film interactions.

Because of this contamination problem, many research groups, ourselves included, have been investigating different metal and metal oxide films that might be introduced between the substrate and HTS film to reduce interdiffusion. Noble metal layers such as Ag, Au, and Pt are attractive for high field magnetic applications but are impractical for high frequency thin film devices because of conduction losses. Among the various oxides with low microwave loss, MgO is a good candidate[3]. MgO has the additional advantage of minimal reaction with YBCO on the one hand, and the reaction with sapphire to form the stable spinel phase ($MgAl_2O_4$) on the other. The spinel interface layer forms a barrier to Al diffusion which increases with thickness and is thereby controllable through a choice of barrier layer-substrate prefiring conditions. To produce MgO and $MgAl_2O_4$ layer thicknesses that are optimal for epitaxy as well as reducing interdiffusion requires as knowledge of the kinetics of spinel formation. In this study we have determined the reaction rate by depositing MgO on sapphire at a temperature below the interdiffusion range followed by rapid heating to a range of isothermal annealing conditions.

EXPERIMENTAL METHODS

The MgO films were produced by ion beam sputtering in a vacuum of 10^{-5} torr using argon as the carrier gas. The test samples were cut from a larger substrate of polished (1̄102) sapphire with 2000Å of MgO deposited at 80°C. The annealing runs were carried out in a silica glass tube furnace where a relatively fast rise time (~10 seconds at 1100°C) was obtained by sliding a silica sample holder plus samples into a preheated region of the tube.

X-ray diffraction (XRD) was used to evaluate the crystallinity and the degree of preferred orientation in the recrystallized MgO and spinel layers. The single crystal substrates were tilted slightly (~1.5°) in a plane normal to the spectrometer rotation to reduce the magnitude of the (01̄12) and (02̄24) reflections of sapphire at 2θ~26° and 2θ~52°. Composition analysis of the films versus depth was achieved using a Perkin-Elmer model 570 Auger spectrometer in conjunction with ion beam sputtering to produce a depth

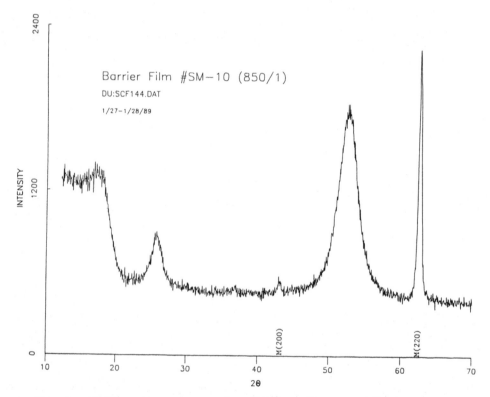

Fig. 1 Diffraction pattern of a 2000Å MgO film on (1̄102) sapphire after an 850°C one hour anneal.

profile of the elemental composition. This was combined with a profilometer measurement of the crater depth to determine sputtering rates for MgO and $MgAl_2O_4$.

EXPERIMENTAL RESULTS

The MgO films as-deposited on sapphire at 80°C were non-crystalline to X-rays. Auger analysis indicated a sputter removal rate of 88Å per minute for the MgO layer. After heat treatment for one hour at 850°C, the XRD pattern is shown in Figure 1 where a weak (200) and stronger (220) MgO reflection are seen. The broad peaks at 26° and 52° are from the substrate described previously.

Figure 2 is the Auger profile versus sputter time for the elements Mg, O, Al and C (assumed to be surface contamination) taken on the sample shown in Figure 1. The Mg and Al signals show abrupt changes in intensity at 24 minutes. From the measured sputtering rate of 88 Å/min on as-deposited MgO, we conclude the MgO thickness on the annealed sample is 2100Å. This implies zero thickness for the spinel layer although other data at 850° to be presented later suggests the spinel layer should be 500Å for these annealing conditions.

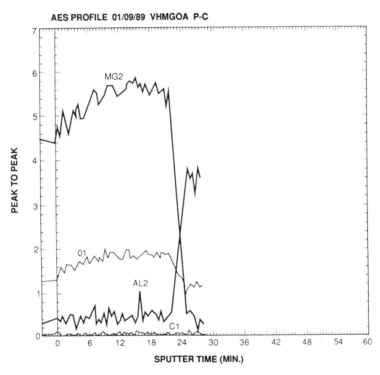

Fig. 2   Auger scan of element concentration versus sputter time for the sample in Figure 1.

In Figure 3 is shown the diffraction scan of a 2000Å MgO film annealed for one hour at 1000°C. The (220) peak for recrystallized MgO is again evident, but there is an additional (111) reflection due to the growth of a spinel interlayer.

Figure 4 is the Auger pattern of the sample whose diffraction pattern is shown in Figure 3. The MgO peak-to-peak intensity shows an inflection point at 12 minutes (MgO-$MgAl_2O_4$ interface) and another at 40 minutes ($MgAl_2O_4$-Al_2O_3 interface). From separate profilometer measurements at 12 and 40 minutes, the thickness of MgO (1000Å) and of $MgAl_2O_4$ (2200Å) are deduced. The sputtering rate of spinel is about 10% slower than that for MgO.

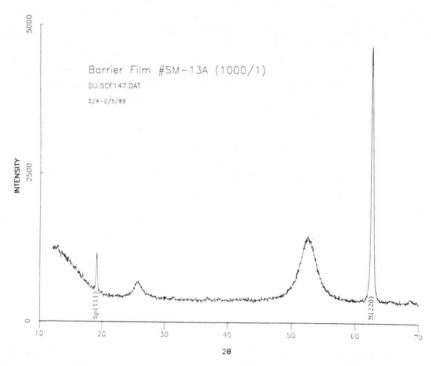

Fig. 3   Diffraction pattern of a 2000Å MgO film on (1̄102) sapphire after one hour at 1000°C.

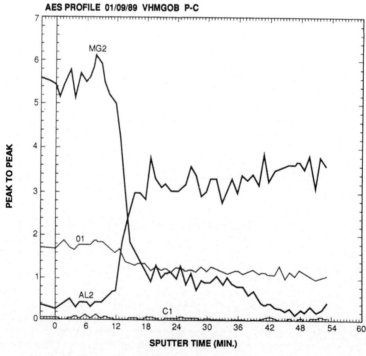

Fig. 4   Auger scan of element concentration versus sputter time for the sample in Figure 3.

Results similar to those shown in Figures 1-4 were obtained on seven separate anneals between 850°C and 1100°C for time periods between 5 minutes and 20 hours. The data shown in Table 1 were determined by combining Auger analysis with profile measurements at appropriate sputtering times. In the last column of the table the rate constant (κ) for spinel formation in $cm^2 sec^{-1}$ is calculated using the relation $\Delta x = (2\kappa t)^{-1/2}$ which applies for the planar geometry with Δx as the spinel layer thickness in cm and t equal to time in seconds.

Table 1. Auger data on 2000Å MgO films on sapphire after various anneals

Temp. (°C)	Time (Sec)	$MgAl_2O_4$ Thickness (Å)	MgO Thickness (Å)	κ cm^2-sec^{-1}
850	3600	0	2100	—
850	36000	2025	1000	5.7×10^{-15}
990	7200	1450	1400	1.5×10^{-14}
1000	3600	2200	1000	6.7×10^{-14}
1100	300	1900	1100	6.0×10^{-13}
1100	1020	3400	300	5.7×10^{-13}
1100	3600	4200	0	—

Apart from the first and last entries to the table where the spinel formation was either not detectable or had reached completion, the data show a spinel layer thickness twice that of the MgO consumed by interdiffusion. For example, at 990°C 600Å of the original 2000Å MgO has been converted to spinel forming a layer about twice that in thickness (1450Å). This relationship between spinel and MgO layer thicknesses, which follows from the similar densities of the reactants is a convenient check on the accuracy of the Auger results.

DISCUSSION

The kinetics of spinel formation from the reaction of single crystal MgO and Al_2O_3 was previously determined by Whitney and Stubican[4] who found rate constants (κ) of 10^{-7} to 10^{-9} $cm^2 sec^{-1}$ in the 1500°-1800°C range. Their results are shown in Figure 5 plotted as an Arrhenius-type reaction with an activation energy $\phi = -88000$ cal-$mole^{-1}$ above 1750°C and $\phi = -125000$ cal-$mole^{-1}$ at lower temperatures. Figure 5 also shows data from Kingery et al.[5] on Mg ion diffusion in MgO and also Al ion diffusion in Al_2O_3. The results of the present study are shown as triangular symbols.

The experimental data on spinel formation are consistent with an extrapolation of the rate data of Whitney and Stubican or with the results on Mg/MgO diffusion reported by Kingery et al.[5]. The reaction rate is orders of magnitude faster than one would predict based on Al/Al_2O_3 diffusion at 800°C. This suggests that the defect structure of the cubic spinel lattice is closer to the MgO host structure than to sapphire. For a deposition rate of 1Å/sec and at a substrate temperature of 1000°C, the reaction rate is such that 30 percent of the MgO film is converted to spinel during deposition. At 1049°C the conversion of MgO to spinel would be complete at the same deposition rate. The orientation of the recrystallized MgO is dependent on method and deposition temperature as well since Talvacchio et al.[6] found (100) MgO on (1102) sapphire by e-beam evaporation at 700°C. The spinel reaction kinetics, however, are known to be independent of orientation in this system[4].

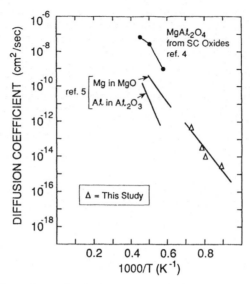

Fig. 5 Arrhenius plot of the reaction to form spinel compared with the bulk diffusion coefficients of the constituent ions.

REFERENCES

1. J. M. Tarascon, L. H. Greene, W. R. McKinnon, and G. W. Hull, Superconductivity in rare-earth-doped oxygen deficient perovskites, Solid State Commun. 63:499 (1987).

2. S. Witanachchi, H. S. Kwok, X. W. Wang, and D. T. Shaw, Deposition of superconducting Y-Ba-Cu-O films at 400°C without post-annealing, Appl. Phys. Lett. 53:234 (1988).

3. J. S. Thorp and N. Enayati-Rad, The dielectric behavior of single crystal MgO, Fe/MgO and Cr/MgO, J. Mater. Sci. 16:255 (1981).

4. W. P. Whitney and V. S. Stubican, Interdiffusion studies in the system $MgO-Al_2O_3$, J. Phys. Chem. Solids 32:305 (1971).

5. W. D. Kingery, H. C. Bowen, and D. R. Uhlmann, Atom Mobility, in: "Introduction to Ceramics," Wiley Interscience (1976).

6. J. Talvacchio, G. R. Wagner, and H. C. Pohl, $YBa_2Cu_3O_7$ films grown on epitaxial MgO buffer layers on sapphire, submitted to Physica C (July, 1989).

PROCESSING AND SUBSTRATE EFFECTS ON YBaCuO THIN FILMS FORMED BY RAPID THERMAL ANNEALING OF Cu/BaO/Y_2O_3 LAYERED STRUCTURES

Q. Y. Ma, M. T. Schmidt, T. J. Licata, D. V. Rossi,
E. S. Yang, Chin-An Chang[*], and C. E. Farrell[*]

Microelectronics Sciences Laboratories and Center for
Telecommunications Research, 1312 S. W. Mudd
Columbia University, New York, NY 10027
[*] IBM T. J. Watson Research Center, Yorktown Heights
NY 10598

ABSTRACT

YBaCuO superconducting thin films have been formed by rapid thermal annealing (RTA) of Cu/BaO/Y_2O_3 layered structures. The films were deposited on various substrates (including MgO, Al_2O_3 and Si) by electron-beam evaporation. The film properties strongly depend on the annealing conditions as well as the substrate used. We found that with 950 - 990 °C anneal for 30 - 90 seconds, the layers intermixed completely and a $YBa_2Cu_3O_x$ polycrystalline film was produced. The best film was obtained on MgO with T_c above 84 K and J_c of 4×10^3 A/cm^2 at 77 K. However, using the same RTA conditions, films on Si showed semiconducting behavior caused by silicon diffusion into the film as indicated by Auger depth profiling. In general, we found that using the RTA process, the film has less film-substrate interaction in comparison with furnace annealing due to a short anneal time. Using the RTA processing with the incorporation of a metal barrier layer, superconductivity of the film on Si was observed. Compared with other techniques, the RTA processing of layered structures provides a simple and rapid method for fabrication of high T_c superconducting thin films.

INTRODUCTION

Over the past two and one half years, many techniques have been developed for the fabrication of the high T_c superconducting thin films including electron-beam evaporation,[1-3] sputtering,[4-5] laser deposition,[6-8] and molecular-beam epitaxy.[9-10] Most as-deposited films were amorphous and required a post-anneal to make the films crystallize. A conventional furnace annealing process was carried out at high-temperature followed by a long cool down period. During a several hour high temperature treatment, interdiffusion and reaction between the film and the substrate occurred which degraded the superconductivity of the film.[11-12] Attempts to avoid the annealing has been made by low temperature processes such as laser deposition[13-14] in which a bulk superconducting material has to be produced as a target. The control of the stoichiometry during the transfer from a bulk material to a film was crucial.

We have developed a simple and rapid method of preparing the high T_c superconducting thin films using rapid thermal annealing (RTA) of multilayered structures. The films were deposited by a single e-beam evaporation with direct metal sources. The stoichiometry of the film was easily achieved by controlling the thickness of each layer. Due to a very short time (few minutes) heating process in a rapid thermal annealing system, the film-substrate interaction was reduced. We have made YBaCuO superconducting thin films on various substrates (including MgO, Si, and Al_2O_3). The film properties strongly depend on the anneal conditions as well as the substrate used. We found that with a 950 - 990°C anneal for 30 - 90 seconds, three layers intermixed completely and a $YBa_2Cu_3O_x$ polycrystalline film was produced. The best film was obtained on MgO with T_c above 84 K and J_c of 4×10^3 A/cm^2 at 77 K.

In this paper, we report studies of YBaCuO superconducting thin films formed by RTA of $Cu/BaO/Y_2O_3$ layered structures on various substrates. First, we will look at the processing effects on the superconducting phase formation of the films. The dependence of the film stoichiometry on the thickness ratio of layers and the optimal annealing conditions will be presented. Then the film-substrate interaction issue will be addressed. The surface morphology and polycrystalline structures of the film are examined by scanning electron microscopy (SEM) and x-ray diffraction. The interdiffusion at the film-substrate interface is investigated through the Auger electron spectroscopy (AES) depth profiling technique. Finally, the relation of the electrical properties of the films to these effects will be discussed.

PROCESSING OF THE FILMS

All films were deposited by multilayer evaporation in a single electron-beam system. The $Cu/BaO/Y_2O_3$ layered structures were prepared by evaporating Y, Ba, and Cu in sequence, with a base vacuum of low 10^{-7} Torr. The first two layers of Y_2O_3 and BaO were deposited in the presence of an oxygen partial pressure of 10^{-4} Torr. To study the stoichiometry of the films, two sets of thickness ratio were used. One had 1000, 2000, and 3000 Å and the other had 1000, 2400, and 900 Å for Y_2O_3, BaO, and Cu, respectively, as shown in Fig. 1. If the Y and Ba layers are fully oxidized, the compositions of two films would be off-stoichiometric $YBa_{1.7}Cu_{10.4}O_x$ and stoichiometric $YBa_2Cu_3O_x$, respectively.[2,15] The substrates used were MgO(100), Al_2O_3(100), and Si(100). During the evaporation the substrates were kept at room temperature.

The annealing was performed in a Heatpulse rapid thermal annealer. The samples were placed on a 4 inch silicon wafer inside a quartz chamber. The Si wafer absorbed infrared radiation from high-intensity, tungsten-halogen lamps and was heated rapidly. A typical RTA process consists of a 30s heating ramp, a constant high temperature period of 1 - 2 min, and a cooling down to 200°C in 5 - 7 min. All RTA cycles were carried out in flowing helium or oxygen gas at atmospheric pressure. The RTA conditions which strongly affect the quality of a film were annealing temperature, duration, and the switching time between helium and oxygen gas flow.

In general, the superconducting films were achieved on MgO and sapphire substrates with a 1-2 min, 970 - 1000°C RTA. For a silicon substrate with a barrier layer, a lower (950 - 970°C) and shorter time (10 - 60s) was needed. This is because the silicon substrate itself will absorb the infrared radiation during a RTA process. When the annealing temperatures were lower than the values mentioned above, three layers of the $Cu/BaO/Y_2O_3$ structures mixed incompletely, resulting in some non-crystalline metallic phases. With a higher temperature RTA the interaction at the film-substrate interface became severe and the film turned out to be insulating. Another important parameter was the switching time from He to oxygen gas. If annealing was performed entirely with flowing oxygen, the top layer of Cu in the structure oxidized first and formed a Cu-oxide layer prior to the intermixing of the three layers, resulting in a insulating film. It was found that flowing He gas during the heating up and the first half period

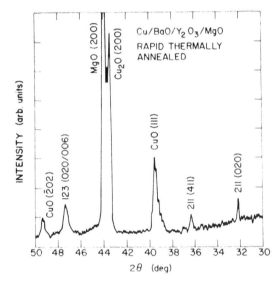

Figure 1. Schematic view of the Cu/BaO/Y_2O_3 layered structures of (a) off-stoichiometry and (b) stoichiometry.

Fig. 2 (a). X-ray diffraction pattern of an YBaCuO off-stoichiometric film.

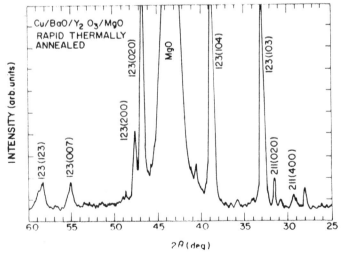

Fig. 2 (b). X-ray diffraction pattern of a stoichiometric film.

of the constant temperature regime, followed by oxygen flowing for the rest of the process, yielded the best superconducting films.

The structure and stoichiometry of the films were studied using x-ray diffraction and Auger electron spectroscopy (AES) analysis. Fig. 2 shows the x-ray diffraction patterns from both the off-stoichiometric and the stoichiometric samples. Both samples were deposited on MgO substrates and

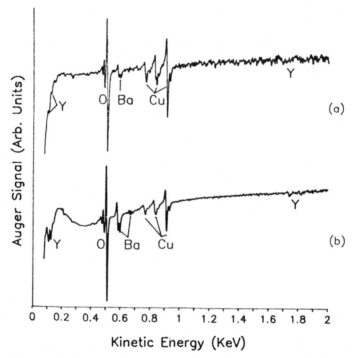

Figure 3. AES spectra of the YBaCuO films on MgO annealed at 980°C for 60s showed (a) off-stoichiometric film, and (b) stoichiometric film.

annealed at the same temperature of 980°C for 90s. The off-stoichiometric film has several large CuO phases and a relatively small 123 superconducting phase, as shown in Fig. 2 (a). This is attributed to the very thick layer of Cu used here. This also was indicated by AES spectra, as shown in Fig. 3 (a), in which the Cu signal intensity was qualitatively large compared to the signal of the stoichiometric film in Fig. 3(b). In contrast, the stoichiometric film contained mainly the 123 phases with no CuO phase as seen in the x-ray pattern of Fig. 2 (b). From Fig. 2, some 211 phases were also seen for both samples. The films on other substrates were deposited with a thickness ratio of 1000:2400:900 Å for Y_2O_3/BaO/Cu layers to obtain the 123 stoichiometric phase.

SUBSTRATE EFFECTS

1. Films on MgO

The YBaCuO superconducting films were formed reproducibly on a MgO substrate with a 980°C RTA for 30 - 90 seconds. The surface morphology of the film is shown by a SEM micrograph in Fig 4. As depicted by the figure, the film was uniform with an approximate 2 - 4 μm average grain size. The electrical properties were measured by a four-point probe technique. Fig. 5 (a) shows a typical temperature-dependent resistance curve for a stoichiometric film on MgO with a 980°C, 45s RTA. The superconducting transition in such a film occurred between 79 and 84 K. For an off-stoichiometric film the highest T_c observed was 84 K.[16] The resistivity at room temperature for both stoichiometric and off-stoichiometric films was in the range of 3 - 15 mΩcm.

The critical current density (J_c) was determined by a transport measurement.[17] For the film with a 980°C, 90s RTA, the J_c obtained was 4×10^3 A/cm². In general, the film on MgO has a larger critical current density than the film on other substrates. This is because the interaction between a film and a MgO substrate is insignificant. During a RTA process the three layers of Y_2O_3, BaO, and Cu intermixed completely as indicated by the Auger depth profile of a film with 980°C 90s RTA in Fig. 6 (a). From the surface to the interface, the composition ratio for the four elements remains uniform. At the film-substrate interface only a small amount of magnesium diffused into the film. The top portion of the film contained only the 123 superconducting phase, which is also indicated by the AES spectra in Fig. 3 (b).

2 Films on Al_2O_3

The superconducting film grown on a sapphire substrate was achieved under the same RTA conditions as on a MgO substrate. However, the film quality was not as good as the film on MgO most likely due to the Al-film interaction.[11-12] The room temperature resistivity of the film on sapphire was one order of magnitude higher than that of the film on MgO. Fig 5 (b) shows the resistance vs temperature measurement of a film on a sapphire substrate with a 980°C, 30s RTA. The T_c was 77 K, two degrees lower than the film on MgO. The critical current density of such a film was 300 A/cm² at 77 K. The aluminum was not seen at the surface of the film from the AES spectra. Again the film composition remained constant over a large portion of the film, as shown in the Auger depth profile of Fig. 6 (b). However, the Al diffusion into the film was a major drawback in this case. When a longer annealing time (>60s) was used, the film was no longer superconducting.

3 Films on Si

The film deposited directly on a Si substrate showed a semiconducting behavior due to the silicon outdiffusion.[18-19] A superconducting transition with an onset temperature at 82 K was observed from a film undergoing a 950°C, 30s RTA. But, the resistance did not drop to zero at lower temperatures. To inhibit the silicon diffusion, a thin (2000Å) layer of Au has been evaporated prior to the YBaCuO film deposition as a diffusion barrier. The results were successful, and superconductivity with T_c of 74 K was obtained as shown in Fig. 5, curve (c) for the film with a 950°C, 30s RTA. The silicon was not seen in the surface portion of the film. This was confirmed by Auger depth profiling as illustrated in Fig. 6 (c). As evidenced by the figure, the Au barrier layer effectively blocked the silicon diffusion. The room-temperature resistivity of such a film was about 1 - 3 mΩcm. The addition of Au into the film improved the electrical conductivity of the film. However, when the annealing temperature was higher than 970°C, even with a Au barrier, the silicon diffused into the film and formed silicon oxides which degraded the superconductivity of the film.

Figure 4. Scanning electron micrograph of an off-stoichiometric YBaCuO film on MgO with a 980°C, 90s RTA.

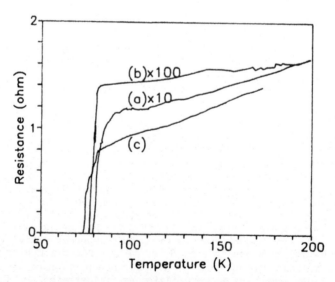

Figure 5. Temperature-dependent resistance measurement for YBaCuO films on (a) MgO with a 980°C, 45s RTA, (b) Al_2O_3 with 980°C, 30s RTA, and (c) Si with a Au barrier layer, annealed at 950°C for 30 seconds.

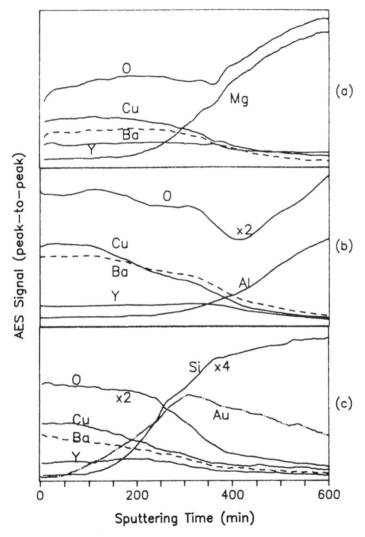

Figure 6. Auger depth profiles of the YBaCuO films on (a) MgO, (b) Al_2O_3, and (c) Si with a Au barrier layer.

CONCLUSIONS

We have demonstrated a simple and rapid technique of making high T_c superconducting thin films. The films were deposited by e-beam evaporation as $Cu/BaO/Y_2O_3$ layered structures and annealed in a rapid thermal annealing system. During a 30 - 90 s high temperature RTA, three layers of the structures intermixed and formed a uniform polycrystalline film. Superconductivity with T_c above 77 K was obtained from a film on MgO and Al_2O_3 substrates. A critical current density of 4×10^3 A/cm^2 was achieved in a film on a MgO substrate. Using a metal barrier layer, the superconducting YBaCuO film was made on a silicon substrate. The interaction at the film-substrate interface has been studied by AES depth profiling. We found that outdiffusion of Al from a sapphire substrate and of Si from a silicon substrate caused severe degradation of superconductivity in the film. Employing a short time RTA and a proper barrier, the outdiffusion was greatly reduced and the superconducting properties restored.

ACKNOWLEDGMENTS

The authors would like to thank X. Wu and J. L. Chow for their assistance. This work is supported by National Science Foundation through the Center for Telecommunications Research, by the Joint Services Electronics Program, and by the ONR-URI program at Columbia.

REFERENCES

1. R. B. Laibowitz, R. H. Koch, P. Chaudhari, and R. J. Gambino, Thin Superconducting Oxide Films, Phys. Rev. B 35, 8821 (1987).

2. Chin-An Chang, C. C. Tsuei, C. C. Chi, and T. R. McGuire, Thin-film YBaCuO superconductors formed by $Cu/BaO/Y_2O_3$ layer structures, Appl. Phys. Lett. 52, 72 (1988).

3. D. K. Lanthrop, S. E. Russek, and R. A. Buhrman, Production of $YBa_2Cu_3O_{7-x}$ superconducting thin films in situ by high-pressure reactive evaporation and rapid thermal annealing, Appl. Phys. Lett. 51, 1554 (1987).

4. M. Hong, S. H. Liou, J. Kwo, and B.A. Davidson, Superconducting Y-Ba-Cu-O oxide films by sputtering, Appl. Phys. Lett. 51, 694 (1987).

5. M. Scheuermann, C. C. Chi, C. C. Tsuei, D. S. Yee, J. J. Guomo, R. B. Laibowitz, R. H. Koch, B. Braren, R. Srinivasan, and M. M. Plechaty, Magnetron sputtering and laser patterning of high transition temperature Cu oxide films, Appl. Phys. Lett. 51, 1951 (1987).

6. D. Dijkkamp, T. Venkatesan, X. D. Wu, S. A. Shaheen, N. Jisrawi, Y.H. Min-Lee, W. L. McLean, and M. Croft, Preparation of Y-Ba-Cu oxide superconductor thin films using pulsed laser evaporation from high T_c bulk material, Appl. Phys. Lett. 51, 619 (1987).

7. J. Narayan, N. Biunno, R. Singh, O. W. Holland, and O. Auciello, Formation of thin superconducting films by the laser processing method, Appl. Phys. Lett., 51, 1845 (1987).

8. G. Koren, A. Gupta, E. A. Giess, A Segmuller, and R. B. Laibowitz, Epitaxial films of $YBa_2Cu_3O_{7-x}$ on $NdGaO_3$, $LaGaO_3$, and $SrTiO_3$ substrates deposited by laser ablation, Appl. Phys. Lett., 54, 1054 (1989).

9. J. Kwo, T. C. Hsieh, R. M. Fleming, M. Hong, S. H. Liou, B. A. Davidson, and L. C. Feldman, Structural and Superconducting Properties of Orientation-Ordered $Y_1Ba_2Cu_3O_{7-x}$ Films Prepared by Molecular-Beam Epitaxy, Phys. Rev. B, 36, 4039 (1987).

10. C. Webb, S. L. Weng, J. N. Eckstein, N. Missert, K. Char, D. G. Schiom, E. S. Hellman, M. R. Beaslay, A. Kapitulnik, and J. S. Harris, Jr., Growth of the high T_c superconducting thin films using molecular beam epitaxy techniques, Appl. Phys. Lett., 51, 1191 (1987).

11. T. Venkatesan, C. C. Chang, D. Dijkkamp, S. B. Ogale, E. W. Chase, L. A. Farrow, D. M. Hwang, P. F. Miceli, S. A. Schwarz, J. M. Tarascon, X. D. Wu, and A. Inam, Substrate effects on the properties of Y-Ba-Cu-O superconducting films prepared by laser deposition, J. Appl. Phys. 63, 4591 (1988).

12. C. T. Cheung and E. Ruckenstein, Superconductor-Substrate Interactions of the Y-Ba-Cu Oxide, J. Mater. Res., 4, 1 (1989).

13. X. D. Wu, A. Inam, T. Venkatesan, C. C. Chang, E. W. Chase, P. Barboux, J. M. Tarascon, and B. Wilkens, Low-temperature preparation of high T_c superconducting thin films, Appl. Phys. Lett., 52, 754 (1988).

14. S. Witanachchi, H. S. Kwok, X. W. Wang, and D. T. Shaw, Depoaition of superconducting Y-Ba-Cu-O films at 400°C without post-annealing, Appl. Phys. Lett., 53, 234 (1988).

15. Chin-An Chang, C. C. Tsuei, T. R. McGuire, D. S. Yee, J. P. Boresh, H. R. Lilienthal, and C. E. Farrell, YBaCuO superconducting films on SiO_2 substrates formed from $Cu/BaO/Y_2Y_3$ layer structures, Appl. Phys. Lett., 53, 916 (1988).

16. Q. Y. Ma, T. J. Licata, X. Wu, E. S. Yang, and Chin-An Chang, High T_c superconducting thin films by rapid thermal annealing of $Cu/BaO/Y_2O_3$ layered structures, Appl. Phys. Lett., 53, 2229 (1988).

17. P. Chaudhari, R. H. Koch, R. B. Laibowitz, T. R. McGuire, and R. J. Gambino, Critical current measurements in epitaxial films of $YBa_2Cu_3O_{7-x}$ compound, Phys. Rev. Lett., 58, 2684 (1987).

18. Q. Y. Ma, E. S. Yang, and Chin-An Chang, Rapid thermal annealing of YBaCuO thin films deposited on SiO_2 substrates, J. Appl. Phys., 66, 1866 (1989).

19. Q. Y. Ma, X. Wu, M. T. Schmidt, E. S. Yang, and Chin-An Chang, Interdiffusion between Si substrates and YBaCuO films, presented in International M²S-HTSC Conference, Stanford, July 1989, and to be published as a special issue of PHYSICA C.

LOW TEMPERATURE METAL-ORGANIC CHEMICAL VAPOR DEPOSITION
(LTMOCVD) ROUTE TO THE FABRICATION OF THIN FILMS OF HIGH
TEMPERATURE OXIDE SUPERCONDUCTORS

Alain E. Kaloyeros, Marianne Holma and   Wendell S. Williams
Aiguo Feng, and   Kenneth C. Brooks
Jonathan Garhart

Physics Department   Physics Department,   Department of
State University of   Chemistry Department   Materials Science &
New York at Albany   University of Illinois Engineering
Albany, NY 12222   at Urbana-Champaign   Case Western Reserve
                      Urbana, IL 61801   University
                                             Cleveland, OH 44106

INTRODUCTION

Although the recent discovery of high T_C superconductivity in bulk oxide ceramic samples has excited the scientific community, the technological potential of this new class of superconductors, especially in device-oriented applications, will not be fully realized until a relatively straightforward and easily reproducible technique can be applied to the synthesis of high quality superconducting thin films.[1,2] There has been considerable progress in the fabrication of superconductor thin films using a variety of techniques, e.g., reactive ion beam deposition, cosputtering from separate sources, dc and rf magnetron sputtering from a single source, sequential evaporation, plasma-assisted laser beam deposition, and chemical vapor deposition.[3-8] In the present paper, we report on the successful preparation of thin films of the Y-Ba-Cu-O system, showing a sharp transition at 90K, by a novel metal-organic chemical vapor method, followed by in-situ post-deposition annealing. This method has the advantages of relative simplicity and controllability, good film adherence, high film uniformity over a large area and reduced susceptibility to interfacial mixing and cross-contamination. In addition, it produces superconductor films on substrates of complex shape and with high growth rates. Because deposition can be achieved at temperatures as low as 300 °C, the method is called LTMOCVD.

EXPERIMENTAL TECHNIQUE

Three metal β-diketonate precursors, namely, bis(acetylacetonato)copper(II), $Cu(acac)_2$, bis(5,5,6,6,7,7,7-heptafluoro-2,2-dimethyl-3,5-octanedionato)barium, $Ba(fod)_2$, and yttrium tris(acetylacetonato)yttrium(III), $Y(acac)_3$, were used as the source compounds for the elemental constituents of the superconducting compound. The metal-organic complexes were synthesized by standard methods.[11,12] A detailed description of the experimental apparatus has already been reported by the present authors.[13]

The solid precursors were heated, in separate bubblers, to 65°C, 85°C, and 105°C, respectively, to sublime the molecular species required for deposition by LTMOCVD--Cu(acac)$_2$, Y(acac)$_3$, and Ba(fod)$_2$. For this purpose, electronic high temperature liquid- and solid-source mass flow controllers were employed to meter the flow of the organometallic precursors directly with high precision (+1% of full scale). This set-up eliminates the need for carrier gases/transport agents and makes the gas handling system less involved and more cost efficient than in standard CVD and MOCVD systems. It also allows precision determination of the amounts of precursors being introduced in the system and better control of stoichiometry.

The chemical vapor deposition was carried out at temperatures as low as 300°C and on a variety of substrates that included (100) MgO, SrTiO$_3$, and ZrO$_2$-coated silicon wafers. The substrates were heated by infrared radiation from quartz-halogen lamps. The deposition was followed by in-situ annealing in oxygen.

METHODS OF CHARACTERIZATION

The techniques used to characterize the films included four-point resistivity, Auger electron spectroscopy (AES), Rutherford backscattering (RBS), scanning electron microscopy (SEM) and energy-dispersive x-ray spectroscopy (EDXS). Auger spectra were recorded using a Physical Electronics PE595 instrument comprising a two-stage, retarding field/cylindrical mirror electron-energy analyzer, a coaxially contained electron gun with a beam normally incident on the sample surface and an electron multiplier. These components are contained in a demountable vacuum system evacuated to below a 10^{-10} torr. The analyzer transmission and resolution were 10% and ~0.6% respectively for all data. A beam energy of 5 keV was used with a beam current of 50nA, adjusted to provide convenient measurement conditions, but kept low enough to avoid specimen damage. Scanning electron microscopy measurements employing a 20 keV primary electron beam were performed using a Hitachi S800 scanning electron miccroscope combined with a LINK energy dispersive x-ray analysis unit.

RESULTS AND DISCUSSION

Deposition Parameters

A typical deposition rate, for films produced at temperatures ranging between 300 and 400°C, and at a working pressure of 10^{-6}-10^{-5} torr, was 2-3 nm/s. Typical film thickness, as measured by Rutherford backscattering, was 1-2 μm. The thickness variation across a typical substrate was found to be <5%/cm over the whole substrate.

Effects of Post-Deposition Annealing

In order to study the effect of post-deposition annealing on the superconducting properties, six Y-Ba-Cu-O films, denoted I-VI, were deposited by LTMOCVD on MgO substrates at 350°C. The films were subsequently annealed in-situ at different temperatures and ramp rates, and in atmospheres of oxygen and air. A four-point resistivity probe was then employed to measure the resistance of the samples versus temperature. The resistance data were normalized to the resistance at 200K. Film I was annealed using a two-step process: (a) a quick anneal in 100% O$_2$, from 25°C to 850°C in 20 mins, remaining at 850°C for 15 mins, and then cooling down to 25°C in 20 mins, followed by (b) same process in air. Film II was annealed using only step (a), while film III was processed according to method (b). Film IV was slowly annealed in air, from 25°C to 850°C in 60 mins, remaining at 850°C for 60 mins, and then cooling down to 25°C in one hour. Film V was processed according to the method used for film IV but in an atmosphere of oxygen instead of air. Finally, film VI was annealed using the process employed for film V, but up to 900°C instead of 850°C, and was kept there for two hours. All annealed films were black and shiny and exhibited metallic behavior at room temperature.

The electrical properties of the films were quite sensitive to the temperature and duration of the high-temperature annealing steps (see figure 1). Too low annealing temperatures led to incomplete reactions,

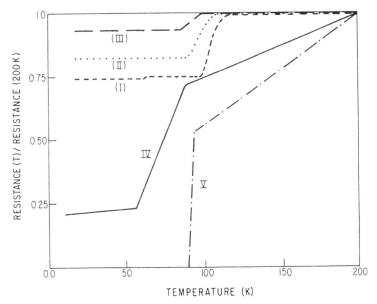

Figure 1 Resistance, normalized to the resistance at 200K, versus temperature is shown for films denoted I-VI. All films were deposited at 350°C and annealed at different temperatures and ramp rates in atmospheres of oxygen and dry air (see text). The best results (film V) were obtained with a slow anneal at 850°C in oxygen. The resistance curve for film VI, annealed at 900°C, is not displayed because of degradation due to heat damage and interaction with the substrate.

resulting in only resistance drops and no superconducting transitions (films I-III). Too high annealing temperatures or too long annealing times led to degradation caused by heat damage and extensive interaction with the substrate, as was the case for film VI. Similarly, slow annealing in air alone led to an appreciable resistance drop but no superconducting transition (film IV). The best transition temperature of ~90K, and the sharpest transition, were obtained for film V using the one-step annealing process described above. The resistance versus temperature curve had a steep slope and an approximately linear dependence on temperature above T~95K, which, if extrapolated, had a

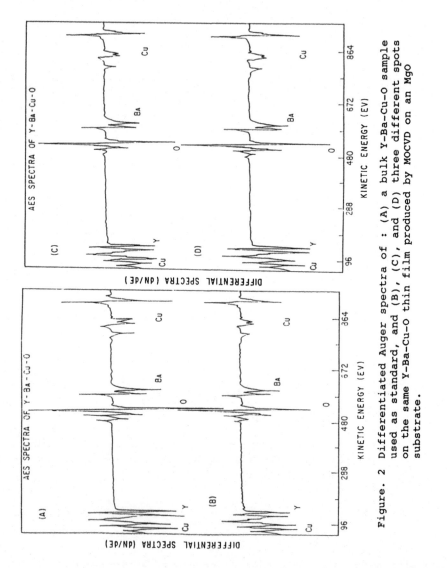

Figure. 2 Differentiated Auger spectra of : (A) a bulk Y-Ba-Cu-O sample used as standard, and (B), (C), and (D) three different spots on the same Y-Ba-Cu-O thin film produced by MOCVD on an MgO substrate.

projected intercept close to zero at T~0K. Subsequent measurements of the critical current gave an average value of 10^5 A/cm^2.

AES Results

Next, the composition, uniformity, and microstructure of the LTMOCVD-produced Y-Ba-Cu-O films were studied on films deposited on MgO substrates at 350°C and annealed according to the step used for film V. The films were characterized by Auger electron spectroscopy (AES), scanning electron microscopy (SEM), and energy-dispersive x-ray spectroscopy (EDXS).

The AES results were standardized using a bulk Y-Ba-Cu-O sample with a T_c~92K. All samples were sputter cleaned with a xenon ion beam before Auger data collection to remove surface contaminants. The choice of a standard of composition and chemical environment and bonding similar to that of the sputtered films allowed us to achieve high accuracy in AES quantitative analysis. (Our results are based on the expectation that the chemical and structural changes, if any, induced during the sputter cleaning process are basically the same in the bulk sample and sputtered films). In addition, all samples were analyzed under the same experimental conditions, using a primary electron beam energy of 5 keV and an electron beam density of 50nA/(0.3mm)^2 to reduce sample decomposition effects.

Figure 2 shows the Y, Ba, Cu and O Auger signals corresponding to a bulk standard sample (fig. 2(A)) and to three different locations on a Y-Ba-Cu-O thin film (figs. 2(B), 2(C) and 2(D)).

AES depth profiling and lateral (x-y) quantitative analyses show that film composition, within the detection limits of AES, was practically constant over the entire film and independent of lateral or depth position. the composition can be expressed as $Y_{1.0}Ba_{2.1}Cu_{3.2}O_x$.

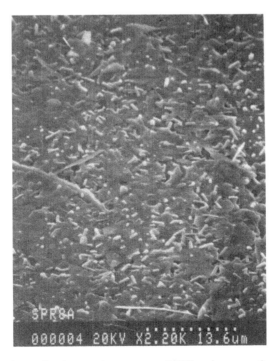

Figure 3 Scanning electron microscope (SEM) micrograph, using a magnification of 2200X, of a Y-Ba-Cu-O thin film produced by MOCVD. A pattern of oriented needles is visible.

EDXS Results

To confirm our AES findings, energy dispersive x-ray spectroscopy (EDXS) was carried out using a Hitachi S800 scanning electron microscope coupled to a LINK multichannel analyzer-minicomputer, equipped with acquisition and display facilities. The EDXS analysis gave results in good agreement with the AES quantitative analysis and yielded a composition of $Y_{1.0}Ba_{2.0}Cu_{3.15}O_x$.

SEM Results

The scanning electron micrograph of the $Y_{1.0}Ba_{2.1}Cu_{3.18}O_x$ film, taken at a magnification of 2200X, is shown in figure 3. The microstructure of the film consists of a typical pattern of oriented needles, with an extremely smooth surface and a very small concentration of voids. Oriented grains are highly desirable for high current capacity.

CONCLUSIONS

In conclusion, high-quality, high purity Y-Ba-Cu-O thin films were deposited by LTMOCVD and showed a transition temperature in the range 78-90K and a critical current of 10^5 A/cm^2. The best films were deposited at 350-400°C on MgO and SrTiO$_3$ substrates followed by a one-step in-situ annealing process in 100% O$_2$ at 850°C. Preliminary microstructural and microchemical studies showed that the films had a constant composition over the whole substrate area. This study suggests that LTMOCVD-produced films should be excellent candidates for photolithography and patterning applications. Efforts are underway to grow Y-Ba-Cu-O and Bi-Sr-Ca-Cu-O thin films with improved current characteristics.

ACKNOWLEDGMENTS

This work was supported by the New York State Energy Resources and Development Authority grant#88F042A (AEK, KCB, AF, and JG), SUNY-Albany grant#450300 (AEK, AF), and National Science Foundation grant#MSM-8617318 (MH, KCB, and WSW). The authors would like to thank SUNY-Albany professional staff member Mario Prividera for his technical advice and for his important contributions to various aspects of this project.

REFERENCES

1. C. W. Chu, P. H. Hor, R.L. Meng, L. Gao, Z. J. Huang, and Y. Q. Wang, "Evidence for Superconductivity above 40K in the La-Ba-Cu-O Compound System," Phys. Rev. Lett. **58** :405 (1987).
2. L. R. Testardi, W. G. Moulton, H. Mathias, H. K. Ng, and C. M. Rey, "Superconducting and Nonsuperconducting phases of Y-Ba-Cu-O: Modifications at the High-Temperature Phase Transition," Phys. Rev. **B16**: 8816 (1987).
3. A. B. Harker, P. H. Kobrin, P. E. D. Morgan, J. F. DeNatale, J. J. Ratto, I. S. Gergis, and D. G. Howitt, "Superconductor Thin Films by Reactive Ion Beam Deposition,"Appl. Phys. Lett. 52:2180 (1988).
4. M. Fukutomi, J. Machida, Y. Tanaka, T. Asano, T. Yamamoto, and H. Maeda, "New Technique for Preparation of Bi-Sr-Ca-Cu-O Thin Films with T$_c$ of 100K and Above," Jpn. J. Appl. Phys. 27: L1484 (1988).
5. B. T. Sullivan, N. R. Osborne, W. N. Hardy, J. F. Carolan, B. X. Yang, P. J. Michael, and R. R. Parsons, "Growth of Y-Ba-Cu-O Superconductor Thin Films by DC Magnetron Sputtering," Appl. Phys. Lett. **52**:1992 (1988).
6. B. F. Kim, J. Bohandy, T. E. Phillips, W. J. Green, E. Agostinelli, F. J. Adrian, K. Moorjani, L. J. Swartzendruber, R. D. Shull, L. H. Bennett. and J. S. Wallace, "Superconducting Thin Films of Bi-Sr-Ca-Cu-O obtained by Laser Ablation Processing," Appl. Phys. Lett. **53**:321 (1988).

7. M. Levinson, S. S. P. Shah, and N. Naito, "Superconducting Films by Magnetron Sputtering of Single Bi_2O_3-SrF_2-CaF_2-CuO Targets," Appl. Phys. Lett. **53**:922 (1988).
8. T. Venkatesam, X. D. Wu, B. Dutta, A. Inam, M. S. Hedge, D. M. Hwang, C. C. Chang, L. Nazar and B. Wilkens, "High Temperature Superconductivity in Ultrathin Films of Y-Ba-Cu-O," Appl. Phys. Lett. **54**: 581 (1989).
9. A. D. Berry, D. K. Gaskill, R. T. Holm, E. J. Cukauskas, R. Kaplan, and R. L. Henry, "Formation Of High T_c Superconducting Films by Organometallic Chemical Vapor Deposition," Appl. Phys. Lett. **52**: 1743 (1988).
10. J. Zhao, K. H. Dahmen, H. O. Marcy, L. M. Tonge, B. W. Wessels, T. J. Marks, and C. R. Kannewurf, "Low Pressure Organometallic Chemical Vapor Deposition of high T Superconducting $YBa_2Cu_3O_{7-d}$ Films," Solid State Comm. **69**: 187 (1989).
11. R. Belcher, C.R. Cranly, J.R. Majer, W.I. Stephen, and P.C. Uden, "Volatile Alkaline Earth Chelates of Fluorinated Alkanoylpivalyl Methanes," Anal. Chim. Acta **60**:109 (1972).
12. R.C. Mehrotra, R. Bohra, and D.P. Gaur, **Metal β-Diketonates and Allied Derivatives**, Academic Press, London (1978).
13. A. E. Kaloyeros, W.S. Williams, and G. Constant, " Method for the Preparation of Protective Coatings by Low-Temperature Metal-Organic Chemical Vapor Deposition (MOCVD)," Rev. Sci Instrum. **59**:1209 (1988).

JOSEPHSON AND QUASIPARTICLE TUNNELING IN HIGH-T_c SUPERCONDUCTORS

M. Gurvitch*

AT&T Bell Labs
Murray Hill, N.J. 07974

INTRODUCTION

The subject of tunneling in high temperature superconductors (HTSC) is too wide for a comprehensive review a few pages long. This brief account is by necessity imcomplete and subjective - a frame taken out of a movie.

Some of the facts have been established; many are still in dispute. Reliable results are show to come out in a tunneling field. This has been the case with Nb and A-15's a decage ago. With HTSC the problems are further compounded at least for three reasons: a) extremely short anisotropic coherence lengths, with the higher values in the range $\xi_{ab} \sim (12 \div 30)$Å [1] (Fig. 1), b) structural complexity, with 4-5 different elements per chemical formula and 13-15 atoms per unit cell, and c) ease of oxygen desorption, at least in a 1:2:3 structure.

From (a) it follows that one needs to prepare tunnel junctions with good material within the first few angstroms from the interface; (b) and (c) make this difficult.

Most arrangements involving Josephson and quasiparticle tunneling in HTSC can be divided into the following seven categories sketched in Fig. 2:

1. Intrinsic junctions across grain boundaries or microcracks.
2. Junctions between a needle made of a conventional superconductor (CS, typically Nb) or a normal metal (NM, typically W,Pt) and HTSC.

---

*From Jan. 1990: Dept. of Physics, SUNY at Stony Brook, Stony Brook, N.Y. 11794

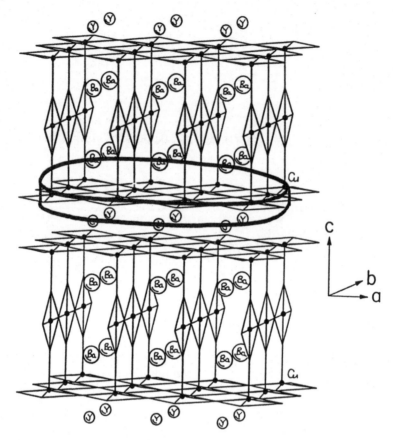

Fig. 1 A sketch of a pancake-shaped Cooper pair in $YBa_2Cu_3O_7$; $\xi_{ab} \sim 15$ Å; $\xi_c \sim 2$ Å.

3. Junctions between ceramic samples or films and pressed soft metal contacts, Pb or In.
4. Coplanar microbridges HTSC-noble metal-HTSC.
5. Sandwich heterostructures HTSC-epitaxial barrier - HTSC.
6. Sandwich structures HTSC-noble metal-(optional tunnel barrier)-CS.
7. Sandwich structures HTSC - CS or NM (evaporated), utilizing a natural barrier formed in air.

JOSEPHSON TUNNELING

So far, the junctions of the first category are the only ones with useful Josephson properties. By this we mean the size of a Josephson current i_c and its sensitivity to the magnetic field and microwave radiation. The size of a current

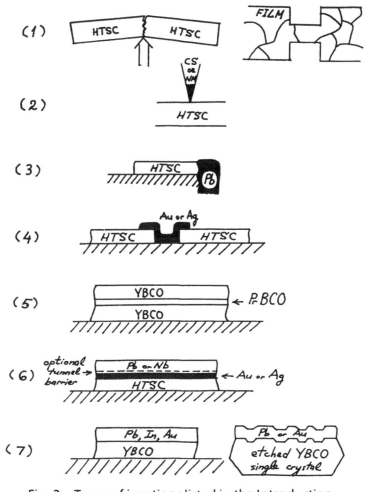

Fig. 2 Types of junctions listed in the Introduction.

is determined with respect to the normal state resistance of a junction R_n. For a symmetrical ($\Delta_1 = \Delta_2 = \Delta$) BCS tunnel junction there exists a theoretical relationship between the product of i_c and R_n and the size of the gap Δ at temperature T:[2]

$$i_c R_n = \frac{\pi \Delta(T)}{2e} \tanh \frac{\Delta(T)}{2K_B T} \qquad (1)$$

From this relationship, at T = 0, i_c equals the tunneling current that would have existed in the absence of superconductivity at a voltage of about 3/4 of the gap voltage $V_g = 2\Delta$. In the best Nb-Al-oxide-Nb junctions at 4.2 K, $i_c R_n \simeq 2$ mV, which is close to the theoretical limit (1) for Nb gap of 1.55 meV.[3] In a high-T_c oxide such as YBCO, the gap is expected to be at least an order of magnitude higher than in Nb. We shall see below, in sec. 3, what experiment has to say on that matter. The question is not fully settled; however, most people would agree on a value of the gap around 20 meV. Substituting this into (1), we obtain $i_c R_n \simeq 31$ mV for a symmetrical YBCO tunnel junction at T = 4.2 K. Experimental values achieved so far are at least an order of magnitude - more often two to four orders of magnitude - lower than this theoretical estimate.

Josephson currents through microcracks in bulk ceramic samples ("break junctions") and across natural grain boundaries in films were observed early on.[4] The IBM group pioneered SQUID's based on natural grain boundary junctions in films, and recently reported very good results using such SQUID's made of thallium compound films.[5] The SQUID loop is patterned so as to incorporate these natural Josephson junctions. Another encouraging results has been reported on natural junctions in films of YBCO deposited by MOCVD.[6] Good quality single Josephson junctions and SQUIDs were produced by patterning ~1 × 1 µm constrictions into these films. The $i_c R_n$ was ~1 mV; in the magnetic field SQUID's produced modulation voltage of up to 200 µV at 4.2 K and up to 15 µV at 60 K.[6] It is interesting, that, according to Prof. Yamashita, the grain boundaries were not visible, so the nature of Josephson junctions in these MOCVD films remains unknown.

Junctions of the 2nd kind, i.e. between a sharp tip made of Nb and a bulk sample of HTSC, with the tip typically driven into the HTSC, have also played their role in the HTSC research. In particular, the Naples-Salerno group [7] may have been the first to observe both the Josephson current and the gap-like structure (albeit rather imperfect) in the I-V's of Nb-YBCO point contacts of that sort. We will show some of the quasiparticle tunneling results obtained with this technique in sec. 3.

The activity towards preparation of useful "man-made" Josephson devices is only beginning, and the characteristics of such devices still fall behind those of the natural grain boundary ones.

A group at Bell Labs, Holmdel, prepared YBCO-Au-YBCO microbridges [8] (type 4 in Fig. 2). The epitaxial YBCO films on $SrTiO_3$ were patterned to have a separation ~1 µm which was in turn bridged with an Au film. Weak Josephson current with $i_c R_n$ ~5 µV was observed. While such low $i_c R_n$ junctions would be of little practical use, they prove in principle the possibility of Josephson coupling between HTSC using normal metals.

Another step in an interesting direction was taken by a Bellcore-Rutgers group which reported work on epitaxial heterostructures $YBa_2Cu_3O_7$-

PrBa$_2$Cu$_3$O$_7$ (500 Å) - YBa$_2$Cu$_3$O$_7$ [9] (type 5 in Fig. 2). PrBa$_2$Cu$_3$O$_7$ (PrBCO) is unique among the lattice - matched rare earth perovskites with 1:2:3 structure in that it is a semiconductor or an insulator at low temperatures. The i_cR_n~4 mV observed in these structures is higher than in most other recent reports. However, judging by the weak manifestations of the Josephson effect reported in Ref. 9, it appears that only part of the zero-voltage current is of Josephson nature; the rest of the current may be flowing through shorts. (The difference between the Josephson junction and a superconducting short is essentially in the strength of coupling between the two sides.)

The heterostructure approach of Ref. 9 reminds us of the successful technology based on the lattice matched GaInAs/GaAlAs system, and thus has a strong appeal. In the present implementation it is not without problems. One problem - possible shorts across PrBCO - have been mentioned above. What is more serious, PrBCO is not a good insulator, especially at the elevated temperatures approaching 77 K. This objection can be overrun by using epitaxially matched insulators such as LaAlO$_3$ in place of PrBCO, as being presently attempted by the Bellcore-Rutgers group as well as by others. Finally, the $C_\perp$ orientation commonly found in epitaxial 1:2:3 films is unfavorable for tunneling. This objection is hard to overcome in the sandwich geometry; it applies to all sandwich structures, with the possible exception of chemically etched nonplanar junctions which will be discussed in the next section.

Junctions of the 6th category in Fig. 2 were described in a number of recent papers.[10] They are mainly serving two purposes: a) as an intermediate stage and a learning ground for future development of all HTSC junctions, and b) to prove the existence of Josephson effect between CS and HTSC. The best, most convincing proof of that is presented in the work of H. Akoh et al.[10] of Electrotechnical Lab in Japan.

It is, however, unclear what physics implications can be drawn from this proof for the question of the wavefunction symmetry in HTSC. We discuss this interesting question very briefly here. In the first order, Josephson effect should be suppressed between superconductors of different Cooper pair symmetry However, it has been argued that 2nd order (relativistic) effects can be significant in superconductors with short coherence length $\xi \sim a_0$ (a_0 - lattice spacing), and can lead to finite Joshepson and proximity effects between superconductors of different symmetry.[11] Now, since the value of Josephson current in experiments described in Ref. 10 is very low, i.e. i_cR_n~7 µV in the work of Akoh et al., it is hard to be sure at this time that we are not observing those 2nd order effects.

Finally, the 7th kind of a junction: sandwich structure between either chemically etched YBCO and Pb,[12] or freshly prepared in-situ YBCO film and Pb[13] have been studied recently at Bell Labs as well as by other groups listed in the next section. In our work (Ref. 12) we have not been able to detect any Josephson current; the work concentrated mainly on the quasiparticle I-V's. In contrast, J.R. Kwo et al.[13] observed Josephson current with $i_c R_n$ up to 0.5 mV. This value may be large enough to have a bearing on the symmetry question discussed above. It is hard to settle this question in the absence of a detailed theory.

QUASIPARTICLE TUNNELING IN HTSC

A more detailed account with an emphasis on vacuum and point-contact tunneling can be found in a recent review by Kirtley.[14] The theory of tunneling shows that the derivative of the tunneling current with respect to voltage across the junction, di/dV is proportional to the density of quasiparticle states in a superconductor. Hence di/dV in an ideal S-I-N junction will essentially reproduce the density of states in a superconductor. In addition, details of di/dV at higher energies $V > V_g$ contain important information on the nature and strength of coupling responsible for the formation of Cooper pairs. That is why tunneling has been traditionally considered the best probe of a superconducting state, and has been vigorously pursued in the field of HTSC. Initial tunneling results, obtained mainly on ceramic samples, were widely different from each other, with gap values differing by as much as a factor of ten.[15] More recently, thanks to persistant work at improving sample surface quality, several groups started to converge at a unique picture.[15] In particular, the data on type 7 junctions from Karlsruhe [16] and from Bell [12,13] can claim sample-to-sample reproducibility which gives confidence in the reported results.

At Bell we learned how to clean surfaces of single crystals of YBCO using chemical etchants. We used two different preparations: perchloric acid water-based etch and bromine in methanol etch.[12] The tunneling results were however the same. It is interesting to note that the surface of an etched crystal is dotted with crystallographically oriented square etch pits some 10-20 μm on the side and a few thousand angstroms deep (Fig. 3). Thus the etching exposes a-b plane terminations to the surface which, in virtue of Fig. 1, should be beneficial for tunneling.

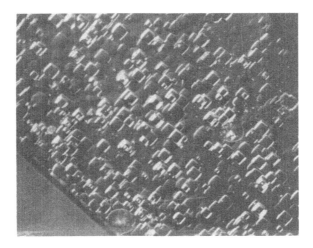

Fig. 3 Photo of a surface of YBCO etched in 10 mM $HClO_4$/ 1 M $NaClO_4$ in water for 5 min, showing etch pits.

The nature of the natural tunnel barrier formed in air and used in Ref. 12, 13 and 16 is unknown. The fact that this barrier may be formed at the expense of near-surface stoichiometry is worrisome and calls for future work on artificial tunnel barriers. The barrier itself, however, is of high quality, which is experimentally proven by observing negligible leakage at low temperatures for voltages smaller then the Pb gap.[12]

Once the reproducible data have been obtained, one can look at the literature and see which results coincide with the reproducible ones. To give several examples, data from NEC Corporation [17] in Japan on ceramic YBCO samples with pressed Pb counterelectrodes (type 3 in Fig. 2) (an unlikely arrangement to give a reliable result, at a first glance) happened to coinside very well with the Bell Labs result (except, possibly for the amplitude of di/dV features which has not been indicated in Ref. 17). Tunneling data from air-cleaved YBCO with evaporated Pb counterelectrode reported in Ref. 18 also has many features is common with Ref. 12, 13, 16, 17, except that it shows a larger gap at ~30 meV. This larger gap at 30-35 meV has been also observed on junctions with an artificial amorphous YBCO tunnel barrier by the Karlsruhe group.[19] Some point contact and vacuum tunneling data [20] also has strong similarities to these results. The three sets of data (Ref. 12, 16, 17) are plotted in Fig. 4 taken from a recent invited presentation of Jochen Geerk.[21] Let us talk about the meaning of these data.

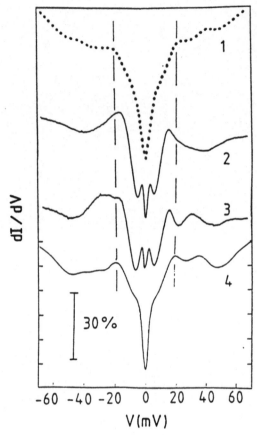

Fig. 4 Comparison of di/dV data on YBCO tunnel junctions from different groups: 1: Takeuchi et al.; 2,3: Geerk et al; 4: Gurvitch et al. The Figure is taken from Ref. 21.

In Fig. 5 we superimposed the tunneling conductance di/dV of an etched YBCO-Pb junction measured at T = 10 K[12] (i.e. with Pb in the normal state) on that of an ideal BCS S-I-N junction with $\Delta = 19$ meV, at $\Delta/kT = 15$, i.e. at approximately 14 K.[19] (The di/dV of an ideal S-I-N junction essentially reproduces the BCS density of states at a given temperature.) The dramatic difference between the two curves is too obvious. Looking at the experimental curve (2) in Fig. 5 we can only talk about "gap-like features" rather than the true clean gaps. It is our guess at this time[12] that the peak at about 19 mV represents the energy of the gap in YBCO. (This gives the ratio $2\Delta/kT_c = 5.3$). The density of states below this energy is reduced in the superconducting state, but the reduction extends only to about 50% at V = 0. There is a change in slope at ~4 mV which may represent a second smaller gap. In contrast, the classical superconductor has no states in the gap, as shown in Fig. 5, curve (1). In practical terms this means that an I-V of a junction with YBCO has only weak nonlinearity at $V<V_g$, with the slope changing by about 50% at $V=0$ compared to $V>V_g$. This, of course, spells doom for applications such as in electronic switching, where the strong nonlinearity of the I-V curve is essential. The important

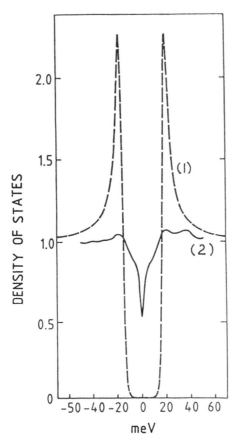

Fig. 5 Tunneling conductance of the YBCO-Pb junction at T = 10 K from Ref. 12 plotted on top of the BCS conductance of an ideal S-I-N junction with Δ = 19 meV, at T = Δ/15 K_B.

question is: are we observing the intrinsic behavior of YBCO, or is this result still reflecting the imperfections at the YBCO surface. Let us not forget that we are dealing with the superconductor which requires near-barrier perfection on a scale of only a few Å, and that the barrier is probably formed at the expense of the material adjacent to the surface.

The fact that the curves shown in Fig. 4 and 5 are reproducible, including the amount of "normal states" in the gap, seem to argue in favor of the intrinsic behavior. Some other experiments, such as Raman and infrared reflectivity measurements, seem to suggest states in the gap of YBCO as well. On the other hand, there exist evidence obtained with the use of point contact

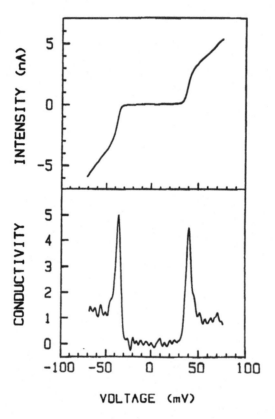

Fig. 6 I-V and di/dV for a point-contact (S-I-S) junction with ceramic $Tl_2Ba_2Ca_2Cu_3O_{10+\delta}$ from Ref. 22.

tunneling which shows essentially clean classical gaps in related high-T_c superconductors (but not in YBCO, interestingly!). Specific heat measurements seem to also argue in favor of clean gaps in all high-T_c materials. In Fig. 6 we reproduce a nice gap obtained on $Tl_2Ba_2Ca_2Cu_3O_{10+\delta}$ with a point contact (STM) measurement by the Madrid-Stockholm collaboration.[22] The authors speculated that this type of a curve results from S-I-S tunneling between two near-surface grains, one of which has been contacted by the tip. Rather clean BCS-type gaps of the S-I-N type were reproducibly observed with Au point contact measurements on $Bi_{1.7}Pb_{0.3}Sr_2CaCu_2O_x$ single crystals by the Argonne National Laboratory group.[23] The spread in values of Δ in these measurements was from 16 to 30 meV. If this and similar STM data are correct, than, presumably, our observations of the states in the gap must be tied to surface imperfection in our junctions. It should be remembered, however, that point-contact experiments a) lack a final proof of tunneling obtained only with the use of superconducting counterelectrodes like Pb (in other words, structures observed in a point-contact measurement can in principle come from processes other than tunneling); b) to a large degree lack a means of checking the temperature and magnetic field dependencies of the observed structures due to a technical difficulty of performing identical low-noise point-contact measurements at different temperatures and in magnetic fields; c) do often show structures due to charging effects which can look very similar to tunneling structures,[24] d) have somewhat irreproducible nature in terms of not always showing the gap structure when the tip is brought to the surface, or showing a spread in gap values for different runs.

One obvious difference between an STM measurement and a classical tunnel junction is in the local character of the former, with an area of only a few tens of square angstroms covered, as compared to fraction of a square mm. It is possible that the local measurement such as in Ref. 22,23 finds occasional good material with the clean gap, while the large area measurement such as in Ref. 12,13,16-18 averages over normal and superconducting regions near the surface. In any case, curves like those in Fig. 6 give at least some hope for the future electronics applications of HTSC in nonlinear tunneling devices.

We left out a number of important issues: the question of gap anisotropy; temperature and magnetic field dependence of the observed structures; asymmetry in the di/dV; normal state tunneling conductance, etc. These issues are covered in a number of listed references.

The author wishes to thank numerous colleagues for valuable discussions. I am particularly grateful to Dr. Jochen Geerk for good discussions and his hospitality in Karlsruhe where parts of this article were written.

REFERENCES

1. B. Batlogg et al., Physica C **153-155**, 1062 (1988); Y. Matsuda et al, Solid State Commun. **68**, 103 (1988); Y. Tajima et al., Phys. Rev. B **37**, 7956 (1988).
2. V. Ambegaokar and A. Baratoff, Phys. Rev. Lett. **11**, 104 (1963).
3. M. Gurvitch and J.R. Kwo, Advances in Cryo. Engineering **30**, 509 (1984), ed. by A.F. Clark and R.P. Reed.
4. R.H. Koch, C.P. Umbach, G.J. Clark, P. Chaudhari, and R.B. Laibowitz, Appl. Phys. Lett. **51**, 200 (1987); H.Nakane, Y. Tarutani, T. Nishino, H. Yamada, and U. Kawabe, Jpn. J. Appl. Phys. **26**, 70 (1987), J.E. Zimmerman, J.A. Beall, M.W. Cromar, and R.H. Ono, Appl. Phys. Lett. **51**, 617 (1987).
5. R.H. Koch et al., to appear in the Proc. of SPIE Symposium on Processing of Films for High-T_c Supercond. Electronics, Santa Clara, CA, Oct. 10-12, 1989.
6. T. Yamashita, ibid.
7. A. Barone, A. DiChiara, G. Peluso, A.M. Cucolo, R. Vaglio, F.C. Matacotta, and E. Olzi, in Novel Superconductivity, Edited by S.A. Wolf and V.Z. Kresin, p. 1003, Plenum Press, N.Y. and London, 1987.
8. P. Mankiewich, D.B. Schwartz, R.E. Howard, L.D. Jackel, B.L. Straughn, E.G. Burkhardt, A.H. Dayem, SPIE **948**, High-T_c Supercond. Thin Films and Devices, p. 37 (1988).
9. C.T. Rogers, A. Inam, M.S. Hegde, B. Dutta, X.D. Wu, T. Venkatesan, preprint.
10. A. Akoh, F. Shinoki, M. Takahashi, and S. Takada, Jpn. J. Appl. Phys. **27**, L519, 1988; also IEEE Trans. Magn. **25**, 795 (1989). K. Mizushima, M. Sagoi, T. Miura, J. Yoshida, Jpn. J. Appl. Phys. **27**, L1489 (1988); L.H. Greene, J.B. Barner, W.L. Feldmann, L.A. Farrow, P.F. Miceli, R. Ramesh, B.J. Wilkens, B.G. Bagley, J.M. Tarascon, J.H. Wernick, M. Giroud, and J.M. Rowell, Proc. of the Int. M^2S-HTSC Conf., Stanford, CA., July 23-28 (1989).
11. E.W. Fenton, Solid State Commun. **60**, 347 (1986); V.N.Kostur and S.E. Shafraniuk, Ukranian Physics Journal **34**, 224 (1989) (in Russian).
12. M. Gurvitch, J.M. Valles, A.M. Cucolo, R.C. Dynes, J.P. Garno, L.F. Schneemeyer, and J.V. Waszczak, Phys. Rev. Lett. **63**, 1008 (1989).
13. J.R. Kwo, T.A. Fulton, M. Hong, P.L. Gammel, and J.P. Mannaerts, to appear in the Proc. of SPIE Symposium on Processing of Films for High-T_c Supercond. Electronics, Santa Clara, CA, Oct. 10-12, 1989.
14. J.R. Kirtley, to appear in the Intern. Journ. of Modern Physics B.
15. For a review of the early work on tunneling see K.E. Gray, M.E. Hawley, and E.R. Moog, in Novel Superconductivity, p. 611, Edited by S.A. Wolf and V.Z. Kresin, Plenum 1987.
16. J. Geerk, X.X. Xi, and G. Linker, Z. Phys. B **73**, 329 (1988).

17. I. Takeuchi, J.S. Tsai, Y. Shimakawa, T. Manako, and Y. Kubo, Physica C **158**, 83 (1989).
18. A. Fournel, I. Oujia, J.P. Sorbier, H. Noel, J.C. Levet, M. Potel, and R. Gougeon, Europhys. Lett. **6**, 653 (1988).
19. J. Geerk, G. Linker, and O. Meyer, Kernforschungszentrum Karlsruhe report #4601, August 1989.
20. J.R. Kirtley, R.T. Collins, Z. Schlesinger, W.J. Gallagher, R.L. Sandstrom, T.R. Dinger, D.A. Chance, Phys. Rev. B **35**, 8846 (1987).
21. J. Geerk, G. Linker, O. Meyer, Q. Li, R.-L. Wang, and X.X. Xi, Physica C - Proc. of the Int.M^{2}S-HTSC Conference, Stanford, CA, July 23-28, 1989.
22. S. Vieira, J.G. Rodrigo, M.A. Ramos, K.V. Rao, and Y. Makino, accepted in Phys. Rev. B: Rapid Comm. (1989).
23. Q. Huang, J.F. Zasadzinski, K.E. Gray, J.Z. Liu, and H. Claus, subm. to Phys. Rev. B - Rapid Comm.
24. S.T. Ruggiero and J.B. Barner, Phys. Rev. B **36**, 8870 (1987).

ELECTRODYNAMICS OF SUPERCONDUCTING $Y_1Ba_2Cu_3O_y$

S. Sridhar

Department of Physics, Northeastern University

360 Huntington Ave., Boston, MA 02115.

The response of superconductors to time-varying electromagnetic fields can yield detailed information on the nature of the superconducting state[1]. Experiments carried out at frequencies in the rf to mm-wave spectral region directly probe the collective response of the condensed electrons and of the quasiparticles, and in conventional superconductors, have been important in determining the gap, the density of states and confirming the nature of the pairing.

The basic parameter that is measured at these frequencies is the complex surface impedance $Z_s = R_s + i\, X_s$. The surface resistance R_s is a measure of the power absorbed, while the surface reactance, which is proportional to the penetration depth λ, is a measure of the reactive power exchanged between the field and the superconductors. These quantities can be computed, from the fundamental constitutive relation which relates current and vector potential for the superconductor, $\vec{J} = -Q\vec{A}$, along with Maxwell's equation and suitable boundary conditions. Experiments can then be compared with detailed predictions for the kernel Q from microscopic theories.

Significant conclusions concerning the intrinsic electrodynamic response require reliable samples. Recent advances in materials processing have yielded high quality samples and experiments indicate that a limiting intrinsic behaviour is observed in these samples. In this paper we summarize measurements of R_s and λ carried out at rf and microwave frequencies on $Y_1Ba_2Cu_3O_y$ crystals. The results are compared[2] with calculations based on the BCS theory, for which detailed comparison is possible at present. Results for the lower critical field H_{c1}, along with temperature dependences of $\lambda(T)$ and $R_s(T)$ and the field dependence of λ, are also discussed. A key feature of the high T_c superconductors is the importance of granularity, and attempts are described to compare the results on polycrystalline samples to models of the granular response in terms of Josephson junction networks.

In addition to the fundamental interest in the high frequency response, a prime motvation is the potential for device applications. Results on two devices that have been fabricated , a microwave superconducting cavity and a filter are discussed briefly.

Measurements on the high T_c superconductors pose new challenges, ironically because of the high temperatures involved. For instance sensitive measurements of R_s require microwave structures whose background absorption is much less than that of the sample. This is difficult to achieve with normal metals like Copper. At 10 GHz, a method was developed[3] which utilizes a high Q ($\sim 10^7$) superconducting cavity made of Pb-plated Cu or Nb and maintained at 4.2K. The sample mounted on a sapphire rod, was placed at the center of the cavity at a maximum of the microwave magnetic field. The sample temperature could be varied from 4.2K to 200K independent of the cavity temperature, thus taking advantage of the low background absorption in Pb or Nb. R_s is determined from the cavity Q using the relation $R_s = \Gamma [Q^{-1}(T) - Q_0^{-1}(T)]$, and changes in the penetration depth are determined from the resonant frequency using $\delta\lambda = -\zeta \delta f$, where Q_0 is the background Q, and Γ and ζ are geometric factors. This method has one of the highest sensitivites currently available for small samples. The technique also circumvents the background thermal contraction (due to the high temperatures) which can severely limit measurements of λ. In addition to the superconducting cavity, Cu cavities at 10, 17 and 35 GHz are also available, in which the sample forms the bottom plate of the cavity. These latter methods are more appropriate for larger samples (≥ 1 cm. dia) and have the advantage of rapid turnaround, although at the expense of sensitivity.

The radio frequency measurements are carried out with an ultra stable oscillator utilizing a tunnel diode. The sample is placed in a small coil (typically 2 mm. dia, 20 turns) which forms the inductive part of the resonant circuit. In order to achieve high stabilites (1 Hz in 10^7 Hz), the tunnel diode was maintained at 77K, close (10 inches away) to the sample whose temperature could be varied from 4.2K to well above 100K. Changes in the resonant frequency are directly related to changes in the penetration depth via the relation $\delta\lambda = -G \delta f/f$, where G is a geometric factor. The high stability enables measurements of changes of 20Å in λ. The temperature dependence of the resonant frequency yields the temperature dependence of λ.

An additional experiment using the rf coil, which has proved to be very informative, is to study the variation of λ with an applied dc magnetic field at fixed temperature. This experiment yields a sharp signature of H_{c1} and has enabled the determination of the complete temperature dependence of H_{c1} in $Y_1Ba_2Cu_3O_y$. The experiment is particularly sensitive to low field effects, both *below and above* H_{c1}, a region not easily accessible by other methods, but of great interest in the high T_c superconductors, because of the possibility of novel flux states[4].

Recent results have shown that spectacularly sharp microwave repsonse can be obtained in well prepared single crystals[5] and thin films[6,7], in contrast to earlier data[8] on polycrystals. These results have raised the hopes of microwave devices with performance far surpassing those made from conventional materials such as Copper. It is apparent however that sample preparation is crucial to obtaining high quality microwave response. The preparation of the single crystals is discussed in ref.5. Below we summarize several electrodynamic properties of these crystals.

PENETRATION DEPTH

The temperature dependence of the penetration depth in the best crystals which possess sharp transitions, are in excellent agreement with calculations based on the BCS theory, down to temperatures of about 0.7 T_c. (Higher sensitivites are required for lower temperatures). The actual quantity that is measured is the change in penetration depth referred to the value at a low temperature T_0 ($<< T_c$) i.e. $\delta\lambda(T) \equiv \delta\lambda(T) - \delta\lambda(T_0)$. The data are then compared to the calculations, using the zero temperature penetration depth $\lambda(0)$ as a fitting parameter. This is typical of all electrodynamic measurements, in which $\lambda(0)$ cannot be measured absolutely, unless sample dimensions are known to accuracies comparable to $\lambda(0)$ - however the same restriction does not apply to changes $\delta\lambda(T)$. Near T_c ($T > 0.9$ T_c) the data can be equally well described by a simple two-fluid model or by the BCS theory. As the temperature range is extended to lower temperatures, the characteristic signature of the gap becomes apparent in the BCS model and the two theories begin to deviate. We have compared the data introducing the gap ratio Δ/kT_c as an additional parameter (see Fig. 1). Best fits to the data are found for $\lambda(0) = 1400$Å and $\Delta/kT_c = 2.15$, with neither the two fluid model nor a weak coupling value of 1.76 providing good fits[2].

The penetration depth measurements are themselves very sensitive tests of sample quality. Using fits to the two-fluid model to extract $\lambda(0)$, we find that $\lambda(0)$ varies from 7.5μm in polycrystals to the lowest[9] value 1400 Å in best crystals, with intermediate values of 0.5 to 1μm in "nominal" crystals.

FIELD DEPENDENCE OF PENETRATION DEPTH $\lambda(H,T)$

London first pointed out on thermodynamic grounds that the penetration depth should vary with magnetic field in the Meissner state. Subsequent experiments by Pippard[10] on the elemental superconductors played an important historical role, since they were the basis of Pippard's concept of a coherence length. Although these variations are small, we have observed them in the rf coil experiment by applying a dc magnetic field, and the results have yielded important information on the relation of the electrodynamic response to granularity.

For $H < H_{c1}$, we find experimentally that the penetration depth increases quadratically with H, i.e. $\delta\lambda(T,H) = k(T) H^2$. This is exactly what one might expect on simple thermodynamic grounds - the magnetic field depresses the order parameter, making all thermodynamic parameters, such as λ, dependent on the magnetic field. To first order the variation should be quadratic (for a T-invariant superconducting state). The coefficent k(T) was determined experimentally for several $Y_1Ba_2Cu_3O_y$ crystalline and polycrystalline samples.

In the best crystals, $k(4.2K) = 10^{-3} Å/G^2$ and increases strongly with temperature by a factor of 10^5 at T_c. The magnitudes and temperature dependence are in good agreement with a Ginzburg-Landau model for the order parameter depression which yields $k(T) = \frac{3}{4}\lambda(T)/H_{c0}^2(T)$. Using BCS temperature dependences for λ and the thermodynamic critical field H_{c0}, we find excellent agreement[2] with data over the entire temperature variation of 5 orders of magnitude.

What is remarkable is the extreme sensitivity of the absolute values of k to sample quality and microstructure. For instance, k(4.2K) varies from $10^{-3} Å/G^2$ in the best crystals to $160 Å/G^2$ in polycrystalline samples, with intermediate values of 10^{-2} to 10^{-1} $Å/G^2$ in nominal crystals. The very large values in the polycrystals are in clear disagreement with the Ginzburg-Landau calculation for the pure bulk superconductor. However, we have shown earlier[11] that these large values can be understood on the basis of a Josephson junction picture. For a single junction, the Josephson penetration depth λ_J

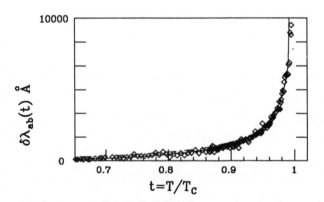

Fig. 1. Temperature dependence of changes in the penetration depth $\delta\lambda(T)$ at 6 MHz with $H_{rf} \perp$ c-axis. The solid line represents a BCS calculation.

will vary quadratically with external magnetic field as $\delta\lambda_J(H) = k_J(T) H^2$, with $k_J = \pi^2 1/24 \lambda_J \Sigma^2/\phi_0^2$, where Σ is the effective junction area, and ϕ_0 the flux quantum. This relation yields $\Sigma \sim 4~\mu m^2$ in the polycrystals. As the samples improve, the sharp decrease in $k(0)$ reflects the decrease of the junction contribution, until the intrinsic properties dominate in the best crystals.

LOWER CRITICAL FIELD H_{c1}

In the above experiment, in which a dc magnetic field is applied in the presence of a probe rf magnetic field, flux entry manifests itself as a sharp deviation from the quadratic dependence of $\delta\lambda$. The feature at H_{c1} is much sharper than is typically seen in magnetization measurements, and the present experiment is a new method to measure H_{c1}. To our knowledge these results constitute the first complete determination of the temperature dependence of H_{c1}. The data for the case of $H \perp$ c-axis are in very good agreement with BCS calculations[2], assuming $H_{c1} \propto 1/\lambda^2(T)$. At 4.2K the measured value of $H_{c1}(0) = 250$ G.

The H_{c1} values also vary considerably with sample quality, providing a plausible explanation for the wide variation reported in the literature. In polycrystals, typical values are < 10 G. This is consistent with the Josephson junction model, since for fields $H_{c1}^* \sim \phi_0/\Sigma \sim 5$ to 10 G, flux entry occurs into the junction. The weakest junctions dominate the observed response, and as the junction contribution is minimised with improved sample quality, H_{c1} appears to saturate at 250 G.

MICROWAVE SURFACE RESISTANCE R_s

The parameter of greatest interest for high frequency applications is the surface resistance R_s. This quantity is typically measured at frequencies > 500 MHz, where the absorption is appreciable.

In contrast to the polycrystals, R_s drops extremely rapidly below T_c in the best crystals. At 10 GHz the drop is nearly 3 orders of magnitude within 4 K below the transition temperature. Comparable results have also been obtained in thin films prepared by laser ablation[6,7].

The temperature dependence near T_c was compared with calculations based on the BCS theory. Such comparisons can be made at several levels, determined primarily by the number of input parameters that one is comfortable with. A complete calculation was carried out using the numerical program of Halbritter[12], which includes the detailed ω and q dependent kernel. Theory and experiment were consistent with the following parameters : $\lambda(0) = 1400$ Å, $\Delta/kT_c = 2.15$, mean free path $l = 70$ Å and coherence length $\xi_0 = 31$ Å, although good fits were also achieved over a region of parameter space surrounding the above values. A wider temperature range is necessary to limit the uncertainty in these parameters.

It is important to realize that near T_c, the T dependence of R_s is dominated by $\lambda(T)$. This is exemplified by the good fit[2] of the data to a simple local limit approximation: $Z_s = R_n \{ 2i/[1 + i\, \delta_n^2/2\lambda^2(T)]\}^{1/2}$. Here quasiparticles are ignored, and all parameters are directly determined from experiment, viz. $\lambda(T)$ as described before, and δ_n and R_n from the measured normal state values. The details of the quasiparticle spectum become important at lower temperatures. Because of the sensitivity limitations due to sample size and high quality, the characterstic signature of an s- wave state, viz. $R_s \propto \exp(-\Delta/kT)$ remains to be verified.

INTRINSIC VS. GRANULAR RESPONSE

A close look at the magnitudes of electrodynamic parameters of several samples reveals that the wide variability in these parameters cannot be dismissed as "junk effects", but rather contain an important piece of physics characterstic of the high T_c superconductors. As one goes from polycrystalline samples to the best crystals, the parameters $\lambda(0)$, $k(0)$ and R_s progressively decrease over several orders of magnitudes, while H_{c1} increases. In the best crystals the behaviour of these parameters is consistent with BCS calculations. On the other extreme, some of the properties of the polycrystals, particularly $k(0)$, can be well described by a single Josephson junction analysis[11], with appropriate large values of the junction area Σ. There have been several attempts[13] to describe the surface impedance of granular material on the basis of junctions.

A realistic network model by Clem[14] provides some insight into the electrodynamic response in the intermediate case where both intrinsic and granular contibutions are present. The penetration depth of a regular network is calculated to be $\lambda_{eff}^2 = \lambda_J^2 + \lambda_{int}^2$. (The inductive impedances of the junction and the superconductor appear in series, leading to a quadrature sum for the penetration depths). Similarly, the field dependent effects represented by k can be written $k_{eff} = k_J + k_{int}$, where k_J and k_{int} are the junction and intrinsic G-L contributions discussed earlier. The observed H_{c1} will be the lower of H_{c1J} and the intrinsic H_{c1}. A detailed model for the surface impedance of networks and quantitative comparison to experiments, remains to be carried out. In any such model however, the junction dissipation can only be included ad hoc, whereas the reactive contributions (such as k and λ) are determined by geometrical factors, such as junction size.

In polycrystals, the normal state R_n is high due to the large resistivity, and is followed by a rather broad transition below T_c with very high residual values measured at 4.2K. Although the data near T_c can be described in terms of a modified Mattis-Bardeen model[8], the low temperature data do not show[15] an $\exp(-\Delta/kT)$ behaviour, even with the residual resistance subtracted. It appears that the temperature dependence of R_s may be almost entirely due to the temperature dependence of λ, which is very large in the polycrystals as stated earlier. Whether a junction model will be able to account in detail for the $R_s(T)$ and $\lambda(T)$ in polycrystals remains to be seen.

A characteristic feature of a superconductor is the $f^{n \sim 2}$ dependence of R_s. This is best demonstrated by measurements on the same sample, as in ref. 15 on a polycrystalline sample. This feature is also present in the junction models, since it arises from the presence of an inductive supercurrent branch.

SUPERCONDUCTING MICROWAVE DEVICES - CAVITIES AND FILTERS

In parallel with the effort aimed at understanding the basic electrodynamic properties of the high T_c superconductors, we have also been involved in the developement of superconducting microwave devices fabricated out of these materials. A microwave cavity is a fundamental device with potential applications in particle accelerator structures[16] and ultra stable clocks[17]. A cylindrical cavity was fabricated[18] from ceramic $Y_1Ba_2Cu_3O_y$. The two pieces were formed using a cold isostatic press. Coupling holes were mechanically drilled into the material. The cavity was resonant in the TE_{011} mode at 8 GHz. Starting from a Q of 10^3 at 100K, the Q improved to 10^4 at 77K and nearly 10^5 at 4.2K. The Q(T) data correlated very well with independent measurements of R_s using the technique mentioned earlier. This device, the first of its kind to be fabricated out of the high T_c superconductors, demonstrated that practical devices are feasible, in which the performance limitations were due to the material and not due to extraneous factors such as joint losses, etc. Based on the single crystal results discussed earlier, Q values exceeding 10^6 at 77K are achievable in principle using high quality material.

Another device of great practical importance in communication systems is a superconducting filter. A typical use of these filters is as a front end to low noise amplifiers. The primary requirement is extremely low insertion loss, so as to avoid added noise to the following amplifier. Although a high Q cavity could be used as a filter, the large number of nearby spurious modes can lead to amplifier saturation. We[19] have designed an interdigital stripline filter with a center frequency of 7.8 GHz and a passband of about 0.3 GHz. The higher order modes, which are all above 20 GHz, are well separated. The filter was machined out of Ag, and coated with YBCO using a process developed at Argonne. At 4.2K the insertion loss was 2 dB, and improvement by a 13 dB from 100K (see Fig. 2). It appears that the limitation in this case is not due to the material, but rather to a temperature independent loss mechanism, probably due to the joints. Insertion losses of 0.1 dB, limited primarily by the connector mismatch, should be achievable with improvements in design.

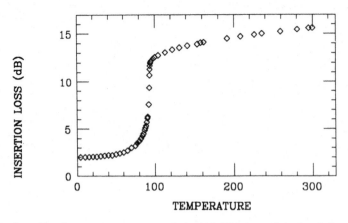

Fig. 2. Insertion loss vs. temperature of the 7.8 GHz interdigital stripline filter.

CONCLUSIONS

Since the discovery of the high critical temperature superconductors, significant progress has been achieved in understanding the electrodynamic response of these materials. Advances in materials fabrication, leading to high quality samples has been an important factor. In $Y_1Ba_2Cu_3O_y$ the electrodynamic parameters display temperature dependences which are in good agreement with calculations based on the BCS theory. Since the central parameter common to the calculations is the gap, the results strongly suggest a mean field temperature dependence for the gap parameter, and a value of $\Delta(0)/kT_c = 2.15$. Further confirmation of the quasiparticle spectrum, viz. an $\exp(-\Delta/kT)$ dependence of $\delta\lambda$ and R_s at low temperatures, requires substantially more sensitivity than is currently available.

From a technological view point, the results on carefully prepared single crystals and thin films offer great promise for device applications at microwave frequencies. Orders of magnitude better performance than conventional devices using Cu are achievable in principle at temperatures exceeding 77K, provided high sample quality can be retained in realistic device structures.

Collaborations with Dong-Ho Wu and W.L.Kennedy, and useful discussions with John Clem are acknowledged. This work was supported by the National Science Foundation under grant ECS-88-11254, the Center for Electromagnetic research at Northeastern University, and NASA/JPL.

REFERENCES

1. M.Tinkham, "Introduction to Superconductivity", Mc-Graw Hill (NY, 1979).
2. S.Sridhar, Dong Ho Wu and W.Kennedy, Phys. Rev. Lett., 63, 1873 (1989)

3. S.Sridhar and W.Kennedy, Rev. Sci. Instr., 59, 531 (1988).

4. David R. Nelson, Phys. Rev. Lett., 60, 1973 (1988).

5. Dong-Ho Wu, W.Kennedy, C. Zahopoulos and S.Sridhar, Appl. Phys. Lett., 55, 696 (1989).

6. L. Drabeck, et. al., Phys. Rev. B, 40, 7350 (1989).

7. N. Klein, et. al., Appl. Phys. Lett., 54, 757 (1989).

8. S.Sridhar, C.A.Shifman and H.Hamdeh, Phys. Rev.B, 36, 2301 (1987)

9. D. Harshman, et. al., Phys. Rev. B, 36, 2386 (1987), A. Fiory, et. al., Phys. Rev. Lett., 61, 1419 (1988), L. Krusin-Elbaum, et. al., Phys. Rev. Lett., 62, 217 (1989).

10. A.B.Pippard, Proc. Roy. Soc. A, 203, 210 (1950).

11. Dong-Ho Wu, C.A.Shiffman and S.Sridhar, Phys. Rev. B, 38, 9311 (1988).

12. J. Halbritter, Z. Phys., 266, 209 (1974).

13. T. Hylton, et. al., App. Phys. Lett., 53, 1343 (1988), A.M. Portis et. al., (Physica C, to be published).

14. J. Clem, Proc. of International M^2SC-HTSC, Stanford, CA, 1989 (Physica C, to be published)

15. W. Kennedy and S.Sridhar, Solid. St. Comm., 68, 71 (1988), W.Kennedy, C.Zahopoulos and S.Sridhar, Sol. St. Comm., (accepted)

16. H. Padamsee, Proc. IEEE, Cat # 87CH2387-9, Part. Accel. Conf., Washington, D.C.(1989)

17. S. Thakur, D.M.Strayer, G.J.Dick and J.E.Mercereau, J. Appl. Phys., 59, 854 (1986)

18. C. Zahopoulos, W. Kennedy and S.Sridhar, Appl. Phys. Lett., 52, 2168 (1988).

19. In collaboration with C. Zahopoulos (NU), J. Bautista (JPL) and M. Flanagan (ANL).

MECHANISMS OF HEAT CONDUCTION IN HIGH-T_c SUPERCONDUCTORS

Ctirad Uher

Physics Department
University of Michigan
Ann Arbor, MI 48109 USA

INTRODUCTION

Much excitement has been generated by the discovery of high-T_c superconductivity by Bednorz and Müller[1] and by the subsequent rapid development culminating in the synthesis of novel superconducting structures with transition temperatures at 90K and beyond.[2-4] Suddenly a whole spectrum of perovskite-related structures has become available upon which one can model theoretical ideas spanning from ordinary to exotic forms of superconductivity. At the same time, material scientists and physicists have been working hard to find ways to make these fascinating materials more mechanically amenable and to improve their superconducting properties.

One of the important parameters of the new high-T_c superconductors is their ability to conduct heat. There is not only an obvious technological interest in how efficiently and by what means the heat flows in these solids but there is also a deep theoretical desire to understand the electronic and vibrational properties of these materials. The magnitude of the thermal conductivity and its temperature dependence are parameters which have an impact on a broad spectrum of devices. For instance, the responsivity of radiation detectors such as bolometers depends sensitively on the thermal conductivity of their thin-film superconducting elements.[5] Likewise, the thermal stability of thin-film microbridge structures[6] owes much to their ability to minimize local hot spots, as does the stable operation of superconducting wires in magnets. From a theoretical viewpoint, the thermal conductivity of superconductors offers important clues about the nature of the carrier and phonon spectra and scattering processes between them. In high-T_c superconductors, such information is even more valuable due to the fact that the traditional galvanomagnetic probes such as resistivity or Hall effect are restricted to a (now) narrow temperature range above T_c. No such limitation exists for the thermal conductivity and one has available an entire temperature range for gathering important transport data. In this overview,

I hope to illustrate the main features of heat transport as they apply to the spectrum of high-T_c superconductors. I shall point out how the free carriers and lattice vibrations contribute to the transport of heat, note similarities in the behavior of ceramic samples, and highlight important characteristics of heat transport in single crystalline specimens. An overview of the thermal conductivity, as part of a broader picture of the thermal behavior of high-T_c superconductors, has been the subject of two prior works.[7,8] Much progress has been made since, including studies on single crystal specimens and on a wider range of high-T_c structures. It is thus appropriate to provide a snapshot of the development as it stands today.

SAMPLES AND EXPERIMENTAL TECHNIQUE

In dealing with high-T_c superconductors, it is important to keep in mind that there are three basic structural forms available: bulk ceramics, single crystals, and oriented epitaxial thin films. The majority of the studies, and thermal conductivity measurements are no exception, have been done on ceramics, i.e., on sintered samples. While such studies are important to establish the magnitude of the thermal conductivity and the relative contributions of the charge carriers and lattice phonons, the "average" structure of the polycrystalline medium precludes investigations of anisotropy in the thermal transport. Anisotropy is one of the key parameters that has an impact on the theoretical understanding of high-T_c materials. Its importance is magnified by the fact that several proposed models of a superconducting mechanism depend sensitively on the degree of anisotropy in the coupling constant. Furthermore, sintered structures frequently contain traces of a secondary phase that can segregate at the grain boundaries and may mask the intrinsic properties of the parent material. Thus, single crystal specimens are indispensable for gaining insight into the intricacies of thermal transport in high-T_c superconductors and in helping to indentify their intrinsic properties. While efforts have been made to measure thermal conductivity on single crystals, the size-limitation of the available samples has been a serious impediment. It is heartening that over the past couple of months more reports are emerging that describe attempts to broaden the data base for thermal conductivity of high-T_c single crystals.

For both of these bulk forms of high-T_c's, the technique most frequently used for thermal conductivity investigations is the well-known longitudinal steady-state method. In the usual configuration, a heater at the free end of a sample generates a temperature gradient that is measured with a pair of thermometers or a differential thermocouple. Alternatively, one can opt for the so-called 2-heater, one-thermometer variant of the steady-state method. In this case, it is the thermometer that is attached at the free end of the sample while the power is switched between two heaters spaced along the length of the sample. Since heaters can usually be made much smaller than thermometers, this is the preferred technique for the very small samples which are typical when dealing with single crystals of high-T_c superconductors. In either case, the thermal conductivity κ is then calculated from the experimentally determined variables using

$$\kappa = \frac{QL}{A\Delta T} \qquad [1]$$

where Q is the power dissipated in the heater, L is the separation between the thermometers (or heaters) and A is the cross-sectional area of the sample.

In using the steady-state technique, care must be exercised to minimize heat flow by means other than through the sample. This is particularly important in the case of high-T_c materials where, as it turns out, the thermal conductivity is rather low. The most serious error usually comes from radiation losses and it may, if not corrected for, easily amount to a substantial overestimate of the thermal conductivity above 250K.

Thin films, the third structural form of high-T_c perovskites, have undergone a spectacular development and in many regards have emerged as the structurally superior form of the material. In particular, high critical current densities on the order of 10^6 Acm^{-2} at 77K achieved[9] on the epitaxially oriented films synthesized at low substrate temperatures fuel hope for a wide range of technological applications. Clearly, heat flow mechanisms in such films will become of great interest to design engineers. Experimentally, thin films pose a difficult problem when attempting to measure their thermal conductivity. Because the film's substrate acts as a heat short, the convenience of the steady-state method is lost. Fortunately, novel techniques based on the reflectivity and refractive index variations that utilize very sensitive laser detection schemes are being developed[10] and, undoubtedly, a way will be found to deal with the thermal transport in thin film structures.

THEORETICAL ASPECTS OF HEAT TRANSPORT IN SUPERCONDUCTORS

It is well known that heat in a solid is carried by two distinct entities, free carriers and quantized lattice vibrations called phonons. While the phonon contribution to the thermal conductivity, κ_p, is always present, the magnitude of the carrier contribution, κ_c, depends on the type of solid because it is directly proportional to the free carrier density. Thus, it is zero in insulators, small in semiconductors, and overwhelmingly dominates heat transport in metals. Since superconductivity inevitably implies some finite carrier density, both κ_p and κ_c are expected to contribute to the thermal conductivity of superconductors. Conventional superconductors have a high free carrier density, therefore the carrier thermal conductivity is usually the dominant contribution. In the high-T_c superconductors, where the Hall effect and other experimental evidence suggest a significant reduction in the carrier density, the relative weight of the two thermal conductivity terms may be entirely different and, as we shall see, the phonon contribution is, by far, the more important one.

All superconductors undergo a phase transition represented by the onset of charge carrier condensation as the temperature is swept through the superconducting transition point T_c. This transition is responsible for a sharp change in the electromagnetic and kinetic properties of the material and, among other things, leads to a drastic modification in the heat flow pattern of the superconductor. Thus, while the conductivity of a superconductor is, in general, more complex than that of a normal metal, the temperature dependent changes in κ as the sample is cooled past and below T_c contain important information on the state of the condensate and the interaction of phonons with the free carriers. Since heat conduction does not cease to exist below T_c while the electrical resistivity vanishes, thermal conductivity is, in essence, a more useful probe of the superconducting state. Being recognized as such, a theoretical description of the thermal conductivity came not long after the development of the macroscopic theory of Bardeen, Cooper and Schrieffer (BCS). Starting with the original paper[11] by Bardeen, Rickhyazen and Tewordt (BRT), a unified picture of the thermal transport in superconductors emerged, initially limited to the case of weak-coupling but later extended[12] to strongly-coupled superconductors.

There are two fundamental properties of the superconducting condensate that have an overriding influence on the thermal transport of superconductors:

1. Cooper pairs carry no entropy,
2. Cooper pairs do not scatter phonons.

The first condition means that the carrier thermal conductivity vanishes rapidly below T_c. This is immediately obvious from a kinetic theory formulation of κ_c

$$\kappa_c = 1/3 c v_F \ell \qquad [2]$$

where c is the specific heat of the carriers, v_F is their Fermi velocity, and ℓ their mean-free path. While v_F is essentially constant and ℓ varies as a power law of temperature, the dominant effect comes from the specific heat which, at $T < T_c$, decreases approximately exponentially.

The second condition has a more subtle effect on the thermal conductivity of superconductors. Provided the mean-free path of phonons at $T > T_c$ is limited by scattering on the charge carriers, on passing into the superconducting domain the phonon thermal conductivity will rise because the number of quasiparticle excitations rapidly decreases leading to an enhancement in the mean-free path of phonons. A competition between the rapidly diminishing κ_c on one hand and increasing κ_p on the other will determine the overall dependence of the thermal conductivity. As already noted, in a vast majority of cases (all pure conventional superconductors and most of their alloys) κ falls rapidly as the material goes superconducting. In a very few cases, where the alloys are sufficiently disordered so that κ_c is small and where κ_p accounts for a large fraction of the normal-state thermal conductivity, one may observe a rise in the total thermal conductivity as the sample enters into the superconducting state. A classic example of this is the data on lead-30% bismuth alloys.[13] Of course, at some lower temperature the thermal conductivity must turn over and start decreasing because the rapidly "freezing out" phonon population will become the determining factor in the transport process. The dominant scattering mode of the phonons determines what power law the temperature dependence of the thermal conductivity will obey. In crystalline materials (including polycrystals as well as single crystals) one would, in principle, expect a T^3-variation of κ arising from phonon scattering on the boundaries of the crystallites. If, however, the crystalline structure contains defects on a scale smaller than the grain size and if these defects scatter phonons strongly at low temperatures, then the temperature dependence of the thermal conductivity may deviate significantly from the T^3 variation. Thus, the power law temperature dependence of κ gives an indication of the scattering mechanism of the phonons which dominate at that particular temperature range. As we shall see, low temperature thermal conductivity of high-T_c perovskites, as a rule, deviates strongly from a T^3-dependence and the intrinsic behavior of the high-T_c materials is a somewhat contentious issue at this time.

With this brief background, let us now look at the characteristic features of the thermal conductivity in high-T_c superconductors. In the space available I cannot give an exhaustive account of all existing thermal transport measurements on this class of materials. Rather, I will try to illustrate typical properties of both the sintered and single crystal samples and how they influence our understanding of high-T_c superconductors.

THERMAL CONDUCTIVITY OF HIGH-T_c SUPERCONDUCTORS

A characteristic feature of high-T_c perovskites is the nearly 2-dimensional Cu-O sheets that are believed to contain all the essential ingredients for superconductivity. The first two families of high-T_c materials, namely the La-based 40K superconductors and the so-called 1-2-3 compounds typified by $YBa_2Cu_3O_{7-\delta}$ that superconduct near 92K are quite sensitive to oxygen stoichiometry. This is reflected in their strong dependence on the conditions of synthesis as well as in their sensitivity to environmental factors such as air humidity. This sensitivity to oxygen can be advantageous when studying the thermal conductivity. By an appropriate annealing treatment in oxygen or vacuum one can drive the material from a superconducting orthorhombic state into a tetragonal insulator and back again while monitoring the ensuing changes in the thermal conductivity. On the other end of the spectrum are the quinary compounds such as the Bi-family of high-T_c superconductors with the general formula $Bi_2Sr_2Ca_{n-1}Cu_nO_{2n+4}$ and the corresponding Tℓ-based superconductors with the highest T_c of about 125K. While far less sensitive to oxygen content, large differences in the volatility and reactivity of their individual chemical constituents make the task of synthesizing single-phase structures of these materials very difficult. The minority phases may affect the magnitude of the thermal conductivity by enhancing phonon-defect and carrier-defect scattering. As I have already noted, to extract the intrinsic behavior, one should resort to measurements on single-crystalline specimens whenever possible.

I. $YBa_2Cu_3O_{7-\delta}$ Compounds

For lack of suitable single crystals, the early measurements were carried out on sintered samples. Because of the nature of the sintering process, the resulting samples have less than the theoretical density and this affects the thermal conductivity in two ways: the voids in the structure scatter phonons, and they also reduce the volume fraction of the conducting medium. Thus, the magnitude of the thermal conductivity of a sample reflects its particular preparation conditions. This is clearly evident in Fig. 1 which shows representative data on the thermal conductivity of sintered $YBa_2Cu_3O_{7-\delta}$ superconductors.[14-18] For comparison, the behavior of a sintered insulator, $YBa_2Cu_3O_6$, is also shown[19] in the same figure. One can see an order of magnitude spread in the value of the conductivity, a direct consequence of the varying degree of compactness of the samples. In particular, the sample of Ref. 15 must have had a very high porosity (which also follows from its unusually high electrical resistivity of 19 m$\Omega-cm$ at 300K). A contributing factor to the wide range in the observed magnitude of κ may also be a variation in oxygen content of the samples that in early studies was not always rigorously monitored.

Oxygen deficiency indeed has a dramatic effect on the magnitude and the temperature dependence of the thermal conductivity of $YBa_2Cu_3O_{7-\delta}$ as is evidenced by measurements of Zavaritskii et al.[21] (Fig. 2), and of others.[22,23] According to Ref. 21, an increase in oxygen deficiency δ creates vacancies that strongly scatter short-wavelength phonons. Of particular interest is their observation that the sample with $\delta=1$, i.e., the O_6 stoichiometry (insulating material), has a thermal conductivity above 120K of nearly a factor of two larger than all the other samples. Because the

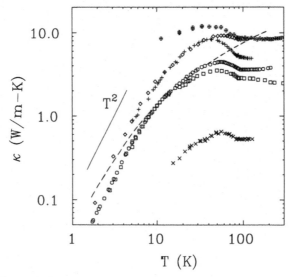

Fig. 1. Thermal Conductivity of $YBa_2Cu_3O_{7-\delta}$
+ Jezowski et al.[14], x Morelli et al.[15], ○ Uher and Kaiser[16]
□ Bayot et al.[17], ◇ Gottwick et al.[18], all measurements refer to superconducting ceramics;
Broken curve—sintered $YBa_2Cu_3O_6$ insulator of Salce et al.[19];
⊞ Hagen et al.[20]—single crystal in the a-b plane.

heat capacities of the materials with the O_6 and O_7 stoichiometries are approximately the same and the sound velocity of $YBa_2Cu_3O_6$ is only larger than that of $YBa_2Cu_3O_7$ by about 9%, the orthorhombic-tetragonal transition cannot account for such a large enhancement in κ on $YBa_2Cu_3O_6$. Instead, Zavaritskii et al. propose that the thermal conductivity is large due to the absence of both oxygen vacancies and free carriers, the two entities strongly dissipating phonon momentum and/or energy.

Assuming the validity of the Wiedemann-Franz law and using the experimental values for the electrical resistivity, one may estimate the maximum possible contribution of the charge carriers to the thermal transport. The result is surprising: typically less than about 10-15% of the heat is being carried by the carriers in the sintered high-T_c samples. This is the first major difference between high-T_c's and conventional superconductors. The next surprising feature is a sudden rise and a pronounced maximum observed on the thermal conductivity curves of the superconducting samples below T_c. Note that the insulating material (depicted by the broken curve) shows a monotonic temperature dependence without any hint of anomaly. Rising thermal conductivity below T_c is a consequence of an enhancement in the phonon mean-free path due to carrier condensation and it is a clear indicator of the importance of phonon-carrier interaction in the normal state of the material. An increase in κ below T_c is precisely the effect predicted theoretically by the BRT-theory for the phonon conductivity limited by carrier scattering. This effect has been difficult to observe in conventional superconductors because their large carrier-contribution masks the changes in the phonon heat conductivity. By fitting the BRT-theory and taking into account the scattering of phonons by structural defects, Tewordt and Wölkhausen[24] have recently estimated that the carrier-phonon coupling constant λ lies within the weak-coupling range.

Turning our attention to temperatures below T_c, see Fig. 1, we observe a maximum in the conductivity near $T_c/2$ after which the thermal conductivity decreases rapidly. How quickly it decreases, i.e. the characteristic temperature dependence of $\kappa(T)$, is not yet a fully settled issue. Over the limited temperature range displayed in Fig. 1 it would appear that κ approaches a quadratic temperature dependence as

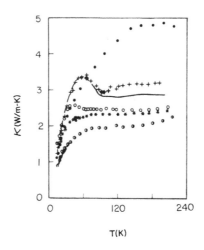

Fig. 2. Thermal conductivity of $YBa_2Cu_3O_{7-\delta}$ as a function of oxygen deficiency δ.
+ original sample, $\delta=0$
○ after annealing for 6h at 435°C, $\delta=0.3$
* further annealing for 6h at 455°C, $\delta=0.47$
◐ after annealing for 6h at 470°C, $\delta=0.69$
● after vacuum annealing for 40h at 675°C, $\delta=1$
Solid line—oxygen annealed for 21h at 540°C, $\delta=0.23$
Adapted from Ref. 21

its limiting behavior. This, again, is unexpected since for a polycrystalline medium (and also for single crystals) at low temperatures one would assume that phonons would scatter predominantly on the grain boundaries resulting in a characteristic T^3-dependence, the well known boundary scattering regime. A comparison of the phonon mean-free path, ℓ_p, calculated from Eq. 1 (now for phonons rather than carriers) using the experimental data for the specific heat and sound velocity, with the typical grain-size in the samples (1-20μm) indicates that the intragrain scattering is important and in some cases even dominates. This would seem to explain why, down to 4K, there is no indication that the conductivity is boundary-limited. The important question is what scatters phonons within the grains?

Regueiro et al.[25] and others[26] have argued that the thermal properties of high-T_c perovskites indicate an amorphous system with two-level tunneling states (TS). According to this picture, the necessary disorder originates from oxygen vacancies and the randomness in their distribution is at the heart of the tunneling states that scatter phonons. Provided the distribution of the relaxation times is sufficiently broad, the total relaxation time for phonon scattering on TS is inversely proportional to phonon frequency, hence the process leads to a T^2-dependence of the thermal conductivity at low temperatures. Sound velocity and internal friction measurements[27] that mimic some but not all characteristic features of the amorphous materials add a measure of credibility to the picture of high-T_c perovskites as glasses. Moreover, the well known but mysterious non-zero γ-term in the specific heat of $YBa_2Cu_3O_{7-\delta}$ and $La_{2-x}Sr_xCuO_4$ at low temperatures could also be explained as arising from the presence of tunneling states. The TS model indeed seems to have a strong appeal. The problem is that on a closer look, the similarities between the high-T_c's and the classical amorphous materials may be more intuitive than substantive. For instance, the TS-phonon interaction is expected to lead to a T^3-dependence of the internal friction rather than the observed $T^{1.8}$ power law, and the γ-term in the specific heat of the La- or Y-based superconductors is some one or two orders of magnitude larger than the values observed in typical amorphous metals or insulators.[28] Furthermore, the most prominent feature of the thermal conductivity in glassy systems, the existence of a plateau, is entirely missing in high-T_c materials. On the other hand, there is at least one disordered crystalline system, KBr_{1-x}-KCN_x mixed single crystals,[29] that is believed to behave as a glassy material but shows no plateau. It is possible, then, that the existence of TS does not assure the presence of a plateau.

Effort in two areas of thermal transport research should be helpful in resolving the issue of phonon scattering in high-T_c superconductors: the extension of measurements to temperatures below 1K, and studies on single crystals. Much progress has been made on both fronts and I will discuss the results of each in turn.

All investigations[18,19,30-34] of sintered samples of superconducting $YBa_2Cu_3O_{7-\delta}$ at temperatures below 1K are showing that the temperature dependence of the thermal conductivity exhibits a remarkable weakening below 0.3K and approaches a limiting T-linear variation, see Fig. 3. In fact all of the measurements can be fitted assuming a conductivity of the form

$$\kappa(T) = aT + bT^3 \qquad [3]$$

with the coefficients a and b given in Table 1.

Table 1. Coefficients a and b of Eq. 3 for Sintered $YBa_2Cu_3O_{7-\delta}$

Reference	18	30	31	32	33	34*
$a[mWK^{-2}m^{-1}]$	1.6	3.6	2.2	3.9	1.4	0.54
$b[mWK^{-4}m^{-1}]$	4.7	36	20	29	8.7	6.9

* Refers to a hot-pressed sample measured parallel to the direction of pressing.

As Cohn et al. showed (see the inset in Fig. 3), the linear term vanishes when a sample is driven into its insulating state by vacuum annealing at 600°C for 5 hours, and reappears with a subsequent annealing treatment at 600°C in flowing oxygen and a slow cool to room temperature. While the experimental situation is very comfortable, interpretations of the data are diverse, and several different mechanisms are proposed. For instance, the authors of Refs. 18, 32, 33 advance the picture of uncondensed free carriers that account for the γ-term in the specific heat and, because of their scattering on defects, also contribute a linear term to the thermal

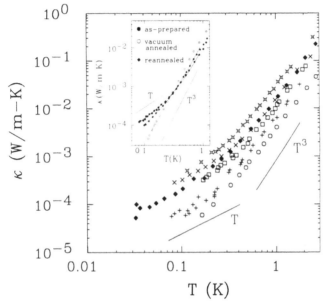

Fig. 3. Thermal conductivity of sintered $YBa_2Cu_3O_{7-\delta}$ at low temperatures.
□ Gottwick et al.[18], ✗ Salce et al.[19], ○ Freeman et al.[30],
◆ Graebner et al.[31], ⋈ Sparn et al.[32], + Kirk et al.[34].
The inset shows the data of Cohn et al.[33].

conductivity. The question of the origin of these carriers is left open; they might be an integral part of some exotic mechanism of superconductivity, arising from a highly anisotropic and gapless Fermi surface, or they could be of extrinsic origin, perhaps due to a minute amount of an impurity phase. In contrast to this explanation, in Refs. 19 and 31 it is proposed that structural defects are responsible for the T-linear thermal conductivity at the lowest temperatures. In Ref. 19, 2-dimensional imperfections such as twins are thought to be the cause, while in Ref. 31, Rayleigh scattering on the microstructure is hypothesized. In fairness it should be said that the situation is far from clear and none of the models have been embraced as the definitive answer. It is hoped that low temperature measurements on single crystals will shed more light on the problem. Unfortunately, at this moment, there is only one set of such data available in the literature and, as far as $YBa_2Cu_3O_{7-\delta}$ is concerned, the results are inconclusive. Graebner et al.[31] measured the thermal conductivity in the a-b plane of single crystals of $YBa_2Cu_3O_{7-\delta}$ and of $HoBa_2Cu_3O_{7-\delta}$ of typical dimensions $0.8 \times 2 \times 0.07$ mm^3 down to about 55 mK, see Fig. 4. While an approximately T^2-dependence seems a good representation of the data on the Ho-substituted single crystal, there is a distinct and continuous curvature on the $YBa_2Cu_3O_{7-\delta}$ sample that pulls the thermal conductivity to a weaker temperature dependence, and the T^2 variation is not a convincing fit over the low temperature range covered. A quadratic power law dependence of κ would, of course, add additional support to the TS model of the thermal conductivity. I cannot emphasize too strongly the need to provide a broader data base so that the mechanisms governing low temperature transport in single crys-

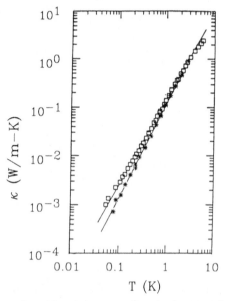

Fig. 4. Thermal conductivity, κ_{ab}, for single crystal samples of
□ $YBa_2Cu_3O_{7-\delta}$ and ✽ $HoBa_2Cu_3O_{7-\delta}$.
Adapted from Ref. 31.

tals of 1-2-3 compounds can be clarified. Obviously, the very small size of available single crystals of $YBa_2Cu_3O_{7-\delta}$ is the major impediment to a greater proliferation of thermal conductivity measurements. This is true not only for studies at subkelvin temperatures but at all temperatures. It is only recently that the first attempt has been made[20] to examine anisotropy in the thermal transport of the 1-2-3 compounds. The data from these measurements are shown in Fig. 5a (conductivity in the a-b plane) and Fig. 5b (conductivity along the c-axis). The in-plane conductivity, κ_{ab}, has a marginally larger magnitude than the conductivity of the sintered samples of $YBa_2Cu_3O_{7-\delta}$ shown in Fig. 1, but otherwise exhibits similar features. To facilitate a comparison, the curve representing sample B of Fig. 5a is replotted on a log-log scale in Fig. 1. The c-axis conductivity, κ_c, is, however, entirely different. Depending on the crystal quality, it is some 5-10 times smaller than κ_{ab}. It slowly increases with decreasing temperature, reaches a broad maximum in the range 50-80K and, eventually, falls sharply with decreasing temperature. There is certainly no hint of any anomalous

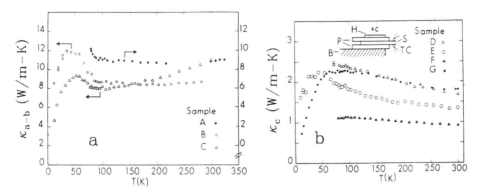

Fig. 5. Thermal conductivity of $YBa_2Cu_3O_{7-\delta}$ single crystals
a) measured in the a-b plane, κ_{a-b}
b) measured along the c-direction, κ_c.
Adapted from Ref. 20.

behavior at or near the transition temperature, T_c. Using the Wiedemann-Franz ratio and the value of the electrical resistivity at 300K ($\sim 150\mu\Omega - cm$), Hagen et al. estimate that above T_c, about 55% of the total heat flow is due to free carriers, a value substantially larger than in the case of sintered specimens. From the fact that κ_{ab} has approximately the same magnitude in superconducting and insulating (not shown here) crystals, they deduce that scattering of phonons on charge carriers in the a-b plane halves the lattice contribution and makes the carrier-limited phonon conductivity about equal to the umklapp-phonon-limited contribution. Since the authors observe no anomaly in κ_c near T_c, they set an upper bound on the carrier-limited lattice resistivity in the c-direction of less than 5% of the total, with 95% due to phonon-umklapp processes. A comparison of the carrier-limited phonon thermal resistivity in the two principal directions reveals an extreme anisotropy in the carrier-phonon coupling and the authors argue that this is not a favorable situation for a phonon-mediated mechanism of superconductivity. While this conclusion is widely accepted, the actual estimate of the carrier contribution to κ_{ab} should be taken with some caution as it is likely to be too high. The point is that it is unlikely that the carrier scattering is perfectly elastic (i.e., one can use the Sommerfeld value $L_o = 2.44 \times 10^{-8} V^2 K^{-2}$ in the Wiedemann-Franz ratio) when, at the same time, the carriers are strongly interacting with phonons.

It is well known that one can substitute for Y other elements of the lanthanide family without, in most cases, seriously affecting the superconducting properties of the material. This, of course, suggests that the Y-site does not play a special role in superconductivity; it is there to stabilize the structure. The effect of substitution on the thermal conductivity can, depending on the nature of the substitution, be rather large. In particular, impurity scattering may be enhanced if substitution leads to the formation of a multiphase structure or if only a partial substitution is achieved. Both the carrier and lattice thermal conductivities can be altered if the substitution leads to changes in the carrier density or to some degree of carrier localization. Finally, magnetic ordering may interfere if magnetic ions are substituted. In practice, it is

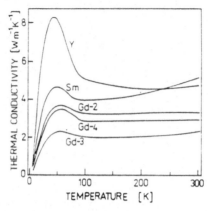

Fig. 6. Thermal conductivity of Sm and Gd-substituted 1-2-3 superconducting ceramics.
Data of Jezowski et al. from Ref. 35.

difficult to resolve specific substitution-related changes in the thermal conductivity unless one makes measurements on single crystal substituted samples. The variability in the structure of sintered specimens is simply too large and its influence on the thermal conductivity overwhelms the changes arising upon substitution. To illustrate this point, in Figs. 6 and 7 are reproduced the data of Jezowski et al.[35] where Y is replaced by Sm or Gd, and the data of Heremans et al.[36] on a series of lanthanide-substituted 1-2-3 compounds. The samples of Heremans et al. have a rather porous structure and, consequently, a thermal conductivity of nearly an order of magnitude smaller than those of Jezowski et al. $GdBa_2Cu_3O_{7-\delta}$ is not only a high-T_c superconductor, but also an antiferromagnet with a Neèl temperature $T_N=2.2K$. Magnetic ordering, in this case, leads[37] to a change in slope of the temperature dependence of the thermal conductivity. Another lanthanide- substituted system on which the thermal conductivity was studied[38] is $EuBa_2Cu_3O_{7-\delta}$; the data are, again, very similar to those on sintered $YBa_2Cu_3O_{7-\delta}$.

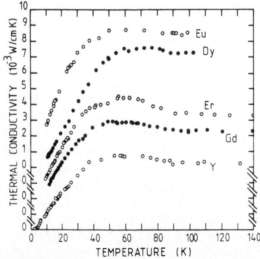

Fig. 7. Thermal conductivity of several rare-earth-substituted superconducting ceramics. For clarity, the vertical axis for each sample is displaced one unit.
Data of Heremans et al. from Ref. 36.

Finally, I consider the changes in the thermal conductivity of 1-2-3 compounds which arise from structural damage created by energetic particles such as fast neutrons. The subject is of technological interest, and, due to changes in the oxygen vacancy network upon irradiation, may also shed some light on the structure-superconductivity relationship. Radiation-induced defects such as extra oxygen vacancies and extended defect cascades alter phonon and carrier distribution functions by enhancing defect scattering. One would thus expect a change in both the magnitude and temperature dependence of the conductivity with increasing neutron fluence. Such changes were indeed observed in the measurements of Uher and Huang[39] on fast neutron-irradiated sintered samples of $YBa_2Cu_3O_{7-\delta}$ shown in Fig. 8. Degradation of thermal conductivity is a direct consequence of strong phonon-defect scattering. With increasing neutron fluence, the phonon-defect scattering rate rises sharply and the phonon-carrier interaction responsible for the peak in the conductivity below T_c is diminished. At fluences of $6 \times 10^{18} ncm^{-2}$ the peak is totally washed out, yet the material still possesses a transition temperature in excess of 80K. Again, this is consistent with the notion that the conventional phonon mediated mechanism cannot account for superconductivity in the 90K range.

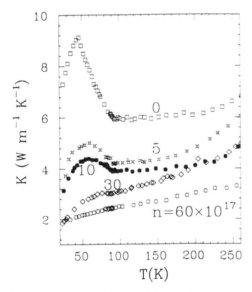

Fig. 8. Thermal conductivity of sintered $YBa_2Cu_3O_{7-\delta}$ at various levels of irradiation. Fast neutron fluence is indicated on each curve. Data of Uher and Huang from Ref. 39.

II. $La_{2-x}Sr_xCuO_4$ Superconductors

The La-family of high-T_c superconductors has a layered K_2NiF_4 structure with quasi-2-dimensional sheets of Cu-O. The parent material, La_2CuO_4, is an orthorhombic insulator that undergoes a phase transition to the tetragonal configuration near 530K. It exhibits rather strong antiferromagnetism with a Neèl temperature near 230K. The existence of long range magnetic ordering in these materials has stimulated considerable theoretical interest and numerous attempts have been made to link superconductivity with the magnetic state of the system. Doping of La_2CuO_4

with 2+ ions such as Sr or Ba introduces a hole into the Cu-O band. As the amount of dopand increases, the interplanar magnetic coupling weakens dramatically, leading to a sharp drop in the Neèl temperature, and a vanishing of the 3-dimensional magnetic ordering. At the same time, the material shows progressively stronger metallic character and, near x=0.5, superconductivity sets in. The maximum transition temperature (~40K) is observed at about x=0.15. Doping to levels of x~0.15-0.2 has little effect on the oxygen stoichiometry, but at higher divalent ion concentrations, the oxygen content is altered and the appropriate representation of these structures is $La_{2-x}Sr_xCuO_{4-y}$, i.e., as oxygen deficient perovskites.

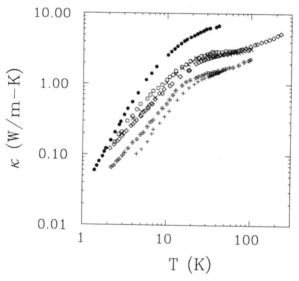

Fig. 9. Thermal conductivity of sintered $La_{2-x}Sr_xCuO_4$.
◇ Uher and Kaiser[16], x=0.2,
+ Bartkowski et al.[40], x=0.2
× Steglich et al.[41], x=0.15
⌗ Regueiro et al.[25], x=0.15
○ Regueiro et al.[25], x=0
● Bernasconi et al.[42], x=0.

As in the case of $YBa_2Cu_3O_{7-\delta}$, most studies of the thermal transport of these materials were made on ceramic forms of $La_{2-x}Sr_xCuO_4$. The magnitude of the thermal conductivity, because of its dependence on density and porosity, reflects the particular sample preparation process used. This is evident in Fig. 9, where the existing thermal conductivity data are collected. The temperature dependence of the conductivity is remarkably similar for all superconducting samples of $La_{2-x}Sr_xCuO_4$, regardless of the actual Sr content. However, measurements of samples with nominally the same Sr concentration differ in magnitude by up to a factor of 3. Comparing the data for a pure La_2CuO_4 sample (open circles) with the $La_{1.85}Sr_{.15}CuO_4$ sample (⌗ symbol) prepared under identical conditions, one can also detect the effect of additional scattering as the solid solution is formed. Overall, the magnitude of the thermal conductivity is low and similar to that of the 1-2-3 compounds. Using the

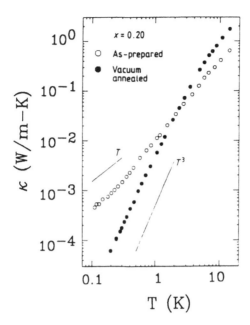

Fig. 10. Thermal conductivity of sintered $La_{1.8}Sr_{0.2}CuO_4$ before and after vacuum annealing.
Data of Uher and Cohn from Ref. 43.

Wiedemann-Franz ratio and values for the electrical resistivity we recognize that the maximum carrier contribution to the thermal conductivity cannot exceed about 10% of the total conductivity. This also is in good accord with an estimate of the carrier contribution in the 1-2-3 compounds. Noting the common features in the heat transport of the 40K and 92K superconductors, one would expect to see a highly anomalous behavior at and below T_c, similar to the one characterizing thermal transport in $YBa_2Cu_3O_{7-\delta}$. No such behavior is observed in sintered $La_{2-x}Sr_xCuO_4$. At most, one detects a change of slope in the neighborhood of T_c. In ceramic samples of the La-family of superconductors, the anomalous behavior near T_c is not well developed for the following reasons: at these relatively low temperatures (T<40K), the rapidly diminishing phonon population compensates for the fact that phonon mean-free path, ℓ_p, increases as the carriers condense, and strong phonon-defect scattering in ceramic structures sets a limit on the possible enhancement in ℓ_p. It should be noted that when the phonon-defect scattering rate is small, such as in the case of high quality single crystals of superconducting $La_{2-x}Sr_xCuO_4$, an increase followed by a peak in the thermal conductivity below T_c is visible even in this kind of high-T_c superconductor (see Fig. 12).

The temperature dependence of the thermal conductivity below T_c (in Fig. 9) obeys an approximately $T^{1.4}$ variation which is a rather weak power law dependence. As in the case of 1-2-3 compounds, speculation centers[25] on tunneling states that might scatter phonons resonantly. By extending the measurements to temperatures well below 1K, Uher and Cohn showed[43] that the sub-quadratic temperature dependence of κ persists down to about 0.5K and, at still lower temperatures, the thermal conductivity approaches a T-linear variation, Fig. 10, much like the situation in sintered 1-2-3 compounds. On destroying superconductivity in the x=0.2 sample by

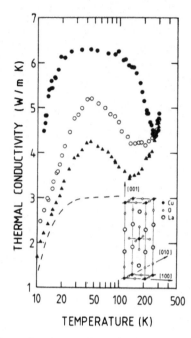

Fig. 11. Thermal conductivity of La_2CuO_4 single crystals.
● Heat flow parallel to [001];
○ Heat flow parallel to [221];
▲ Heat flow parallel to [110].
The dashed line is the data of Ref. 25 for sintered La_2CuO_4.

vacuum annealing, the temperature dependence of the conductivity is drastically changed and a T^3 power law is observed at all temperatures below 3K.

Unlike the 1-2-3 compounds, it is relatively easy to grow fairly large single crystals of both the parent and doped La-based perovskites and such crystals have been used recently by Morelli et al.[44] in investigations of the thermal conductivity. The data for insulating single crystals of La_2CuO_4 with various orientations of their crystallographic axes relative to the heat flow are shown in Fig. 11. The thermal conductivity of all the samples exhibits a peak near 40K which, the authors believe, indicates that the samples display a crystalline character in their thermal transport, contrary to the idea of tunneling states and a glassy behavior. They suggest that the amorphous-like behavior of the sintered materials is an artifact of their poor crystalline state which masks the intrinsic effects displayed in Fig. 11. A remarkable feature of these data is a minimum in the thermal conductivity at higher temperatures that appears to depend strongly on the orientation of the sample. Morelli et al. associate this minimum with magnetic order-disorder transitions in the lattice whereby the heat-carrying phonons are predominantly scattered by spin waves. In their interpretation, above T_N where the spins are disordered and phonon scattering on such spins is temperature independent, the thermal conductivity increases with the temperature because the phonon density is still increasing. However, as one cools below T_N, the spin disorder disappears and the phonon-spin scattering scales with the magnetic order parameter which increases with decreasing temperature. Hence, the thermal conductivity rises as the temperature falls. The position and sharpness of the minimum reflects the nature of the magnetic ordering in each antiferromagnetic

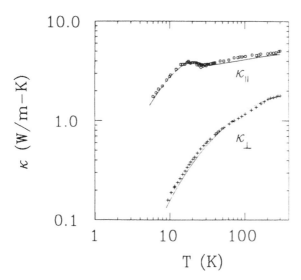

Fig. 12. Thermal conductivity of a $La_{1.96}Sr_{0.04}CuO_4$ single crystal measured parallel ($\kappa_\parallel$) and perpendicular ($\kappa_\perp$) to the CuO_2 planes. Solid lines are fits to the BRT-theory.
Adapted from Ref. 45.

sublattice. It is somewhat surprising that the thermal conductivity is found to be significantly higher when the heat flow is parallel to the c-axis of the structure (across the planes) than in the case where the flow is in-plane. One would expect that the lower rigidity and the partially layered character of the structure would result in a smaller thermal conductivity in the c-direction. This certainly seems to be the case in measurements currently underway in my laboratory in collaboration with Dr. Morelli that aim to extend the thermal transport studies on similar La_2CuO_4 single crystals to subkelvin temperatures.

A study of anisotropy in the heat conduction of superconducting single crystals of Sr-doped lanthanum cuprates has been completed by Morelli et al.[45] quite recently. The data for parallel, $\kappa_\parallel$, and perpendicular, $\kappa_\perp$, orientations of the heat flow relative to the CuO_2 planes of a single crystal with a nominal composition $La_{1.96}Sr_{0.04}CuO_4$ are shown in Fig. 12. Resistance measurements on this sample indicate an onset of superconductivity near 25K with the transition complete at 10K. The most important aspect of the measurements is a large anisotropy in the thermal conductivity that at room temperature stands at 3 and increases to 16 at 10K. Note also that, as one would expect, the heat flows much better in-plane than across the planes, contrary to the data on La_2CuO_4 single crystals displayed in Fig. 11. Morelli et al. fitted the data to a modified BRT-theory (the curves are shown in Fig. 12) and arrived at the following conclusions: for the in-plane transport, the carrier-phonon coupling constant, λ, is approximately 0.6, a value very similar to the one obtained in Ref. 24 for sintered $YBa_2Cu_3O_{7-\delta}$ compounds. In the direction across the planes of CuO_2, phonons scatter very strongly on the stacking faults and this process masks all other dissipative channels. In particular, the phonon-carrier scattering term has virtually

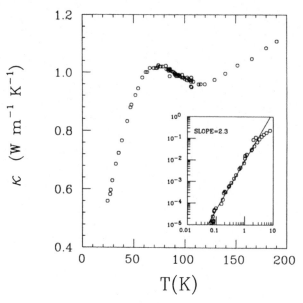

Fig. 13. Thermal conductivity of sintered $Bi_2Sr_2Ca_2Cu_3O_{10}$. Inset shows the conductivity at low temperatures. Data of Peacor and Uher from Ref. 46.

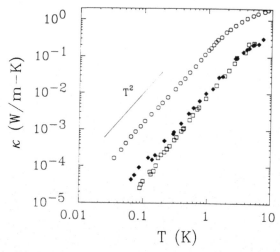

Fig. 14. Thermal conductivity of Bi-Sr-Ca-Cu-O superconductors.
○ single crystal of $Bi_2Sr_2Ca_1Cu_2O_8$ measured parallel to the a-b plane. Data of Zhu et al. from Ref. 47;
◆ hot-pressed $Bi_2Sr_2Ca_1Cu_2O_8$ measured perpendicular and
□ hot-pressed $Bi_2Sr_2Ca_1Cu_2O_8$ measured parallel to the pressing direction.
Data of Peacor and Uher from Ref. 46.

no effect on the fit even if its strength is allowed to vary over a wide range. From their fit, the authors find a stacking fault spacing of some 100Å which, if associated with misaligned CuO_2 sheets, would mean that every tenth CuO_2 atomic plane is not in complete registry with the crystal structure. Such strong defect scattering creates a very unfortunate situation since it prevents one from estimating the anisotropy of the carrier-phonon coupling constant. Perhaps with further improvements in crystal quality one might be able to extract this important parameter.

II. $Bi_2Sr_2Ca_{n-1}Cu_nO_{2n+4}$ Superconductors

Following an intensive search for superconductors with still higher T_c, Maeda et al.[3] reported in early 1988 the discovery of superconductivity in a quinary system based on Bi-Sr-Ca-Cu-O, the first high-T_c material synthesized without a rare-earth element. As a result of a thorough structural analysis, two compositions (and subsequently a third, lower T_c phase) were identified as being responsible for superconductivity: $Bi_2Sr_2CaCu_2O_8$, the so-called (2212) phase with a T_c near 85K, and $Bi_2Sr_2Ca_2Cu_3O_{10}$, the (2223) phase with $T_c \sim 110K$. In general, it is difficult to prepare these materials as single phases. As a consequence, the resistance of the (2223) structure does not vanish following a sharp drop near 110K but, instead, gradually diminishes down to 85K where it becomes zero. It is only because of success with partial substitution of Pb for Bi (a process which promotes the connectivity of the 110K phase), that the zero resistance state has been realized near 110K. There are several notable features of the structure of Bi-Sr-Ca-Cu-O superconductors that distinguish it from the 40K and 92K materials and that are considered important from a theoretical and practical viewpoint: The structure does not support Cu-O chains such as occur in the orthorhombic phase of the 1-2-3 compounds, and, because there is no tetragonal-orthorhombic transition in the Bi-family of superconductors, there are no mirror-type (110) twins. For potential technological applications, it is of interest that the Bi-system is relatively insensitive to oxygen treatment. Finally, a very important feature of these materials is their apparent lack of a γ-term in the low temperature specific heat. As indicated below, this property may have an impact on the behavior of the thermal conductivity at low temperatures.

Currently, two reports exist in the literature describing measurements of the thermal conductivity on Bi-based high-T_c superconductors, one pertaining to polycrystalline samples and the other to the in-plane conductivity of single crystals. Peacor and Uher[46] investigated thermal transport from room temperature down to 70 mK on a sintered sample of nominally (2223) material (which inevitably contained some (2212) phase), and two hot-pressed specimens of the (2212) compound, see Figs. 13 and 14. The hot-pressed samples were cut from the same ingot parallel (open squares) and perpendicular (solid diamonds), respectively, to the pressing direction. As already noted, hot-pressing promotes alignment of the crystallites during the processing stage and this gives rise to a certain degree of structural anisotropy. While the processing-induced anisotropy is reflected in the resistivity (the anisotropy ratio of the resistance measured parallel and perpendicular to the pressing direction is about 4 at 300K), the effect on the thermal conductivity is rather small. The data of Fig. 13 reveal all the characteristic features found in high-T_c superconductors: a small value of κ, a dominant phonon contribution (87% in this particular case), and a peak in κ below T_c. Since the Bi-based samples exhibit these common features, one would expect that their transport behavior at low temperatures would be similar to that of the 40K and 92K superconductors. However, at least for the ceramic structures, this is not so. As the data of Fig. 14 and the inset in Fig. 13 indicate, from about 2K down to the lowest temperatures, the data follow an approximately T^2-dependence. This contrasts sharply with the behavior of sintered samples of $La_{2-x}Sr_xCuO_4$ and $YBa_2Cu_3O_{7-\delta}$ where the thermal conductivity always displays a distinct T-linear limiting behavior. I have already noted that this linear "pull" in κ occurs in ceramic samples that show a large γ-term in their specific heat. Since all attempts to detect a γ-term in the Bi-based superconductors have failed, the lack of a T-linear term in the thermal conductivity is consistent with a one-to-one correlation between T-linear contributions to the heat transport and the specific heat. Therefore, just as the presence of a linear specific heat at low temperatures is not a universal feature of high-T_c superconductors, so the limiting T-linear dependence of the thermal conductivity is not an intrinsic feature of the thermal transport in these ceramics. The sintered Bi-Sr-Ca-Cu-O material simply bucks the trend. It is possible that linear terms in the specific heat and the thermal conductivity (even allowing for an exception in the case of the Bi-based materials) might follow naturally from a better understanding of the transport processes in some of the more exotic models of superconductivity. Since we lack such an understanding at this time, it is reasonable to assume that the linear contributions are of extrinsic origin. An obvious candidate in this regard is a minute amount of a normal phase dispersed in the high-T_c matrix. This would yield both the γ-term and, at sufficiently low temperatures, the T-linear carrier contribution to the thermal conductivity. Clearly, this does not happen in Bi-Sr-Ca-Cu-O but perhaps among the Ba- or La-containing compounds there exists a minority phase that affects their superconducting forms. Such an impurity phase would be primarily squeezed out of high quality single crystals and this might be the reason why the linear term in the thermal conductivity of $YBa_2Cu_3O_{7-\delta}$ single crystals is much suppressed, but not completely eliminated. This line of argument points to a scenario of an approximately quadratic temperature dependence as the intrinsic power law variation for the thermal conductivity of high-T_c superconductors at low temperatures. Indeed, the predominantly T^2-dependence observed in the measurements of Zhu et al.[47] on single crystals of $Bi_2Sr_2Ca_1Cu_2O_8$ would also support this point of view, (see Fig. 13). However, whether the T^2 variation of the thermal conductivity necessarily implies the existence of two-level tunneling states such as those which dominate the thermal transport in amorphous materials is, I believe, not yet proven conclusively. Because of the enhanced layered character of the Bi-Sr-Ca-Cu-O structure, it is also possible that dimensional effects play a role in the thermal transport. Obviously, much work remains to be done before a coherent picture of the low temperature transport properties of high-T_c superconductors is developed.

CONCLUSIONS

Thermal conductivity is an important transport parameter with a significant impact on the possible technological applications of a given material. In addition, the magnitude and temperature dependence of the thermal conductivity are powerful probes of the fundamental interaction processes taking place in a solid since they provide information on scattering events. Unlike galvanomagnetic properties, the thermal response of superconductors remains finite and useful throughout the entire temperature range, not just above T_c

Considerable progress has been made over the past 30 months in exploring the essential features of heat transport in the family of high-T_c superconductors. It has been shown that the thermal conductivity of all investigated high-T_c's is rather poor, that the heat is carried chiefly by phonons, and that structural defects greatly affect the conductivity. From the fact that the thermal conductivity increases below T_c and has a peak near $T_c/2$, it has been deduced that the phonon-carrier interaction is an important relaxation process. By applying the existing theory (BRT), an estimate of the carrier-phonon coupling constant has been made which falls within the domain of weak coupling. We have also learned that the thermal transport is highly anisotropic, but the quality of existing crystals is not yet good enough to allow an estimate of the anisotropy in the coupling constant. Studies at low temperatures all agree that the behavior of the thermal conductivity is unusual and depends sensitively on the quality of the material and on whether the structure is superconducting or insulating. There seems to exist a remarkably good correlation between a γ-term in the specific heat and a linear limiting temperature dependence of the thermal conductivity. Models put forward to explain the low temperature behavior are controversial and none have yet been accepted as the final answer. While some pioneering measurements have been done on single crystals, there is a great need to continue in this effort. In particular, the data on 1-2-3 compounds are sparse and need to be augmented. At present, there is a dearth of information regarding the thermal transport in thin films of high-T_c superconductors. As novel experimental techniques, tailored specifically for this problem, become more mature, these technologically important structures will, undoubtedly, become the focus of intensive research. Among the new superconducting materials not yet scrutinized for their thermal transport properties are the cubic, copperless superconductors based on $Ba_{1-x}K_xBiO_3$. It is widely believed that this family of superconductors will be a bridge between the traditional, phonon-mediated superconductors and the new, exotic high-T_c materials. A detailed study of the thermal transport properties here would be quite interesting, particularly if single crystals of suitable size become available.

In writing an overview on a subject that is undergoing a rapid evolution, one runs the risk of being immediately out of date or of inadvertantly ommiting relevant papers. While I have tried to cover the most important developments, I apologize to those colleagues whose work may have been overlooked. It is a challenge to keep abreast of the progress in this exciting field.

ACKNOWLEDGEMENTS

It is my great pleasure to thank my students, Ray Richardson and Scott Peacor, for their assistance in writing this paper. I have also benefitted from discussions with my previous students, Drs. Donald Morelli and Joshua Cohn. I thank all my colleagues for permission to use their published results and, in particular, to Professor Ong and Dr. Morelli for allowing me to use their data prior to publication. This work was supported in part by U.S. Army Research Office Contract No. DAAL-03-87-K-0007 and by a grant from the Kellogg Foundation.

REFERENCES

1. J.G. Bednorz and K.A. Müller, Z. Phys., B **64**: 189 (1986).
2. M.K. Wu, J.R. Asburn, C.J. Torng, P.H. Hor, R. L. Meng, L. Gao, Z. J. Huang, Y.Q. Wang, and C.W. Chu, Phys. Rev. Lett. **58**: 908 (1987).
3. H. Maeda, Y. Tanaka, M. Fukutomi, and T. Asano, Jap. J. Appl. Phys. **27**: L209 (1988).

4. Z.Z. Sheng and A.M. Herman, Nature **332**: 55 (1988).

5. P.L. Richards, J. Clarke, R. Leoni, Ph. Lerch, B. Verghese, M.R. Beasley, T.H. Geballe, R.H. Hammond, P. Rosenthal, and S.R. Spielmann, Appl. Phys. Lett. **54**: 283 (1989).

6. M.I. Flik and C.L. Tien, to appear in Journal of Heat Transfer.

7. H.E. Fischer, S.K. Watson, and D.G. Cahill, Comments Cond. Mat. Phys. **14**: 65 (1988).

8. E. Gmelin, in "Studies of High Temperature Superconductors", ed. A. Narlikar, Nova Science Publishers, New York, Vol. 2, Ch. 4, p. 95 (1989).

9. A. Inam, M.S. Hegde, X.D. Wu, T. Venkatesan, D. England, P.F. Miceli, E.W. Chase, C.C. Chang, J.M. Tarascon, and Y.B. Wachtman, Appl. Phys. Lett. **53**: 908 (1988).

10. J.T. Fanton, A. Kapitulnik, D.B. Mitzi, B.T. Khuri-Yakub, and G.S. Kino, Appl. Phys. Lett. in press.

11. J. Bardeen, G. Rickhayzen, and L. Tewordt, Phys,. Rev. **113**: 982 (1959).

12. B.T. Geilikman and V.Z. Kresin, "Kinetic and Nonsteady-State Effects in Superconductors", John Wiley & Sons, New York (1974).

13. J.L. Olsen, Proc. Phys. Sec., **A65**: 518 (1952).

14. A. Jezowski, J. Mucha, K. Rogacki, R. Horyn, Z. Bukowski, M. Horobiowski, J. Rafalowicz, J. Stepien-Damm, C. Sulkowski, E. Trojnar, A.J. Zaleski, and J. Klamut, Phys. Lett. **A122**: 431 (1987).

15. D.T. Morelli, J. Heremans, and D.E. Swets, Phys. Rev. B **36**: 3917 (1987).

16. C. Uher and A.B. Kaiser, Phys. Rev. B **36**: 5680 (1987).

17. V. Bayot, F. Delannay, C. Dewitte, J.-P. Erauw, X. Gonze, J.-P. Issi, A. Jonas, M. Kinany-Alaoui, M. Lambricht, J.-P. Michenand, J.-P. Minet, and L. Piraux, Solid State Commun. **63**: 983 (1987).

18. U. Gottwick, R. Held, G. Sparn, F. Steglich, H. Rietschel, D. Ewert, B. Renker, W. Bauhofer, S. von Molnar, M. Wilhelm, and H.E. Hoenig, Europhysics Lett. **4**: 1183 (1987).

19. B. Salse, R. Calemczuk, C. Ayache, E. Bonjour, J.Y. Henry, M. Raki, L. Forro, M. Couach, A.F. Khoder, B. Barbara, P. Burlet, M.J.M. Jurgens, and J. Rossat-Mignod, Physica, **C153-155**: 1014 (1988).

20. S.J. Hagen, Z.Z. Wang, and N.P. Ong, Phys. Rev., submitted.

21. N.V. Zavaritskii, A.V. Samoilov, and A.A. Yurgens, Sov. Phys. JETP Lett. **48**: 242 (1988).

22. A. Jezowski, J. Klamut, R. Horyn, and K. Rogacki, Supercond. Sci. & Technol., to be published.

23. V.B. Yefimov, A.A. Levchenko, L.P. Mezhov-Deglin, R.K. Nikolaev, N.S. Sidorov, Sov. Journal Superconductivity: Physics, Chemistry, Engineering **2**: 16 (1989).

24. L. Tewordt and Th. Wölkhausen, Solid State Commun. **70**: 839 (1989).

25. M. Núñez Regueiro, D. Castello, M.A. Izbizky, D. Esparza, and C.D'Ovidio, Phys. Rev. B **36**: 8813 (1987).

26. B. Golding, N.O. Birge, W.H. Haemmerle, R.J. Cava, and E. Rietman, Phys. Rev. B **36**: 5606 (1987).

27. see, e.g., R. Srinivasan in "Studies of High Temperature Superconductors", ed. A. Narlikar, Nova Science Publishers, New York, Vol. 1, p. 267 (1989).

28. S. Hunklinger and A.K. Raychaudhuri, in "Progress in Low Temperature Physics", ed. by D.F. Brewer, Elsevier, Amsterdam, Vol. 9 (1986).

29. J.J. DeYoreo, W. Knaak, M. Meissner, R.O. Pohl, Phys. Rev. B **34**: 8828 (1986).

30. J.J. Freeman, T.A. Friedmann, D.M. Ginsberg, J. Chen, and A. Zangvil, Phys. Rev. B **36**: 8786 (1987).

31. J.E. Graebner, L.F. Schneemeyer, R.J. Cava, J.V. Waszczak, and E.A. Rietman, Symp. Proc. Mat. Res. Soc. **99**: 745 (1988).

32. G. Sparn, W. Schiebeling, M. Lang, R. Held, U. Gottwick, F. Steglich, and H. Rietschel, Physica **C153-155**: 1010 (1988).

33. J.L. Cohn, S.D. Peacor, and C. Uher, Phys. Rev. B **38**: 2892 (1988).

34. W.P. Kirk, P.S. Kobiela, R.N. Tsumura, and R.K. Pandey, Ferroelectrics **92**: 151 (1989).

35. A. Jezowski, A.J. Zaleski, M. Ciszek, J. Mucha, T. Olejniczak, E. Trojnar, and J. Klamut, Helvetica Physica Acta **61**: 438 (1988).

36. J. Heremans, D.T. Morelli, G.W. Smith, and S.C. Strite III, Phys. Rev. B **37**: 1604 (1988).

37. K. Mori, K. Noto, M. Sasakawa, Y. Isikawa, K. Sato, N. Kobayashi, and Y. Muto, Physica **C153-155**: 1515 (1988).

38. M.A. Izbizky, M. Núñez Regueiro, P. Esquinazi, and C. Fainstein, Phys. Rev. B **38**: 9220 (1988).

39. C. Uher and W.-N. Huang, Phys. Rev. B **40**: 2694 (1989).

40. K. Bartkowski, R. Horyn, A.J. Zaleski, Z. Bukowski, M. Horobiowski, C. Marucha, J. Rafalowicz, K. Rogacki, A. Stepien-Damm, C. Sulkowski, E. Trojnar, and J. Klamut, Phys. Stat. Solidi (a) **103**: K37 (1987).

41. F. Steglich, U. Ahlheim, D. Ewert, U. Gottwick, R. Held, H. Kneissel, M. Lang, U. Rauchschwalbe, B. Renker, H. Rietschel, G. Sparn and H. Spille, Physica Scripta **37**: 901 (1988).

42. A. Bernasconi, E. Felder, F. Hulliger, H.R. Ott, Z. Fisk, F. Greuter, and C. Schueler, Physica **C153-155**: 1034 (1988).

43. C. Uher and J.L. Cohn, J. Phys. **C21**: L957 (1988).

44. D.T. Morelli, J. Heremans, G. Doll, P.J. Picone, H.P. Jenssen, and M.S. Dresselhaus, Phys. Rev. B **39**: 804 (1989).

45. D.T. Morelli, G.L. Doll, J. Heremans, M.S. Dresselhaus, A. Cassanho, D.R. Gabbe, and H.P. Jenssen, to be published.

46. S.D. Peacor and C. Uher, Phys. Rev. B **39**: 11559 (1989).

47. Da-Ming Zhu, A.C. Anderson, E.D. Bukowski, and D.M. Ginsberg, Phys. Rev. B **40**: 841 (1989).

ACOUSTIC RESPONSE OF $YBa_2Cu_3O_x$ FILMS

Soon-Gul Lee, C.C. Chi, G. Koren, A. Gupta, and Armin Segmüller

IBM, Thomas J. Watson Research Center
P.O. Box 218
Yorktown Heights, New York 10598.

INTRODUCTION

Since the discovery of high temperature oxide superconductors many different experiments have been conducted to find some clues to the understanding of the basic mechanism. Acoustic attenuation measurement was one of the techniques applied to the conventional metal superconductors to verify the existence of a superconducting gap (Morse and Bohm, 1957). It would be natural to apply the standard acoustic echo technique to the oxide superconductors. However, unlike the conventional metal superconductors, large single crystals of the oxide superconductors suitable for the bulk acoustic measurement is still not available. Therefore most reported acoustic measurements were done using ceramic samples (Horie et al., 1987; Bhattacharya et al., 1988; Xu et al., 1989). Unfortunately, the reported data on ceramic samples showed subtantial variations from samples to samples. It is not surprising because the porosity and the process-dependent intergrain materials of the ceramic sample can cause a large scattering background as well as sample-to-sample variations. Nevertheless several broad peaks of the acoutic attenuation as a function of temperature have been observed more than once by different groups. There have been two reported acoustic measurements on single-crystal $YBa_2Cu_3O_7$ (Saint-Paul et al., 1988; Shi et al., 1989). Saint Paul et al. used a conventional echo technique with a quartz rod buffer between the transducer and a small single crystal. They reported a gradual reduction of attenuation with decreasing temperature without any broad peaks often seen in ceramic samples. On the other hand, Shi et.al. reported such peaks of a single crystal sample using a vibrating reed technique. None of the experiments on either ceramics or single crystals have indicated a superconducting gap below the superconducting transition temperature.

In this paper, we report the results of acoustic attenuation measurements on high quality of $YBa_2Cu_3O_7$ films epitaxially grown on $LiNbO_3$ substrates (Höhler et al., 1989; Lee et al., 1989) using the surface acoustic wave (SAW) technique. Similar technique has been previously used to study superconducting Pb and Sn films (Akao, 1969). We have also carried out a systematic study of the effects of oxygen deficiency on dc resistivity and acoustic attenuation.

EXPERIMENTAL DETAILS

It is clear that the SAW attenuation of high T_c oxides is meaningful only if high quality films can be grown on the SAW substrates. Otherwise, the interfacial coupling between the film and substrate is going to dominate the attenuation. We have demonstrated our capability to fabricate high quality epitaxial $YBa_2Cu_3O_x$ thin films on the Y-cut $LiNbO_3$ substrates (Lee et al., 1989), one of the commonly used SAW media. Fully oxygenated superconducting films are deposited *in situ* using a standard laser ablation technique. Typical thickness of the film is 1 μm. X-ray diffraction studies were conducted to find the orientation of the film. The c-axis of the film is normal to the substrate plane and (110) orientation is parallel to the [00.1] orientation of the hexagonal lattice or the Z-direction of the substrate. Films have two domains with the [11.0] acting as a mirror plane. $LiNbO_3$ crystal has a trigonal symmetry with the lattice constants of a hexagonal unit cell with $a = 5.148$ Å and $c = 13.863$ Å (Weis and Gaylord, 1985). The plane of Y-cut crystal has a rectangular unit cell with lattice parameters 5.148 Å and 13.863 Å. The shorter side is closer to the diagonal of the a-b plane of $YBa_2Cu_3O_x$ ($a = 3.822$ Å, $b = 3.891$ Å), but the lattice match is still relatively poor in comparison with the well-known lattice matching substrates, $SrTiO_3$ or gallates (Koren et al., 1989). The room temperature resistivity of the films ranges from 200 to 700 $\mu\Omega$-cm, and the zero-resistance T_c varies from 88 to 92 K with a transition width of 0.5 to 2 K. The critical current density of one thinner film was measured to be $2\times10^5 A/cm^2$ at 77 K. Such a high value is a good indicator of good crystalline orientation of the film and the absence of non-stoichiometric intergranular materials.

A schematic diagram of the experimental setup for the acoustic measurement is shown in Figure 1. Prior to the deposition of $YBa_2Cu_3O_x$ films three identical gold or platinum transducers are patterned on the Y-cut $LiNbO_3$ substrates with equal spacing, 6 mm. Each transducer is made of 10 pairs of interdigital fingers and is designed for 30 MHz SAWs. Width of the fingers is 30 μm and gap between fingers is also 30 μm. The center transducer is used as a surface wave generator, and the end sets are the receivers. The high T_c film, typically 1 μm thick and 3-4 mm wide, is deposited between the generator and one of the receivers. Thus one receiver detects the SAW transmitted through the film and substrate, and the other one detects SAW transmitted through the substrate only. The latter is used as a reference to normalize out any possible effects due to the bare substrate.

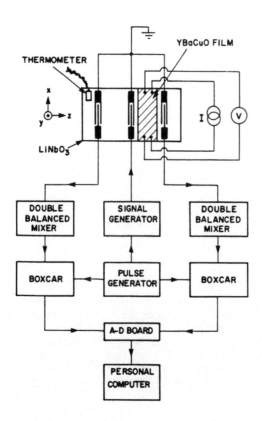

Fig. 1. Schematic diagram of experimental setup for the acoustic measurement. x, y, and z coordinates indicate the crystallographic axes of the $LiNbO_3$ substrate.

A cold finger cryostat is used with the sample in a vacuum environment. The cryostat is equipped with three thin gold braided coaxial cables for rf signals and several pairs of twisted wires for dc resistance measurements and thermometry. To generate SAW a 30 MHz sine wave gated by a pulse generator with a pulse width of 0.5 to 1 μ sec and a period of 100 μ sec is applied to the center transducer. A pulse train of surface acoustic waves is generated and propagates in both directions. To avoid reflection of the waves by the edges of the substrate, surfaces near both edges of the substrate are crosshatched by a diamond scriber. In addition, sufficiently lengthening the pulse period helps to avoid the interference effect caused by edge reflections. The signal detected by each finger set is self mixed, then amplified by a boxcar, and finally digitized and recorded by a personal computer. Acoustic attenuation of the film is obtained by dividing the signal through the film side by that through the bare side. Sample temperature is monitored by a Si diode thermometer attached to the substrate. Both acoustic signals and the dc resistance of the film are measured simultaneously as the temperature of substrate is decreased slowly from room temperature to about 4 K, lowest obtainable temperature.

Before presenting our results it should be pointed out that there is a difference between the SAW technique and the usual pulse echo technique. Unlike the usual echo technique in which the bulk sound waves can be purely longitudinal or transverse, SAW is inherently a combination of both. Furthermore, for our case, the surface acoustic wave propagates along the diagonal direction of the ab plane of the high T_c film. As a result, the observed attenuation is an average over all three axes of the $YBa_2Cu_3O_x$ film, similar to the case of bulk acoustic attenuation in randomly oriented ceramics, but without the complications of voids and intergranular materials.

We have studied a total of eight different samples. In this paper we focus on two of them, labelled S1 and S2. Judged by their room temperature resistivity and T_c, these two are the best in this group. Figs. 2 and 3 show the results of S1 and S2. They have very similar resistance and resistive transition, and yet the temperature dependences of their acoustic attenuation are very different. S1 has two broad peaks, one at 135 K and the other near 250 K, while S2 shows only a monotonic decrease. Roughly speaking, half of our samples showed a behaviour similar to S1 with variations in the peak height but not the peak temperatures. And the other half are similar to S2. The latter have in general a larger attenuation than the former as illustrated in Figs. 2 and 3 for S1 and S2. This implies that peaks might be overshadowed by a larger monotonic background. The only common feature for all the samples is the overall attenuation reduction with T.
At the present moment, we do not know the reason for these two kinds of behaviors. There does not seem to be any correlations with dc resistivity or T_c. Our results can be considered as a miniature scope of what have been published in the literature about the acoustic attenuation of $YBa_2Cu_3O_7$ ceramics and single crystals. Most likely the variations are caused by some subtle crystalline deffects which does not affect T_c or resistivity. Once this is understood, the acoustic attenuation can be used as a sensitive probe for sample quality.

Consistent with other reports, our acoustic attenuation data do not show any evidence of superconducting gap near T_c. Near the low T_c of the usual metal superconductors, electron-phonon interaction is the dominant mechanism for acoustic attenuation. A sharp decrease of quasiparticle density due to the gap opening results in an abrupt drop of the attenuation at T_c (Schrieffer, 1964). For the high T_c superconductor a rough estimate using $\alpha \simeq (mv_F/Mv_s)q^2\ell$ (Callaway, 1974) indicates that the attenuation due to electron-phonon interaction is several orders of magnitude smaller than that actually observed. Here m and M are effective electron mass and atomic mass respectively, v_F and v_s are fermi velocity and sound velocity respectively, q is phonon momentum, and ℓ is electron mean free path. Moreover, the attenuation due to a simple electron-phonon interaction increases as temperature decreases as predicted from the above equation. Our results and others suggest that the electron-phonon interaction is simply dominated by some other attenuation mechanisms for high T_c superconductors.

An important advantage of using thin film samples over bulk single crystals is that the annealing time required to change the oxygen content is much shorter. It takes typically hours for thin films versus weeks for single crystals. To study the role of oxy-

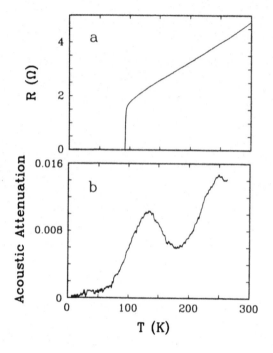

Fig. 2. Results for sample S1 in full oxygen state. (a) Resistance versus temperature. (b) Normalized acoustic attenuation versus temperature.

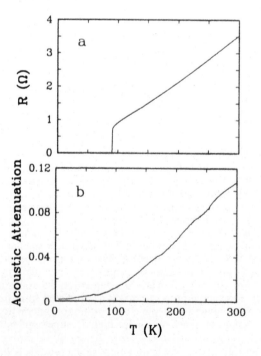

Fig. 3. Results for sample S2 in full oxygen state. (a) Resistance versus temperature. (b) Normalized acoustic attenuation versus temperature.

gen atoms in the acoustic response we annealed oxygen out gradually and repeated the measurement at each step. The oxygen content was controlled by annealing the sample in the quartz tube furnace with a small amount of oxygen gas flowing. Annealing temperature was 500 °C. For sample S1, the oxygen was gradually reduced by increasing annealing time in vaccuum. For sample S2, the oxygen content was controlled by the oxygen flow rate. During annealing the sample resistance was indirectly monitored *in situ* by measuring the resistance of a nearby control film. In the metallic regime the annealing time required to reach an equilibrium resistance for a given condition is typically 10 minutes, while in semiconducting and insulating regimes it takes about an hour. After each annealing, the c-axis lattice constant of the film was determined by X-ray diffraction spectrum. Its oxygen content was then inferred from the relation of oxygen content and the c axis lattice constant for the powdered ceramics (Cava et al., 1987; Specht et al., 1988; Koren et al., 1989). The estimated oxygen content for S2 is shown in Table I. In Table I the lattice constant for the fully oxygenated film is 11.71 Å, which is substantially larger than 11.68 Å for power samples. It indicates that the film may be strained by the lattice mismatch.

Table I. Parameters for sample 2 in different oxygen contents.

	$R_{300K}(\Omega)$	c(Å)	x^a
A	4.8	11.707	6.83
B	82	11.811	6.14
C	1.7×10^3	11.845	5.90
D	$> 10^7$	11.866	5.76

[a]Calculated by using the results of powdered sample.

The oxygen extraction process is reversible. The oxygen in the sample can be completely replenished by anealing at 500 C in oxygen atmosphere. Both the resistance and the acoustic attenuation versus temperature curves of the re-oxygenated sample are nearly identical to those of the virgin sample.

Figs. 4 and 5 show the acoustic attenuation data for four different oxygen contents and their corresponding resistance curves for the samples S1 and S2 respectively. Although the acoustic attenuation curves are quite different when both samples are in full oxygen content, they become similar to each other when the sample oxygen depletion is significant. Note that the resistance curves for S1 and S2 are quite different when both samples are in the oxygen depleted states. It is probably due to the inhomogeneity of oxygen distribution in S1 caused by the time-controlled vacuum anealing process. The oxygen-flow-rate-controlled anealing process produced a more homogeneous oxygen depletion for the sample S2, which was subtantiated by the line width of its X-ray diffraction spectrum. It is a fact that the resistance is largely determined by the most conducting path in the sample, while the acoustic attenuation is averaged over the whole sample.

It is interesting to point out that the linear slopes of the resistance curves A and B in Fig. 5 are almost identical. Several resistance curves, not shown in Fig. 5, with oxygen contents between those of A nd B all have the same linear slope while the room temperature resistance increases with decreasing oxygen content. Similar behaviors have also been observed for $YBa_2Cu_3O_7$ films on $SrTiO_3$ substrates (Koren et al., 1989). On the other hand, the resistance as a function of temperature is very different for oxygen deficient bulk ceramic samples (Cava et al., 1987). This indicates that the grain boundary material of ceramic samples behaves differently when oxygen is taken out.

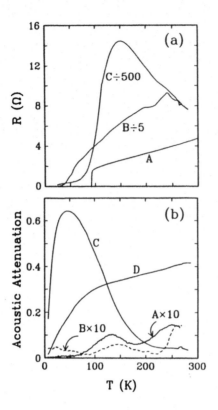

Fig. 4. Results for various oxygen states of sample S1. (a) Resistance versus temperature. R(T) for the insulating state D has a room temperature resistance of $> 20 M\Omega$ and is not shown here. (b) Normalized acoustic attenuation versus temperature.

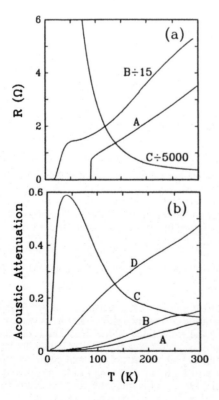

Fig. 5. Results for various oxygen states of sample S2. (a) Resistance versus temperature. R(T) for the insulating state D is not shown here. (b) Normalized acoustic attenuation versus temperature.

From both resistive transition and acoustic attenuation data, the results can be categorized into three regimes: metallic, semiconducting and insulating regimes. In the metallic or superconducting regime, A and B, the acoustic attenuation does not change much in absolute magnitude. In the semiconducting regime there is a drastic change. The attenuation has a large broad peak at about 40 K. It is independent of a small amount of inhomogeneity because both samples S1 and S2 showed the same behavior. In the insulating regime the attenuation decreases monotonically as the temperature decreases but the change is more than an order of magnitude larger than that of the metallic regime. The acoustic attenuation curves in the non-metallic regimes are remarkably reproducible for all the samples we have tried. The acoustic attenuations for all three regimes approach each other as the temperature decreases toward zero.

For the entire range of oxygen content we have studied, the lattice constants change only about 1%, so elastic properties are not expected to differ much. Nevertheless, the acoustic attenuation changes drastically in semiconducting and insulating regimes. We suspect that the origin of the acoustic attenuation in non-metallic regimes is not due to ordinary phonon relaxation processes. In other words, the dominant attenuation process is not electron-phonon interaction. One of the possibilities is two-level tunneling of the mobile oxygen atoms, which has been seen by McKenna et al. (1989) as a relatively weak phonon attenuation mechanism for fully oxygenated samples. It is expected to become more important when more oxygen vacancies are created. But at the present moment, we do not have a quantitative theoretical calculation to compare with.

CONCLUSION

We have studied acoustic responce of high quality epitaxial $YBa_2Cu_3O_x$ films deposited on Y-cut $LiNbO_3$ substrates using 30 MHz surface acoustic waves. Three interdigital transducers are patterned on the same substrates prior to the deposition of the high T_c film to fill in one of the two spacings. Acoustic attenuation is obtained by simultaneously measuring both signals through the high T_c film and through the bare substrate. Results of acoustic measurements are categorized into three regimes which are metallic, semiconducting, and insulating. In the metallic regime, the results showed a relatively large sample-dependent variation. Out of the eight samples studied some show two broad peaks near 150 K and 250 K and others show a monotonic decrease with temperature. As oxygen is depleted, a very large broad peak appears at about 40 K in the semiconducting regime, and finally the attenuation decreases monotonically in the insulating regime. In both semiconducting and insulating regimes the magnitude of the change in the attenuation is at least an order of magnitude larger than that of metalic regime. We believe that the attenuation is not due to ordinary phonon relaxation processes. Two level tunnelling of mobile oxygen is a possible candidate.

REFERENCES

Akao, F., 1969, Attenuation of elastic surface waves in thin film superconducting Pb and In at 316 MHz, *Phys. Lett.*, 30A:409.

Bhattacharya, S., Higgins, M.J., Johnston, D.C., Jacobson, A.J., Stokes, J.P., Lewandowski, and Goshorn, D.P., 1988, Anomalous ultrasound propagation in high-T_c superconductors: $La_{1.8}Sr_{0.2}CuO_{4-y}$ and $YBa_2Cu_3O_{7-\delta}$, *Phys. Rev.*, B 37:5901.

Brewster, J.L., Levy, M., and Rudnick, I., 1963, Ultrasonic determination of the superconducting energy gap in vanadium, *Phys. Rev.*, 132:1062.

Callaway, J., 1974, "Quantum Theory of the Solid State," Academic Press, New York.

Cava, R.J., Batlogg, B., Chen, C.H., Rietman, E.A., Zahurak, S.M., and Werder, D., 1987, Single-phase 60-K bulk superconductor in annealed $Ba_2YCu_3O_{7-\delta}(0.3 < \delta < 0.4)$ with correlatedoxygen vacancies in the Cu-O chains, *Phys. Rev.*, B 36:5719.

Höhler, A., Guggi, D., Neeb, H., and Heiden, C., 1989, Fully textured growth of $Y_1Ba_2Cu_3O_{7-\delta}$ films by sputtering on $LiNbO_3$ substrates, *Appl. Phys. Lett.*, 54:1066.

Horie, Y., Terashi, Y., Fukuda, H., Fukami, T., and Mase, S., 1987, Ultrasonic studies of the high T_c superconductor $Y_2Ba_4Cu_6O_{14}$, *Solid State Commun.*, 64:501.

Koren, G., Gupta, A., Giess, E.A., Segmüller, A., and Laibowitz, .B., 1989, Epitaxial films of $YBa_2Cu_3O_{7-\delta}$ on $NdGaO_3$, and $SrTiO_3$ substrates deposited by laser ablation, *Appl. Phys. Lett.*, 54:1989.

Koren, G., Gupta, A., and Segmüller, A., 1989, *in Proceedings of International Conference on Materials and Mechanisms of Superconductivity: High-Temperature Supercondcutivity, Stanford, CA, July 23-28, 1989.* To be published in *Physica C.*

Lee, S.G., Koren, G., Gupta, A., Segmüller, A., and Chi, C.C., 1989, Epitaxial growth of $YBa_2Cu_3O_{7-\delta}$ thin films on $LiNbO_3$ substrates, *Appl. Phys. Lett.*, 55:1261.

Mckenna, M.J., Hikada, A., Takeuchi, J., Elbaum, C., Kershaw, R., and Wold, A., 1989, Electron and phonon interactions with two-level-tunneling systems in the High-T_c superconductor $YBa_2Cu_3O_{7-\delta}$ and in Niobium, *Phys. Rev. Lett.*, 62:1556.

Morse, R., and Bohm, H., 1957, Superconducting energy gap from ultrasonic attenuation measurements, *Phys. Rev.*, 108:1094.

Saint-Paul, M., Tholence, J.L., Monceau, P., Noel, H., Levet, J.C., Potel, M., Gougeon, P., and Capponi, J.J., 1988, Ultrasound study of $YBa_2Cu_3O_{7-\delta}$ single crystals, *Solid State Commun.*, 66:641.

Schrieffer, J.R., 1964, "Theory of Superconductivity," W.A. Benjamin, Inc., Reading, MA.

Shi, X.D., Yu, R.C., Wang, Z.Z., Ong, N.P., and CHaikin, P.M., 1989, Sound velocity and attenuation in single-crystal $YBa_2Cu_3O_{7-\delta}$, *Phys. Rev.*, B 39:827.

Specht, E.D., Sparks, C.J., Dhere, A.G., Brynestad, J., Cavin, O.B., Kroeger, D.M., and Oye, H.A., 1988, Effect of oxygen pressure on the orthorhombic-tetragonal transition in the high-temperature superconductor $YBa_2Cu_3O_x$, *Phys. Rev.*, B 37:7426.

Weis, R.S., and Gaylord, T.K., 1985, Lithium niobate: summary of physical properties and crystal structure, *Appl. Phys.*, A 37:191.

Xu, X-F., Bein, D., Wiegert, R.F., Sarma, B.K., Levy, M., Zhao, Z., Adenwalla, S., Moreau, A., Robinson, Q., Johnson, D.L., Hwu, S.-J., Poeppelmeier, K.R., and Ketterson, J.B., 1989, Ultrasonic attenuation measurements in sinter-forged $YBa_2Cu_3O_{7-\delta}$, *Phys. Rev.*, B 39:843.

MAGNETIZATION STUDY OF HIGH T_c SUPERCONDUCTORS

Wei-Yan Guan

Texas Center for Superconductivity
University of Houston
Houston, TX 77204

Yun-Hui Xu

Institut fur Schicht und Ionentechnik der
Kernforschungsanlage Julich, Postfach 1913
D-5170 Julich, Federal Republic of Germany

K. Zeibig

Institut fur Angewandte Physik, Universitat
Giessen, D-6300 Giessen, Federal Republic of Germany

INTRODUCTION

We measured the magnetization curves of sintered samples $RBa_2Cu_3O_{7-y}$ (R=Y, Eu, Dy, Er, Gd). Our result gives strong support to the idea that there are two different types of superconductivity, bulk superconductivity of grains and weak link superconductivity between grains, existing in ceramic superconductors.[1,2] Using the critical state model critical current densities were derived from the hysteresis loops. There are indications of enhanced flux pinning by a quenched second superconducting phase, since the field dependence of critical current density was found to exhibit a maximum at intermediate temperatures.[3] We present the results of hysteretic ac losses of sintered samples calculated from our magnetization data graphically.[4] We find that the relation of ac loss w versus amplitude of sweeping field H_m follows a power law: $w \propto H_m^n$, where n varies from 7 to 1 in various regions of H_m. We find that at a fixed field amplitude $H_m >> H_p$ (H_p is the field of full penetration), the dependence of ac losses w on temperature T follows an exponential law: $w = \alpha e^{-\beta T}$, which can not be treated within a mechanism controlled by flux creep. We report detailed magnetic susceptibility and magnetization studies for the high T_c oxide superconductor $ErBa_2Cu_3O_{7-y}$[5]. The magnetic susceptibility χ of $ErBa_2Cu_3O_{7-y}$ follows a Curie-Weiss law both above and below T_c. The saturation of M in high fields at low temperature can be explained by the theory of paramagnetizm with Brillouin function. We find that superconductivity and paramagnetizm in $ErBa_2Cu_3O_{7-y}$ exists largely independent of one another.

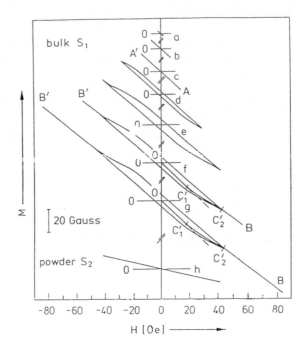

Fig. 1. Magnetization curves of $EuBa_2Cu_3O_{7-y}$ at 4.2K: for the bulk sample S_1, $H_{max}=\pm 1.4$ Oe (a), ± 7 Oe (b), ± 14 Oe (c), ± 28 Oe (d), ± 42 Oe (e), ± 56 Oe (f) and ± 84 Oe (g); and for the powder sample S_2, $H_{max}=\pm 1.4$ Oe (h).

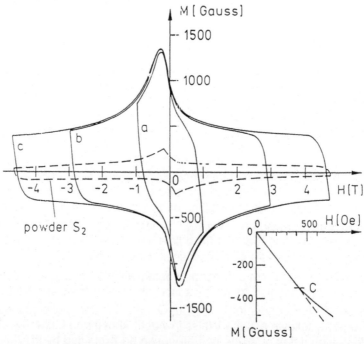

Fig. 2. Magnetization curves of $EuBa_2Cu_3O_{7-y}$ at 4.2K: for the bulk sample S_1, $H_{max} = \pm 1.$ T (a), ± 3 T (b), ± 4.7 T (c) (solid lines); for the powder sample S_2, $H_{max} = \pm 4.7$ Tesla (broken line).

EXPERIMENTAL DETAIL

The ceramic samples were prepared standard powder-metallurgical method. The powders of R_2O_3 (R = Eu, Dy, Er, Gd, and Y), CuO and $BaCO_3$ were mixed at the ratio of atoms: R : Ba : Cu = 1: 2: 3, pressed into pellets (6.4 ton/cm^2) and heated at 800°C for 12 hours. Then the pellets were reground into very fine powders and pressed (13.3 ton/cm^2) into pellets again, then sintered in flowing oxygen at 940°C for 38.5 hours, at 700°C for 5 hours and at 500°C for 22 hours. After that the pellets were cooled slowly to room temperature in the furnace.

X-ray diffraction revealed that all samples are of single phase which is now well known as an orthorhombic, oxygen-deficient layered perovskite-like structure. T_c was measured with the standard four probes method. All samples have T_c (R=0) ≥ 90K. Magnetization measurements were performed using a low-drift high-sensititivy magnetometer with a fast SQUID system for the low field range (≤ 84 Oe) and a vibrating sample magnetometer again with a SQUID system as detector for the high field rang (up to 4.7 T). The magnetization measurements have ben performed on samples cut from pellets in a suitable shape and size.

RESULTS AND DISCUSSIONS

Two Types of Superconductivity

Fig. 1 shows magnetization curves with maximum external field H_{max} ±1.4, ±7, ±14, ±28, ±42, ±56, and ±84 Oe at 4.2K for the bulk sample S_1 of $EuBa_2Cu_3O_{7-y}$ with T_c (R=0) = 93K and the size 1.0x2x2.7 mm^3. When $|H_{max}|$ ≤ 14 Oe (Fig. 1 a,b,c) the M-H curve was a reversible straight line with a slope χ_1 and no hysteresis at all. Magnetic hysteresis can be seen for H_{max} = ± 28 Oe (Fig. 1,d). When H_{max} = ± 42 Oe, a closed hysteresis loop with twofold rotational symmetry about the origin formed. When H_{max} reached ±56 and ±84 Oe, the shape and the size of the loops remained exactly the same having a reversible straight line at the ends with a slope χ_2.

To compare the low field behaviour with that a high fields, Fig. 2 shows the magnetization curves of the same $EuBa_2Cu_3O_{7-y}$ bulk sample S_1 for H_{max} = ±1, ±3, and ±4.7 Tesla at 4.2K. For applied fields H < 400 Oe the part OC of the curve is a straight line with the slope χ_3 which is equal to χ_2 (insert of Fig. 2). When H_{max} is high enough (H_{max} > 3T, Fig. 2), the frozen in moment M(H) at lower fields assume a constant.

The prepared high-T_c oxide superconductor is a kind of granular superconductor. It seems that there are two parts of materials which coexist in this kind of superconductor: superconducting grains, and the boundary materials between the grains which usually are not superconductors, however because of tunneling or proximity effect, the Cooper pairs can pass through these regions. It can be called weak link granular superconductivity. Because there are two different sorts of superconductivity existing in our samples, we observed two different types of hysteresis loops appearing in different ranges of magnetic fields (Fig. 1 and Fig. 2). In the lower field range (Fig. 1) the applied fields being lower than H_{c1} of the grains (≈400 Oe), the contribution of magnetization of grains to the total magnetization of a reversible straight line BB' with a slope χ_2 (Fig. 3,a,b). We assume that the other part of the magnetization comes from the granular behavior of the superconductor.

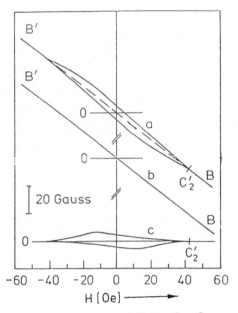

Fig. 3. Magnetization curves of EuBa$_2$Cu$_3$O$_{7-y}$ sample S$_1$ at 4.2K and H$_{max}$=±4.2 Oe: (a) total magnetization, (b) reversible straight line BB', (c) the resulting magnetization after subtracting (b) from (a).

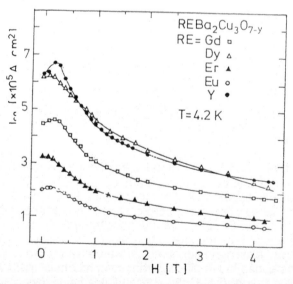

Fig. 4. Intragranular critical current densities J$_{cg}$ versus applied field at 4.2K for bulk REBa$_2$Cu$_3$O$_{7-y}$, Re = Gd (□), Dy (△), Er (▲), Eu (○), Y (●).

The magnetization contribution of weak link superconductivity to total magnetization of the whole sample can be obtained by subtracting the straight line from the total hysteresis loop (Fig. 3,c). The resulting curve seems like the typical magnetization curve of a non-ideal type II superconductor with pinning.

We may conceive such a physical picture. When H_{max} is lower than ≈ 17 Oe, there is a shielding current around the surface of the whole sample which is in the Meissner state. The magnetic penetration depth is the London penetration depth, and the supercurrent shielding effects results in the negative slope χ_2 of the lines BB' (Fig. 1, f,g). When 17 Oe<$|H_{max}|$< 42 Oe, magnetic flux begins to penetrate into the sample, the higher the field, the deeper the penetration, but only into the boundary material between the grains, not inside the grains themselves. There is a shielding current around the surface of each grain itself which is inside the penetration region. When H_{max} reaches 42 Oe, the flux lines penetrate into the whole sample (but not inside the grains), and the frozen in magnetic moment remains constant independent of H_{max}. When 42 Oe < $|H_{max}|$ < 400 Oe, the high intensity of the field destroys the weak coupling between superconducting grains, and the supercurrent around the whole sample disappears. However, there is shielding current around each grain, which contributes to the total magnetization. In the case of 400 Oe < $|H_{max}|$ < 3T, a similar situation as described above exists inside each grain. H_{c1} = 400 Oe is the lower critical field of the grains.

It can be deduced that for powdered samples there should be no magnetic hysteresis at low fields. To verify this, one part of the bulk sample S_1 of $EuBa_2Cu_3O_{7-y}$ was reground into very fine powder, mixed with non-magnetic grease for protection against particle coupling, and shaped to a small sphere S_2. The M-H curve of S_2 is shown in Fig. 1h, it is a reversible straight line at the fields $H_{max} = \pm 42$ Oe at 4.2K is expected.

Similar behaviour was observed at 4.2K and 77K for $RBa_2Cu_3O_{7-y}$ samples (R = Y, Dy, Er, Gd).[1,2]

Magnetic Hysteresis and Critical Current Densities.

If we assume the specimen to be a large flat plate with an applied field parallel to its surface, using Bean's Model, the critical current density Jc(H) and Me(H) can be determined by the formula:

$$4 \Pi (M^+ + M^-) = 8 \Pi Me \qquad (1)$$

$$4 \Pi (M^+ + M^-) = KWJc(H) \qquad (2)$$

where Me is the reversible magnetization of the sample, 2W is the specimen thickness, the value of K is $4\Pi/10$ for M in gauss and Jc in a cm^{-2} and M^+ and M^- are the magnetization of the decreasing and increasing field branches.

The intragranular critical current Jc versus H curves at 4.2K deduced from magnetization hysteresis curves in high field range (for example, Fig. 2) using Equation (2) for some of the samples $RBa_2Cu_3O_{7-y}$ (R = Y, Eu, Dy, Er, Gd) are plotted in Fig. 4.

We now turn to those features which are the anomalous field dependence of intragranular critical current density at intermediate temperatures for some of our $GdBa_2Cu_3O_{7-y}$ and $YBa_2Cu_3O_{7-y}$ samples.

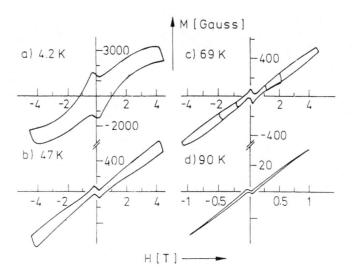

Fig. 5. Magnetization curves of GdBa$_2$Cu$_3$O$_{7-y}$ bulk sample at different temperatures.

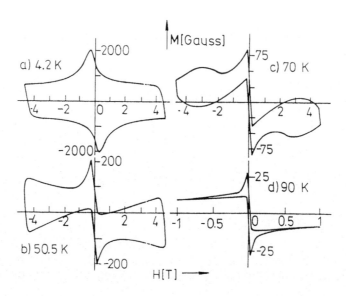

Fig. 6. Magnetization curves of YBa$_2$Cu$_3$O$_{7-y}$ bulk sample at different temperatures.

In Fig. 5 and 6 we present M(H) curves of $GdBa_2Cu_3O_{7-y}$ and $YBa_2Cu_3O_{7-y}$ at four typical temperatures. At 4.2K and temperature close to T_c the difference $M^+ - M^-$ monotonically decreased with increasing applied fields (Fig. 5 a,d and 6 a,d). At intermediate temperature however, $M^+ - M^-$ is observed to go through a maximum with increasing applied field (Fig. 5 b,c and 6 b,c).

The corresponding anomalous behaviour in Jc(H) is plotted in Fig 7 and 8 for these two materials.

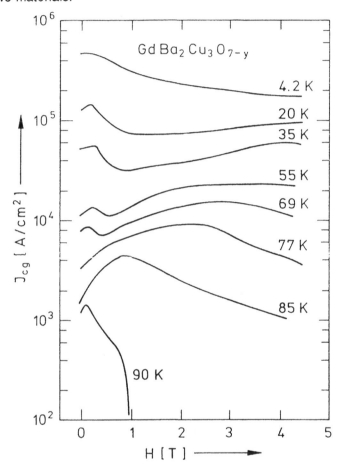

Fig. 7. Critical current densities versus applied field at different temperatures for $GdBa_2Cu_3O_{7-y}$.

If we ignore the extra dip near the origin, we can affirm: The J_C monotonically decreases with increasing applied field from 0 to 4.7T at very low temperatures (4.2K) and at higher temperature close to the T_c; the J_c(H) curves express a maximum at some intermediate temperatures. The maximum in J_c(H) curves gradually shifts to lower fields with increasing temperature. We believe that the corresponding maximum of J_c(H) curves for lower temperatures (Fig. 7, 20K and Fig. 8, 31 and 51K) occurs at magnetic fields higher than the maximum field in our measurements (4.7T).

We have also observed similar anomalous features of J_c versus H for samples with R = Er, Dy, Eu[3]. One interpretation of the anonalous behaviour in field dependence of J_c may be based on the assumption that the sintered samples

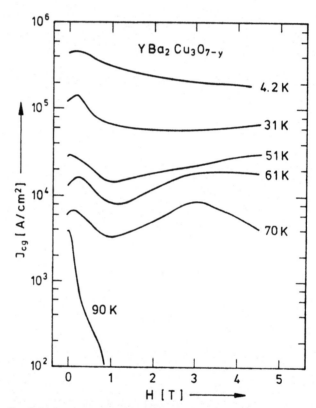

Fig. 8. Critical current densities versus applied field at different temperatures for $YBa_2Cu_3O_{7-y}$.

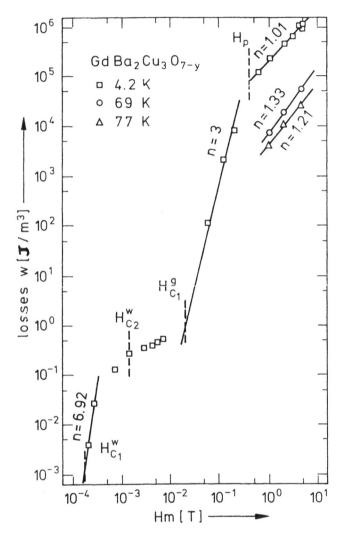

Fig. 9. ac losses w vs amplitude of sweeping field H_m at 4.2, 69, and 77K for $GdBa_2Cu_3O_{7-y}$. The plot at 4.2K can be divided into four regions.

are not strictly single phase. Our results suggest that there are at least two superconducting phases existing in our sintered samples. When the superconductivity of one of them (phase 1) is quenched in a magnetic field, the normal granules of phase 1 may act as flux-pinning centers dispersed in the matrix that forms the second superconducting phase (phase 2). A distribution of T_c and H_{c2} of phase 1 then will lead to a gradual increase of the critical current of phase 2, with increasing field for a certain region of the magnetic field at some intermediate temperatures.[6]

As the critical field of phase 1 increases with decreasing temperature, the maximum in the $J_c(H)$ curves shifts to higher fields at lower temperatures as we observed.

ac Losses in $RBa_2Cu_3O_{7-y}$ Superconductors

It has been now well established that the dominant ac loss mechanism in type II superconductors above H_{c1} is hysteretic in nature and is due to the interaction of fluxoids with the lattice as they move in the superconductor. The frequency-dependent eddy current loss in the type II superconductor is small and can be omitted compared to the hysteretic one. Magnetic hysteresis measurements provide a method of measurement of ac loss which in each cycle is frequency independent and proportional to the area of magnetization hysteresis loop.[7,8]

The dependence of the ac losses per $1m^3$ and 1 cycle w on the amplitude of the sweeping field H_m for $GdBa_2Cu_3O_{7-y}$ at 4.2, 69 and 77K is illustrated in Fig. 9. The plot at 4.2K can be divided into four regions. In the first region ($H < 5 \times 10^{-4}T$), the losses increase approximately with the seven power of H_m. This is the region that the amplitude of sweeping field exceeds the lower critical field H_{c1} of weak link superconductivity between grains. At a certain value of H_m the losses start to saturate and remain nearly constant for the interval of $1.4 \times 10^{-3}T - 2 \times 10^{-2}T$. In the third region ($H_{c1} < H_m < H_p$) in Fig. 9, the losses exhibit a field dependence approximately as $w \propto H_m^3$, where H_p is the field of full penetration in the critical state model. For $H_m >> H_p$, the losses are almost proportional to the amplitude of field: $w \propto H_m$. At field near H_p, the dependence of losses on H_m was performed complicatedly.

The energy dissipated at 4.2K in one cycle per $1m^3$ of samples $RBa_2Cu_3O_{7-y}$ (R = Gd, Dy, Er, Eu, Y) is point plotted against the amplitude of field H_m in Fig. 10, where the low-field losses have been omitted for the sake of clarity. All the data points taken from different samples lie close to a common line over most of the field range. The relation between w and H_m follows a power law: $w = aH_m^n$, for small H_m, $n \approx 3$; for $H_m >> H_p$, n is well close to 1. This is consistent with a hysteretic loss mechanism.

It can be seen in Fig. 11 that at a fixed value of field amplitude H_m ($H_m >> H_p$), the ac loss of $GdBa_2Cu_3O_{7-y}$ and $YBa_2Cu_3O_{7-y}$ decrease with increasing temperature exponentially: $w = \alpha e^{-\beta T}$, were α and β are constants. Our results w(T) cannot be explained by the flux creep model which predicts a power law.[9]

Coexisting of Superconductivity and Paramagnetizm

The substitution of most of the lanthanide rare-earth ions for Y^{3+} in $YBa_2Cu_3O_{7-y}$ compounds shows no harmful effects on superconducting

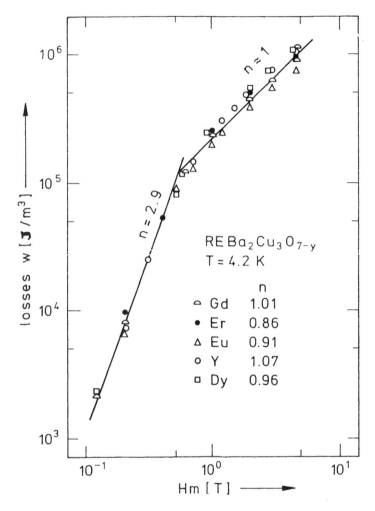

Fig. 10. ac losses w vs H_m at 4.2K for $REBa_2Cu_3O_{7-y}$. (RE = Gd, Dy, Er, Eu, Y). The relation is $w = aH_m^n$ for $H_m < H_p$ n = 3, for $H_m \gg H_p$ n=1.

transitions; T_C remains near 90K.[10] This finding is unexpected. Among these lanthanide elements, there are some ions having large intrinsic magnetic moments at the ground state such as Gd^{3+}, Dy^{3+}, Ho^{3+}, and Er^{3+} ions. The interaction between the magnetic moment of the ion and the superconducting electrons breaks the Cooper pairs responsible for superconductivity thereby suppressing T_C. The depression of T_C upon the introduction of magnetic impurities is described by the theory of Abrikosov and Gor'kov.[11]

Magnetic measurements reveal that the susceptibility of some rare-earth ions substituted high T_C superconducting compounds, for example, $DyBa_2Cu_3O_{7-y}$, follows a Curie-Weiss law above T_C.[10]

These experimental facts indicate clearly that the effect of the magnetic interactions on the superconductivity is very weak in these compounds and magnetic interactions between rare-earth ions mediated by conducting electrons (RKKY interactions) are also unimportant.

How about the situation at the temperature below T_C? In this case the Meissner diamagnetism, or diamagnetic screening, introduces a reduction in the local magnetic field. The susceptibility data, therefore, could deviate from the Curie-Weiss law.

Our measurements above T_C confirm observations[10] that the susceptibility obeys a Curie-Weiss law with the Curie constant scaling with the effective moment for the isolated rare-earth ions. Our measurements below T_C show that the Curie susceptibility is basically unchanged by the superconducting transition provided that the reduction in the local magnetic field is properly allowed for.

The magnetic susceptibility χ above T_C can be described by the sum of a constant Pauli-like contribution χ_0 and a Curie-Weiss contribution $C/(T - \theta)$; i.e.,

$$\chi(T) = \chi_0 + C/(T - \theta) \qquad (3)$$

where χ is the dimensionless (volume) susceptibility. In Fig. 12, we plot the reverse of the susceptibility $1/(\chi - \chi_0)$ as a function of temperature. The linear relation of $1/(\chi - \chi_0)$ vs T is well coincident with formula (3). From the slope, an effective number of paramagnetic Bohr magnetons of Er^{3t} of $9.2\mu_B$ are obtained, which is close to that of $9.58\mu_B$ deduced from the ground state of Er^{3+} ion ($^4I_{15/2}$).

At temperature $T < T_C$, the data acquired as isothermal magnetization curves M(H). Fig. 13 shows the magnetization curves of $ErBa_2Cu_3O_{7-y}$ in the superconducting state with magnetic field varying between -4.7 and 4.7 T.

The hysteresis loops at 4.2K (Fig. 13) and Me(H) deduced from formula (1) and Fig. 13 look like a superpostion of a superconducting loop and a paramagnetic magnetization curve. The magnetization of the sample changes its sign from negative to positive at a certain field intensity.

The magnetic hysteresis loop at 4.2K tends to saturation in high fields. The saturation is greatest at low temperature and high fields and gradually disappeared with increasing temperature.

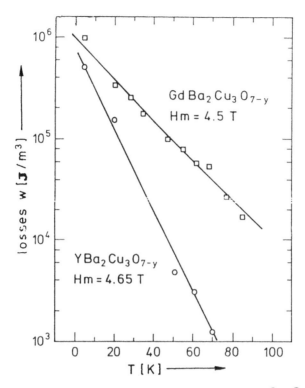

Fig. 11. ac losses w vs temperature T for GdBa$_2$Cu$_3$O$_{7-y}$ (H$_m$=4.5T) and YBa$_2$Cu$_3$O$_{7-y}$ (H$_m$=4.65T). The dependence of w on T follows an exponential law: w=αe$^{-\beta T}$.

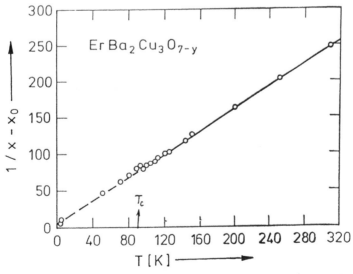

Fig. 12. Inverse of the Curie-Weiss susceptibility $1/(\chi - \chi_0)$ vs T, for ErBa$_2$Cu$_3$O$_{7-y}$.

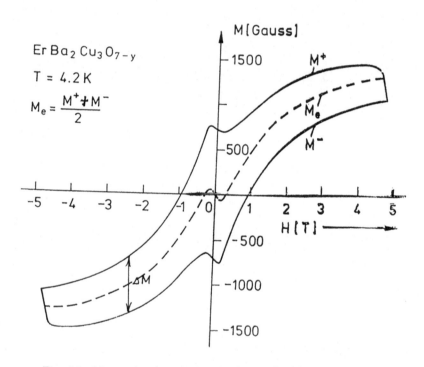

Fig. 13. Magnetization curves for ErBa$_2$Cu$_3$O$_{7-y}$ at T = 4.2K.

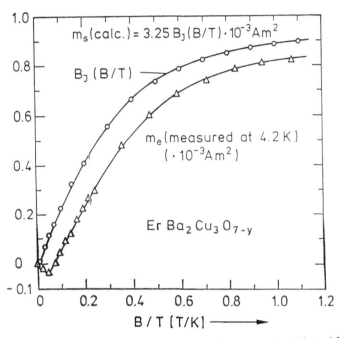

Fig. 14. M_{cal} [magnetization calculated using formula (4)] and M_{meas} [at 4.2K, $M_{meas} = M_e$ in Fig.2(b)] vs B/T for $ErBa_2Cu_3O_{7-y}$.

For nointeracting moments with angular momentum J, the variation of M(H,T) is given by the Brillonin function B_J,

$$M(H,T) = Ng\, J\mu_B\, B_J(gJ\mu_B\, H/K_BT) \qquad (4)$$

For comparison, the theoretical values of magnetization M_{cal} deduced from relation (4) are shown in Fig. 14 together with the measured value M_{meas}. These two curves have similar tendency in their dependence on argument H/T. Both curves tend to a saturation value at highest H/T.

As seen in Fig. 12, the Curie-Weiss behavior observed above T_C continues in the temperature region well below T_C. This shows that the material's superconductivity exists independently of the rare-earth paramagnetizm to a high degree. Thus we deduce that the layered magnetic ions have almost no effect on the mechanism responsible for superconductivity, and that the rare-earth ions retain their local moments and are readily magnetized by an applied magnetic field, even in the superconducting state.

References

1. Xu Yunhui, Guan Weiyan and K. Zeibig, Physica C 153-155, 1657 (1988).
2. Yunhui Xu, Weiyan Guan and K. Zeibig, Solid State Communications 68, 47 (1988).
3. Yunhui Xu, Weiyan Guan, K. Zeibig and C. Heiden, Cryogencies 29 March supplement, 281 (1989).
4. Yunhui Xu, Weiyan Guan and K. Zeibig, Appl. Phys. Lett. 54, 1699 (1989).
5. Yunhui Xu and Weiyan Guan, Appl. Phys. Lett. 53, 334, (1988).
6. Kuan Wei-yan (Guan Weiyan), Chen Sy-sen, Yi Sun-sheng, Wang Zu-lun, Cheng Wu, and Pierre Garoche, Jour. of Low Temp. Phys. 46, 237 (1982).
7. A.M. Campbell and J.E. Evetts, Adv. Phys. 21, 199 (1972).
8. P.H. Melville, Adv. Phys. 21, 647 (1972).
9. Y. Yeshuran and A.P. Malozemoff, Phys. Rev. Lett. 60, 2202 (1988), M. Tinkham, Halv. Phys. Acta 61, 443 (1988).
10. Cao Ning, Duan Zhangno, Shao Xiuyu, Zheng Jiaqi, Ran Qize, Liu Jinxiang, Chang Yinchuan, Hou Deshen, Fan Hui, Chen Xichen and Guan Weiyan, Solid State Commun. 63, 965 (1987); 63, 1125 (1987).
11. A.A. Abrikosov and L.P. Gor'kov, Zh.Eksp.Teor.Fiz. 39, 1781 (1960) (Sov. Phys. - JETP 12, 1243 (1961)).

MICROSTRUCTURE AND ELECTRICAL PROPERTIES OF BULK HIGH-T_c SUPERCONDUCTORS

U. Balachandran, M. J. McGuire, K. C. Goretta, C. A. Youngdahl,
Donglu Shi, R. B. Poeppel, and S. Danyluk*

Argonne National Laboratory, Argonne, IL 60439
*University of Illinois at Chicago, Chicago, IL 60680

INTRODUCTION

   Bulk forms of the superconductors Y-Ba-Cu-O and Bi-Sr-Ca-Cu-O generally have a low critical current density (J_c). At 77 K in zero magnetic field, values greater than about 10^3 A cm^{-2} have seldom been obtained. The most notable exceptions are for melt-textured rods and filaments,[1-3] for which J_c values often exceed 10^4 A cm^{-2}. Values of J_c approaching 10^4 A cm^{-2} have also been reported for wires and tapes processed in Ag tubes.[4-6] Little information about the microstructures of these composite conductors has emerged. This work was undertaken to examine the relationships between processing, microstructure, and electrical properties for high-temperature superconductors processed as pellets or Ag-clad tapes.

EXPERIMENTAL METHODS

   Powders of $YBa_2Cu_3O_x$ (YBCO), $Bi_2Sr_2CaCu_2O_y$ (BSCCO), and $Bi_{0.7}Pb_{0.3}SrCaCu_{1.8}O_z$ (Pb-BSCCO) were synthesized by solid-state reaction.[7-9] The powders were characterized by scanning electron microscopy, X-ray diffraction, and differential thermal analysis. The YBCO powder was phase-pure, nearly equiaxed, and about 1.5 µm in average diameter. Both Bi-based powders were plate-shaped and about 1-5 µm in size (Fig. 1). The BSCCO contained a small amount of the $Bi_2Sr_2CuO_6$ phase, and the Pb-BSCCO consisted of the 110 K phase, a smaller amount of the 80 K phase and a trace of unreacted CuO. Onsets of incongruent melting in oxygen for the three powders were 1020°C for YBCO, 870°C for BSCCO, and 845°C for Pb-BSCCO.

   Each powder was either cold-pressed into pellets or loaded into Ag tubes and rolled into a tape. The pellets, 1.5 g in mass and 12.7 mm in diameter, were pressed at 140 MPa pressure. Average green densities were about 60% of theoretical. To form the tapes, powders were placed in vibrating, closed-end tubes that were 50 mm long, 6.4 mm in outer diameter, and 4.4 mm in inner diameter. Vibration enabled green densities to reach 60% of theoretical. The Ag tubes were then evacuated, sealed mechanically, and rolled. Reductions per rolling pass varied from 25 to 250 µm. Final core thicknesses were 300-400 µm.

 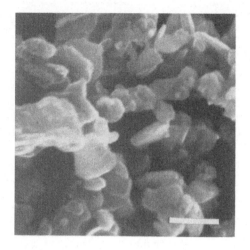

Fig. 1. Starting powders: YBCO (left, bar = 2 µm) and BSCCO (right, bar = 2 µm)

For each type of specimen, sintering was conducted below and above the temperature of melting. The YBCO specimens were given an oxygenation anneal at 450°C. Bars were cut from the final specimens, and T_C and J_C measurements were obtained. J_C of the tape was defined as the current divided by the cross-sectional area of the superconducting core. A criterion of 1 µV cm^{-1} was used to define J_C.

RESULTS AND DISCUSSION

For YBCO, sintering at sufficiently high temperatures to form a liquid phase resulted in fully dense specimens with relatively low J_C values. For sintering in the solid state from 915-980°C in oxygen, J_C was about 10^3 A cm^{-2} and was nearly independent of sintering temperature.[10] For the BSCCO pellets, sintering was very poor in the absence of a liquid phase and J_C was very low. The highest J_C of about 300 A cm^{-2} was obtained by sintering in a region of partial melting, at 910°C, for 0.3 h, followed by annealing at 825°C for 4 h. Liquid-phase sintering was not, however, successful for sintering the Pb-BSCCO because of Pb volatilization. Results for the pellets are summarized in Table 1. The T_C values shown are zero-resistance temperatures.

Table 1. Data for Sintered Pellets

Material	Sintering Temp. (°C)	% Theor. Density	T_c (°C)	J_c (A cm^{-2})
YBCO	915-980	84-98	92	700-1000
YBCO	1020-1030	98-99	90	<350
BSCCO	865	50-55	88	~0
BSCCO	910	65-70	89	300
Pb-BSCCO	845	~65	106	200

In the absence of a liquid phase, the YBCO grains were generally smaller than 40 μm in maximum dimension, and all of the grain boundaries were clean (Fig. 2). If a liquid phase was present during sintering, growth resulted in grains that could exceed 200 μm in the a-b plane. Transmission electron microscopy revealed the presence of phases such as $BaCuO_2$ and CuO at the boundaries of these grains. The lower J_c values for specimens sintered in the presence of a liquid appear to be attributable to the presence of the grain boundary phases.[10,11]

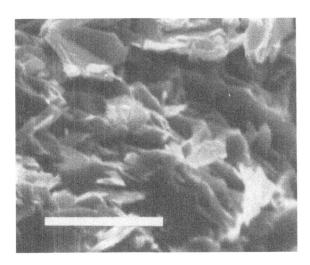

Fig. 2. YBCO sintered in oxygen at 940°C (bar = 10 μm)

For BSCCO, solid-state sintering induced no densification because of extensive grain growth in the a-b planes of the grains. Liquid-phase sintering produced still larger grains, but the a-b planes were aligned preferentially and the connectivity between grains was good. Second phases such as CuO were present on some grain boundaries. The Pb-BSCCO specimens possessed similar microstructures to those of the BSCCO. The grains were large, platelike, and partially textured. As was found for YBCO sintered with a liquid phase, J_c values for the BSCCO pellets are limited by the solidification products of the liquid.[8]

Processing superconductor powders in Ag tubes had few effects on microstructure or properties. The YBCO/Ag tapes has higher green densities than did the pellets (80% versus 60%), but the fired densities were similar. Little texturing was observed in the YBCO/Ag-clad tapes because the starting powder was equiaxed. T_c was reduced to 88 K, and J_c values were about 50% lower than those of the pellets. The lower T_c is attributable to reduced oxygen levels in the clad tapes: as revealed by iodometric titration, the pellets had average stoichiometries of $YBa_2Cu_3O_{6.92}$, whereas the tapes had an average of $YBa_2Cu_3O_{6.84}$. The reduction in J_c resulted, at least in part, from the reduction in T_c.

The interface between the YBCO and the Ag was good, and the J_c was measured through the Ag. J_c in tapes is generally measured in this fashion.[4-6,12] In these measurements, the Ag carried a small amount of the total current. If the YBCO core is thin relative to the sheath, a substantial fraction of the current may be transported through the sheath. In addition, as cross-sectional areas are reduced, magnetic fields decrease and J_c rises.[13] In one study, a YBCO core of 300 μm yielded a J_c of 300 A cm^{-2} at 77 K, but the same roll-processing to a

30-μm core produced a Jc of 800 A cm^{-2} at 77 K and 3000 A cm^{-2} at 4.2 K.[12] In that and other studies, J_c values increased further when the microstructure became more highly textured.[4-6,12]

The BSCCO/Ag tapes were very similar to the pressed pellets: T_c was 88 K and the highest J_c was 200 A cm^{-2}. Microstructural developments were similar. Sintering in the absence of a liquid induced substantial grain growth. The tapes were only modestly textured, and because of this and the constraint imparted by the Ag sheath, plastic deformation and fracture of the grains occurred during sintering (Figs. 3 and 4). Connectivity between grains was poor in the resultant structure. Liquid-phase sintering produced very large, highly textured grains (Fig. 5). J_c values were still modest, however, probably because of extraneous grain-boundary phases. The best combination of processing appears to be solid-state sintering of a textured green body. The resultant microstructure should be free from deleterious second phases and relatively easy to sinter.

The Pb-BSCCO sintered similarly to the BSCCO. Critical current densities were very low in the tapes (less than 15 A cm^{-2}). One cause for the low J_c appears to have been a reaction between the Ag and the Pb-BSCCO. It was reported recently that this reaction can be mitigated by processing the Ag/Pb-BSCCO in reduced oxygen pressures of 300-670 Pa.[14] Work to confirm this contention is now underway.

ASSESSMENT

Processing $YBa_2Cu_3O_x$ and Bi-based superconductors as pellets and as Ag-clad tapes has revealed several considerations for obtaining optimal properties. In the absence of texture, processing in Ag tubes offers little advantage over other bulk forms. YBCO and Ag do, however, form a continuous, low-resistance interface. The interfaces between BSCCO and

Fig. 3. Microstructure of BSCCO sintered in Ag in the solid state

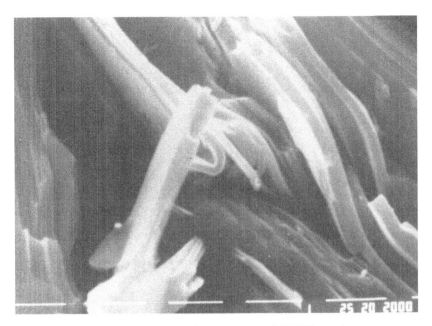

Fig. 4. Bending and fracture of BSCCO grains

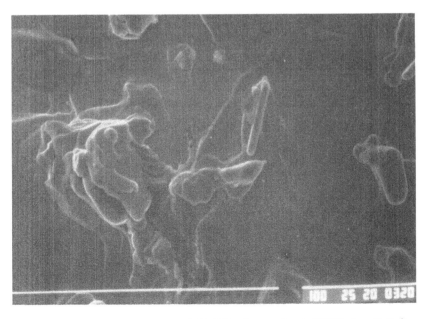

Fig. 5. Microstructure of BSCCO sintered at 900°C for 0.3 h

Pb-BSCCO and Ag appear to be of lower quality. All of the high-T_c superconductors melt incongruently, and in no case was it possible to obtain a J_c value greater than about 500 A cm^{-2} when a liquid was present during sintering. This statement refers to conventional sintering[10] of bulk forms and not to processes such as melt texturing in which very large, highly aligned grains with clean grain boundaries are produced.[1-3]

Solid-state sintering, or post-sintering anneals capable of removing the effects of liquid phase,[15] appear to be essential in producing good bulk superconductors. Because tape rolling is a directional process, use of powders with aspect ratios should enhance texture. Improvements in superconducting properties should result.[12]

ACKNOWLEDGMENTS

Work was supported by the U. S. Department of Energy, Offices of Energy Storage and Distribution, Conservation and Renewable Energy, and Basic Energy Sciences -- Materials Science, under contract number W-31-109-Eng-38. The work of M. J. McGuire was in partial fulfilment of the requirements of the M.S. degree at the University of Illinois at Chicago and was performed while Mr. McGuire was a thesis participant in a program administered by the Argonne Division of Educational Programs and funded by the U. S. Department of Energy.

REFERENCES

1. S. Jin, T. H. Tiefel, R. C. Sherwood, R. B. van Dover, M. E. Davis, G. W. Kammlott, and R. A. Fastnacht, Melt-textured growth of polycrystalline YBa$_2$Cu$_3$O$_{7-\partial}$ with high transport J_c at 77 K, Phys. Rev. B 37:7850 (1988).

2. K. Salama, V. Selvamanickam, L. Gao, and K. Sun, High current density in bulk YBa$_2$Cu$_3$O$_x$ superconductor, Appl. Phys. Lett. 54:2352 (1989).

3. R. S. Feigelson, D. Gazit, D. K. Fork, and T. H. Geballe, Superconducting Bi-Ca-Sr-Cu-O fibers grown by the laser-heated pedestal growth bethod, Science 240:1642 (1988).

4. M. Okada, A. Okayama, T. Morimoto, T. Matsumoto, K. Aihara, and S. Matsuda, Fabrication of Ag-sheathed Ba-Y-Cu oxide superconductor tape, Jpn. J. Appl. Phys. 27:L185 (1988).

5. K. Togano, H. Kumakura, H. Maeda, E. Yanagisawa, N. Irisawa, J. Shimoyama, and T. Morimoto, Fabrication of flexible ribbons of High-T_c Bi(Pb)-Sr-Ca-Cu-O superconductors, Jpn. J. Appl. Phys. 28:L95 (1989).

6. T. Hikata, K. Sato, and H. Hitotsuyanagi, Ag-Sheathed Bi-Pb-Sr-Ca-Cu-O superconducting wires with high critical current density, Jpn. J. Appl. Phys. 28:L82 (1989).

7. U. Balachandran, R. B. Poeppel, J. E. Emerson, S. A. Johnson, M. T. Lanagan, C. A. Youngdahl, D. Shi, K. C. Goretta, and N. G. Eror, Synthesis of phase-pure orthorhombic YBa$_2$Cu$_3$O$_x$ under low oxygen pressure, Mater. Lett. (in press).

8. U. Balachandran, D. I. Dos Santos, A. W. von Stumberg, S. W. Graham, J. P. Singh, C. A. Youngdahl, K. C. Goretta, D. Shi, and R. B. Poeppel, Processing Y- and Bi-based superconductors, in "Proceedings of the Xth Winter Meeting on Low Temperature Physics - High-Tc Superconductors," World Scientific, Singapore (1989, in press).

9. U. Balachandran, D. Shi, D. I. Dos Santos, S. W. Graham, M. A. Patel, B. Tani, K. Vandervoort, H. Claus, and R. B. Poeppel, 120 K superconductivity in the (Bi,Pb)-Sr-Ca-Cu-O system, *Physica C* 156:649 (1988).

10. D. Shi, J. G. Chen, M. Xu, A. L. Cornelius, U. Balachandran, and K. C. Goretta, Transport critical current behavior and grain boundary microstructure in $YBa_2Cu_3O_x$, *J. Appl. Phys.* (in press).

11. D. Shi, D. W. Capone II, G. T. Goudey, J. P. Singh, N. J. Zaluzec, and K. C. Goretta, Sintering of $YBa_2Cu_3O_{7-x}$ compacts, *Mater. Lett.* 6:217 (1988).

12. K. Osamura, T. Takayama, and S. Ochiai, Effect of cold-working on the critical current density of Ag-sheathed $Ba_2YCu_3O_{6+x}$ tapes, *Supercond. Sci. Technol.* 2:111 (1989).

13. L. Chen and Y. Zhang, Theoretical issues for critical currents of bulk polycrystalline $YBa_2Cu_3O_{7-\partial}$ superconductors, *J. Appl. Phys.* 66:1886 (1989).

14. E. O. Feinberg, ed., High Tc Update, Ames Laboratory, 3:(14):1 (1988).

15. I. Bloom, B. S. Tani, M. C. Hash, D. Shi, M. A. Patel, K. C. Goretta, N. Chen, and D. W. Capone II, Effect of heat treatment time and temperature on the properties of $YBa_2Cu_3O_{7-x}$, *J. Mater. Res.* 4:1093 (1989).

ELECTRONIC STRUCTURE STUDIES OF CUPRATE SUPERCONDUCTORS

N. Nücker, H. Romberg, M. Alexander, S. Nakai, P. Adelmann, and J. Fink

Kernforschungszentrum Karlsruhe
Institut für Nukleare Festkörperphysik
P.O.B. 3640, D-7500 Karlsruhe, FRG

INTRODUCTION

Since the discovery of the cuprate superconductors [1] three years have gone and still the mechanism for superconductivity in this new class of materials is unclear. Strongly linked to this problem is the knowledge of the electronic structure of these materials, mainly at energies near the Fermi level. In this field a lot of experimental and theoretical work has been done bringing some insight into the problem. It seems to be established now that (maybe excluding the so-called electron conducting cuprates) superconductivity is connected with hole states on oxygen most probably in the CuO_2 planes or ribbons. But there is a strong debate whether the cuprates can be described within a Fermi liquid theory or whether due to the strong onsite Coulomb interaction ($U \sim 8$ eV) of the Cu d electrons a more local description is required. Band-structure calculations in the local density approximation (LDA)[2] cannot describe the electronic structure of $La_{2-x}Sr_xCuO_4$: The band-structure calculation predicts a half-filled band and hence a metal, even in the case of the undoped La_2CuO_4 which is an antiferromagnetic insulator. XPS investigations [3] showed that the density-of-states (DOS) of the valence band is at lower energy by about 2 eV compared to LDA band-structure calculations. On the other hand, band-structure calculations predict rather well the topography of the Fermi surface as found by angle resolved photoemission (UPS) investigations on $Bi_2Sr_2CaCu_2O_8$ [4]. While the tools for band-structure calculations are rather elaborated giving excellent agreement with experimental results in "normal" cases it is not yet possible to calculate the electronic structure for the highly correlated systems such as La_2CuO_4. For the discussion of our results we may use a "band" picture [5] (Fig. 1) in which, due to the Coulomb interaction, the copper-oxygen band splits into two Cu3d Hubbard bands with an oxygen 2p band in between. In the case of the undoped La_2CuO_4 or $YBa_2Cu_3O_6$ the oxygen band is filled. We will argue in this paper that upon doping holes are formed in this band. Another model for the electronic structure in the doped case may be a description where impurity states are created upon doping at or near the Fermi level in between the filled oxygen band and the upper Hubbard band.

EXPERIMENTAL

The measurements were performed at the Karlsruhe electron energy-loss spectrometer (EELS) described elsewhere[6]. This spectrometer operates with high energy electrons (170 keV) in transmission. Samples were about 100 nm thick and therefore bulk properties are probed. The energy resolution is adjustable between 0.1 eV and 0.7 eV. By means of 4 pairs of deflection plates electrons scattered with a momentum transfer perpendicular to the beam line can be realigned to the optical axis and thereby the momentum transfer perpendicular to the beam direction can be measured. This is used to orient single crystals, to analyze the crystal structure, and to compare excitation spectra in different crystal symmetry directions. For example, for the measurement of the O1s excitation spectrum with momentum transfer parallel or perpendicular to the **a,b** plane the sample is oriented with its **c**-axis in the scattering plane and tilted by 45° to the beam axis. If now the measurements are performed with a momentum transfer perpendicular to the beam direction $q_\perp = \pm q_\parallel = \pm 0.4$ Å^{-1} where $q_\parallel = 0.4$ Å^{-1} is the momentum transfer associated with the energy-loss $\Delta E \sim 530$ eV of the O1s excitation, the total momentum transfer will either be in the **c** direction or in the **a,b**-plane, respectively, depending on the sign of $q_\perp$. A direct comparison of scattered intensities is possible since the scattering angle $\Theta = q_\perp/q_0 \sim 0.4$ Å$^{-1}/228$ Å^{-1} $\sim 0.1°$ is very small, and hence the electron path through the sample is almost identical.

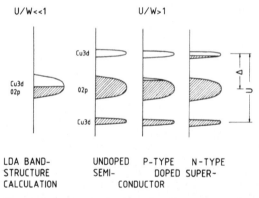

Fig. 1. Schematical band structure for cuprate superconductors.

In most cases thin films were cut from ceramic samples or from single crystals with a diamond knife using an ultramicrotome. $Bi_2Sr_2CaCu_2O_8$ thin films were produced by peeling them off from a single crystal with a tape. Suitable films of ~100 nm thickness were selected by observing interference colors under a microscope. Some $YBa_2Cu_3O_7$ and $GdBa_2Cu_3O_7$ films were sputtered on freshly cleaved surfaces of CaO single crystals. The films were removed from the substrate and mounted on electron-microscope grids. $YBa_2Cu_3O_6$ samples were obtained from $YBa_2Cu_3O_7$ thin films in situ by annealing in ultrahigh vacuum. By on-line measurements of low energy excitations monitoring the plasmon near 1 eV and the sharp Cu-O excitation near 4 eV the oxygen concentration was controlled. For all samples the structure was checked in situ by elastic electron scattering.

RESULTS

Fig. 2 shows the O1s absorption spectra of doped and undoped $La_{2-x}Sr_xCuO_4$. The O1s binding energy as defined by XPS is marked by a broken line. The excitation spectrum of undoped La_2CuO_4 shows no intensity near the binding energy which is in agreement with the semiconducting behavior of this cuprate. Upon replacing some of the trivalent La by divalent Sr [7] a triangular-shaped peak emerges near the binding energy the intensity of which is roughly proportional to the Sr doping up to x~0.25. Since no

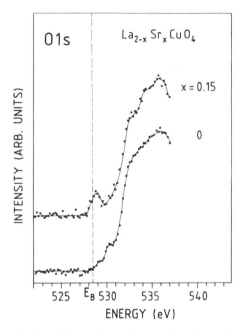

Fig. 2. O1s absorption edges of undoped and p-type (Sr) doped La_2CuO_4.

corresponding intensity was observed in valence band XPS spectra [3] where the Cu3d states have a large cross section, the intensity observed near the binding energy has predominantly O2p character and hence holes on oxygen are created upon doping. At higher energies beginning with ~3 eV above the binding energy La5d and 4f states hybridized with O2p states are observed in the absorption spectra of the doped and undoped compound. XPS results [7] on the same system showed that no trivalent Cu was formed upon doping. Similarly holes in the $YBa_2Cu_3O_6$ compounds can be created and superconductivity can be induced by oxygen doping or in the case of $Bi_2Sr_2Ca_{1-x}Y_xCu_2O_8$ when replacing trivalent Y by divalent Ca [8].

The replacement of trivalent Y by trivalent Pr[9] in $YBa_2Cu_3O_7$ leads to a continuous reduction of T_c until superconductivity disappears near the composition $Y_{0.4}Pr_{0.6}Ba_2Cu_3O_7$ [10]. On the other hand, O1s absorption spectra do not show drastic changes, and holes on oxygen are observed even for the pure Pr-compound [11]. This indicates that T_c reduction and suppression may be due to a localization of holes.

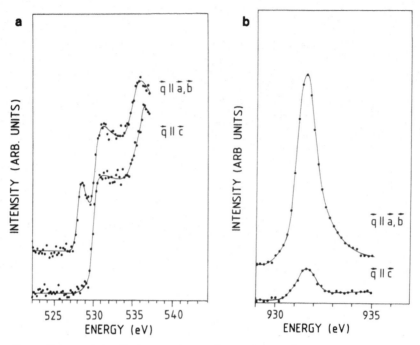

Fig. 3 a. O1s and b. Cu2p$_{3/2}$ absorption edges of Bi$_2$Sr$_2$CaCu$_2$O$_8$ with q∥a,b and q∥c.

More insight in the electronic states near E_F can be extracted from measurements on single crystals [12]. Fig. 3a shows the O1s absorption spectra of Bi$_2$Sr$_2$CaCu$_2$O$_8$ for momentum transfer parallel to the a,b-plane (q∥a,b), compared to that with q∥c. We observe a sharp edge at the O1s binding energy $E_B = 528.7$ eV as determined by XPS. According to LDA band-structure calculations [13] and to UPS results reproducing the calculated Fermi surface [4] these states are assigned to O2p$_{x,y}$ states of a (probably renormalized) dpσ* band in the CuO$_2$ planes and to O2p$_{x,y}$ states in the BiO$_2$ planes. There are no O2p$_z$ holes within ~1.5 eV from the threshold as can be seen in Fig. 3a for q∥c. Above 530 eV we observe nearly identical density-of-states ascribed to flat Bi6p-O2p bands as well for both polarizations in Bi$_2$Sr$_2$CaCu$_2$O$_8$ and as in polycrystalline samples of insulating Bi$_2$Sr$_2$YCu$_2$O$_8$ [8]. The Cu2p$_{3/2}$ absorption spectra for Bi$_2$Sr$_2$CaCu$_2$O$_8$ are shown in Fig. 3b. For q∥a,b we observe an asymmetric line. The main absorption line is ascribed to final states with 3d$_{x2-y2}$ character and the shoulder is due to the influence of holes on neighboring O sites [14]. There is almost no such asymmetry for q∥c which is in line with the absence of holes with 2p$_z$ symmetry in the O1s spectrum in Fig. 3a. The intensity of the absorption line for q∥c is about 10-15% of that for q∥a,b and may be explained by an admixture of 3d$_{3z2-r2}$ states to the empty 3d$_{x2-y2}$ bands.

Recently similar results were obtained for Tl$_2$Ba$_2$CaCu$_2$O$_8$ [15]. Again we observe in the O1s absorption spectra (Fig. 4) holes in the CuO$_2$ planes. But for q∥c there is a strong edge starting about 1 eV above threshold for q∥a,b. Band-structure calculations [16] predict two antibonding Tl-O bands crossing the Fermi level near Γ and Z. At these symmetry points O2p$_z$ orbitals from Tl-O and Ba-O layers hybridized with Tl6s and Tl5d$_{3z2-r2}$ orbitals determine the band character. The shift of the onset between q∥a,b, and q∥c absorptions may be

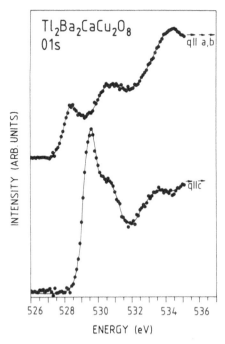

Fig. 4. O1s absorption edges of $Tl_2Ba_2CaCu_2O_8$ with **q∥a,b** and **q∥c**.

explained by a chemical shift which is close to that deduced from LDA band-structure calculations. The much higher intensity of the absorption for **q∥c** compared to that for **q∥a,b** is consistent with band-structure calculations. As in $Bi_2Sr_2CaCu_2O_8$, the $Cu2p_{3/2}$ absorption spectra of $Tl_2Ba_2CaCu_2O_8$ indicate that there is an admixture of about 10% $Cu3d_{3z^2-r^2}$ to the empty $3d_{x^2-y^2}$ band observed. In contrast to XAS results [17] we do not observe any shift between the absorption lines for **q∥a,b** and **q∥c** neither in $Tl_2Ba_2CaCu_2O_8$ nor in $Bi_2Sr_2CaCu_2O_8$ nor in $YBa_2Cu_3O_7$ within 50 meV.

Fig. 5a shows measured orientation dependent O1s absorption spectra for $YBa_2Cu_3O_7$ [12] (points) compared to calculated absorption edges [18] (solid line). The O1s edge shows for **q∥a,b** a double peak structure near the O1s binding energy measured by XPS. This may be due to a chemical shift between the oxygen atoms in the CuO_2 plane and those in the ribbons. The apex oxygen exhibits an even stronger shift to lower energies. The LDA band-structure calculations of the absorption edges including the dipole matrix elements are not far from the experimental values. This indicates that in doped cuprates the electronic structure is not too far from that derived from LDA band-structure calculations. For the undoped system $YBa_2Cu_3O_6$ no states at the Fermi level are observed [12] which is in strong contrast to LDA band-structure calculations. In Fig. 5b and 5c the $Cu2p_{3/2}$ spectra of $YBa_2Cu_3O_7$ and $YBa_2Cu_3O_6$ are shown. In contrast to the symmetric $Cu2p_{3/2}$ line for **q∥c** of $Bi_2Sr_2CaCu_2O_8$ discussed above the absorption spectra for $YBa_2Cu_3O_7$ show extra intensity near 933 eV for **q∥a,b** and **q∥c**. Furtheron, the intensity ratio of the main line between the **q∥c** and the **q∥a,b** results is increased compared to that in the $Bi_2Sr_2CaCu_2O_8$ spectrum. While the absorption in $YBa_2Cu_3O_7$ for **q∥a,b** is caused predominantly by $3d_{x^2-y^2}$ final states and to a lesser extent by $3d_{z^2-y^2}$ final states from the ribbons, for **q∥c** mainly $3d_{z^2-y^2}$ states from the ribbons are reached. Both the **q∥a,b** and the **q∥c** $Cu2p_{3/2}$ absorption spectra show a second

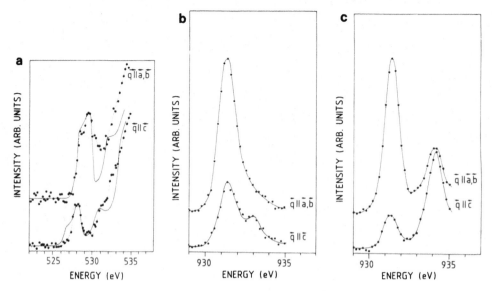

Fig. 5. a. O1s absorption edges of $YBa_2Cu_3O_7$ for $q\|a,b$ and $q\|c$: dots experimental results, solid line LDA band-structure calculation by Zaanen et al.[17]
b. $Cu2p_{3/2}$ absorption edges of $YBa_2Cu_3O_7$ for $q\|a,b$ and $q\|c$.
c. same as b) for $YBa_2Cu_3O_6$.

contribution assigned to the influence of holes on O in the neighborhood of the excited Cu atoms [14]. Since the distance between the apex O and the copper in the ribbons is shorter than all other Cu-O distances the energy shift due to a hole on an apex O may be largest as observed in the spectrum. In the insulating $YBa_2Cu_3O_6$ the lines at 931 eV are symmetric and show no extra intensity near 933 eV confirming the given assignment. The intensity for $q\|c$ near 931 eV may be interpreted as in the case of $Bi_2Sr_2CaCu_2O_8$ by ~10% admixture of $3d_{3z^2-r^2}$ states to the empty $3d_{x^2-y^2}$ band. The resonance-like absorptions at 934 eV are due to monovalent Cu in the ribbons produced by removing the O atoms from the chains. The unoccupied states are expected to have mainly $3d_{3z^2-r^2}$ character and indeed much stronger absorption is observed for $q\|c$.

With the discovery of the new class of electron conducting cuprate superconductors there was hope that superconductivity of high T_c cuprates could be understood better, since charge symmetry would reduce the number of possible models considerably. However, O1s absorption measurements [19] (Fig. 6a) on $Nd_{2-x}Ce_xCuO_{4-\delta}$ showed the existence of holes at the Fermi level already for the undoped compound and a slightly increasing number of holes upon n-type doping. We carefully investigated most of the known n-type doped superconducting cuprates including the F-doped compound $Nd_2CuO_{3.7}F_{0.3}$ [20] and all of them exhibit a prepeak in the O1s spectra near E_F. Orientation dependent measurements on a Nd_2CuO_4 single crystal showed holes in O2p states parallel to the a,b-plane. Since the oxygens in the Nd_2O_2 layers have no planar symmetry the observed holes are likely to be in $O2p_{x,y}$ orbitals of the CuO_2 plane. The $Cu2p_{3/2}$ absorption edge of $Nd_{1.85}Ce_{0.15}CuO_{4-\delta}$ is compared to that of undoped Nd_2CuO_4 in Fig. 6b. The observed lines are narrower than those found for the p-type doped cuprates, but the width is comparable to that of $YBa_2Cu_3O_6$. They do not show an asymmetry as observed in the p-doped cuprates. This may be an indication that there are no holes on

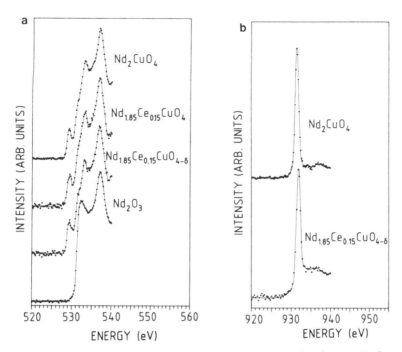

Fig. 6. a. O1s and b. Cu2p$_{3/2}$ absorption edges of Nd$_{2-x}$Ce$_x$CuO$_{4-\delta}$ for $x=0$, $\delta=0$ and $x=0.15$, $\delta>0$. In addition the O1s absorption edge is shown for unreduced, non-superconducting Nd$_{1.85}$Ce$_{0.15}$CuO$_4$ and for Nd$_2$O$_3$.

O-sites in CuO$_2$ planes hybridized to Cu, contrary to the results above from the O1s spectra.

On the other hand we would expect in the simple model, outlined in the introduction an n-type doped cuprate to show some Cu$^+$ in the Cu 2p$_{3/2}$ absorption edge at about 934 eV as was observed for YBa$_2$Cu$_3$O$_6$. This is not observed but instead there is some extra intensity on the high energy side of the main line and there is a shift towards higher energy of about 0.2 eV upon Ce doping. We tried to measure whether there is a reduction in the intensity of the Cu^{2+} line at 931 eV upon Ce doping by measuring the ratio of the Cu2p$_{3/2}$ to the Nd3d$_{5/2}$ intensities and correcting for the reduced Nd content. But due to problems in correcting the background influenced by the Ce doping the error in the intensity ratios is almost of the same order as the expected effect.

By the comparison of Ce3d$_{5/2}$ edges the number of f-electrons on Ce in Nd$_{1.85}$Ce$_{0.15}$CuO$_{4-\delta}$ was found to be the same as in CeO$_2$. Similarly, the observation of the Nd3d$_{5/2}$ absorption edge of the cuprate superconductor was very close to that of trivalent Nd in Nd$_2$O$_3$. By these measurements of the valency of the rare earth ions in these systems n-doping would be expected.

CONCLUSIONS

While in p-doped cuprate superconductors the charge carriers are probably holes on oxygen, the electronic structure of the n-doped cuprates is completely unclear. The observation of holes on oxygen in n-type superconductors cannot be understood in terms of an n-type doped charge-transfer insulator. For Nd$_{2-x}$Ce$_x$CuO$_{4-\delta}$ recently disproportionation into two different domains ~50 Å wide was observed [21]. Possibly one of these phases may be an ordinary p-type superconductor responsible for the holes on oxygen observed in the EELS experiment.

ACKNOWLEDGEMENTS

We are grateful to D. Ewert, D. Hahn, B. Maple, B. Nick, C. Politis, K.F. Renk, B. Renker, S. Uchida, T. Wolf, X.X. Xi, Z.X. Zhao, and T. Zetterer for providing us samples.

REFERENCES

1. J.G. Bednorz and K.A. Müller,Z. Phys. B - Condensed Matter 64:189 (1986).
2. L.F. Mattheiss, Phys. Rev. Lett. 58:1028 (1987).
3. J.C. Fuggle, P.J.W. Weijs, R. Schoorl, G.A. Sawatzky, J. Fink, N. Nücker, P.J. Durham, and W.M. Temmerman, Phys. Rev. B 37:123 (1988).
4. R. Manzke, T. Buslaps, R. Claessen, M. Skibowski and J. Fink, submitted to Physica Scripta
5. J. Zaanen, G.A. Sawatzky, and J.W. Allen, Phys. Rev. Lett. 55: 418 (1985).
6. J. Fink, Adv. Electron. Electron Phys. 75:121 (1989).
7. N. Nücker, J. Fink, B. Renker, D. Ewert, C. Politis, P.J.W. Weijs, and J.C. Fuggle, Z. Phys. B, 67:9 (1987)
8. J. Fink, N. Nücker, H. Romberg, M. Alexander, S. Nakai, B. Scheerer, P. Adelmann, and D. Ewert, to be published in Physica C.
9. U. Neukirch, C.T. Simmons, P. Sladeczek, C. Laubschat, O. Strebel, G. Kaindl, and D.D. Sarma, Europhys. Lett. 5:567 (1988).
10. C.-S. Jee, A. Kebede, D. Nichols, J.E. Crow, T. Mihalisin, G.H. Myer, I. Perez, R.E. Salomon, and P. Schlottmann, Solid State Commun. 69:379 (1989) and references therein
11. N. Nücker et al., to be published
12. N. Nücker, H. Romberg, X.X. Xi, J. Fink, B. Gegenheimer, and Z.X. Zhao, Phys. Rev. B 39:6619 (1989).
13. M.S. Hybertsen and L.F. Mattheiss, Phys. Rev. Lett. 60:1661 (1988)
    H. Krakauer and W.E. Pickett, Phys. Rev. Lett. 60:1665 (1988).
14. A. Bianconi, A. Congiu Castellano, M. de Santis, P. Rudolf, P. Lagarde, A.M. Flank, and A. Marcelli, Solid State Commun. 63:1009 (1987).
15. H. Romberg et al., to be published
16. J. Yu, S. Massidda, and A.J. Freeman, Physica C 152:273 (1988).
    R.V. Kasowski and W.Y. Hsu, Phys. Rev. B 38:6470 (1988).
17. A. Bianconi, P. Castrucci, M. De Santis, A. Di Cicco, A. Fabrizi, A.M. Flank, P. Lagarde, K. Katayama-Yoshida, A. Kotani, A. Marcelli, Z.X. Zhao, and C. Politis, Modern Phys. Lett. B 11:1313 (1988)
18. J. Zaanen, M. Alouani, and O. Jepsen: Phys. Rev. B 40:837 (1989).
19. N. Nücker, P. Adelmann, M. Alexander, H. Romberg, S. Nakai, J. Fink, H. Rietschel, G. Roth, H. Schmidt, and H. Spille, Z. Phys. B 75:421 (1989).
20. M. Alexander et al., to be published
21. C.H. Chen, D.J. Werder, A.C.W.P. James, D.W. Murphy, S. Zahurak, R.M. Fleming, B. Batlogg, and L.F. Schneemeyer, preprint

STUDIES OF MICROSTRUCTURES IN HIGH-Tc SUPERCONDUCTORS

BY X-RAY ABSORPTION TECHNIQUES

A. Krol, C.J. Sher, Z.H. Ming, C.S. Lin, L.W. Song
and Y.H. Kao

Department of Physics and Astronomy
and New York State Institute on Superconductivity
State University of New York at Buffalo
Buffalo, New York 14260

G.C. Smith

Instrumentation Division
Brookhaven National Laboratory
Upton, New York 11973

Y.Z. Zhu and D.T. Shaw

New York State Institute on Superconductivity
State University of New York at Buffalo
Buffalo, New York 14260

Fluorescence emission and total-electron-yield vs. grazing angle of incidence at fixed energy of incoming soft x-ray radiation, around O 1s absorption edge, for $YBa_2Cu_3O_{7-\delta}$ and $Bi_2Sr_2CaCu_2O_{8+x}$ thin films on MgO and ZrO_2 substrates were investigated. Surface rms roughness and oxygen depth distribution were estimated. We have found that in these materials the information depth of microstructures derived from fluorescence yield and total-electron-yield is around 2500 Å and 100 Å, respectively.

INTRODUCTION

Thin films of high-T_c superconductors are a very important and promising class of new superconducting materials for potential device applications. The performance of these films is controlled by some critical structural parameters. X-ray absorption is one of the convenient and nondestructive methods allowing us to obtain information on the quality of interfaces and surface, and depth distribution of selected atomic species.

In this contribution we will discuss employment of angular x-ray absorption spectroscopy in order to extract information on thin film microstructures. In this technique, the energy of incoming x-ray radiation is kept fixed and absorption is measured vs. grazing angle of incidence. This method should be contrasted with the spectroscopic approach where x-ray absorption is monitored vs. energy of incoming radiation while the angle of incidence is kept constant, giving rise to so called NEXAFS and/or (S)EXAFS spectra. The latter spectroscopy provides us with information on local density of empty states and local (usually not more than third next-nearest neighbor distance) ordering around selected atomic species.

MODEL

The total number of primary photoelectrons and hence primary core-holes created at a depth z in the small volume dV within the sample is proportional to the total number of x-ray photons photoabsorbed in this volume, i.e., to the loss of the radiant flux. Dynamical absorption in thin film can be obtained from optical theory of x-ray interaction with stratified media[1]. The number of primary core-holes of type q created per unit time in the small volume dV at a depth z normalized to incoming flux, is

$$\frac{dN_q(z,\Theta)}{F_0} = Cn_q(z)s_q \frac{\sin\phi_1(z_0)}{\sin\phi_0(z_0)} \frac{dP_1(z,\Theta)}{P_0}, \tag{1}$$

where C is a constant, Θ the grazing angle of incidence, $n_q(z)$ the concentration of atoms giving rise to q-type core-holes at depth z, s_q the partial photo-ionization cross section, F_0 incoming flux, $P_1(z,\Theta)$ and P_0 are the Poynting vectors below and above the surface of the sample (at $Z=Z_0$), and ϕ_0 and ϕ_1 are the directions of Poynting vectors incident and at a depth z. The probability that a fluorescence photon created at depth z will reach the surface is given by

$$P_q(\vec{r}) = P_q(z) = \frac{1}{2\pi} \int_\Omega e^{-\beta z/\cos\zeta} d\Omega, \tag{2}$$

where β is absorption coefficient of characteristic q-type fluorescence emisssion and ζ is the polar angle of emission.

Fluorescence yield, due to radiative filling of q-type core-hole normalized to incoming flux (neglecting secondary fluorescence events) can be written in the form

$$FY_q(\Theta) = \int_0^L w_q dN_q(z) P_q(z) dz, \tag{3}$$

where L is the incoming x-ray attenuation length, w_q is the probability of this process.

Total-electron-yield consists mainly of low energy (below 10 eV) secondary electrons produced in cascades of inelastic collisions by primary photo- and Auger- electrons. The escape probability of secondary electrons created at depth z, with average energy a is given by

$$P_a(\vec{r}) = P_a(z) = \frac{1}{2\pi} \int_\Omega e^{-z/A\cos(\xi)} d\Omega, \tag{4}$$

where A is secondary electron attenuation length and ξ is the polar angle of emission.

Each of the primary processes is described by its characteristic angular distribution of emission $D_q(\vec{r},z)$. In our model of secondary electron production we approximate real, intricate (due to elastic and inelastic collisions) trajectories of primary electrons by average linear radial paths traversed by electrons from the source (i.e., the primary core-hole at depth z) to the point given by the effective penetration range of the electron in the considered direction. We employ Bethe's continuous slowing down approximation to calculate the ranges and rate in which energy is lost by primary electron along its trajectory[2] (so called stopping power $S_q(r,z)$). We assume that the number of secondary electrons created per unit length of the primary electron path at a given point r is proportional to the stopping power at this point. The effective penetration range of the q-type primary electrons with initial energy E_q is given by

$$R(E_q) = \int_{E_q}^{0} S_q(\vec{r},z)dE, \qquad (5)$$

Secondary electron current due to primary electron excited from shell q can be written as follows:

$$i_q = \frac{q}{a} \int_0^{E_q} \int_{sample}^{L} a_q P_a(z) dN_q(z) D_q(\vec{r},z) S_q(\vec{r},z) d^3r \, dz, \qquad (6)$$

where a_q is the probability of emission.

In the case of primary photoelectron $a_q = 1$, and $D_q(r,z)$ is given by[3]

$$D_q(r,z) = \frac{\sin^2 u}{(1 - v/c \cos u)^4}, \qquad (7)$$

where u is the angle between k vector of incoming x-ray photon and the direction of emission of the primary photoelectron. In the case of the Auger qrs process when two secondary holes are created in the r and s shells $a_q = a_{qrs}$, $E_q = E_{qrs}$ and $D_q(r,z) = 1$.

Total-electron-yield is the following sum

$$i(\theta) = \sum_q [i_q(\theta) + \sum_{rs} i_{qrs}(\theta)]. \qquad (8)$$

EXPERIMENTAL

The experimental set-up utilized in the angular x-ray absorption method consists of monochromatic x-ray source, radiation intensity monitor, collimator, goniometer and detectors of electrons and fluorescence photons emitted from the sample. Our experiment was carried out at the U15 beamline at NSLS. Total-electron-yield and fluorescence emission around O K-edge were measured. A current amplifier was used to measure the photocurrent of the negatively biased sample mounted on an insulating plate. The characterisic oxygen K fluorescence photons were detected by a low pressure parallel plate avalanche chamber[4] placed perpendicular to the plane of incidence.

The $YBa_2Cu_3O_{7-\delta}$ and $Bi_2Sr_2CaCu_2O_{8+x}$ thin films on MgO and ZrO_2 substrates were investigated. Laser ablation method was used in order to obtain good quality overlayers of superconductors.

The angular variations of x-ray absorption spectra were obtained simultaneously in the fluorescence and total-electron-yield mode for selected energies of incoming photons around the oxygen K-edge. The grazing angle of incidence was varied in the range 0-200 mrad.

RESULTS AND DISCUSSION

The stopping power and effective penetration range were calculated based on the data published by Sugiyama[2]. For both types of the superconductors investigated it was found that ranges of primary electrons created due to excitation of O 1s electron are slightly smaller than 100 Å. Based on the data published in Ref. 5 the attenuation length for secondary electrons was estimated to be 5 Å. The experimental absorption curves were fitted to the model described above. Surface rms roughness parameters were found to be in the range of 70 - 100Å, which indicates that no smoothing processes occurred. The thickness of the oxygen depleted and/or surface contaminated layer was estimated to be of the order of 200 Å. The optical constants were found, with extinction coefficient approximately 50% larger than the values expected from theoretical calculations which place a limit on soft x-ray penetration depth to approximately 2500 Å at the oxygen K-edge energy.

SUMMARY

In summary, a grazing incidence angular x-ray absorption technique was applied to the investigation of two types of thin films of superconductors. Surface rms roughness and oxygen depth distribution were estimated. We have found that the information depth of fluorescence yield and total-electron-yield is of order of 2500 Å and 100 Å, respectively. These results indicate that fluorescence emission is a much better tool than total-electron-yield for the investigation of microstructures in superconducting films with thickness greater than 100 Å.

ACKNOWLEDGEMENTS

This research is sponsored by AFOSR under grant No. AFSOR-88-0095, and by DOE under grant No. DOE-FG02-87ER45283.

REFERENCES

1. A. Krol, C.J. Sher, and Y.H. Kao, Phys. Rev. B39: 8579 (1988).
2. H. Sugiyama, Jap. Bull. Electr. Lab. 38: 351 (1974).
3. B.K. Agrawal "X-Ray Spectroscopy" ed., Springer-Verlag, New York (1979).
4. G.C. Smith, A. Krol, and Y.H. Kao, to be published in Nucl. Instr. Meth.
5. M.P. Seah and W.A. Dench, Surface and Interface Analysis 1: 2 (1979).

STRUCTURAL COMPLICATIONS IN SUPERCONDUCTING SOLIDS: CHEMICAL DISORDER, STRUCTURAL MODULATION, AND LOW TEMPERATURE PHASE TRANSITIONS IN THE Bi-Sr-Ca CUPRATES

Philip Coppens, Yan Gao, Peter Lee, Heinz Graafsma, James Ye and Peter Bush

Chemistry Department and Institute on Superconductivity
State University of New York at Buffalo
Buffalo, NY 14214

INTRODUCTION

The superconducting inorganic cuprates are unusually complex solids. They are non-stoichiometric with certain atom types being delocalized over different crystallographic sites. In particular the bismuth-based solids are structurally modulated, i.e. they do not have the regular three-dimensional periodicity characterisitic for most crystalline solids. Furthermore subtle, not fully understood phase transitions on cooling are evident in the diffraction patterns. The superconducting properties are sensitive to doping with other elements, which may occupy various sites in the crystals.

While technological applications of the high T_c compounds may make use of thin-films and perhaps polycrystalline materials, single crystal studies are an essential prerequisite to the understanding of the properties of these materials, and the optimization of such properties by variation of the structural parameters through an astute choice of preparation conditions.

We describe here the application of a number of single crystal diffraction techniques to the characterization of several BiSrCaCuO phases. Using the SUNY X3 beamline at Brookhaven National Laboratory we have applied anomalous scattering techniques, in which the scattering power of specific atoms is varied by tuning the X-ray wavelength to values at and near an absorption edge. Such "selective atom diffraction" allows determination of the distribution of one particular atom type over the sites in the crystal. We have used formalisms for the scattering of modulated crystals in terms of super-spacegroup theory to analyse the modulation that occurs in almost all known phases in the series. Finally we have applied low temperature (down to about 15K) crystallographic techniques to study phase transitions which occur on cooling of the single crystals.

We make the point that the properties of high T_c solids can not be fully understood from calculations based on an idealised average of the true structure. They depend to a considerable extent on the structural details, which are the subject of our studies.

DESCRIPTION OF THE SAMPLES

Crystals of the 2212 phase were obtained from Dupont Central Research Station (Dr. Subramanian, this sample is referred to below as crystal 1), and the Academia Sinica (Dr. Zhao, sample referred to as crystal 2). A slight difference in T_c, reported as 95 and

87K respectively, is attributed to a larger Ca content in the second crystal, in accordance with a slight difference in the c-axis cell dimensions of the two specimens. All experiments on 2212 described below used crystal 1, except one of the studies of the phase transitions occuring on cooling, as mentioned specifically in the text.

Crystals of the of the 221 and Pb-doped 221 phase were prepared from stoichiometric mixtures of the oxides using a flux method. The temperature variation during single crystal preparation is illustrated in fig. 1. The composition and homogeneity of all single crystals used in the experiments was checked with Scanning Electron Microscopy (SEM) and Energy-Dispersive X-ray Analysis (EDX). The results of the EDX analysis was used in the calculation of X-ray absorption coefficients.

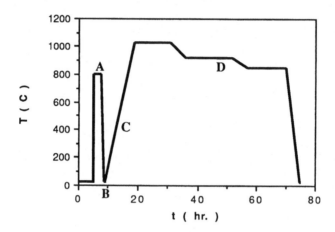

A: Calcining   B: Pulverizing   C: Sintering   D: Crystallization

Fig. 1. Temperature variation during single crystal preparation.

ANOMALOUS SCATTERING STUDY OF THE Bi DISTRIBUTION IN THE 2212 SUPERCONDUCTOR

Substitutional disorder in $Bi_2Sr_2Ca_1Cu_2O_8$

The 2212 Bi-Sr-Ca-Cu-O superconductor contains CuO_2 layers separated by a cation layer of mixed composition, as well as layers consisting of bismuth and oxygen, and strontium and oxygen atoms (Fig. 2). The mixed layer has been described as containing Sr and Ca in about equal amounts (Subramanian, 1988, Gao et al, 1988, von Schnering et al, 1988, Tallon, et al, 1988, Eibl, 1988), though other authors have described the layer's composition as either Sr/Ca or Bi/Ca (Bordet, et al, 1988), and as containing 20% bismuth (Sunshine et al, 1988). The highest Bi percentage is listed by Sastry et al, (1988), who estimate that there is a substantial amount (>50%) of Bi/Sr at the Ca sites, in addition to 30% Bi at the Sr sites. The exact composition of a site in the complex crystals is not easily determined from single wavelength X-ray diffraction data as the occupancies correlate with the atomic temperature factors, which have to be refined simultaneously. Nevertheless, the chemical occupancy is essential for an understanding of the properties. For example, substitution of a bivalent atom (Ca or Sr) by a trivalent atom (Bi) changes the

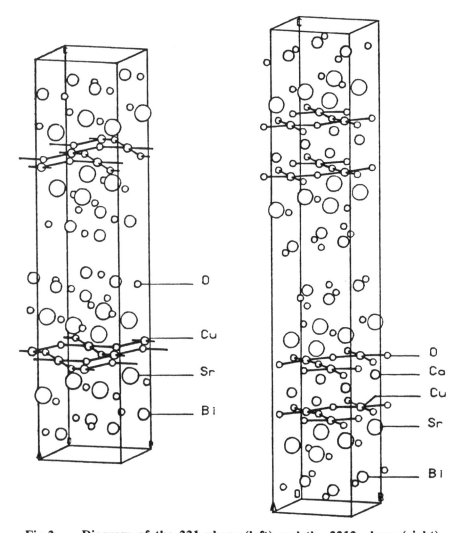

Fig.2. Diagram of the 221 phase (left) and the 2212 phase (right)

hole concentration in the CuO$_2$ planes, as does a deficiency in the occupancy of the cation layers (Sleight, 1988). As pointed out by Cheetham et al, (1988), a deficiency of the Sr-site occupancy, according to the general formula $Bi_2Sr_{2-x}CaCu_2O_8$, may be the principal mechanism for controlling Cu valency in the Bi-cuprate superconductors.

Description of the method

The strong variation of the bismuth scattering factor in the immediate vicinity of the Bi L$_{III}$ edge at 0.924 Å (13.42 keV) makes it possible to do Bi-specific diffraction experiments by means of tunable synchrotron radiation. The real part of the anomalous scattering factor f' (Bi) varies from -21.15 at 0.924 Å (the edge) to -9.51 at 0.9600 Å (Fig. 3). The variation of the imaginary part f'' (Bi) is of less importance in this almost centrosymmetric structure, which has commonly been described in the space group Amaa.. The variation of f'' can be minimized by a careful choice of the wavelengths λ_1 and λ_2 at the lower part of the edge and well below the edge respectively. Combining the data sets at the two wavelengths after appropriate scaling leads to:

$$\Delta F = F(\lambda_1) - F(\lambda_2) = \sum_\nu \Delta(f_\nu + f'_\nu + if''_\nu) \exp(2\pi i\, \mathbf{H}.\mathbf{r}_\nu) T_\nu,$$

where the sum is over all atoms ν and T is the temperature factor. Or, for a centrosymmetric structure:

$$\Delta F = \sum_\nu \Delta(f_\nu + f'_\nu + if''_\nu) \cos(2\pi \mathbf{H}.\mathbf{r}_\nu) T_\nu$$

With the choice of wavelengths made, the only significant variation is for the f' of the Bi atoms. Thus

$$\Delta F = \sum_{Bi} \Delta f'(Bi) \cos(2\pi \mathbf{H}.\mathbf{r}_{Bi}) T_{Bi}$$

In other words, the signal modified in this way selectively gives information on the Bismuth atoms only. This is the principle of the Selective Atom Diffraction Method.

Description of the experiment

With a (220) perfect Si double-crystal monochromator at the SUNY X3 beamline at the National Synchrotron Light Source, an energy resolution of about 5.4eV was obtained. Using new software (Restori, 1988) 154 pairs of reflections with $\sin\theta/\lambda < 0.54$ Å^{-1} and $I(\lambda_2)-I(\lambda_1) > 3\,\sigma(\Delta I)$, were collected at wavelengths of 0.9243 and 0.9600 Å, where λ, θ and I are the wavelength, the Bragg angle and the intensity of the diffracted beam respectively. Data were corrected for absorption by means of an analytical integration procedure (Templeton and Templeton, 1978), and values of the absorption coefficient of 377.7 cm^{-1} and 369.4 cm^{-1} at the two wavelengths respectively. These values were obtained by calculation (Cromer and Liberman, 1970) and subsequent interpolation for the on-edge-wavelength, using our experimental EXAFS curve (Fig. 3). Values of f' of -18.56 and -9.51 for $\lambda = 0.9243$ and 0.9600 Å respectively were obtained, while f'' equals 3.96 and 4.21 at the two wavelengths. The values of f' at each of the wavelengths were averaged over the experimental band width for use in the subsequent analysis (Fig. 3).

X-ray scale factors at each wavelength were obtained by refinement of each of the data sets, keeping the other parameters fixed. The relative scale of the two sets was checked by calculating the ΔF ($=F(\lambda_2)-F(\lambda_1)$) Fourier map, which showed peaks in the Bi and Sr and Ca planes, but no significant features in the Cu-O plane. As the scattering factors of the other atoms show very little variation (< 0.15e), the ΔF values are only dependent on the scattering of the Bi atoms.

The Bi occupancies of the metal atom sites in the Bi, Sr and Ca planes were refined, keeping positional and temperature parameters constant at the values from the conventional refinement of the data on the same crystal specimen. The structural modulation was included in the model, as described fully in the next section. The modulation included a significant **c** axis amplitude of the Ca atoms, which was not taken into account in our earlier work (Gao et al, 1988).

Results (Table 1) indicate significant Bi occupancies of about 5 and 6% respectively at the Sr sites in the 'pure' Sr layer at $z = 0.14$, and the Ca site in the 'mixed' layer at $z = 0.25$. The agreement factors obtained in this refinement, listed in the table, are compatible with the ratio of about 10 between the ΔF values and their standard deviations derived from the counting statistics of each of the measurements. After completion of the analysis of the synchrotron data, the conventional data set was reanalysed, keeping the Bi occupancies at the values obtained in the anomalous scattering analysis. The fact that agreement factors are identical to those of the original refinement demonstrates the insensitivity of conventional X-ray data to the detailed chemical occupancy of the metal sites. The mixed plane contains 17(7) % Sr, in addition to Bi and Ca. While this amount is somewhat below 3σ, a partial occupancy of this site by strontium atoms is likely.

The average valency of Cu

Taking into account the different multiplicities of the sites, the occupancies listed in the table correspond to the formula $Bi_{2.03}Sr_{1.85}Ca_{0.77}Cu_2O_8$. This leads to an average valency of 2.34(6) for the Cu atoms, which confirms the existence of holes in the CuO_2 layers. The stoichiometry may be compared with the composition $Bi_2Sr_{1.67}Ca_{0.99}Cu_2O_8$ given by Chippindale et al, (1988). Except for a perhaps significant difference in Ca content, which is common between different preparations (Chippindale et al, 1988, Niu et al, 1988), the two results are in good agreement, in particular regarding the (Sr + Ca)/Bi ratio of 1.3, rather than 1.5 in the idealized composition. The value of 2.34(6) for the average Cu valency is lower than the value of 2.62 we obtained previously in the analysis in which the Ca atom modulation was omitted. The corrected value is in better agreement with values obtained by other techniques, such as chemical titration: <2.30 (Gopalakrishnan et al, 1989), Hall effect measurements: 2.42 (Clayhold et al, 1989), and the disappearance of superconductivity on Y^{3+} substitution: 2.3 (Manthiram and Goodenough, 1988). The Bi occupancy of the Sr and Ca sites, and the resulting Sr deficiency, strongly support the conclusion that Cu valency is controlled by the Sr, Ca deficiency as first proposed by Cheetham et al, (1988).

The deficiency (i.e. less than three for Ca+Sr as in the idealized formula $Bi_2Sr_2Ca_1Cu_2O_8$) accounts for the increase in Cu valency and the corresponding creation of electron holes in the CuO_2 layers. Neither extra oxygen, nor overlap of the Bi $6p$ and Cu $3d$ bands (Hybertsen and Mattheiss, 1988) is needed to create the holes, which are the charge carriers in these compounds.

STRUCTURAL MODULATIONS

What is a modulated solid ?

The common definition of a crystal is based on its three-dimensional periodicity, with a unit pattern being repeated at constant intervals in each of three non-coplanar

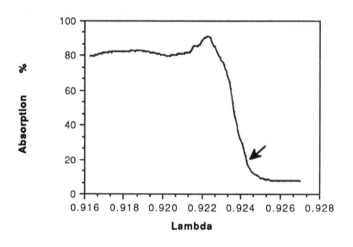

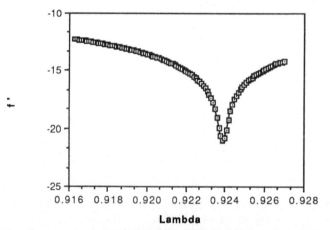

Fig. 3. TOP: Extended X-ray absorption fine structure curve of the 2212 Bi-Sr-Ca-Cu-O sample, indicating the wavelength used to collect the "on edge" diffraction data. BOTTOM: Variation of f' (Bi) with λ in the same interval, averaged over the 5.4-eV bandwidth of the Si (220) monochromator.

directions, much like the two-dimensional analogue of a wallpaper pattern. Modulated crystals do not quite conform to this elegant definition, because the periodicity is removed, at least in one direction, by the superposition of a different periodicity with a repeat which is generally not a simple multiple of one of the three original translation distances. The different periodicity may be due to a displacement of the atoms (a 'displacive modulation'), or to non-random substitution by a different atomic species (a 'substitutional modulation').

Table 1. Site occupancies and agreement factors from the least squares refinement. Numbers in parentheses represent statistical standard deviations.

	Synchrotron	Conventional	
Occupancies	ΔF	Gao et. al	this work
Bi/Bi	0.93(1)	0.94(2)	0.93
Bi/Sr	0.05(1)	--	0.054
Sr/Sr		0.98(3)	0.84(3)
Bi/Ca	0.06(1)	--	0.063
Sr/Ca		0.41(7)	0.17(7)
Ca/Ca		0.59	0.767
R factors			
R	15.73	10.32	10.32
R_w	18.04	11.09	11.09
Number of reflections	154	597[†]	597[†]

[†] /Bi: z=0.0522, /Sr: z=0.1409 /Ca: z=0.25.
[††] Including 108 first-order satellite reflections.

In either case the distortion can be described by a plane wave, with wave vector $\mathbf{q}$, such that points at the position defined by $\mathbf{r}$ have identical displacements when $\mathbf{q}\cdot\mathbf{r}$ is the same. For a displacive modulation with a sinusoidal displacement wave, the displacement U at $\mathbf{r}$ is the sum of the contributions of two waves with 0 and 90° phase respectively:

$$u_v = U_v^x \sin(2\pi \mathbf{q}\cdot\mathbf{r}_v) + U_v^y \cos(2\pi \mathbf{q}\cdot\mathbf{r}_v)$$

In addition to such a sinusoidal wave higher harmonic waves with a smaller repeat distance are often necessary to describe a modulation. An illustration of a one-dimensional modulation wave with a longitudinal displacement parallel to the modulation wave vector q is given in fig. 4.

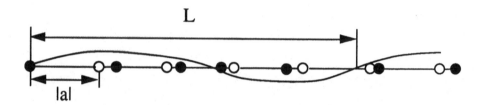

● Real atomic position   ○ Ideal atomic position

Fig. 4. An illustration of displacive modulation in one dimensional lattice. L is the modulation wave length, |a| the length of the lattice vector. |L|/|a| is generally non-integral.

For the 2212 BiSrCaCuO phase the q-vector equals $0.21a^*$, which means that an identical point along the distortion wave is reached after $1/0.21 \simeq 4.8$ unit cells. But, since this is not an integer number there is no identical atom at that position. Thus, the incommensurateness of the modulation means that no two unit cells along the direction of the wave vector q are distorted exactly in the same way, thereby introducing a much less regular pattern than would be the case without the modulation. (The regularity can be retrieved, however, by the mathematical concept of introducing of a fourth dimension, see De Wolff et al, (1981) and references therein.)

In the diffraction pattern the modulation gives rise to a whole set of additional reflections, the so-called satellite reflections, which are displaced from the 'main' diffraction maxima. The atomic displacement amplitudes and their directions can be derived from the intensities of these satellite reflections, using formalisms described elsewhere (Petricek et al, 1985). This leads to the description of the modulations in the superconducting crystals discussed in the following sections.

Dependence of the modulation on the number of CuO_2 layers and on chemical substitution

The modulations in the Bi-superconductors are dependent on n, the number of CuO_2 layers and are strongly influenced by Pb substitution. The q-vectors for a number of compounds are given in Table 2. Whereas in the 2212 phase the q-vector is along the a-axis, it points in an oblique direction in the one-layer 221 BiSrCaO crystals, thus lowering the symmetry from orthorhombic to monoclinic. Unlike the multilayer n=2 and 3 phases the BiSrO 221 phase can also be prepared without Ca. But this last phase does not show any modulation! The effect of Pb substitution is equally dramatic. It causes a set of additional satellite reflections to occur with a different periodicity, and therefore a different

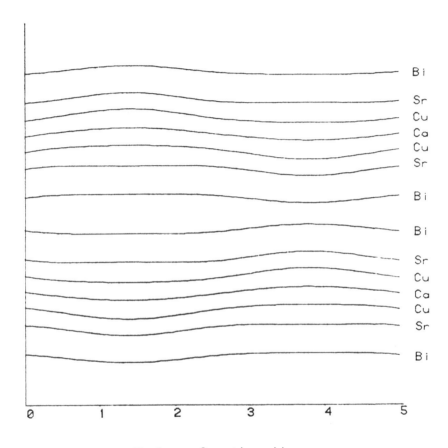

Fig.5. Atomic modulations of the Bi 2212 phase along the c-axis direction. The horizontal axis represents the number of unit cells along the a-axis, the vertical axis the relative position of the successive layers along the c-axis.

q-vector, which equals 0.144 a^*, compared with 0.234 a^* for the original "2212-like" wave. However, our anomalous scattering experiments (Lee et al, 1989) show that they are not solely due to a displacement of the Pb atoms: both the 'original' satellite reflections and the 'new' satellites contain Bi and Pb contributions.

Comparison of modulation amplitudes in the different phases

There are considerable similarities between the modulations in the three phases discussed here, the structures of which are illustrated in fig. 2. In all three phases we have studied so far, the 2212, the BiSrCaCuO and the lead doped BiSrCaCuO 221 phases, the displacements of the bismuth atoms are pronounced in the crystallographic **a**- and **c**- directions, while displacements along **b**- are small or absent (Table 3). The Bi displacements are much larger in the **a**- direction, while for Sr both **a** and **c** amplitudes are of equal magnitude. On the other hand, the Cu atoms are strongly constrained in the CuO_2 planes by the covalent Cu-O bonding, and show only a displacive modulation in the c- direction. The wavy curves in fig. 5 illustrate the **c**-axis displacement in the successive layers of the 2212 phase. The amplitude of the curve at any position represents the displacement an atom at that position in the **c**-axis direction.

Table 2. Modulation wave vector for (Bi- , Pb-doped Bi-) 2212 and 221 superconductors

Phase	Wave vector
Bi-2212	0.210a*
Bi-221	0.213a* + 0.61c*
Pb-doped Bi-2212	0.220a* (q_1)
	0.136a* (q_2)
Pb-doped Bi-221	0.234a* (q_1)
	0.144a* (q_2)

The amplitudes of the modulations increase when the number of the more rigid Cu-O planes decreases. Thus the 221 phase shows larger amplitudes than the 2212 material. This is also true for the Cu atoms, which have a maximum displacement (which is the largest value of the sum of the first and second harmonic displacements) of 0.22(2)Å in the 2212 and 0.43(1)Å in the 221 compound. Lead-doping of 221 leads to the two displacement waves with maximum displacements for Cu of 0.43(2) and 0.32(2)Å. If the waves occur in the same microdomain this would lead to a total displacement of about 0.70Å at its maximum. It is possible, however, that the two waves occur in different microdomains, sufficiently small to give a coherent diffraction pattern including the observed "combination" satellites at the positions $q_1 - q_2$.

The crystal-chemical description of the modulation requires information on the displacement of the oxygen atoms. Though the oxygen atom modulations can be refined from the X-ray data, the results are inaccurate because of the relatively small X-ray scattering power of the oxygen atoms. Neutron diffraction data on single crystals or homogeneous powders are required for a full analysis of the bismuth cuprates.

Table 3. Atomic modulation amplitudes (Å) in the 2212, 221, and Pb-doped 221 phases

A) $Bi_2Sr_2Ca_1Cu_2O_8$

Atom	sin components			cos components		
	U_a	U_b	U_c	U_a	U_b	U_c
Bi						
q	-0.39(1)	----	----	----	0.04(1)	0.18(1)
2q	0.15(1)	----	----	----	-0.02(2)	-0.05(1)
Sr/Bi						
q	-0.23(1)	----	----	----	-0.03(3)	0.20(1)
2q	0.14(2)	----	----	----	-0.07(6)	-0.08(2)
Ca/Sr/Bi						
q	----	----	----	----	0.03(4)	0.26(2)
2q	-0.04(6)	----	----	----	----	----
Cu						
q	-0.06(1)	----	----	----	0.01(2)	0.27(1)
2q	0.03(3)	----	----	----	0.11(8)	-0.05(2)

B) $Bi_2(Sr,Ca)_2CuO_6$

Atoms	sin components			cos components		
	U_a	U_b	U_c	U_a	U_b	U_c
Bi(1)						
q	----	0.05(1)	-0.23(1)	-0.46(1)	0.02(1)	----
2q	-0.08(1)	0.01(1)	0.03(1)	-0.04(1)	-0.05(1)	-0.03(1)
Bi(2)						
q	0.39(1)	0.00(1)	-0.10(1)	-0.11(1)	0.01(1)	0.29(1)
2q	0.10(1)	0.02(2)	-0.02(1)	-0.03(1)	-0.02(2)	0.06(1)
Sr/Ca/Bi(1)						
q	-0.07(2)	-0.04(2)	-0.17(2)	-0.12(1)	-0.01(2)	0.12(2)
2q	-0.18(3)	0.03(4)	-0.12(4)	0.06(4)	-0.02(4)	0.23(4)
Sr/Ca/Bi(2)						
q	0.29(2)	-0.01(2)	-0.23(2)	0.02(2)	-0.03(2)	0.35(2)
2q	0.04(2)	0.00(3)	-0.02(3)	0.07(2)	-0.02(4)	-0.16(2)
Cu						
q	0.05(1)	-0.01(2)	-0.48(1)	-0.00(1)	-0.00(2)	0.03(2)
2q	0.04(3)	0.01(4)	-0.03(2)	0.16(3)	-0.01(3)	0.06(4)

(continued)

Table 3. Continued

C) $(Bi,Pb)_2(Sr,Ca)_2CuO_6$

Atoms	sin components			cos components		
	U_a	U_b	U_c	U_a	U_b	U_c
Bi/Pb(1)						
q_1	----	0.04(1)	-0.11(1)	-0.27(1)	-0.01(1)	----
$2q_1$	-0.14(1)	-0.02(2)	-0.04(2)	-0.04(2)	-0.19(1)	-0.01(1)
q_2	---	0.04(1)	-0.31(1)	-0.21(1)	0.01(1)	----
$2q_2$	0.00(1)	0.04(2)	-0.02(1)	0.05(1)	-0.05(2)	-0.06(1)
Bi/Pb(2)						
q_1	-0.00(1)	-0.00(1)	-0.08(1)	0.15(1)	-0.01(1)	-0.02(1)
$2q_1$	-0.16(2)	0.09(3)	0.02(2)	0.05(3)	-0.07(4)	0.04(2)
q_2	0.02(1)	-0.01(1)	-0.22(1)	0.13(1)	0.01(1)	0.01(1)
$2q_2$	-0.17(2)	-0.04(2)	0.04(1)	-0.05(2)	-0.04(2)	-0.11(2)
Sr/Ca/Bi/Pb(1)						
q_1	-0.01(1)	0.02(1)	-0.14(1)	-0.17(1)	-0.04(1)	-0.01(1)
$2q_1$	0.09(3)	-0.08(3)	-0.16(3)	-0.09(2)	0.17(3)	-0.20(2)
q_2	0.11(1)	0.03(1)	-0.22(1)	-0.03(1)	-0.03(1)	-0.09(2)
$2q_2$	-0.10(3)	0.08(4)	-0.02(3)	-0.05(3)	-0.20(3)	-0.12(2)
Sr/Ca/Bi/Pb(2)						
q_1	-0.01(1)	0.06(1)	-0.14(1)	0.08(1)	0.00(1)	0.00(1)
$2q_1$	-0.04(3)	0.10(3)	0.22(3)	-0.05(3)	0.14(2)	-0.16(3)
q_2	-0.11(1)	0.01(1)	-0.31(1)	0.21(1)	0.00(1)	-0.08(2)
$2q_2$	-0.04(2)	0.04(2)	0.09(2)	-0.19(2)	-0.05(2)	-0.02(2)
Cu						
q_1	-0.03(2)	-0.03(2)	-0.19(1)	0.03(2)	-0.02(2)	-0.03(2)
$2q_1$	0.01(4)	-0.15(4)	0.04(4)	-0.10(4)	-0.01(5)	-0.12(3)
q_2	0.06(1)	0.02(2)	-0.34(1)	0.15(2)	-0.01(2)	-0.18(3)
$2q_2$	-0.04(4)	0.01(5)	0.05(2)	0.00(5)	-0.05(5)	0.00(3)

The Cu displacements are relevant to the superconducting properties which are likely affected by variations in the Cu-O-Cu angles. Though the oxygen displacements cannot be derived accurately from the X-ray data, our results indicate large fluctuations of about 10° and sometimes as large as 20° about the average values.

EVIDENCE FOR LOW TEMPERATURE PHASE TRANSITIONS

Splitting of the diffraction peaks

Though structural phase transitions have been described for the Ba-doped $La_{2-x}M_xCuO_4$ superconductor (Axe et al, 1989), to our knowledge no phase transitions in the Bi compounds have been reported. However, we have found that on cooling of both 2212 crystals, a splitting of the diffraction peaks occurs. The two crystals, obtained respectively from the Dupont Central Research Station and the Physical Science Institue of the Academia Sinica, differ slightly in Ca content, leading to somewhat different c-axis cell dimensions. On cooling of the samples in a DISPLEX cryostat mounted on a four-circle diffractometer, first a broadening and subsequently a splitting of the peaks is observed, as illustrated in fig. 6 for the 0 2 10 reflection of crystal 1. The splitting is first observed just below room temperature, and increases as the temperature is lowered further. Thus the onset of the phase transition does not coincide with the superconducting transition temperature. Below 70K an even more complicated behavior is observed. As illustrated in fig. 7 for the 0 2 10 reflection of crystal 2, at 60K the main peak shows an additional splitting. On further cooling to 30K one of the two equal components is reduced in height, while the overall splitting increases. Below 30K one of the three peaks seems to disappear completely. This behavior is not understood at present.

Table 4.   Cell Dimensions of Bi 2212 Crystal

	a(Å)	b	c	$\alpha°$	β	γ
Room temperature	5.403(1)	5.412(2)	30.800(15)	89.97(3)	90.01(2)	89.99(2)
140K Subcell I	5.402(3)Å	5.407(2)	30.79(2)	90.48(4)°	89.95(5)	89.93(3)
Subcell II	5.401(3)Å	5.407(1)	30.76(2)	89.86(3)°	89.93(4)	89.91(3)

A set of cell dimensions obtained by careful centering of one of the two components of each peak of crystal 1, mounted in a gas-flow stream at 140K on a CAD-4 diffractometer shows the splitting at that temperature to be due to a difference in the α angle of about $0.6°$ (Table 4). This difference is the result of a tilt of the c-axis, which would correspond to a relative shift of successive layers in the superconductor. It may be accompanied by distortions, such as canting of the Cu-O octahedra. However such an effect can only be confirmed by a full structural analysis, which is complicated by the overlap of the diffraction peaks of the twin components.

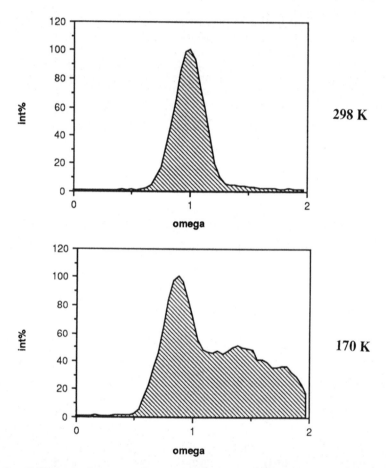

Fig. 6. Splitting of the profile of the 0 2 10 reflection of crystal 1 (2212) on cooling from room temperature to 170K. The vertical axis represents intensity on an arbitrary scale, the horizontal axis rotation of the crystal around an axis perpendicular to the diffraction plane.

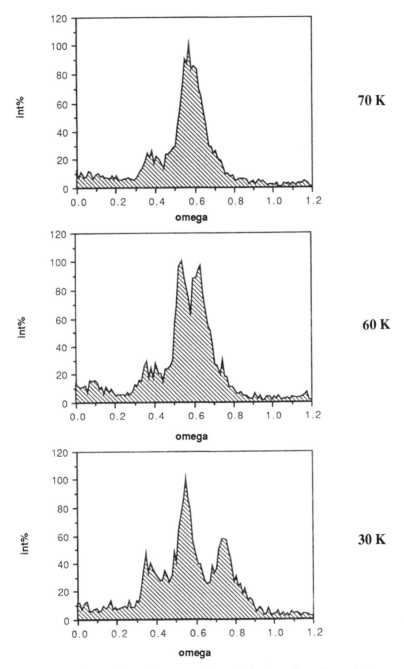

Fig. 7. The (0210) profile of crystal 2 (2212) as a function of temperature. Axes as in figure 6

The temperature dependence of the modulation

Ramesh et al (1988) reported a disappearance of the satellite reflections in the electron diffraction pattern in the lead-doped 2223 phase on cooling below 88K. In order to investigate if a similar behavior occurs in the Pb-doped 221 phase, a crystal was mounted in the DISPLEX cryostat and cooled from room temperature to 20K at a rate of 1K/min. The temperature was maintained for one hour after each decrease of 10 degrees. During this period a main reflection and two strong satellites of each of the two modulations were recorded. Except for an increase in intensity attributable to a reduction in the temperature factor, no significant change in intensity was observed. The experiment was repeated for a Pb-doped 2212 crystal, with a similarly negative result.

CONCLUSIONS

The superconducting cuprates are unusually complex phases, in which structural modulations, chemical disorder and phase transitions occur in the superconductivity temperature range. Band structure calculations of these solids are by necessity approximate if only an average structure is considered. As the properties are dependent on each of these effects, extension of the single crystal work to oriented thin films is necessary if the properties of thin films and their dependence on the method of preparation are to be understood. This will require careful control of thin-film preparation conditions coupled with diffraction experiments at intense X-ray sources.

ACKNOWLEDGEMENTS

Support of this work by the New York State Institute for Science and Technology (NYSIS 88F088) and the National Science Foundation (CHE8711736) is gratefully acknowledged. The SUNY X3 Beamline at NSLS is supported by the U.S. Department of Energy (DEFG0286ER45231).

REFERENCES

Axe, J.D., Moudden, A.H., Hohlwein, D., Cox, D.E., Mohanty, K.M., Moodenbaugh, A.R., and Xu, Y., 1989, Phys. Rev. Lett., 62:2751.

Bordet, P., Capponi, J.J., Chaillout, C., Chenavas, J., Hewat, A.W., Hewat, E.A., Hodeau, J.L., Marezio, M., Tholence, J.L., Tranqui, D., Feb., 1988, Phys.C., Proc. Interlaken Conf. Superconductivity, Interlaken, Switzerland, North-Holland, Amsterdam.

Cheetham, A.K., Chippindale, A. M., Hibble, S.J., 1988, Nature, 333:21.

Chippindale, A.M., Hibble, S.J., Hriljac, J.A., Cowey, L., Bagguley, D.M.S., Day, P., Cheetham, A.K., 1988, Phys. C., 152:154.

Clayhold, J., Hagen, S.J., Ong, N.P., Tarascon, J.N., and Barboux, P., 1989 Phys.Rev. B, 39:7320.

Cromer, D.T. and Liberman, D., 1970, J. Chem. Phys., 53:1891.

DeWolff, P.M., Janssen, T., Janner, A., 1981, Acta Cryst., A37:625

Eibl, O., 1988, Solid State Comm, 67:703.

Gao, Y., Lee, P., Coppens, P., Subramanian, M.A., Sleight, A.W., 1988, Science, 241:954.

Gopalakrishnan, J., Subramanian, M.A., and Sleight, A.W., 1989, J. Solid State Chem., 80:156.

Hybertsen, M.S., and Mattheiss, L.F., 1988, Phys. Rev. Lett., 60:1661. H. Krakauer and W.E. Pickett, ibid., p. 1665.

Lee, P., Sheu, H.-S., Darovskikh. A., Coppens, P., To Be Published, 1989.

Manthiram, A. and Goodenough, J.B., 1988, Appl. Phys. Lett., 53:420.

Niu, H., Fukushima, N., Ando, K., 1988, Jap. J. Appl. Phys., 27:L1442.

Petricek, I.V., Coppens, P., Becker, P.J., 1985, Acta Cryst., A41:478.

Ramesh, R., van Tendeloo, G., Thomas, G., Green, S.M., Luo, H.L., 1988, Appl. Phys. Lett., 53:2220.

Restori, R., Diffractometer Control Program ZACK, State University of New York at Buffalo (1988).

Sastry, P.V.P.S.S., Gopalakrishnan, I.K. Sequeira, A., Rajagopal, H., Gangadharan, K., Phatak, G.M., Iyer, R.M., 1988, Phys. C, 156:230.

Sleight, A.W., 1988, Science, 242:1519.

Subramanian, M.A., Torardi, C.C., Calabrese, J.C., Gopalakrishnan, J., Morrissey, K.J., Askew, T.R., Flippen, R.B., Chowdhry, U., Sleight, A.W., 1988, Science, 239:1015.

Sunshine, S.A., Siegrist, T., Schneemeyer, L.F., Murphy, D.W., Cava, R.J., Batlogg, B., van Dover, R.B., Fleming, R.M., Glarum, S.H., Nakahara, S., Farrow, R., Krajewski, J.J., Zahurak, S.M., Waszczak, J.V., Marshall, J.H., Marsh, P., Rupp, L.W., Jr., Peck, W.F., 1988, Phys. Rev. B, 38:893.

Tallon, J.L., Buckley, R.G., Gilberd, P.W., Presland, M.R., Brown, I.W.M., Bowden, M.E., Christian, L. A., Goguel, R., 1988, Nature, 333:153.

Templeton, D. H. and Templeton, L. K., Crystallographic Computer Program AGNOSTC, University of California at Berkeley (1978).

von Schnering, H.G., Walz, L., Schwarz, M., Becker, W., Hartweg, M., Popp, T., Hettich, B., Muller, P., Kampf, G., 1988, Angew. Chem. Int. Ed. Engl. 2: 574.

LOCAL STRUCTURE AND DISTORTIONS IN PURE AND DOPED $Y_1Ba_2Cu_3O_{7-\delta}$: X-RAY ABSORPTION STUDIES

J. B. Boyce[a], F. Bridges[a,b] and T. Claeson[c]

[a]Xerox Palo Alto Research Center, Palo Alto, CA 94304
[b]Dept. of Physics, University of California, Santa Cruz, CA 95064
[c]Physics Dept., Chalmers Univ. of Techn., Gothenburg, Sweden

INTRODUCTION

The superconducting properties of $Y_1Ba_2Cu_3O_{7-\delta}$ (YBCO) are strongly effected by small changes in the structure and composition. To achieve a better understanding of the interactions between the structure of YBCO and its superconducting properties, we have made an extensive local-structure investigation of pure YBCO, both oxygen-rich ($y = 7-\delta \approx 7$) and oxygen-deficient ($y = 7-\delta \approx 6$), and YBCO doped with Co, Fe and Ni, using x-ray absorption spectroscopy.

The structure of YBCO is a distorted, oxygen-deficient, trilayer perovskite[1-4], which as a 90K superconductor ($\delta < 0.2$) is orthorhombic with two distinct Cu sites: two Cu(2) sites and one Cu(1) site, corresponding to two-dimensional Cu-O planes and one-dimensional Cu-O chains, respectively. Since our discussion centers around the location of the atoms in this material, we show this much-discussed structure in Fig. 1. The notation for the atoms is that given in Ref. 3 for both the orthorhombic and tetragonal phases. When oxygen is removed from the sample, it is the O(1) site in the Cu(1)-O chains that is depleted. As the oxygen content of the sample decreases, the oxygen defects that drive the formation of the Cu-O chains become disordered and T_c decreases. For δ greater than about 0.5 (depending on the thermal treatment), an orthorhombic-to-tetragonal transition takes place, and superconductivity is lost. For the transition-metal-substituted YBCO, T_c is strongly suppressed[5]. The orthorhombic distortion vanishes at very low concentrations of the dopant, 2.5-3%, for Co and Fe but not for Ni[5,6]. The weight of evidence on the location of the dopants is as follows: Co occupies the Cu(1) chain sites[5,7,8]; Fe primarily resides on Cu(1) sites with about 10-20% on the Cu(2) plane site, depending on the concentrations[5,7,9]; for Ni there is still some uncertainty[5,10,11]. Our results indicate that Ni occupies both sites[12].

EXPERIMENTAL DETAILS

Three pure YBCO samples with different oxygen content were studied: $y = 6.98$, 6.87, and 6.15[13]. Samples of YBCO doped with Co, Fe and Ni were prepared by a solid-state reaction. For these $YBa_2(Cu_{3-x}M_x)O_{7-\delta}$ samples the following compositions were investigated: For M=Co, x=0.033, 0.067; 0.17, and 0.3; for Fe, x=0.033, 0.10, 0.17; for Ni, x=0.033, 0.067, and 0.10. Resistive and magnetic measurements of their

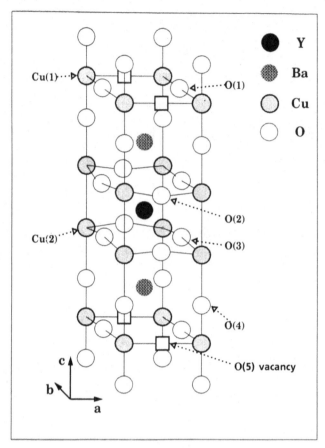

Fig. 1. The structure of YBCO for the fully oxygenated case using the notation of Jorgenson, et al[3]. In the ideal structure the O(5) sites are vacant, and in the low-oxygen case, the O(1) sites are depleted.

superconducting properties are reported in Ref. 5. Diffraction measurements indicated that all the samples are single-phase within a few percent. The two high-O samples are orthorhombic, whereas the low-O sample is tetragonal, as are the doped samples that were studied. Note that a slightly orthorhombic compound, twinned on a scale of ≈10Å, could not have been distinguished from a tetragonal compound in the x-ray diffraction. Both orthorhombic samples have a superconducting transition temperature of approximately 90K, whereas the oxygen-deficient tetragonal material is not superconducting down to 4.2K and the doped samples have a depressed T_c.[5]

The samples were ground into fine powders and brushed onto tapes which were stacked together to give approximately two x-ray absorption lengths at the Cu K-edge. The x-ray absorption data were collected at the Cu K-edge (8980eV) for all the samples and on the Co, Fe, and Ni K-edges for the doped samples. The data were collected on wiggler beamline VII-3 at the Stanford Synchrotron Radiation Laboratory using a Si(400) double monochromator with an exit slit height of 1mm. All absorption measurements were taken in the transmission mode at 80K and, in some cases, as a function of temperature from 4K to 600K.

The reduction and analysis of the XAFS data were performed in the usual way[7,13,14]. After background removal, the energy-space data were converted to k-space data, $k\chi(k)$ and then Fourier transformed to complex, r-space data, $[FT(k\chi(k))]$. For a quantitative comparison, data in real space were fit to structural standards in order to determine the number of neighbors at specific distances and spreads in distance with temperature.

We divide the Cu near-neighbor region out to 4Å into two parts: (1) the first neighbor region consisting of O atoms at distances of < 2.5Å and (2) a second neighbor region between 3Å and 4Å spacing, dominated by the metal atoms. For the purposes of the XAFS discussion we combine into a single peak those peaks of the same kind of neighbor that are too close for the XAFS to resolve in this complex structure, i.e., those within about 0.05Å of each other. This reduces the number of peaks for r<4Å from 15 to 9 for the orthorhombic materials and from 10 to 9 for the low-O tetragonal samples. Weighted averages must be used since XAFS, being a bulk probe, sees a superposition of all pairs in an unoriented powdered sample. This weighing must include not only the number of neighboring atoms but also the number of Cu central atoms since there are two inequivalent Cu sites: one Cu(1) and two Cu(2). Since the Cu XAFS is normalized to the number of Cu central atoms, the number of neighbors must be normalized in a similar fashion.

XAFS RESULTS for PURE YBCO

A thorough analysis was made of the XAFS data on the Cu edge in each sample. The Fourier transform of the Cu K-edge XAFS, $k\chi(k)$, to real space is shown in Fig. 2 for $y=6.98$ and 6.15, both at 5K. The vertical bars on the r axis indicate the position and relative amplitude of each of the neighbors to the Cu according to the diffraction

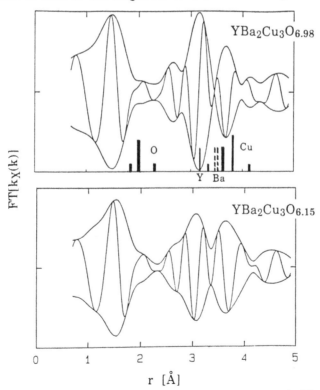

Fig. 2. The Fourier transform of the Cu K-edge XAFS, $k\chi(k)$, to real space for $y=6.98$ and 6.15, at 5K. The envelope curve is the magnitude of the transform and the oscillatory curve is the real part of the transform.

results. Note that there is a shift of the XAFS peaks to lower r due to the phase shifts. The two orthorhombic samples, y=6.87 and y=6.98, yielded essentially identical spectra over the entire temperature range.

The changes in the near-neighbor environment on going from y=6.98 to 6.15 are readily apparent. In the region of the first-neighbor oxygen peaks, the low-O sample has fewer O neighbors. The Cu(2)–O(4) bond length lengthens due to the removal of the O(1) atoms. Since the O(1) atoms contribute to the peak at 1.94Å, the weighted average number of neighbors in this peak is reduced in the y=6.15 sample.

As seen in Fig. 2, there are dramatic differences in the spectra between y=6.15 and y=6.98 for the second neighbor region, even though the major structural change on going from y=7 to y=6 is the removal of a first neighbor to a Cu, namely, the O(1) atoms in the chains. With the removal of the O(1) atoms, the a and b axes become equivalent, and the orthorhombic structure becomes tetragonal. This change in symmetry does not alter the number of Cu–metal second neighbor pairs nor does it substantially change their separations. Nonetheless, the second neighbor peaks in the XAFS spectrum do change substantially due to a modification of the interference between the various Cu–metal peaks due to a broadening and a movement of the Ba atoms, as discussed below.

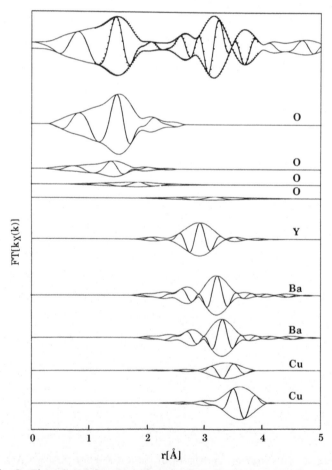

Fig. 3. The nine component peaks that fit the spectra for the high-O sample, y=6.98, at 5K listed below the data (solid line) and fit (dotted curve).

The nine component peaks that fit the spectra for the high–O sample, y=6.98, at 5K are shown in Fig. 3. It is seen that the O first–neighbor region is dominated by the O(1), O(2) and O(3) neighbors in the a–b plane, all at about 1.94Å. The short and long Cu–O(4) peaks, the ones due to O atoms along the c–axis, make a smaller contribution. In the second–neighbor region, the O makes a negligible contribution compared with the metal atoms. The three peaked second–neighbor feature observed in the data corresponds roughly to the three metal atoms: the Cu–Y, Cu–Ba and Cu–Cu constituents. These components do, however, strongly overlap and interfere with one another.

An analogous fit to the low–O spectra shows that the three–peaked second–neighbor structure for y=6.98 becomes two–peaked for y=6.15 (Fig. 2) largely due to a reduction in the Cu–Ba peak. This reduction is due, in part, to a slightly larger disorder for the two Ba peaks in the low–O material, but, more importantly, to destructive interference from the larger separation of the two Cu–Ba peaks. The result of both these effects is a two-peak second–neighbor structure in the data despite the fact that it is composed of several metal-atom neighbors.

The temperature dependence of the Cu–X distances are presented in Fig. 4 for $YBa_2Cu_3O_{6.87}$. They increase only slightly from 5K to 600K, except for the shorter of the two Cu–Ba distances. This Cu(2)–Ba spacing decreases with increasing temperature, while the Cu(1)–Ba distance increases. This indicates that, as the temperature increases, the Ba atoms move toward the planes containing the Cu(2) atoms and away from the chains containing the Cu(1) atoms. There is a normal expansion of the lattice, giving rise

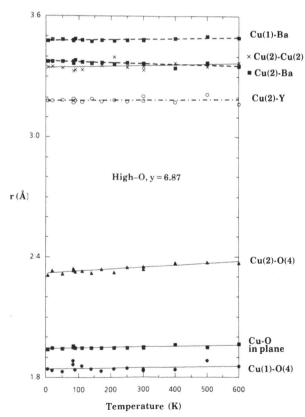

Fig. 4. The temperature dependence of the Cu–X distances for $YBa_2Cu_3O_{6.87}$.

to the small increases in atomic spacing; but, in addition, there is a distortion of the Ba position within the unit cell as it moves toward the planes. This result is qualitatively consistent with the structural measurements at higher temperature[15]. Similar features are observed for the Cu-X distances in the low-O (y=6.15) sample. However, there is a larger separation between the Cu-Ba peaks and a larger temperature variation for the low-O material. This indicates that the low-O sample has a larger motion of the Ba atoms toward the Cu(2)-O planes with increasing temperature.

XAFS RESULTS for YBCO:Co

Our XAFS study of Co-doped YBCO indicates that the Co primarily replaces the Cu(1) atoms, in agreement with other investigations. This conclusion is arrived at by a detailed analysis (similar to the one described above for pure YBCO) of the second neighbor environment around the Co atoms. The results of such an analysis are shown in Fig. 5 for YBCO:Co$_{0.07}$. Here it is seen that no Y is contained in the second neighbor shells, ruling out significant occupation of the planes. In addition, we find that the Co does not just simply substitute for the Cu(1) atoms, but rather significant distortions occur. The near neighbor O peak is composed of about 5 O atoms, ~ 3.5 O atoms at 1.8Å

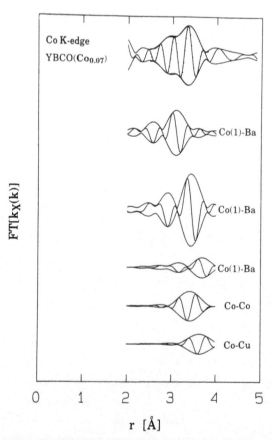

Fig. 5. Second neighbor environment around the Co atoms for YBCO:Co$_{0.07}$.

and ~1.3 O atom at 2.4Å. For comparison, the weighted number of O neighbors in YBCO are 2/3 neighbor at 1.85Å, 10/3 neighbors at 1.94Å, and 2/3 neighbor at 2.3Å. This means that both long and short bonds must be present within the plane containing the Cu(1) sites. The Co second neighbor peak is unexpectedly low in amplitude, but has considerable structure that is inconsistent with a simple Gaussian broadening of the expected Co-Ba and Co-Cu(Co) bond distances. Measurements of the **Cu** environment in the highly doped $Co_{0.3}$ sample show a well defined second neighbor peak composed of a (larger than in normal YBCO) Co-Y peak, two Co-Ba peaks of unequal amplitude, plus the Co-Cu(Co) contribution. These results indicate that most of the remaining Cu in the YBCO:$Co_{0.3}$ sample is on the Cu(2) site, and that, viewed from the Cu(2) site, the Y, Ba, and Cu(Co) atoms are at their expected distances. The amplitudes are consistent with a small amount of Co (~11% best fit, 5-20% reasonable fits) on the Cu(2) site in this highly doped sample. The most important point of this analysis is that the Ba atom positions are *not* strongly *disordered*, and therefore cannot account for the small second neighbor amplitude observed for the Co K-edge data. Therefore, some of the Co atoms must be *displaced considerably* from the normal Cu(1) site, resulting in several different Co-Ba distances. The interference of these Co-Ba peaks produces the small overall second neighbor peak amplitude.

We propose a simple local structure model, shown in Fig. 6, that can account for the different Co-O bond lengths in the Cu(1) layers, provides a good fit to the Co second neighbor peak for the low concentration data, and suggests an explanation for the apparent tetragonal structure of the Co doped samples. Zigzag chains of three-Co-atom segments can be formed along the <110> direction, that (1) have Co-Co distances that are shorter than the usual Cu-Cu planer distance, (2) provide a combination of long and short Co-O bonds as are observed in the first neighbor peak, (3) provide a good fit and a simple explanation for the reduced amplitude of the second neighbor Co multi-peak structure (some of the Co(1) atoms have an off-center

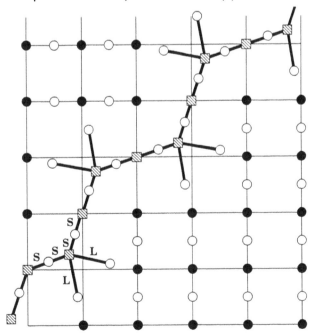

Fig. 6. Chain model for YBCO:Co showing layers containing the the Cu(1) and Co atoms.

displacement), (4) have an increased number of O atoms (one additional O for every two Co in the single chain model of Fig. 6), and (5) fit easily into the square lattice obtained using the lattice constant from diffraction experiments. In this model, some of the O(5) sites near a Co atom are occupied; consequently, linear Cu-O chains can leave the zigzag chain in either the x- or y-direction. We think this promotes twinning on a microscopic scale[8,9] and thus leads to the observed tetragonal structure. Even at very low concentrations, a tendency for Co to aggregate into chains would lead to a tetragonal component[16] in the crystal structure.

A few experiments have hinted at the distortions discussed here[8,9,17,18]. Neutron scattering measurements on Co substituted samples indicate that both magnetic and structural disorder exist within the plane containing the Co(1) atoms. Electron diffraction studies of Co and Fe doped material have indicated chains of dopant atoms along the <110> directions. The large thermal parameters observed for the Co(1) and O(1) atoms in the neutron measurements can be understood in terms of the displaced atoms of the chain models proposed here.

XAFS RESULTS for YBCO:Fe and YBCO:Ni

The data for the Fe doped samples are very similar to the Co substituted samples suggesting a distorted site for the Fe atoms also. However, there appears to be a non-negligible fraction of the Fe on the Cu(2) site (~10% of the Fe), even at the lower concentrations, which makes detailed modeling inappropriate. The small, second-neighbor peak clearly indicates that the second neighbor Fe-X distances are badly disordered. We think this is a result of an off-center displacement of some of the Fe on the Cu sites. Consequently, the Fe(1) sites are not equivalent to the Cu(1) sites. This feature must be taken into account in assigning the various quadrupole splittings, observed in the Mössbauer data, to specific structures.

By contrast, the Ni second neighbor peak for YBCO:$Ni_{0.033}$ is comparable to the Cu data with little broadening of the peaks and a uniform distribution on the two Cu sites. This similarity to pure YBCO is seen in Fig. 7 where the component peaks that fit the second neighbor region around the Ni atoms is shown. It is similar to Fig. 3 for pure YBCO, indicating that the Ni atoms equally occupy the chain and plane sites. Higher Ni concentration samples have a significant fraction of undissolved Ni in the form of NiO.

SUMMARY

We find that, within the experimental error, the XAFS-determined local structure for $Y_1Ba_2Cu_3O_y$ agrees well with the long-range order given by diffraction results for both oxygen-rich (y=6.98 and 6.87) and oxygen-deficient (y=6.15) compounds. The temperature dependence of the parameters shows only a smooth variation. No significant anomalies are observed versus temperature. The Cu-X distances have a negligible to a small positive change with temperature, consistent with the lattice expansion. The exception is the Cu-Ba distances which change substantially; the Cu(1)-Ba distance increases and the Cu(2)-Ba distance decreases. This indicates that the Ba moves away from the Cu(1)-O chains and toward the Cu(2)-O planes with increasing temperature and that anharmonicity plays a role. This motion is larger for the oxygen-depleted compound than for the fully-oxygenated material.

For YBCO:Co, a zigzag chain model accounts for the different Co-O bond lengths in the Cu(1) layer, provides a good fit to the Co second neighbor peak for the low

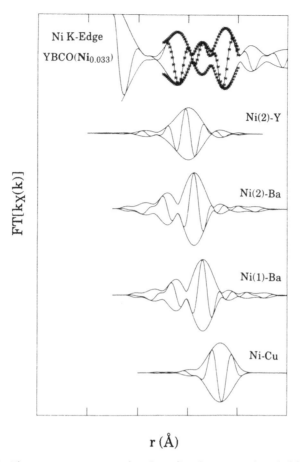

Fig. 7. The component peaks that fit the second neighbor region around the Ni atoms in YBCO:$Ni_{0.033}$.

concentration data, and suggests twinning on a microscopic scale as an explanation for the apparent tetragonal structure of the Co doped samples

The data for the Fe doped samples are very similar to the Co substituted samples suggesting a distorted site for the Fe atoms also. However, there appears to be a non-negligible fraction (~10%) of the Fe on the Cu(2) site, even at the lower concentrations. Consequently, the Fe(1) sites are not equivalent to the Cu(1) sites. This feature must be taken into account in assigning the various quadrupole splittings, observed in the Mössbauer data, to specific structures. Ni is different from Co and Fe in that it occupies both chains and planes.

We think that the observed structural distortions, induced by Co or Fe substitution, play an important role in the suppression of T_c by changing the electron–electron coupling within the Cu(1) layer and also the coupling between the Cu(2) planes and the Cu(1) layers. We suggest that these distortions are more important that the orthorhombic to tetragonal transition.

ACKNOWLEDGMENT

We would like to thank T. H. Geballe for helpful discussions, M. Nygren for the pure YBCO and J. M. Tarascon for the doped samples. The experiments were performed at SSRL, which is funded by the Department of Energy under contract DE-AC03-82ER-13000, Office of Basic Energy Sciences, Division of Chemical Sciences, and the National Institutes of Health, Biotechnology Resource Program, Division of Research Resources. This research was supported in part by NSF Grant No. DMR 85-05549 and the Swedish Natural Science Research Council.

REFERENCES

1. M. A. Beno, L. Soderholm, D. W. Capone II, D. G. Hinks, J. D. Jorgensen, J. D. Grace, I. K. Schuller, C. U. Segre and K. Zhang, Appl. Phys. Lett. **51**, 57 (1987).
2. T. Siegrist, S. Sunshine, D. W. Murphy, R. J. Cava, and S. M. Zahurak, Phys. Rev. B **35**, 7137(1987).
3. J. D. Jorgensen, B. W. Veal, W. K. Kwok, G. W. Crabtree, A. Umezawa, L. J. Nowicki and A. P. Paulikas, Phys. Rev. B **36**, 5731 (1987).
4. C. C. Torardi, E. M. McCarron, P. E. Bierstedt, A. W. Sleight, and D. E. Cox, Solid State Comm. **64**, 497 (1987).
5. J. M. Tarascon, P. Barboux, P. F. Miceli, L. H. Greene, G. W. Hull, M. Eibschutz and S. A. Sunshine, Phys. Rev B **37**, 7458 (1988).
6. Y. Maeno, T. Tomita, M. Kyogoku, S. Awaji, Y. Aoki, K. Hoshino, A. Minami, and T Fujita, Nature **328**, 512 (1987).
7. F. Bridges, J. B. Boyce, T. Claeson, T. H. Geballe, and J. M. Tarascon, Phys. Rev. B **39**, 11603 (1989).
8. H. Renevier, J. L. Hodeau, P. Bordet, J. J. Capponi, M. Marezio, J. C. Martinez, and J. J. Prejean, in Proceedings of the International Conference on Materials and Mechanisms of Superconductivity, Stanford, CA, July, 1989, to be published.
9. P. Bordet, J. L. Hodeau, P. Strobel, M. Marezio, and A. Santoro, Solid State Comm. **66**, 435 (1988).
10. R. S. Howland, T. H. Geballe, S. S. Laderman, A. Fischer-Colbrie, M. Scott, J. M. Tarascon, and P. Barboux, Phys. Rev. B. **39**, 9017 (1989).
11. M. Qian, E. A. Stern, Y. Ma, R. Ingalls, M. Sarikaya, B. Thiel, R. Kurosky, C. Han, L. Hutter, and I. Aksay, Phys. Rev. B **39**, 9192 (1989).
12. F. Bridges, J. B. Boyce, T. Claeson, T. H. Geballe, and J. M. Tarascon, in Proceedings of the International Conference on Materials and Mechanisms of Superconductivity, Stanford, CA, July, 1989, to be published.
13. J. B. Boyce, F. Bridges, T. Claeson, and M. Nygren, Phys. Rev. B **39**, 6555 (1989).
14. J. B. Boyce, F. Bridges, T. Claeson, R. S., Howland and T. H. Geballe, Phys. Rev. B **36**, 5251 (1987).
15. P. Marsh, T. Siegrist, R. M. Fleming, L. F. Schneemeyer and J. V. Waszczak, Phys. Rev. B **38**, 874 (1988).
16. M.Eibschutz, M. E. Lines, J. M. Tarascon, and P. Barboux, Phys. Rev. B **36**, 2896 (1988).
17. J. Hodeau, P. Bordet, J. Capponi, C. Chaillout, and M. Marezio, Physica C **153-155**, 582 (1988).
18. Y. K. Tao, J. S. Swinnea, A. Manthiram, J. S. Kim, J. B. Goodenough and H. Steinfink, J. Mater. Res. **3**, 248 (1988); Mat. Res. Soc. Symp. Proc. **99**, 519, (1988).

Probing Charge on Cu in Oxide Superconductors

E. E. Alp, S. M. Mini, M. Ramanathan, G. L. Goodman
Argonne National Laboratory, Argonne, Illinois 60439

O.B. Hyun
Iowa State University, Ames, Iowa 50011-3020

ABSTRACT

The non-destructive measurement of charge on an ion in a solid has always been a source of great controversy. In this chapter, we will review how x-ray absorption spectroscopy addresses this issue. We will describe the underlying principles of the technique, list some recent experiments, and try to explain the controversial aspects. A new method of evaluation of the position of the absorption edge will be introduced, and its relation to the Mulliken populations determined by discrete variational method for calculating orbital states of a molecular cluster will be discussed. We will then show how x-ray absorption spectroscopy can probe variations in effective charge on atoms as a result of chemical doping. Finally, we will point out some newer developments as a result of using polarized x-rays and oriented crystals.

INTRODUCTION

Ever since the discovery of metallic conductivity in oxides (1), as well as induction of metallic conductivity in cuprates and in other transition metal oxides by chemical doping (2-5), the question of change in valency as a result of non-isovalent doping has been a hot topic of discussion. There are many spectroscopic techniques that addresses the issue of valency, and mixed valency. For example, Mossbauer Spectroscopy is one such non-destructive technique which via isomer shift scale and systematic studies can provide an answer (6). The main shortcoming of this technique is the limitation imposed by the availability of the Mossbauer active isotopes. Nuclear Magnetic Resonance (NMR) is another technique which once promised to be an alternative via systematic "chemical shift" measurements. However, the question of long-range correlations made the interpretation difficult, and sometimes impossible like in the case of magnetic systems. Photoemission spectroscopy and Auger spectroscopy, on the other hand, have suffered from their extreme sensitivity to surfaces. Especially in the case of oxides synthesized via solid state sintering techniques, samples may contain impurity phases, as well as oxygen non-stoichiometry at the surface. What is needed is a bulk sensitive local probe which can cover most of the elements in the periodic table. The advent of synchrotron radiation has helped the situation considerably. Synchrotron radiation sources with x-rays ranging from 1-100 keV are now routinely available. As a result of this, there is a revival in one of the oldest methods of x-ray techniques, namely the X-Ray Absorption Spectroscopy (XAS). There are by now excellent reviews (7), and books (8) which covers the field. Here, we will try to explain how XAS technique so far handled the situation in the case of oxide superconductors.

THE PRINCIPLES OF CHARGE DETERMINATION BY XAS

The X-Ray absorption process involves electronic transitions between specific initial and final states. This is due to the fact that the core levels of an atom in a solid is still not too far away from the free atomic levels, and hence known within few electron volts, and the fact that transition is governed by a dipole operator, allowing a change in angular momentum quantum number of ± 1. The latter property defines what type of final states are accessible. However, in the case of an atom in a lower symmetry configuration, quadrupole component of the electric field may couple two states with a different, but known selection rules.

The basic premise of XAS to measure effective charge on an ion is that it will require more energetic photons to excite an electron away from the core potential as the positive charge on the atom that absorbs the photon is increased. This effect manifests itself as a shift in the overall energy position of the absorption edge to higher energies with increasing net charge. The magnitude of this shift in transition metals is a few electron volts every time the net charge on the absorbing atom is increased by one. However, the width of the absorption edge is 10 to 40 eV. The details, also, may be much more complicated due to solid-state effects which determine the energy widths of the final states. Nevertheless, systematic studies on transition metals so far have shown that the higher the valence of a cation in a solid, the higher the position of the absorption edge on the energy scale.

An example is given in Fig. 1, where normalized x-ray absorption cross-sections of Cu_2O, CuO, and $NaCuO_2$ are shown. These compounds are chosen to represent nominally +1, +2 and +3 Cu. The shift in the position of the first peak from Cu_2O to CuO is 3 eV, and to $NaCuO_2$ is 5 eV. However, the same numbers for the main peaks are 2.5 and 4.7 eV, respectively. There are then two questions: i) What is the origin of the peaks, and ii) where should one take the difference in order to provide a quantitative evaluation. The answer to the first question is given by Guo et. al (9). They have shown that in highly anisotropic environments, the first peak of CuO, and $KCuO_2$ (which is isostructural to $NaCuO_2$) is due to 1s->$4p_\pi$ (out-of-plane) transitions, where as the main peak corresponds to 1s->$4p_\sigma$ (in-plane) transitions where Cu is planar, four-fold coordinated by oxygen atoms. For Cu_2O, on the other hand, the first peak corresponds to transitions into orbitals which are perpendicular to the Cu-O bond, where as the main peak is due to transitions into the orbitals along the Cu-O bond. Note that Cu is two-fold coordinated by oxygen in a linear chain form in Cu_2O.

The answer to the second question requires some elaboration. Since we are treating the atom as a whole, a shift evaluation at either of the peak positions would be wrong. We believe the proper approach is as follows: The assignment of an energy position for the overall absorption edge is described in a previous paper (10). The method is based on the ratio of the first moment to the area of the normalized absorption cross-section as a function of energy within an energy window of -5 to 40 eV around the absorption edge. We take

$$M^{(n)} = \int_{L_0}^{L_1} E^{(n)} \mu(E) \, dE \quad \ldots (1)$$

as the nth energy moment of the normalized absorption cross-section $\mu(E)$. The mean energy for this portion of the cross-section is then

$$\langle E \rangle = M^{(1)} / M^{(0)} \quad \ldots (2)$$

For a step function model of the absorption edge, with a step height of unity, located at position S in the range between L_0 and L_1, it was shown that

$$S = 2 \langle E \rangle - L_1 \quad \ldots (3)$$

to be the "<u>characteristic energy</u>" for an absorption edge with a mean energy of <E>. Here, E is the energy expressed in electron volts relative to a reference point. This is chosen to be the first inflection point of Cu-metal absorption spectrum for Cu K-edges. This procedure for assigning the overall position of an X-Ray absorption edge has no adjustable parameter, and provides the first reference scale of its kind, similar to Mossbauer isomer shifts (6).

Based on such a quantitative evaluation scheme, one can attempt to do relative comparisons between samples with different chemical doping rates. However, what is further needed for a meaningful scale is some theoretical charge calculations for some of these compounds. One can, then, accurately associate the absorption edge position to the effective charge on an atom as a whole. Goodman et.al. has recently calculated Mulliken charge populations for several copper compounds including Cu_2O, CuO, $KCuO_2$, La_2CuO_4, and $YBa_2Cu_3O_x$ (11). It is based on using the discrete variational method for calculating orbital states of a molecular cluster. A linear fit gives a good correlation between the calculated Cu charges and the experimentally measured <u>characteristic energies.</u> The following equation expresses this linear relationship:

$$S = 3.33 \, q + 1.68 \quad \ldots (4)$$

where q is the effective positive charge on copper atom. In other words, for every electron removed from copper atom, there is a shift of 3.33 eV in the absorption edge position to higher energies. Since the linear relation in Eq. (4) seems to be followed for compounds with a wide range of formal valences, we have inverted it to obtain a proposed method to assign an effective charge for the Cu atoms in a sample based on the observed characteristic energy of the Cu K-edge absorption :

$$q = 0.30 \, S - 0.50 \qquad \ldots (5)$$

Based on this scale, we can go back to the compounds whose XAS are shown in Fig. 1. The characteristic energies for Cu_2O, CuO, and $NaCuO_2$ are 3.58, 6.49 and 8.35 eV, which correspond to a net charge of + 0.57, +1.44 and +2, respectively. Before discussing the meaning of these numbers, we would like to point out that the controversial aspect of the valence determination is in the spectroscopic assignment of the ground state. Cu atom has $(Ar)3d^{10}4s^1$ configuration. Typically, there seems to be no problem assigning $|3d^{10}4s^0>$ configuration for Cu^{+1}, and $|3d^94s^0>$ for Cu^{+2}. However, for energetic reasons i.e. high cost of 3d-3d repulsion of two holes in the 3d band $|3d^84s^0>$ configuration is not favored for Cu^{+3}. Instead, the extra hole is associated to the ligands by writing $|3d^9 \underline{L}>$ to represent the ground state. The degree with which the Cu atoms "effectively" lose charge depends on the degree of hybridization between Cu 3d and oxygen 2p orbitals. In fact, one can measure such changes in the O 2p occupation by O K-edge XAS (19,20). The interesting point is that such holes in the O 2p levels are not exclusive to high temperature superconductors. CuO, for example, has similar electron holes in the O 2p levels (21). So, going back to the experimentally determined numbers, we believe that the changes in the measured absorption edge position, no matter how they are evaluated, indicate removal of electrons from the absorbing atom. Therefore, copper in $NaCuO_2$ should be considered as Cu^{+3}, whose spectroscopic ground state should be reconciled with photoelectron spectroscopy (22), EELS, and Cu L_3-edge XAS (23) measurements.

The accuracy with which characteristic absorption edge can be determined is limited by the calibration procedure for the energy scale as well as by the accuracy of Mulliken population calculations. We estimate the accuracy of the energy scale to be around ± 0.15 eV, which limits the ability of this method finding effective charge transfers at the level of ± 0.05 electron. However, the same level of sensitivity cannot be expected of the theoretical calculations, yet.

THE EFFECT OF CHEMICAL DOPING ON CHARGE ON Cu in $Nd_{2-x}Ce_xCuO_4$

In the following, we will discuss how the method described above can help to understand the effect of Ce doping in RE_2CuO_4 (RE : Pr, Nd, Sm, Eu, Gd) system. This group of materials have been shown to become metallic and superconducting upon Ce or Th doping (12-14). The crystallographic structure is designated as T'-phase, which is tetragonal with a= 3.943 and c= 12.15 Å for Nd_2CuO_4. The question of whether this kind

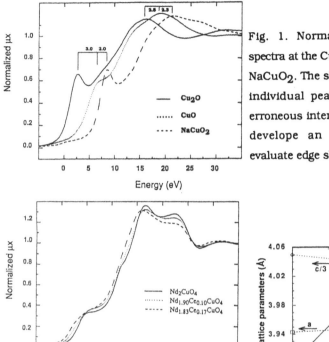

Fig. 1. Normalized X-Ray Absorption spectra at the Cu K-edge of Cu_2O, CuO, and $NaCuO_2$. The shifts in the positions of the individual peaks may cause somewhat erroneous interpretation, which led us to develope an integrated approach to evaluate edge shifts, as shown in the text.

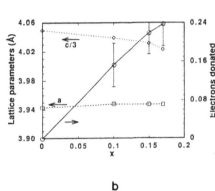

a b

Fig. 2. The effect of Ce doping onto Cu charge in $Nd_{2-x}Ce_xCuO_4$ system can be measured from the Cu K-edge XAS as shown in part (a), and quantified by the procedure described in the text, as shown in part (b).

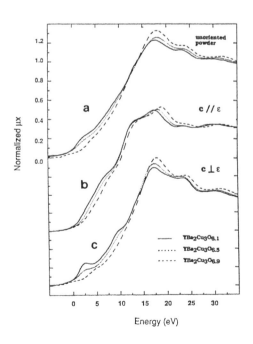

Fig. 3. The Cu K-edge XAS of $YBa_2Cu_3O_x$ system (a) measured using unoriented powders, (b) when the polarization of the photon beam is along the c-axis of the crystal, and (c) when it is along the ab-planes. Due to existence of two Cu sites, the change caused by removing two oxygens around Cu (1) atoms as the oxygen content is reduced from x=7 to x=6 becomes more visible, only when the polarization direction is set parallel to ab-planes.

of non-isovalent doping i.e. substituting Ce^{+4} into Nd^{+3} sites, has any effect on Cu charge cannot easily be measured by any other spectroscopic technique except X-ray absorption spectroscopy. One reason is very small change in lattice parameters as a result of Ce doping (0.65 % reduction in c lattice parameter, and 0.12% increase in a lattice parameter with x=0.17 in $Nd_{2-x}Ce_xCuO_4$). Cu atoms are four-fold planar coordinated by oxygen with a Cu-O bond length of 1.972 Å, typical of divalent copper.

The effect of Ce doping can be seen in Fig. 2 (a). As the Ce content is increased, there is a gradual shift in the overall absorption edge position, indicating donation of electrons onto all of the copper atoms (14), instead of being localized on some Cu atoms, and converting them into +1 Cu. In Fig. 2 (b), the magnitude of this charge donation is quantitatively estimated, using the procedure described above. Based on the shifts of characteristic energy, and using Eq. (5) given above, we conclude that for a Ce doping of x= 0.17 in $Nd_{2-x}Ce_xCuO_4$, 0.28 ± 0.05 electrons are donated into Cu 3d-derived bands.

POLARIZED X-RAY ABSORPTION SPECTROSCOPY

The radiation coming out of a synchrotron is linearly polarized in the plane of the ring This property can be beneficially exploited in understanding the relative energy positions of those orbitals lying parallel to the polarization direction if single crystals or oriented powders are used. In the case of oxide superconductors, it was shown (15) that the anisotropy of the electronic structure is large enough to be observed, when magnetically oriented crystals are used for XAS measurements. Below we will discuss the results of some polarized XAS measurements we have done on the oxide superconductors (16).

i) $YBa_2Cu_3O_x$ system

The crystallographic structure of $YBa_2Cu_3O_{7-x}$ is perhaps one of the most complicated one when compared to the other superconducting cuprates. One reason for this is the large oxygen solid solubility with x ranging from zero to one. There is by now a vast literature on this subject, specifically studying the oxygen ordering as the oxygen content is changed (17,18). There seems to be a difference in the resulting oxygen ordering as a function of the oxygen removal mode or speed. The samples measured in this study are prepared according to the procedure given in Ref. (18). In Fig. 3 (a), the Cu K-edge XAS of unoriented powders of $YBa_2Cu_3O_{7-x}$ with x=0.1, 0.5, and 0.9 are shown. In Fig. 3 (b) and (c), the spectra obtained from the magnetically oriented powders are shown. Here, the polarized nature of the synchrotron radiation is advantageously used to probe the anisotropic nature of the charge removal process, as the oxygen content is reduced. Due to the electric dipole operator coupling the initial and final states, one can bring planar or axial empty 4p orbitals of Cu atoms in-and-out of resonance using single crystals or oriented powders.

The energy splitting of Cu 4p orbitals along the Cu-O bond and perpendicular to it can be understood in terms of the local anisotropic coordination of oxygen ions. Oxygen ions have negative charge, and therefore, Cu 4p orbitals pointing in the direction of oxygen ions are pushed up in energy due to Coulomb repulsion compared to the states perpendicular to the bonding axis (9). This analysis forms the basis for studying the details of charge removal or addition to particular orbitals as a result of doping by polarized x-ray absorption spectroscopy. Note that there are two inequivalent Cu sites in $YBa_2Cu_3O_{7-x}$. It is the Cu(1) sites in the chains which are being reduced as the oxygen is removed from $YBa_2Cu_3O_7$. We see the evidence of this reduction process by the rising peaks at 1.78 and 6.28 eV, in Fig. 3 (c). Our previous Mulliken population calculations (11) indicate that the net charge on Cu(1) decreases from 1.81 to 0.52 as the oxygen content is reduced from 7 to 6, while for the planar Cu(2) this reduction is from 1.85 to 1.74. The same type of cluster calculations yield 0.65 for Cu in Cu_2O, 1.57 for Cu in CuO, and 1.87 for Cu in $KCuO_2$. Therefore, we conclude that copper in $YBa_2Cu_3O_7$ is highly oxidized compared to Cu in CuO. Furthermore, the simulations of near edge absorption structure by multiple scattering calculations based on the potential derived from from these charges agree very well with the experimental results (9).

ii) La_2CuO_4-Nd_2CuO_4 systems

The discovery of superconductivity in RE_2CuO_4 by Ce doping (12) showed that highly oxidized copper may not be necessary for superconductivity. In Fig. 4 (a), we compare the Cu K-edge x-ray absorption spectra of Cu_2O, CuO, Nd_2CuO_4 and La_2CuO_4. Based on the relation described in Eq. 5, we estimate the "effective charge" on Cu in this series of compounds as 0.57, 1.44, 1.53 and 1.87. These numbers can be made more plausible if one considers the coordination numbers, and Cu-O bond distances in these compounds, which are summarized in Table 1.

Table 1. Coordination numbers, interatomic distances, characteristic energies and effective charges on several Cu compounds of interest to high temperature superconductors :

Compound	Coordination Number	Cu-O bond distance (Å)	Characteristic Energy (eV)[*]	Effective charge
Cu_2O	2	1.849	3.57	0.57
CuO	4	1.956	6.49	1.44
Nd_2CuO_4	4	1.97	6.77	1.53
La_2CuO_4	6	1.89, 2.42	7.94	1.88
$NaCuO_2$	4	1.89	8.35	2.0

[*] measured with respect to the first inflection point in the K-edge absorption spectrum of copper metal, which corresponds to 8979 eV.

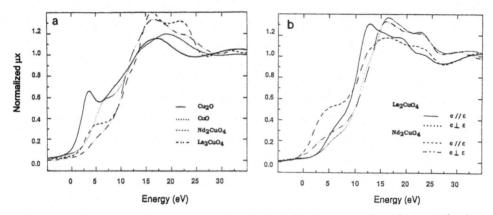

Fig. 4. (a) X-Ray absorption spectra of Cu_2O, CuO, Nd_2CuO_4, and La_2CuO_4 at the Cu K-edge. Note that how the existence of two apical oxygens push the rising part of La_2CuO_4 absorption edge to a higher energy. (b) Polarized Cu K-edge spectra of La_2CuO_4 and Nd_2CuO_4. The relative shift of spectra of Nd_2CuO_4 when polarization is along the c-direction indicates the effect of two apical oxygens that exist in La_2CuO_4, but not in Nd_2CuO_4.

Such an analysis of polarized XAS measurements may also form a basis to measure the changes in the energy level positions of different 4p orbitals as the coordination of Cu changes. An example is given in Fig.4, where we compare polarized XAS results measured on oriented samples of La_2CuO_4 and Nd_2CuO_4. La_2CuO_4 has T structure with a distorted six-fold coordination of oxygen atoms. Four of these oxygens are at a distance of at 1.894 Å, where as two of them are at 2.428 Å. On the other hand, Nd_2CuO_4 has T' phase in which copper is four-fold planar coordinated with oxygens at a distance of 1.97 Å. One should then expect that transitions into out-of-plane 4p orbitals of Nd_2CuO_4 would have a lower energy than the ones in La_2CuO_4 because of lack of Coulombic repulsion of apex oxygens. It is gratifying to see that this is what is observed in the experimental results, as shown in Fig. 4 (b).

What would be needed at this point is an improved simulation procedure for near edge structures to incorporate small changes in the interatomic distances, and compare them to the measured values. One can almost tune the interatomic distance by replacing Nd with the rest of the rare-earth series of elements, yet maintain the same crystal structure. Such a set of systematic measurements may provide a good control for the accuracy of theoretical calculations. We should also mention that using ground powders may complicate these measurements, since each individual grain may not be a single crystal. What would be needed is an independent characterization of the degree of orientation, like rocking curve widths of selected diffraction peaks. Grinding single crystals, and reorienting them may be another solution.

CONCLUSIONS

We have discussed how the relative shifts in the Cu K x-ray absorption edge positions can be converted into effective charge estimations for Cu compounds. For Ce - doped electron superconductor $Nd_{2-x}Ce_xCuO_4$, it was found that, indeed, replacing Nd^{+3} with Ce^{+4} causes donation of electrons onto Cu-derived 3d bands. It was shown that for $YBa_2Cu_3O_x$ system, using polarized XAS, one can monitor the change in the charge of Cu(1), and Cu(2) atoms as x changes form 7 to 6. Finally, a comparison is given between T and T' phases of $RE2CuO4$ system, and the effect of the existence of apex oxygens in La_2CuO_4 versus Nd_2CuO_4 is discussed within the context of polarized XAS.

ACKNOWLEDGMENTS

We acknowledge the help of Dr. A. Bommanavvar during the XAS measurements at beamline X-18-B at NSLS. We thank Drs. D.E. Ellis and J. Guo for useful discussions., and Drs. B. Dabrowski and D.G. Hinks for most of the samples. Part of the work is done at beamline X-11A at NSLS, which is supported by US DOE-BES Division of Materials Sciences under contract No : DE-AS05 -80ER10742, and at SSRL. This work is supported by US DOE-BES under contract No W-3-109-ENG-38. One of us (SMM) is supported by NSF-Office of Science and Technology Centers under contract No: #STC-8809854

REFERENCES

1. G.Hagg, A. Magnelli, Arkiv Kemi Mineral. Geo. **19A** (1944) 2.
2. N.Nguyen, J. Choisnet, M. Hervieu, B. Raveau, J. Solid State Chem., **39** (1981) 120.
3. C. Michel, B. Raveau, Rev. Chim. Miner., **21** (1981) 407.
4. K.K. Singh, P. Ganguly, J. B. Goodenough, J. Solid. State Chem., **52** (1984) 254.
5. J.G. Bednorz, K.A. Muller, Z. Phys.B, **64** (1986) 189.
6. "Mossbauer Isomer Shifts", ed. G.K. Shenoy, F.E. Wagner, North-Holland , (1978).
7. J. Röhler, "X-ray Absorption and Emission Spectra" Chapter 71, Handbook on Physics and Chemistry of Rare-Earths, Vol. **10** Eds. K.A. Gschneider, Jr., L. Eyring, S. Hüfner, North-Holland (1987).
8. B. K.Teo, EXAFS: Basic Principles and Data Analysis, Springer-Verlag , Berlin (1986).
9. J. Guo, D. E. Ellis, G. L. Goodman, E. E. Alp, G.K. Shenoy, Phys. Rev. B (submitted).
10. E. E. Alp, G. L. Goodman, L. Soderholm, S. M. Mini, M. Ramanathan, G.K. Shenoy, A. Bommanavar, J. Phys. : Condensed Matter (in print).
11. G.L. Goodman, D. E. Ellis, E. E. Alp, L. Soderholm, J. Chem. Phys (in print).
12. H. Tagaki, S. Uchida, Y. Tokura, Phys. Rev. Lett. **62** (1989) 1197.
13. J.T. Markert, E.A. Early, T. Bjørnholm, S. Ghamaty, B.W. Lee, J.J. Neumeier, R.D> Price, C.L. Seeman, and B.M. Maple, Physica C **158** (1989) 178.

14. E.E. Alp, S.M. Mini, M. Ramanathan, B. Dabrowski, D.R. Richards, D.G. Hinks, Phys. Rev. B: Rapid Communications **40** (1989) 2617.
15. S. Heald, J. Tranquada, Phys. Rev. B **38** (1988) 761.
16. E. E. Alp, S.M. Mini, M. Ramanathan, B.W. Veal, L. Soderholm, G. L. Goodman, B. Dabrowski, G.K. Shenoy, J. Guo, D.E. Ellis, A. Bommannavar, O.B. Hyun, Mat. Res. Soc. Proc. vol **143** (1989) 97.
17. R. Beyers, B.T. Ahn, G. Gorman, V.Y. Lee, S.S. Parkin, M.L. Ramirez, K.P. Roche, J.E. Vazquez, T.M. Gür, R.A. Huggins, Nature (London) **340** (1989)619.
18. M.A. Beno, L. Soderholm, D.W. Capone, D.G. Hinks, J.D. Jorgensen, J.D. Grace, I.K. Schuller, C.U. Segre, K. Zhang, Appl. Phys. Lett. **51** (1987) 57.
19. N. Nücker, H. Romberg, X.X. Xi, J. Fink, B. Gegenheimer, Z.X. Zhao, Phys. Rev. B (in print).
20. E. E. Alp, J.C. Campuzano, G. Jennings, J. Guo, D.E. Ellis, L. Beaulaigue, S.M. Mini, M. Faiz, Y. Zhou, B.W. Veal, J.Z. Liu, Phys. Rev. B Rapid Communication (in Print)
21. S. Nakai et.al. Phys. Rev. B **36** (1987) 9241.
22. P. Steiner, S. Hüfner, V. Kinsinger, I. Sander, B. Siegwart, H. Schmitt, R. Schulz, S. Junk, G. Schwitzgebel, A. Gold, C.Politis, H.P. Müller, R. Hoppe, S. Kemmler-Sack, C. Kunz, Z. Phys. B **69** (1988) 449.
23. A. Bianconi, J. Budnick, G. Demazeau, A.M. Flank, A. Fontaine, P. Lagarde, J. Jegoduez, A. Revcolevski, A. Marcelli, M. Verdauguer, Physica C, **153-155** (1988) 117.

Andreev reflection and tunneling results on $YBa_2Cu_3O_7$ and $Nd_{2-x}Ce_xCuO_{4-y}$ single crystals

T.W. Jing, Z.Z. Wang, and N.P. Ong
Department of Physics, Princeton University, Princeton, NJ 08544
J.M. Tarascon and E. Wang
Bell Communications Research, Redbank, NJ 07701

1. Introduction

We will describe some results from recent experiments on Andreev reflection and tunneling in the superconductors $YBa_2Cu_3O_7$ ("*123*") and $Nd_{2-x}Ce_xCuO_{4-y}$ ("*NCCO*")[1,2]. The interface resistance between gold and *123* crystals has been greatly reduced using a technique[3] which does not involve high temperature annealing. Gold is evaporated *in situ* onto a crystal cleaved in high vacuum. Using the specific contact resistivity $\rho_{[]}$ as a gauge of the interface quality, we find that $\rho_{[]}$ in these junctions approaches the ideal value $\rho_{[]}^0$ derived by counting the number of conduction channels in *123*. ($\rho_{[]}$ is the product of the contact resistance R_c and the contact area.) With these junctions, we have studied the decrease of the interface resistance near T_c. We have also observed[4] Andreev reflection of injected carriers at low temperatures T in these junctions. The T depedence of the Andreev signal is studied from 4.2 to 80 K. In the "T' phase" superconductor[1,2] *NCCO* we have used the same cleaving technique to make Giaever tunneling junctions with Pb as the counter electrode. We observe rather clearly the gap in the tunneling spectrum[5]. The dependence of the spectrum on temperature and magnetic field has been studied. A number of anomalous features are observed. These are described below.

2. The Interface problem

Although it was previously demonstrated that careful high-temperature annealing of Au pads evaporated on *123* ceramics can yield contacts with very low values of $\rho_{[]}$, very little work has been done on sharp interfaces between Au and any of the high-T_c oxides. Ekin et al[6] reported that $\rho_{[]}$ as low as 10^{-10} Ωcm^2 can be attained using annealing temperatures near 600 C. Because significant diffusion of Au into *123* occurs at such high temperatures, it is quite unclear if the measured impedance is that of the Au-*123* interface or a complicated graded Au-*123* composition. (Au is known to substitute for Cu in the chain sites and become trivalent.) Previous work on single crystals obtain values of $\rho_{[]}$ which are 3 or 4 orders of magnitude larger[7].

In our work, high quality crystals of *123* are cleaved in a vacuum of 10^{-6} torr. (The flux-grown crystals have a T_c between 90 and 92 K as-grown, and an in-plane resistivities ρ_{ab} = 150 µΩcm at 290 K.) Au is then evaporated *in situ* to a thickness between 1,000 and 3,000 Å onto the exposed surface. The resulting Au-*123* interface is found to have very low contact resitivities. The plane of the interface is perpendicular to the CuO_2 planes and parallel to the c axis. The exposed surface is very reactive. If dry oxygen is readmitted into the evaporator before Au is evaporated the junction resistance increases by 3 to 4 orders of magnitude. The junction is also very sensitive to the presence of moisture in the residual gases. To minimize absorption of residual gasses the Au evaporation is started just before the

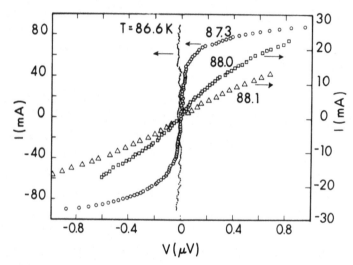

FIG 1 *I-V* curves of Au-*123* junctions near the limit of our measurement resolution. R_{meas} at zero-bias equals 39, 20.5, and 0.85 µΩ at 88.1, 88.0, and 87.3 K respectively (Ref. 3).

crystal is cleaved. At no point is the crystal heated above 320 K. Current-voltage (I-V) curves at some temperatures are shown in Fig. 1.

The value of the contact resistance obtained is so low that accurate measurements of $\rho_{[]}$ becomes impractical as soon as T falls 2 or 3 K below T_c, due to the effect of spreading resistance[3]. We show in Fig. 1 the roughly exponential decrease in the ratio (V/I which we call R_{meas}, denoted by open triangles) as T decreases from 90 K to 87 K. When R_{meas} reaches about 100 µΩ, the effect of spreading resistance artificially suppresses V, so that R_{meas} appears to decrease even faster. We have used a technique borrowed from Logan, Wick and Rice[8] to compute the electrostatic distribution in the Au pad, in order to correct for the potential suppression. The corrected resistance R (open circles) is seen to fall on the extrapolated line in Fig. 2. It may also be seen that if R continues to fall at the same rate, $\rho_{[]}$

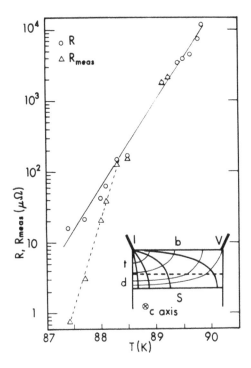

FIG 2 Decrease of the contact resistance with temperature in Au-*123* junction. The observed R_{meas} has been corrected for spreading resistance to R. Insert shows a schematic of the potential contours in the spreading resistance regime (Ref. 3).

reaches the value $\rho_{[]}^0 = 2 \times 10^{-11}$ Ωcm^2 near 85 K. This is the value for the ideal interface obtained by associating a conductance of $e^2/\hbar$ (where e is the electronic charge, and $\hbar$ Planck's constant/2π) to each conduction channel on the *123* side[9]. ($\rho_{[]}$ is essentially determined by the number of channels on the *123* side, since the corresponding number of channels on the Au side is very large.) If we associate each unit cell of *123* with one channel, we have $\rho_{[]}^0 = a'c' / (e^2/\hbar)$, where a', c' are the unit-cell parameters of *123* along the axes **a** and **c**, respectively.

3. Andreev reflection spectrum

Andreev reflection[10,11] involves the reflection of an electron with change of character (into a hole) at the interface between a normal metal N and a superconductor S. We consider an electron of energy ε_k and momentum **k** incident on the interface from the left (the N side). The electron either becomes a quasi-particle (qp) in S if $\xi_k \gg \Delta$, or enters the condensate in S if $\xi_k \leq \Delta$. ($\xi_k = \varepsilon_k - \mu_s$ is the energy referred to the chemical potential μ_s in S.) In the first case, the incident flux is unchanged for a zero-impedance interface, so that the junction conductance G (= dI/dV) equals the value G_N it attains when S is normal. By contrast, each electron that enters the Cooper Sea in the second case results in two electrons which appear

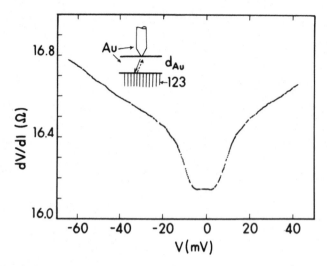

FIG 3 The differential resistance vs. bias of point contact junction showing Andreev reflection spectrum in Au-123 interface. The insert shows the experimental geometry (Ref. 4).

bound as a Cooper pair in S. To conserve charge, a hole is Andreev reflected back into N with momentum $-\mathbf{k}$. Since a left-going hole current is equivalent to a right-going electron current, the total current is enhanced over the injected value by the factor $(1 + A)$, where A is, roughly speaking, the fraction of the incident electrons that are Andreev reflected as holes. Hence, for $\xi_\mathbf{k} \leq \Delta$, the junction conductance is *increased* by a fraction A. By decreasing $\xi_\mathbf{k}$ starting from a value much larger than Δ, we should observe G increasing abruptly from G_N to $(1 + A)G_N$ as $\xi_\mathbf{k}$ decreases below Δ. [If a fraction A of the incident flux is Andreev reflected, the magnitude of the supercurrent in S must be $2A$, in units of the incident current. The remaining electrons enter S as quasi-particles. To conserve probability flux, we must have

$$1 = A + C, \qquad (1)$$

with A and C (the quasi-particle flux) defined as

$$A = |a|^2, \qquad C = |c|^2 v_G. \qquad (2)$$

The quantity a is the probability amplitude for the electron to end up as a hole reflected back to N, while c is the amplitude for transmission into S as a qp. v_G is the group velocity in the qp branch in units of the Fermi velocity v_F in N. By including the qp current, we find that the total current in S is $C + 2A$. This is seen (from Eq. 1) to be equal to the current in N, $(1 + A)$. Thus, electric current is conserved at the interface. From the foregoing, both A and C vary strongly with $\xi_\mathbf{k}$ near $\pm \Delta$. See Blonder, Tinkham and Klapwijk[11] (BTK) and Griffin and Demers[12] (GD).]

Hoevers et al[13] have reported observing at 1.2 K Andreev reflection off the Ag-*123* interface, using a thin-film sample. We have extended their work using our Au-*123* junctions

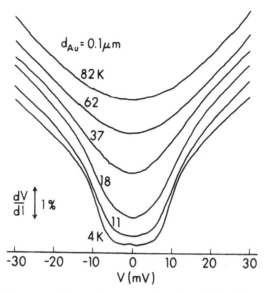

FIG 4 The I-V curves of the point contact junction in Fig. 3 at various T (Ref. 4).

to investigate the T dependence of the Andreev signal[4]. As in their work, we use a point contact (Au) to inject current into the junction. The insert in Fig. 3 shows the experimental geometry. By tuning the voltage bias V across the point contact, we change the energy of the injected electrons relative to the Fermi energy ε_F in the Au film (which is equal to μ_s in the *123*), and measure the junction differential resistance $R = dV/dI$. (In the figure, note that R is the quantity plotted instead of G.) We identify the abrupt decrease in R at $|V| \sim 10$ mV with the gap at the Au-*123* interface. This value is similar to that obtained by Hoevers et al.

At a finite bias, the electrons are injected "hot" into the Au film. They undergo multiple scattering (mostly elastic at low T) within the Au film before they reach the Au-*123* interface. If Andreev reflection occurs at the interface, the reflected hole retraces the path of the incident electron, undergoing the same scattering events in reversed order. However, if some of these scattering events are inelastic (phonon emission or absorption) the phase memory of the electron or hole is disrupted, and the reflected hole fails to return to the point of injection. Therefore, we expect the Andreev signal to be strongly suppressed as T is increased. Figure 4 shows this thermal smearing quite clearly. As T increases from 4.2 K, the gap feature broadens until it is barely resolved at 62 K. The fractional decrease of the junction resistance $-\Delta R(0)/R_{117}$ (normalized to the value at 117 K) is shown in Fig. 5. Even at 4.2 K, the value of $-\Delta R(0)/R_{117}$ is only 3 %. It is possible that this small signal may be due to residual inelastic events such as phonon emission or to the large anisotropy of the normal state conductivity in *123*. This determines the anisotropy of the quasi-particle conductivity. The two-dimensional dispersion may lead to a strong decrease of the Andreev signal because of incompatibility with the isotropic dispersion of gold[13].

Finally, we comment on the shape of the Andreev spectrum. From Fig. 3 and 4, we observe that the spectrum has a rather "flat" bottom, i.e. R (or G) is independent of V for $|V|$

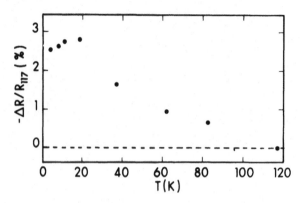

FIG 5 The temperature dependence of the fractional decrease in the zero-bias resistance referenced to the resistance at 117 K (Ref. 4).

$\leq \Delta$. In BTK's theory[11], this implies that the strength of the interface barrier (measured by the dimensionless parameter Z) is zero. If Z is non-zero, $G(0)$ is suppressed while its value at the gap, $G(\Delta)$, remains unaffected. The resulting spectrum has a conductance minimum at zero-bias and sharp conductance maxima at $V = \pm \Delta$. [This may be undertood as follows. A finite Z has two effects: Some fraction of the incident electrons are classically reflected back into N (with amplitude b), and some transmitted into the second qp branch (with amplitude d). Instead of Eq. 1, we generalize to the conservation relationship

$$1 = A + B + C + D, \qquad (3)$$

where $B = |b|^2$ is the reflected flux, and $D = |d|^2 v_G$ is the flux in the second qp branch. Let us consider what happens when a transmitted electron has just sufficient energy to enter the qp branch at its lowest energy ($V = \pm \Delta$). At the bottom of the qp branch the group velocity v_G vanishes. Hence, at this energy both qp contributions, C and D, vanish in Eq. 3, regardless of the magnitude of Z. At precisely this energy, the incident electrons either enter the Cooper Sea, or are classically reflected back into N. Moreover, at the bottom of the qp branch, the Bogolyubov amplitudes u_k, v_k are equal, i.e. the qp wave function in S is a combination of *equal* amounts of electron and hole plane waves. The boundary conditions can be satisfied only if, on the N side, the incident electron wave has exactly the same amplitude as the reflected hole wave, or $a = 1$ and $b = 0$. This argument shows that $A = 1$ when $V = \pm \Delta$, regardless of Z's magnitude[11]. At zero-bias, in contrast, a finite Z implies that some fraction of the electrons are classically reflected back into N ($B \neq 0$). To conserve probability flux, A therefore must be less than 1.]

The absence of minima in R at $\pm \Delta$ in Figs. 3 and 4 suggests that the interface at which the Andreev reflection occurs has a vanishing (classical) barrier, i.e. the barrier strength Z at

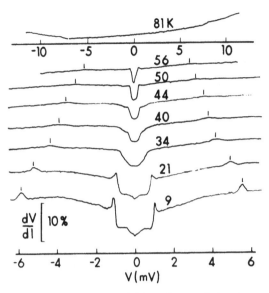

FIG 6 I-V characteristics of a Au-123 junction made by rf sputtering the *a-b* face of a 123 crystal. Gaps associated with the Andreev reflection are much smaller than in the cleaved crystal junctions (Ref. 4).

the Au-*123* interface is close to zero. The Andreev results were actually obtained before the contact impedance work described in Sec. 2. Thus, direct measurement of the contact impedance confirms this aspect of the Andreev spectrum. Electrons are transmitted through the Au-*123* interface with negligible scattering.

The data in Figs. 3-5 are for a junction in which the thickness of the Au film d_N equals 1,000 Å. We have also performed similar studies for two junctions with d_N = 3,000 Å. The spectra and the gap values obtained are similar to that in the 1000 Å sample (~10 mV). However, the magnitude of $\Delta G(0)/G_N$ is reduced from ~3 % to 1 %. Although this part of the research is preliminary, the similarity suggests that the Au film in our junction is not being made superconducting by proximity[14-16] with the *123* crystal. In a proximity junction comprised of an *N-S* sandwich, the rapid tunneling of electrons between *N* and *S* enables the electrons in *N* to benefit from the pairing potential in *S*. As a result a gap $\Delta_N < \Delta$ exists in *N*. The gap Δ_N decays exponentially into *N* with a decay length ξ_N. For $d_N < \xi_N$ (thin-film limit) Δ_N is approximately uniform in *N* [14]. In an Andreev reflection measurement the spectrum measured is dominated by Δ_N. On the other hand, for $d_N \gg \xi_N$, Andreev reflection occurs mostly off Δ in *S*. Only near zero-bias does reflection occur from the small Δ_N in *N*. This produces a small conductance peak near zero-bias[17]. Our failure to observe this characteristic change in the spectrum as d_N is changed by a factor of 3 argues against any proximity effect in these junctions. However, more study is required to investigate the possibility of proximity effect in Au.

Aside from cleaving crystals in vacuum we have also experimented with preparing

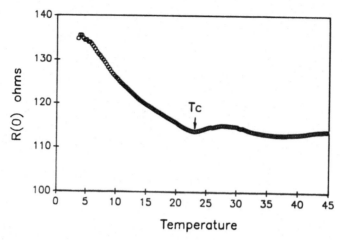

FIG 7 The temperature dependence of the zero-bias resistance in a Pb-*NCCO* junction, showing the sharp upturn at T_c. (Ref. 5).

junctions by *rf* sputtering the *a-b* face with Ar ions (at an rf power of 40 watts). Au is then evaporated onto the *a-b* face. (It is much harder to expose the *a-b* face by cleaving.) The value of $\rho_{[]}$ is much higher in such junctions (typically 1-2 x 10^{-3} Ωcm^2). The large junction impedance allows the *I-V* curves to be measured without using a point contact. Figure 6 shows the *I-V* curves at various temperatures. In contrast to Figs. 3 and 4, the values of the gap are much smaller (typically 1 mV at low *T*, and vanishing near 60 K). The small gap value suggests that rf sputtering leads to significant lattice disorder and possibly serious oxygen loss near the interface.

4. Tunneling spectrum in $Nd_{2-x}Ce_xCuO_{4-y}$

Recently, single crystals of the "*T'* phase" superconductor $Nd_{2-x}Ce_xCuO_{4-y}$ have been grown at Bellcore[18]. Much interest has been aroused by initial reports that the carriers may be electrons rather than holes[1,2]. In our crystals, the Hall effect shows that the carriers appear to be hole-like in the superconducting phase[19]. However, this interesting question remains open. Here, we wish to focus on the tunneling results. To obtain the superconducting phase, it is necessary to subject the crystals to a rigorous reducing procedure with rather stringent temperature requirements. It is our experience that many crystals show a sharp transition in the ac susceptibility, but no trace of superconductivity in resistance (or at best a very broad transition). In a particular batch which we call JW2 most of the crystals showed sharp resistive transitions with T_c as high as 22 K. Zero-resistance is attained between 15 and 20 K, depending on sample. In these crystals, we have carried out

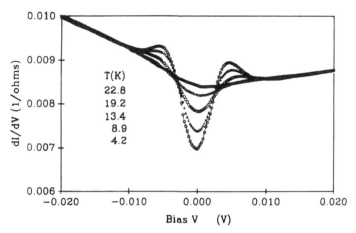

FIG 8 The tunneling spectra of a Pb-*NCCO* junction at various temperatures below T_c (unnormalized). The gap inferred appears to be independent of T. Note the asymmetry of the magnitude of G between positive and negative bias. Positive V means that the *NCCO* side has a higher potential (Ref. 5).

tunneling studies, using Pb as the counter-electrode[5]. The crystals are cleaved at ambient pressure and then placed in an evaporator where Pb is deposited onto the cleaved face to a thickness of 2,000 Å. In three crystals we observe clear evidence of tunneling. In Fig. 7, the zero-bias resistance shows a sharp upturn at T_c, increasing roughly linearly with the reduced temperature $(T_c - T)/T_c$. This provides evidence that the decrease in the tunneling current is associated with changes in the electronic spectrum starting at T_c. Figure 8 shows the dependence of the conductance dI/dV on the bias V at a number of temperatures. Clearly, states that are removed from below the gap Δ, approximately 2.5 mV in size, are transferred to above the gap.

The present results provide convincing evidence for tunneling into the *NCCO* crystals. However, there are a number of anomalous features which we do not understand at present. First, no trace of the Pb gap is observed. In conventional S-I-S' junctions, structures are observed at $\Delta \pm \Delta'$ where Δ (Δ') is the gap in S (S'). The complete absence of such gap structures is puzzling in Fig. 8. In all 3 crystals investigated, we failed to observe the Pb gap (For bulk Pb, $2\Delta \approx 2.7$ mV). We speculate that very strong depairing processes such as magnetic impurity scattering may be strongly suppressing the Pb gap. Although we do not have evidence to support this conjecture, we have found in Pb-*123* junctions made in the same way the Pb gap is clearly observed, but significantly broadened by some as-yet unknown pair-breaking process. No trace of the *123* gap is observed, although we observe the broad bumps and shoulders often cited as evidence for the *123* gap.

The second anomalous feature is the absence of any T dependence in the inferred gap value as T_c (~22 K) is approached from below. Because the gap is strongly smeared, its value is difficult to estimate without specific models. However, if we take the energy at which $G(V)$ intersects the conductance curve $G_N(V)$ in the normal state as a rough measure of the gap, we infer that Δ shows very little dependence on T even close to 22 K. (However, the minimum in G at zero bias rapidly approaches G_N, as is clear from Fig. 8.) The insensitivity of Δ to $(T_c - T)$ may reflect a fundamental difference between the high T_c superconductivity and conventional BCS behavior. The slight asymmetry in Fig. 8 is also observed in all crystals. When the crystal is at a higher potential than the Pb electrode the conductance G is lower, i.e. *it is harder to inject electrons into the high T_c oxide than to withdraw electrons*. It would be interesting to investigate this point further in other systems.

The field dependence of the tunneling spectra up to 1 Tesla has also been studied with **H** parallel to the CuO_2 planes (i.e. normal to the junction interface). In this geometry, the field has a weak effect on the superconductivity since H_{c2} is estimated to exceed 40 T. Nevertheless, we observe a significant effect of the field on the states above the gap. In one crystal (Sample 3), the conductance above the gap shows a prominent feature which resembles a mesa. This feature, which exists between 5 and 10 mV, is suppressed in relatively weak fields (~0.1 T). The mesa feature is also apparent in some of the tunneling spectra recently obtained by Gurvitch et al[20] on *123*. In their work the mesa remains unaffected by fields in excess of 10 T. In our Sample 1 it is completely absent, whereas Sample 2 shows a smeared feature at the same energy. At present, we do not know if it is due to the tunneling of electrons along the **c** axis, or associated with the Pb superconductivity (H_{c2} for bulk Pb is ~0.08 T).

We have benefitted from discussions with P.W. Anderson and R.C. Dynes. The research at Princeton is supported by the Office of Naval Research (Contract No. N00014-88-K-0283).

References

1. T. Tokura, H. Takagi, and S. Uchida, Nature **337**, 345 (1989).
2. H. Takagi, S. Uchida, and Y. Tokura, Phys. Rev. Lett. **62**, 1197 (1989).
3. T.W. Jing, Z.Z. Wang and N.P. Ong, Appl. Phys. Lett., Nov. 1st 1989
4. T.W. Jing, N.P. Ong, Z.Z. Wang, and P.W. Anderson, unpublished.
5. T.W. Jing, N.P. Ong, Z.Z. Wang, J.M. Tarascon and E. Wang, unpublished.
6. J.W. Ekin, T.M. Larson, N.F. Bergren, A.J. Nelson, A.B. Swartzlander, L.L. Kazmerski, A.J. Panson, and B.A. Blankenship, Appl. Phys. Lett. **52**, 1819 (1988).
7. S.W. Tozer, A.W. Kleinsasser, T. Penney, D. Kaiser, and F. Holtzberg, Phys. Rev. Lett. **59**, 1768 (1987).
8. B.F. Logan, S.O. Rice, and R.F. Wick, J. Appl. Phys. **42**, 2975 (1971).
9. P.W. Anderson, private communication.
10. A.F. Andreev, Zh. Eksp. Teor. Fiz. **46**, 1823 (1964) [Sov. Phys. JETP **19**, 1228 (1964)].
11. G.E. Blonder, M. Tinkham and T.M. Klapwijk, Phys. Rev. B **25**, 4515 (1982).

12. A. Griffin and J. Demers, Phys. Rev. B 4, 2202 (1971). This work does not obtain the Andreev enhancement in G because of a different treatment of the qp flux from Ref. 11.

13. W.A. Little, private communication.

14. W.L. McMillan, Phys. Rev. **175**, 537 (1968).

15. J. Vrba and S.B. Woods, Phys. Rev. B **3**, 2243 (1971); C.J. Adkins and B.W. Kington, Phys. Rev. **177**, 777 (1969).

16. For a review see E.L. Wolf and G.B. Arnold, Phys. Rept. **91**, 33 (1982).

17. P.C. van Son, H. van Kempen, and P. Wyder, Phys. Rev. Lett. **59**, 2226 (1987).

18. J.M. Tarascon et al, Phys. Rev. B **40**, 4494 (1989).

19. Z.Z. Wang, D.A. Brawner, T.R. Chien, N.P. Ong, J.M. Tarascon and E. Wang, unpublished.

20. M. Gurvitch et al, AT&T preprint 1989, and Proceedings of this conference.

POSITRON ANNIHILATION STUDIES OF HIGH TEMPERATURE

SUPERCONDUCTORS

C.S.Sundar, A.Bharathi, Liyang Hao, and Y.C.Jean

University of Missouri-Kansas City, Kansas City
Missouri 64110

P.H.Hor, R.L.Meng, Z.J.Huang and C.W.Chu

Texas Centre for Superconductivity,
University of Houston, Houston, Texas 77204-5506

INTRODUCTION

The discovery[1] of high temperature superconductivity in Cu-O based perovskite materials has initiated a large effort in the study of the various physical properties of these materials with a view to obtain a basic understanding of the mechanism of superconductivity in these systems[2]. In this paper we provide an overview of the positron annihilation studies on high temperature superconductors (HTS). After a brief introduction to the experimental method, the results of positron annihilation studies on high temperature superconductors are presented in three major parts: (1) Studies on the temperature dependence of the annihilation characteristics across T_c, (2) Investigation of the Fermi surface in these materials, and (3) Studies on the structure and defect properties.

POSITRON ANNIHILATION SPECTROSCOPY

Detailed reviews on positron annihilation spectroscopy (PAS) and its applications to the study of electronic structure and defect properties of solids can be found in Ref 3. The measured annihilation characteristics are the lifetime of the positron, the angular correlation of the annihilation photons and the Doppler broadened energy spectra of the annihilation radiation. The lifetime of the positron in the medium is determined by the electron density at the site of the positron. The angular correlation of the annihilation photons provides information on the electron momentum distribution in a solid. The electron momentum distribution also results in the Doppler broadening of the annihilation radiation.

In a perfect crystalline material, the positron exists in a delocalised Bloch state, whereas in the presence of vacancy type defects, the positrons are localised at these defects. The annihilation characteristics of the positron trapped at defects are different from those in the Bloch state and this can be used to characterise the defects. The different modes of annihilation of a positron in a crystal appear as different components in the lifetime spectrum. From an analysis of the lifetime spectra, the various lifetime components and their relative intensities can be extracted. In addition one can obtain the average lifetime of the positron in the medium and the bulk lifetime corresponding to the annihilation from the Bloch state. The Doppler broadened energy spectra are in general analysed in terms of a lineshape parameter S, defined as a ratio of the counts in the central low momentum region to the total counts in the entire photopeak of the 511 keV annihilation radiation.

TEMPERATURE DEPENDENCE OF ANNIHILATION CHARACTERISTICS ACROSS THE SUPERCONDUCTING TRANSITION

The annihilation characteristics of a positron in a medium are dependent on the local electronic properties at the site of the positron. There has been a large number of PAS studies to investigate changes in the electronic properties across the superconducting transition. In this section, we present the results in a variety of HTS materials, and discuss the results in Y-Ba-Cu-O system in some detail.

In the conventional superconductors, no change in the annihilation characteristics is observed below T_c. As an example, Fig.1 shows the results of lifetime and S parameter measurements in polcrystalline Pb[4]. Similar results have also been obtained in Nb_3Sn and V_3Si[4]. The absence of any temperature dependence of annihilation characteristics at or below T_c is

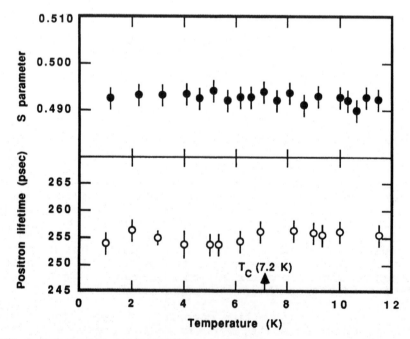

Fig.1 Variation of positron lifetime and S parameter as a function of temperature in polycrystalline Pb.

rationalised as arising due to the fact that the superconducting pairing affects only a small portion of electrons near the Fermi surface. The smearing of the Fermi surface in these BCS superconductors is too small to be detected by angular correlation measurements.

With the advent of the HTS, the first measurement of the temperature dependence of annihilation characteristics across T_c was carried out in $Y_1Ba_2Cu_3O_{7-x}$ (YBCO) by Jean et al[5]. In the ceramic superconducting compound, the lifetime component having a value of 220ps in the normal state was observed to decrease at T_c, whereas no such change was seen in the non-superconducting tetragonal phase. Figure 2 shows the results of a recent measurement of the temperature dependence of lifetime in a well characterised high quality single crystal sample of YBCO having a T_c of 92K.

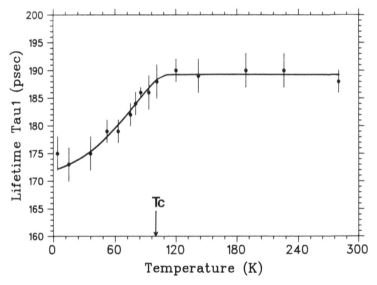

Fig.2 Variation of positron lifetime as a function of temperature in a single crystal of $Y_1Ba_2Cu_3O_7$.

It is clearly seen that the lifetime decreases below T_c. There have been several experiments in YBCO by various groups[4-12]. While most studies indicated a decrease in lifetime and lineshape parameter below T_c, there were a few studies that indicated other kind of temperature dependencies. For example, Harshman et al[9] reported an increase in lifetime below T_c in a single crystal sample of YBCO. Corbel et al[12] observed a variety of temperature dependencies in different samples. This naturally raises the question as to what are the various factors that influence the temperature dependence of annihilation characteristics and what is the efficacy of PAS for the study of high temperature superconductivity?

In order to obtain a better understanding of the temperature dependence of annihilation characteristics in YBCO, detailed positron annihilation experiments have been carried out[13] in YBCO doped with Zn and Ga. It is well known that these dopants[14], in particular Zn, causes a large suppression of T_c. The results of positron lifetime measurements in the Zn doped YBCO are shown in Fig.3. The main features of the results are (1) a large decrease in the bulk lifetime in the normal state with increasing Zn concentration, (2) a dramatic change in the nature of temperature dependence from a decrease in bulk lifetime below T_c in undoped YBCO to an increase in bulk lifetime in the Zn doped sample. In all cases the changes in lifetime with temperature are well correlated with T_c.

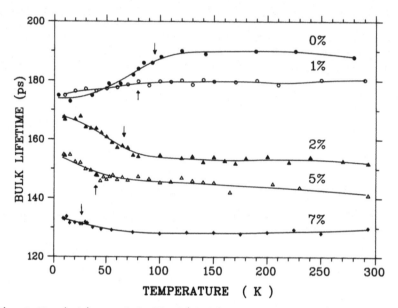

Fig.3 Variation of bulk lifetime as a function of temperature in undoped and Zn doped YBCO. Arrows are marked at T_c.

In order to understand the positron annihilation characteristics, it is necessary to have a knowledge of the positron density distribution. Calculations[11,15,16] of the positron distribution in orthorhombic $Y_1Ba_2Cu_3O_7$ indicates that the positron density is mainly distributed in between the Cu-O chains in the basal plane. Fig.4 shows the positron distribution in the basal plane of YBCO. Calculation[13] of the positron distribution in the case of Zn doped YBCO indicates, that with Zn substitution at Cu(2) site and in particular Y site, the weight of the positron distribution shifts from the Cu-O chains to the CuO_2 planes (see Fig.5).This suggests that the different temperature dependence of annihilation parameters in undoped and Zn doped YBCO (cf.Fig 3) are related to the different positron distributions within the unit cell of YBCO, in particular with respect to the superconducting CuO_2 layers. With the knowledge of the positron distribution in Zn-doped and undoped YBCO, a consistent model to explain the observed temperature dependence of annihilation parameters below T_c is to invoke that there is a local transfer of electron density between the CuO_2 planes and the Cu-O chains below T_c. In the

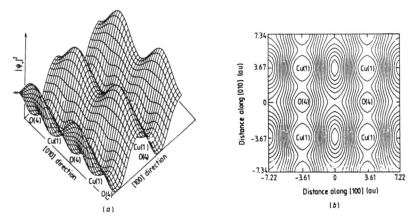

Fig.4 (a) Positron density distribution in the basal plane of $Y_1Ba_2Cu_3O_7$. (b) contour plot of the positron density distribution.

undoped YBCO, wherein the positrons are located in the chains, the transfer of electrons from the planes to the chains leads to a decrease in lifetime below T_c. In the case of Zn-doped YBCO, such a transfer of electrons from the planes to the chains leads to an increase in lifetime since the positrons are located predominantly in the Cu-O planes. The above mentioned picture is also consistent with the results in Ga doped YBCO[13], wherein experimentally a decrease in lifetime is observed below T_c, and the calculations indicate that the positron are distributed in between the chains as in undoped YBCO. The above mentioned positron annihilation studies on undoped and controlled doped YBCO lllustrates the importance of positron distribution with respect to the CuO_2 planes in determining the temperature dependence of annihilation characteristics below T_c. In addition to the explanation based on a local charge transfer, the temperature dependence of annihilation parameters below T_c has also been discussed in terms of (1) local changes in structure such as the motion of O atoms and (2) changes in the electron-positron correlation arising due to the pairing of holes in the CuO_2 planes[17].

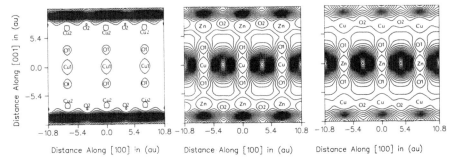

Fig.5 Contour plot of the positron density distribution in the (010) plane in Zn doped YBCO. The maximum in the positron density is found to be at the centre of the Cu-O chains for Zn replacing Cu(1) (right), Cu(2) (Centre), and in the space between the CuO_2 planes for Zn replacing Y site (left).

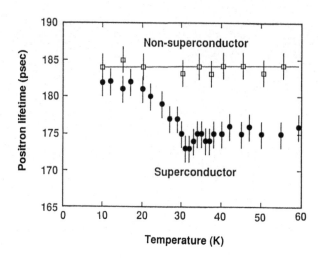

Fig.6 Positron lifetime Vs temperature in La-Sr-Cu-O.

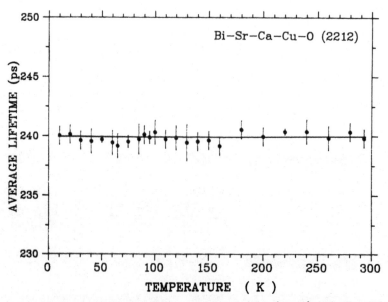

Fig.7 Positron lifetime Vs temperature in Bi-Sr-Ca-Cu-O.

Positron Studies on New HTS Materials

The temperature dependence of lifetime in the La-Sr-Cu-O system[7] is shown in Fig 6. In the non-superconducting sample, there is no change in lifetime, while the superconducting sample shows an increase in lifetime correlated with T_c. The calculation of positron distribution in this system indicates that the positrons are mainly distributed in the CuO_2 plane.

It is seen from Fig.7, that in the Bi-Sr-Ca-Cu-O (2212) superconducting compound, no change in lifetime is seen across T_c. A calculation of the positron distribution in this system indicates that the positron density is mainly distributed in the region between the Bi-O planes (see Fig.8). Similar positron distribution has been reported by Singh et al [18]. In view of the complete absence of positron density at the superconducting CuO_2 plane, it is not surprising that no changes in annihilation characteristics are seen at T_c.

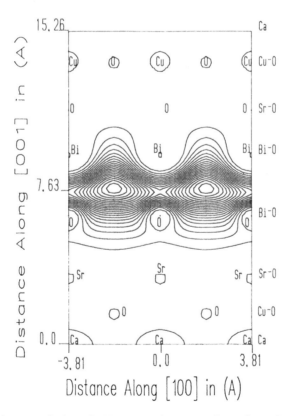

Fig.8 Contour plot of the positron distribution in the (010) plane of Bi-Sr-Ca-Cu-O.

The superconducting system Tl-Ba-Ca-Cu-O forms a heirarchy of structures[19] with differing number of CuO_2 layers having different transition temperatures. The variation of positron lifetime as a function of temperature in the Tl(2201), Tl(2212) and Tl(2223) systems are compared in Fig.9. The Tl (2223) system shows a dramatic increase[20] in lifetime at T_c (124K). In the Tl (2212), a decrease in lifetime is seen below T_c

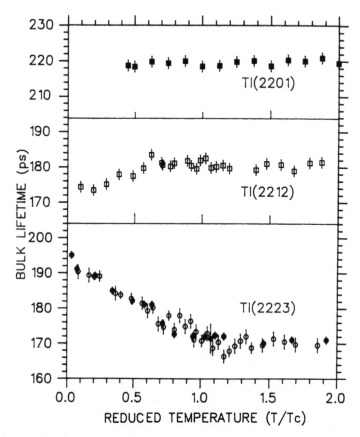

Fig.9. Positron lifetime plotted as function of reduced temperature in Tl-Ba-Ca-Cu-O superconductors containing one (Tl2201), two (Tl2212) and three (Tl2223) CuO_2 layers.

(110K), wheras in Tl(2201) no change in lifetime is seen below T_c (20K). These contrasting temperature dependencies may be related to the differences in the positron distribution with respect to the CuO_2 planes. Calculation[18] of the positron distribution in Tl(2212) system indicates that the positrons probe the CuO_2 layers in contrast to the Bi(2212) system.

The discovery[21] of superconductivity in the $Ba_{.6}K_{.4}BiO_3$ has generated considerable excitement in view of the contrasting properties when compared to the cuprate superconductors. The bulk of the experimental evidences indicate that this system is a phononic superconductor[22]. The results[23] of positron lifetime measurements in this system are shown in Fig 10. It is seen that there is a large decrease in lifetime in the normal state with no discernable change below T_c (25K). The decrease in lifetime in the normal state, which is much larger than that can be accounted for by thermal contraction, may be an indication of (1) an increase in the concentration in the O^- species and/or (2) phonon softening.

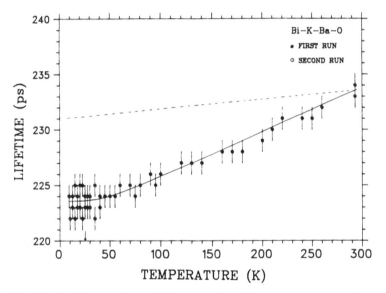

Fig.10 Variation of positron lifetime as a function of temperature in $Ba_{.6}K_{.4}BiO_3$. The dashed line indicates the decrease in lifetime expected on the basis of volume contraction at low temperatures.

The results of positron lifetime measurements as a function of temperature in the electron doped superconductor Nd-Ce-Cu-O is shown in Fig.11 along with the results in the corresponding non-superconducting compound. It is well known[24] that in these systems, the oxygen vacancies play an important role in the occurrance of superconductivity. It is seen from Fig.10 that in the superconducting compound, dominant changes in lifetime occur in the range of 100K whereas no such change is seen in the non-superconducting compound. These results may be related to the detrapping of positrons from the oxygen vacancies which act as shallow traps.

FERMI SURFACE MEASUREMENTS

Positron annihilation spectroscopy, in particular the two dimensional angular correlation (2D-ACAR) technique, has been sucessfully used to observe the Fermi surface in many conventional superconducting elements and there is significant agreement with detailed band structure calculations. One of the important questions in the field of high temperature superconductivity is, "Is there a Fermi surface in these materials ? and if so what is the magnitude of the discontinuity at E_f? In an attempt to obtain an answer to this question, there have been several 2D-ACAR measurements in both the La-Sr-Cu-O and YBCO systems. Unfortunately, there is no unequivocal answer to date, despite the fact measurements are done on good quality single crystals.

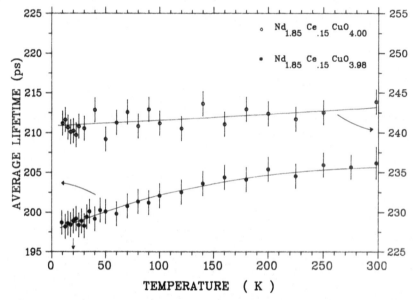

Fig. 11 Positron lifetime Vs temperature in superconducting and non-superconducting Nd-Ce-Cu-O.

In YBCO, measurements have been done by several groups. Hoffman et al[25] report the presence of a Fermi surface and their results are found to be good agreement with the band structure calculations. Smedskjaer et al[26] have measured the 2D-ACAR in the ab plane of YBCO and have compared their results with detailed band structure calculations of Bansil et al[27]. Recently, Haghighi et al[28] have carried out measurements with higher statistics, and they do not find a unique signal corresponding to the presence of the Fermi surface. Fig 12 shows their results. These authors have stressed the importance of the wave function effects, and have cautioned against attributing all the observed structures in the momentum distribution, to Fermi surface features.

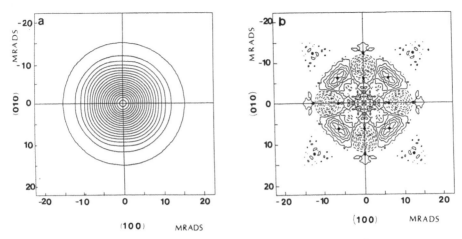

Fig.12 (a) Contour plot of the 2D electron-positron momentum density distribution in $Y_1Ba_2Cu_3O_{7-x}$ after correcting for 12% background contribution from sample mount, correcting for momentum sampling function of the spectrometer and folding with he symmetry operation of the projected reciprocal lattice of the effective tetragonal structure of the twinned specimen. (b) Contour plot of the anisitropy of the spectrum in (a).

STUDIES ON THE STRUCTURAL AND DEFECT PROPERTIES

One of the fruitful applications of PAS in metals and alloys has been in the study of defects such as vacancies, dislocations, voids etc. In the HTS materials, it is well known that there exist a variety of defects and these defects in particular, the oxygen vacancies play an important role such as doping. The oxygen vacancies in YBCO has been investigated by Bharathi et al[29] and Smedskjaer et al[30]. Fig 13 shows the variation of positron annihilation parameters as a function of

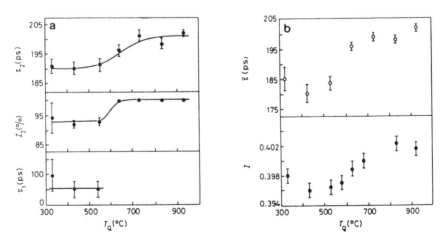

Fig.13 (a) Variation of positron lifetime parameters as a function of quench tempratue (T_Q) in YBCO. (b) Variation of meanliftime and S parameter as a function of T_Q. The changes in annihilation parameters are correlated with the orthorhombic to tetragonal transformation.

quench temperature, which varies the oxygen stoichiometry. The change in the lifetime and lineshape parameter in the temperature range of 500°C correlates with the orthorhombic to tetragonal structural transition. The variation in annihilation parammeters with quench temperature has been qualitatively understood as an indication of the change in the nature (charge state) of the oxygen vacancies. Smedsjkaer et al have analysed their results in terms of the trapping model and have concluded that oxygen vacancies act as weak trapping centres with a very small trapping rate.

It is seen from Fig 13 that the lifetime and lineshape parameters in the tetragonal phase is larger than in the orthorhombic phase. This is substantiated by the calculation[15] of annihilation characteristics in the orthohombic and tetragonal phases of YBCO. The difference in the annihilation characteristics in the orthorhombic and tetragonal phases has been used in the study[31] of phase separation in off-stoichiometric YBCO subjected to heat treatment. Fig 14 shows that the average lifetime increase as a function of ageing time at 200°C, while the overall oxygen stoichiometry is observed to be constant. The increase in lifetime from a value characteristic of orthorhombic phase to a value characteristic of the tetragonal phase is taken as an evidence of the presence of the tetragonal phase due to phase separation.

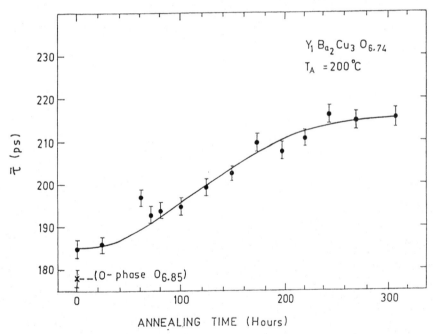

Fig.14 Variation of lifetime as a function of annealing time at 200°C.

SUMMARY AND CONCLUSIONS

In the present paper, the result of positron studies on the electronic and defect properties of the high temperature superconductors were discussed. Extensive studies on the temperature dependence of annihilation characteristics indicate that the variation in annihilation parameters is correlated with T_c. The different kinds of temperature dependences can be understood if we take into account the distribution of positrons with respect to the superconducting CuO_2 planes. The changes in annihilation characteristics with temperature points to a local change in the electron distribution below T_c. With regard to obtaining a definitive answer to the question of the existance of Fermi surface, further experiments on high quality samples with higher statistics are required. The application of PAS to the detailed study of electronic structure and defect properties of the HTS systems appears very promising.

ACKNOWLEDGMENTS

We would like to acknowledge our fruitful collaboration and and discussions with Prof. W.Y.Ching, Prof. Z.Z.Wang, Drs.R.H.Howell, P.E.A.Turchi, A.L.Wachs, M.J.Fluss, Prof. D.M.Schrader, Prof. P.K.Tseng, Prof.J.Kaiser, Prof.R.N.West, Dr.H.Nakanishi and Prof.S.J.Wang. The research has been supported by the Weldon Spring Endowment Funds, University of Missouri,, Prime Grant MDA 972-88-G002 to the University of Houston from DARPA and a subcontract to the University of Missouri-Kansas City from the Lawrence Livermore National Laboratory.

REFERENCES

1. J.G.Bednorz and K.A.Muller, Possible high Tc in Ba-La-Cu-O system, Z.Phys. B 64, 189 (1986)

2. J Muller and J.L.Olsen, eds.," High Temperature Superconductors and Materials and Mechanisms of Superconductivity," North-Holland, Amsterdam (1988)

3. W.Brandt and A.Dupasquier, eds., " Positron Solid State Physics," North- Holland, Amsterdam (1984): D.M.Schrader and Y.C.Jean, eds., " Positron and Positronium Chemistry, Elsevier, Amsterdam (1988)

4. Y.C.Jean, H.Nakanishi, S.J.Wang, J.Kyle, P.E.A.Turchi, R.H.Howell, A.L.Wachs, M.J.Fluss, R.L.Meng, P.H.Hor, Z.J.Huang and C.W.Chu, A systematic study of high temperature superconductivity by positron annihilation spectroscopy, " Proc. 8th Intl. Conf. Positron annihilation", L.Dorikens-Vanpraet, M.Dorikens and D.Segers ed., World Scientific Publishers, Singapore (1989)

5. Y.C.Jean, S.J.Wang, H.Nakanishi, W.N.Hardy, M.Y.Hayden, R.F.Kiefl, R.L.Meng, P.H.Hor, Z.J.Huang and C.W.Chu, Positron Annihilation in the High Temperature Superconductor $YBa_2Cu_3O_7$, Phys. Rev. B 36, 3994 (1987)

6. S.G.Usmar, P.Sferlazzo, K.G.Lynn and A.R.Moodenbaugh, Temperature dependence of positron-annihilation parameters in $Y_1Ba_2Cu_3O_7$ above and below the superconducting transition, Phys. Rev. B36, 8854 (1987)

7. Y.C.Jean, J.Kyle, H.Nakanishi, P.E.A.Turchi, R.H.Howell, A.L.Wachs, M.J.Fluss, R.L.Meng, P.H.Hor, Z.J.Huang and C.W.Chu, Evidence for a common high-temperature superconducting effect in $La_{1.85}Sr_{0.15}CuO_4$, Phys. Rev. Lett. 60, 1069 (1988)

8. C.S.Sundar, A.K.Sood, A.Bharathi and Y.Hariharan, Positron annihilation measurement across the superconducting transition in $Y_1Ba_2Cu_3O_7$, Physica C 153-155, 155 (1988)

9. D.R.Harshman, L.F.Schneemeyer, Y.V.Waszzak, Y.C.Jean, M.J.Fluss, R.H.Howell and A.L.Wachs, Temperature dependence of the positron lifetime in single-crystal $YBa_2Cu_3O_7$, Phys. Rev. B 38, 848 (1988)

10. S.G.Usmar, K.G.Lynn, A.R.Moodenbaugh, M.Suenga and R.L.Sabatini, The influence of twinning in $YBa_2Cu_3O_7$ on the positron annihilation parameters, Phys. Rev. B 38, 5126 (1988)

11. E.C.von Stetten, S.Berko, S.S.Li, R.R.Lee, J.Brynestad, D.Singh, H.Krakauer, W.E.Pickett and R.E.Cohen, High sensitivity of positrons to oxygen vacancies and to copper-oxygen chain disorder in $Y_1Ba_2Cu_3O_7$, Phys. Rev. Lett. 60, 2198 (1988)

12. C.Corbel, P.Bernede, H.Pascard, F.Rullier-Alberque, R.Korman and J.F.Marucco Appl. Phys. A 48, 335 (1989)

13. Y.C.Jean, C.S.Sundar, A.Bharathi, J.Kyle, H.Nakanishi, P.K.Tseng, P.H.Hor, R.L.Meng, Z.J.Huang, C.W.Chu, Z.Z.Wang, P.E.A.Turchi, R.H.Howell, A.L.Wachs and M.J.Fluss, Local charge density change and superconductivity: A positron study, Phys.Rev.Lett.(26 March 1990)

14. G.Xiao, M.Z.Cieplak, A.Garvin, F.H.Streitz, A.Bakshshai, and C.L.Chien, High temperature superconductivity in tetragonal perovskite structures: is oxygen vacancy order important?, Phys. Rev. Lett. 60, 1446 (1988)

15. A.Bharathi, C.S.Sundar and Y.Hariharan, A study of positron distribution and annihilation characteristics in $Y_1Ba_2Cu_3O_{7-x}$, J.Phys. Condens. Matter 1, 1467 (1989)

16. K.Jensen, R.Nieminen, M.Puska, Positron states in $YBa_2Cu_3O_{7-x}$, "Proc. 8th Intl. Conf Positron Annihilation", L.Dorikens-Vanpraet, M.Dorikens and D.Segers ed., World Scientific, Singapore (1989)

17. B.Chakraborty, A novel probe of pairing in high temperature superconductors, Phys. Rev B 39, 215 (1989)

18. D.Singh, W.E.Pickett, R.E.Cohen, Henry Krakauer and Stephan Berko, Positron annihilation in high T_c superconductors, Phys.Rev.B 39, 9667 (1989)

19. R.Beyers, S.S.P.Parkin, V.Y.Lee, A.I.Nazzal, R.J.Savoy, G.L.Gorman, T.C.Huang and S,J. La Placa, Tl-Ca-Ba-Cu-O Superconducting oxides, IBM Jr Res. Develop. 33, 228 (1989)

20. Y.C.Jean, H.Nakanishi, M.J.Fluss, A.L.Wachs, P.E.A.Turchi, R.H.Howell, Z.Z Wang R.L.Meng, P.H.Hor, Z.J.Huang and C.W.Chu, A comparison of the temperature dependence of the electron-positron momentum density characteristics in Tl(2223), Y(123) and La(214) superconductors, J.Phys: Condens Mater Phys. $\underline{1}$, 2696 (1989)

21. R.J.Cava, B.Batlogg, J.J.Krajewski, R.Farrow, L.W.Rupp Jr, A.E.White, K.Short, W.F.Peck, and T.Kometani, Superconductivity near 30 K without copper: the $Ba_{0.6}K_{0.4}BiO_3$ perovskite, Nature $\underline{332}$, 814 (1988)

22. C.K.Loong, P.Vasishta, R.K.Kalia, M.H.Degani, D.L.Price, J.D.Jorgensen, D.G.Hinks, B.Dabrowski, A.W.Mitchell, D.R.Richards, and Y.Sheng, High energy phonon modes and superconductivity in $Ba_{1-x}K_xBiO_3$: an inelastic-neutron scattering experiment and molecular dynamics simulation, Phys.Rev.Lett. $\underline{62}$, 2628 (1989)

23. C.S.Sundar, A.Bharathi, Y.C.Jean, D.G.Hinks, B.Dabrowski, Y.Zheng, A.W.Mitchell, J.C.Ho, R.H.Howell, A.L.Wachs, P.E.A.Turchi, M.J.Fluss, R.L.Meng, P.H.Hor, Z.J.Huang, and C.W.Chu, The electronic properties of high T_c superconductors probed by positron annihilation, Physica C $\underline{162-164}$, 1379 (1989)

24. H.Takagi, S.Uchida, and Y.Tokura, Superconductivity produced by electron doping in CuO_2 -layered compounds, Phys.Rev.Lett. $\underline{62}$, 1197 (1989)

25. L.Hoffman, A.A.Manuel, M.Peter, E.Walker and A.Damento, Electronic structure of $YBa_2Cu_3O_7$ by positron annihilation, Europhys. Lett. $\underline{6}$, 61 (1988)

26. L.C.Smedskjaer, J.Z.Liu, R.Benedek, D.G.Legini, D.J.Lam, M.D.Stahulek and H.Claus, Fermi surface by 2D-ACAR, Physica C $\underline{156}$, 269 (1988)

27. A.Bansil, R.Pankaluoto, R.S.Rao, P.E.Mijnarends, W.Dlugosz, R.Prasad and L.C.Smedskjaer, Fermi surface, ground state electronic structure and positron experiments in $YBa_2Cu_3O_7$, Phys. Rev. Lett. $\underline{61}$, 2480 (1988)

28. H.Haghighi, J.H.Kaiser, S.Rayner, R.N.West, M.J.Fluss, R.H.Howell, P.E.A.Turchi, A.L.Wachs, Y.C.Jean and Z.Z.Wang J.Phys. Conden. Matter (submitted)

29. A.Bharathi, Y.Hariharan, A.K.Sood, V.Sankara Sastry, M.P.Janawadkar and C.S.Sundar, Positron annihilation study of oxygen vacancies in $Y_1Ba_2Cu_3O_{7-x}$, Europhys. lett., $\underline{6}$, 369 (1988)

30. L.C.Smedskjaer, B.W.Veal, D.G.Legini, A.P.Paulikas and L.J.Nowicki, Positron Trapping in the superconductor $YBa_2Cu_3O_7$, Physica B+C $\underline{150}$, 56 (1988)

31. D.Vasumathi, C.S.Sundar, A,Bharathi, Y.Hariharan and A.K.Sood, A positron annihilation study of decomposition of $Y_1Ba_2Cu_3O_{7-x}$, Physica C (in print)

WEAK LINKS AND PLANAR DEFECTS IN SUPERCONDUCTING CUPRATES

J. Halbritter
Kernforschungszentrum, Postfach 3640
75 Karlsruhe, WEST GERMANY

Applied Physics, Stanford University
Stanford, California 94305

Abstract

In the new cuprate superconductors weak links are the reasons for deterioration of the -super- conducting properties. These weak links are classified according to their conductivity indicating that insulating cuprate surface layers are the origin of weak links.

1. INTRODUCTION

"Defects define properties of materials", this statement is especially true for the superconducting cuprates, especially for the conduction, because the cuprates are at the verge of a metal -insulator-transition.[1,2] Thus most defects will be insulating hindering the conduction, yielding not only percolative enhanced dc resistivities, penetration depths, and surface resistances R, but drastically lowered critical currents j_c or enhanced rf residual losses R_{res}, also.[3] This will be discussed in this paper. But before starting this, the structure and the conductivity is summarized. This analysis concentrates on $YBa_2Cu_3O_x$, but most results presented here hold also for the other cuprate superconductors.

The new cuprate superconductors are layered compounds with electric -hole- conduction confined to CuO_2-planes.[2] The -hole-concentration depends on oxygenation[2] and oxygen ordering.[3] Ordering - or disordering - relate to charge transfer via directional d-bonds up to distances $d_u \geq 1nm$, i.e. the lattice constant. That is, there exist no "point defects" or atomically sharp surfaces, i.e. every inhomogeniety may be spread out over ≥ 1nm by the directional d-bonds. Thus structural measurements, without a measure of the local charge transfer, e.g. by XPS[3], don't give a valid picture of defects or weak links in cuprate superconductors. But, detailed XPS measurements on weak links don't exist, yet. Thus we characterize weak links by their electric performance in the normal and superconducting state outlined in Parts 2 and 3. In Part 4 the small amount structural and stoichiometric information on weak links, i.e. on external and internal surfaces, is summarized.

2. ELECTRIC CONDUCTION

The normal- and super-conduction in the cuprates is dominated by the layered crystal structure.[1] This layering yields the planar electric conduction in CuO_2 double planes with negligible impurity scattering as shown by a negligible residual resistance.[2,3] This negligible impurity scattering seems a consequence of the verge of the metal-insulator transition (correlated electron motion) and of the layered structure dominated by nonlocal d-bonds smearing out charge transfer. Most defects reducing the conductivity seem to be organized in planes, as, e.g., stacking faults parallel to the CuO_2-planes (a - b) or as small or large angle grain boundary perpendicular to the a - b plane. The latter, if <u>insulating</u>, force the electric current to meander through the cuprate which in the most simple fashion is depicted in Fig. 1 and Eq. (1):[2,3]

$$\rho(T) = \Sigma R_b/d_b + p(\rho^i_{OL} + \alpha^i T) \qquad (1)$$

In Eq. (1), R_b describes the grain boundary resistance, d_b the effective crystallite size and $\rho^i_{OL} + \alpha^i T = \rho^i(T)$ the intrinsic resistance of weak link free crystallites. $p = L/L_o C_o/C \geq 1$ describes a mean percolative lengthening of conduction paths from L_o to L and $C/C_o < 1$ describes the mean shrinking of the current cross section C_o. Fig. 1 and Eq. (1) describe the resistivity of small blocks. In macroscopic samples of cuprates such blocks occur in parallel and in series, which we still approximate by Eq. (1) as shown in Fig. 2. There $\alpha = d\rho/dT = p\alpha^i$ is plotted against ρ_{100} for poly-(O) and "single-crystalline" (□)[4] and postannealed epitaxial films (△)[5]. This plot shows a qualitative correlation between α and ρ_{100}. Connecting epitaxial films and single crystals a line is defined as border to the left. This line going through zero proves with Eq. (1)

$$\rho_b + p\rho_{OL}^i \sim 0 \text{ , i.e. } \rho_b \sim 0 \text{ and } \rho_{OL}^i \sim 0 \qquad 1.1$$

for good material and shows that our model allows a simple description of different samples. Using fluctuation analysis one obtains $\alpha^i = 0.5\mu\Omega cm/K$ and $\rho_{100} = 50\ \mu\Omega cm$.[5]

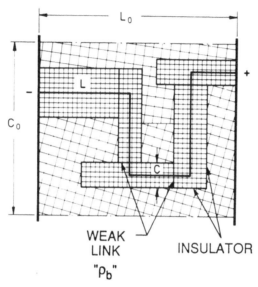

Fig. 1. Sketch of current path lengthening (L/Lo > 1) and cross section shrinking (C/Co < 1) by planar defects in a cuprate block. This sketch reflects the case typical for post annealed epitaxial films or single crystals.

In summary, the path systems <u>meandering</u> through the cuprates are defined by <u>insulators</u> or <u>cracks</u>. Grain boundary resistances R_b change the path system for $R_b \approx d_b p(\rho^i{}_{OL} + \alpha^i T)$ only. This may occur at higher temperatures for not so good material causing also activated conduction often observed, e.g., for $\rho_\perp$. As summarized in Ref. 3, c-axis epitaxial films, single crystals and sintered material cluster at $p \sim 3$. This is a strong hint, that the percolation is 2-dimensional due to the low resistance in the a-b-planes forcing the coupling to parallel a-b-planes via weak links. Also, as summarized in Ref. 3, the pronounced zero B-field peak in $j_c(B)$ with $j_c(0)/j_c(>B_1) > 50$ evidences this a-b- a-b--plane coupling. The clustering at $p \approx 3$ for all preparations is consistent with the small $R_b = \rho_b/d_b \leq 10^{-9}\Omega cm^2$ and randomization. In this connection it should be stated, that the observed conduction in c-direction can still be explained by a zig-zag percolation path. For example, the observed metallic conduction[4] with $\rho_\perp/\rho_{||} \geq 30$ fits to $p \approx$

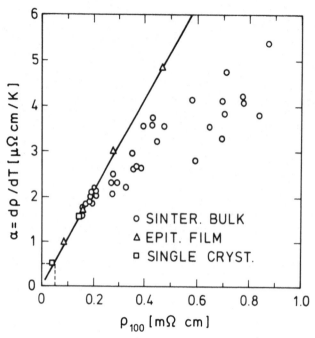

Fig. 2. The increase of $\alpha = d\rho/dT$ with ρ_{100} indicates one common source for the $d\rho/dT$ and ρ_{100}. The straight line through zero is obtained as left boundary by connecting $d\rho/dT$ and ρ_{100} of epitaxial films and of single crystals. The "intrinsic slope" $\alpha^i \approx 0.5$ μΩcm/K and intrinsic 100 K resistivity $\rho^i_{100} \sim 50$ μΩcm of $YBa_2Cu_3O_x$ is indicated on this line.

40, and this can be explained by zig-zag conduction in a-b-planes, which are connected by weak links.

To evaluate ρ_b (Eq. (1)) quantitatively for intra- and intergrain weak links, normal conduction measurements are not suited because of $\rho_b \ll \rho_{100}$ (Eq. (1.1)). But in the superconducting state $\rho^i = 0$ holds and thus R_b becomes prominent defining the <u>superconducting critical current</u>[3]

$$j_c R_b \sim c \Delta(T) \tanh(\Delta/2kT) . \qquad (2)$$

This equation holds for SIS and SNS junctions or microbridges and for 2 and 3 dimensional networks of identical junctions. Taking the above result $R_b \ll d_b \rho^i_{100}$ the experimental j_c's are between 1 and 4 orders of magnitude below expectation. This discrepancy and the steep $j_c(T)$-decrease with T is explained[3] by <u>"normal conductors" coating</u> the <u>cuprate interfaces as depicted</u> in Fig. 3 and discussed below. With increasing B-field, $j_c(B)$ decreases as shown in Fig. 4 up to the first

JOSEPHSON JUNCTION

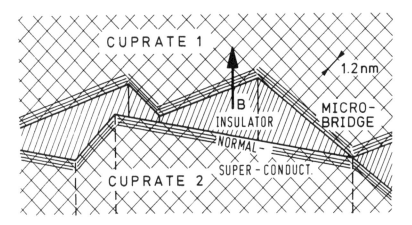

Fig. 3. Sketch of a small part of a planar intergrain defect with microbridges. As discussed in Part 2, the banks of the cuprate seems to act "normal conducting" as resolved in Part 3. For intragrain planar defects, as, e.g., small angle grain boundaries, the microbridges may be related to the dislocation lines occurring in regular distances.[3]

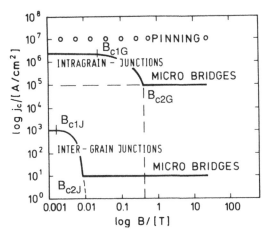

Fig. 4. Summary of typical $j_c(B, 77\ K)$ results obtained for polycrystalline and oriented $YBa_2Cu_3O_x$. Depending on preparation the indicated absolute values on inter- and intragrain weak links may vary by an order of magnitude.

interference minimum at B_1. The large ratio $j_c(0)/j_c(B>B_1) \geq 50$, as compared to theory ~ 5, and the missing secondary maxima are explained by <u>microbridges</u> carrying the main part of the current - see Fig. 3. These conclusions hold not only for intergrain weak links but also for intragrain weak links in so-called, "single crystals" or "epitaxial films". The intragrain weak links show a more regular $j_c(B)$ behavior and show a larger $j_c R_b$ value than intergrain weak links[3].

In extended SNS or SNINS junctions flux penetrates being described for intergrain weak links by:

$$j_{cJ} \approx 10^2 A/cm^2 : H_{c1J} \leq 1 \text{ Oe}$$
$$H_{c2J} \geq 50 \text{ Oe} \quad (3.1)$$

$$j_{cJ} \approx j_{cJ}(0) \frac{H_o}{|H| + H_o} ; H_o \approx H_{c1J} \quad (3.2)$$

and for intragrain weak links by

$$j_{cG} \geq 10^4 \text{ A/cm}^2 : H_{c1G} \geq 500 \text{ Oe} \quad (3.3)$$

These fluxoids get pinned between the grains as obvious from Fig. 3 and described by Eq. (3.2). The fluxoids are nucleated fast and move fast along the weak links allowing critical state behavior even above 10 GHz.[3,6] This corresponds to nonlinear behavior, i.e. a surface resistance and penetration depths increasing with H_{rf}. To this hysteretic behavior flux flow has to be added, as worked out in Ref. 6.

The two types of inter- and intragrain- weak link networks, described by Eq. (3.1) - (3.3) dominate the ac or rf response and limit the critical transport current. They are easily separable because of their vastly different current densities - see Fig. 4. In contrast, intragrain weak link effects (3.3) cannot be separated from pinning with certainty yet. They have shown up in decoration experiments[7a], for annealed samples or in TEM measurements as subgrain boundaries for insitu films.[7b] These intragrain weak links show a higher density of fluxoids than the grain interior. This is caused by a depressed order parameter in insulating weak link material. Thus flux creep and pinning in the 1 Tesla range is still dominated by the correlated motion of fluxoids and Josephson type behavior at intragrain weak links[8]. Only above 2T > Hc1 (bulk) ≈ 1.5T the $YBa_2Cu_3O_x$ bulk pinning may dominate. Below about 500 Oe the creep and pinning is dominated by intergrain weak links.

3. ELECTRONIC MAKE-UP OF WEAK LINKS

The crucial question, what is actually the "normal conduction" occurring at inter- or intragrain surfaces is discussed in the following. First, one should state, that "normal conduction" is inferred from a small Cooper pair current as compared to the normal current across the weak links. Second, the low density of -hole-states in the narrow, two dimensional conduction band of cuprates[1] is due to the overlap of O2p levels mediated via directional states. Thus disorder in position or in energy, e.g., due to strain or O disorder, changes this delicate hybridization yielding localized states being more electron like[1]. Thus, any external ("inter") or internal ("intra") surface yields a layer of localized electron states because of the disorder or strain and of the missing hybridization partners (Fig. 5). This yields in transport[9,10] a correlation gap Δ_{corr} diminishing in approaching the cuprate because the hybridization strengthen and Δ_{corr} is reduced to Δ^*_{corr}. For the cuprate becoming superconducting the hybridization is reduced, i.e. Δ^*_{corr} is approaching Δ_{corr}, especially in the superconducting energy gap. But this situation is even more complex for our cuprate surfaces: not only we are dealing with Cu and O sites, but also two electrons (holes) may be localized at one site causing Coulomb repulsion U_{Cu-Cu} and U_{O-O}[1] adding to the correlation gap. But because the cuprate is a hole conductor whereas the surface layer show localized electrons I propose as additional tunnel channel the transfer of electron - hole pairs showing no correlation gap.

The localized states carry the current across weak links by resonant tunneling because of the slower attenuation with distance as compared to direct tunneling.[11] This one particle resonant tunneling shows also a correlation gap (Coulomb blockade) and so resonant tunneling of electron - hole pairs without such a gap will dominate at low voltages increasing with the thickness of the layer housing these localized electron. This electron - hole resonant tunnel channel will not carry Cooper pairs, which are carried by "one particle" resonant tunneling being subject to pair weakening in the localized state. Thus the large difference between normal and supercurrent being usually explained by a "normal conductor" seems simply a consequence of the supression of Cooper pair tunneling by pair weakening and by the Coulomb gap as compared to the electron - hole tunneling -or creation- process occurring at cuprate - localized state interfaces. In summary, localized states forming weak links can explain the "normal conducting" banks and the correlation gap by electron - hole and one particle tunneling occuring at superconducting cuprate - insulating cuprate interfaces. The transition from the hole - superconductor to electron like localized state can explain the "normal conducting" channel by correlated electron - hole tunneling and the correlation gap by one particle tunneling. Thus our model can explain experimental details observed in tunneling[10,13], like, e.g., the correlation gap $\Delta_{corr} = 0.1 eV$

in wide area junction showing also a linear increase in the tunnel conductivity up to 0.5V indicative for a uniform distribution of localized states up to ± 0.5eV around E_F.

4. STRUCTURE AND CHEMISTRY

In Part 2 and 3 we concluded, that weak links are due to planes containing weak links. These planes behave insulating for thicknesses

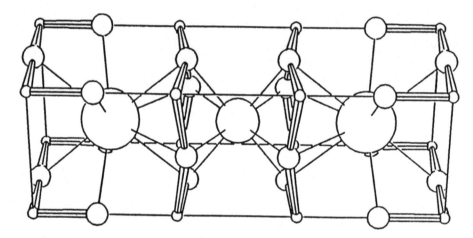

Fig. 5. Perspective drawing of one unit cell (supercell) of the structure of $YBa_2Cu_3O_{7-\delta}$. The two largest circles denote Ba and Y, respectively; the smallest circles denote Cu; the medium-size circles represent O-atoms. It is obvious from this figure that a - b surfaces have to relax and that BaO_2 formation is a very likely process. In contrast, c-axis surfaces may show a stable Ba-oxide surface.

above about 4nm but allow resonant tunneling as thin layers. This tunneling is more isotropic due to the localized states, i.e. allows charge transfer in a - b and c - direction. This resonant tunneling carries a small Cooper pair current but a large correlated electron - hole current simulating the normal conducting banks deduced in Part 2 (Fig. 3). Thus chemistry wise we have to explain why the cuprate becomes insulating at surfaces and what are the localized states at E_F and for which surfaces are those electron states occuring.

The new cuprate superconductors show a layered structure with electric conduction confined to the CuO_2 (double) planes. The Cu planes are conducting by electron transfer to the Cu - chain Ba plane region. This electron transfer is due to O^{2-} ions in the Cu chain regions being reduced by O loss or peroxy ion (O_2^{2-}) formation yielding an insulator. There are two types of planar defects: parallel to the Cu planes (‖ a - b) and perpendicular (⊥). The latter interrupt the electric conduction as discussed above and are classified as small and large angle grain boundaries, e.g., twin planes (90°).

The first type (‖ a - b) is most often studied because it is easily attainable because cleaving prefers to follow these "inert" insulating planes along Ba[14],-Bi, or Tl-oxides. As summarized, e.g., in Ref. 14 for $YBa_2Cu_3O_x$ these surfaces "are" often a specific Ba-O layer which is inert and stable. The second type of surface is naturally occurring between grains, as intergranular weak link, or as intragranular weak link at "internal surfaces" being small or large angle grain boundaries. Studies on these surfaces are scarce,[14,15] because breaks are irregular and are thus not well defined and because such surfaces deteriorate.[3] But despite their irregularity, they show more properties relevant to weak link superconductivity than the commonly studied -dead- c-axis surfaces.

To study the above mentioned surfaces and hence the weak links, structure measurements and profiling is not appropriate because of the following reasons: The cuprate superconductors are built by directional d-bonds allowing nonlocal stoichiometry, i.e. stoichiometry may be achieved over distances of 1nm, i.e. roughly the unit cell.[14] This information defining the electric conduction is difficult to obtain by structure measurements. Because of the d-bonds and the neighborhood to the metal insulator transition, cuprates are easily damaged by irradiation, especially the "insulating planes", ruling out profiling.

To measure the local chemistry, XPS is suited best in its ARXPS form because by measuring as function angle the whole "surface transition region" is analized and separated from the bulk. This is even more important for the layered cuprates, because these exist with different surfaces (‖ or ⊥ - see above), which have often residues from cleavage or growth. We studied epitaxial films and cleaved polycrystalline samples with the following results: The XPS signature of c-axis textured $YBa_2Cu_3O_x$ surfaces show often a specific "surface Ba (Ba-2)" coating of the cuprate if grown Cu poor (see - Fig. 6). The "superconducting Ba (Ba-1)" underneath is differing from Ba-2 by more than 1eV level shift, but is transforming to Ba-2 for driving the cuprates insulating.[14] This Ba-2 seems to be accompanied by a peroxidic O line (O - 3 in Table 1). In contrast Y - and Cu - XPS lines

Table 1. Binding energy and assignments of fitted Ba 4d5/2-, Y3d3/2-, C 1s- and O 1s- components.[14,16]

B.E. [eV]	Assignment	
87.2	Ba-1	Superconducting Cuprate
88.2	Ba-2	Cuprate Surface, Ins. Cupr.
89.6	Ba-3	BaO_2, $BaCO_x$, $BaCO_3$, BaO
91.0	Ba-4	$Ba(OH)_2$
156.2	Y-1	Superconducting Cuprate
157.3	Y-2	Y_2O_3, ...
158.6	Y-3	$Y(OH)_3$, ...
284.0	C-1	Graphite, Hydrocarbons
285.0	C-2	Graphite oxide
286	C-3	Alcohol
288.1	C-4	Carboxyle
289.1	C-5	Carbonate
528	O-0	Cuprate
529.1	O-1	Cuprate, Y_2O_3, BaO
530.0	O-2	CO_x
531.4	O-3	Peroxide (Cuprate, BaO_2)
532.4	O-4	CO_x
534	O-5	Hydroxide, H_2O, alcohol

don't show drastic changes for the cuprate becoming insulating. Thus only the Ba - line is changing from going from the superconducting to the insulating state and so we propose Ba - 1 as a signature of the superconducting state.[14,15] For a - b - surfaces the transition to Ba-2 and O-3 yielding insulating surfaces are caused by a loss of hybridization in the Cu-chain region causing peroxidic ions, i.e. less holes at the Cu planes - see Fig. 5. Less holes means more electrons which are now localized in the CuO_2 planes at E_F because lack of overlap and less electron transfer to the Ba-planes Cu chains ($2O^{2-} \rightarrow O_2^{2-} + 2e^-$). The Ba-2 line occurs also for c - surfaces, but there it seems an indication of Ba-peroxidic stacking fault being insulating and not involved in localized states, i.e., in current carrying.

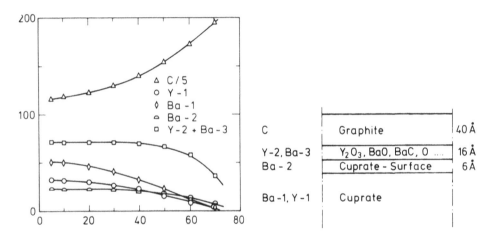

Fig. 6. Sketch of a cuprate surface with the corresponding XPS-intensities F (θ) changing with photo electron angle $\theta \leq 80°$. The intensities are normalized to the carbon intensity.

Using the Ba-1 signature we found[14-16] that c-axis surfaces are insulating and that the cut parallel to the c-axis shows more Ba -1 than perpendicular. This assignment is confirmed further by the observation that such cuts show a higher density of states at the Fermi energy.[17] The observation, that this density of states diminishes with time depends strongly on strain in cutting and seems strongest for a - b surfaces, which have to relax their dangling bonds - see Fig. 5. c - axis surfaces are more inert and thus stable and the observed instabilities may be caused by other reasons, e.g. to contamination. Aside from the above results on intrinsic weak links by insulating cuprate layers contaminations were found by ARXPS by fracturing sintered $YBa_2Cu_3O_x$ showing, that at intergrain surface C -, Ba - 2, 3 and Y - 2 layers exists.[14-16]

References

1. e.g.: K. C. Hass in: Solid State Physics 42 (H. Ehrenreich and D. Turnbull eds., Acad. Press, New York, 1989), p. 213.

2. R. Byers and T. G. Shaw, ibid., p. 135.

3. J. Halbritter, Int. J. Mod. Physics B3, 719 (1989).

4. G. Weigang, K. Winzer, to be published.

5. B. Oh, K. Char, A. D. Kent, M. Naito, M. R. Beasley, T. H. Geballe, R. H. Hammond, A. Kapitulnik, J. M. Graybeal, Phys. Rev. B37, 7861 (1988), K. Char, thesis (Stanford, 1989).

6. J. Halbritter, submitted to J. Appl. Phys.

7a. G. J. Dolan, G. V. Chandreshekhar, T. R. Dinger, C. Feild, F. Holtzberg, Phys. Rev. Lett. 62, 8271 (1989).

7b. J. Chang, M. Nakajima, K. Yamamoto, A. Sayama, Appl. Phys. Lett. 54, 2349 (1989).

8. K. Osamura, T. Takayama, S. Ochiai, Appl. Phys. Lett. 55, 396 (1989).

9. J. H. Davies, P. A. Lee, T. M. Rice, Phys. Rev. B29, 4260 (1984).

10. M. Lee, M. Naito, A. Kapitulnik, M. R. Beasley, Sol. State Comm. 70, 449 (1989).

11. J. Halbritter, Surface Sci. 122, 80 (1982) and J. Appl. Phys. 58, 1320 (1985).

12. A. B. Kaiser, J. Phys. 3, 410 (1970) and J. Halbritter, Sol. State Comm. 34, 675 (1980).

13. M. Gurvoitch, J. M. Valles, A. M. Cucolo, R. C. Dynes, J. P. Farno, L. F. Schneemeyer, J. V. Wasrcrak, to be published.

14. J. Halbritter, P. Walk, H.-J. Mathes, W. Harwink, I. Apfelstedt, AIP Conference Proc. 182 (N.Y., 1989), p. 208.

15. F. Stucki, P. Bruesch, Th. Baumann, Physica C156, 461 (1988).

16. J. Halbritter, P. Walk, H.-J. Mathes, B. Haeuser, H. Rogalla, Z. Phys. B73, 277 (1988).

17. D. Fowler, D. Brundle, private communication.

MAGNETIC FIELD DEPENDENCE OF CRITICAL CURRENT DENSITY

IN SILVER-DOPED Y-Ba-Cu-O SUPERCONDUCTORS

L.W. Song and Y.H. Kao

Department of Physics and
New York State Institute on Superconductivity
State University of New York at Buffalo
Buffalo, NY 14260

Effects of silver doping on critical current density in the bulk superconductor Y-Ba-Cu-O system are investigated by measuring the variation of critical current density in low magnetic fields. Below a newly defined characteristic field value H_c', the critical current is mainly controlled by intergranular effects due to Josephson weak-link coupling. This field H_c', which signifies a crossover between regions dominated by inter- and intra-granular effects, is found to vary with the silver content. An optimum enhancement of both the critical current density and H_c' is obtained with silver doping around 5-10%. These results lend direct support to the interpretation that the Josephson coupling is enhanced due to the presence of silver atoms along the grain boundaries in the material.

INTRODUCTION

The influence of chemical doping on superconducting properties has been a subject under extensive investigation in recent years. Of considerable interest is the effect of silver doping in the bulk high-Tc superconductor system Y-Ba-Cu-O (YBCO). It has been found that the critical current density of ceramic YBCO can be enhanced by more than an order of magnitude at 77K by silver doping[1]. There are reports on improved mechanical properties of composite YBCO tapes prepared on a silver substrate[2]. Silver-doping of YBCO thin films also proves to be effective in protecting the films against moisture-induced degradation[3,4]. The mechanism of Ag-doping in YBCO is therefore worth studying in light of these fundamental as well as technical interests.

It has been suggested that all the improved superconducting and mechanical properties can be largely attributed to the formation of Ag clusters which tend to fill in the voids in the ceramic material[5-7]. The Ag-impregnated YBCO ceramics thus show a lower porosity, making the material less permeable to atmospheric contaminants[6]. To explain the enhancement in critical current density, it is generally believed that the silver dopants must reside on the grain boundaries, which leads to an increased Josephson weak-link coupling between the adjacent superconducting grains[5-10].

In this communication, we report our recent measurements of the magnetic field dependence of critical current density in Ag-doped YBCO. A characteristic field H_c' is defined which serves as an indication of crossover between regions dominated by inter- and intra-granular effects[11]. Below H_c', the critical current density J_c is predominantly controlled by the intergranular weak-link coupling, the field dependence of J_c therefore can be understood in terms of the Josephson current averaged over the randomly oriented grains. For field values above H_c', the Josephson current is largely suppressed, and the field-dependence of J_c is mainly affected by the penetration of magnetic field into the superconducting domains and depairing caused by the field. Hence, an increase in the value of H_c' should also signify an increased coupling strength between adjacent grains.

The connection between H_c' and weak-link coupling can be further investigated by examining the response of critical current to microwave radiation. It is expected that the material should show an averaged microwave response similar to a bulk granular superconductor for $H < H_c'$ while it should become insensitive to microwave radiation for $H > H_c'$ when the weak-link coupling is suppressed. This observation is borne out in our experiment.

The variation of H_c' with Ag content is investigated by measuring the field dependence of J_c in several Ag-doped YBCO samples. For Ag content below 10% where an enhanced J_c was observed, the critical current enhancement is indeed accompanied by an increase in H_c', thus providing further support to the identification of H_c' as the point of crossover between the intergrain (weak-link coupled) and intragrain effects.

EXPERIMENTAL

Silver-doped YBCO samples used in this study were prepared by solid state reaction. High purity powders of Y_2O_3, $BaCO_3$, CuO and Ag_2O were mixed in accordance with the formula $YBa_2(Cu_{1-x}Ag_x)_3O_y$ with x=0 (pure $YBa_2Cu_3O_y$ material), 0.05, 0.10, and 0.15, respectively. Fine mixed powders were then pressed into pellets under 1-1.5 ton/in^2 pressure and sintered at 900°C for 10 hrs. in flowing oxygen. Grinding and pressing were repeated again, and the pellets were calcined at 950°C for over 10 hrs. in flowing oxygen. The typical size of samples for transport measurements was 0.5mm×0.5mm×10mm. Small size of "bridge" was made first by gluing a pellet onto a quartz substrate and then cut into the shape of a bridge with a blade.

Electrical resistivity was measured as a function of temperature between 10K and 300K using the standard four-probe method. Critical current density J_c as a function of magnetic field and microwave power

Table I. Variations of transition temperature T_c (R = 0), zero-field critical density $J_c(0)$, and the characteristic field H_c' in $YBa_2(Cu_{1-x}Ag_x)_3O_y$.

X	T_c(K)	$J_c(0)$ (A/cm^2)	H_c'(Oe)
0	92.0	81	39
0.05	91.3	480	76
0.10	90.2	170	60
0.15	87.6	3	40

was also determined by a four-probe measurement at 77K. The sample was mounted at the end of a waveguide connected to a microwave spectrometer operating at 8.95 GHz. J_c was determined at the onset voltage of 2μV and the noise level was about 1μV. Up to 1 T magnetic field was applied transverse to the current direction. The characteristic superconducting properties of the samples are listed in Table I.

All the samples were examined by x-ray powder diffraction. The well-known $YBa_2Cu_3O_y$ structure[12] was the major phase in the silver-free sample, and remained as the dominant phase even when 15% Ag was added. Small quantities of impurity phases were identified as due to Ba_2CuO_2, Y_2O_3, and Y_2BaCuO_5. Scanning electron micrograph (SEM) showed granular structure of the sample on a 10μm scale, while energy dispersion x-ray analysis (EDXA) indicated the samples were very uniform in composition. A commerical SQUID magnetometer was employed for the DC magnetization measurements.

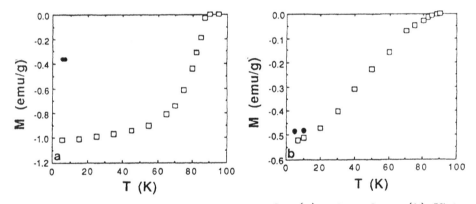

Fig. 1 DC magnetization vs. temperature for (a) silver-free, (b) 5% Ag-doped sample. □ - cooled from above T_c with zero field; ● - cooled from above T_c with field (~100 Oe) on. The measurements were done at a field ~100 Oe.

The field dependence of J_c can be used to estimate H_{c1} of the samples[13]. J_c was determined by applying a field up to a trial value H_m and then back to zero. If $H_m < H_{c1}$, J_c will remain at J_{co}, the value of J_c before any field was applied. However, for $H_m > H_{c1}$, J_c will deviate from J_{co}. Based on these assumptions, we have estimated that $H_{c1} \approx 150$ Oe for our samples.

RESULTS AND DISCUSSIONS

To see the effects of silver doping on sample quality, we first compare the silver-free and 5% Ag-doped samples. The grain size of these two samples is about the same (~10μm) based on our SEM observations. The density of the 5% Ag-doped sample, however, is increased (from 55% to 70% with respect to the theoretical value of 6.4 g/cm^3). Therefore, silver does make the ceramic material less porous by filling the voids. This was also observed directly under SEM. These metallic silver clusters can be detected by EDXA in 5% Ag-doped sample. Due to the limitation of our instrument, only silver clusters with size greater than 1μm can be resolved. We were unable to detect any silver clusters along the grain boundaries.

Figure 1 shows the static magnetization curves of 0% Ag and 5% Ag-doped samples. Magnetization measurements were performed by first cooling the sample from above the transition temperature down to ~5K in zero field, then applying a magnetic field (~100 Oe) and taking data while the sample was warmed up. Flux expulsion data were obtained by cooling from above the transition temperature down to ~5K in a constant field (~100 Oe). Figure 1(a) indicates that the 0% Ag sample has only ~36% while the 5% Ag-doped sample has ~93% flux expulsion effect [Fig. 1(b)].

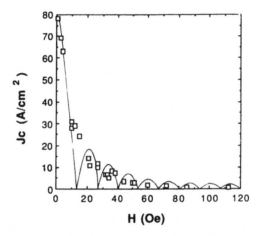

Fig. 2 Magnetic field dependence of the critical current density for a silver-free YBCO sample on a linear scale. The solid curve is a fit to the behavior of a Josephson junction in a magnetic field.

The behavior of critical current density in YBCO ceramic material in low fields has been studied by many others[9,10,13-15]. Figure 2 shows the data from our silver-free sample, which can be described by an average over a random distribution of Josephson weak links formed between adjacent superconducting grains as discussed in Refs. 13,15,16. The field dependence of the Josephson critical current is given by $J_c(H) = J_c(0) |\frac{\sin \pi x}{\pi x}|$, where $x = H/H_o$[17]. The data in Fig. 2 can be fitted quite satisfactorily with an average of $J_c(H)$, from which the parameter H_o can be deduced. In this model, H_o is the field corresponding to the presence of a flux quantum Φ_o inside the weak link area Ld, or $\mu L d H_o = \Phi_o$, where μ is the permeability, $d = 2\lambda+t$ the effective junction thickness, λ the London penetration depth, t the average barrier thickness, and L is the average junction length taken to be the average grain size. By assuming an effective permeability of $\mu = 0.3$, and d = $0.5\mu m^{15}$, we obtained $H_o = 12$ Oe and an average grain size of 10μm. This value is in agreement with the typical grain size observed directly under SEM.

The variation of J_c with H can be more usefully displayed by plotting J_c on a logarithmic scale, as shown in Fig. 3. In this semilog plot, a clear change of slope at a low characteristic field is revealed!! This characteristic field is now called H_c'. It is suggestive to identify H_c' as the point of crossover from a region dominated by the intergranular effects to that by the intragranular effects. The value of H_c' in Fig. 3 corresponds approximately to the presence of 3 flux quanta inside the weak link area. At this field value, the Josephson critical current $J_c(H)$ is only about 7% of its zero-field value and it is reasonable to assume that the intergranular coupling effects have become negligibly small after averaging over all the random junctions.

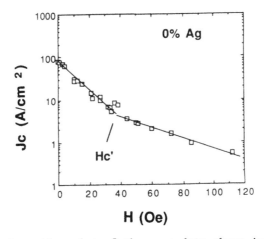

Fig. 3 A semilog plot of the same data shown in Fig. 2.

To provide a test of this interpretation, the response of $J_c(H)$ to microwave radiation was measured, as shown in Fig. 4. The zero-field J_c of a silver-free sample decreases with increasing power of incident microwave radiation (curve a). This behavior is similar to that found in a bulk granular superconductor formed by random packing of oxidized Nb powders in which Josephson effects arising from an array of weak-links were observed[18]. In a magnetic field higher than H_c' (e.g., at 48 Oe, curve b), however, J_c becomes independent of the microwave radiation. This is consistent with our interpretation of H_c' since for $H > H_c'$ the Josephson effects in the intergranular region are practically all averaged out, and the microwave energy is insufficient to alter the superconducting order parameters inside the grains; neither the inter- nor the intra-granular behavior can be influenced by the microwave radiation.

The more interesting result is the variation of H_c' in Ag-doped samples. This result is summarized in Fig. 5 as well as in Table I. As noted before, $J_c(0)$ is enhanced by adding a small amount of silver (5-10 atomic percent) while T_c remains practically unchanged. This enhancement of J_c has been interpreted as due to the presence of silver on the grain boundaries, thereby leading to an enhanced Josephson coupling. It is interesting that H_c' also shifts to a higher value (curve b) along with the enhanced J_c. It indicates, therefore, that an extension to higher field of the regime dominated by the intergranular (weak-link coupled) effect. This observation is consistent with the presence of the Josephson weak links as well as silver clusters on the grain boundaries, therefore lends more support to the previous interpretations.

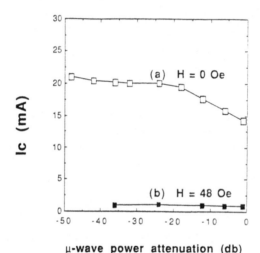

Fig. 4 Variations of the critical current density with microwave power for a silver-free sample. (a) H = 0; (b) H = 48 Oe.

When silver content is further increased, however, both $J_c(0)$ and H_c' start to decrease (Fig. 5, curves c and d). This is believed to be caused by local defects and a decrease in the superconducting order parameter due to the presence of Ag impurities in the Cu-O planes in the material[1,19]. The substitution of Ag for Cu in the YBCO structure is supported by our data on extended x-ray absorption fine structure (EXAFS) near the Ag K edge. When the short-range-order structure in YBCO is severely perturbed by the impurity atoms, the superconducting order parameter is expected to diminish. This indicates an extension to lower field of the regime dominated by the intragranular effects.

SUMMARY

The low field behavior of the critical current in ceramic YBCO material is dominated by Josephson weak-link coupling between superconducting grains. By adding a small amount of silver (5-10 atomic percent), the Josephson weak-link (intergranular) coupling can be largely enhanced; the critical current density J_c is therefore increased. We have introduced a new physical quantity H_c', which is defined as an indication of crossover between the regimes dominated by intergranular and intragranular effects; it can be used as a measure of enhancement of the Josephson weak-link coupling. The variation of H_c' in our Ag-doped samples supports the interpretation that the presence of silver along the grain boundaries will improve the weak-link coupling in ceramic YBCO material.

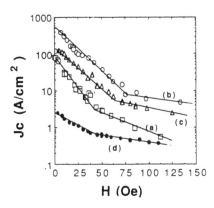

Fig. 5. Semilog plots of the critical current density as a function of external magnetic field H. (a) silver-free; (b) 5% Ag-doped; (c) 10% Ag-doped; (d) 15% Ag-doped.

ACKNOWLEDGEMENTS

This research is supported by AFOSR under grant No. AFOSR-88-0095, and by DOE under grant No. DE-FG02-87ER45283.

REFERENCES

1. Y.D. Yao, A. Krol, Y.H. Kao, C. Walters, S. Spagna, J. Althoff, S. Woronick, L.Y. Jang, and F. Xu, Mat. Res. Soc. Symp. Proc. 99: 407 (1988).
2. J.P. Singh, D. Shi and D.W. Capone II, Appl. Phys. Lett. 53: 237 (1988).
3. Chin-An Chang, Appl. Phys. Lett. 53: 1113 (1988).
4. Chin-An Chang and Julie A. Tsai, Appl. Phys. Lett. 53: 1976 (1988).
5. S. Jin, R.C. Sherwood, R.B. van Dover, T.H. Tiefel, and D.W. Johnson, Jr., Appl. Phys. Lett. 51: 203 (1987).
6. G.G. Peterson, B.R. Weinberger, L. Lynds, and H.A. Krasinski, J. Mater. Res. 3: 605 (1988).
7. Y.H. Kao, Y.D. Yao, L.Y. Jang, F. Xu, A. Krol, L.W. Song, C.J. Sher, A. Darovsky, J.C. Phillips, J.J. Simmins, and R.L. Synder, to be published in J. Appl. Phys.
8. D.E. Farrell and B.S. Chandrasekhar, Phys. Rev. B36: 4025 (1987).
9. W. Ekin, A.J. Panson, A.I. Braginski, M.A. Janocko, M. Hong, J. Kwo, H. Liou, D.W. Capone II, and B. Flandermeyer, Mat. Res. Soc. Sympt. Proc. EA-11: 223 (1987).

10. D.W. Capone II and B. Flandermeyer, in Ref. 9, p. 181.
11. L.W. Song and Y.H. Kao, B. Am. Phys. Soc. 34: 602 (1989).
12. For example, T. Siegrist, S. Sunshine, D.W. Murphy, R.J. Cava, and S.M. Zahurak, Phys. Rev. B35: 7137 (1987); M.A. Beno, L. Soderholm, D.W. Capone II, D.G. Hinks, J.D. Jorgensen, J.D. Grace, I.K Schuller, C.U. Segre, and K. Zhang, Appl. Phys. Lett. 51: 57 (1987).
13. J.F. Kwak, E.L. Venturini, D.S. Ginley, and W. Fu, in "Novel Superconductivity", S.A. Wolf and V.Z. Kresin, ed., Plenum, New York (1987) p. 983.
14. J.W. Ekin, Adv. Ceram. Mater. 2: 586 (1987).
15. R.L. Peterson and J.W. Ekin, Phys. Rev. B37: 9848 (1988).
16. J.R. Clem, Physica C153-155: 50 (1988).
17. M. Tinkham, "Introduction to Superconductivity", McGraw-Hill, New York (1975) p. 199.
18. J. Warman, M.T. John, and Y.H. Kao, J. Appl. Phys. 42: 5194 (1971).
19. Y.H. Kao, Y.D. Yao, and L.W. Song, Int. J. Mod. Phys. B3: 573 (1989).

DESCRIPTION OF THE Tl, Pb AND Bi CUPRATE HIGH-T_c SUPERCONDUCTORS (HTSC'S) AS INTERCALATION COMPOUNDS AND CLASSIFICATION OF ALL HTSC'S ACCORDING TO DOPING MECHANISM

B.C. Giessen[a,c] and R.S. Markiewicz,[b,c]

Departments of Chemistry (a) and Physics (b) and Materials Science Division, Barnett Institute (c), Northeastern University Boston, MA 02115

INTRODUCTION

After an initial period of rapid growth during 1987 and 1988, the number of known ceramic HTSC phases is presently increasing only slowly and may be close to having reached a stable plateau. Cataloguing these phases at this time is useful for their understanding and to find directions for further development.

One approach to classification can be based on crystal structure, and there are compilations of HTSC's based on this principle.[1] Making use of the description of many HTSC's as intercalation phases which was presented by us previously[2] and is briefly repeated here, we give specifics of a classification of HTSC's based on doping type, doping charge sign and the chemistry of the doping functional group. We note that there is evidence for the possibility of a unified characterization of the T_c values of cuprate layer HTCS's in terms of a density-of-states (DOS) model recently developed and published by us.[3,4]

TYPES OF DOPING

<u>Intercalation Layer Doping:</u> In analogy to other intercalation compounds such as graphite intercalation compounds,[5] we visualize the HTSC cuprates as consisting of single layers or multilayers of the (semiconducting) "parent" cuprate layer compound $(Ca,Sr)CuO_2$[6] and intercalated layers of "doping units" such as $TlBaO_{2.5}$ which contain

partially unfilled oxygen sites. These intercalant sites can accept additional oxygen atoms which, upon ionizing to O^{2-} (and changing the doping group composition, e.g., to $TlBaO_{2.5+\delta}$), oxidize the parent compound layers and, by introducing enough holes into the network of CuO_2^{2-} squares, can produce conduction. (The ratio of parent cuprate layers (host units) to intercalant units is referred to as the "stage number".)

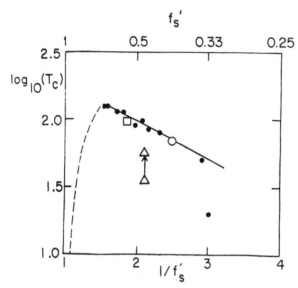

Fig. 1 Scaling of T_c with superconducting fraction f_s' in cuprate HTSC's. Filled circles = Tl- and Bi-intercalant classes; open circle = $Pb_2CuSr_2Y_{0.5}Ca_{0.5}Cu_2O_8$ (Ref. 11); open square = YBCO; open triangles = LSCO (arrow showing pressure dependence). See Ref. 2 for further discussion of f_s'.

Without addressing the mechanism of oxidation or the preferential location of the holes in the electron configuration of the cuprate layer here further, we mention our recent demonstration[2] that this concept qualitatively explains a maximum in the values of T_c seen in a plot of T_c versus the fraction of the unit cell occupied by CuO_2^{2-} sheets, represented by f_s' in Fig. 1 (adapted from Ref. 2). T_c then increases with f_s' according to $\ln T_c \sim const - 1/f_s'$ up to a maximum value of f_s', at which the total charge transferred from the CuO_2^{2-} sheets to the oxidizing intercalant group is no longer sufficient to provide an adequate hole concentration, causing T_c to drop again. A closer discussion of the nature of this limiting factor can be found in Ref. 2.

Counter-Ion Doping: In $(La_{1-x}Sr_x)_2CuO_4$ (LSCO) and in the copper-free cubic perovskite HTSC phase $(Ba_{1-x}K_x)BiO_3$ there is no intercalation group. Instead, the doping is provided by heterovalent substitution of cations in the counter-ion sites (La and Ba, respectively). This substitution can be either hole doping (as in the two cases listed) or electron doping, as in the more recently discovered T' phase $(Nd_{.92}Ce_{0.08})CuO_{3.93}$.[7] In principle, it is possible to have both types of doping present simultaneously, and indeed the presence of counter-ion doping in intercalation-doped compounds is suggested by the fact that the optimal T_c in such phases tends to occur at compositions producing mixed-site occupation in, e.g., $(Tl,Bi)Sr_2Ca_{n-1}Cu_nO_x$.[8]

REPRESENTATION OF HTSC's ACCORDING TO FUNCTIONAL GROUPS

Structural Classification based on Dopant and Intercalant Type: Using "pure" counter-ion doping and intercalation layer doping as prime classification criteria, we propose a taxonomy of the essential HTSC oxide phases as given in Table I, classifying these oxides into categories

Table I. Structural Families of High T_o Superconductor Oxides

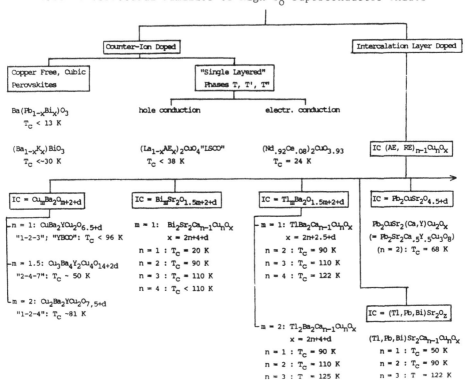

(determined by the type of doping), classes, families and (family) members; the latter three groupings are defined below.

Within the counter-ion doped category, we subdivide into two classes, the copper-free perovskites and the LSCO-related T, T', and T"(T*) type phases;[7] the latter class we subdivide further into families according to the sign of the doping charge, as shown in Table I.

The intercalation layer doped phases follow the general formula $(IC)(AE,RE)_{n-1}Cu_nO_x$, where IC is the intercalant group and (AE,RE) designates the alkaline earths/rare earths mixed counter-ions. Within this category, there is a subdivision into 5 classes according to the intercalant groups, as defined broadly by the "functional" metallic elemental constituent of the intercalant (Cu,Bi,Tl,Pb); where necessary, each class is subdivided into families according to the stoichiometry of these "functional" elements, designated by the subscript m as listed in the following. The IC groups are:

Cu_mBa_2 oxide (with m = 1, 1.5 and 2): $Cu_mBa_2O_{m+2+\delta}$;

Bi_mSr_2 oxide (with m = 2): $Bi_2Sr_2O_{5+\delta}$;

Tl_mBa_2 oxide (with m = 1 and 2): $Tl_mBa_2O_{1.5m+2+\delta}$;

$(Tl,Pb,Bi)Sr_2$ oxide (with m = 1): $(Tl,Pb,Bi)Sr_2O_x$;

Pb_2CuSr_2 oxide: $Pb_2CuSr_2O_{5+\delta}$.

Within each family, thus defined, members are listed according to the number of successive CuO_2^{2-} layers (the "stage number" in the language of intercalation compounds[2]). We now review the classes, families and their members in some detail. (The order of listing for the intercalant groups in Table I is essentially historical, following the order of discovery, i.e., they are listed in the order Cu, Bi, Tl and Pb_2Cu.)

The first class is characterized by the intercalant groups given by $Cu_mBa_2O_{m+2+\delta}$, forming three families. Of these, the first family $CuBa_2O_{3+\delta}$; (m = 1) has as its sole member the "workhorse" HTSC $CuBa_2YCu_2O_{6.5+\delta}$ (commonly known as $YBa_2Cu_3O_{6.5+\delta}$; the generically proper description given here is preferred in the present context).

The family based on the Cu-double-layer intercalant $Cu_2Ba_2O_{4+\delta}$ (m = 2) contains as its sole member the cuprate $Cu_2Ba_2YCu_2O_{7.5+\delta}$ (known as $YBa_2Cu_4O_{7.5+\delta}$); last, the oxide $(Cu_2Ba_2)_{0.5}(CuBa_2)_{0.5}YCu_2O_{7+\delta}$ (or $YBa_2Cu_{3.5}O_{7+\delta}$) corresponds to m = 1.5 and contains both of the intercalants with m = 1 and m = 2 in equal proportion.

The second class listed in Table I is based on the intercalant $Bi_mSr_2O_{1.5m+\delta}$. This class consists of a single family with m = 2, i.e., with the intercalant $Bi_2Sr_2O_{5+\delta}$ and the composition $Bi_2Sr_2Ca_{n-1}Cu_nO_{2n+4+\delta}$. It has members with n = 1 to 4; of these, the ones with n = 2 and 3 constitute the technically important members of the bismuth cuprate class.

The class of cuprates containing the Tl-Ba oxide intercalant $Tl_mBa_2O_{1.5m+2+\delta}$ is made up of two families with m = 1 and 2, each with members ranging up to n = 4. These families, with compositions $Tl_mBa_2Ca_{n-1}Cu_nO_{1.5m+2n+1+\delta}$, contain two compounds with T_c > 120 K, of these, one (with m = 2 and n = 3) has the highest known T_c of any stable HTSC and the other (with m = 1 and n = 4) has the highest T_c known for a compound with four consecutive copper oxide network layers.[9,10]

Next, we list the mixed-composition intercalant group $(Tl,Pb,Bi)Sr_2O_x$ which forms a class with a single family having three members containing cuprate layers $Ca_{n-1}Cu_nO_{2n-1}$ with n = 1 to 3. Conceptually, one might lump together the three last named intercalant groups (and the three classes of HTCS's based on them) and consider them as special cases of an oxide with the overall formula $(Tl,Pb,Bi)_m(Sr,Ba)_2$ with m = 1 or 2. The exploration of the cuprate phases in the complex six-component system formed by this generalized intercalant with $Ca_{n-1}Cu_nO_{2n-1}$ as a function of composition, temperature and oxygen pressure is doubtlessly a most challenging task.

The last (and latest to be discovered) intercalant group has the formula $Pb_2CuSr_2O_x$, with the Cu ion being located between two Pb atom layers.[11,12] For this class and the single family based on it, only one member (with n = 2) has been described, with the overall formula $Pb_2Sr_2(Ca_{0.5}Y_{0.5})Cu_3O_8$. It should be noted that here Ca must be partially (about half) replaced by a trivalent rare earth metal, just as a complete replacement of Ca by Y was required in the families based on the Cu_mBa_2 oxide intercalant.

375

This descriptive review exhausts the list of the currently known HTSC phases.

Classification by Dopant Type and Dopant Charge Sign: The explicit listing in Table I is primarily based on dopant type, with the sign of the dopant charge as a subordinate consideration that is applicable only to one specific, "single-layered" class of counter-ion doped cuprates (as far as known today). It is a natural extension of the concepts presented here to make a list of the HTSC phases both in terms of doping type <u>and</u> charge carrier sign, with intercalant group and composition (m, n as used above) as subsidiary considerations. This approach is implemented in Table II, which contains the categories and classes of Table I without listing individual families (with different m) and specific family members (n = 1 to 4) in detail.

Table II. HTSC Families Arranges According to Doping Type and Charge Carrier Sign.

Doping Type \ Charge Carriers	HOLES		ELECTRONS
COUNTER-ION DOPING	Perovskites $Ba(Pb_xBi_{1-x})O_3$ $(Ba_{1-x}K_x)BiO_3$ $T_c < 30$ K	Single Layer Phase (T) $(La_{1-x}Sr_x)_2CuO_4$ (LSCO) $T_c < 38$ K	Single Layer Phase (T') $(Nd_{0.92}Ce_{0.08})CuO_{3.93}$ $T_c = 24$ K
INTERCALATION DOPING	(IC) (AE,RE)$_{n-1}$Cu$_n$O$_x$		No Known Examples
	IC Group	Compounds	
	$Cu_mBa_2O_{m+2+d}$ m = 1; 1.5; 2	$YBa_2Cu_{2+m}O_{6.5+d}$	
	$Bi_2Sr_2O_{5+d}$	$Bi_2Sr_2Ca_{n-1}Cu_nO_{2n+4+d}$	
	$Tl_mBa_2O_{1.5m+2+d}$ m = 1; 2	$Tl_mBa_2Ca_{n-1}Cu_nO_x$ x = 1.5m+2n+1+d	
	$(Tl,Pb,Bi)Sr_2O_z$	$(Tl,Pb,Bi)Sr_2Ca_{n-1}CuO_x$	
	$Pb_2CuSr_2O_{4.5+d}$	$Pb_2CuSr_2(Ca_{0.5}Y_{0.5})CuO_8$	

CONCLUDING REMARKS

The two tabular presentations of the data given here lead to several observations.

The intercalant groups fall into two chemically distinct sets, one with intercalants containing Cu (accomodating excess oxygen along Cu chains) and the other with single or close-packed double layers consisting of the large metal atoms Tl, Pb and Bi (containing the excess oxygen in or between the large metal atom planes). It is still unclear how the intercalant groups in these two, chemically distinctly different sets are related and whether there can be other, functionally similar groups.

Further it is not clear why counter-ion doping does not seem to play as strong a role in the intercalation-doped phases as it does in the non-intercalation oxides, i.e., why there seem to be no true mixed-doping-type cuprates. As observed above, there can be counter-ion mixing, e.g., Ca for Y in $PbCuSr_2(Y_{0.5}Ca_{0.5})Cu_2O_x$, but this seems more to be required for phase formation than to affect the conduction properties via the doping level.

We note, parenthetically, that the prototypical counter-ion doped compound LSCO may also be intercalation-doped by adding oxygen without insertion of another cation group. It is not clear whether a part of the La double layer (with the extra oxygen) should be considered as an intercalation group, e.g., $La_{1.33}O_{2+\delta}$.

Last, we note that the absence of electron-conducting, intercalation-doped superconductors is striking. There is currently no answer why such phases have not been seen to form; appropriate further searches may be fruitful.

ACKNOWLEDGEMENT

We thank I.E. DuPont de Nemours and Co., Inc. for financial support of this work. We are also pleased to acknowledge collection of data by Dr. J. Sigalovsky.

REFERENCES

1. R. Beyers and T.M. Shaw, in "Solid State Physics", ed. by H. Ehrenreich and D. Turnbull, 42: 135 (1989).
2. R.S. Markiewicz and B.C. Giessen, Mod. Phys. Lett. B 3: 723 (1989).
3. R.S. Markiewicz and B.C. Giessen, Physica C_2 160: 497 (1989).
4. R.S. Markiewicz and B.C. Giessen, Proc. M^2S-$HTSC_{II}$ Conf., Stanford, CA, Physica C (1989), (in print).
5. M.S. Dresselhaus and G. Dresselhaus, Adv. Phys. 30: 139 (1988).
6. T. Siegrist, S.M. Zahurak, D.W. Murphy and R.S. Roth, Nature 334: 231 (1988).
7. Y. Tokura, H. Takagi and S. Uchida, Nature 337: 345 (1989).
8. P. Haldar, A. Roig-Janicki, S. Sridhar, R. Markiewicz, and B.C. Giessen, J. of Superconductivity, 1: 211 (1988).
9. P. Haldar, K. Chen, B. Maheswaran, A. Roig-Janicki, N. Jaggi, R. Markiewicz, and B.C. Giessen, Science 24: 1198 (1988).
10. H. Ihara, R. Sugise, M. Hirabayashi, N. Terada, M. Jo, K. Hayashi, A. Negishi, M. Tokumoto, Y. Kimura and T. Shimomura, Nature 334: 510 (1988).
11. R.J. Cava, B. Batlogg, J.J. Krajewski, L. W. Cupp, L.F. Schneemeyer, T. Siegrist, R.B. van Dover, P. Marsh, W. F. Peck, Jr., P.K. Gallagher, S.H. Glarum, J.H. Marshall, R.C. Farrow, J.V. Waszczak, R. Hull and P. Trevor, Nature 336: 211 (1988).
12. M.A. Subramanian, J. Gopalakrishnan, C.C. Torardi, P.L. Gai, E.D. Boyes, T.R. Askew, R.B. Flippen, W.E. Farneth and A.W. Sleight, Physica C 157: 124 (1989).

$La_2MO_{4+\delta}$ (M=Cu, Ni, Co): PHASE SEPARATION AND SUPERCONDUCTIVITY RESULTING FROM EXCESS OXYGEN DEFECTS.*

B. Dabrowski, J.D. Jorgensen, D.G. Hinks, Shiyou Pei, D.R. Richards, K.G. Vandervoort, G.W. Crabtree, H.B. Vanfleet[#] and D.L. Decker[#]

Materials Science Division, Argonne National Laboratory, Argonne, IL 60439 and [#]Department of Physics and Astronomy, Brigham Young University, Provo, Utah 84602

ABSTRACT

The structural, superconducting and normal-state properties of $La_2MO_{4+\delta}$ (M=Cu, Ni, Co) were studied by neutron powder diffraction, thermogravimetric analysis and resistivity measurements. $La_2CuO_{4+\delta}$ was found to undergo phase separation near room temperature into antiferromagnetic La_2CuO_4 and superconducting $La_2CuO_{4.08}$. Single-phase, metallic, superconducting $La_2CuO_{4.08}$ was prepared under 25 kbar of O_2 at 500°C. The excess oxygen present as $O^=$ ions provides a doping mechanism analogous to alkaline-earth metal ion substitution.

$La_2NiO_{4+\delta}$ and $La_2CoO_{4+\delta}$ undergo a similar phase separation for $0.02<\delta<0.12$ and $0<\delta<0.16$, respectively. The structure of the interstitial excess oxygen defect was determined for a single-phase $La_2NiO_{4.18}$ sample. A different excess oxygen defect forms for material with $0\leq\delta\leq0.02$. All stoichiometric ($\delta=0$) compounds have the same Cmca orthorhombic structure and are semiconducting and antiferromagnetic.

INTRODUCTION

Copper, nickel and cobalt are the only known transition metal oxides forming the K_2NiF_4 structure with lanthanum. All three compounds ($La_2MO_{4+\delta}$, M=Cu, Ni, Co) display a large range of oxygen nonstoichiometry[1]. The nature of this nonstoichiometry is not fully understood. Several reports attribute the observed oxygen nonstoichiometry to a variable metal ion ratio[2-4]. For example, Davis and Tilley[2] reported the presence of Ruddlesden-Popper intergrowths of the type $La_{1+n}Cu_nO_{1+3n}$ with n=2, 3, 4 and 5 for the Cu system. Such intergrowths with n=2, 3 and 4 had been seen earlier for the Ni system[3]. Conversely, a large concentration of La vacancies was reported for the Co system, suggesting a composition of $La_{1.83}CoO_{4-\delta}$[4]. More recent studies indicate that it is possible to prepare these compounds with stoichiometric metal ion ratios by carefully controlling the oxygen partial pressure during the high temperature firing[5-7]. Thus, it seems that the introduction of metal ion nonstoichiometry may only be possible under specific preparation conditions. In particular it is known that the perovskite, $LaMO_3$, (which may be considered as the n=∞ member of the Ruddlesden-Popper intergrowth family) forms for M=Ni and Co by firing in O_2 and for M=Cu by firing in 25 kbar of O_2, at appropriate temperatures[8,9].

The discovery of high temperature superconductivity in oxides led to the need for a better understanding of the effects of oxygen stoichiometry on the structural and transport properties of these materials. $La_2CuO_{4+\delta}$ was initially thought to support only oxygen deficiency ($\delta<0$)[10]. However, Schirber et al.[11] demonstrated the presence of excess oxygen ($\delta>0$) for samples annealed in 3 kbar oxygen. By comparing weight loss and iodometric titration measurements they concluded that the excess oxygen is incorporated in the host structure as a superoxide ion, O_2^-. Similarly, recent density measurements on a single crystal of $La_2NiO_{4+\delta}$ showed a large concentration of excess lattice oxygen[6]. Ganguly at al.[12] and Buttery et al.[6] proposed that the excess oxygen enters the lattice as a superoxide or peroxide ion. Ram et al.[7] prepared polycrystalline $La_2CoO_{4+\delta}$ at 1200°C in a CO_2 atmosphere and measured an excess oxygen concentration $\delta=0.11$ by iodometric titration.

We have recently performed a detailed study of the structural, superconducting and magnetic properties of $La_2CuO_{4+\delta}$ and $La_2NiO_{4+\delta}$ using neutron powder diffraction, thermogravimetric analysis, resistivity, and magnetic susceptibility measurements[1,12]. $La_2CuO_{4+\delta}$ was found to phase separate near room temperature into antiferromagnetic La_2CuO_4 and superconducting $La_2CuO_{4.08}$. Similarly, $La_2NiO_{4+\delta}$ for $0.02 \leq \delta \leq 0.13$ separates into two phases with different defect concentrations and structures. The structure of the interstitial oxygen defect was determined for $La_2NiO_{4.18}$[12].

In this paper we compare the structural and transport properties of the $La_2CuO_{4+\delta}$ and $La_2NiO_{4+\delta}$ systems with our preliminary data for the $La_2CoO_{4+\delta}$ system. New data is presented for single-phase superconducting $La_2CuO_{4.08}$ prepared using very high oxygen pressure. All compounds studied contain excess oxygen ($\delta>0$) when annealed in oxygen. They have remarkably similar structural properties, all displaying phase separation. The stoichiometric ($\delta=0$) compounds have the same Cmca orthorhombic structure and are semiconducting and antiferromagnetic. Only oxygen-rich $La_2CuO_{4.08}$ is metallic and superconducting.

EXPERIMENTAL PROCEDURE

Samples were prepared from high purity La_2O_3 (oxygen prefired at 800°C to remove hydroxides and carbonates), CuO, NiO and Co_3O_4. The oxide powders were thoroughly wet-ball-milled in agate for 12h, dried in air, and dry-ball-milled to remove any possible inhomogenity occurring during the drying process. The $La_2CuO_{4+\delta}$ was fired in flowing oxygen at 975°C for 12 hours and furnace cooled (about 10 hours) to room temperature. The high-oxygen-content samples were annealed in 0.1 and 25 kbar of oxygen at 500°C for 48 hours followed by a 24-hour cool to room temperature. The details of the 25 kbar anneal are described later. The stoichiometric sample ($\delta=0$) was prepared in nitrogen at 600°C for 12 hours. The $La_2NiO_{4+\delta}$ and $La_2CoO_{4+\delta}$ samples were fired in flowing nitrogen at 1035 and 1215°C, respectively. The high-oxygen-content samples were annealed in flowing oxygen at 450°C and slowly cooled to room temperature. The stoichiometric La_2NiO_4 and La_2CoO_4 samples were prepared by partially reducing the oxygen-annealed samples in a flowing 15 mol % hydrogen-in-nitrogen gas mixture at 360 and 380°C, respectively.

Neutron powder diffraction data were obtained using the Special Environment Powder Diffractometer (SEPD) at Argonne National Laboratory's Intense Pulsed Neutron Source (IPNS). Resistivities were measured by the standard four-probe technique in a closed-cycle refrigerator from 300 to 10K. The oxygen content of the samples was determined using a Perkin Elmer TGA 7 system.

THERMOGRAVIMETRIC ANALYSIS

The overall oxygen stoichiometry of the various samples was determined by in situ hydrogen reduction on a thermobalance. Small (~100 mg) samples were heated on a platinum pan in flowing hydrogen (0.1 L/min) to 700°C (800°C for Co) at 2°C/min.

Fig. 1 shows the weight loss on heating in N_2 for both 0.1 kbar O_2-annealed and as-prepared (1 bar of O_2) $La_2CuO_{4+\delta}$ samples which were stored in air for long periods. The weight-loss curve has two well defined regions, the first starting around 150°C and the second around 300°C. Using low temperature in-situ O_2-N_2 cycling it was determined that the initial weight loss is related to a surface species (superoxide, hydroxide, carbonate or others which have been seen in x-ray photoemission experiments[14]) and the excess interstitial oxygen is evolved in the second region. A similar conclusion was drawn by Zhou et al.[15] for samples made by a different process.

Fig. 2 shows the weight loss of the N_2-annealed La_2CuO_4 sample on heating in H_2. Assuming that the sample is reduced to La_2O_3 and Cu-metal, the oxygen content is 3.998±0.004. The oxygen contents of the 0.1-kbar O_2-annealed and as-prepared samples determined in the same way are 4.023±0.004 and 4.009±0.004, respectively. The weight-loss curve in H_2 for the Cu compound shows two-plateau character. The first plateau observed on heating corresponds to the stoichiometric La_2CuO_4 material. The second, poorly defined, plateau corresponds to material of the approximate oxygen stoichiometry 3.67, i.e. material containing a mixture of Cu^{+1} and Cu^{+2} ions. The X-ray diffraction patterns for material obtained by quenching from the second plateau region show broad diffraction peaks corresponding to the K_2NiF_4 structure. Since the impurity level for that material is small, this suggests that large concentrations of oxygen vacancies can also be introduced into the La_2CuO_4 structure. This is consistent with the neutron diffraction observation that samples prepared in air at 1100°C and quenched to room temperature contain oxygen vacancies in the CuO planes[16].

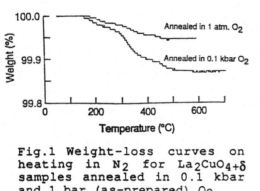

Fig.1 Weight-loss curves on heating in N_2 for $La_2CuO_{4+\delta}$ samples annealed in 0.1 kbar and 1 bar (as-prepared) O_2.

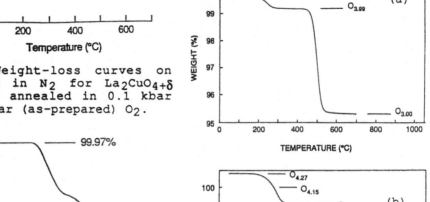

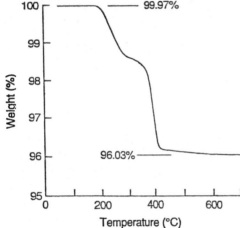

Fig.2 Weight loss curve on heating in H_2 for La_2CuO_4 sample annealed in N_2.

Fig.3 Weight loss curves on heating in H_2 for (a) $La_2NiO_{4+\delta}$ and (b) $La_2CoO_{4+\delta}$ samples annealed in 1 bar O_2.

The weight loss curves in H_2 for oxygen-rich $La_2NiO_{4+\delta}$ and $La_2CoO_{4+\delta}$ samples are shown in Fig. 3. Assuming that the resulting products from the hydrogen reduction are Ni or Co metal and La_2O_3, the overall oxygen stoichiometry of the oxygen-annealed samples is 4.18±0.01 and 4.27±0.01 for Ni and Co, respectively. A similar analysis for nitrogen-prepared samples gives oxygen stoichiometries of 4.07±0.01 and 4.15±0.01 for the Ni and Co compounds, respectively. These weight-loss curves also show well-defined broad plateaus near stoichiometries of 4.00 extending over a wide temperature range. The existence of these plateaus makes possible the straightforward preparation of stiochiometric ($\delta=0$) samples.

NEUTRON POWDER DIFFRACTION

The structure of $La_2CuO_{4+\delta}$ was studied as a function of oxygen content and temperature[1]. The data for high pressure oxygen-annealed samples taken at 10K revealed the presence of new Bragg peaks appearing as shoulders on some of the peaks of the known orthorhombic Cmca structure. The intensities of the new peaks scaled with the amount of superconducting phase and the oxygen pressure under which the samples were prepared (i.e.

the oxygen excess). This observation led to the conclusion that two different phases are present: the first one is nearly stoichiometric La_2CuO_4 and the second is an oxygen-rich phase. The structure of the new phase (most probably orthorhombic Fmmm) is closely related to the orthorhombic La_2CuO_4 structure. It was not possible to uniquely determine the structure of the new phase due to substantial overlap of the diffraction peaks. However, clear evidence for the phase separation behavior was provided by the temperature dependence of the positions and intensities of the new peaks. The phase separation was found to occur reversibly near room temperature. The stoichiometry of the new phase, $La_2CuO_{4.08}$, was determined using the phase fractions determined from Rietveld refinements and the total oxygen content determined from thermogravimetric measurements[17].

These initial studies suggested that the synthesis of single-phase, superconducting $La_2CuO_{4+\delta}$ ($\delta \geq 0.08$) would require very high oxygen pressures. Following a well-known high-pressure technique, first applied to the $La_2CuO_{4+\delta}$ system by Zhou et al.[15], we succeeded in making single-phase, bulk superconducting samples[17]. Stoichiometric La_2CuO_4 powder was sealed in an inconel can along with Ag_2O and pressurized to 25 kbar in a 400-ton hexahedral anvil press at Brigham Young University. High oxygen pressures were then achieved by passing current through the inconel to heat the sample and decompose the Ag_2O. In this way, by annealing in 25 kbar of oxygen at 500-600^{0}C, a bulk superconducting sample of $La_2CuO_{4+\delta}$ was made. Neutron diffraction measurements were somewhat hindered by the small sample size (~1 g) and the occurrence of a second oxide impurity phase which forms if the anneal temperature is too high (>550^{0}C). However, the neutron diffraction data were consistent with the oxygen-rich phase of the two-phase samples and no two-phase behavior could be seen at room temperature.

Since it was not possible to determine the structure of the oxygen defect for $La_2CuO_{4+\delta}$ due to the low concentration of defects we next examined samples of the isostructural $La_2NiO_{4+\delta}$ compound, which exhibits a much wider range of oxygen stoichiometry[13]. By studying three samples with different oxygen contents (δ=0.00, 0.07, 0.18) at room temperature and at 10K we found that the $La_2NiO_{4+\delta}$ system has structural properties remarkably similar to those of the $La_2CuO_{4+\delta}$ system. In particular, we found that $La_2NiO_{4.07}$ is separated into $La_2NiO_{4.02}$ and $La_2NiO_{4.13}$ phases at room temperature. Thus, a miscibility gap exists for $0.02<\delta<0.13$ at room temperature and below. The structure of the interstitial oxygen defect was determined for the orthorhombic Fmmm phase which is stable for $\delta>0.13$. The defect is located at the $(1/4,1/4,z\approx0.23)$ site within the double La-O layer. Oxygen enters the lattice as an $O^=$ ion coordinated to four La ions and forcing four neighboring oxygen ions to displace by about 0.5Å from their normal lattice sites. A different oxygen defect (whose structure was not determined) exists for the orthorhombic Cmca phase for $0.00<\delta<0.02$. By examining the Ni-O distances, which are the most sensitive to electronic changes, as a function of oxygen content it was concluded that this low-oxygen-concentration defect is different from the interstitial oxygen defect present in the Fmmm phase.

The $La_2CoO_{4+\delta}$ system displays the largest solubility for excess oxygen. Three samples with δ=0.00, 0.15 and 0.27 were

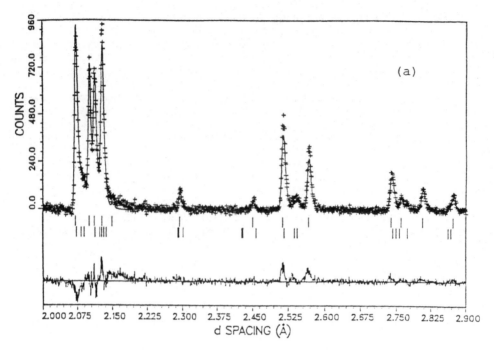

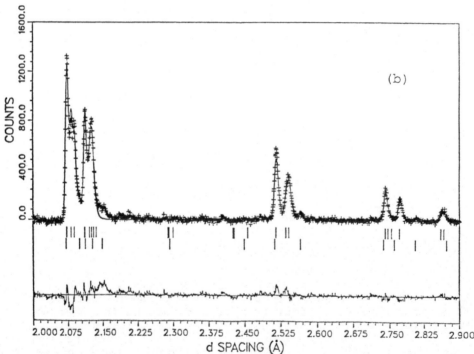

Fig.4 Portion of the Rietveld refinement profiles for the (a) H_2-reduced and (b) N_2-prepared $La_2CoO_{4+\delta}$ samples. Plus marks (+) are the raw data. The solid line is the calculated profile. Thick marks below the diffraction profile indicate the positions of allowed reflections in the model. For the H_2-reduced sample the upper row of thick marks is for the orthorhombic Cmca phase; the second row is for the monoclinic C2/m phase. The order of the thick marks is inverted for the N_2-prepared sample.

studied at room temperature and at 10K. Fig.4 shows room temperature Rietveld refinement profiles for La_2CoO_4 and $La_2CoO_{4.15}$. Both these compositions show two-phase behavior similar to that in the $La_2CuO_{4+\delta}$ and $La_2NiO_{4+\delta}$ systems. However, unlike those cases in which the phase separation was observable only with high instrumental resolution, the splittings of many peaks are now clearly resolved. The majority phase (~90%) of the $\delta=0.00$ sample has the orthorhombic Cmca structure. The minor phase of the $\delta=0.00$ sample has identical peak positions and intensities as the majority phase (~95%) of the $\delta=0.15$ sample. This oxygen-rich phase has a monclinic C2/m structure. The $\delta=0.00$ and 0.15 samples are, thus, close to the borders of the miscibility gap and we see that our initial synthesis methods did not produce a sample in the middle of the gap. The $\delta=0.27$ sample is single phase and has an orthorhombic structure, implying yet another phase line in the region $0.15<\delta<0.27$. The full determination of the structures of these oxygen-rich phases, including the structures of the oxygen defects, is in progress.

TRANSPORT PROPERTIES

The temperature dependence of the resistivity depends strongly on oxygen content for $La_2CuO_{4+\delta}$. Fig. 5 shows the resistivities for samples annealed in N_2, 0.1 kbar O_2 and 25 kbar O_2. The single-phase N_2-annealed sample is semiconducting. The 25-kbar O_2-annealed sample is metallic and superconducting with a sharp transition at 38K. The two-phase 0.1-kbar sample displays a broad resistive anomaly in the 150-300^{0}C range and a sharp superconducting drop below 38K. The resistive anomaly is due to the phase separation process which develops in the same temperature range and is present only for two-phase samples[17]. In contrast, all $La_2NiO_{4+\delta}$ samples were semiconducting in the 100-300K range. We were unable to measure resistivities for these samples below 100K due to their very high resistivity. $La_2CoO_{4+\delta}$ samples also have very high room temperature resistivities.

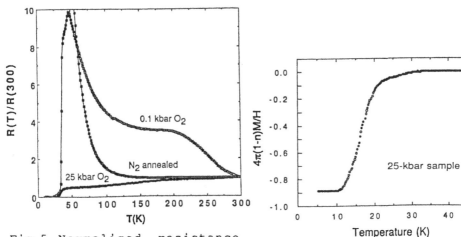

Fig.5 Normalized resistance for the N_2-, 0.1-kbar O_2-, and 25-kbar O_2-annealed $La_2CuO_{4+\delta}$ samples.

Fig.6 Normalized diamagnetic shielding for the 25-kbar O_2-annealed $La_2CuO_{4+\delta}$ sample.

Magnetic susceptibility measurements confirmed bulk superconductivity in the $La_2CuO_{4+\delta}$ samples annealed at high oxygen pressures. Fig. 6 shows diamagnetic shielding data for the 25 kbar oxygen-annealed sample. The superconducting volume fraction is near 90%. The stoichiometric compound ($\delta=0.00$) is antiferromagnetic with $T_N \approx 310K$. The two-phase sample displays both the onset of antiferromagnetism at $T_N \approx 260K$ due to the presence of the nearly-stoichiometric phase and superconductivity due to the presence of the oxygen-rich $La_2CuO_{4.08}$ phase. The decrease of T_N from 310K for the stoichiometric compound to 260K for the two-phase sample indicates that the topology of the miscibility gap allows the incorporation of a small amount of excess oxygen into the Cmca structure. We found no evidence for superconductivity in the $La_2NiO_{4+\delta}$ and $La_2CoO_{4+\delta}$ samples.

CONCLUSION

We have studied the structural and transport properties of the isostructural $La_2MO_{4+\delta}$ (M=Cu, Ni, Co) compounds as a function of the excess oxygen content δ. The stoichiometric ($\delta=0$) compounds have the same Cmca orthorhombic structure and are semiconducting and antiferromagnetic. Very small concentrations of excess oxygen can be introduced into this phase. The nature of this oxygen defect is presently unknown. For larger concentrations of excess oxygen the compounds separate into a nearly-stoichiometric La_2MO_4 phase and an oxygen-rich $La_2MO_{4+\delta}$ phase with $\delta \approx 0.08$, 0.13, 0.16 for the Cu, Ni, Co compounds, respectively. Only the oxygen-rich $La_2CuO_{4.08}$ phase is metallic and superconducting. For the Ni and Co compounds, annealing in one 1 bar of oxygen produces samples with excess oxygen concentrations beyond the miscibility gap and the properties of the oxygen-rich phases can be easily studied. The Cu compound, however, has a much lower solubility for excess oxygen and very high (kbar-range) oxygen pressures are required to produce single-phase oxygen-rich samples.

* Work supported by National Science Foundation-Office of Science and Technology Centers under contract #STC-8809854 (BD, SP, KGV) and U.S Department of Energy, BES, under contract #W-31-109-ENG-38 (DGH, JDJ, GWC). One of us (DRR) would like to thank American Air Liquide for financial support.

REFERENCES

1. J.D. Jorgensen, B. Dabrowski, Shiyou Pei, D.G. Hinks, L. Soderholm, B. Morosin, L.E. Schiber, E.L. Venturini and D.S. Ginley, Phys. Rev. B 38, 11337 (1988) and references therein
2. A.H. Davies and R.J.D. Tilley, Nature 326, 859 (1987).
3. R.A.M. Ram, L. Ganapathi, P. Ganguly and C.N.R. Rao, J. Solid State Chem. 63, 139 (1986).
4. J.T. Lewandowski, R.A. Beyerlein, J.M. Longo and R.A. McCauley, J. Amer. Ceram. Soc. 69, 699 (1986).
5. B. Dabrowski, D.G. Hinks, J.D. Jorgensen, K. Zhang and C.U. Segre, Bull. Amer. Phys. Soc. 33, 557 (1988).
6. D.J. Butterey, P. Ganguly, J.M. Honig, C.N.R. Rao, R.R. Schartman and G.N. Subbanna, J. Solid State Chem. 74, 233

(1988).
7. R.A.M. Ram, P. Ganguly and C.N.R. Rao, Mat. Res. Bull. 23, 501 (1988).
8. G. Demazeau, C. Parent, M. Pouchard and P. Hagenmuller, Mat. Res. Bull. 7, 913 (1972).
9. T. Nakamura, G. Petzow and L.J. Gauckler, Mat. Res. Bull. 14, 649 (1979).
10. D.C. Johnston, J.P. Stokes, D.P. Goshorn and J.T. Lewandowski, Phys. Rev. B 36, 4007 (1987).
11. J.E. Schirber, E.L. Venturini, B. Morosin, J.F. Kwak, D.S. Ginley, and R.J. Baughman, Mater. Res. Soc. Symp. Proc. Vol. 99, 479 (1988).
12. P. Ganguly, in "Advances in Solid State Chemistry" (C.N.R. Rao, Ed.). p. 135ff, Indian National Scince Academy, New Delhi (1986).
13. J.D. Jorgensen, B. Dabrowski, S. Pei, D.R. Richards and D.G. Hinks, Phys. Rev. B 40, 2187 (1989).
14. J.W. Rogers, Jr., N.D. Shinn, J.E. Schirber, E.L. Venturini, D.S. Ginley and B. Morosin, Phys. Rev. B 38, 5021 (1988).
15. J. Zhou, S. Sinha and J.B. Goodenough, Phys. Rev. B 39, 12331 (1989).
16. J.D. Jorgensen, H.B. Schuttler, D.G. Hinks, D.W. Capone II, K. Zhang, M.B. Brodsky, and D.J. Scalapino, Phys. Rev. Lett. 58, 1024 (1987).
17. B. Dabrowski, D.G. Hinks, J.D. Jorgensen and D.R. Richards (to be published in Mater. Res. Soc. Symp. Proc.)

COMPOSITIONAL DEPENDENCE AND CHARACTERISTICS OF THE SUPERCONDUCTIVE Bi-Sr-Ca-Cu-O SYSTEM*

H. L. Luo, S. M. Green †, Yu Mei and A. E. Manzi

Department of Electrical and Computer Engineering, R-007
University of California, San Diego
La Jolla, California 92093-0407

ABSTRACT

The results of a systematic study on compositional dependence of the Bi-Sr-Ca-Cu-oxide superconductors are presented. The optimal range of Pb substitution for Bi in the stability of $Bi_{2-x}Pb_xSr_2Ca_2Cu_3O_\delta$ is determined to be $0.20 \leq x \leq 0.35$. There is a limited range of solid solution for Sr replacing Ca. But Bi substituting Ca can lead to the instability of 2223 structure. Once the 2223 is formed, its oxygen content is quite stable.

INTRODUCTION

There are at least three superconducting polytypoids in the high transition-temperature (T_c) Bi-Sr-Ca-Cu oxide (BSCCO) system [1,2] which can be described by the general formula $Bi_2Sr_2Ca_{n-1}Cu_nO_\delta$ (n = 1, 2 and 3). [3-6] They are often referred by their cation ratio as 2201, 2212 and 2223; and when properly prepared, their T_c are in the ranges of 10-20, 70-80 and 105-115 K respectively. The three, although very closely related in structure, are quite distinct in their formation and stability. The 2212 type appears to be the most stable. Under normal circumstance, the 2212 and 2223 often coexist in most grains with the former in the extensive and the later in the core region of the grains. [7,8] The superconductivity of the 2201 type which was first reported with T_c at ~ 20 K [1] has been quite elusive. More often than not, polycrystalline samples of 2201 which were structurally correct are not superconductive down to 4 K. [1,9]

The 2223 polytypoid can be stabilized and promoted through partially replacing Pb by Bi [7,10] and zero resistance above 105 K can be attained readily in ceramic samples. [11,12] Recently, through careful preparation virtually single-phase material of 2223 has been reported. [13]

---

* This work is supported by the California MICRO Program and Hughes Aircraft Company
† Currently at the Superconductivity Center, University of Maryland, College Park, Maryland

According to the literature, there appeared to exist a range of random substitution between Sr and Ca which prompt the expression $Bi_2(Sr,Ca)_3Cu_2O_\delta$.[4,6,14,15] High resolution transmission electron microscopy reveals random occupation of Ca sites by Bi ions in both 2212 and 2223 polytypoids.[16]

The role of oxygen in the Cu-O based high-T_c superconductors is a fascinating question. In $RBa_2Cu_3O_{7-y}$ (RBCO, where R represents any one of the rare-earth elements except Ce, Pr, Tb),[17-20] it has been established that the oxygen parameter y can vary between 0 and 1. For y = 0, the normal-state resistivity is metallic and the compound is superconductive with $T_c \approx 95\,K$. For y = 1, the general character of the crystalline structure does not change substantially, but the resistivity becomes semiconductor-like and the superconductivity disappears altogether.[21] The absorption and desorption of oxygen by as much as one atom per formula unit is completely reversible. However, in BSCCO, the effect of oxygen content is less clear.

In this study, a systematic approach on some of these questions is attempted. The work on the 2223 type is emphasized.

EXPERIMENTAL PROCEDURES

For the study of different compositional variants in 2223, three groups of samples were prepared with the following nominal compositions:

(A) $Bi_{2-x}Pb_xSr_2Ca_2Cu_3O_\delta$
    x = 0, 0.2, 0.25, 0.3, 0.35, 0.4, 0.6, 0.8, 1.0, 1.5, 2.0;

(B) $Bi_{1.75}Pb_{0.25}Sr_{2-x}Ca_{2+x}Cu_3O_\delta$
    x = −0.50, −0.25, 0.0, 0.25, 0.50;

(C) $Bi_{1.75-x}Pb_{0.25}Sr_2Ca_{2+x}Cu_3O_\delta$
    x = −0.2, −0.1, 0.0, 0.1, 0.2.

Samples within each group were processed simultaneously and are assumed to share identical thermal histories.

To investigate the effect of oxygen, a master sample of fixed composition $Bi_{1.65}Pb_{0.35}Sr_2Ca_2Cu_3O_\delta$ was employed. For reference purpose, samples of 2212 and 2201 were also prepared.

Approximate amounts of Bi_2O_3, PbO, $SrCO_3$, $CaCO_3$ and CuO (all at least 99.9% purity) were mixed and ground together until visually homogeneous. After prefiring the mixtures at 750–800°C overnight, the powders were thoroughly reground and pressed into 1-cm diameter pellets. The samples were then sintered at 845–850°C for 60 - 120 hours with at least one intermediate regrinding. Except oxygen-controlled experiments, all heat treatments were conducted in air and were allowed to furnace-cool to room temperature. The controlled oxygenation experiments were carried out in the form of post-annealing at

740°C under various oxygen partial pressure. Oxygen desorption was attempted by heating samples at ~ 700°C in pure argon atmosphere.

Powder x-ray diffraction patterns which were obtained using Ni-filtered CuK$_\alpha$ radiation over the interval of $2\theta = 20$–$50°$ were used as the first screening for phase identification. Resistivity data was collected on bar-shaped samples approximately $10\times2\times1$ mm^3 from room temperature to 77 K. Electrical contacts consisted of spring-loaded Rh-plated steel needles pressed against small silver point pads in the standard four-point configuration. About 1 - mA rms current was supplied at 40 Hg with the sample's voltage drop monitored through a lock-in amplifier. Zero resistance point was defined as the temperature at which the lock-in reached its signal detection limit of ~ 0.5 $\mu\Omega$–cm. DC magnetic susceptibility was measured using a SQUID magnetometer (Quantum Design MPMS) at 20 Oe over the temperature range 5-300 K on the same bar samples used for resistivity measurements.

RESULTS AND DISCUSSION

(1) The Replacement of Bi by Pb

The purpose of studying the series of samples $Bi_{2-x}Pb_xSr_2Ca_2Cu_3O_\delta$ was to determine the optimal Pb content for the stability of 2223 polytypoid. The x-ray diffraction patterns are sufficient to differentiate the phases involved. For $x = 2$, the diffraction pattern, though not yet completely identified, is totally different from any of the BSCCO family. This phase begins to appear for $x = 0.4$ and becomes more intense as x increases. On the other hand, mixed patterns of 2212 and 2223 with various strengths were always observed in the range $x < 0.20$. With repeated regrinding and prolonged annealing, single phase material of 2223 has been achieved between $x = 0.20$ and 0.35 as shown in Fig. 1 and listed in Table I.

(2) The Variation of the Sr/Ca Ratio

The effect of Sr-Ca substitution is best illustrated by their magnetic susceptibilities (Fig. 2). The resistivity data is also shown in Fig. 3. The sample most deficient in Ca ($x = -0.5$) displays the predominance of the 2212 polytypoid. Its normal state resistivity is much lower than others which is characteristic of the 2212 type as compared to 2223.

Two of the samples ($x = -0.25$ and $x = 0$) attained zero resistance above 100 K, indicating that a significant fraction of the superconducting grains are not separated from each other by a lower-T_c material. Both samples demonstrate a sharp diamagnetic onset at ~ 110 K with an additional step at 70-75 K which signals the 2212 presence. The similar behavior in resistivity and susceptibility of these two samples might be interpreted as indicating a range of solid solution between Sr and Ca within the 2223 polytypoid.

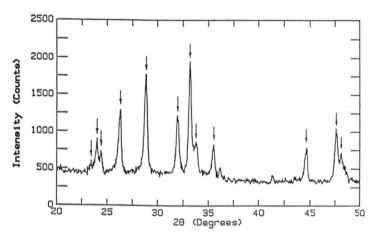

Figure 1. Powder x-ray diffraction pattern of the 2223 polytypoid, using CuK$_\alpha$ radiation. The arrows indicate the peaks used to compute the latter parameters.

Table I. Powder x-ray diffraction data for $Bi_{1.7}Pb_{0.3}Sr_2Ca_2Cu_3O_y$.

h k l	2θ	I_{obs}/I_o	d_{obs} (Å)	d_{calc} (Å)
008	19.17	6	4.630	4.637
111	23.43	10	3.797	3.799
00 10	23.98	29	3.711	3.710
113	24.40	19	3.648	3.649
115	26.25	57	3.395	3.395
117 \} 00 12	28.83	88	3.097	3.098 \} 3.092
119	31.94	53	2.802	2.801
200	33.17	100	2.701	2.700
00 14	33.77	28	2.654	2.650
11 11	35.52	25	2.527	2.528
206	36.25	8	2.478	2.474
20 10	41.33	7	2.184	2.183
20 12	44.60	23	2.032	2.034
220	47.61	40	1.910	1.909
11 17 \} 12 14	48.06	19	1.893	1.895 \} 1.892

Lattice parameters a = 5.400Å, b = 5.401Å, c = 37.101Å.

On the other hand, the samples deficient in Sr are severely affected with considerably higher normal-state resistivity and lower zero-resistance temperature. The diamagnetic transitions are also much broader. This experimental evidence points out that Ca enrichment at the expense of Sr is detrimental to the high-T_c superconductivity in the BSCCO system.

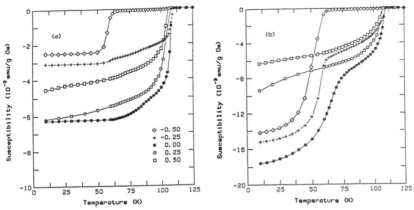

Figure 2. (a) Meissner and (b) shielding data for samples with normal composition $Bi_{1.75}Pb_{0.25}Sr_{2-x}Ca_{2+x}Cu_3O_\delta$ (x as indicated).

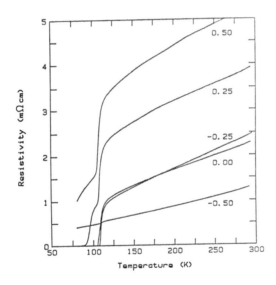

Figure 3. Temperature-dependence of resistivity for samples with normal composition. $Bi_{1.75}Pb_{0.25}Sr_{2-x}Ca_{2+x}Cu_3O_\delta$ (x as indicated).

(3) The Substitution Between Bi and Ca

The susceptibility and resistivity data are shown in Figs. 4 and 5. The latter have been normalized to the room-temperature values which are in the range of 2.5-3.0 $\mu\Omega$-cm for all samples in this group. Bi-rich samples display stepped resistive transition. The height of the step and the zero-resistance temperature show systematic variation. The Meissner data also demonstrate a similar trend: A shoulder at ~ 70 K which corresponds to the 2212 presence becomes progressively sharper with decreasing x. Thus, increasing the Bi content at the expense of Ca promotes the 2212 formation. This view is reasonable since, relatively speaking, the 2212 type is rich in Bi and deficient in Ca as compared to 2223. Bi occupying Ca sites may induce the instability of the 2223 structure in favor of 2212.

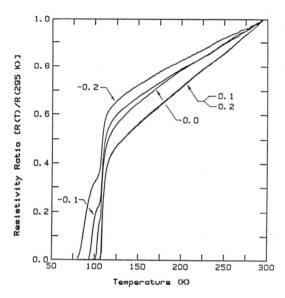

Figure 4. Temperature dependence of normalized resistivity for samples with normal composition $Bi_{1.75-x}Pb_{0.25}Sr_2Ca_{2+x}Cu_3O_\delta$ (x as indicated).

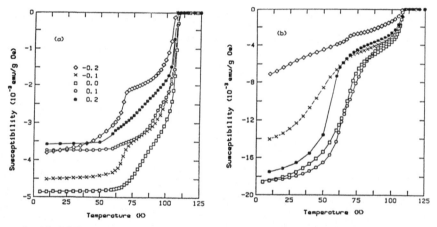

Figure 5. (a) Meissner and (b) shielding parts of susceptibility for samples with nominal composition $Bi_{1.75-x}Pb_{0.25}Sr_2Ca_{2+x}Cu_3O_\delta$ (x as indicated).

The resistive transitions of Ca-rich samples are step-free with zero-resistance at ~ 107 K. The situation is totally different from that in the Ca/Sr replacement series. The direct contrast simply means that, if the Ca enrichment is at the expense of Sr, intergranular connectivity is disrupted, while if it is at the expense of Bi, then the 2223 material is contiguous. At this juncture, an important remark must be made. For Ca-rich samples, the shielding curves show a two-step diamagnetic response in the Bi/Ca series, but a single transition in the Sr/Ca replacements. This observation serves to emphasize that the nature of the resistive and magnetic signals for the same superconductive material does not correlate in this complicated oxide system. A two-stage diamagnetic response does not imply a stepped resistive transition, and a step-free susceptibility curve does not necessarily correspond to a single resistive transition. The magnetic data provide information regarding the relative amounts of the various superconducting phases present together with the effects due to flux trapping and expulsion. The resistive signals simply indicate how well the highest- T_c regions connect to each other and manifest to a certain degree the proximity effect and tunneling phenomenon.

(4) <u>The Issue Regarding Oxygen</u>

The effect due to oxygen content is very intriguing. Obviously the issue is not as clear-cut and drastic as in the RBCO superconductors. The post-annealing at various partial pressures indeed raised T_c, but only to a minor degree, from 107 K at 1 bar to 110 K at 10 mbar of oxygen partial pressure (Fig. 6). The complete resistivity and susceptibility data for the sample post-annealing at the optimal oxygen pressure (10 mbar) are presented in Fig. 7. Note that the zero-resistance occurs at 109 K which is among the highest reported in the BSCCO system. The present result gives no indication that over-oxygenation destabilizes the 2223 structure once it has formed. The slight diminution of T_c is in distinct contrast to the 2212 polytypoid whose T_c changes by as much as 20% over a similar range of oxygen partial pressure. [22,23]

The oxygen desorption experiments yield results which are also consistent. Annealing at ~ 700°C in pure oxygen does not cause any change in the 2223 structure and the T_c remains nearly the same. This part of the work is continuing.

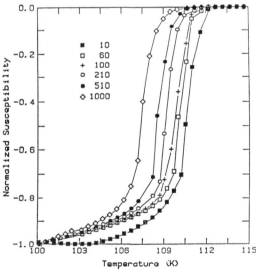

Figure 6. The effect of partial oxygen pressure on the T_c of single-phase 2223 samples.

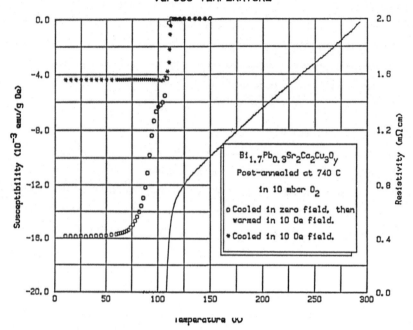

Figure 7. Susceptibility and resistivity of a 2223 sample after post-annealing at the optimal oxygen partial pressure.

REFERENCES

1. C. Michel, M. Hervieu, M. M. Borel, A. Grandin, F. Deslandes, J. Provost, and B. Raveau, *Z. Phys. B* **68**: 421 (1987).

2. M. Maeda, Y. Tanaka, M. Fukutomi, and T. Asano, *Jpn. J. Appl. Phys. Lett.* **27**: L209 (1988).

3. H. W. Zandbergen, Y. K. Huang, M. J. V. Menken, J. N. Li, K. Kadowaki, A. A. Menovsky, G. van Tendeloo, and S. Amelinckx, *Nature* **332**: 620 (1988).

4. M. A. Subramanian, C. C. Torardi, J. C. Calabrese, J. Gopalakrishnan, K. J. Morrissey, T. R. Askew, R. B. Flippen, U. Chowdhry, and A. W. Sleight, *Science* **239**: 1015 (1988).

5. S. M. Green, Yu Mei, C. Jiang, H. L. Luo, and C. Politis, *Mod. Phys. Lett.* **B2**: 915 (1988).

6. J. M. Tarascon, Y. LePage, P. Barboux, B. G. Bagley, L. H. Greene, W. R. McKinnon, G. W. Hull, M. Giroud, and D. M. Hwang, *Phys. Rev.* **B37**: 9382 (1988).

7. R. Ramesh, G. Thomas, S. M. Green, C. Jiang, Yu Mei, M. L. Rudee, and H. L. Luo, *Phys. Rev. B* **B38**: 7070 (1988).

8. R. Ramesh, G. Thomas, S. M. Green, Yu Mei, C. Jiang, and H. L. Luo, *Appl. Phys. Lett.* **53**: 1759 (1988).

9. J. M. Tarascon, W. R. McKinnon, P. Barboux, D. M. Hwang, B. G. Bagley, L. H. Greene, G. W. Hull, Y. LePage, N. Stoffel, and M. Giroud, *Phys. Rev.* **B38**: 8885 (1988).

10. R. J. Cava, B. Batlogg, S. A. Sunshine, T. Siegrist, R. M. Fleming, K. Rabe, L. F. Schneemeyer, D. W. Murphy, R. B. van Dover, P. K. Gallagher, S. H. Glarum, S. Nakahara, R. C. Farrow, J. J. Krajewski, S. M. Zahurak, J. V. Waszczak, J. H. Marshall, P. Marsh, L. W. Rupp, Jr., W. F. Peck, and E. A. Rietman, *Physica C* **153-155:** 560 (1988).

11. S. M. Green, C. Jiang, Yu Mei, H. L. Luo, and C. Politis, *Phys. Rev.* **B38:** 5016 (1988).

12. B. W. Statt, Z. Wang, M. J. G. Lee, J. V. Yakhmi, P. C. deCamargo, J. F. Major, and J. W. Rutter, *Physica C* **156:** 251 (1988).

13. Yu Mei, S. M. Green, C. Jiang, and H. L. Luo, *J. Appl. Phys.* **66:** 1777 (1989).

14. Z. Iqbal, H. Eckhardt, F. Reidinger, A. Bose, J. C. Barry, and B. L. Ramakrishna, *Phys. Rev.* **B38:** 859 (1988).

15. G. S. Grader, E. M. Gyorgy, P. K. Gallagher, H. M. O'Bryan, D. W. Johnson, Jr., S. Sunshine, S. M. Zahurak, S. Jin, and R. C. Sherwood, *Phys. Rev.* **B38:** 757 (1988).

16. C. J. D. Hetherington, R. Ramesh, M. A. O'Keefe, R. Kilaas, G. Thomas, S. M. Green, and H. L. Luo, *Appl. Phys. Lett.* **53:** 1016 (1988).

17. H. L. Luo, S. M. Green, C. Jiang, Yu Mei, and C. Politis, *Metal. Trans.* **19A:** 734 (1988).

18. E. M. Engler, V. Y. Lee, A. I. Nazzal, R. B. Beyers, G. Lim, P. M. Grant, S. S. P. Parkin, M. L. Ramirez, J. E. Vazquez, and R. J. Savoy, *J. Am. Chem. Soc.* **109:** 2848 (1987).

19. Z. Fisk, J. D. Thompson, E. Zirngiebl, J. L. Smith, and S. W. Cheong, *Solid State Commun.* **62:** 743 (1987).

20. S. Tsurumi, M. Hitkita, T. Iwata, K. Semba, and S. Kurihara, *Jpn. J. Appl. Phys.* **26:** L856 (1987).

21. J. D. Jorgensen, M. A. Beno, D. G. Hinks, L. Soderholm, K. J. Volin, R. L. Hitterman, J. D. Grace, I. K. Schuller, C. V. Segre, K. Zhang, and M. S. Kleefisch, *Phys. Rev.* **B36:** 3608 (1987).

22. D. E. Morris, C. T. Hultgren, A. M. Markelz, J. Y. T. Wei, N. G. Asmar, and J. H. Nickel, *Phys. Rev.* **B39:** 6612 (1989).

23. R. G. Buckley, J. L. Tallon, I. W. M. Brown, M. R. Presland, N. E. Flower, P. W. Gilberd, M. Bowden, and N. B. Midestone, *Physica* **C156:** 629 (1988).

GRAIN BOUNDARY DIFFUSION AND SUPERCONDUCTIVITY

IN Bi-Pb-Sr-Ca-Cu-O COMPOUND

T.K. Chaki and Shiaw C. Tseng
Department of Mechanical and Aerospace Engineering
State University of New York
Buffalo, NY 14260

INTRODUCTION

The discovery [1] of high temperature oxide superconductors has caused tremendous excitement due to great technological importance. The critical current in polycrystalline oxide superconductors is low. Cava et al. [2] first measured critical current density, J_c in $Y_1Ba_2Cu_3O_7$ by transport method and reported J_c at 77K and zero magnetic field to be only 1100 A/cm^2, compared to 10^5 A/cm^2 reported [3] for epitaxial films. Moreover, the critical current in the polycrystalline material decreases drastically [4] at a low magnetic field. Similar low critical currents have also been observed [5] in polycrystalline $Bi_2CaSr_2Cu_2O_8$ and $Tl_2Ca_2Ba_2Cu_3O_{10}$ compounds. Weak linking at the grain boundaries is thought [6] to be the cause for low J_c in polycrystalline high-T_c materials. By measuring J_c across single grain boundaries between [001] oriented grains of $Y_1Ba_2Cu_3O_7$, Mannhart et al. [7] have shown that high angle tilt grain boundaries behave like normal layers in SNS-type Josephson junctions.

The critical current is likely to change if the coupling between the grains is altered. Introduction of foreign atoms by grain boundary diffusion [8] can possibly change the coupling between the grains. Grain boundary diffusion can occur at temperatures below those at which lattice diffusion is predominant and thus, can dope the grain boundaries selectively without affecting the interior of the grains. In this paper, we report the effects of grain boundary diffusion of Zn and Sb on the superconducting properties in $Bi_{1.5}Pb_{0.5}Sr_2Ca_2Cu_3O_{10}$ compounds. Zinc has low melting point and high vapor pressure and can easily diffuse from the vapor phase. The effects of bulk doping of Zn on superconductivity in $Y_1Ba_2Cu_3O_7$ [9] and Bi-Sr-Ca-Cu-O system [10] have been studied. In both materials, the transition temperature, T_c is depressed, but the lattice parameters change only slightly. Thus, Zn has destructive effect on superconductivity in oxide materials. On the other hand, Sb is known [11] to stabilize 110 K phase in

Bi-Pb-Sr-Ca-Cu-O system. The present study shows that grain boundary diffusion of both Zn and Sb decreases coupling between grains.

EXPERIMENTAL

The specimens of $Bi_{1.5}Pb_{0.5}Sr_2Ca_2Cu_3O_{10}$ were prepared by solid state sintering. A mixture of $SrCO_3$, $CaCO_3$, CuO, Bi_2O_3 and PbO in proper proportions is ground and annealed at 840°C for 48 hours. The material was reground and compressed into pellets. The pellets were annealed at 860°C for 120 hours and then quenched into liquid nitrogen. The pellets of original size 9mm x 6mm x 4mm were ground by a file to make bridge-type specimens. The cross section of the neck was 2mm x 2mm and the length was 3mm. T_c and J_c were measured on the bridge-type specimens by transport technique at zero magnetic field. During T_c measurement, a current of 3mA was sent through the specimens.

The bridge specimens were encapsuled in pyrex glass tubes under vacuum. In some of the capsules, zinc chips along with the high-T_c specimens, were introduced. The specimens in the capsules without zinc, were used as control specimens for comparison. Different batches of the capsules were annealed at 260° and 450°C for 150 hours. Some specimens encapsuled with Sb chips were annealed at 300°C for 170 hours. The capsules were taken out of the furnace and cooled. The capsules were broken, and T_c and J_c were measured both in control specimens and in specimens with Zn and Sb diffusion. A few specimens were prepared by bulk replacement of a few at % Cu by Zn. T_c and J_c of these specimens were also measured.

The microstructure of sintered specimens was observed by examining polished surfaces by a scanning electron microscope. A few specimens after Zn diffusion were fractured in the neck region. The fracture surface was examined by an Auger electron microscope to detect the presence of Zn.

RESULT

Fig. 1 shows a secondary electron image of sintered superconducting $Bi_{1.5}Pb_{0.5}Sr_2Ca_2Cu_3O_{10}$ compound. Long plate-like grains about 10 μm long and 1 μm wide, were observed. Pores and second phase were also present. Typical resistivity vs. temperature plot is shown in curve (a) of Fig. 2. T_c at zero resistance was (105 ± 1) K and transport J_c at 40 K was $(12.5 \pm 0.5) A/cm^2$. A bulk replacement of 2.5 mol.% of Cu by Zn depressed T_c drastically, as shown in curve (b) of Fig. 2. T_c was reduced to 64K and J_c at 40 K was only 1.3 A/cm^2. The zinc-doped samples behaved like a semiconductor as the resistivity increased upon cooling from room temperature to 105 K. The step at 90 K indicates that Zn affects 110 K phase in Bi-Sr-Ca-Cu-O system more adversely than 90 K phase. The specimens with 5 mol.% of Cu replaced by Zn did not show superconductivity during cooling up to 20 K.

Fig. 3 shows resistivity vs. temperature plots for specimens with grain boundary diffusion of Zn. The specimens annealed at 260 and 450°C for 150 hrs without any Zn diffusion have T_c of (105 ± 1) K and transport J_c at 40 K equal to (12.5 ± 0.5) A/cm^2. For specimens subject

Fig.1. Scanning electron micrograph of sintered $Bi_{1.5}Pb_{0.5}Sr_2Ca_2Cu_3O_{10}$ compound

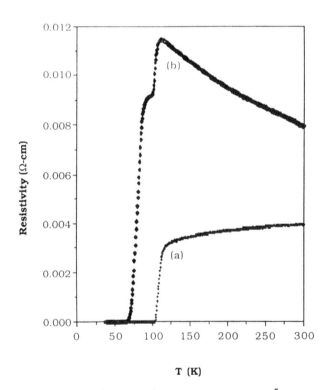

Fig.2. Resistivity vs. temperature curves for $Bi_{1.5}Pb_{0.5}Sr_2Ca_2Cu_3O_{10}$ specimens with bulk replacement of Cu by Zn. (a) No zinc (b) 2.5 mole % of Cu is replaced by Zn.

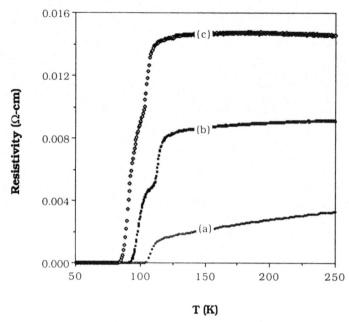

Fig.3. Resistivity vs. temperature curves for sintered $Bi_{1.5}Pb_{0.5}Sr_2Ca_2Cu_3O_{10}$ specimens, showing depression of T_c due to grain boundary diffusion of Zn. (a) Annealed at 260°C for 150 hrs. without Zn. (b) Zn diffusion at 260°C for 150 hrs. (c) Zn diffusion at 450°C for 150 hrs.

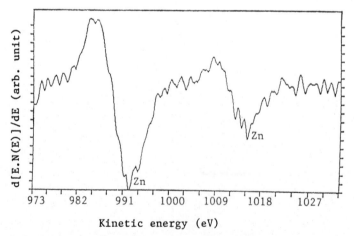

Fig.4. Auger electron spectrum of fractured surface of $Bi_{1.5}Pb_{0.5}Sr_2Ca_2Cu_3O_{10}$ specimens, showing presence of Zn, upon diffusion at 260°C for 150 Hrs.

to Zn diffusion at 260°C for 150 hrs, T_c dropped to (92 ± 1) K and transport J_c at 40 K decreased to (4.2 ± 0.3) A/cm^2. T_c and J_c values are average of 4 specimens. At temperatures as low as 260°C, lattice diffusion of Zn in $Bi_{1.5}Pb_{0.5}Sr_2Ca_2Cu_3O_{10}$ is inappreciable. Thus, grain boundary diffusion which can occur at relatively low temperatures decreases T_c and J_c. Due to diffusion at 450°C for 150 hours, T_c dropped further to (83 ± 1) K and J_c at 40 K to (3.3 ± 0.3) A/cm^2.

The normal state resistivity in $Bi_{1.5}Pb_{0.5}Sr_2Ca_2Cu_3O_{10}$ increased due to Zn diffusion. The resistivity at room temperature increased from 3.7 to 7.0 milliohm-cm due to Zn diffusion at 260°C for 150 hrs. It became 14.5 milliohm-cm due to Zn diffusion at 450°C for 150 hrs. Thus, grain boundary diffusion of Zn makes grain boundaries in $Bi_{1.5}Pb_{0.5}Sr_2Ca_2Cu_3O_{10}$ more insulating. Auger electron microscopic study detected presence of Zn (Fig. 4) on the fracture surface of the specimens subject to Zn diffusion at 260 and 450°C. On polished surfaces, the presence of Zn could not be detected. The fracture in high-T_c superconductors is predominantly intergranular [12]. Thus, the fracture surface comprises mostly of grain boundary interfaces which contain higher Zn concentration due to grain boundary diffusion.

$Bi_{1.5}Pb_{0.5}Sr_2Ca_2Cu_3O_{10}$ specimens which underwent Sb diffusion at 300°C for 170 hrs. became semiconducting and the resistivity increased with decreasing temperatures. They did not show any superconducting transition during cooling up to 20 K. This result is very surprising, since bulk substitution of Sb is beneficial [11]. Grain boundary region is in disordered state [13]. It is not clear how doping of foreign elements is affecting the electronic properties of grain boundaries.

CONCLUSION

Grain boundary diffusion of Zn and Sb in $Bi_{1.5}Pb_{0.5}Sr_2Ca_2Cu_3O_{10}$ at relatively low temperatures has detrimental effects on superconducting properties. Due to diffusion of Zn at 260°C for 150 hrs., T_c at zero resistance decreases from 105 to 92 K and transport J_c at 40 K decreases from 12.5 to 4.2 A/cm². The specimens which underwent Sb diffusion at 300°C for 170 hrs. did not show any superconducting transition during cooling up to 20 K. Grain boundary diffusion decreases coupling between grains and T_c and J_c in polycrystalline ceramic superconductors are reduced.

ACKNOWLEDGEMENT

The work was supported by a grant from New York State Institute on Superconductivity.

REFERENCES

1. J.G. Bednorz and K.A. Muller, Z. Phys. B64:189 (1986).
2. R.J. Cava, B. Batlogg, R.B. VanDover, D.W. Murphy, S. Sunshine, T. Siegrist, J.P. Remeika, E.A. Rietman, S. Zahurak and G.P. Espinosa, Phys. Rev. Lett. 58:1676 (1987).
3. P. Chaudhari, R.H. Koch, R.B. Laibowitz, T.R. McGuire and R.J. Gambino, Phys. Rev. Lett. 58:2684 (1987).
4. G. Paterno, C. Alvani, S. Casadio, U. Gambardella and I. Maritato, Appl. Phys. Lett. 53:609 (1988).

5. H. Kupfer, S.M. Green, C. Jiang, Yu Mei, H.L. Luo, R. Meier-Hirmer and C. Politis, Z. Phys. 71:63 (1988).
6. R.L. Peterson and J.W. Ekin, Phys. Rev. B 37:9848 (1988).
7. J. Mannhart, P. Chaudhari, D. Dimos, C.C. Tsuei and T.R. McGuire, Phys. Rev. Lett. 61:276 (1988).
8. P.G. Shewmon, "Diffusion in Solids", McGraw-Hill Book Company, New York, (1964), p. 164.
9. B. Jayaram, S.K. Agarwal, C.V. Narasimha Rao and A.V. Narlikar, Phys. Rev. B. 38:2903 (1988).
10. T. Kanai, T. Kamo and S-P Matsuda, Jap. J. Appl. Phys. 28:L551 (1989).
11. Z.J. Yang, H. Bratsberg, T.H. Johansen, N. Norman, J. Tafto, G. Helgesen, I. Lorentzen and A.T. Skjeltorp, to be published in Physica C. (1989).
12. J.D. Verhoeven, A.J. Bevolo, R.W. McCallum, E.D. Gibson and M.A. Noack, Appl. Phys. Lett. 52:745 (1988).
13. H. Gleiter, Phys. Stat. Sol. B45:9 (1971).

DISSIPATIVE STRUCTURES IN THIN FILM SUPERCONDUCTING Tl-Ca-Ba-Cu-O

M. L. Chu, H. L. Chang, C. Wang, T. C. Wang‡,
T. M. Uen‡ and Y. S. Gou‡

Institute of Electro-optical Engineering
‡Institute of Electrophysics
National Chiao Tung Univ., Hsinchu, Taiwan, R. O. C.

ABSTRACT

We have measured the temperature dependences of the current-voltage characteristics of Tl-Ca-Ba-Cu-O superconducting thin film. There are direct evidences that the thin film goes through a vortex glass transition well below the superconducting transition temperature at $T_m \simeq 32.5K$. At lower temperatures below T_m, the dissipations are due to the motion of a fixed number of vortices with a vortex-glass structure, while at temperatures above T_m it increases with temperature in proportional to the number of thermally activated vortices in correspondence with Kosterlitz-Thouless (K-T) transition.

Defect and disorder play crucial roles of physical properties in the new high temperature oxide-superconductors. It is well known that defect forms the pinning site for the flux to take seriously affect on the fluctuation of critical current and temperatures, and disorder destroys the translational long-range order of the flux lattice. Furthermore, the oxide-superconductors have layered structure and large anisotropy[1,2], it shows two dimensional behavior manifested by the Kosterlitz-Thouless (K-T) transition[3,4] in the ab-plane. It is already understood that in two dimensional superconductor, it will be in favor of the formation of thermally activated flux vortices. There is a melting temperature (T_m) below the K-T transition temperature (T_c), the superconductor enters a glass state[4] at temperatures below (T_m). Indeed, Fisher[5] has currently formulated a theory to describe the glassy state and predict a supercurrent decay time. Nelson and Seung[6] has formulated a theory of the melting fluid state as well. Mueller[7] etc. first experimentally reported the superconductive glassy state in La_2BaCuO_{4-y} powder samples. Gregory[8] etc. showed evidence of vortex glass state in $Y_1Ba_2Cu_3O_{7-x}$ thin film. Artemenko[9] etc. measured the Kosterlitz-Thouless transition and demonstrated the vortex unbinding in superocnducting single crystal $Bi_2Sr_2CaCu_2O_x$. Yeh and Tsuei[10] measured the K-T transition in $Y_1B_2Cu_3O_7$ single crystals. Although a considerable effort has been made in these topics, all these results are fragmentary with a global description of vortex dynamics interrelated with these defect and disorder systems in tne low dimensionality yet to be given.

In this work, we described our experimental results on the vortex-glass state in thin film of Tl-Ca-Ba-Cu-O with the formation of quasi-particle tunneling line (Q T L). The glass state is manifested as a hysteresis in the current-voltage characteristics of the structure below the melting temperature. At higher temperature, the number of Abrikosov vortices increases following the simple Boltzman factor and indicates its thermally activated nature. In our work the number of vortices freezes below the melting temperature at $T_m=32.5K$.

The Tl-Ca-Ba-Cu-O superconducting thin film was fabricated by dc sputtering followed by a post annealing at ~895°c for 3 minutes. The dimension of the sample is 1cm by 0.5cm and the film thickness is about 1μm. Detailed preparation process of the sample has been published elsewhere.[11] Scanning electron microscopy and x-ray diffraction patterns show that although the thin film is totally c-axis oriented, it exhibits a highly disordered permutation of the grains in the ab plane. Electrical contacts to the films were made with a conductive silver paint. The resulting sample has a superconducting transition temperature about 98K and the resistance is about one percent of R (the resistance at 98K) at 90K, then the resistance vanishes at 72K. All the measurements were made in an Air Products 10K closed-cycle regrigerator system and temperature was controlled to within ±0.1K.

Fig.1 shows that characteristics of current-voltage at 12.7K. The film remains in the superconducting state at current lower than I_{c1}, it was forced into the first dissipative state with further increase of applied current. Then it enters the phase slippage at a constant current level and cause the voltage jump. After the first jump, the second dissipative state will be followed when the applied current is increased further. In the return path of lowering the current, there are some subgap structures in the dissipative state which may be due to Andreev reflection in the S-N interface of the film[12] and/or the

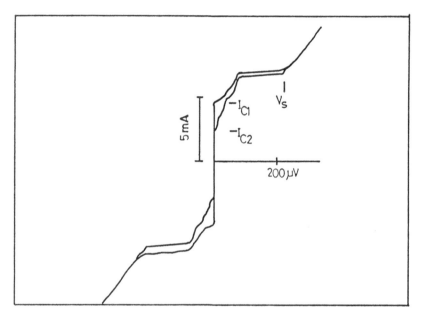

Fig.1   Current-voltage characteristic curve of Tl-Ca-Ba-Cu-O Thin Film at 12.7K.

interaction of the system of microjunctions with Josephson radiations. The film, however, reenters the superconducting state at current lower than I_{c2}. Fig.2 shows the temperature dependence of the two critical current I_{c1} and I_{c2} on the I-V curves. The hysteresis of the I-V curves shows up at temperature below 32.5K. The hysteresis can be manifested or interpreted as disorder pinning effect of the vortex lattice in the glass state and so the temperature at 32.5K can be designated as the freezing temperature (T_m) of vortices. In Fig.1, moreover, the sudden increase of voltage before the onset of the second dissipative current

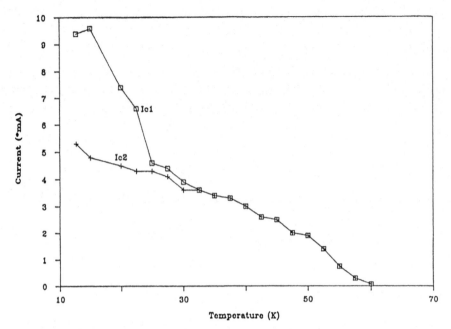

Fig.2 Temperature dependence of critical current. The difference I_{c1} and I_{c2} is caused by the hysteresis.

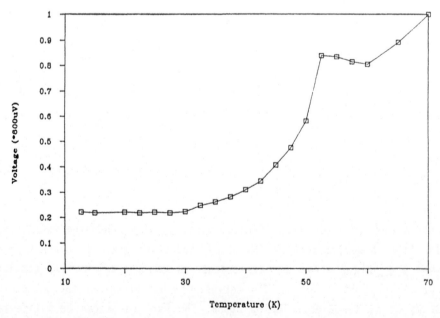

Fig. 3 The voltage V_s after the first voltage jump as a function of temperaure.

state is due to the phase slippage. Fig.3 shows the temperature dependences of the amplitude of the voltage V_s at the end of the first jump. It also shows a characteristic temperature at ~32.5°K, below which the voltage jump remains at a constant value. In what follows we will explain that this is also a typical behavior in the vortex-glassy state. Because of the Josephson ac effect, the magnitude of the voltage jump is dependent on the rate of the transverse vortex flow across the film. Accordingly the constant jump at temperatures below 32.5K can be

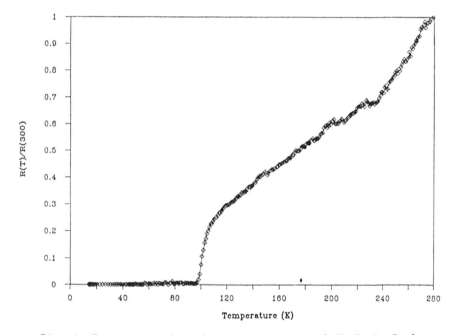

Fig. 4 Temperature dependence resistivity of Tl-Ca-Ba-Cu-O.

illustrated as the rate of the induced fixed number of vortices remained a constant. On the other hand, it is well known that all the physical properties keep unchange in the glassy state. Naviely we believe that the constant jump is a typical behavior for the vortex-glassy state of the Tl-based film at temperatures below T_m. Whereas, it increases at higher temperatures above T_m in seemingly agreement with the Boltzmann factor $e^{-\nu/kT}$, where ν is a measure of the vortex unbinding energy. We note that the jump feature of Fig.3 at temperatures above ~50K could be corresponded to K-T behavior in a quasi-two dimensional system.

We have also measured the temperature dependences of the film resistance as shown in Fig.4. The curve is in qualitative agreement with the observation of Asadov and Mikheenko,[13] who predicts that the form of the curve favors formation of quasi-particle tunneling, and thus all the results presented here can be interpreted as the formation of quasiparticle tunneling lines in the granular films as discussed by the above authors.

To be concluded the thermally activated vortex picture indicates the two dimensional nature of the Tl-Ca-Ba-Cu-O thin films. The disorder permutation of the sueprconduting grains forms the quasiparticle tunneling lines which allow the flow of the vortices in the films results in the dissipative state in the I-V curves. Therefore, some experimental evidences of the temperature dependence of the vortex (dissipative) structure in the Tl-based system can be elucidated.

REFERENCES

1. S. W. Tozer, A. W. Kleinsasser, T. Penney, D. Kaiser and F. Holtzberg, Phys. Rev. Lett. $\underline{59}$, 1768 (1987).
2. L. Ferro, J. Y. Henry, C. Ayache and P. Stamp, Phys. Lett. $\underline{A128}$, 283 (1988).
3. T. M. Kosterlitz and D. J. Thouless, J. Phys. $\underline{C6}$, 1181 (1972).
4. P. Minnhagen, Rev. Mod. Phys. $\underline{59}$, 1001 (1987).
5. M. P. A. Fisher, Phys. Rev. Lett. $\underline{62}$, 1415 (1989).
6. D. R. Nelson and H. S. Seung, Phys. Rev. B $\underline{39}$, 9153 (1989).
7. K. A. Mueller, M. Takashige and J. G. Bednorz, Phys. Rev. Lett. $\underline{58}$, 408 (1987).
8. S. Gregory, C. T. Rogers, T. Venkatesan, X. D. Wu, A. Inam and B. Dutta, Phys. Rev. Lett. $\underline{62}$, 1548 (1989).
9. S. N. Artemenko, I. G. Gorlova and Y. I. Latyshev, Phys. Lett. A, $\underline{138}$, 428 (1989).
10. N. C. Yeh and C. C. Tsuei, Quasi Two-Dimensional Phase Fluctuations in Bulk superconducting $YBa_2Cu_3O_7$ Single Crystals, Phys. Rev. B39:9708 (1989).
11. H. L. Chang, C. Wang, M. L. Chu, T. M. Uen and Y. S. Gou, Japanese J. Appl. Phys. $\underline{28}$, L631 (1989).
12. A. F. Andreev. Soviet Phys. JETP, $\underline{19}$, 1228 (1964).
13. A. K. Asadov and P. N. Mikheenko, Sov. Phys. Solid State $\underline{28}$, 2083 (1986).

GRAIN ORIENTATION IN HIGH T_c CERAMIC SUPERCONDUCTORS

Sudhakar Gopalakrishnan and Walter A. Schulze

NYS College of Ceramics

Alfred University, Alfred, NY-14802

ABSTRACT

Molten salt or the 'flux' method has been a popular technique for growing cystals and particles with anisotropy morphology and has been utilized in this study. Preformed $Ba_2YCu_3O_{7-x}$ powders were reacted with the molten salts of Na, K, and Li belonging to the chloride, sulfate and iodide systems. The $Ba_2YCu_3O_{7-x}$ phase degraded when heat treated in the presence of these salt systems. $Ba_2YCu_3O_{7-x}$ powders prepared in the presence of NaCl-KCl salt system does not appear feasible in the temperature range of 400°C- 900°C. In another method the salt prepared grains were used as seed crystals in the formation of $Ba_2YCu_3O_{7-x}$. This method proved more promising permitting easy removal of the salts from BaY_2CuO_5 by water treatment and formation of phase pure 213 phase. X-ray diffraction analysis and Scanning Electron Microscopy were used to determine the compostion and morphology of the particles.

INTRODUCTION

Grain oriented fabrication of ceramics utilizes the presence of some form of anisotropy in the particles of the starting material to obtain textured microstructures. Since the discovery of $Ba_2YCu_3O_{7-x}$ ceramic superconductors[1] it has been realised that problems with its stability, its fabrication, its brittle nature and its degradation with time and atmospheric conditions are the obstacles in its applications. The existing anisotropy in the current density of $Ba_2YCu_3O_{7-x}$[2,3] necessitates a fabrication process of polycrystalline ceramics which would yield net single crystal-like properties in certain physical direction with the purpose of enhancing the directional current density of the ceramic. Grain oriented processing which yields 'textured' microstructures, is a solution to such a fabrication challenge. This microstructural engineering enhances the favorable properties by controlling the crystallographic direction of each grain in the ceramic. Holmes[4] developed a simpler fabrication process for layered structures from which small particles with platelike morphology are easily grown and orinted during green body formation. The process involved molten salt synthesis of the powder which is composed of small anisotropic single crystals, tape casting, and lamination to orient the powder and form a monolith and controlled heat treatment of the ceramic. This study evaluates the feasibility of producing $Ba_2YCu_3O_{7-x}$ powders with anisotropic particle morphology by molten salt synthesis.

EXPERIMENTAL PROCEEDURE

In the initial attempt reagent grade $BaCO_3$, Y_2O_3 and CuO were mixed in the molar proportions of $Ba_2YCu_3O_{7-x}$. This mixture was ball milled for 24 hours in a polyethylene jar with distilled water and zirconia media. The suspension was dried at 90°C to form a homogeneous powder and then calcined at 930°C for 12 hours and annealed at 530°C for 6 hours. The calcination was carried out in oxygen atmosphere in a muffle furnace. The resulting powder was characterized by x-ray diffraction (XRD) which indicated the presence of the 213 phase. Equal weights of this 213 powder and the salts were mixed using pestle and mortar and heat treated. The salt systems investigated here are the Chlorides, Sulfates, and Iodides of Na, K and Li. All of the molten salt synthesis experiments were carried out in covered magnesia crucibles. The reaction temperature depended on the salt system investigated and varied from 750°C to 1000°C depending on their melting point.

In another attempt the $Ba_2YCu_3O_{7-x}$ - salt equilibrium was approached from the other side by combining reagent grade $BaCO_3$, Y_2O_3 and CuO in molar proportions of 213 with an equal weight of the eutectic mixture of NaCl-KCl salt system. Here the ball milling was carried out in ethanol media because of its limited solubility of the salts. The heat treatment was carried out in closed magnesia crucibles at 900°C with times varying from 2-4 hours.

It has been quite established[5] that the 213 phase is instable in water. It is this fact that led us to approach grain oriented 213 superconductor from an entirely new direction. It has been proved that the BaY_2CuO_5 phase is highly stable with respect to water. Moreover it is known[6] that additions of $BaCO_3$ and CuO to the 121 phase results in the formation of the 213 phase. So we now proceeded in the molten salt synthesis of BaY_2CuO_5. As was the case with 213, we proceeded to approach the equilibrium here also from two directions.

In the initial attempt reagent grade $BaCO_3$, Y_2O_3 and CuO were mixed in molar proportions of BaY_2CuO_5 This mixture was ball milled for 24 hours in a polyethylene jar with distilled water and zirconia media. The suspension was dried at 90°C to form a homogeneous powder and then calcined at 925°C for 12 hours and annealed at 530°C for 6 hours. The calcination was carried out in an oxygen atmosphere in a muffle furnace. The resulting powder was characterized by x-ray diffraction (XRD) which indicated the presence of the 121 phase. Equal weights of this 121 powder and the salts were mixed using pestle and mortar and reacted. The salt systems investigated here are the Chlorides, Sulfates, and Iodides of Na, K and Li. All of the molten salt synthesis experiments were carried out in covered magnesia crucibles. The reacting temperature depended on the salt system investigated and was 825°C for Chloride systems, 925°C for the sulfates and 700°C for the iodides.

In another attempt the BaY_2CuO_5 - salt equilibrium was approached from the other side by combining the reagent grade $BaCO_3$, Y_2O_3 and CuO in molar proportions of 121 with an equal weight of the eutectic mixture of NaCl-KCl salt system. Here the ball milling was carried out in ethanol media because of its limited solubility to the salts. The reaction was carried out in closed magnesia crucibles at 900°C with times varying from 2-4 hours.

The salts from the resulting mixtures obtained from experiments with the 121 phase were removed by water treatment. The mixture with water was boiled and filtered. The residue from the filter paper was collected after drying. XRD analysis of this powder confirmed the absence of the salts. To the resultant pure 121 powders, $BaCO_3$ and CuO in the desired molar ratios were added and the mixture was blended well using a pestle and mortar. The batch was then calcined at 980°C for 12 hrs in an oxygen atmosphere and then annealed at 530°C for 12 hours. The resulting sample was characterized by XRD to detect the presence of 213 phase. These samples were then pressed into pellets and hot forged. The hot forging was carried out at 950°C for 4 hours. The pressure was applied in increments of 0.5 mPa. For hot forging, an initial pressure of 0.5 mPa was applied which was raised by increments of 0.5mPa whenever the deformation rate reached a constant value. The samples obtained from hot forging were characterized using XRD and SEM.

RESULTS AND DISCUSSION

It was noted from our experiments with the 213 powders, that 213 phase degraded with the salt synthesis. The formation of $Ba_2YCu_3O_{7-x}$ in the presence of molten salts of Na, K and Li does not appear feasible in the temperature range of 400°C - 900°C. In our experiments with 121 powders we had better success. It was observed from our experiments that 121 was a highly stable phase and proved to be stable against most of the salt systems investigated. The XRD results confirmed the stable nature of 121 phase. Figure 1 shows the XRD analysis of one of the samples of 121 powder heat treated with KCl system. Peaks of KCl are marked on the graph. The other peaks belongs to the 121 phase.

Removal of salts was made easy by water treatment. The 121 phase remained stable even after the water treatment. Figure 2 gives the XRD analysis of the sample after water treatment. The absence of the KCl peaks are clearly seen from this graph. The remaining peaks represent the pure 121 phase.

Formation of 213 from the 121 powders was easily obtained and XRD results confirmed an almost phase pure 213 sample. Figure 3 clearly gives credence to this fact. The peaks shown represents the 213 phase. The heat treatment was carried out as explained earlier. Results of our hot forging experiments are yet to be fully analysed but the initial results show encouraging results. Further work is needed in this area to find an optimum flux and reaction conditions and to facilitate the growth of 121 grains which will then act as seed crystals in the formation of oriented crystals of 213 superconductor.

CONCLUSION

A study of the molten salt synthesis of $Ba_2YCu_3O_{7-x}$ was performed. The 213 phase was found to degrade with most of the salt systems investigated. For this reason processing of 121 phase to act as seed crystals in the formation of grain oriented 213 was carried out. It has been noted that 121 is a very stable phase and

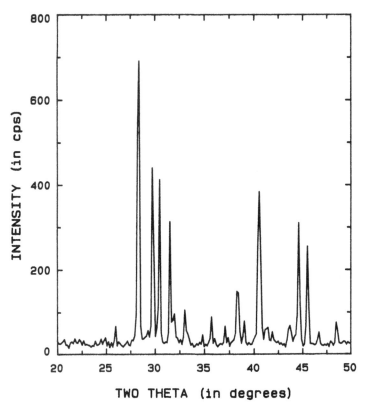

Figure 1. Preformed 121 powder heat treated with KCl at 825°C for 3 hours in O_2 atmosphere

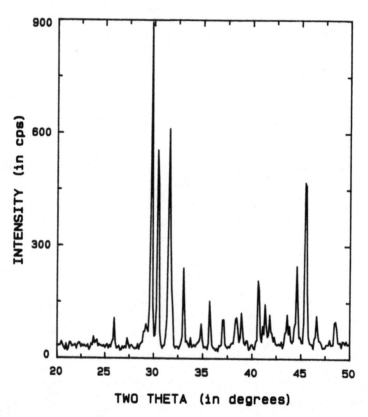

Figure 2: Preformed 121 powder with KCl after water treatment

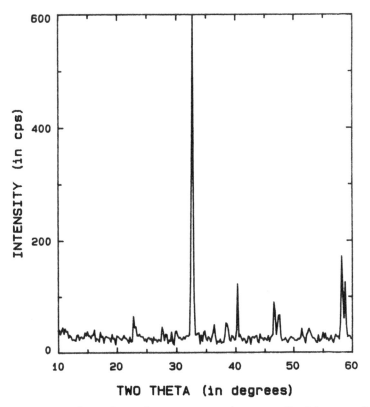

Figure 3: Formation of **213** from **121** powders + 3 $BaCO_3$ + 5 CuO

proved to be stable against most of the salts. The 121 phase being water stable allows the easy removal of the salt fluxes. Formation of 213 from the 121 phase and mixed oxides yielded a phase pure material.

REFERENCES

1. M.K. Wu, J.R. Ashburn, C.J. Torng, P.H. Hor, R.L. Meng, L. Gao, Z.L. Huang, Q. Wang and C.W. Chu, "Superconductivity of 93K in a new mixed phase Y-Ba-Cu-O compound system at ambient pressure," Phys. Rev. Lett., 58 988-89 (1987)

2. T.R. Dinger, T.K. Worthington, W.J. Gallagher and R.L. Sandstrom, "Direct observation of Electronic Anisotropy in single crystal $Ba_2YCu_3O_{7-x}$ " Phys. Rev. Lett., 58 [25] 2687-90 (1987)

3. T.K. Worthington, W.J. Gallagher and T. R. Dinger, " Anisotropic nature of high temperature superconductivity in single crystal $Ba_2YCu_3O_{7-x}$," Phys. Rev. Lett., 59 [10] 1160-63 (1987)

4. M. Holmes, R.E. Newnham and L.E. Cross,"Grain oriented ferroelectric ceramics," Am. Ceram. Soc. Bull., 58 [9] 872 (1979)

5. R.L. Burns and R.A. Laudise,"Stability of superconducting $Ba_2YCu_3O_{7-x}$ in the presence of water",Appl. Phys. Lett. 51 (17) 1373-75 (1987)

6. Eli Ruckenstein, Sanjeev Narain and Nae-lih Wu,"Reaction pathways for the formation of the $Ba_2YCu_3O_{7-x}$ compound," J. Mater. Res., Vol.4, No.2, 267-272 Mar/Apr 1989

MOSSBAUER EFFECT STUDIES IN $YBa_2Cu_{3-y}Sn_yO_{7-x}$

M. DeMarco, G. Trbovich

Physics Department
Buffalo State College
Buffalo, NY 14222

X. Wang

Department of Electrical Engineering
Alfred University
Alfred, NY 14802

P.G. Mattocks, M. Naughton

Physics Department
SUNY at Buffalo
Buffalo, NY 14260

ABSTRACT

We have measured the Sn Mossbauer Effect in $YBa_2Cu_{3-y}Sn_yO_{7-x}$ samples. The samples have been analyzed using thermogravimetric and magnetometer analysis. The Mossbauer measurements show a broadened line that can be fit with two sets of Quadrupoles that indicate two different oxygen vacancy sites in the 1-2-3 compounds.

INTRODUCTION

Many Iron Mossbauer effect studies have been done on $YBa_2Cu_3O_{7-x}$ but the details of the oxygen vacancies and ionization states of Cu and O are still uncertain. It is therefore important to study these compounds doped with Sn to coroborate and further investigate these properties because this element does not substantially alter the Tc[5]. We have produced $YBa_2Cu_{3-y}Sn_yO_{6+x}$ to study the effects of Sn and Oxygen concentration.

EXPERIMENTS

We have prepared Sn119 doped samples with the stoichiometry $YBa_2Cu_{3-y}Sn_yO_{7-x}$. The y value was varied in steps of .05 from .05 to 0.2 while x was either 0.9, 1.0 or unknown. The samples were prepared by solid state reaction. In some samples Nitrogen was used in the annealing process instead of oxygen. Thermogravimetric analysis was used to determine the oxygen content.

The Sn119 Mossbauer Effect absorption measurements were performed using a Ranger Mossbauer Spectrometer, Germanium Intrinsic detector and a Nuclear Data Multichannel Analyzer. The source was $Sn_{119m}O_2$ (annealed in air) prepared at the Buffalo Materials Research Reactor. The full width linewidth was 1.4mm/s using a $BaSnO_3$ absorber with 1 mg/cm^2 of Sn_{119}. All measurements were performed in the same geometry with about 1 mg/cm^2 Sn_{119} in the 1-2-3 doped compounds.

In addition samples were analyzed using a SQUID magnetometer. The superconducting volume percentage was found along with the Tc of the materials. Table 1 shows these values.

Table 1

COMPOUND	Tc(K) (±1)	MEISSNER/SHIELDING EFFECT % B (applied)=100 gauss
$YBa_2Cu_{2.85}Sn_{.15}O_{6.9}$	90	33
$YBa_2Cu_{2.9}Sn_{.1}O_{7-X}$	90	33
$YBa_2Cu_{2.95}Sn_{.05}O_7$	90	20
$YBa_2Cu_{2.85}Sn_{.15}O_7$	90	25
$YBa_2Cu_{2.85}Sn_{.15}O_{7-X}$	90	45
$YBa_2Cu_{2.8}Sn_{.2}O_{6.9}$	80	10

RESULTS

The results of the experiments performed at room temperature are listed in Table II. All the experimental results showed a single broadened line which was centered on the zero velocity channel. This is consistent with Sn in the +4 state. Each broadened line was fit with a single or double pair of lines using a commercially prepared computer program by Ranger Scientific. The linewidths used in the fitting process were those that were found experimentally using the $BaSnO_3$ absorber.

Table 2

COMPOUND	CS_1	Q_1	I_1	I_2	Q_2	CS_2	S	CS
$Sn_{.15}O_{6.9}$	269	2	1000	9200	12	256.5		
$Sn_{.15}O_7$				8550	17	257.5		
$Sn_{.05}O_7$	266.5	7	2000	3000	12	254		
$Sn_{.2}O_{6.9}$	268	4	3200	4800	12	254		
$Sn_{.1}O_{7-x}$				12,225	15	257.5		
SnO_2							19.5	255.7
$Sn_{.1}O_{7-x}$	263.5	5	500	4600	19	256.5		
$Sn_{.15}O_{7-x}(N_2)$	251.5	13	1500	3900	23	260.5		
$BaSnO_3$							17	255.8
$Sn_{.05}O_{7-x}(N_2)$				19,150	18	260		

CS   center shift channel

Q   Quadrupole splitting in channels

I   absorption intensity

S   single line width in channels

(N_2) designates Nitrogen annealing, all the rest are oxygen annealed.

Sn_yO_{7-x} is short for $YBa_2Cu_{3-y}Sn_yO_{7-x}$

1 channel = .042 mm/s

1, 2 denote different sites

The experimental results for all the oxygen annealed samples consistently show a large intensity doublet formed for the same isomer or center shift. Another doublet of smaller intensity is also present. It has a different range of isomer shifts than the large intensity doublet. The Nitrogen annealed samples show a different behavior.

Mossbauer Spectra for $YBa_2Cu_{2.95}Sn_{.15}O_{6.9}$ and $YBa_2Cu_{2.85}Sn_{.15}O_7$ are shown in Figure 1. These results show the contrast in lineshape between the $Sn_{.15}O_{6.9}$ and $Sn_{.15}O_7$ samples. A Nitrogen annealed sample of $Sn_{.15}O_{7-x}$ is also shown.

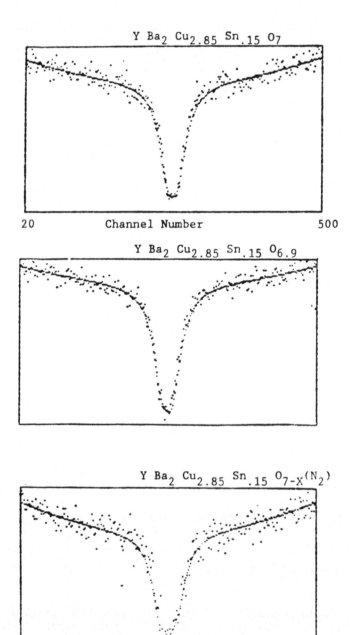

Figure 1. Mossbauer Effect $Sn_{119m}O_2$ Source -vs- 1-2-3 Absorbers.

DISCUSSION

The Sn ion is thought to be replacing the Cu ions in the 1-2-3 compounds [6,7]. This being the case, all the experimental results and particularly those for the samples of $Sn_{.15}O_{6.9}$ and $Sn_{.15}O_7$ suggest that there are two different types of Sn sites surrounded by oxygen ions. The dominant site has a large splitting (about 15 channels) which indicates a large asymmetry in the number of near neighbor oxygen ions. The large absorption intensity indicates a high population of this type of site relative to the other site. The second or less intensely populated site has a smaller Quadrupole splitting. This may be explained by a non-symmetric distribution of

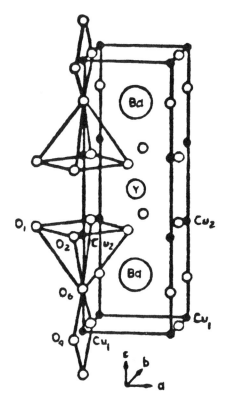

Figure 2. Crystal Structure of $YBa_2Cu_3O_{7-x}$.

oxygen ion neighbors more distant from the Sn ion. If one considers the crystal structure shown in Figure 2, the separation distances[8] O_6 and O_4 from Cu(1) are 1.785A and 1.930A. In the same figure we see that separations of oxygen neighbors from Cu(2) are 1.946 A and 2.382 A.[9] Therefore we see a correspondence between the Oxygen and Cu ion separations, the Sn Quadrupole splittings and the Cu(1) and Cu(2) sites. The interpretation of two types of Sn vacancy sites is consistent with Blue's[3] results for Fe substituting at the copper sites.

The Nitrogen annealed samples present different center or isomer shifts for each Quadrupole splitting and thus, are different types of sites. This may be due to Nitrogen filling the Oxygen vacancies.

It is interesting to note that small changes in Sn concentration have a small effect on the Tc and Mossbauer spectra.

ACKNOWLEDGEMENTS

Support by CACT, No. 12-12-58, DOE and Buffalo Materials Research Corporation. SQUID measurements performed by M. White, SUNY at Buffalo.

REFERENCES

1. I. Felner, Y. Wolfus, G. Hilscher, N. Pillmayr, Susceptibility, crystal structure and specific heat studies of the high Tc superconductor and quenched $YBa_2(Cu_{.91}Fe_{.09})_3O_x$ Phy Rev B 39. 225 (1989)
2. S. Suharan, C.E. Johnson, D.H. Jones, M.F. Thomas, R. Driver, The Charge State of ^{57}Fe doped into $YBa_2Cu_3O_{7-\delta}$, Solid State Comm 67, 125 (1988)
3. C. Blue, K. Elgaid, I Zitkovsky, P. Boolchand, D. McDaniel, W.C. Joiner, J. Oosters, W. Huff, Oxygen-Vacancy-formation in $YBa_2(Cu_{.985}Fe_{.15})_3O_{7-\delta}$, Phy Rev B 37, 5905 (1988)
4. Q.A. Pankhurst, A.H. Moorish, X.F. Zhou, I. Martense, Magnetic ordering below 10K in the high Tc superconductor $YBa_2(Cu_{.985}Fe_{.015})_3O_{7-\delta}$, Hyperfine Interactions 42, 1235 (1988)
5. B.R. Zhao, Y.H. Shi, Y.Y. Zhao, Lin Li, Effect of Fe on the superconductivity of the Ba-Y-Cu system, Phy Rev B 38, 2486 (1988)
6. P. Boolchand, R.N. Enzweiler, Ivan Zitkovsky, J. Wells, W. Bresser, D. McDaniel, R.L. Meng, P.H. Has, C.W. Chu, Softening of Cu-O vibrational modes as a precursor to onset of superconductivity in $EuBa_2Cu_3O_{7-\delta}$, Phy Rev B, 37, 3766 (1988)
   T. Yuen, C.L. Lin, J.E. Crow, G.N. Meyer, R.E. Saloman, P. Schlottman, N. Bykovetz, W.N. Herman, Mossbauer study of the lattice dynamics in ^{119}Sn doped superconductivity and non superconductivity $YBa_2Cu_3O_{7-\delta}$, Phy Rev B, 37, 3770 (1988)
7. X. Wang, M. DeMarco, G. Trbovich, P.G. Mattocks, Mossbauer Effect Measurements in Sn, Fe doped $YBa_2Cu_3O_{6+x}$, Bulletin of the American Physical Society Vol 34, No 3, 744 (1989)
8. T. Siegrist, S. Sunshine, D.W. Murphy, R.J. Cava, S.M. Zahurak, Crystal Structure of the high Tc superconductor $Ba_2YCu_3O_{9-\delta}$, Phy Rev B, 35, 7137 (1987)
9. R.P. Gupta and M. Gupta, Role of Cu-O linear chains in high temperature superconductivity and antiferromagnetism in $YBa_2Cu_3O_{7-\delta}$, Solid State Comm 67, 129 (1988)

ENERGETICS OF SUPERSTRUCTURES IN THE Bi-SERIES OF SUPERCONDUCTING COMPOUNDS

Anurag Dwivedi and A. N. Cormack
NYS College of Ceramics
Alfred University, Alfred, NY 14802

ABSTRACT

The existence of structural patterns in Bi and Tl-containing superconducting compounds has been well established. In this work, we concentrate on Bi-containing compounds whose stoichiometries are described by the homologous series $3AO$-$nA'BO_3$ ($3A=Bi_2Sr$ and $nA'=SrCa_{n-1}$, $B=Cu$). The first three compounds of this series commonly known as 2021 ($Bi_2Sr_2CuO_6$), 2122 ($Bi_2CaSr_2Cu_2O_8$), and 2223 ($Bi_2Ca_2Sr_2Cu_3O_{10}$) (corresponding to n = 1, 2, and 3 respectively) have well been studied and the T_c of these compounds has been found to increase with n (at least for $n \leq 3$) in the general formula. In other words, T_c apparently is correlated with the number of Cu-O planes present in the perovskite layer of the unit cell.

Structural information on higher order members is, however, incomplete. Based on the structural patterns of the homologous series, model structures and stoichiometries of the first five members (n=1 to 5) were setup. We have investigated the energetics of both model and XRD (where known) structures using atomistic simulation techniques. Preliminary results, using a pairwise potential model, revealed large differences between X-ray and model structures which indicate that distortions and buckling in the planes seen by XRD and which have been ignored in the model structures, are associated with large relaxation energies. These may contribute to the limited phase stability seen in this system. Our results also highlight the need for very accurate interatomic potentials, which we are in the process of developing. An explanation of observed buckling in the Cu-O planes of higher order compounds is also suggested.

INTRODUCTION

The complex crystal chemistry of newly discovered high temperature superconductors makes it really difficult, if not impossible, to study their structural and defect chemistry in detail. In addition to the original $La_{2-x}Sr_xCuO_4$, and the subsequently discovered $YBa_2Cu_3O_{6+x}$, there have been added Bi-containing and Tl containing compounds. Apart from these there have been reported some non-copper containing ceramic compounds which also show superconductivity; some examples are $Ba_{1-x}K_xBiO_3$ and $La_{2-x}Sr_xNiO_4$. Interestingly enough all of these compounds, irrespective of whether or not they contain copper ions, have similar structural features. Copper-containing superconductors, however, have till now shown higher T_c than their non-copper containing counterparts.

Most of the copper-containing superconducting systems form a Ruddlesden-Popper (R-P) type of homologous series of compounds with structures that have alternate stacking of rocksalt and perovskite structured blocks. As the number of Cu-O planes increases in the perovskite blocks (forming higher order members of these R-P type compounds), the size of unit cell grows exceedingly bigger and the crystal chemistry becomes rather complicated. Notwithstanding observations that higher order members may have interesting properties (as is evident from the studies of the first few members of Bi and Tl-containing compounds where T_c is reported to increase as the number of Cu-O planes increase in the perovskite blocks; see Table-1) it is not at all easy to synthesize them phase pure and, thus, to study their defect and structural chemistry. One of the goals of this project is to understand why it should be difficult to prepare phase pure samples.

Due to a combination of experimental complications there have not been many studies on the higher order members of Bi and Tl-containing compounds and only the first few members are reported to form. Detailed structural information beyond the 4th member is missing from the literature. In the somewhat simpler La-Cu-O system, however, indications of the existence of up to the 7th member of the R-P type series have been found by Davies and Tilley [1] through HREM studies (although not necessarily phase pure).

At this stage, it becomes crucial to determine more about the thermodynamical stability, crystal and defect chemistry of these new ceramic superconductors. This is possible using careful theoretical methods. In theoretical studies, also, it is difficult to deal with such complex systems with extremely large unit cells, especially in the case of the higher order compounds. But modern powerful supercomputing resources (such as those available to us at the Cornell National Supercomputing Facility) make it possible to perform a detailed atomistic computer simulation study of these high T_c superconducting oxides. A major problem in the atomistic simulation studies of these systems is the lack of a reliable interatomic potential for all the interactions taking place in the solid. We have been constantly working on the

Table 1. Few members of R-P homologous series of Bi-containing compounds.

Compound	Ideal Prototype $mAO-nA'BO_3$	m:n	% Missing Oxygen	T_c
$Bi_2Sr_2CuO_6$	$3AO-A'BO_3$ $3A=Bi_2Sr$; $A'=Sr$	3:1	0	12 K
$Bi_2Sr_2CaCu_2O_8$	$3AO-2A'BO_3$ $3A=Bi_2Sr$; $2A'=SrCa$	3:2	11	85 K
$Bi_2Sr_2Ca_2Cu_3O_{10}$	$3AO-3A'BO_3$ $3A=Bi_2Sr$; $3A'=SrCa_2$	3:3	17	110 K
$Bi_2Sr_2Ca_3Cu_4O_{12}$	$3AO-4A'BO_3$ $3A=Bi_2Sr$ $4A'=SrCa_3$	3:4	20	90 K (Thin film)
$Bi_2Sr_2Ca_4Cu_5O_{14}$ (Not reported in literature)	$3AO-5A'BO_3$ $3A=Bi_2Sr$ $5A'=SrCa_4$	3:5	22	?

modification of existing interatomic potentials and further, aim to develop much more accurate potentials.

COMPUTATIONAL TECHNIQUES

The calculations are based on the Born model of the solid, which treats the solid as a collection of point ions with short range repulsive forces acting between them, an approach which has enjoyed a wide range of success, although it has been found that the reliability of the simulations ultimately depends on the validity of the potential model used in the calculations. Detailed discussions on the simulation techniques have been presented by Catlow [2], and in the monograph edited by Catlow and Mackrodt [3]. Brief descriptions of interatomic potentials and lattice energy calculations are given below.

Potential Models

In the present study, the potentials used are described by a simple analytical function, the Born-Mayer central-force potential, supplemented by an attractive r^{-6} term, i. e.

$$V_{ij}(r_{ij}) = A_{ij}exp\left(-r_{ij}/\rho\right) - C_{ij}r_{ij}^{-6} \tag{1}$$

The polarisability of individual ions is included by the Shell Model of Dick and Overhauser [4] in which the outer valence electron cloud of the ion is simulated by a massless shell of charge Y and the nucleus and inner electrons by a core of charge X. The total charge of the ion is, thus, X+Y which equals the oxidation state of the ion. The interaction between core and shell of any ion is harmonic with a spring constant, k, and is given by:

$$V_i(r_i) = \frac{1}{2} k_i d_i^2 \qquad (2)$$

where d is the relative displacement of core and shell of ion i.

For the Shell Model, the value of the free-ion electronic polarizability is given by:

$$\alpha_i = Y_i^2/k_i \qquad (3)$$

The potential parameters A, ρ, and C in Equation-1 and the shell charges, Y and spring constant, k, associated with the shell-model description of polarizability have been derived by the procedure of 'empirical fitting', i.e. these parameters were adjusted by a least-squares fitting routine, so as to achieve the best possible agreement between calculated and experimental crystal properties, in this case, the structural features of the binary and superconducting compounds. The oxygen-oxygen interaction was taken from the earlier work of Catlow [5]. Lewis and Catlow [6] discuss in greater detail the derivation of the potential parameters for oxides.

Lattice Energy Calculations

The lattice energy is the binding or cohesive energy of the perfect crystal (per unit cell or per formula unit). It is of central importance in treating thermochemical properties of solids and in assessing the relative stabilities of different structures. Moreover, its derivatives with respect to elastic strain and displacement are related to dielectric, piezoelectric and elastic constants and phonon dispersion curves.

The lattice energy is calculated in the Born model of the solid by the relation

$$U = \frac{1}{2} \sum_i \sum_j V_{ij} \qquad (4)$$

where the total pairwise interatomic potential, V_{ij}, is given by:

$$V_{ij}(r_{ij}) = q_i q_j/r_{ij} + A_{ij}\,exp(-r_{ij}/\rho_{ij}) - C_{ij} r_{ij}^{-6} \qquad (5)$$

with the first term representing the Coulombic interactions between species i and j, and the last two the noncoulombic short range contributions discussed above. The lattice energy is thus calculated exactly and the only limitations in the procedure arise from a lack of precise knowledge of the interatomic potentials.

Calculation of the equilibrium atomic configuration involves adjusting the coordinates until the internal basis strains (i.e. the net forces acting on a species) are totally removed. For complete structural equilibration (what is generally termed the "constant pressure condition"), the lattice vectors are also relaxed, using elasticity theory. From the bulk lattice strains, e_{ij}, (i, j, k, being the three orthogonal basis vectors), obtained from the derivatives of the lattice energy, a new set of basis vectors may be defined

$$\begin{bmatrix} i' \\ j' \\ k' \end{bmatrix} = \begin{bmatrix} 1 + e_{ii} & e_{ij} & e_{ik} \\ e_{ij} & 1 + e_{jj} & e_{jk} \\ e_{ik} & e_{jk} & 1 + e_{kk} \end{bmatrix} \begin{bmatrix} i \\ j \\ k \end{bmatrix}$$

The new lattice vector matrix can thus be expressed directly in terms of the original basis vectors and the bulk lattice strains. The atomic coordinates are then re-equilibrated with the new lattice vectors and the procedure repeated until the bulk lattice strains are completely eliminated. Details of the procedure have been outlined by Cormack [7] and Parker [8].

SUMMARY OF THE WORK

Development of a Suitable Interatomic Potential Model

The first requirement for any atomistic simulation is a reliable potential model. All known high T_c ceramic superconductors have a complex structure with variable stoichiometry and we believe that a potential model, if reliable, should describe the structure and properties of the superconducting complex oxides as well as the constituent binary oxides. It has been found, through experience, that for complex oxides such as these, potential parameters may be transferred from the appropriate binary oxide potential model, if one exists. In the present case, this is possible as we discuss below.

a) Bi-O Interactions: Bi-O potentials are needed for the simulation of both $BaBiO_3$, the non-copper containing compound, and the copper-containing Bi-series of compounds. Bi-O potentials have been reported in the literature by Jacobs and MacDonaill [9] for their atomistic simulation study of Bi_2O_3 where the valence of bismuth ion is +3. Note that they used an unusual double exponential form. One problem that we anticipate is that the difference in the coordination of the bismuth between Bi_2O_3 and superconductors may affect the potential. Initially we assumed that it does not.

b) Cu-O Interactions: Copper ions play an important role in all copper-containing ceramic superconductors, both in exhibiting superconductivity and in governing the stoichiometry of the compounds. We have obtained Cu-O pair potential parameters by fitting to the CuO structure. We found it very difficult to derive an adequate pairwise potential model for this system; however, the best possible pairwise Cu-O

Table 2. Cu-O Potential Parameters Derived at Alfred.

Potential form: $V(r) = A\ exp(-r/\rho) - C\ r^{-6}$

Interaction	A	ρ	C
Cu^{2+}–O^{2-}	202.4567	0.4333	0.000
O^{2-}–O^{2-}	22764.200	0.1491	17.890

Table 3. Shell-Model Parameters Derived at Alfred.

Interaction	Shell Charge (Y)	Spring constant (k)
Cu(Core)–Cu(Shell)	0.000	∞ (Rigid Ion)
O(Core)–O(Shell)	-2.07	27.290

potential parameters which we have developed are reported below in Tables-2 and 3. Central force models being inadequate, the next consideration is the inclusion of three body potentials, as in the model that Islam et al used [10]. As will be seen below, the double exponential form of O-O interaction did not perform well for CuO. However, we retained it in the potential model for the high T_c Bi-Ca-Sr-Cu oxides.

We had to modify our programs to include three body potentials. This allowed us to compare the results of our two body potentials with those from three body potentials reported by Islam et al[10]. In our systematic study of potentials for CuO, we examined potential models that neglected ion polarisability and whether the double exponential form of the oxygen-oxygen potential would be suitable. Our conclusion is that, for CuO, oxygen ion polarizability is necessary and that only the single exponential form of the oxygen-oxygen interaction is practicable. The double exponential form derived by Jacobs and MacDonaill was determined not to be suitable for CuO itself.

With the Cu-O pair potentials derived by us and the bond bending potentials derived by Islam et al we were able to model both for CuO and La_2CuO_4. It is worth noting that the angle Ox(2)-Ox(1)-Cu is about 38° both in CuO and in La_2CuO_4. In the perovskite block of Bi-containing compounds it is 40° which is quite close to the value found in CuO and La_2CuO_3 (Fig-1). This provided us with confidence in using the same potentials which were derived for CuO and La_2CuO_4 in simulating Bi-Sr-Ca-Cu oxides. Ca-O and Sr-O potentials were taken from the study of Lewis and Catlow [6].

Studies on Ruddlesden-Popper Type Compounds

It is important to note any structural patterns that exist in these materials. Such patterns can assist in the search of additional superconducting oxides, and may even contribute to the development of a better theoretical understanding. Almost all of the presently known high T_c superconductors fit either the basic perovskite formula ABO_3 or are large-cation-rich relative to this composition [11]. In the present study we have concentrated on the system Bi-Sr-Ca-Cu-O in which T_c has been reported to increase as the number of Cu-O layers increases in the compounds (at least for n=1 to 3).

Compounds in the system Bi-Sr-Ca-Cu-O have very complicated crystal chemistry as is evident from their crystal structures. It has been noticed that the ideal prototype stoichiometry of these compounds is of the form mAO-$nA'BO_3$ and that the structures of these materials can be related to a generalised form of the Ruddlesden-Popper (R-P) series of structures, in which two dimensional perovskite like slabs are separated by layers of the large-cation (e.g. Bi) oxide with the NaCl stucture. There are minor distortions and displacements in the actual systems which produce slight deviations from the regular patterns.

It is further interesting to note that the T_c of the compound is found to go up as 'n' in the stoichiometry $mAO.nA'BO_3$ increases. The apparent correlation of T_c with the number of Cu-O planes led us to include in our study the higher member ($n=4,5$) of the series. Subsequently thin film fabrication of the $n=4$ member was reported [12] with a T_c of 90 K. Bulk preparation and T_c has still not been achieved.

In the literature, there are reported the first three members, corresponding to n=1, 2 and 3, of the homologous series. Higher members can also be modeled for theoretical investigations, but the structural information and evidence of their existence is lacking in the literature. We produced model stoichiometries of the next two members of the series and also their model structures. These are reported in Figures 2-6. This has helped the experimentalists in their attempt to produce phase pure materials with the stoichiometries suggested. Matheis *et al* [13] used this information to calculate XRD patterns of these ideal model compounds.

DISCUSSION OF RESULTS
Buckling in the Cu-O Planes in the Bi-Sr-Ca-Cu-Oxides

We have used various types of Cu-O potentials to investigate the energetics and relaxed structures of the first five members of the series. Results of three body potentials were almost same as those using pair potentials. After relaxing the structure we found that Bi ions moved substantially from their initial position. The Cu-O planes also showed buckling, except in the first member $Bi_2Sr_2CuO_6$.

Table 4. Results of Three-Body Potential Model.

Compound	Lattice Energy of		Relaxation Energy (eV)
	Unrelaxed Structure (eV)	Relaxed Structure (eV)	
2021	-368	-448	80
2122	-525	-601	76
2223	-664	-744	80
2324	-811	-892	81
2425	-958	-1039	81

It has been noticed by many of those working in this area that the homologous series of Bi-containing compounds is structurally very similar to the strontium titanate R-P series. However, we have observed that there are some significant differences; for example, the rocksalt and perovskite blocks in the new superconducting compounds are not electrically neutral, unlike in the Sr-Ti-O system, and may well lead to significant differences in their structural chemistry. The Rocksalt block has a net positive charge which remains constant throughout the homologous series since the size and stoichiometry remains the same. When additional Cu-O planes are inserted, increasing the size of the perovskite blocks of the higher order members, it becomes necessary to create oxygen vacancies in the basic perovskite structure in order to maintain the electrical neutrality. Our calculations suggest that it is these missing oxygen ions that allow the Cu-O planes to buckle. This is also supported by the absence of buckling in the first member of Bi-containing compound (Fig-2) in which there are no missing oxygen ions, and in the Sr-Ti-O series of compounds. Thus one of the major differences between the Cu containing superconductors and the strontium titanates, the buckling of Cu-O planes (and "non-buckling" of the comparable Ti-O planes) can be explained in terms of the stoichiometry and crystal chemistry. Note that these vacant oxygen sites are with respect to the ideal structure of the perovskite-like slabs. They are not really vacancies with respect to the superconductor structure. Any non-stoichiometry, such as oxygen deficiency, in these coumpounds will produce additional point defects.

These non-neutral perovskite and rocksalt blocks, apart from providing the cause of the buckling of Cu-O planes, also render an additional coulombic bonding energy between two different types of blocks. This may provide a different trend of energetics from that seen in the Sr-Ti-O system [14]. Some results of our work on the

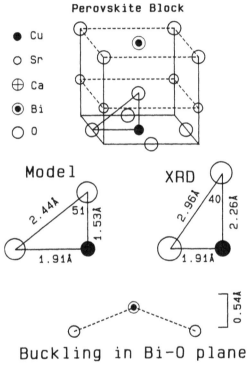

Fig 1. O-O-Cu angles in perovskite blocks and buckling of Bi-O planes after relaxation.

energetics of R-P type of Bi-containing compounds are reported in Table-4 and 5.

Our model structures were highly idealized and in the initial structures interplanar distances were same in perovskite and rocksalt blocks and the planes were not buckled. We expected significant distortions after relaxation, and we do see these (Table- 4 and 5), producing the buckling in the Cu-O planes, as seen in those structures which have been solved so far.

Table 5: Energetics of Model and XRD Structures.

Compound	$E_{Lattice}$		Relaxation Energy (eV)
	Unrelaxed (eV)	Relaxed (eV)	
2021 (XRD)	-420	-448	28
2021 (Model)	-368	-448	80
2122 (XRD)	-572	-597	25
2122 (Model)	-525	-601	76

Progress in Modeling the Bi-Sr-Ca-Cu-O System

Although we have made substantial progress in modelling these complex structures, there are still some problems that need to be worked out. Chief amongst them is the observation that our models apparently overestimate the degree of lattice relaxation. This problem appears to be associated with the extent of Bi ion relaxation and the buckling of the Bi-O layers (Fig-1). Whilst the difficulty probably lies with the Bi-O interatomic potential, it may also reflect the constraints imposed on the model structures during the calculations. We used the basic unit cell, although there are indications that the true structure is a modulated structure with a superlattice cell parameters of the order of 5 times the $\underline{a}$ parameters we used. The origin of this superlattice ordering is apparently related to the structure of the Bi-O planes, although this has not been demonstrated conclusively. Our results may be reflecting this desire of the Bi ions not to be constrained to the smaller periodicity. Further work is needed to clarify this.

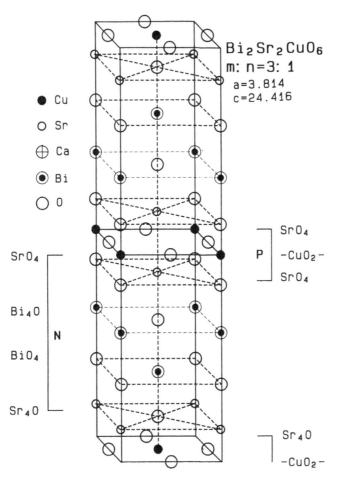

Fig 2. Model Structure for 2201 (n=1).

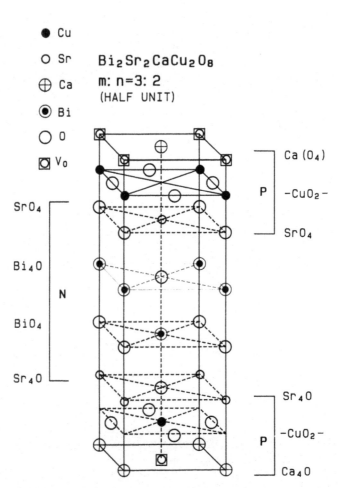

Fig 3. Model Structure for 2212 ($n=2$).

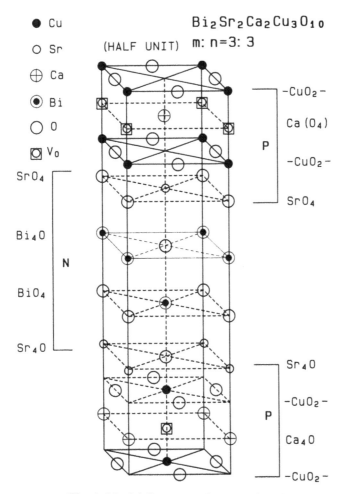

Fig 4. Model Structure for 2223 ($n=3$).

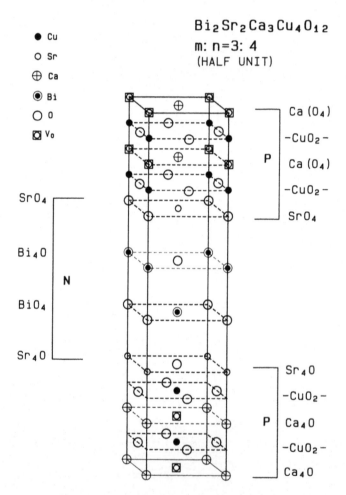

Fig 5. Model Structure for 2234 ($n=4$).

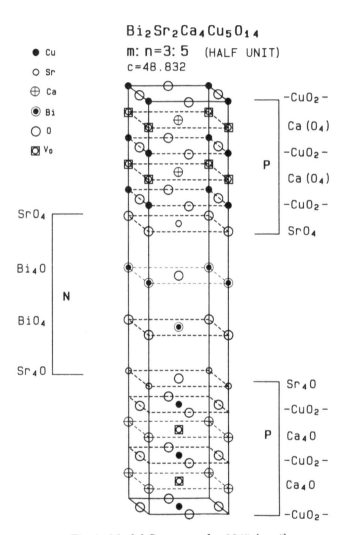

Fig 6. Model Structure for 2245 ($n=5$).

References

[1] Davies, A. H. and Tilley, J. D., Nature, Vol. 326, 1987.

[2] Catlow, C. R. A., in Mass Transport in Solids, edited by F. Beniere and C. R. A. Catlow, p1, Plenum (1983).

[3] Catlow, C. R. A. and Mackrodt, W. C., eds. Computer Simulation of Solids, Lecture Notes in Physics, 166, Springer, (1982).

[4] Dick, B. G. and Overhouser, A. W., Phys. Rev., 112[1], 90 (1958).

[5] Catlow, C. R. A., Proc. Roy. Soc., A, **353**, 533(1977).

[6] Lewis, G. V. and Catlow, C. R. A., 1985, J. Phys. C:Solid State Phys., **18**, 1149(1985).

[7] Cormack, A. N., Solid State Ionics, **8**, 187 (1983).

[8] Parker, S. C., Computer Modeling of Minerals, Ph. D. Thesis, Univ. of London, London, UK, (1983).

[9] Jacobs, P. W. M., and MacDonaill, D. A., Solid State Ionics, **23**, 279(1987).

[10] Islam, M. S., Leslie, M., Tomlinson, S. M., and Catlow, C. R. A., J. Phys. C: Solid State Phys. **21**(1988), L109-L117.

[11] Smyth, D., M., "Structural Patterns in High T_c Superconductors", Preprint (To be published in the Proceedings of the Superconductor Symposium, 90th Annual Meeting of the American Ceramic Society, Cincinnati, May 1-5, 1988).

[12] Sleight, A. W., Subramanian, M. A., and Torardi, C. C., Mat. Res. Bulletin, Vol XIV, No. 1, pp 45-47, (1989).

[13] Matheis, D. P., McIntyre, P., Snyder, R. L., "X-ray Powder Diffraction Standards for Copper Oxide Based Superconductors", Paper presented in the 91st annual meeting of Amer. Cer. Soc., Indianapolis, IN, April 23-27, 1989.

[14] Udaykumar, K. R., and Cormack, A. N., J. Amer. Cer. Soc. **71**, C-469(1988).

MICROSTRUCTURAL EFFECTS IN OXIDE SUPERCONDUCTORS

J.G. Darab, R. Garcia, R.K. Macgrone and K. Rajan
Materials Engineering Department
Rensselaer Polytechnic Institute
Troy, NY 12180-3590

ABSTRACT

Mechanical deformation (ball milling) has been found to degrade the superconducting properties of $YBa_2Cu_3O_{7-x}$. The degraded samples have been found to contain a large density of non-intersecting (110) twins and at the same time exhibit a large EPR signal. A clear correlation betwen the degree of degradation (as measured by the relative susceptibility), the twin density and the EPR intensity was established. A detailed analysis of the EPR parameters shows conslusively that the copper(II) complexes involved cannot lie in perfect defect-free $YBa_2Cu_3O_{7-x}$. Rather, they have very different environments most probably associated with twin boundaries. The twin structure and associated EPR spectra are very sensitive to modest thermal and mechanical stress, and a new "ribbon-like" defect is subsequently observed. On the other hand, annealing at 500°C, produces the "tweed-like" twin structure, with no associated EPR spectra and with restored superconducting properties.

INTRODUCTION

It has long been recognized that structural defects have a significant influence on the superconducting properties of the high T_c oxides. This paper is concerned with the relation between the density of localized Cu(II) ions as monitored by EPR, the density of twin boundaries, and some of the superconducting parameters, such as the ac susceptibility and H_{C1}.

In most work on this subject, the defects invoked are not explicitely identified, nor the mechanism of their influence discussed. For example Couach et al[1] deduced from their pioneering ac magnetic measurements that losses resulted from both inter-grain and intra-grain effects, but did not identify the responsible defect within grains. Zwicknagel et al[2] have pointed out that the broadening of the resistive transition is due to structural inhomogeneties, while Ekin[3] ascribes magnetic effects on the resistive transition to regions with low H_{C2} values. The specific imhomogeneities and low H_{C2} regions have not been specified, e.g.[4].

Structural defects in the oxides have also been studied. For example Beyers et al[5] have studied various microstructures and twin boundaries in $YBa_2Cu_3O_{7-x}$. There is evidence[6] that the oxygen concentration effects the twin density, it being reported that the inter-twin separation decreases from 1000 Å to a few hundred angstroms as x changes from 7.0 to 6.5. Super-lattices and stacking faults have been studied by Tan et al[7].

None of the above studies have related specific defects to specific superconducting effects directly. Recently however, Shi et al[8] have indirectly related the influence of oxygen deficiency on the width of the transition to the influence of twin boundaries and the presence of the tetragonal phase. Their conclusion was based on previous structural studies, specifically references [2] and [6].

In the context of point defects, EPR studies of the homologous oxides are relevant. The general consensus is that pure single phase $YBa_2Cu_3O_{7-x}$ is EPR "silent" above the superconducting transition. Any observed well defined EPR spectrum in this temperature regime is rather ascribed to the presence of small amounts of the "green" phase, Y_2BaCuO_5, and the "brown" phase, $YBa_3Cu_2O_y$ [9,10]. On the other hand, a weak line with g=2.03 and width of about 250 gauss sometimes observed is ascribed to various causes. For example Shalteil et al[11] have ascribed this resonance to conduction electrons.

This paper is concerned with the relation between the degradation of the superconducting properties and structural defects introduced by ball-milling. Specifically the defects involved are twin and other faults and localized holes on Cu(II) sites, which are most probably associated with (110) twin boundaries.

EXPERIMENTAL RESULTS

Carefully prepared superconducting powder was ball milled in hexane for various times and the superconducting transition measured magnetically using an ac inductance bridge operating at 5kHz. The behavior of the superconducting transition was also determined using a VSM at 40 K. Figure 1 shows the B-H curves for samples milled for zero, one and two hours. As can be seen, both H_{C1} and the diamagnetism decrease as the milling time is increased.

In addition, the corresponding EPR spectra of the powders were obtained at 300 K and at 4 K. The 300 K spectra are shown in Figure 2. The relatively small spectrum in the as-prepared powder changes to a large and different absorption after ball milling. We find that there is a very close correlation between the EPR intensity and the ac susceptibility. Taking X_r as the relative susceptibility at 77 K, ($X_r = X_t/X_0$ where X_t is the susceptibility after t hours of milling and X_0 is the susceptibility after 0 hours of milling), we find a smooth monotonic relation between $1/X_r$ and the intensity of the EPR resonance, as shown in Figure 3. The g_o values of the ball milled material systematically change with milling time and are summarized in Table I. This change in g_o indicates that the EPR active environments of the copper(II) ions before and after milling are different.

As a control, a portion of the starting $YBa_2Cu_3O_{7-x}$ material was soaked in hexane for two hours. The EPR spectra before and after soaking show no significant differences in intensity or shape, thus ruling out corrosive effects.

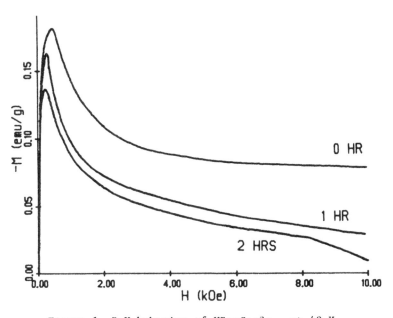

Figure 1 B-H behavior of $YBa_2Cu_3O_{7-x}$ at 40 K after various milling times.

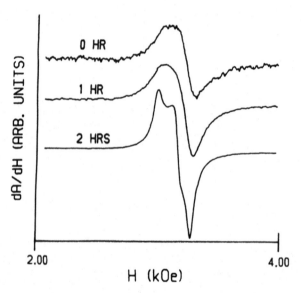

Figure 2 Room temperature EPR spectra of $YBa_2Cu_3O_{7-x}$ after various milling times.

Table I. Summary of EPR Parameters of $YBa_2Cu_3O_{7-x}$ after various milling times.

	MILLING TIME (HOURS)		
	0	1	2
EXPERIMENTAL VALUES:			
g_o	2.067	2.089	2.095
Resonance width (G)	180	150	75
SIMULATED VALUES:			
g_X		2.106	2.106
g_Y		2.055	2.025
g_Z		2.226	2.226
$A_X (\times 10^{-4}\ cm^{-1})$		-97	-23
A_Y		-29	-4
A_Z		-76	-16

After ball milling, the specimens were examined using transmission electron microscopy, and a typical micro-structures is shown in Figure 4. A dramatic increase in twin density and twin structure with progressive ball milling is evident.

The Cu(II) spectra observed in the $YBa_2Cu_3O_{7-x}$ compound after one and two hours of ball milling have been carefully simulated. The principle g and A values used for the simulations are also summarized in Table I. These simulated spectra are indistinguishable from the observed spectra. We note that the values of g used in the simulations are commonly observed. The values of the hyperfine terms are however unusually low, and as we will discuss later, indicate that a significant degree of spin exchange is involved.

We find that the twinned structure and the associated EPR spectra are extremely sensitive to even modest subsequent mechanical and thermal stresses. When the ball milled powder is mixed with polyethylene and extruded at 100°C through a die with a reduction of 10:1, both the twin density and the EPR signal decrease, see Figure 5. Also note that a new "ribbon-like" defect has appeared. At the same time the superconducting properties remain degraded. Additional experiments are needed to determine whether mechanical or thermal stress alone is sufficient to produce the effect.

On annealing at 500°C in air for several hours, the EPR signal is further reduced. In contrast to the previous observation, the twin structure is now completely different, namely the so-called "tweed structure" has arisen as can be seen in Figure 6. At the same time the sharp superconducting transition has been restored.

DISCUSSION

The EPR Spectra

The as-prepared material used in this work is relatively EPR "silent", and we take this to indicate that our starting material is of high quality and is essentially single phased $YBa_2Cu_3O_{7-x}$.

The possibility exists that the increase in EPR intensity could be related to the new surface area generated by the ball milling

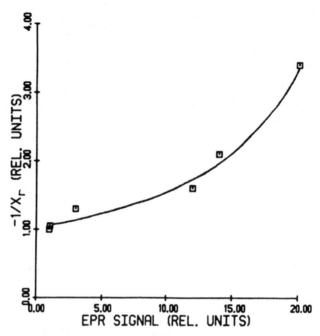

Figure 3   Plot of $1/X_r$ vs intensity of EPR resonance, showing the close correlation between superconducting degradation and EPR active defect content.

Figure 4   Transmission electron micrograph (dark field image) of $YBa_2Cu_3O_{7-x}$ ball milled for two hours showing a typical twinned microstructure.

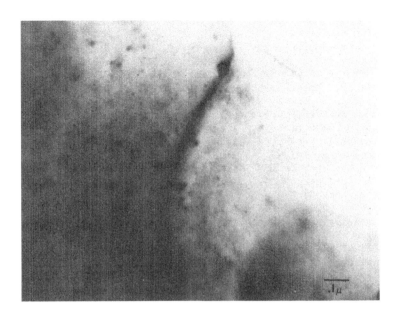

Figure 5 Transmission electron micrograph of $YBa_2Cu_3O_{7-x}$ ball milled for two hours and extruded showing "ribbon-like" defect and noticeable absence of twins.

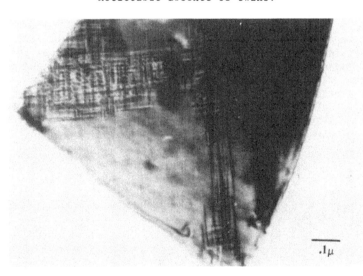

Figure 6 Transmission electron micrograph of $YBa_2Cu_3O_{7-x}$ ball milled for two hours, extruded, then annealed in air at 500°C. The "tweed-like" microstructure is clearly evident.

as the particles fracture, in view of the results of low temperature photo-emission studies reported by Los Alamos, Sandia, Argonne and Ames Laboratories[12]. This work showed that oxygen is lost from the surface, transforming the intrinsic metallic material into a semi-conductor. To test for this possibility, we show a plot of EPR intensity vs particle surface area in Figure 7. As can be seen, there is little evidence that the EPR signal scales with surface area, but strongly suggests a bulk origin for the spectra.

The low temperature EPR behavior observed here can also be used to rule out that the impurity phases are responsible for the EPR activity after ball milling. It is well established that the EPR signal from the "green" phase undergoes a catastrophic decrease in intensity at 15 K, due to an anti-ferromagnetic transition[9,10,13]. On the other hand the "brown" phase displays an intense but poorly resolved spectrum between 300 K and 4 K, supposedly due to copper in a tetragonally distorted octahedral environment (D_{4h}) with g_Z = 2.23 and g_X = g_Y = 2.09[9]. The ball milled samples here show no catastrophic drop in EPR intensity down to 4 K, ruling out the presence of the "green" phase. In addition, there is no significant EPR spectral component characteristic of D_{4h} symmetry with g_Z = 2.23 and g_X = g_Y = 2.09 at these and higher temperatures, thus ruling out the presence of the "brown" phase. We thus conclude that the EPR spectrum above T_c seen after ball milling is not due to the presence of these other phases, but rather is due to copper(II) complexes at internal defects of some sort in the $YBa_2Cu_3O_{7-x}$ phase.

The EPR "silence" of good superconducting oxide has been suggested as being due to super-exchange coupling of the copper ions through the oxygen p-orbitals[11]. Our observations indicate that upon ball milling, a strong EPR signal appears which correlated quantitatively in intensity with the density of twin boundaries. At the same time, the relative susceptibility decreases, see Figure 3. Thus it is natural to ascribe the observed EPR and superconducting changes to copper(II) complexes which are associated with the twin boundaries in some way, and not due to the formation of the "green" phase. We suppose that the local oxygen ligation and structure near the twin boundary changes the local and translational symmetry of the copper ions associated with the boundaries in such a manner and extent that isolated copper(II) states occur. The EPR signal then arises from these especial copper ions in the twin boundaries. As shown below, using a straight foreward structural model of the twin boundary

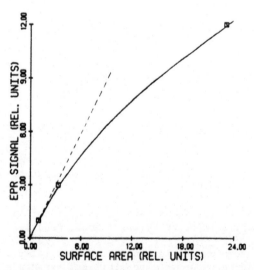

Figure 7 Plot of EPR intensity vs particle surface area. Note that the EPR signal does not scale with surface area.

and the copper(II) environment, we are able to quantitatively simulate
the EPR spectra observed here. At the same time we assert, with
others[14], that the twin boundaries are the weak links in question
responsible for the widening of the transition.

The Copper(II) Environment

Localization of holes onto distinct Cu(II) sites requires that
the site symmetry be very different from those in the host crystal
for hole diffusion not to occur. The Cu(II) sites associated with
(110) twin boundary are of different symmetry and are thus natural
candidates where hole localization can be expected. The structure
local to copper(II) ions in un-twinned material, and twinned material
are shown in Figure 8. These diagrams represent the structures appropriate
to (110) twinning, which is commonly reported, e.g.[18,19] and which
serve as a first order structural basis for understanding the EPR
spectra.

The spin-Hamiltonian which is of relevance to these structures
may be written as

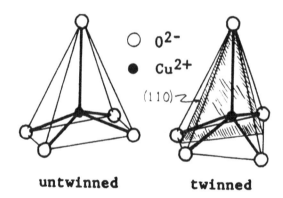

Figure 8 Structure of copper(II) ion
environments in untwinned and
twinned $YBa_2Cu_3O_{7-x}$.

$$H = \beta \mathbf{H} \cdot \mathbf{g} \cdot \mathbf{S} + \mathbf{I} \cdot \mathbf{A} \cdot \mathbf{S} \qquad [1]$$

where β is the Bohr magneton, **g** and **A** are the system g and nuclear
hyperfine coupling tensors respectively, and **S**, **I**, and **H** are the
electronic spin, nuclear spin, and magnetic field vectors respectively.
Expanding out the terms for the diagonalized tensors then,

$$H = \beta[g_X S_X H_X + g_Y S_Y H_Y + g_Z S_Z H_Z]$$
$$+ [A_X S_X I_X + A_Y S_Y I_Y + A_Z S_Z I_Z] \qquad [2]$$

The values of g_i represent the degree to which the electronic spin
moment is coupled to the orbital angular momentum.

The most striking feature of the EPR spectra of the milled samples is that the observed hyperfine coupling constants, $A_i{}^{OBS}$, are relatively small for copper(II) complexes approaching square pyramidal symmetry, and in particular such as those found in pure YBa_2CuO_{7-x}. For complexes such as these, $|A_z|$ is expected to be $\sim 120\text{-}150\times 10^{-4}$ cm^{-1} with $|A_X|$, $|A_Y|$ $\ll |A_Z|$ [20]; whereas, experimentally we find that $|A_Z{}^{OBS}| \simeq 16\text{-}76\times 10^{-4}$ cm^{-1} and that $|A_X{}^{OBS}| > |A_Z{}^{OBS}|$, see Table I. We believe that the small values of $|A_i{}^{OBS}|$ are due to spin exchange between the complexes in question which narrows the hyperfine spacing[21]. Indeed, for the crystallographic phases of concern here, namely $YBa_2Cu_3O_{7-x}$ and Y_2BaCuO_5, there exists a large degree of super-exchange between neighboring Cu^{2+} cations through intervening O^{2-} anions[9]. For the case of Y_2BaCuO_5, this superexchange gives rise to antiferromagnetism below 15 K[10]. Therefore, the above mechanism for the small values of $|A_i{}^{OBS}|$ is certainly reasonable.

The EPR spectra of copper(II) complexes are well understood: expressions for pure and distorted octahedral coordination have been given by Harrowfield et al[15], for pure and distorted square pyramidal coordination by Bencini et al[16] and for pure and distorted trigonal bipyramidal coordination by Bertini et al[17]. We have accordingly performed a search for the possible structures for copper(II) complexes compatible with the observed EPR spectra subject to the following assumptions:

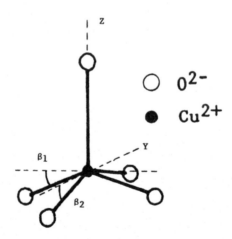

Figure 9 Structural model used to evaluate the EPR parameters of the observed spectra. The bond angles β_1 and β_2 are indicated.

1. Since the copper(II) complexes found in pure $YBa_2Cu_3O_{7-x}$ are of five-fold coordination and of C_{2v} symmetry, we assume that the complexes which give rise to the observed EPR spectra are also of C_{2v} symmetry. This coordination scheme will be generally described by the distortion of bond angles β_1 and β_2 as shown in Figure 9. This is a more general case than that described by Bertini et al[17] in which only one bond angle is allowed to be distorted. The relevant

expressions for g_i and A_i are intricate and space precludes listing them here.

2. Since the effect of spin exchange is to narrow the hyperfine spacing, only calculated values of $|A_i|$ which are greater than or equal to $|A_i^{OBS}|$ will be allowed.

For copper(II) complexes in the pure $YBa_2Cu_3O_{7-x}$ phase, β_1 and β_2 correspond to $\sim 7.5°$ and $\sim 8.0°$ respectively. Our structural search shows that for the EPR parameters describing both one- and two-hour milled samples, no such C_{2v} complex satisfies the assumptions made above. In other words, the observed EPR spectra cannot be due to copper(II) complexes in symmetries found in perfect $YBa_2Cu_3O_{7-x}$ crystals. The allowed site structures, which cover a range of β_1 and β_2 values, but not including those of the perfect crystal, must lie elsewhere, viz. at twin boundaries.

Since the experimentally determined EPR parameters may not strictly arise in the C_{2v} symmetry of the copper(II) complexes as assumed, the analysis should be extended to investigate lower symmetry complexes. This will involve asymmetric bending of bonds and a revised set of relations for A_i and g_i. The corresponding quantum mechanical derivations are presently being worked out. These lower symmetry complexes can be expected at twin boundaries, and other such defects as illustrated in Figure 8.

TEM Studies

The extruded sample which shows a significant smaller EPR signal than the ball milled samples also exhibited a remarkable change in microstructure, see Figure 5. It was found that the defect density decreased sharply and in many instances the material showed no planar defects when viewed under a variety of diffraction contrast imaging conditions. However using lattice imaging one is able to clearly observe planar faulting at the nanometric scale. Unlike the other samples, the length of these faults extends only a few tens of nanometers. Another significant observation is the presence of what we wish to term as "ribbon-like" defects. These defects neither exhibit contrast behavior of dislocations nor do they have the straight appearance of twin boundaries. Initial studies on the nature of these defects have been conducted using convergent beam diffraction. The initial findings indicate a slight displacement of Higher Order Laue Zone (HOLZ) lines in the disc of the diffracted beam, when using two-beam condition. While preliminary results do not show complimentary contrast of HOLZ lines on either side of the defect, typical of inversion domain boundaries, it is clear that some rigid body displacement exists across the defect. Further work is in progress to study the nature of this defect; however it is significant that this new type of defect appears in the material exhibiting a noticeably different EPR intensity.

The sample that has been annealed at 500°C shows a very high twin density, see Figure 6. Unlike the other samples this material contained a high density of intersecting twins. The microstructure is identical to that observed by Shaw et al[23] in material given a slow cooling oxygen heat treatment. Strong bands of strain contrast are observed to run diagonally across each cell formed by the intersecting twins. It has been suggested that the strain contrast is due to the buckling of the foil caused by the relaxation of the stresses associated with the intersecting twins. We propose to be more specific, by noting that this buckling can be caused by the formation of out-of-plane secondary twinning occuring at the twin intersections. This type of phenomena has been well established in twin-twin interactions in fcc alloys[24,25]. While the presence of twins has been interpreted in terms of localized changes in chemical environment, it is suggested here that secondary twins associated with transformation strains may also have an important effect. This is presently being examined in more detail.

CONCLUSIONS

Using ball milling as a method of introducing defects into optimally prepared $Y_2Ba_2Cu_3O_{7-x}$, we have shown that deformation induced twins are formed which degrade the superconducting properties. At the same time the copper(II) complexes associated with the boundaries give rise to a strong EPR signal. The structural environment of these associated copper(II) complexes has been shown on the basis of the EPR spectra to be different from those in bulk material. This very different coordination of the copper(II) ions associated with the planar defect boundaries provides a natural explanation for their influence on the superconducting properties.

Our future work will involve more quantitative methods of introducing planar defects into these materials. Further study of the new "ribbon-like" defects, the symmetry of the EPR active copper(II) complexes, and the superconducting properties is envisaged.

ACKNOWLEDGMENTS

We acknowledge G. Hamlin for the development of the AC bridge and careful susceptibility measurements, and the assistance of R. Dove with the electron microscopy. This research was partially supported by the National Science Foundation under Contract No. DMR-8510617.

REFERENCES

1. M. Couach, A.F. Khoder, B. Barbara, J.Y. Henry, C. Ayache, E. Bonjour, R. Calemczuk and B. Salce, Physica C, 153-155, 844 (1988).
2. G.G. Zwicknagel and J.W. Wilkins, Phys. Rev. Lett. 53(13), 1276 (1984).
3. J.W. Ekin, Advanced Ceramic Materials (Special Supplementary Issue), vol. 2, p. 586 (1987).
4. M.M. Fang, V.G. Jogan, D.K. Finnemore, J.R. Clem, L.S. Chumbley, and D. Farrel, Phys. Rev. B 37, 4, 2334 (1988).
5. R. Beyers, G. Im, E.M. Engler, R.J. Savoy, T.M. Shaw, T.R. Dinger, W.J. Gallagher and R.L. Sandstrom, Appl. Phys. Lett. 50(26) 191B (1987).
6. G. Van Tendeloo, H.W. Zandbergen and S. Amelinckx, Solid State Comm., 63(5), 389 (1987).
7. N.X. Tan, A.J. Bourdillon, S.X. Don, H.K. Kiu and C.C. Sorrel, Phil. Mag. Lett., 58(3), 1988.
8. D. Shi, M. Patel and P. Tashishta, to be published in J. Appl. Phys.
9. W.R. McKinnon, J.R. Morton, K.F. Preston and L.S. Selwyn, Solid State Commun., 65 No. 8, 855 (1988).
10. E.W. Ong, B.L. Ramakrishna and Z. Iqbal, to be published in Solid State Comm.
11. D. Shalteil, J. Genossar, A. Grayevsky, Z.H. Kalman, B. Fisher and N. Kaplan, Solid State Commun., 63(11), 987 (1987).
12. Sandia Science News 2/89 as reported in Ceram. Bull., 68 No. 5, 966 (1989).
13. J.Z. Sun et al., Phys. Rev. Lett., 58, 1574 (1987).

14. R. Flukiger, T. Muller, W. Goldacker, T. Wolf, E. Seibt, I. Apfelstedt, H. Kupfer and W. Schauer, Physica C 153-155, 1574 (1988).
15. B.V. Harrowfield, A.T. Dempster, T.E. Freeman, and J.R. Pilbrow, Solid State Phys., 6 2058 (1973).
16. A. Bencini, I. Bertini, D. Gatteschi and A. Scozzafava, Inorg. Chem., 17, 3194 (1979).
17. I. Bertini, G. Canti, R. Grassi and A. Scozzafava, Inorg. Chem., 19, 2198 (1980).
18. M. Herview, B. Domengas, C. Michel, and B. Raveau, Europhys. Lett., 4, 211 (1987).
19. H. You, J.D. Axe, X.B. Kan, S.C. Moss, J.Z. Liu, and D.J. Lam, in High Temperature Superconductors, Eds. M.B. Brodsky, R.C. Dynes, K. Kitazawa, and H.L. Tuller, MRS Symposium Proceedings, vol. 99, 1988, p. 83.
20. C. Michel and B. Raveau, J. Solid State Chem. 49, 150 (1983).
21. Y.N. Molin, K.M. Salikhov, and K.I. Zamaraev, "Spin Exchange Principles and Applications in Chemistry and Biology", Springer-Verlag, New York, 1980, Chap. 1-3.
22. D.W. Murphy, S.A. Sunshine, P.K. Gallagher, H.M. O'Bryan, R.J. Cava, B. Batlogg, R.B. Van Dover, L.F. Schneemeyer and S.M. Zahurak in "Chemistry of High-Temperature Superconductors", eds. D.L. Nelson, M.S. Whittingham and T.F. George, ACS Symposium Series 351, 1987, Chapter 18.
23. T.M. Shaw, S.L. Shinde, D. Dimos, R.F. Cook, P.R. Duncombe and C. Knoll, J. Mat. Res. 4, 248 (1989).
24. S. Mahajan and G.Y. Chin, Acta Met 22, 1113 (1974).
25. K. Rajan, J. Mat. Sci. Lett. 1, 482 (1982).

THEORY OF STRONGLY FLUCTUATING SUPERCONDUCTIVITY

S. H. Liu

Solid State Division, Oak Ridge National Laboratory

Oak Ridge, Tennessee 37831-6032

ABSTRACT

In superconductors with short coherence lengths, such as the copper oxides, the order parameter is prone to phase fluctuations. Josephson showed that phase fluctuations are coupled to the density of Cooper pairs so that they are always accompanied by space and time varying supercurrents. In this paper we show how the variation of the order parameter, its phase, and the charge density can be mapped onto a magnetic problem by a real space extension of the pseudo-spin representation of Anderson. It is shown that at long wavelengths the charge fluctuation is asymptotically decoupled from the order parameter fluctuation so that the problem is isomorphic to the XY model. In particular, at low temperatures there exist collective excitations, the so-called phase waves, which are analogous to spin waves in magnetic systems. Two experiments are proposed to test the theoretical predictions.

I. INTRODUCTION

Specific heat measurements on high-T_c superconductors have revealed strong critical fluctuations like what is seen in magnetic systems.[1] In contrast, the critical region in low temperature superconductors is usually so narrow that no critical effects can be observed. The difference between the two types of superconductors lies in their pair coherence lengths, typically 1000 Å for low-T_c superconductors but 10–20 Å for high-T_c superconductors. Short coherence length implies that the order parameter can vary rapidly in space. On the other hand, Josephson showed that spatial variations of the phase must be accompanied by flow of supercurrents.[2] Therefore, it is now important to include the Josephson effect in the fluctuation problem and reexamine the importance of charge fluctuation. In this paper we show that a complete theory of superconducting fluctuation can be formulated by an extension of the pseudo-spin formalism of Anderson.[3] In the mean-field approximation the theory is equivalent to the BCS theory.[4] In the critical region where long wavelength fluctuations dominate, the charge degree of freedom is asymptotically decoupled. This means that the Ginzburg theory,[5] which considers only the fluctuations in amplitude and phase of the order parameter, is an adequate starting point for studying the critical phenomena in superconductors. At low temperatures, the same consideration yields a collective excitation analogous to the spin waves in magnetic systems. The mode has a linear dispersion relation, and is distinct from the zeroth sound discussed by Bogoliubov[6] and the plasmon mode discussed by Anderson and others.[7-9] The mode is common to all singlet paired superconductors, so they are perhaps more readily observable in ordinary superconductors. Methods for detecting these excitations will be described.

II. PSEUDO-SPIN FORMALISM

In a 1958 paper Anderson demonstrated that the BCS operator $b_{\vec{k}}^\dagger = c_{\vec{k}\uparrow}^\dagger c_{\vec{k}\downarrow}^\dagger$, its hermitian conjugate $b_{\vec{k}}$, and the operator $\frac{1}{2}(c_{\vec{k}\uparrow}^\dagger c_{\vec{k}\uparrow} + c_{-\vec{k}\downarrow}^\dagger c_{-\vec{k}\downarrow} - 1)$ commute with one another like S^+, S^-, and S^z components of the spin operator.[3] In the following we show that pseudo-spin operators can be extended to real space. We divide the crystal into identical cells such that the linear dimension of each cell is much larger than the Coulomb screening length but much smaller than the pair coherence length. The two lengths differ by orders of magnitude, so there is considerable flexibility in the choice of cell size. In superconducting compounds with many atoms in a unit cell, such as A15 and the copper-oxide family, the cell can be simply the unit cell of the crystal structure. In elemental superconductors with one atom per unit crystal cell, we need to coarse grain the system by combining a few unit cells into every supercell. We redefine the electron states in the Brillouin zone of the new cells and label them by the band index n and wavevector $\vec{k}$. Next we construct Wannier operators in the cells by

$$c_{n\ell s} = N^{-\frac{1}{2}} \sum_{\vec{k}} c_{n\vec{k}s} e^{i\vec{k}\cdot\vec{R}_\ell} \tag{1}$$

where $\vec{R}_\ell$ is a convenient reference point for the ℓ-th cell, N is the total number of cells, and s labels the spin. The band index n can be interpreted as the orbital index in each cell. The real space pair operators are defined in terms of the Wannier operators:

$$b_\ell^\dagger = \sum_n c_{n\ell\uparrow}^\dagger c_{n\ell\downarrow}^\dagger \ . \tag{2}$$

Following Anderson we define the pseudo-spin operators by

$$\begin{aligned}
\xi_\ell &= \frac{1}{2}(b_\ell^\dagger + b_\ell) \ , \\
\eta_\ell &= \frac{1}{2i}(b_\ell^\dagger - b_\ell) \ , \\
\zeta_\ell &= \frac{1}{2}\sum_n (c_{n\ell\uparrow}^\dagger c_{n\ell\uparrow} + c_{n\ell\downarrow}^\dagger c_{n\ell\downarrow} - 1) \ .
\end{aligned} \tag{3}$$

It is straightforward to verify that these operators commute with one another like the x, y, z components of the spin operator respectively. The necessity for coarse graining becomes clear in the definition of $b_\ell^\dagger$. It costs a large Coulomb energy to put two electrons in the same atomic orbital, but the energy is much reduced by screening when they are put in one cell orbital defined above.

In the ground state of a uniform superconductor we can choose the phase of the order parameter to be zero. This means that the expectation values $<b_\ell^\dagger>$ and $<b_\ell>$ are both equal to Δ/V, where Δ is the gap function and V is the interaction potential. The expectation values of the pseudo-spin operators are: $<\xi_\ell> = \Delta/V$, $<\eta_\ell> = 0$, and $<\zeta_\ell> = \frac{1}{2}\sum_n(<c_{n\ell\uparrow}^\dagger c_{n\ell\uparrow}> + <c_{n\ell\downarrow}^\dagger c_{n\ell\downarrow}> - 1)$. The last quantity, which is a measure of the electron density, has the same value in the normal state. Alternatively, we may regard the superconducting transition as a magnetic ordering problem of the pseudo-spins. The expectation value of the ζ component remains unchanged at the transition, but the ξ component develops a long range order below T_c. Furthermore, the ordered moment may be defined in an arbitrary direction in the $\xi\eta$ plane. Such a magnetic system is the XY model. In the mean-field approximation we treat $<\xi_\ell>$ as a c-number and $<\eta_\ell> = 0$, we find that $<b_\ell^\dagger> = <b_\ell>$ is a c-number. This is precisely the starting assumption of the BCS theory, and a self-consistent calculation of the order parameter yields the BCS gap equation.[4]

If the system is perturbed by a weak and slowly varying electromagnetic field, the order parameter acquires a phase ϕ_ℓ, then $\zeta_\ell = (\Delta/V)\cos\phi_\ell$ and $\eta_\ell = -(\Delta/V)\sin\phi_\ell$. Thus, when ϕ_ℓ is small, ξ_ℓ remains the order parameter but η_ℓ represents phase variations. The ζ_ℓ component can be readily related to the number of pairs ρ_ℓ in the cell, $\zeta_\ell = \rho_\ell - n_b/2$, where $\rho_\ell = \frac{1}{2}\sum_n(c^\dagger_{n\ell\uparrow}c_{n\ell\uparrow} + c^\dagger_{n\ell\downarrow}c_{n\ell\downarrow})$ and n_b is the number of orbitals in a cell. Putting all these results into the pseudo-spin commutation relation $[\eta_\ell, \zeta_{\ell'}] = i\delta_{\ell\ell'}\xi_\ell$, we obtain

$$[\phi_\ell, \rho_{\ell'}] = -i\delta_{\ell\ell'} \ . \tag{4}$$

This is the discrete lattice version of the Anderson relation[10]

$$[\phi(\vec{r}), \rho(\vec{r}\,')] = -i\delta(\vec{r}-\vec{r}\,') \ , \tag{5}$$

which is the mathematical foundation of the Josephson effect. In the pseudo-spin space, the Josephson effect can be regarded as the result of an infinitesimal rotation around the ξ axis.

In a similar manner we may carry out an infinitesimal rotation around the phase axis η. This transformation mixes the order parameter with the charge degree of freedom so that the two must vary together. A spatial variation of the order parameter without phase change can be realized at the junction between a superconductor and a normal metal. As a result of the pseudo-spin commutation relation, there must be charge flow across the junction when proximity effect causes the normal metal to be superconducting near the junction.

The familiar superconducting fluctuation corresponds to local rotations around the ζ axis. The charge is conserved, but the real and imaginary parts of the order parameter undergo local variations. In the next Section we will develop a dynamical theory for this type of pseudo-spin excitation.

III. CALCULATION OF EXCITATION ENERGY

We describe the electron system by the BCS Hamiltonian

$$H = H_0 + H_1 \ , \tag{6}$$

where H_0 is the band energy

$$H_0 = \sum_{nks} \epsilon_{n\vec{k}} c^\dagger_{n\vec{k}s} c_{n\vec{k}s} \ , \tag{7}$$

and H_1 is the pair interaction energy

$$H_1 = -V \sum_{n\vec{k}} \sum_{n'\vec{k}'} c^\dagger_{n\vec{k}\uparrow} c^\dagger_{n,-\vec{k}\downarrow} c_{n',-\vec{k}'\downarrow} c_{n',\vec{k}'\uparrow} \ . \tag{8}$$

The band energies are measured from the Fermi level. The origin of the interaction V need not concern us here. In the ground state all pseudo-spins are aligned in the ξ direction, and the ground state energy is entirely determined by the order parameter Δ. We now investigate the dynamical properties by pulling the pseudo-spins slightly away from the ξ direction. We define the Fourier components of the pseudo-spins in the usual way and represent the deviations in terms of the Fourier components $\xi_{\vec{q}}$ etc. Next we expand the Hamiltonian around the ground state configuration and find that, to the lowest order of the spin deviations

$$H = E_0 + \sum_{\vec{q}} \left(\frac{\delta H}{\delta \xi_{\vec{q}}} \xi_{\vec{q}} + \frac{\delta H}{\delta \eta_{\vec{q}}} \eta_{\vec{q}} + \frac{\delta H}{\delta \zeta_{\vec{q}}} \zeta_{\vec{q}} \right) \ , \tag{9}$$

where E_0 is the ground state energy and the functional derivatives are the effective interactions of the electron system with the pseudo-spins. It is straightforward to determine that

$$\frac{\delta H}{\delta \xi_{\vec{q}}} = -V(b^\dagger_{-\vec{q}} + b_{\vec{q}}) \; , \tag{10}$$

$$\frac{\delta H}{\delta \eta_{\vec{q}}} = iV(b^\dagger_{-\vec{q}} - b_{\vec{q}}) \; , \tag{11}$$

where

$$b_{\vec{q}} = N^{-\frac{1}{2}} \sum_q b_\ell e^{i\vec{q}\cdot\vec{R}_\ell} \; . \tag{12}$$

The quantities in Eqs. (10) and (11) can be identified as the longitudinal and transverse components of the pairing field, and the divergence of the susceptibilities of the electron system to this field marks the onset of superconductivity. In contrast, the interaction with charge fluctuation has the expression

$$\frac{\delta H}{\delta \zeta_{\vec{q}}} = \frac{V}{\sqrt{N}} \sum_{\vec{q}} [c^\dagger_{n\vec{k}\uparrow} c_{n,\vec{k}+\vec{q},\uparrow} + c^\dagger_{n\vec{k}\downarrow} c_{n,\vec{k}+\vec{q}\downarrow}] \; , \tag{13}$$

which has the form of an internal electric field.

The contributions of the fluctuations to the total energy are calculated by second order perturbation theory. The calculation is standard,[11] and it is found that the increase in energy is

$$\delta E = -\frac{1}{2} V \Delta \sum_{\vec{q}} \{I_1(\vec{q})[\xi_{\vec{q}} \xi_{-\vec{q}} + \eta_{\vec{q}} \eta_{-\vec{q}}] + I_2(\vec{q}) \zeta_{\vec{q}} \zeta_{-\vec{q}}\} \; , \tag{14}$$

where $I_1(\vec{q})$ and $I_2(\vec{q})$ denote two integrals

$$I_1(\vec{q}) = \frac{1}{2N} \sum_{n\vec{k}} [1 + \frac{\epsilon_{n\vec{k}} \epsilon_{n,\vec{k}+\vec{q}} + \Delta^2}{E_{n\vec{k}} E_{n,\vec{k}+\vec{q}}}] \frac{1}{E_{n\vec{k}} + E_{n,\vec{k}+\vec{q}}} \; , \tag{15}$$

which is the pair susceptibility of the non-interacting electrons, and

$$I_2(\vec{q}) = \frac{1}{2N} \sum_{n\vec{k}} [1 - \frac{\epsilon_{n\vec{k}} \epsilon_{n,\vec{k}+\vec{q}} + \Delta^2}{E_{n\vec{k}} E_{n,\vec{k}+\vec{q}}}] \frac{1}{E_{n\vec{k}} + E_{n,\vec{k}+\vec{q}}} \; , \tag{16}$$

which is the paramagnetic part of the Meissner kernel.[4,11] In the above expressions $E_{n\vec{k}} = [\epsilon^2_{n\vec{k}} + \Delta^2]^{1/2}$ is the quasi-particle energy.

The right-hand side of Eq. (14) may be interpreted as an effective Hamiltonian for pseudo-spins written in terms of Fourier transforms of spins. The integrals $I_1(\vec{q})$ and $I_2(\vec{q})$ are proportional to the Fourier transforms of pair exchange interactions, and they determine the dynamical properties of the spin system. It is of particular interest to study the long wavelength behavior of the interaction integrals because long wavelength fluctuations determine both the critical behavior near the critical temperature and the collective excitations at low temperatures. Without making any assumptions about the band structure we can determine in the limit $q \to 0$ that $I_1(\vec{q}) = 1/V - O(q^2)$ and $I_2(\vec{q}) = O(q^2)$. This shows that the spin system has highly anisotropic interactions such that in the long wavelength limit the third component, the charge fluctuation, becomes asymptotically decoupled from the other two components for the order parameters. This result assures that the Ginzburg approach to the critical phenomenon in superconductors, which leaves out the Josephson coupling, is basically sound. For short wavelength fluctuations, however, a varying pair density, i.e. a supercurrent, is an integral part of the dynamics. The consequences of this effect remains to be explored.

The isomorphism of the superconductivity problem to the XY model implies that there must exist collective modes in superconductors analogous to spin waves. To find these modes we quantize the operators $\xi_{\vec{q}}, \eta_{\vec{q}}, \zeta_{\vec{q}}$ according to their commutation relations:

$$[\xi_{\vec{q}}, \eta_{\vec{q}'}] = i\zeta_{\vec{q}+\vec{q}'} \tag{17}$$

and others. The well-known Holstein-Primakoff transformation accomplishes this task:[12]

$$\xi_{\vec{q}} = \frac{\Delta}{V}\delta_{\vec{q}0} - \sum_{\vec{q}'} a^\dagger_{\vec{q}+\vec{q}'} a_{\vec{q}'}$$

$$\eta_{\vec{q}} = \sqrt{\frac{\Delta}{2V}} (a^\dagger_{\vec{q}} + a_{-\vec{q}}) ,$$

$$\zeta_{\vec{q}} = \frac{1}{i}\sqrt{\frac{\Delta}{2V}} (a^\dagger_{\vec{q}} - a_{-\vec{q}}) , \tag{18}$$

where $a^\dagger_{\vec{q}}$ and $a_{\vec{q}}$ are boson operators. One can easily verify that the commutations relations in Eq. (17) are satisfied to the lowest order in the $a_{\vec{q}}$ operators.[13] The effective Hamiltonian in Eq. (14) can be expressed in terms of the boson operators up to the second order as

$$\delta E = \Delta \sum_{\vec{q}} \{2a^\dagger_{\vec{q}} a_{\vec{q}} + \frac{1}{2}VI_1(\vec{q})(a^\dagger_{-\vec{q}} + a_{\vec{q}})(a^\dagger_{\vec{q}} + a_{-\vec{q}}) \\ - \frac{1}{2}VI_2(\vec{q})(a^\dagger_{-\vec{q}} - a_{\vec{q}})(a^\dagger_{\vec{q}} - a_{-\vec{q}})\} . \tag{19}$$

Diagonalization of this Hamiltonian yields the energy spectrum:

$$\omega_{\vec{q}} = 2\Delta\{[1 - VI_1(\vec{q})][1 - VI_2(\vec{q})]\}^{\frac{1}{2}} . \tag{20}$$

In the long wavelength limit the first factor $1 - VI_1(\vec{q})$ is or the order q^2 and the second factor is essentially unity. Thus, the mode has a linear dispersion relation, and the velocity depends in the details of the electronic structure. To make contact with previous theories we make the simplifying assumption that there is only one parabolic band that crosses the Fermi level. The integrals can be carried out, and the dispersion relation has the expression

$$\omega_{\vec{q}} = [N(0)V]^{\frac{1}{2}} v_F q/\sqrt{3} , \tag{21}$$

where v_F is the Fermi velocity and $N(0)$ is the density of states at the Fermi level.

The search for collective modes, the Goldstone boson, in superconductors has a long history. Bogoliubov studied the fluctuation of pair density and obtained a sound-like mode with velocity $v_F/\sqrt{3}$.[6] Physically this mode is the analog of the zeroth sound in liquid ^{3}He. Anderson argued that since Cooper pairs are charged, any fluctuation of their density must be identical to the plasma mode, which are very high energy excitations.[7,8] Ambegaokar and Kadanoff proposed that the collective mode should be a phase wave, and they searched for it by finding the pole in the response function of the superconductor to a weak electromagnetic field.[9] Their result confirmed the strong coupling to the charge such that mode must be the plasma oscillation. The mode found in this paper is physically very different from all of the above. We have already shown in deriving the phase wave that the charge is asymptotically decoupled, so the mode is the one sought by Ambegaokar and Kadanoff. The difference between our calculation and that in Ref. 9 is that the new mode is a pole in the response function of the system to a weak and varying pairing field. We believe this is the more physical way to approach the problem. The result in Eq. (21) indicates that the wave velocity contains the BCS coupling parameter $N(0)V$ such that the mode does not exist unless the system is superconducting.

The phase wave becomes overdamped once its energy overlaps with the continuum of electron spectrum, and this occurs when $\omega_q = 2\Delta$. The maximum wave vector can be estimated from $q_{max} \simeq \Delta/v_F$, which is the inverse of the pair coherence length. Typical low temperature superconductors have coherence lengths of a few thousand Å, so the phase mode exists only in a negligibly small part of the Brillouin zone and makes no detectable contribution to the thermodynamic and critical properties. For high-T_c materials, however, the coherence length is comparable to the lattice parameter. We expect the phase wave to exist over a sizable region of the Brillouin zone.

Finally we discuss how the phase wave may be detected. Consider a superconducting film whose thickness d is 2-3 times the coherence length. At the film surfaces the normal gradient of the phase vanishes. This confines the phase waves to a set of discrete energy levels given by $\omega_n = cn\pi/d$, where c is the velocity and n is an integer. We put one Josephson junction at one side of the film so that the film can be biased relative to the outer electrode. Under zero bias a dc Josephson current can flow through the entire junction. When the film is biased, one should observe a dc Josephson current again whenever the bias voltage satisfies $V_n = \omega_n/2e$. The magnitude of the current should be proportional to the Boltzmann factor $\exp(-\omega_n/kT)$ because this measures the probability that the system is in the n-th excited state. Alternatively, one can put a Josephson junction on one side of the film, bias the film sufficiently, and observe ac tunneling current signals at a series of frequencies $2eV - \omega_n$ for $n = 0, 1, 2$, etc. Again, the signal strength should be proportional to the Boltzmann factor for the n-th state.

IV. CONCLUSION

We have shown that in strongly fluctuating superconductors the Josephson effect is an integral part of the fluctuation dynamics. In the long wavelength limit, which determines the critical phenomena, the charge fluctuation becomes asymptotically decoupled and the problem is isomorphic to the XY model of magnetism. At low temperatures there should exist a collective mode analogous to the spin waves, and two experiments are proposed to verify this prediction.

V. ACKNOWLEDGMENTS

The author wishes to thank R. S. Fishman, G. M. Stocks, Y. H. Kao, and D. C. Mattis for illuminating discussions and constructive criticisms. This research was supported by the Division of Materials Sciences, U.S. Department of Energy, under Contract No. DE-AC05-84OR21400 with Martin Marietta Energy Systems. Inc.

REFERENCES

1. S. E. Inderhees, M. B. Salamon, N. Goldenfeld, J. P. Rice, B. G. Pazol, D. M. Ginsberg, J. Z. Liu, and G. W. Crabtree, *Phys. Rev. Lett.* **60**, 1178 (1988).
2. B. D. Josephson, *Phys. Lett.* 1, 251 (1962).
3. P. W. Anderson, *Phys. Rev.* **112**, 1900 (1958).
4. J. Bardeen, L. Cooper, and J. R. Schrieffer, *Phys. Rev.* **108**, 1175 (1957).

5. V. L. Ginzburg, *Fiz. Tverd. Tela* **2**, 2031 (1960) [*Soviet Physics, Solid State* **2**, 1824 (1960)].
6. N. N. Bogoliubov, V. V. Tomachev, and D. V. Shirkov, *A New Method in the Theory of Superconductivity* (Academy of Sciences of the U.S.S.R., Moscow, 1958 and Consultants Bureau, Inc., New York, 1959), Chap. 4.
7. P. W. Anderson, *Phys. Rev.* **110,** 827 (1958).
8. G. Rickayzen, *Phys. Rev.* **115**, 795 (1959).
9. V. Ambegaokar and L. P. Kadanoff, *Nuovo Cimento* **22**, 914 (1961).
10. P. W. Anderson, *Rev. Mod. Phys.* **38**, 298 (1966).
11. J. R. Schrieffer, *Theory of Superconductivity* (Benjamin, New York, 1964).
12. T. Holstein and H. Primakoff, *Phys. Rev.* **58**, 1098 (1940).
13. The boson representation does not reproduce the $q = 0$ component of the charge component ζ, but this component is a constant throughout and is not a part of the superconducting fluctuation.

CORRELATION IN METALLIC COPPER OXIDE SUPERCONDUCTORS:

HOW LARGE IS IT?

W.E. Pickett,[a] H. Krakauer,[b] R.E. Cohen,[a]
D. Singh,[a] and D.A. Papaconstantopoulos[a]

[a]Complex Systems Theory Branch, Code 4690
Naval Research Laboratory, Washington, DC 20375-5000

[b]Department of Physics, College of William and Mary
Williamsburg, VA 23185

INTRODUCTION

The remarkably high superconducting transition temperatures (T_c) of the copper oxide superconductors suggest some unusual property is necessary to account for high T_c. Suggestions for such a characteristic property include (1) the layered, highly two-dimensional (2D) structure itself, (2) the proximity to metal/insulator, magnetic/nonmagnetic or structural transitions, (3) large anharmonicity, and (4) very large electron-phonon coupling, among others. However, the increasing number of distinct compounds[1] which display T_c values ranging quasicontinuously from low temperature to above 100 K can be interpreted as indicating there is some property (related to pairing, of course) which is a rather general property of layered cuprates and which is variable. Moreover, the question of whether the layered cuprates are special is raised by the bismuthates, since $(Ba,K)BiO_3$ (containing no Cu) superconducts above[2] 30 K.

The number of speculations about the pairing mechanism for high T_c has been enormous, and after nearly three years of such speculation no convincing picture is in sight. A more systematic approach is to learn what picture is required to describe the properties of the normal, metallic phase above T_c, and thereby to discover the possible, and indeed likely, instabilities which may lead to superconductivity. We have applied the density functional (DF) method (in the local density approximation (LDA)) to calculate a variety of properties. The properties, which we review in this paper, are of both the ground state variety (unit cell volume, shape and internal parameters, and phonon frequencies) where the DF approach is rigorous, as well as excited state properties which are approximately correct as long as many-body correlations are not too strong.

A central question centers on strong correlation effects in the copper oxide materials. In fact, recent practice is to define "strong correlation" in terms of the inability of LDA to describe the desired properties. Since we will not review the density functional (and LDA) formalism here, we will just comment that such "strong correlation" effects for ground state properties implies an inadequacy of the local density approximation (true density functional theory would give the exact properties) while "strong correlation" effects in excited states more likely indicates large dynamic self-energies rather than any explicit problem with LDA (see, for example, Ref. 3 for discussion and references).

As with many strongly correlated systems, a popular approach[4] has been to try to identify a model Hamiltonian with the appropriate terms to describe both an antiferromagnetic insulating phase and a metallic, nonmagnetic phase, determine the parameters, and then to begin from the insulating regime (which usually is used to fix the parameters) and describe the alloying ("doping") through the insulator-to-metal transition. This approach, which tends to center on magnetic interactions, may well be an unnecessarily complex line of reasoning from which to describe the metallic phase, especially if it is simply a Fermi liquid (FL) phase. Thus as we discuss the various properties and the apparent "strong correlations" or lack thereof, we will also indicate whether the properties appear to fit a FL picture.

SPECTROSCOPY

A wide variety of spectroscopic measurements have been reported and reviewed. From the point of view of comparing the gross characteristics of the insulating and metallic phases, electron energy loss spectroscopy (EELS) has been exceedingly important. Unlike photoemission, which is very surface sensitive and as a result was misleading for the first two years, EELS is a bulk measurement which gives a comparitatively unambiguous picture of the unoccupied states above the Fermi level E_F. The results, reported by Fink et al.[5], dispute the picture of the metallic phase as doped insulators (whether Mott insulators or band insulators). The various pictures are shown schematically in Fig. 1. For the insulating phase (i.e., La_2CuO_4, $YBa_2Cu_3O_6$, etc.) the picture should be as in the upper panel, with no spectral density at E_F and a gap above, and such spectra are seen.

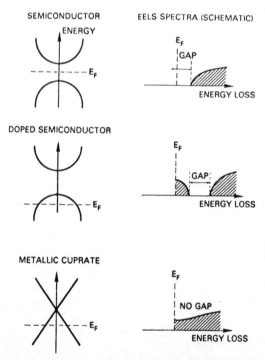

Figure 1. Schematic diagram of the band structure (left side) and the resulting EELS spectrum (right side, hashed regions) for a semiconductor, a hole-doped semiconductor, and a metal. The EELS data of Fink et al.[5] in the metallic phase of several layered cuprates resemble the bottom panel, with no gap occurring in the unoccupied region.

If the metallic phase is a (heavily) doped semiconducting phase, E_F should be moved down into the valence bands (for holes), and the hole states just above E_F will be seen and there will be a gap above. Such spectra are not seen, rather a non-zero density is seen at E_F and a <u>continuous</u> spectrum occurs above E_F. Evidently the alloying/doping has **completely destroyed the gap** (2-4 eV in La_2CuO_4 by various estimates) and a continuous spectrum without apparent band edges appears. Thus the evidence from EELS is that the gap actually collapses in the metallic phase, consistent with and highly suggestive of a FL-like state, and in agreement with LDA calculations.

Recent photoemission (PE) data has provided clear evidence of FL behavior. Earlier PE data on the $(La,M)_2CuO_4$ and $YBa_2Cu_3O_x$ systems had indicated an occupied spectral density in considerable disagreement with LDA predictions. Arko et al.[6] have shown, by cleaving crystals <u>in situ</u> (i.e. in the high vacuum necessary for PE experiments) at 20 K, that the previous data on $YBa_2Cu_3O_x$ materials represented data from oxygen-poor surfaces, and PE is an extremely surface-sensitive technique. Their data on samples in which oxygen had not been depleted from the surfaces were found to exhibit six peaks which could be aligned very convincingly, both in position and strength, with LDA predictions of Redinger et al.,[7] if an overall 0.5 eV shift to high binding energy was taken into account. A peak at very low binding energy also is consistent with a similar, chain-derived peak in our calculations.[8] The copper valence band satellite at 12.5 eV binding energy seen in earlier experiments was still present, indicating this feature is not due to oxygen deficiency of the surface.

The implications from this work are as follows: (1) LDA band theory gives a very realistic picture of the relative placement of oxygen and copper states and the large hybridization which occurs. This is a drastic change from the earlier indications (based on bad surfaces) which had suggested that LDA bands were considerably different from the spectral density. (2) The (apparently uniform) shift to 0.5 eV higher binding energy demonstrates both the strength and character of the self-energy corrections to the band structure. The size of the shift is significant but not much different than in copper metal itself, and it appears to be strongly band-like in the sense that it shifts all bands uniformly. This does not however rule out the possibility, which has been widely discussed and modelled, that the main corrections result from correlation processes associated with the Cu d states, since these states are strongly mixed into all of the valence band states. Costa-Quintana et al.[9], for example, have shown that one implementation of Cu d related self-energy provides a shift of E_F (and nearby bands) relative to valence band peaks which are nearly unaffected. (3) The valence band satellite also reflects self-energy effects, similar to those seen in Ni and Pd metal,[10] which also have a small unoccupied density of d states. The Ni bands and the satellite have been modelled successfully in terms of self-energy corrections to the LDA band structure.

Angle-resolved PE (ARPES) studies, which are capable of measuring the quasiparticle (QP) band structure (bulk and surface) of the material, are becoming available, although neither a thorough picture nor any consensus has arisen as yet. Manzke et al.[11] used both direct and inverse ARPES for momentum transfers in the basal plane of $Bi_2Sr_2CaCu_2O_8$, and found a band crossing E_F precisely where our Linearized Augmented Plane Wave (LAPW) calculation based on LDA predicted it to be. The gradient of the band, which is a measure of the inverse mass, appeared to be 3-4 times smaller than predicted by LDA. If this behavior is found to hold more generally, the situation will be reminiscent of the heavy electron compound UPt_3, whose very complicated Fermi surface was predicted remarkably well[12] by LDA calculations but whose effective masses at very low temperature are enhanced by roughly a factor of 20 by dynamic self-energy effects. The effective mass enhancements in the cuprates are not expected to be particularly large, however, since various estimates of the QP density of states (DOS) $N^*(E_F)$ (from specific heat, critical field slope, etc.) are similar to the LDA band value of $N(E_F)$.

If the mass enhancement measured by ARPES however does turn out to be sizable (such as even a factor of 3-4) it will have profound implications for the theoretical picture of the QPs. The mass enhancement of the heavy fermion metals[13] apparently arises from a frequency dependence of the QP self-energy, since it "evaporates" rapidly upon raising the temperature. It appears to arise from very low energy spin-fluctuations, and also becomes ineffective for excitations above a few tens of meV. In both respects it is similar to the mass enhancement from the electron-phonon interaction, except that it can be much larger. Conversely, since the mass enhancement of this one band of $Bi_2Sr_2CaCu_2O_8$ extends to high energy (of the order of 1 eV) and at least to room temperature, it either reflects an origin in a wavevector dependence of the self-energy or it originates from a frequency dependence arising from some excitation with high energy (of the order of 1 eV). In either case, it is expedient to verify this result and to determine whether this apparent mass enhancement is general for the high T_c cuprates.

Although other ARPES studies[14] have generally reflected well-defined QP bands near E_F, including crossings leading to Fermi surfaces, the correspondence with the calculated LDA bands has not been close enough to compare Fermi surfaces and mass enhancements. Whether this is an experimental difficulty (with oxygen-poor surfaces, confusion caused by surface bands, etc.) or a true discrepancy seems now to be resolvable, however. Finally, several high resolution PES measurements have shown sharp Fermi edges in the high T_c cuprates, and have even detected and estimated the magnitude[15] of the gap.

To summarize this section, EELS and PE spectroscopies are strongly supportive of not only a band-like FL metallic phase, but one which has much in common with LDA predictions. The Cu satellite, and perhaps the bands near E_F, will require self-energy/correlation effects for their description, and it is crucial to determine the character and magnitude of these many-body effects.

DISORDERED PHASES; RELATED OXIDES

LDA predictions of the electronic structure, bonding, densities of states, and charge densities have been reviewed elsewhere.[16] Here we briefly comment on recent work which should be taken into account when evaluating the magnitude of many-body effects.

<u>Electronic Structure for disordered phases</u>. Since many of the high T_c superconductors are either not well ordered or are alloyed, it is important to understand how the band structure is modified due to the effects of chemical disorder and vacancies. To address this problem we have first constructed tight-binding Hamiltonians by performing very accurate fits to our LAPW results, and then using these Hamiltonians in the coherent potential approximation (CPA) to obtain the electronic DOS for the alloyed or substoichiometric compounds. For $La_{2-x}Ba_xCuO_4$ the CPA results[17] support the hypothesis of rigid-band-like lowering of E_F (once in the metallic phase). This trend strongly increases $N(E_F)$ near x=0.15, revealing a striking correlation between $N(E_F)$ and T_c. On the other hand, we have found that deoxygenation does not raise E_F, contrary to simple doping arguments. For small amounts of vacancies, E_F remains pinned while, somewhat surprisingly, for larger vacancy concentration the Fermi level moves towards lower energies, creating an effect on $N(E_F)$ similar to

alloying with Ba. For $YBa_2Cu_3O_{7-y}$ our preliminary study of reducing the oxygen content in the chains also shows a strong correlation of the calculated $N(E_F)$ with the measured T_c.

Non-Cu-based perovskites. Because of the absence of Cu and the much lower T_c (<40 K), the $(Ba,K)(Pb,Bi)O_3$ system is considered by many to be in a different category of superconductor. There is much less concern about many-body correlations in this system, and there is strong evidence that the electron-phonon interaction is responsible for T_c. Our electronic structure calculations[18] for both stoichiometric and disordered phases of this system show that, while the position of E_F moves qualitatively according to the rigid band model, the oxygen-dominated bands widen as a result of alloying. The evaluation of the electron-phonon coupling strength λ in this system indicates a strong enhancement provided by Bi, and indicates that the standard theory to evaluate this coupling is adequate to explain the measured T_c.

STRUCTURAL ENERGIES AND PHONON FREQUENCIES

LDA calculations have been exceedingly successful in predicting relative structural energies between phases and in reproducing and predicting phonon frequencies, which are given by energy changes versus the amplitude of a phonon which is frozen into the lattice. Using the Linearized Augmented Plane Wave (LAPW) method, we have calculated[19] the lattice constants a and c and the internal structural parameters $z(O_z)$ and $z(La)$ of La_2CuO_4, obtaining excellent agreement with experimental values. The calculated cell volume showed a small discrepancy, being about 5% less than experiment, but including zero point motion would improve the comparison somewhat. Even so, the agreement is very satisfactory and better than many "normal," less complicated, compounds.

Encouraged by these results, we have carried through calculations of several zone center and X point phonons, including the X point "tilt" distortion which carries the high T tetragonal phase to the low T orthorhombic phase and therefore corresponds to the structural instability. For the zone center there are two A_{1g} modes (derived from c axis motion of the O_z and La atoms) and the couplings were also calculated, leading to a 2x2 dynamical matrix whose eigenvalues give the mode frequencies and whose eigenvectors give the atomic displacements. At the X point all four displacements with the symmetry of the oxygen breathing mode (breathing, scissors, and La and O_z axial breathing displacements) were calculated as well as their couplings, giving a 4x4 dynamical matrix and four frequencies. More of the details have been published elsewhere.[16]

All of the calculated modes are shown in Table 1, where they are compared with the available data from Raman scattering and inelastic neutron scattering experiments. Overall the agreement is very good. The energy versus displacement curves are shown in Fig. 2. The tilt mode displacement, and only this one, is found to be an instability of the tetragonal structure, in agreement with experiment, and moreover the amplitude of the orthorhombic distortion is predicted well. Similar studies for $YBa_2Cu_3O_7$ are underway.

These results make it clear that, at least for La_2CuO_4, LDA does perfectly well for structural energies and phonon frequencies. Only for the two highly anharmonic modes is there any significant discrepancy with experiment. There is no indication that additional correlation would lead to significant improvement. Some questions have been raised by Pinchovius et al.,[20] who seem to detect an "extra" mode in both the $(La,M)_2CuO_4$ and $YBa_2Cu_3O_7$ materials, which they suggest to be a mode whose spectral density

Table 1. Comparison of calculated phonon frequencies (cm^{-1}) in La_2CuO_4 with experimental data from Raman, ir, and neutron scattering. "Character" denotes the principal characteristic of the eigenvector.

	Mode	Character	Calc.	Exp.
Γ Point	O E_u	Sliding	(78)	110,162,177*
	O B_{2u}	Silent	293	264*
	O E_g	Sliding	233	228,243*
	O A_{1g}	Axial	417	424,429,427*
	La A_{1g}	Axial	220	228,231
X Point	La A_g	Axial	156	147*
	O A_g	Axial	299	317* or 373*
	O A_g	Scissors	475	497*
	O A_g	Breathing	731	714*
	O B_{1g}	Rotational	(87)	157*
	O B_{1g}	Quadrupolar	548	500*

* denotes Pinchovius et al. (Ref. 20); see Ref. 19 for references to other experimental data. Calculated values in parentheses denote highly anharmonic modes.

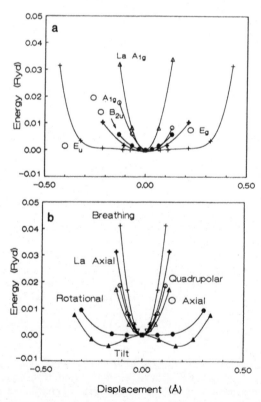

Figure 2. Calculated energies for frozen phonons at the Γ and X points in La_2CuO_4, versus amplitude of the phonon. The double well form of the energy for the tilt distortion reflects a T=0 instability with respect to an orthorhombic distortion, in agreement with experiment.

is split by interaction with some (as yet undetermined) low energy electronic interaction. Such a splitting could also arise from strong anharmonic interactions which do not involve low energy electronic processes.

The one-electron potentials which result from these frozen phonon calculation provide exactly the change in potential which is necessary to evaluate λ. While the calculations of the Fermi surface averages of the matrix elements (which provide λ) are not yet completed, evidence of unusual behavior can already be seen in the charge redistribution and the resulting self-consistently-screened change in potential. An example is given in Fig. 3, which shows these quantities for Ba displacements along the c axis in $YBa_2Cu_3O_7$. In standard metals λ is evaluated within a rigid atomic sphere approximation, in which the potential near an atom is assumed to be displaced rigidly with it. In such a case the contribution of each atom to λ depends on its local state density at E_F, and atoms like Ba which do not contribute at E_F give no coupling. In Fig. 3 it can be seen however that Ba motion causes charge redistribution, and therefore changes in potential, on the metallic atoms within the cell (primarily the Cu for this particular displacement). Analysis indicates that the non-local change in potential is driven by Madelung shifts, arising from the strongly ionic nature of the ions and only weakly screened by the carriers in the Cu-O planes and chains. Calculations in progess will determine the strength of the coupling resulting form these Madelung shifts.

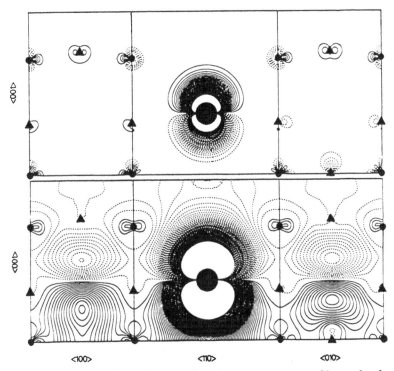

Figure 3. Change in the charge density (upper panel) and the self-consistently screened potential (lower panel) due to a displacement of the Ba ions by 0.004c=0.05 Å. Contours are separated by 1 millielectron/bohr^3 and by 1 mRy, respectively; solid (dashed) contours denote positive (dashed) values. Symbols indicate locations of Ba (large circles), Cu (small circles), and O (triangles).

TRANSPORT PROPERTIES

Using standard Bloch-Gruneisen transport theory and approximating the quasiparticles by the LDA band eigenvalues and eigenfunctions, Allen, Pickett and Krakauer[21] (APK) have calculated the Hall tensor R^H, thermopower S, and quantities related to the resistivity, for $(La,M)_2CuO_4$ and $YBa_2Cu_3O_7$. Since these results have been reported in detail previously, we only provide a synopsis here of the Hall tensor results.

Some rather striking predictions have been made, and subsequently confirmed, for the Hall tensor. For $YBa_2Cu_3O_7$, the predictions were:
1. $\underline{H}$ parallel to $\underline{c}$: $R = +0.2 \times 10^{-9}$ m^3/C (hole-like sign).
2. $\underline{H}$ parallel to $\underline{b}$: $R = -1.1 \times 10^{-9}$ m^3/C (electron-like sign).
3. $\underline{H}$ parallel to $\underline{a}$: $R = -0.4 \times 10^{-9}$ m^3/C (electron-like sign).

Thus, for in-plane motion of the carriers the prediction was for hole-like sign, as seen in ceramic samples and thus expected, while for carrier motion out of the plane an entirely unexpected **electron-like sign** was predicted. These predictions were borne out by measurements on a twinned crystal by Tozer et al.,[22] who reported
1. $\underline{H}$ parallel to $\underline{c}$: $R = +0.2 \times 10^{-9}$ m^3/C (hole-like; T=300 K).
2. $\underline{H}$ in $\underline{a},\underline{b}$ plane: $R = -0.8 \times 10^{-9}$ m^3/C (electron-like).

This last number, because of twinning, should be some average of the theoretical values, and the agreement for both orientations is excellent.

For a metal with a sharp Fermi surface and isotropic scattering (assumed in these calculations), the Hall coefficient is a complicated measure of the curvature of the Fermi surface. The simple relations often quoted which involve "carrier density" and "effective masses" are inappropriate in systems such as these which have Fermi surfaces with regions of both positive and negative curvature, and therefore partially cancelling contributions to R^H. The interpretation in terms of LDA calculations is also borne out in the $La_{2-x}Sr_xCuO_4$ system.

For concentrations within the range 0.05<x<0.20) including where T_c peaks, the ($\underline{H}$ parallel to $\underline{c}$) Hall coefficient is predicted to have hole-like sign with magnitude 5-10 smaller than the prediction of a model based on a single parabolic band containing x holes/cell. However, as x increases, R(x) decreases faster than 1/x and changes sign at x=0.24, beyond which point it is negative (electron-like sign) and increasing in magnitude. This behavior is standard in a nearly half-filled band and, although for x>0 the band is less than half-filled, in the $(La,Sr)_2CuO_4$ system interaction of the band crossing E_F with bands below E_F changes its curvature and moves the point at which R changes sign from x=0 to x=0.24.

Early in 1989 Suzuki[23] reported measurements on crystalline films which track the trend in our calculation from x=0.10 to above x=0.20. However, rather than crossing zero around x=0.24 it remains positive but continues to decrease in magnitude in an asymptotic fashion. More recently, Kitazawa[24] has reported data on single crystals which show the change in sign near x=0.25. If these results are confirmed by other groups, it will be a strong affirmation of the LDA band picture in $(La,M)_2CuO_4$.

ANTIFERROMAGNETIC INSULATING PHASE

For most of the properties of the metallic phase which have been discussed in the previous sections, the considerable agreement with local density predictions suggests that many-body correlation effects are not dominant, although they may be necessary to obtain quantitative agreement with data (the copper satellite is an exception which necessitates a many-body treatment). The most obvious shortcoming of LDA calculations has been the inability to obtain a qualitatively correct description of the antiferromagnetic (AFM) insulating phase. This is indeed a shortcoming of the local spin density approximation (LSDA) itself, for true density functional theory must reproduce the AFM ground state spin density. Hence there are necessarily "correlations beyond the local density approximation" giving rise to the AFM insulating phase. Here we briefly review our studies of this matter.

Several reports of AFM solutions within LSDA have appeared, occasionally with an insulating band structure but often with metallic bands. The evidence, which has been reviewed recently,[16] is that the more accurate studies obtain no AFM solution at all. Using the full potential LAPW method, which makes no shape approximations and a minimum of simplifications in solving the LSDA equations, we have studied La_2CuO_4, $CaCuO_2$, and Sr_2VO_4, all of which are AFM insulators, and have found no AFM solution. In carrying out these studies, we made several efforts to encourage a polarized solution, such as by including the orthorhombic distortion in La_2CuO_4, by lowering the symmetry even below what has been seen experimentally, by removing the symmetry-breaking exchange splitting slowly to encourage the attainment of a metastable solution, and giving extra weight to the Fermi surface region in zone sampling. More of the details of the calculations are presented elsewhere.[25] For Sr_2VO_4 the paramagnetic state was predicted to be unstable to ferromagnetism due to a density of states peak at E_F, but no AFM solution could be found.

For the cuprates we have not pursued the possibility of ferromagnetic solutions, but rather have evaluated the Stoner enhancement[26] $S=1/[1-<I>N(E_F)]$ within LSDA. For neither $YBa_2Cu_3O_7$ nor $La_{1.85}M_{0.15}CuO_4$ is an instability predicted, and the enhancements are modest. For $YBa_2Cu_3O_7$, $<I>N(E_F)=0.38$, with the exchange strength $<I>$ arising 20% from the chain O, 30% from the plane Cu, and 40% from the bridging O, the remainder contributed from other atoms. For $(La,M)_2CuO_4$, $<I>N(E_F)=0.36$, and $<I>$ is dominated (95%) by the Cu contribution. Our results differ from the large value $<I>N(E_F)=1.12$, indicating a ferromagnetic instability, reported by Freeman et al.[27] for $YBa_2Cu_3O_7$. The reason for this discrepancy is not known at present.

Our investigations have resulted in some insight into what may be necessary (within density functional theory) to produce an AFM insulator in these materials. Using an accurate empirical tight-binding representation[28] of the La_2CuO_4 bands, it was discovered that an exchange splitting (between spin up and spin down d states on Cu atoms) which splits primarily the $d(x^2-y^2)$ states will readily induce both a gap opening and a local moment of the order of 0.5 Bohr magneton per Cu, whereas splittings which affect all d orbitals equally induce a weaker moment and do not produce a gap.[29] There are two obvious ways such a splitting could arise. First is via the self-interaction correction (SIC) to LSDA, in which case a single hole in the Cu d shell of $d(x^2-y^2)$ symmetry will not experience an attractive SIC potential seen by all other minority spin electrons and by all majority spin electrons, and therefore becomes split off into an upper unoccupied band. Svane and Gunnarsson[30] have used their discrete lattice LSDA and SIC-LSDA models to study this effect, and indeed find that SIC greatly enhanced the tendency toward the AFM insulating state.

The SIC-LSDA approach is an orbital functional, rather than a density functional, method, and our studies have suggested that density functional theory might produce the AFM insulating state by encouraging stronger anisotropy in the exchange potential than is done in LDA. By applying an entirely anisotropic exchange potential (i.e. having zero spherical average) to $CaCuO_2$ which splits primarily the $d(x^2-y^2)$ states, we have verified[23] that a gap opens up readily in a self-consistent calculation similarly to that seen in the tight-binding calculation. Although the application of this anisotropic, externally imposed symmetry-breaking potential produced a strongly polarized, insulating electronic structure, this AFM insulating state rapidly degenerated to the unpolarized metallic solution upon removal of the imposed potential. The description of the AFM insulator from a density functional viewpoint thus remains an outstanding (in both senses) problem.

SUMMARY

In this overview, which covers much of our work as well as some from other groups, we have described how the local density approach describes quite well several properties of the metallic phase of superconducting cuprates. For the AFM insulating phase the outlook is not so bright, yet even here it is possible that straightforward improvements to LSDA may succeed. Since the superconducting state represents an instability of the metallic phase, its description is most expedient. As Friedel[31] has carefully concluded, "there is no clear cut evidence that treating (correlations) as a perturbation to a band scheme is not enough to describe the physical properties of these oxides."

Of course there are several unusual properties which have not been touched upon here. As examples, we mention that NMR and NQR data are not understood, there are indications from ultrasonics, EXAFS, and radial distribution functions that subtle atomic rearrangements may occur very near T_c, and the positron lifetime and momentum distribution behaves anomalously near T_c. These difficult items require further study to understand their implications.

We wish to thank Howard Lu and Daniel Reich for assistance in calculations, and acknowledge the help of and useful discussions with M. J. DeWeert, L. L. Boyer, B. M. Klein, J. W. Serene and P. B. Allen. The computations involved in the studies described here were carried out at the Cornell National Supercomputing Center. H.K. was supported by NSF Grant No. DMR-87-19535, and work at NRL was supported by the Office of Naval Research.

REFERENCES

1. See, for example, R. J. Cava et al., Nature 336, 211 (1988).
2. L. F. Mattheiss, E. M. Gyorgy, and D. W. Johnson, Jr., Phys. Rev. B37, 3745 (1988); D. G. Hinks et al., Nature 333, 836 (1988).
3. W. E. Pickett, Comments on Solid State Phys. 12, 1 (1985); ibid. 59 (1986), and references therein.
4. A. K. McMahan, R. M. Martin, and S. Satpathy, Phys. Rev. B38, 6650 (1988).
5. J. Fink, N. Nücker, H. Romberg, and J. C. Fuggle, IBM J. Res. Devel. (1989, in press).
6. A. J. Arko et al., Phys. Rev. B40, 2268 (1989).
7. J. Redinger, A. J. Freeman, J. Yu, and S. Massidda, Phys. Lett. A124, 469 (1987).
8. H. Krakauer, W. E. Pickett, and R. E. Cohen, J. Supercon. 1, 111 (1988).
9. J. Costa-Quintana, F. Lopez-Aguilar, S. Balle, and R. Salvador, Phys. Rev. B39, 9675 (1989).
10. See, for example, T. Miyahara et al., J. Phys. Soc. Japan 58, 2160 (1989).
11. R. Manzke, T. Buslaps, R. Claessen, M. Skibowski, and J. Fink, preprint; J. Fink et al., Proc. Intl. Winter School on Elect. Props. of Polymers and Related Compounds, Kirchberg, Austria, 1989 (in press).

12. C. S. Wang, M. R. Norman, R. C. Albers, A. M. Boring, W. E. Pickett, H. Krakauer, and N. E. Christensen, Phys. Rev. B35, 7260 (1987).
13. P. A. Lee, T. M. Rice, J. W. Serene, L. J. Sham, and J. W. Wilkins, Comments on Cond. Matt. Phys. 12, 99 (1986); W. E. Pickett, in <u>Novel Superconductivity</u>, eds. S. A. Wolf and V. Z. Kresin (Plenum, New York, 1987),233.
14. T. Takahashi et al., Phys. Rev. B39, 6636 (1989); F. Minami, T. Kimura, and S. Takekawa, Phys. Rev. B39, 4788 (1989); C. G. Olson et al., Proc. Intl. Conf. on Materials and Mechanisms of Superconductivity, Stanford, 1989.
15. J.-M. Imer et al., Phys. Rev. Lett. 62, 336 (1989); see also Refs. 6, 11, and 14.
16. W. E. Pickett, Rev. Mod. Phys. 61, 433 (1989).
17. D. A. Papaconstantopoulos, W. E. Pickett, and M. J. DeWeert, Phys. Rev. Lett. 61, 211 (1988).
18. D. A. Papaconstantopoulos, A. Pasturel, J. P. Julien, and F. Cyrot-Lackmann, Phys. Rev. B40, (1989) in press.
19. R. E. Cohen, W. E. Pickett, and H. Krakauer, Phys. Rev. Lett. 62, 831 (1989).
20. L. Pinchovius, N. Pyka, W. Reichardt, A. Yu. Rumiantsev, A. Ivanov, and N. Mitrofanov, in Proc. Intl. Symp. on High T_c Superconductivity, Dubna, July 1989 (in press).
21. P. B. Allen, W. E. Pickett, and H. Krakauer, Phys. Rev. B37, 7482 (1988); ibid. B36, 3926 (1987).
22. S. W. Tozer, A. W. Kleinsasser, T. Penney, D. Kaiser, and F. Holtzberg, Phys. Rev. Lett. 59, 1768 (1987).
23. M. Suzuki, Phys. Rev. B39, 2312 (1989).
24. K. Kitazawa, invited paper at the Spring MRS Mtg., San Diego, 1989.
25. D. Singh, W. E. Pickett, R. E. Cohen, D. A. Papaconstantopoulos, and H. Krakauer, Physica B (1989, in press).
26. S. H. Vosko and J. P. Perdew, Can. J. Phys. 53, 1385 (1975).
27. A. J. Freeman, J. Yu, S. Massidda, and D. D. Koelling, Physica 148B, 212 (1987).
28. M. J. DeWeert, D. A. Papaconstantopoulos, and W. E. Pickett, Phys. Rev. B39, 4235 (1989).
29. W. E. Pickett and D. A. Papaconstantopoulos, in <u>Atomic Scale Calculations in Materials Science</u>, eds. J. Tersoff, D. Vanderbilt, and V. Vitek, MRS Symp. Proc. No. 141 (MRS, Pittsburgh, 1989), 109.
30. A. Svane and O. Gunnarsson, Europhys. Lett. 7, 171 (1988).
31. J. Friedel, preprint.

SQUEEZED POLARON MODEL OF HIGH T_c SUPERCONDUCTIVITY

D.L. Lin

Department of Physics and Astronomy
State University of New York at Buffalo
Amherst, New York 14260

Hang Zheng

Center of Theoretical Physics, Chinese Center of Advanced
Science and Technology (World Laboratory),
Beijing 100080, P.R. China and

Department of Applied Physics
Shanghai Jiao Tong University,
Shanghai, P.O. China

Abstract

On the basis of the squeezed polaron model of superconductivity for strong-coupling narrow-band electron phonon interacting systems, we present here a calculation of the critical temperature T_c for simple lattice structures by assuming n = 1 where n is the mean number of carriers per unit cell. It is found that T_c can be higher than 100 K. The isotope effect on T_c is also discussed.

Introduction

There has always been interest in searching for the high T_c superconductivity[1], even before the discovery of new superconductors at the beginning of 1987. Early theoretical work has mainly been concerned with the strong electron-phonon interaction (SEPI) systems. In the strong-coupling theory, which is a natural extension of the weak-coupling theory of Bardeen, Cooper and Schrieffer (BCS)[2], only electrons near the Fermi surface are assumed to be coupled to phonons. As the new superconductors are all of the strong-coupling narrow-band (SCNB) type, the carrier concentration is generally low. The Fermi energy is therefore small and is not necessarily very much larger than the characteristic phonon frequency. Thus the well-known Migdal theorem on which the strong-coupling theory is based, breaks down.[3]

In SCNB systems, all electrons in the Fermi sea are involved in the interaction with phonons. Such strong interactions lead to the formation of small polarons with extremely narrow band.[4] The band narrowing effect is induced by the random motion of phonons in the phonon cloud around each polaron. It should be emphasized here that the renormalized polaron band width reduces to its bare width in the limit of zero temperature, and that the quantum fluctuation can never really be eliminated under any circumstances.

When the electrons per unit cell n is finite, the phonons may form some kind of ordered state of zero momentum to reduce the narrowing effect. It has recently been proposed[5] that the two-phonon coherent state is just such an ordered state, and demonstrated that the phonon-induced attraction between electrons on-site or off-site can be produced. Because of the strong on-site Coulomb repulsion, it is the attractive correlation between the intersite electrons that stabilizes a new superconducting phase. It is also shown that as long as the renormalized off-site correlation is attractive, a nonzero gap function $\Delta_{\vec{k}}$ exists whether the renormalized on-site electron correlation is repulsive or attractive, thus a new type of pairing occurs. In this paper, we calculate the critical temperature T_c from a $\vec{k}$-dependent gap function $\Delta_{\vec{k}}$ for simple lattice structures. It is found that T_c can be higher than 100 K. The mean carrier concentration n is assumed in this calculation to be unity for simplicity. In reality, high T_c materials generally have n smaller than but close to unity. Since we are considering systems in which there does not exist long range order, there can not be a carrier on each site even if n=1. Hence the sites are not all equally deformed. Nevertheless, calculations for $n \lesssim 1$ are being carried out and the results will be reported elsewhere.

The Model

The total Hamiltonian of our SEPI system is

$$H = H_e + H_p + H_{ep}. \tag{1}$$

where we have used the generalized Hubbard model for the electron subsytem

$$H_e = \sum_{\vec{l},\sigma} \epsilon\, n_{\vec{l}\sigma} - \sum_{\vec{l},\vec{\delta}} \sum_{\sigma} T_o\, d^{\dagger}_{\vec{l}\sigma} d_{\vec{l}+\vec{\delta},\sigma} + \sum_{\vec{l}} U\, n_{\vec{l}\uparrow} n_{\vec{l}\downarrow}$$

$$+ \tfrac{1}{2} \sum_{\vec{l},\vec{\delta}} V(n_{\vec{l}\uparrow} + n_{\vec{l}\downarrow})(n_{\vec{l}+\vec{\delta}\uparrow} + n_{\vec{l}+\vec{\delta}\downarrow})$$

$$- \tfrac{1}{2} \sum_{\vec{l},\vec{\delta}} \sum_{\sigma,\sigma'} J\, d^{\dagger}_{\vec{l},\sigma} d^{\dagger}_{\vec{l}+\vec{\delta},\sigma'} d_{\vec{l},\sigma'} d_{\vec{l}+\vec{\delta},\sigma}. \tag{1a}$$

In (1a), ϵ represents the unperturbed site level, T_o is the hopping energy, U and V are the Coulomb repulsion energies between on-site and off-site electrons respectively. $d^{\dagger}_{\vec{l}\sigma}$ creates an electron of spin σ at the site $\vec{l}$ and $n_{\vec{l}\sigma} = d^{\dagger}_{\vec{l}\sigma} d_{\vec{l}\sigma}$. The last term reduces to Ising model when $\sigma' = \sigma$ and to Heisenberg model when $\sigma' = -\sigma$. J stands for the exchange energy and only $J > 0$ is assumed in our treatment to reflect the antiferromagnetic coupling. The phonon subsystem is simply

$$H_p = \sum_{\vec{l}} \hbar\omega\, b^{\dagger}_{\vec{l}} b_{\vec{l}} \tag{1b}$$

where $b^{\dagger}_{\vec{l}}$ ($b_{\vec{l}}$) creates (annihilates) a phonon at the site $\vec{l}$. For simplicity, we assume that the phonon frequency is dispersionless. The electron-phonon interaction is given by

$$H_{ep} = g_1 \sum_{\vec{l},\sigma} n_{\vec{l}\sigma}(b^{+}_{\vec{l}} + b_{\vec{l}}) + g_2 \sum_{\vec{l},\sigma} \sum_{\vec{\delta}} n_{\vec{l}\sigma}(b^{+}_{\vec{l}+\vec{\delta}} + b_{\vec{l}+\vec{\delta}}), \tag{1c}$$

where g_1 and g_2 are the on-site and intersite electron-phonon coupling constants, respectively.

For sufficiently low temperatures, the state of phonon subsystem does not change significantly with the electron motion. In other words, we assume that the phonon occupation numbers remain approximately the same as electrons move. It is therefore possible to separate the electron coordinates by averaging H over thermal equilibrium distribution of the phonon populations. As is well known, this averaging procedure introduces the band narrowing factor[6] of the type $\exp(-g^2/\hbar^2\omega^2)$, where g is the characteristic strength of the electron-phonon interaction and $\hbar\omega$ the characteristic phonon energy. Since we are considering SEPI systems, the situation is quite different especially when the carrier concentration is finite. For $g/\hbar\omega > 1$, the narrowing effect remains fairly strong even at zero temperature and increases the total energy of the system as has been shown previously[7].

The narrow band implies large effective mass of the polaron. As has been pointed out above, it is due to the random motion of phonons in the surrounding cloud of the electron. Consequently, the lattice periodicity in the neighborhood of the polaron is partially destroyed, and the Bloch wave form of the electron can no longer maintain. When two polarons have their wave functions overlapping, their phonon clouds interfere strongly and form some kind of ordered state. The two-phonon coherent state can be regarded as such a state. This long-range phase-ordered state restores partially the Bloch form of the electorn wave function and hence reduces the polaron effective mass. Thus, we assume the ground state of the phonon subsystem to be

$$|\psi_p\rangle = e^{-S}|vac\rangle \tag{2}$$

$$S = \Sigma_i \, \alpha \, (b_i b_i - b_i^\dagger b_i^\dagger) \tag{2a}$$

where α is the variational parameter. This is formally identical to the "squeeze" transformation[8] in quantum optics, and for this reason we call the new type of polaron the squeezed polaron. It is obvious that $|\psi_p\rangle$ goes back to the vacuum as $\alpha = 0$. The value of α is determined by variational principle such that the ground state energy of the interacting system is an absolute minimum.

The effective Hamiltonian of our electron subsystem is then the average of H over the state $|\psi_p\rangle$, namely,

$$\begin{aligned}H_{eff} = & N\hbar\omega \, (\sinh 2\alpha)^2 + \Sigma_{i,\sigma} \, \epsilon_e n_{i\sigma} \\ & - \Sigma_{i,\sigma} \Sigma_\delta \, T_e d_{i\sigma}^\dagger d_{i+\delta,\sigma} + \Sigma_i U_e n_{i\uparrow} n_{i\downarrow} \\ & + \tfrac{1}{2} \Sigma_{i,\delta} \, V_e (n_{i\uparrow} + n_{i\downarrow})(n_{i+\delta\uparrow} + n_{i+\delta\downarrow}) \\ & - \tfrac{1}{2} \Sigma_{i,\delta} \Sigma_{\sigma,\sigma'} \, J \, d_{i\sigma}^\dagger d_{i+\delta\sigma'}^\dagger d_{i\sigma'} d_{i+\delta,\sigma}\end{aligned} \tag{3}$$

$$\epsilon_e = \epsilon - (g_1^2 + z g_2^2)/\hbar\omega \tag{3a}$$

$$T_e = T_0 \rho \tag{3b}$$

$$U_e = U - 2(g_1^2 + z g_2^2)/\hbar\omega \tag{3c}$$

$$V_e = V - 4g_1 g_2/\hbar\omega \tag{3d}$$

$$\rho = \exp\{-[(g_1-g_2)^2 + (z-1)g_2^2]\tau^2/(\hbar\omega)^2\} \tag{3e}$$

$$\tau = e^{-2\alpha} \tag{3f}$$

where we have introduced the coordination number z. It is noted that all the phonon-induced correlations between next-nearest and third-nearest neighboring electrons have been dropped for simplicity. Their inclusion would not change any of our conclusions qualitatively. The energy parameters ϵ_e, T_e, U_e and V_e are already renormalized by the electron-phonon interaction. It is seen from (3e) and (3f) that our band narrowing factor ρ has an extra factor $\tau^2 = e^{-4\alpha}$ in its exponent. This is a consequence of the squeeze transformation (2). It reduces greatly the narrowing effect as well as T_e for $\alpha > 0$. On the other hand, the mean phonon energy represented by the first term of (3) increases with α. These two competing effects of α result in a stable minimum of the ground state energy of our SEPI system for some nonzero α to be determined by the variational method. As has been shown in Ref. 5, the total energy as a function of α indeed possesses a stable minimum at some finite value α_m.

Critical Temperature

In the finite temperature region, the gap equation can be derived from the effective Hamiltonian within the framework of the generalized Hartree-Fock approximation[2,5,7]

$$\Delta_{\vec{k}} = \Delta_o - \Delta_1 \gamma(\vec{k}) \tag{4}$$

$$\Delta_o = \frac{1}{2N} \sum_{\vec{k}} U_e \{[\Delta_1 \gamma(\vec{k}) - \Delta_o]/W(\vec{k})\} \tanh[\tfrac{1}{2}\beta W(\vec{k})] \tag{4a}$$

$$\Delta_1 = \frac{1}{2N} \sum_{\vec{k}} z(|V_e|+J)\gamma(\vec{k}) \cdot$$

$$\{[\Delta_1 \gamma(\vec{k}) - \Delta_o]/W(\vec{k})\} \tanh[\tfrac{1}{2}\beta W(\vec{k})] \tag{4b}$$

where

$$\gamma(\vec{k}) = \frac{1}{z} \sum_{\vec{\delta}} \exp(i\vec{k}\cdot\vec{\delta}) \tag{5}$$

$$W(\vec{k}) = [E^2(\vec{k}) + \Delta_{\vec{k}}^2]^{1/2} \tag{6}$$

$$E(\vec{k}) = \epsilon_e - \mu + \tfrac{1}{2} U_e n - \tfrac{1}{2} z (2|V_e| + J)n$$

$$- zT_e \gamma(\vec{k}) + \tfrac{1}{2} z (|V_e| + 2J) n_{\vec{k}} \tag{7}$$

$$n = \frac{1}{N} \sum_{\vec{k}} \{1 - \frac{E(\vec{k})}{W(\vec{k})} \tanh[\tfrac{1}{2}\beta W(\vec{k})]\} \tag{8}$$

$$n_{\vec{k}} = \frac{1}{n} \sum_{\vec{k}'} \gamma(\vec{k}-\vec{k}')\{1 - \frac{E(\vec{k}')}{W(\vec{k}')} \tanh[\tfrac{1}{2}\beta W(\vec{k}')]\} \qquad (9)$$

$\mu = \mu(T)$ is the chemical potential and $\beta = 1/k_B T$. It can be shown from symmetry considerations that[5]

$$n_{\vec{k}} = 2\,\xi(T)\gamma(\vec{k}) \qquad (10)$$

$$\xi(T) = -\frac{1}{2N} \sum_{\vec{k}'} [\gamma(\vec{k}')E(\vec{k}')/W(\vec{k}')] \tanh[\tfrac{1}{2}\beta W(\vec{k}')] \qquad (10a)$$

In general, the chemical potential μ and the parameters $\xi(T)$, $\Delta_0(T)$, and $\Delta_1(T)$ can only be calculated numerically. We shall discuss in the following their critical behavior.

When $T \to T_c$, $\Delta_{\vec{k}} \to 0$ and $W(\vec{k}) = E(\vec{k})$. Although the parameters $\Delta_0(T_c)$ and $\Delta_1(T_c)$ tend to zero separately, the ratio $x = \Delta_0/\Delta_1$ remains to be finite. Thus, we have

$$x = \frac{1}{2N} \sum_{\vec{k}} U_e \frac{\gamma(\vec{k}) - x}{E_c(\vec{k})} \tanh[\tfrac{1}{2}\beta_c E_c(\vec{k})] \qquad (11a)$$

$$1 = \frac{1}{2N} \sum_{\vec{k}} z(|V_e|+J)\gamma(\vec{k}) \frac{\gamma(\vec{k})-x}{E_c(\vec{k})} \tanh[\tfrac{1}{2}\beta_c E_c(\vec{k})] \qquad (11b)$$

where $E_c(\vec{k})$ is the value of $E(\vec{k})$ at $T = T_c$, and $\beta_c = 1/k_B T_c$. For simplicity, we assume in our actual calculation $n = 1$ for the mean carrier concentration. As we have noted previously, this does not imply that there is actually one electron at every lattice site. All the energy parameters are measured in the unit of zT_o. For the new high T_c superconductors, n is somewhat smaller than, but close to 1 and zT_o is of the order of 1 eV. When $n = 1$, we have $x = 0$ and the chemical potential

$$\mu(T_c) = \epsilon_e + \tfrac{1}{2} U_e - \tfrac{1}{2} z(2|V_e| + J) \qquad (12)$$

$$\xi(T_c) = \rho_m T_o/(2|V_e| + 3J) \qquad (13)$$

where ρ_m is the value of ρ corresponding to r_m (or α_m) for which the ground state energy of the system is at its stable minimum. Equation (11b) then becomes

$$1 = \frac{1}{2N} \sum_{\vec{k}} [(2|V_e|+ 3J)/\rho_m T_o]\gamma(\vec{k}) \cdot \tanh \frac{\beta_c z \rho_m T_o(|V_o|+ J)\gamma(\vec{k})}{4|V_e| + 6J} \qquad (14)$$

from which the critical temperature T_c is determinted numerically.

479

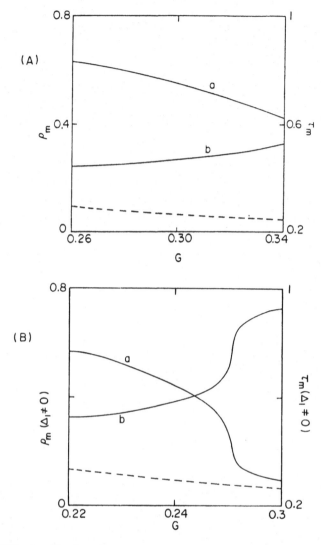

FIG. 1. The band-narrowing parameter ρ_m (curve a) and the parameter τ_m (curve b), both evaluated in the stable ground state, are plotted as functions of the coupling constant G for simple cubic lattice. Other parameters used are $\hbar\omega = 0.08$, $U = 2$, $V = 0.1$, $J = 0.1$. (A) simple cubic lattice, (B) Body-centered cubic lattice.

Results and Discussion

In Fig. 1, we plot for (A) a simple cubic (sc) and (B) a body-centered cubic (bcc) lattices the band narrowing factor ρ_m (curve a) as a function of the coupling constant $G = 4 g_1 g_2 / z T_o \hbar\omega$. The variation of τ_m (curve b) is shown simultaneously on the right had side of the figure while the dashed line represents the narrowing factor for $\tau = 1$. In these calculations we have assumed that $\Delta_1 \neq 0$, and $g_1^2 = z g_2^2$. It is observed in Fig. 1(A) for the sc case that for the range $0.26 < G < 0.34$, the parameter r_m is within the range of $0.4 \sim 0.6$ and ρ_m is within $0.6 \sim 0.8$. Compared to the dashed line for which $r = 1$ and $0.05 < \rho < 0.1$, it is clear that the band narrowing effect is greatly weakened by the two-phonon coherent states. As a consequence, the effective mass of our squeezed polaron is much smaller than what is expected in the literature because the band width is inversely proportional to the polaron effective mass. Therefore the system can undergo Bose-Einstein condensation at high temepratures.

The critical temperature $k_B T_c$ (curve a) and the superconducting order parameter Δ_1 (curve b) are plotted in Fig. 2 versus G, again for (A) a sc

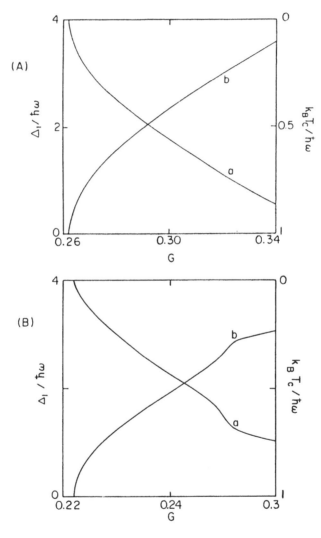

FIG. 2. The critical temperature $k_B T_c/\hbar\omega$ (curve a) and the superconducting order parameter $\Delta_1/\hbar\omega$ versus the coupling constant G. Other parameters are the same as in Fig. 1. Note, however, that the ordinate for $k_B T_c/\hbar\omega$ decreases upward. (A) simple cubic lattice. (B) Body-centered cubic lattice.

and (B) a bcc lattice. Both $k_B T_c$ and Δ_1 are in the unit of the characteristic phonon energy $\hbar\omega$. Notice that the ordinate for $k_B T_c/\hbar\omega$ decreases from 1 to 0. For a crude estimate, we take $\hbar\omega/k_B \approx$ 200 K or $\hbar\omega \approx$ 17 meV, the critical temperature T_c can vary from 0 K ~ 160 K when the coupling strength G changes within the range 0.26 ~ 0.34. Similar results are found for bcc lattice. With the range of the coupling strength $0.22 \leq G \leq 0.3$, we find $0 K \leq T \leq 150 K$. This G actually measures the effective attraction $|V_e| + J$ between intersite squeezed polarons according to (3d). We emphasize that the high T_c occurs because all squeezed polarons are involved in the superconducting process.

Let us now make a few remarks to conclude our discussion. Although the phonon subsystem is approximately decoupled from the electron subsystem in our treatment, they are still interwoven via the paramter α. The value of α determined by the variational method not only specifies the phonon state but also depends on the electron state as well as the electron-phonon coupling strength. This is of course very different from the small polaron system in which the phonon is in the vacuum state.

The $\vec{k}$ dependence of the gap function $\Delta_{\vec{k}}$ has two implications. Firstly, it means that the energy gap is in general anisotropic in nature. This implies then the anisotropic current. Secondly, it means that the gap is not a constant in the Brillouin zone and may even vanish at some points in the zone. Thus, the quasi-particle excitation spectrum in this theroy is quite different from the BCS theory. It may result in totally different thermal and electromagnetic properties of the system under consideration. For instance, the energy gap can not be detected in specific heat measurements but can be seen in the infrared reflection spectrum. This is not difficult to understand. The specific heat measurements involve elementary excitations in the whole Brillouin zone while the infrared measurements invovle only excitations near $\vec{k} = 0$.

Finally, we note that the isotope effect is rather weak in this model. This can be seen qualitatively as follows. Since the carrier concentration is low in narrow-band systems, the Fermi energy may be of the same order as the characteristic phonon energy $\hbar\omega$. Thus all the electrons in the Fermi sea may participate in the superconductivity. This is quite different from the BCS theory in which only electrons with energy between E_F and $E_F - \hbar\omega_D$ are involved, and hence $T_c \sim \hbar\omega_D$ where ω_D is the Debye frequency. This is the origin of the isotope effect since $\omega_D \propto M^{-\frac{1}{2}}$. In the squeezed polaron model, we have $T_c \sim E_F$. However, such qualitiative argument does not rule out the isotope effect completely. More careful calculation is being carrried out for n = 0.85. Our preliminary results from a rough calculation are plotted in Fig. 3 which indicate that T_c is a slowly varying function of ω. It increases with increasing ω in the small ω regime and decreases with increasing ω in the large ω regime. There exists a region around the maximum of ω in which the isotope effect is almost zero. It must be emphasized at this point that in such a crude model calculation, the results are not taken to be quantitatively meaningful. It is, however, interesting to note that qualitatively T_c is not a monotonic function of ω and that the maximum T_c corresponds to zero isotope effect.

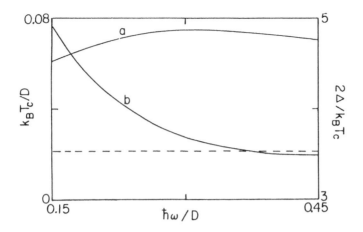

FIG. 3. Isotope effect on T_c. The half-width D of the bare band is taken to be unity and the mean carrier density n = 0.85. The dashed line denotes the BCS value of $\frac{2\Delta}{k_B T_c} = 3.53$ where Δ is the gap at T = 0.
(a) $k_B T_c/D$, (b) $2\Delta/k_B T_c$.

Acknowledgement

One of us (HZ) is partially supported by Fok Ying Dong Education Foundation, Hong Kong.

References

1. See, for example, V. L. Ginzburg and D. A. Kirzhnits, ed., <u>High Temperature Superconductivity</u>. Plenum, New York, 1982, and references cited therein.
2. J. Bardeen, L. N. Cooper and J. R. Schrieffer, Phys. Rev. <u>108</u>, 1175 (1957).
3. A. S. Alexandrov, J. Ranninger, and S. Robszkiewicz, Phys. Rev. Lett. <u>56</u>, 949 (1986); Phys. Rev. B <u>33</u>, 4526 (1986); M.T. Beal-Monod, Phys. Rev. B <u>35</u>, 8788 (1987).
4. Y. Toyazawa, in <u>Polarons and Excitons</u>, ed., G.B. Kruper and G.D. Whitfield (Oliver and Boyd, Edinburgh 1962).
5. H. Zheng, Phys. Rev. B. <u>37</u>, 7419 (1988); D.L. Lin and H.Zheng, J. Appl. Phys. <u>64</u> (10) 5905 (1988).
6. T. Holstein, Ann. Phys. (NY) <u>8</u>, 343 (1959).
7. A. S. Alexandrov and J. Ranninger, Phys. Rev. B 23, 1796 (1981); <u>24</u> 1164 (1981); S. Robaszkiewicz, M Micnas and K.A. Chao, Phys. Rev. B <u>24</u>, 4018 (1981); <u>26</u>, 3915 (1982).
8. H. P. Yuen, Phys. Rev. A <u>13</u>, 2226 (1976).

LOCAL PAIRING AND ANTIFERROMAGNETISM IN HIGH-T_c SUPERCONDUCTORS

H. R. Lee and Thomas F. George

Department of Physics & Astronomy and Chemistry
239 Fronczak Hall
State University of New York at Buffalo
Buffalo, New York 14260

INTRODUCTION

The CuO_2 planes in high-temperature superconductors are studied by a two-dimensional Hubbard Hamiltonian with a new proposed vacuum (d^9p^6). The superexchange energy between Cu sites and the hopping parameters of the carriers estimated in this new vacuum are in good agreement with experimental observations as well as numerical calculations with a finite cluster. Single carriers (electrons for n-type or holes for p-type) are localized due to the rigidity of the antiferromagnetic background. Local pairing in the CuO_2 plane enables the paired states to move around without disturbing the antiferromagnetic background. The binding energy is estimated from the pair hopping energy, which causes the paired state to be lower in energy than the single carrier state.

BACKGROUND

Since the discovery of p-type high-temperature superconductivity in ceramics [1], many theoretical studies [2] have been carried out on these compounds. There is a variety of experimental evidence that the explanation of the phenomenon does not lie within the traditional phonon-mediated mechanism. Neutron and Raman scattering experiments [3] show that strong antiferromagnetic correlations coexist in the superconducting state. Thus, the antiferromagntism is intimately related to superconductivity in these materials, in contrast to magnetism opposing superconductivity in conventional superconductors. Other prominant features, such as short coherent length (12 ~ 20Å) and the direct proportionality of the superconducting phase transition temperature (T_c) to the carrier density observed in many experiments, strongly support the idea of local space pairing, which is very different from the ordinary BCS-type pairing caused by a mean-field excitation.

The hole superconductors [$La_{2-x}(Sr,Ba)_x CuO_4$] [1] are obtained by partially substituting divalent barium or strontium for the trivalent lanthanum in La_2CuO_4. Meanwhile, the n-type superconductors [$Ln_{2-x}Ce_xCuO_4$] (Ln: rare earth) [4] can be obtained as in semiconductors when a tetravalent lanthanide is partially substituted for the trivalent lanthanide in compounds of the form Ln_2CuO_4. The crystal structure of Nd_2CuO_4, the parent material for electron superconductors, is similar to that of the hole superconductors except that each copper atom is bonded to only four in-plane oxygen atoms, possessing no apical oxygen atoms above or below the CuO_2 plane. These two prominent features suggest that: (1) electron-hole symmetry must hold, and (2) the superconducting transition is purely a property of the two-dimensional CuO_2 sheets. The argument for electron-hole symmetry is strengthened by the report of both electron and hole superconductivity in $TlCa_{1-x}Ln_xSr_2Cu_2$-$O_{7+\delta}$ [5].

X-ray absorption spectra show that the charge carriers (electrons) are located on the Cu sites (Cu^{1+}) in n-type superconductors [6], while the charge carriers (holes) are located on O sites (O^{1-}) in p-type superconductors [7]. Clearly, this observation is due to the presence or absence of electrons on the antibonding states of $Cu(d_{x^2-y^2})$ and $O(p_x$ or $p_y)$ orbitals. This kind of behavior is very similar to that of the impurity states in semiconductors. In this case, the choice of vacuum for the system plays a critical role for fixing the electron and hole energy levels, particularly for the two-dimensional Hubbard model.

The hole superconductors have been extensively studied with the Hubbard Hamiltonian [8]. In all papers using this model, the closed shell $Cu(d^{10})$-$O(p^6)$ was assumed to be the vacuum, that is, the other energy levels were well referenced to this closed shell orbit. However, this vacuum is not amenable to describing electron superconductivity. To explain both electron- and hole-state superconductors, we define a new vacuum, $Cu(d^9)$-$O(p^6)$, which exhibits an antiferromagnetic background on the Cu lattice of the parent, undoped materials. We shall study various parameters with the extended Hubbard Hamiltonian in this new vacuum, where the electron and hole creations can be explained as follows: the extra electron which is inserted within the CuO plane by doping with a tetravalent lanthanide will occupy the empty antibonding state $Cu^{1+}(d^{10})$. If a divalent atom is the dopant instead of a tetravalent atom, the CuO_2 plane will lose one electron from the highest-occupied antibonding state $O^{2-}(p^6)$. Then the problem in the theory of high-temperature superconductors narrows down to the best description of the charge carriers (electron or hole) in an antiferromagnetic background, which is consistent with many experiments on high-tempertaure superconductors [9].

THEORY

We write the Hamiltonian as

$$H = t_{pd} \sum_{<i,j>}{}' (d^{h\dagger}_{i\sigma} p_{j\sigma} + d^{e\dagger}_{i\sigma} p^{\dagger}_{j\sigma} + h.c.)$$

$$+ \sum_{i,\sigma} (E_d + U_d) d^{e\dagger}_{i\sigma} d^{e}_{i\sigma} + \sum_{i,\sigma} E_d d^{h\dagger}_{i\sigma} d^{h}_{i\sigma}$$

$$+ \sum_{j,\sigma} (E_p + U_p) p^{\dagger}_{j\sigma} p_{j\sigma} - \sum_{\substack{j \\ \sigma \neq \sigma'}} U_p n_{j\sigma} n_{j\sigma'} + \sum_{\substack{<i,j> \\ \sigma,\sigma'}}{}' U_{pd} n^{e}_{i\sigma} n^{h}_{j\sigma'} \quad , \quad (1)$$

where t_{pd} is the hopping integral between neighboring Cu and O sites, and the $d_{i\sigma}$'s and $p_{j\sigma}$'s are second-quantization operators related to the Cu sites and O sites, respectively. The superscripts h and e on the Cu site operators are related to the holes $Cu^{3+}(d^8)$ and electrons $Cu^{1+}(d^{10})$, respectively, due to the new definition of the vacuum, but the O site operators always represent the hole state and consequently possess the same parameters as in the old vacuum. The prime on a summation indicates $i \neq j$, and $<i,j>$ indicates nearest neighbors between the Cu and O sites. E_d and E_p are the diagonal kinetic energies of the 3d and 2p orbitals, respectively. Notice the on-site Coulomb repulsion (U_d) of the Cu sites appears when an electron is created on a Cu site, while the on-site Coulomb repulsion (U_p) of the O sites is released when a hole is created on an O site instead of two holes being created. The interatomic Coulomb potential U_{pd} in the last term of the Hamiltonian is the electron-hole pair creation potential; it essentially gives the change in Coulomb attraction between neighboring ionic sites due to the creation of an electron-hole pair out of the vacuum.

The electron-hole pair can be created when an electron on an O site moves to a neighboring Cu site, i.e., $Cu^{2+}(d^9)$-$O^{2-}(p^6) \to Cu^{1+}(d^{10})$-$O^{1-}(p^5)$. This changes the valence charge configurations most significantly for later calculations. The oxygen 2p orbitals are assumed to be of the σ type, pointing towards the two neighboring positive Cu ions. The large covalent interaction between the Cu and O neighbors in this band can result in a rapid disappearance of the antiferromagnetic background of the Cu sites for a small amount of doping. This effect can be understood if we consider that for an electron superconductor an electron on a Cu site can hop to neighboring Cu sites only through the O site between them. The relevant energy parameters are given by $t_{pd} \simeq 1.2$ eV, $U_p \sim 5 \sim 7$ eV, $U_d \sim 8 \sim 10$ eV, $U_{pd} \approx 1 \sim 2$ eV, and $\Delta E = E_d - E_p \sim 1 \sim 2$ eV [8,10].

In the absence of doping, the effective Hamiltonian can be written as

$$H_{cm} = J \sum_{<i,j>} \vec{S}_i \cdot \vec{S}_j , \qquad (2)$$

where the sum is over all the nearest-neighbor Cu sites, and $\vec{S}_i$ are spin-$\frac{1}{2}$ operators. Because the superexchange energy (J) involves an oxygen site, it is found by fourth-order perturbation theory:

$$J = \frac{4 t_{pd}^4}{(\Delta E + U_d + U_{pd} - U_p)^2} \left[\frac{1}{U_d} + \frac{2}{2\Delta E + 2U_d + 4U_{pd} - U_p} \right] \qquad (3)$$

This energy is significantly reduced by the on-site Coulomb potential U_d and interatomic Coulomb potential U_{pd}, and its value will be determined later.

Now any holes, produced by doping with divalent atoms or changing the oxygen content, will go onto the O site given the condition $U_p > \Delta E$. These oxygen holes will hop from site to site, becoming the carrier of the supercurrent. The effective hopping Hamiltonian of the single carrier hole can be written as

$$H_h = (t_2 - t_1) \sum_{\substack{\ell \neq \ell' \\ \sigma,\sigma'}} p_{\ell\sigma}^\dagger p_{\ell'\sigma'} + t_2 \sum_{\substack{\ell \neq \ell' \\ \sigma}} p_{\ell\sigma}^\dagger p_{\ell'\sigma} , \qquad (4)$$

where $p^\dagger_{\ell\sigma}(p_{\ell\sigma})$ creates(annihilates) holes of spin σ on the O sites. An oxygen site (ℓ or ℓ') is one of the four nearest neighbors for a given Cu site. The propagation of an oxygen hole by the first term exchanges its spin with the corresponding Cu, leaving behind a string of broken antiferromagnetic bonds among pairs of Cu sites. By means of the second term, an oxygen hole is localized over the four neighboring O sites surrounding a given Cu site. To second order in the Cu-O hopping parameters t_{pd}, we obtain

$$t_1 = \frac{t_{pd}^2}{\Delta E + U_d + 2U_{pd} - U_p} \tag{5}$$

and

$$t_2 = \frac{t_{pd}^2}{U_p - \Delta E} . \tag{6}$$

Equation 3 and these expressions illustrate the dependence of the superexchange energy and the O hole hopping on the parameters of the original Hamiltonian and must be compared with the result obtained from the vacuum defined as the closed shell [8,11]. As we can see from the energy parameters, the hole hopping energy ($t_h = t_2 - t_1$) is greater than the superexchange energy between Cu sites, resulting in the propagation of the hole carriers to lower its energy.

In the case of electron superconductors, any electrons produced by doping with tetravalent atoms will obviously go into the lowest empty states, which is the antibonding state of a Cu site. Since the spins of the neighboring four Cu sites are opposite to the spin of the given Cu site, the propagation of an electron carrier always leaves behind a string of broken antiferromagnetic bonds between the adjacent Cu sites. Similar to the hole hopping, we can write the effective hopping Hamiltonian for the Cu site electron as

$$H_e = t_e \sum_{<i,j>} d^{e\dagger}_{i\sigma} d^e_{j\sigma} , \tag{7}$$

where $<i,j>$ indicates nearest-neighbor Cu sites. To second order in the Cu-O hopping parameter t_{pd}, we have

$$t_e = \frac{t_{pd}^2}{\Delta E + U_d + 2U_{pd} - U_p} \quad . \tag{8}$$

In equation 7, we considered only one channel for the effective hopping of the electron carrier, i.e., the electron carriers are assumed to hop through the $O^{1-}(p^5)$ configuration. The electron hopping will be reduced in equation 8 if we count the hopping through the $O^{3-}(3s^1)$ configuration on an O site, which is a process completely out of phase with equation 8. This calculation is now underway in our laboratory [13]. Using the middle values of the reported parameters given earlier -- $\Delta E = 1.5$ eV, $t_{pd} = 1.2$ eV, $U_d = 9.0$ eV, $U_p = 6.0$ eV, $U_{pd} = 1.5$ eV -- the carrier hopping energy (t_e and t_h) and the superexchange energy can be obtained. The estimated values are $t_e = 0.192$ eV, $t_h = t_2 - t_1 = 0.128$ eV and $J = 48$ meV. These results can be compared with the experimental values of $J = 50 \sim 110$ meV [3], $t_h \approx 0.15$ eV for Y-Ba-Cu-O and $t_h \approx 0.1$ eV for La-Sr-Cu-O [12]. The superexchange energy (J), which is also estimated in a cluster calculation (2 Cu atoms and 4 O atoms) [13], agrees with the above results.

Let us now consider one hole on an O site for hole superconductors and one electron on an Cu site for electron superconductors. As discussed earlier, the single carriers will lower their energies by hopping to their neighbors, but the propagation of these carriers leaves behind a string of broken Cu-Cu antiferromagnetic bonds. The price for doing this is 2J times the number of hopped-to sites. With its maximum kinetic energy of 0.77 eV, it can sweep over approximately seven or eight Cu sites. If the local spin relaxation is long enough compared to the time scale of the carrier motion, the carriers will be localized within seven to eight units cells in the linear direction. In this picture, the Cu-O plane will be fully covered with broken antiferromagnetic bonds for less than 4% concentration of the carriers. This is consistent with the rapid disappearance of the antiferromagnetic fluctuation for a small amount of doping [9,14].

As the concentration of the carriers is increased to twice that of the carrier concentration at which the antiferromagnetic disappears, the neighboring two carriers will lose their origins and make pairs by reducing the number of broken antiferromagnetic bonds between Cu sites. This local pairing in space is due to the rigidity of the antiferromagnetic background of the vacuum and is possible only when the carrier hopping energy is larger than the antiferromagnetic superexchange energy between Cu sites. This paired state can move around freely without breaking the antiferromagnetic background of the vacuum and lowers its energy compared to the localized single electron.

We now consider the propagation of a pair of O site holes in an antiferromagnetic background hole superconductors as shown in figure 1. The propagation of a pair of Cu site electrons is the same as the O site hole propagation except that the carriers are located on Cu sites in electron superconductors as shown in figure 2. In this local pairing in space, we may consider two carriers which are close together in figure 1. The energy differences of the configurations (a)-(d) in figure 1 are due to the number of broken antiferromagnetic bonds and to intersite Coulomb repulsion. The superexchange energy can be slightly changed due to the presence of the carriers nearby [8,11]; however, we ignore this effect because it has little influence in the new vacuum. The paired holes in configuration (a) can hop to (d), which is degenerate with (a), via two intermediate states (b) or (c). By this hopping process, the paired holes can move freely, keeping the possible number of broken antiferromagnetic bonds to a minimum. The number of such bonds is determined by the Coulomb potential (V) between two carriers and the superexchange energy. The configuration (a) in figure 1 has two more broken antiferromagnetic bonds than the configuratin (c), but the Coulomb repulsion between carriers must be considered in the real system. Similar arguments for the one- and two-hole motion are discussed in simple models [15].

In this paper, we estimate the energy gap (Δ) using V as a parameter, since it is very difficult to calculate it in a microscopic domain. The Coulomb repulsions in (a) or (d) will be very small if the dielectric limit is already reached in those configurations. If this is not the case, the paired electrons will have two more broken antiferromagnets between them, but the general descriptions will not be changed. As in the case of single-carrier hopping, the effective hopping of a paired hole can be obtained by degenerate perturbation theory,

$$t_p^{(b)} = \frac{t_e^2}{2J} \tag{9a}$$

and

$$t_p^{(c)} \simeq \frac{t_e^2}{V - 2J}, \tag{9b}$$

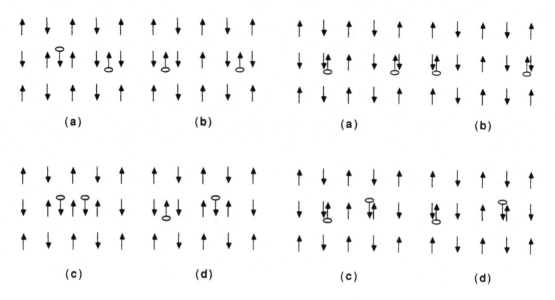

Fig. 1. Paired holes and their propagation: (a) initial state, (b) intermediate state, (c) intermediate state and (d) final state. The carrier holes are located on the O sites.

Fig. 2. Paired electrons and their propagation: (a) initial state, (b) intermediate state, (c) intermediate state and (d) final state. The carrier electrons are located on the Cu sites.

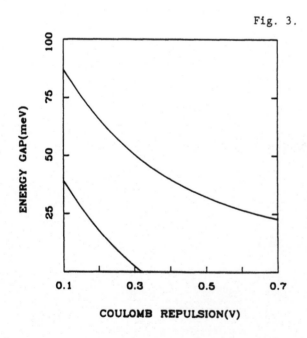

Fig. 3. Superconducting energy gap ($\Delta \sim t_p$) as a function of the Coulomb repulsion V between two carriers of the intermediate state (c) in figure 1 for the upper curve when the paired state has two spin-filled Cu sites between two carriers. V is assumed between two carriers of (a) in figure 1 for the lower curve when the paired state has three spin-flipped Cu sites between them (see the text for details).

where $t_p^{(b)}$ and $t_p^{(c)}$ are the pair hopping energy via two intermediate states (b) and (c) in figure 1, and V is the Coulomb repulsion between two carriers in (c) of figure 1. These expressions are not accurate when the energy denominators are comparable to the one hole hopping energy t_e or t_h. However, more accurate values can be obtained by a modified form of Wigner-Brillouin perturbation between three states: |initial>, |intermediate> and |final> states [8,11]. As discussed earlier, the paired holes can move freely with hopping energy $t_p^{(b)} + t_p^{(c)}$, but some of this kinetic energy is used to keep two spin-flipped Cu sites between them, which is 4J eV in energy. Therefore, the effective pair hopping energy will be $t_p = t_p^{(b)} + t_p^{(c)} - NJ$, where N is the number of spin-flipped Cu sites between two carriers. Then the energy gap, which is the energy difference between the paired state and the single-hole state, will be approximately $\Delta \sim t_p$.

NUMERICAL RESULTS AND CONCLUDING REMARKS

The energy gap versus the Coulomb repulsion of two carriers in (c) of figure 1 is shown by the upper curve in figure 3. If the Coulomb repulsion between two carriers of (a) in figure 1 is less than 0.3 eV, the paired state with three spin-flipped Cu sites between them is also lower in energy than two single carrier states. This result explains the possibility of more than one energy gap in the high-T_c ceramics, which is observed in many experiments [16]. We believe that better results can be obtained if the Coulomb interactions between valence charges in a microscopic domain, which are very difficult to estimate, are included.

Finally, we should point out that this local pairing in space is very different from the ordinary BCS-type pairing caused by a mean-field interaction. By this local pairing, all the carriers in the CuO_2 plane can make pairs and contribute to the superconducting current, which is consistent with experiments [12].

ACKNOWLEGMENTS

This research was supported by an award from the New York State Institute on Superconductivity in conjunction with the New York State Energy Research and Development Authority, and by the Office of Naval Research. We are happy to acknowledge useful conversations with Y. C. Lee and Andreas Langner.

REFERENCES

1. J. G. Bednorz and K. A. Müller, Z. Phys. B $\underline{64}$, 189 (1986).
2. P. W. Anderson, Science $\underline{235}$, 1196 (1987); Z. Zou and P. W. Anderson, Phys. Rev. B $\underline{37}$, 627 (1988); J. E. Hirsch, E. Loh, Jr., D. J. Salapino and S. Tang, Phys. Rev. B $\underline{39}$, 243 (1989).
3. G. Shirane, Y. Endoh, R. J. Birgeneau, M. A. Kastner, Y. Hidaka, M. Oda, M. Suzuki and T. Murakami, Phys. Rev. Lett. $\underline{59}$, 1613 (1987); J. M. Tranquada, D. E. Cox, W. Kunnmann, H. Moudden, G. Shirane, M. Suenaga, P. Zolliker, D. Vaknin, S. K. Sinha, M. S. Alvarez, A. J. Jacobson and D. C. Johnston, Phys. Rev. Lett. $\underline{60}$, 156 (1988); K. B. Lyons, P. A. Fleury, L. F. Schneemeyer and J. V. Waszczak, Phys. Rev. Lett. $\underline{60}$, 732 (1988).
4. Y. Tokura, H. Takagi and S. Uchida, Nature $\underline{337}$, 345 (1989).
5. C. N. R. Rao, A. K. Ganguli, R. Vihayaraghavan, N. Y. Vasanthachara, A. K. Sood and N. Chandrabhas, High-T_c Update $\underline{3(9)}$, 7 (1989).
6. J. M. Tranquada, S. M. Heald, A. R. Moodenbaugh, G. Liang and M. Croft, Nature $\underline{337}$, 720 (1989).
7. J. M. Tranquada, S. M. Heald and A. R. Moodenbaugh, Phys. Rev. B $\underline{36}$, 5263 (1987).
8. E. B. Stechel and D. R. Jennison, Phys. Rev. B $\underline{38}$, 4632 (1988); V. J. Emery and G. Reiter, Phys. Rev. B $\underline{38}$, 4547 (1988).
9. R. L. Greene, H. Maletta, T. S. Plaskett, J. G. Bednorz and K. A. Müller, Solid State Commun. $\underline{63}$, 379 (1987); K. Yamada, E. Kudo, Y. Endoh, Y. Hidaka, M. Oda, M. Suzuki and T. Murakami, Solid State Commun. $\underline{64}$, 753 (1987).
10. A. K. McMahan, R. M. Martin and S. Satpathy, Phys. Rev. B $\underline{38}$, 6650 (1988); D. E. Ramaker, N. H. Turner, J. S. Murday, L. E. Toth, M. Osofsky and F. L. Hutson, Phys. Rev. B $\underline{36}$, 5672 (1987).
11. H. R. Lee and T. F. George, unpublished.
12. Y. J. Uemura, V. J. Emery, A. R. Moodenbaugh, M. Suenaga, D. C. Johnston, A. J. Jacobson, J. T. Lewandowski, J. H. Brewer, R. F. Kiefl, S. R. Kreitzman, G. M. Luke, T. Riseman, C. E. Stronach, W. J. Kossler, J. R. Kempton, X. Y. Yu, D. Opie and H. E. Schone, Phys. Rev. B $\underline{38}$, 909 (1988).
13. H. R. Lee, K. H. Yeon and T. F. George, manuscript in preparation.
14. J. H. Brewer and 34 co-authors, Phys. Rev. Lett. $\underline{60}$, 1073 (1988); D. C. Johnston, J. P. Stokes, D. P. Goshorn and J. T. Lewandowski, Phys. Rev. B $\underline{36}$, 4007 (1987); R. J. Cava, B. Batlogg, C. H. Chen, E. A. Reitman, S. M. Zahurak and D. Werder, Nature $\underline{329}$, 423 (1987).
15. J. E. Hirsch, Phys. Rev. Lett. $\underline{59}$, 228 (1987); S. A. Trugman, Phys. Rev. B $\underline{37}$, 1597 (1988).
16. J. R. Kirtley, R. T. Collins, Z. Schlessinger, W. J. Gallagher, R. L. Sandstorm, T. R. Dinger and D. A. Chance, Phys. Rev. B $\underline{35}$, 8846 (1987).

SUPERCONDUCTIVITY DUE TO THE SOFTENING OF THE BREATHING
MODE IN $Ba_{1-x}K_xBiO_3$ COMPOUNDS

C.S. Ting and Z.Y. Weng

Department of Physics and Texas Center for
Superconductivity
University of Houston, Houston, TX 77204-5504

I. Introduction

Recent experiments[1-3] seem to indicate that the superconductivity in $Ba_{1-x}K_xBiO_3$ compounds occurs in samples with stoichiometry close to the transition from metallic to charge density wave (CDW) state as the value of x decreases. The isotope effect has been measured for $Ba_{0.625}K_{0.375}BiO_3$ and was found to be large ($\alpha=0.41$), indicating that the pairing interaction is predominantly phonon -mediated[4]. Although a later measurement[5] indicates a smaller isotope effect ($\alpha=0.2$). Evidence of phonon softening has also been observed very recently by neutron scattering in this type of material[6]. The crystal structure of $Ba_{1-x}K_xBiO_3$ is similar to that of $BaPb_xBi_{1-x}O_3$[7] except the dopant K atom replaces Ba atom and leave the BiO_3 octahedral network intact. The BiO based compound has perovskite-type structure with one Bi-atom surrounded by six nearest neighboring oxygen atoms in three dimension. The lattice formed by the Bi atoms in the metallic phase is simple cubic. In this article we assume that local breathing mode due to the oxygen atoms around a Bi-atom is responsible for both CDW and superconducting states and study the enhancement of the superconducting transition temperature T_c due to the softening of this mode from the metallic side near the CDW instability as a function of carrier doping. In the next section, we shall adopt the band structure from the tight binding theory to describe the motion of charged carriers on Bi atoms and discuss how the phonon spectrum can be renormalized and softened. In section III, the phonon softening is demonstrated numerically and the superconducting transition temperature T_c is calculated as a function of the doping parameter x. The obtained result yields the essential feature of the experimental measurements[3]. A summary and discussion weill be given by Section IV.

II. Hamiltonian

The model Hamiltonian describing such a system can be written as

$$H = \sum_{k} (\varepsilon_k - \mu) c^+_{k\sigma} c_{k\sigma} + \sum_{q} \omega_o(q) b^+_q b_q + \sum_{kq\sigma} g(q)(b_q + b^+_{-q}) c^+_{k\sigma} c_{k+q\ \sigma} \ , \quad (1)$$

here the single particle energy $\varepsilon_k = -2t(\cos k_x + \cos k_y + \cos k_z)$, with t as the hopping integral, is obtained by using three dimensional tight binding theory for a simple cubic lattice formed by the Bi atoms. $\omega_o(q) = \omega_o$ corresponds to the frequency of the breathing mode. μ is the chemical potential and $g(q) = \gamma/\sqrt{(\omega_o(q))}$, with γ as the coupling constant, is the electron-phonon interaction. $c^+_{k\sigma}$ ($c_{k\sigma}$) and b^+_q (b_q) are respectively the electron and the phonon creation (destruction) operators. For convenience the lattice constant a is chosen to be one unit length. In the presence of electron-phonon interaction $g(q)$, the phonon frequency should be renormalized. In the random phase approximation, the renormalized phonon frequency can be written as

$$\omega^2 = \omega^2(q) = \omega_o^2 [\ 1 - \frac{2\gamma^2}{\omega_o^2} \Pi(q,\omega)\ ] \ . \quad (2)$$

$\Pi(q,\omega)$ is the density-density correlation function and it takes the following form

$$\Pi(q,\omega) = \frac{2}{N} \sum_k \frac{f(k) - f(k+q)}{\varepsilon_{k+q} - \varepsilon_k + \omega + i\delta} \ , \quad (3)$$

here N is the number of lattice site and $f(\mathbf{k}) = 1/[\exp((\varepsilon_\mathbf{k}-\mu)/T)+1]$ is the Fermi-Dirac distribution function at temperature T. Since the magnitude of the carrier energy $\varepsilon_\mathbf{k}$ is of order t (~ 1 electron volt) and the phonon energy $\omega(\mathbf{q})\ll t$, our numerical study shows that the ω-dependence of $\Pi(\mathbf{q},\omega)$ makes almost no contribution to the phonon renormalization. We therefore can replace $\Pi(\mathbf{q},\omega)$ with $\Pi(\mathbf{q},0)$ in Eq.(2). For the half-filled case which corresponds to one electron per each lattice site, the chemical potential can be taken as $\mu=0$ and one has a nested Fermi surface in the sense that the nesting wave vector is $\mathbf{Q} = (\pi,\pi,\pi)$ and $\varepsilon_{\mathbf{k}+\mathbf{Q}} = -\varepsilon_\mathbf{k}$, the density-density correlation function $\Pi(\mathbf{Q},0)$ diverges logarithmically[8] at T=0K

$$\Pi(\mathbf{Q},0) = -\int_{-6t}^{0} \frac{N(\varepsilon)}{\varepsilon} d\varepsilon \to \infty , \qquad (4)$$

$N(\varepsilon)$ is the density of states evaluated from $\varepsilon_\mathbf{k}$ according to the tight binding theory for a simple cubic lattice. If we set t=1, $N(\varepsilon)$ as a function of ε is plotted in Figure 1. For $|\varepsilon|\leq 2.0$, $N(\varepsilon)\approx 0.1425$ is almost ε independent. The divergence of $\Pi(\mathbf{Q},0)$ corresponds to the system attempting to form an insulating CDW state below certain temperature T_{CDW}. T_{CDW} can be determined from the condition $\omega(\mathbf{Q})\equiv 0$ when $T=T_{CDW}$. Namely the $\mathbf{q}=\mathbf{Q}$ breathing mode becomes completely soft at the CDW transition. The CDW state can be destroyed by either rising T or doping more electrons into the system. For $Ba_{1-x}K_xBiO_3$, the half filled case is represented by x=0. The parent compound $BaBiO_3$ is a CDW insulator and an energy gap opens up at $\varepsilon=0$. If the doping parameter x is increased to a certain value x_o, the CDW state disappears and the normal metallic state will show up. Here we wish to mention that $\omega(\mathbf{Q}')\equiv 0$ at T=0K and $x=x_o$ but with $\mathbf{Q}'\neq\mathbf{Q}$. This result implies that the CDW state at finite doping $x\neq 0 (x<x_o)$ may become imcommensurate with the lattice and its structure could be rather complicated that no analytic study has been made. However a general review of the experiments and theories involving with the competition between CDW and superconductivity can be found in Ref.9. In the present paper we shall concentrate our effort on the metallic state of the compound $(x>x_o)$ and study the enhancement of the superconducting temperature T_c due to softening of the breathing mode as $x\to x_o$.

III. Superconductivity

Following the strong coupling theory of superconductivity, T_c should be determined by the linearized Elliashberg equation in the Matsubara representation[10]

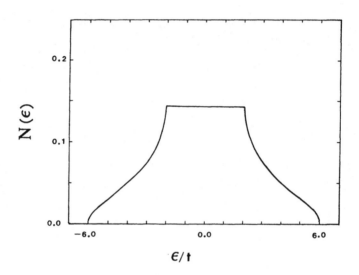

Figure 1. Density of states $N(\varepsilon)$ of the three dimensional tight binding model. The energy ε is in units of t.

$$G_n \Delta_n = \lambda \sum_{m=0}^{\infty} (\bar{\lambda}_{n-m} + \bar{\lambda}_{n+m+1}) \Delta_m - 2\mu^* \sum_{m=0}^{j_c} \Delta_m , \qquad (5)$$

in which

$$G_n = 2n+1+\lambda+2\lambda \sum_{l=1}^{n} \bar{\lambda}_l , \qquad (6)$$

$$\bar{\lambda}_l = \frac{2}{\lambda} \int_0^{\omega_0} d\omega\, \alpha^2 F(\omega) \frac{\omega}{\omega^2 + (2\pi T_c l)^2} \qquad (7)$$

Δ_n is the superconducting order parameter at $\omega_n = 2\pi nT$, j_c is an integer determined by the condition $\omega_0 < (2j_c+1)\pi T_c < \omega_0 + 2\pi T_c$, μ^* is the Coulomb pseudo potential and the dimensionless electron-phonon coupling constant λ is defined by setting $\lambda_0=1$ in Eg.(7). The function $\alpha^2 F(\omega)$ in the above equation is the product of the effective electron phonon coupling and the phonon density of states[11]

$$\alpha^2 F(\omega) = N(0) < |\bar{g}_q|^2 B(q,\omega) >_{FS} , \qquad (8)$$

N(0) is the density of states at the Fermi level. It is noted that in the proper doping regime the variation of the density of states around the Fermi level is negligible according to Figure 1. The phonon spectral function $B(q,\omega)$ is approximately replaced by $\delta(\omega-\omega(q))$ and the effective electron-phonon coupling g_q could be generally expressed as[10]

$$\bar{g}_q = \frac{\gamma}{\sqrt{(\omega('q))}} \qquad (9)$$

$\omega(q)$ here is the renormalized phonon frequency as shown in Eq.(2). $<...>_{FS}$ in Eq.(8) denotes that the average is taken over the Fermi surface with respect to the momenta p and p' ($q = p-p'$). There are several methods[12] to solve Eq.(5). One intuitive way to determine T_c is to set the determinant from Eq.(5) equal to zero. However for $\lambda < 1$, we expect that T_c obtained from Eq.(5) can be approximately represented by the revised McMillan T_c formula[13]

$$T_c = \frac{\bar{\omega}}{1.20} \exp\left[- \frac{1.04(1+\lambda)}{\lambda - \mu^*(1+0.62\lambda)} \right]. \qquad (10)$$

Here phonon frequency moment $\bar{\omega}$ are defined respectively by

$$\bar{\omega} = \frac{2}{\lambda} \int d\omega \frac{\alpha^2 F(\omega)}{\omega} \omega , \qquad (11)$$

In the following study we shall set the Coulomb pseudo potential $\mu^*=0$.

The numerical calculation has been performed on a 76x76x76 finite lattice in which the wave vectors are defined as

$$k_i = \frac{2\pi}{N_i} n_i \quad \text{and} \quad -\frac{N_i}{2} < n_i \leq \frac{N_i}{2} \qquad (12)$$

with $i = x,y,z$, $N_i=76$ and n_i is an integer. For $x=0$, $\Pi(q,0)$ has an infinite maximum at the nesting wave vector $q=Q=(\pi,\pi,\pi)$. At finite doping ($x \neq 0$) the infinite maximum splits into six finite maxima with nesting wave vectors moving away from (π,π,π) but located along the three edges $(\pi,\pi,\pi \pm b)$, $(\pi,\pi \pm b,\pi)$ and $(\pi \pm b,\pi,\pi)$ in the reciprocal lattice space. b satisfies $b \geq 0$ and is a function of x. For example $b=0$ at $x=0$, and $b>0$ when $x \neq 0$. Although the nesting wave vectors $(\pi,\pi,\pi+b)$, $(\pi,\pi+b,\pi)$ and $(\pi+b,\pi,\pi)$ are outside the first Brillouin zone, they can be mapped back to the zone boundary as $(\pi,\pi,-\pi+b)$, $(\pi,\pi \pm b,\pi)$ and $(\pi \pm b,\pi,\pi)$ in the reciprocal lattice space. b satisfies $b \geq 0$ and is a function of x. For example $b=0$ at $x=0$, and $b>0$ when $x \neq 0$. Although the nesting wave vectors $(\pi,\pi,\pi+b)$, $(\pi,\pi+b,\pi)$ and $(\pi+b,\pi,\pi)$ are outside the first Brillouin zone, they can be mapped back to the zone boundary as $(\pi,\pi,-\pi+b)$, $(\pi,-\pi+b,\pi)$ and $(-\pi+b,\pi,\pi)$. In the discrete lattice case, the average over the Fermi surface is defined as

$$\langle f(q) \rangle_{FS} = \frac{\sum_{p,p'} f(q)}{\sum_{p,p'}}, \qquad (13)$$

here $p'=p-q$ and $\varepsilon_F - \Delta < \varepsilon_p(\varepsilon_{p'}) < \varepsilon_F + \Delta$, ε_F is the Fermi energy and $\Delta (<< \varepsilon_F)$ is an energy of the order of magnitude of a phonon frequency. Since the superconductivities in BiO based compounds seem to occur in the metallic phase, we shall study T_c as a function of x with $x>x_o$. When $x>>x_o$ we expect that the phonon frequency $\omega(q)$ should not be strongly renormalized by the electron-phonon interaction g_q or γ. As $x \to x_o$ the softening of $\omega(q)$ at the six nesting reciprocal lattice vectors mentioned after Eq.(12) begins to show up, when $x<x_o$, the system becomes a metallic CDW state. Choosing[14] $\omega_o=819K$ and $2\gamma^2/\omega_o^2=1.77$, the numerical results of $\omega(q)$ as a function of q with $q=(\pi,\pi,0)$ to (π,π,π) are plotted in Figure 2 for several different values of x. We expect $\omega(q)$ vanishes at the nesting wave vectors when $x=x_o=0.27$. T_c as a function of x ($x>x_o$) has also been numerically studied according to Eq.(5).

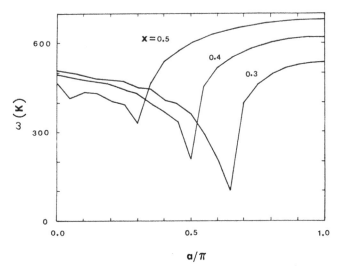

Figure 2. Renormalized phonon frequency $\omega(\mathbf{q})$ as a function of $\mathbf{q}=(\pi,\pi,a)$ for doping x=0.3, 0.4 and 0.5 where $0<a\leq 1$, ω_0=819K and $2\gamma^2/\omega_0^2$=1.77 on a 76x76x76 lattice.

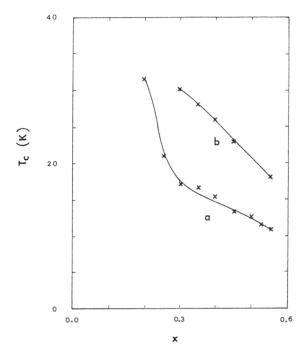

Figure 3. The superconducting transition temperature T_c as a function of doping x on a 76x76x76 lattice. The cross are numerical data points. Curve a and b are respectively for $2\gamma^2/\omega_0^2$=1.53 and 1.77, here ω_0=819K.

The results are shown in Figure 3. It is readily seen that T_c, when x varies from $x \gg x_o$ to $x \to x_o$, can be enhanced by one to two times as $2\gamma^2/\omega_o^2$ is changed from 1.77 to 1.53. It is interesting to note here that the dimensionless coupling constant $\lambda=0.924$ and 0.39 respectively for $x=0.2$ and $x=0.5$ as $2\gamma^2/\omega_o^2=1.53$. Similar calculation has also been performed for a two dimensional tight binding system. The results are shown in Figure 4 for $2\gamma^2/\omega_o^2=1.471$ and 1.266. The enhancement of T_c as a function of x in 2-D system is quite significant and is much more dramatic than that of a 3-D system. Our numerical studies shown in Figure 3 and 4 are confined to the region in which x is larger or slightly larger than x_o. When the doping parameter $x=x_0$ is at the CDW instability, the phonon becomes completely soft at certain wave vectors Q' and we have $\omega \simeq 0$ and $\lambda \simeq \infty$. Under this condition the

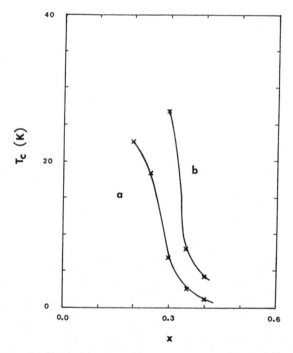

Figure 4. The superconducting transition temperature T_c as a function of doping x on a 100x100 lattice. The cross are numerical data points. Curve a and b are respectively for $2\gamma^2/\omega_o^2=1.266$ and 1.471, here $\omega_o=800K$.

revised McMillan T_c formula, which yields $T_c \approx 0$, is no longer valid and a correct and finite T_c should be obtained from the Eliashberg equation in Eq.(5).

IV Conclusion

In this paper we have studied the enhancement of T_c due to the softening of the breathing modes of oxygen-atoms around Bi atoms in $Ba_{1-x}K_xBiO_3$ compound near the CDW instability in the normal metallic phase as the doping parameter x varies. The superconducting transition temperature T_c can be calculated by either the Elliashberg equation or the revised McMillan formula as long as $\lambda \leq 1$ and $x > x_0$. The magnitude of the enhancement from $x > x_0$ to x near x_0 is quite significant and the result seems to be consistent with the essential feature of the experiments[3]. Moreover our result for a 2-D system with $x > x_0$ agrees qualitatively with that of a recent Monte Carlo simulation performed for a 6x6 lattice[15]. In the region fo large x, the system may become rather disordered and its T_c may be substantially depressed. When the effect due to disorder at large x is taken into account, we would expect that the enhancement of T_c could be more dramatic than those depicted in Figs.3 and 4. When $x < x_0$, CDW will set in. Since the nesting wave vector Q' is no longer at $Q=(\pi,\pi,\pi)$ for finite x, the CDW state may become incommensurate with the lattice and is thus rather difficult to describe. However if one assumes that the nesting still occurs at Q even for $x \neq 0$, T_c can be calculated by the Monte Carlo simulation[15] for a 2-D system and the result shows that T_c suddenly drops to zero as x varies from $x > x_0$ to $x < x_0$. This feature is consistent with the experimental measurements[3] and should be persisting to a 3-D system like $Ba_{1-x}K_xBiO_3$ compound. For $x < x_0$, the charge carriers in the CDW state have been pointed out recently to be bipolarons with very large effective mass[16] and they can be easily pinned near the impurity sites like those of K or oxygen vacancies. This may be the reason why superconductivity has not been observed in $Ba_{1-x}K_xBiO_3$ for $x < x_0$. On the other hand, the superconductivity in $BaPb_{1-x}Bi_xO_3$ may coexist with the CDW state. This is because its parent compound $BaPbO_3$ is a semimetal[17] with $T_c \simeq$ 0.45 K. When $BaPbO_3$ is doped with Bi (x=0.25), superconductivity[7] (T_c =12K) and CDW are both present. Although the CDW state due to the presence of Bi atoms yields a pseudo-gap[14], the superconductivity still comes from the metallic band of the parent compound which possesses no real gap near the Fermi level. One of the reasons for the large enhancement of T_c in $BaPb_{1-x}Bi_xO_3$ with x≈0.25 may due to the hybridyzation of the metallic band and the localized bipolarons brought in by the CDW state[18]. Finally we wish to mention that the main part of the present paper has been accepted for publication[19] and a similar work based upon the band structure calculation[20] has also been done very recently by Shirai et al[21].

Acknowledgement - We wish to thank Profesors W.P. Su, X.X. Yao and J.L. Shen for useful discussions. This work was supported by a grant from the Robert A. Welch Foundation and also by the Texas Center for Superconductivity at the University of Houston under the prime grant No. MDA 972-8-G-0002 from the U.S. Defense Advanced Research Projects Agency and the State of Texas

References

1. T.M. Rice, Nature 332, (1988) 780.
2. R.J. Cava et al. Nature 332, (1988) 814.
3. D.G. Hinks et al. Nature 332, (1988) 836.
4. D.G. Hinks et al. Nature 335, (1988) 419.
5. B. Batlogg et al. Phys. Rev. Lett. (1989).
6. C.K. Loong et al. Phys. Rev. Lett. 62, (1989) 2628.
7. A.W. Sleight, in "Superconductivity: synthesis, properties and processing", edited by W. Hatfield (Marcel Dekker, New York 1988).
8. S. Doniach and E. Sondheimer, "Green's Function for Solid State Physicists", (Benjamin, Reading, Mass, 1982) p165.
9. C.Y. Huang, CRC critical Reviews in Solid State and Materials Science v12 (1984) 75.
10. C.S. Owen and D.J. Scalapino, Physica, 55 (1971) 691.
11. D.J. Scalapino, "Superconductivity Vol. 1, ed. R.D. Parks", (Marcel Dekker, New York 1969) p449.

12. Zheng-yu Weng and Hang-sheng Wu, J. Phys. C$\underline{19}$ (1986) 5459. (See the references there in).
13. R.C. Dynes, Solid State Commun. $\underline{10}$ (1972) 615.
14. S. Tajima et al., Phys. Rev. B$\underline{35}$, (1987) 696.
15. W.P. Su, (Physical Review B, in press)
16. J. Yu, X.Y. Chen and W.P. Su (Physical Review B, in press)
17. V.V.Bogatko and Yu.N. Venevtsev, Sov. Phys. Solid State $\underline{22}$ (1980)705.
18. C.S. Ting and D.Y. Xing in "Novel superconductivity" edited by S.A. Wolf and V.Z. Kresin (Plenum Publishing, 1987) 539.
19. C.S. Ting, X.Y. Chen and Z.Y. Weng (Mordern Phys. Lett. 1989).
20. L.F. Mattheiss and D.R. Hamann Phys. Rev. Lett. $\underline{60}$ (1988) 2681.
21. M. Shirai, N. Suzuki and K. Motizuki (preprint- 1989).

Evidence for Pairing of Semions in Finite Systems

Weikang Wu, C. Kallin and A. Brass[a]

Institute for Materials Research and Department of Physics
McMaster University, Hamilton, Ontario, Canada L8S 4M1

ABSTRACT

We present the results of numerical studies of semions on a lattice. From our studies of the pairing energies, flux quantization and effects of external magnetic fields, we find strong evidence in support of the theory that semions will pair due to their statistical interaction and form a coherent or superfluid ground state.

I. Introduction

Quantum statistics have profound consequences for condensed matter systems at low temperatures, giving rise to the concepts of a Fermi surface for Fermions and a condensate for Bosons. In three dimensions, Fermi and Bose statistics are the only statistics allowed.[1] However, in two-dimensions, fractional statistics, which interpolate between the Fermi and Bose cases, are possible.[2] Fractional statistics arise in several field theory models,[3] as well as in the fractional quantum Hall effect.[4] More recently, Laughlin[5] has proposed a model for high temperature superconductivity in which the carriers obey fractional statistics.

The difference between two and three dimensions lies in the different topologies of the configurations spaces. The configuration space for N identical particles is not simply the Cartesian product of single paricle spaces. One needs to identify those points in this product space which represent the same physical configuration. Hence, two configurations differing by the exchange of particle positions are considered to be identical. The resulting configuration space is singular at points where the positions of two or more particles coincide. The essential difference between two and three dimensions can be understood by considering a path in configuration space in which one particle moves along a loop which encloses a second particle. In three dimensions, this loop can always be deformed to a point without crossing any singularities (positions of other particles). Thus the phase associated with such a loop must be a multiple of 2π. (Encircling a second particle is equivalent to two pairwise exchanges.) This is only consistent with Bose or Fermi statistics. However, in two dimensions, the loop cannot be contracted to a point without crossing a singularity and hence a nontrivial phase can be associated with such a loop. In two dimensions, quantum statistics can be parameterized by a continuous variable α, where the phase acquired by the wave function upon interchange of two identical particles is $\pm i\pi\alpha$. Bosons correspond to $\alpha = 0$ and fermions correspond to $\alpha = 1$. All physical quantities are periodic in α, with periodicity 2.

The possible existence of fractional statistics is unique to two dimensional systems and it is natural to look for their existence in physical systems where the dynamics of the particles are constrained to two dimensions. The electrons in an inversion layer at low temperatures provide an example of such a system, and in a large transverse magnetic field, under certain conditions, are expected to give rise to quasiparticles which obey fractional statistics. The high temperature oxide superconductors also appear to be highly two dimensional, and Laughlin has proposed a theory of superconductivity in which the quasiparticles obey fractional statistics with $\alpha = 1/2$. Particles obeying fractional statistics are referred to as anyons and those obeying half statistics, in particular, are referred to as semions.

Wilczek[6] showed that anyons may be described as charged Fermions or Bosons with a flux tube attached to each particle. If, for example, each flux tube carries α flux quanta, the Bohm-Aharonov phase a Bose particle picks up when moving in the presence of the other Bose particles is precisely the phase that would be acquired due to α-statistics. Fermions carrying flux α correspond to $(1-\alpha)$-statistics. This transformation which changes statistics is a singular gauge transformation which changes the wavefunctions (from allowed anyon wave functions to allowed Bose wave functions, for example) but leaves the energy spectrum unchanged. Arovas et al.[7] calculated the second virial coefficient of a free anyon gas, and found that when anyons are treated as fermions with a gauge interaction, the second virial coefficient is reduced from that of free fermions. This suggests that there is an attractive interaction between anyons when treated in the fermion representation. Furthermore, Laughlin[5] pointed out that pairs of semions form bosons and hence it is possible that semions may pair due to their statistical interaction. Laughlin treated semions in mean field theory and found some support for this idea. More recently, Fetter, Laughlin and Hanna [5], calculated the collective excitations of a semion gas in the random phase approximation and found that it behaves as a charge 2e superfluid. Canright and Girvin[8] have found numerical support for this theory from exact diagonalization studies of semions on a square lattice with cylindrical symmetry (i.e. periodic boundary conditions were applied in one direction).

In this paper we report on some of the numerical results we have obtained for semions on a lattice superimposed on a sphere. The spherical topology was chosen in order to eliminate edge effects which one has on an open topology. A torus, or periodic boundary conditions in both directions, is the usual choice for eliminating edge effects in numerical studies. However, there are some subtleties in applying two-dimensional periodic boundary conditions to a system of particles obeying fractional statistics, whereas the sphere is particularly simple in this regard.

II. Numerical Technique

Anyons with α-statistics may be treated as hard core bosons with flux tubes of strength $\alpha\phi_0$, where $\phi_0 = hc/e$ is the flux quantum. Then the Hamiltonian is

$$\mathcal{H} = \frac{1}{2m}\sum_i \left(\mathbf{p}_i + \frac{e}{hc}\sum_{j<i}\mathbf{A}_{ij}\right)^2 + \sum_{j<i}V(\mathbf{r}_i - \mathbf{r}_j), \qquad (2.1)$$

where V is the interparticle interaction (and is zero for free anyons) and the gauge interaction is

$$\mathbf{A}_{ij} = \frac{\alpha\phi_0(\mathbf{r}_i - \mathbf{r}_j) \times \hat{\mathbf{z}}}{|\mathbf{r}_i - \mathbf{r}_j|^2} = \alpha\phi_0 \nabla\theta_{ij}. \qquad (2.2)$$

Here θ_{ij} is the angle between particles labelled by i and j, with respect to some fixed coordinate system. It is clear from this Hamiltonian that even the problem of free anyons is intrinsically a many-body problem. This makes it difficult to obtain analytic solutions, and exact solutions only exist for the case of two anyons, and for a few special cases of three anyons.

We study the problem of anyons on a finite lattice, since this results in a finite Hilbert space and can be studied numerically. At low densities the results should be similar to those of the continuum problem, but near half filling there will typically be important lattice effects. For anyons confined to lattice sites, the Hamiltonian becomes

$$\mathcal{H} = -t \sum_{<lm>} e^{i\phi_{lm}} a_l^\dagger a_m + V \sum_{<lm>} n_l n_m, \qquad (2.3)$$

where $a_l^\dagger$ creates a boson on lattice site l if the site is initially empty, $n_l = a_l^\dagger a_l$, and we have specialized to the case of nearest neighbour interactions V only. The statistical phase associated with hopping from site l to m is

$$\phi_{lm} = \alpha \phi_0 \int_l^m \mathbf{A} \cdot d\ell, \qquad (2.4)$$

where $\mathbf{A}$ is the statistical vector potential due to all the other particles. Without lose generality, we take $t = 1$. To eliminate edge effects, we study lattices with spherical topology. There are five regular polyhedra whose vertices lie on the surface of a sphere. The two largest are of interest, the icosahedron with 12 vertices which are taken to be the lattice sites, and the dodecahedron with 20 sites. Thus, we study up to eleven anyons on the icosahedron and up to nineteen anyons on the dodecahedron. The largest Hilbert space is for 10 anyons on 20 sites, which has 184,756 states. Using the five-fold rotational symmetry we are able to diagonalize this.

The advantage to working on a regular polyhedron is that each site is equivalent, *i.e.* there are no edges, and fractional statistics are particularly simple to formulate in this topology. However, on the sphere, and, in fact, on any closed surface, there is a constraint on the allowed values of the statistical parameter α:

$$\alpha(N_p - 1) = \text{integer}, \qquad (2.5)$$

where N_p is the total number of anyons. This constraint is easily seen to arise from considering a path where one particle encircles all the other particles. On the surface of a sphere, this path can also be considered to enclose no particles and hence must give rise to a phase which is simply a multiple of 2π. This constraint can also be thought of as arising from the monopole constraint for the total flux through a closed surface, keeping in mind that each particle only sees the flux attached to the other particles and does not see its own flux.

In the thermodynamic limit $N_p \to \infty$ this constraint is insignificant. However, it does restrict the statistics one may consider for a finite system. In particular, for the case of semions, only odd numbers of semions are allowed. In order to study the possible pairing of semions, it is necessary to study even numbers of semions. This can still be done in the spherical geometry in two different ways. One possible way is to introduce a small uniform magnetic field. In the presence of a magnetic field, the constraint becomes

$$\alpha(N_p - 1) + \phi/\phi_0 = \text{integer}, \qquad (2.6)$$

where ϕ is the total flux due to the external field. Thus, one may study an even number of semions in a field of strength $\phi_0/2$. One may think of this as the particle interacting with its own flux tube in a mean field sense. For a large system, this field is very small, since ϕ is the total flux over the entire surface. Alternatively, one can study an odd number of semions, but with the position of one semion fixed, at the north pole, for example, so that it has no dynamics. This is equivalent to an even number of semions in the presence of a defect, which restores the monopole constraint. We will use both of these procedures in order to interpret the results for even numbers of semions.

For anyons confined to the surface of a sphere, the vector potential in the "string gauge" may be written as

$$\mathbf{A}_{ij} = \alpha\phi_0 \delta(\phi_i - \phi_j)[\Theta(\theta_i - \theta_j) - 1/2], \tag{2.7}$$

where Θ is the Heavyside theta function. In this gauge a string emanates from each particle and terminates at the south pole. Every time a particle crosses another particle's string, the wave function, and hence the hopping matrix element, picks up a phase of $\pm i\alpha\pi$. This choice of gauge preserves the azimuthal symmetry which we use to block diagonalize the Hamiltonian matrix.

III. Pairing in Finite Systems

The main purpose of this study is to investigate the correlations due to statistics in systems of semions and in particular to investigate the possible pairing of semions. The questions we focus on are: (i) Do semions pair due to their statistical interaction? and (ii) If so, do these pairs form a coherent state and hence a superfluid? We will discuss the following four signatures of a coherent state formed from pairs of particles:

(1) *Pairing Energy:* As a first check on whether there are correlations which tend to pair the particles, one can calculate the pairing energy defined as

$$\Delta(N) = E(N+2) + E(N) - 2E(N+1). \tag{3.1}$$

If the particles are paired, one finds that $\Delta(N) < 0$ for N even and $\Delta(N) > 0$ for N odd. In the thermodynamic limit, $\Delta(N_{\text{odd}}) = -\Delta(N_{\text{even}})$.

(2) *Periodicity of $E(\phi)$:* If an infinitely thin solenoid is passed through the north and south poles of the sphere and the flux through the solenoid is varied continuously, then, in the thermodynamic limit, all physical quantities, including the energy $E(\phi)$ must be periodic in the flux ϕ with period ϕ_0. However, if the particles are paired, as are the electrons in a BCS superconductor, then the energy will be periodic with period $\phi_0/2$; *i.e.* the flux quanta for a paired state is $\phi_0/2$. In a finite system, the energy will not be precisely periodic. However, if pairing exists, the energy will exhibit minima at integer multiples of $\phi_0/2$, and the difference in energy between the minima at odd integer multiples and the minima at even integer multiples will vanish in the thermodynamic limit.

(3) *Flux Quantization:* If a coherent paired state exists with off diagonal long range order (ODLRO), then the flux through the solenoid is quantized in integer multiples of $\phi_0/2$. Thus, in the thermodynamic limit, the energy barrier between two adjacent minima in $E(\phi)$ is infinite. In a finite system, this energy barrier, $E_{\max}(\phi) - E_{\min}(\phi)$, scales as the number of particles N, if there is ODLRO.

(4) *Critical Field:* If there is ODLRO the flux quantization will persist in the presence of a small uniform magnetic field, but will be destroyed by a sufficiently strong field. For the case of semions, even in the absence of an external magnetic field, the system does not have time reversal symmetry. Therefore the critical field which destroys the ODLRO will, in general, be different for the two different orientations of the field.

The first two signatures are indications of pairing, while the last two are indications of coherence or ODLRO. In any finite size calculations, it is crucial to distinguish between those signatures resulting from finite size effects and those that will survive in the thermodynamic limit. Our strategy is to simultaneously search for several independent signatures of pairing to check for consistency. Therefore, we study all four signatures described above in order to identify pairing.

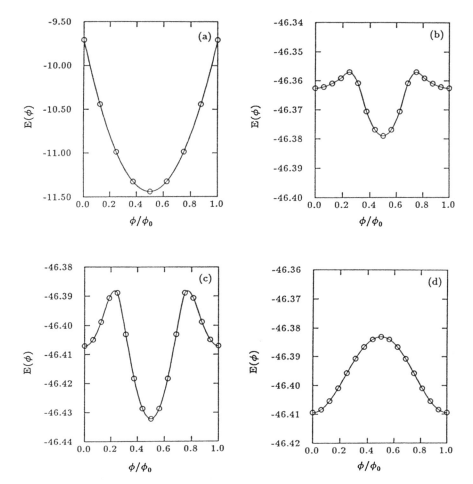

FIG. 1. Ground state energy as function of flux for six fermions on an icosahedron lattice. a) $V = 0, B = 0$; b) $V = -15, B = 0$; c) $V = -15, B = \phi_0$; d) $V = -15, B = 5\phi_0$.

IV. Numerical Studies of Fermion Superconductivity

In order to check that the systems we can study numerically are large enough to give rise to clear signatures of ODLRO, we first study the case of fermions with attractive interactions. We use the Hamiltonian given in Eq. (2.3) with $\alpha = 1$ and the interaction V is negative. Since we are considering spinless fermions, for sufficiently strong attractive interactions, the fermions are expected to pair and form a triplet superconductor.

Fig. 1 shows the results for six fermions on an icosahedron (12 sites). With no nearest neighbor interaction V there is no indication of a second minimum in $E(\phi)$ at $\phi_0/2$. (Fig. 1a) With a sufficiently strong attractive interaction $V \geq V_0$, a second minumum develops at $\phi_0/2$. (Fig. 1b) These two minima persist in a small magnetic field, (Fig. 1c) and the second minima is destroyed by a field of $B \geq 5\phi_0$. (Fig. 1d) Similar results are observed for fermions on the dodecahedron. (If the interaction V is increased sufficiently, the fermions will form a cluster, rather than a coherent paired state, and other minima develop. This results indicate that the finite systems we study are sufficiently large to show evidence of ODLRO.

V. Pairing of Semions

We have studied many semion systems on icoshedral and dodecahedral lattices. Here, we present some numerical results for free semion systems.

Fig. 2 shows the energy as a function of flux (in unites of ϕ_0) for 4 free semions on an icosahedral lattice. Fig. 2a clearly shows a second minimum in the energy at $\phi_0/2$, a signature of pairing. Fig. 2b shows the same system in the presence of a uniform external magnetic field with total flux equal to one flux quanta. In this case, the two energy minima persist. However, when the magnetic field is increased to 2 flux quanta, only one minimum remains. (See Fig. 2c.) This magnetic field dependence suggests that the semions form a coherent or superfluid state.

The above results for a free semion system suggest that the statistical correlation alone is sufficient to provide a pairing interaction between the particles. One would expect this pairing could be enhanced by adding a suitable attractive interaction between semions and that it would be destroyed by the addition of a sufficiently strong repulsive interaction. Fig. 2d shows an example of the effect of a repulsive interaction, where the pairing signature of a second minimum no longer exists.

For odd numbers of semions, no pairing was observed. Fig. 3, for example, shows the energy flux curve for 5 semions. This is a finite size effect, which one would expect to disappear in sufficiently large systems. However, it clearly points out the need to look for multiple signatures of pairing and ideally to study the signatures as a function of system size.

Figs. 4 shows the results for semions systems on a dodecahedral lattice. Fig. 4a shows the results for 4 semions and Fig. 4b shows the case of 12 semions. Both display a second minimum in energy at exactly $\phi_0/2$.

In all of these curves, the boundary conditions for even numbers of semions were handled by localizing an extra "semion" at the north pole. The observed second minimum is deeper using this method rather than adding a small external field.

We have also calculated the pairing energies for semions. It is found that pairing energies for even number semions are negative and those for odd number are positive. This is another evidence that pairing does occur in free semion systems.

VI. Conclusions

We have studied semions on finite lattice by numerically diagonalizing the Hamiltonian matrix. We investigate possible superfluid correlations in semion systems by studying several different signatures of a coherent paired state, namely, the pairing energy, flux quantization and the effects of an external magnetic field. By studying fermions with attractive interactions, we first show that our systems are large enough to display the properties of a superfluid state. However, finite size

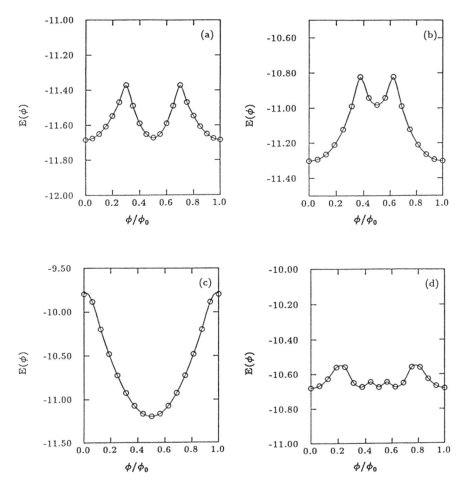

FIG. 2. Ground state energy as function of flux for four semions on an icosahedron lattice. a) $V = 0, B = 0$; b) $V = 0, B = \phi_0$; c) $V = 0, B = 2\phi_0$; d) $V = 1, B = 0$.

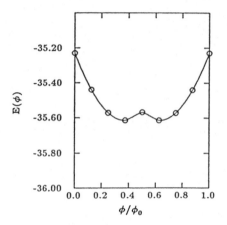

FIG. 3. Ground state energy as function of flux for five semions with $V = 0, B = 0$ on an icosahedron lattice.

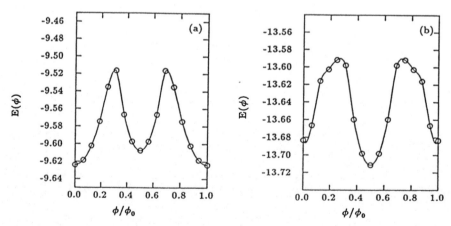

FIG. 4. Ground state energy as function of flux for semions with $V = 0, B = 0$ on a dodechedron lattice. a) 4 semions; b) 12 semions.

effects are relatively large as evidenced by the fact that signatures of pairing are not seen in any system with an odd number of particles. Therefore, we believe it is necessary to study several different signatures simultaneously. Our numerical results lend strong support to the hypothesis that semions pair and form a superfluid due to the effective statistical interaction between the particles. Thus, it seems likely that in the thermodynamic limit, charged semion systems will form a superconducting state. Such a superconductor is expected to exhibit some exotic properties. For example, because time-reversal symmetry is broken in a system with fractional statistics, the magnetization curve for a semion superconductor will depend on the direction of the external magnetic field.

The authors would like to thank A.J. Berlinsky, G. Canright, S.M. Girvin, A.-M.S. Tremblay and X.G. Wen for useful conversations. This research was supported by the Natural Sciences and Engineering Research Council of Canada and by the High Temperature Superconductivity Consortium of the Ontario Centre for Materials Research. In addition, one of the authors (CK) acknowledges support from the Alfred P. Sloan Foundation.

REFERENCES

(a). Present address: Department of Biochemistry and Molecular Biology, Unversity of Manchester, Manchester M13 9PT, UK.

1. Laidlaw and DeWitt, *Phys. Rev. D* **3**, 1375 (1971).

   . J.M. Leinaas and J. Myrheim, *Nuovo cimento* **37B**, 1, (1977).

2. F. Wilczek, *Phys. Rev. Lett.* **48**, 1144 (1982).

   . Y.S. Wu, *Phys. Rev. Lett.* **52**, 2013 (1984).

3. R. Jackiw and C. Rebbi, *Phys. Rev. Lett.* **36**, 1116 (1976).

   . E. Witten, *Phys. Lett.* **86B**, 283 (1979).

   . A. Goldhabar, *Phys. Rev. Lett.* **49**, 905 (1982).

   . G.W. Semenoff, *Phys. Rev. Lett.* **61**, 517 (1988).

4. B.I. Harperin, *Phys. Rev. Lett.* **52**, 1583 (1984).

   . D. Arovas, J.R. Schrieffer, and F. Wilczek, *Phys. Rev. Lett.* **53**, 722 (1984).

5. R.B. Laughlin, *Science* **242**, 525 (1988).

   . V. Kalmeyer and R.B. Laughlin, *Phys. Rev. Lett.* **60**, 1057 (1988).

   . A. Fetter, C. Hanna, and R.B. Laughlin, *Phys. Rev. B* **39**, 9679 (1989).

   . Y.-H. Chen, F. Wilczek, E. Witten, and B.I. Halperin, preprint.

6. F. Wilczek, *Phys. Rev. Lett.* **49**, 975 (1982).

7. D. Arovas, J.R. Schrieffer, F. Wilczek, and A. Zee, *Nucl. Phys.* **B251**, 117 (1985).

8. G.S. Canright, S.M. Girvin, and A. Brass, preprint.

OXYGEN STABILIZED ZERO-RESISTANCE STATES WITH TRANSITION

TEMPERATURES ABOVE 200 K IN YBaCuO

J.T. Chen, L-X Qian, L-Q Wang and L.E. Wenger

Department of Physics and Institute for Manufacturing Research
Wayne State University
Detroit, MI 48202

E.M. Logothetis

Ford Motor Company
Dearborn, MI 48121

The zero-resistance states in mixed-phase $Y_5Ba_6Cu_{11}O_y$ samples with transition temperatures in the range of 235 K to 265 K are found to be thermally recycleable when the samples are enclosed in an oxygen atmosphere. The nonlinear current-voltage characteristics and their sensitivity to small magnetic fields near the transition temperature resemble the behavior of superconducting microbridges.

INTRODUCTION

Since the discovery of the LaBaCuO[1] and YBaCuO[2,3] superconductors, the superconducting transition temperature, T_c, in a stable phase material has been increased to approximately 125 K in the TlBaCaCuO system.[4] However over the past two years, there have been numerous reports and publications of sharp resistive transitions[5-10] (some even with "zero" resistance) and other superconducting-like phenomena[11-14] at much higher temperatures. These higher-temperature, superconducting-like phenomena generally suffer from an inability to survive thermal cycling making it difficult to perform detailed studies and experimental measurements necessary for a better scientific understanding of these phenomena and for the identification of the structure and/or composition responsible for the higher T_c.

In our previous work,[11] similar partial resistive drops were observed in some mixed-phase $Y_{1.8}Ba_{0.2}CuO_x$ and $Y_5Ba_6Cu_{11}O_y$ samples at temperatures above 150 K, also only for a few thermal cycles. During subsequent work, our magnetization studies (unpublished) on several $Y_5Ba_6Cu_{11}O_y$ samples above 200 K indicated a magnetic field-thermal history dependence similar to flux-trapping in a type-II superconductor, albeit the magnetic response had a positive, paramagnetic-like background. Although this magnetic effect became progressively weaker after several thermal cycles in the magnetometer, this flux-trapping-like effect reappeared after the samples had been stored in air for a period of six to nine months, suggesting that oxidation near room temperature may aid in the stabilization of the higher-temperature, superconducting phases. Consequently by utilizing an additional low-temperature oxygenation process and maintaining an oxygen environment during the measurements, the stability of the "zero" resistive states above 200 K have been significantly improved in mixed-phase YBaCuO materials. This improved stability against thermal cycling has correspondingly permitted detailed measurements of the temperature-dependent resistance, contact potentials, bias current dependence, current-voltage (I-V) characteristics in zero and applied magnetic fields, and the temperature-dependent magnetization curves. In this paper, a detailed description of the sample fabrication, the low-temperature oxygenation process, and the experimental results are presented.

EXPERIMENTAL METHODS

Sample preparation

Over twenty samples with nominal compositions of $Y_5Ba_6Cu_{11}O_y$, $Y_1Ba_2Cu_3O_{7-\delta}$, and $Y_1Ba_2Cu_4O_{8-z}$ were synthesized by conventional ceramic techniques from Y_2O_3, $BaCO_3$, and CuO powders. The unreacted powders were calcined at 925°C (±25°C) for 10 hours in either an oxygen atmosphere or in air. This calcination step followed by subsequent grinding of the powder was repeated until a dark black color and a fine crystallite texture was obtained. This reacted black material was ground into a powder, which was then pressed into a 3/4-inch disk-shaped pellet at approximately 50,000 psi. The pellet was sintered between 800°C and 900°C for 24 hours in flowing O_2, cooled to room temperature over a period of 6 hours, and then reground. After pressing into another pellet at 50,000 psi, this pellet was sintered again in O_2 between 800°C and 900°C for another 24 hours and cooled to room temperature over a six hour period.

Low-temperature oxygenation process

The pellets were subsequently cut into several small bar-shaped pieces of approximately 2 mm x 1 mm x 1 mm dimensions and subjected to two different oxygenation processes. (i) Samples (denoted by LP) were placed inside a glass desiccator with a moisture absorbing agent (anhydrous $CaSO_4$) at an oxygen partial pressure of one atmosphere. These samples in the glass desiccator were slightly heated to temperatures of 50°C to 70°C by using an infrared heating lamp for at least one week and only removed from this environment when measurements were to be made at lower temperatures. (ii) Other samples (denoted by HP) prepared by similar steps described in the preparation section were placed in a high pressure vessel filled to an oxygen pressure of 130 atmospheres. After an oxidation anneal for 96 hours at 150°C and cooling back to room temperature over a period of several hours, these samples were removed from the vessel and placed inside the O_2-filled desiccator to prevent possible deterioration in air at room temperature. In addition to the storage of the samples in an oxygen environment (70°C, one atmosphere of oxygen), experimental measurements of the electrical resistance and magnetization were performed with samples surrounded by an oxygen environment. This additional requirement of an oxygen measuring environment not only appears to prevent the degradation of the superconducting properties, but even improves the superconducting properties of sample LP-1 during the first few thermal cycles between 150 K and 300 K.

Material characterization

Although all three compositions show diamagnetic-like deviations from the paramagnetic background in the temperature-dependent magnetization, so far zero resistance phenomenon was observed only in the $Y_5Ba_6Cu_{11}O_y$ samples. The x-ray diffraction pattern for the $Y_5Ba_6Cu_{11}O_y$ samples showed $Y_1Ba_2Cu_3O_{7-\delta}$, Y_2BaCuO_5, and CuO as the predominate phases with smaller unidentified peaks at $2\Theta \approx 37.5°$ and $43.8°$ as shown in Fig. 1. This identification is consistent with energy-dispersive x-ray analysis indicating the presence of these three materials. Unfortunately most microstructural analytical techniques, such as SEM and TEM, are *in-vacuo* techniques resulting in the loss of oxygen that appears to be necessary for the stabilization of the higher-T_c phases. Thus further structural analyses have been handicapped to date. The XRD pattern also shows a high degree of (00*l*) orientation for the $Y_1Ba_2Cu_3O_{7-\delta}$ phase in the direction of the force applied during the formation of the pellet, i.e. perpendicular to the top and bottom surfaces of the pellet. It is unclear at the present time what role the mixed-phases and the (00*l*) orientation play in leading to the formation of the higher-T_c superconducting phase through the oxygenation process.

Measurement techniques

Detailed electrical measurements are made by using six cold-pressed indium contacts and gold leads as shown in the inset of Fig. 2. This multiple-lead arrangement permits resistance and current-voltage (I-V) measurements for several different paths as well as checking the contacts and leads for possible experimental artifacts. For example, the resistance along the sample's surface can be obtained by sending current I_{EF} from E to F and measuring the voltage V_{BC} as in a conventional four-lead method. If a sample is superconducting, the resistance associated with the lead and contact at E can be obtained from a three-lead measurement, such as voltages V_{EB}, V_{EC}, or V_{ED} for a current I_{EA}. By comparing these values, an easy check is made to verify whether the sample has zero resistance or not. The temperature dependence of a contact resistance can also be monitored by using a such three-lead measurement in order to eliminate the possibility of an open or highly resistive contact. In addition, a two-lead measurement, such as V_{EA} and I_{EA}, permits a determination of whether any section of the sample is not probed by the voltage leads in the four-lead arrangement due to an insulating section shunting off the bias current.

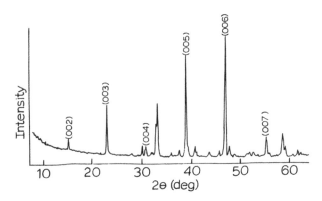

Figure 1. X-ray diffraction pattern of a $Y_5Ba_6Cu_{11}O_y$ sample after oxygenation process (i). The pattern indicates a preferential $(00l)$ orientation of the $Y_1Ba_2Cu_3O_{7-\delta}$ crystallites with the c-axis perpendicular to the surface of the sample.

The total bias current entering the sample is monitored continuously at all temperatures and is maintained at a stability to better than 10^{-4} as determined by the voltage across a standard resistor in series with the sample, control resistors, and a 12-V storage battery. The minimum detectable voltage is 20 nV using a Keithley 180 nanovoltmeter and a X-Y chart recorder. To prevent sample degradation, the resistive probe is inserted inside a metallic dewar filled with one atmosphere of O_2 gas. The probe is then cooled by either liquid nitrogen (LN_2) or a dry ice/acetone mixture surrounding the tail section of the metallic dewar. This cooling process typically took 6 to 24 hours to reach the lowest temperatures in order to minimize the thermal emf fluctuation. The temperature of the sample is measured by a diode sensor thermally mounted on the copper resistance probe with an estimated accuracy of 2 K. However, there could be additional temperature uncertainty up to 10 K because of the physical separation between the thermometer and the sample. In addition, the earth's magnetic field is reduced to less than a few mOe inside the tail section by a mu-metal shield.

RESULTS AND DISCUSSION

One of five $Y_5Ba_6Cu_{11}O_y$ samples (labeled LP-1) following the oxygenation of step (i) consistently showed a zero-resistive T_c in the temperature range of 235 K to 250 K for various electrical paths and different thermal cycles. For current flowing along the surface (in the direction *perpendicular* to the applied force used in pressing the pellets), the *first* four-probe resistance (V_{BC}/I_{EF}) measurement showed sharp partial resistive drops at approximately 90 K, 235 K and 255 K when a dc bias current of 5 µA was applied. These

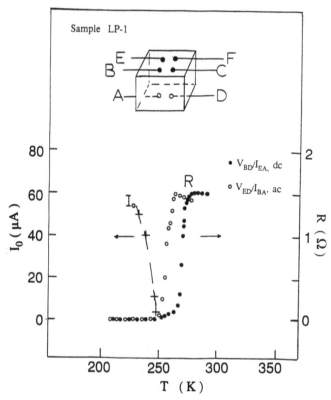

Figure 2. The zero-voltage current I_o (+), left scale, and resistance R, right scale, versus temperature for a $Y_5Ba_6Cu_{11}O_y$ sample (LP-1). The inset shows the electrical contact arrangement for the sample. The solid dots represent resistance (V_{BD}/I_{EA}) data measured during the 6th thermal cycle with a dc bias current of 10 µA and the open circles represent resistance (V_{ED}/I_{BA}) data measured during the 8th thermal cycle with a 7 mHz, 9.3 µA$_{p-p}$ ac current.

resistive drops indicated that several possible superconducting phases might be present in this sample. In addition, the resistance of the sample decreased in each of the first three thermal cycles. In order to measure the resistive path through the sample's body, the electrical lead arrangement was changed so that the current flowed in a direction *parallel* to the applied force used in pressing the sample, e.g. I_{EA}. The resistance measurements for this current direction resulted in zero resistance at ~250 K as shown in Fig. 2 for the lead arrangement of V_{BD} and I_{EA}. It should be noted that the current was slowly increased from zero to a maximum value at each temperature and then decreased back to zero in order to account for thermal emfs, to minimize Joule heating, and to prevent possible flux trapping (and the destruction of a superconducting zero-resistance state) that might arise when a weakly-coupled superconductor is cooled through its transition in the presence of a steady dc current. To minimize the inaccuracy in the thermal emf determination and still measure the dc transition, this procedure was subsequently replaced by a low frequency (7 mHz) ac method utilizing a function generator and monitoring the ac voltage with the nanovoltmeter. As shown in Figs. 3 and 4, the resistive transition was clearly distinguishable from the thermal emf background by this method. The resistance data for the ac measurements (see Figs. 2-4) including for _different_ electrical paths and thermal cycles indicate a sharp resistance transition in the _same_ temperature range of 250 K to 260 K as for the dc mode of measurement. The stability of the ac bias current was also monitored and showed no indication of being reduced at any temperature as seen in Fig. 3(b). To ensure that the temperature was indeed above 200 K, a

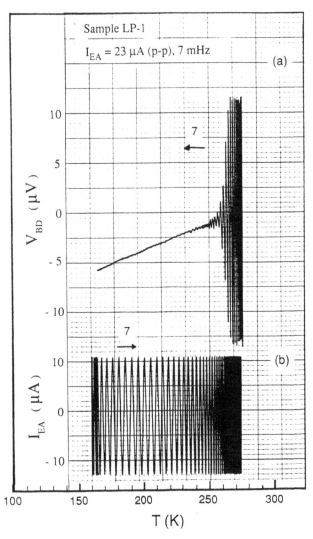

Figure 3. (a) This curve represents the voltage V_{BD} vs temperature using a 7 mHz, 23 μA_{p-p} ac current during the 7th thermal cycle of sample LP-1. Note that the nonzero voltage seen below 250 K is due to the thermal emf. (b) The amplitude of the ac current I_{EA} is displayed in the lower curve during the warming part of this thermal cycle.

bath of acetone with dry ice was used to replace LN_2 as the coolant. The resistive transition from this measurement with the dry ice-acetone bath (lower curve in Fig. 4) occurred in the same temperature region as those using LN_2, even for two different paths. An additional advantage for using an acetone/dry ice bath is that by changing the amount of dry ice in the acetone bath, the temperature can be varied extremely slowly to achieve a nearly perfect thermal equilibrium condition (less than 0.2 K/hr variation) for reliable current-voltage characteristics (Fig. 5) and magnetic field dependent voltage measurements (Figs. 6 and 7) in the temperature range of 220 K to 290 K. The zero-voltage current I_0 using the 20 nV criterion is also shown in Fig. 2 with the maximum I_0 being >45 µA at 240 K, which corresponds to an experimental resistance resolution of 0.44 mΩ (= 20 nV/45 µA) and a

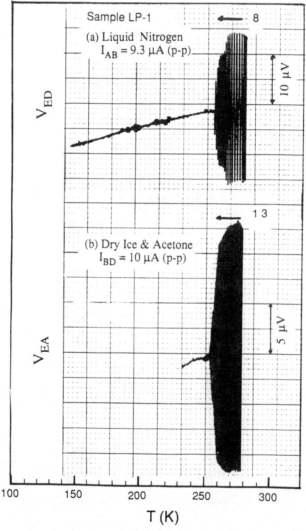

Figure 4. Measurements of the resistance for sample LP-1 by using different coolants: (a) LN_2 with electrical contact arrangement (V_{ED}/I_{BA}) and (b) a dry ice-acetone mixture with electrical contact arrangement (V_{EA}/I_{BD}). Note the large density of voltage traces in the lower curve which indicate an extremely slow cooling rate. These resistance curves were measured during the 8th and 13th thermal cycles, respectively.

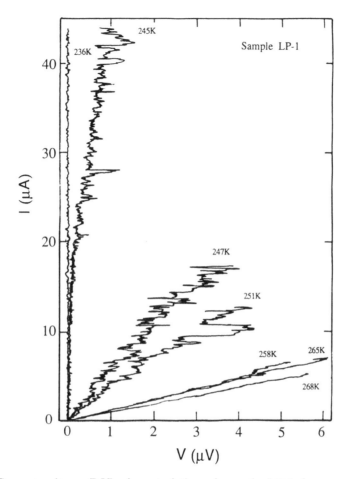

Figure 5. Current-voltage (I-V) characteristics of sample LP-1 for several different temperatures. The voltage structures in the transition region are much larger than the instrumental voltage fluctuation of approximately 20 nV as seen at 236 K.

resistance drop at least three orders of magnitude from the normal-state resistance. The zero-resistance phenomenon in this sample was observable over a period of eleven days and 29 thermal cycles, finally disappearing as a result of excessive Joule heating arising at the contacts and in the sample when the sample was in a normal resistive state.

The current-voltage (I-V) characteristics of sample LP-1 measured along the direction of the applied force during pressing are shown in Fig. 5 for several temperatures near the transition. In the transition region, a temperature-dependent nonlinearity is clearly distinguishable from the ohmic behavior above 258 K. The I-V curves also show considerable voltage fluctuations in the transition region which are larger than the instrumental resolution of 20 nV. These voltage fluctuations reflect instabilities near T_c consistent with the behavior of a network of weakly-coupled superconducting junctions or microbridges. Moreover, the voltage is sensitive to both the bias current I_b and the magnetic field H when I_o is small. Figures 6 and 7 show the voltage change versus applied magnetic field for different dc bias currents. One notes that the voltage changes are more sensitive to the field as the bias

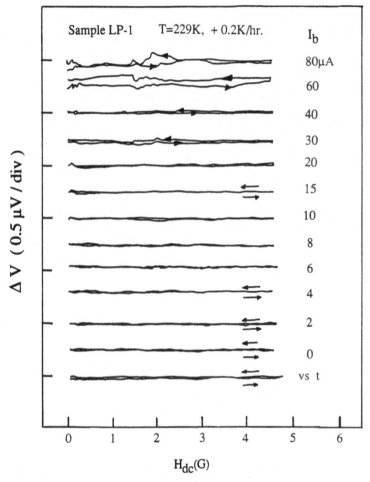

Figure 6. The dc voltage change versus dc magnetic field for several different bias currents at 229 K ($T<T_c$) for sample LP-1. Each voltage-change curve is displaced vertically for clarity. The bottom curve shows the voltage change as a function of time.

current increases and as the temperature approaches T_c. However, no voltage dependence upon magnetic field is detected at higher temperatures where the I-V characteristics are ohmic. Again this magnetic field effect upon the I-V characteristics is consistent with a network of weakly-coupled superconducting junctions or microbridges.

The use of the I-V characteristics measurements is especially beneficial in electrical measurements of very low resistance samples where conventional constant-current resistance measurements may require large bias currents for voltage detection of the superconducting transition. For example, a sample with a normal-state resistance of 5 $\mu\Omega$ and a critical current of 10 mA would have only an equivalent voltage change of 50 nV for a bias current of 10 mA when going from the normal-to-superconducting state. This 50 nV change is comparable to the limit of the sensitivity of even a well-designed electrical circuit and consequently nearly undetectable as a function of temperature. In addition, the thermal voltage fluctuations caused by the temperature controller may further increase the limitations of the voltage sensitivity. As an alternative approach, the I-V characteristics measurements in a quasi-static equilibrium condition using the dry ice-acetone mixture are found to be ideal for determining the superconducting transition of very low resistance samples in the 200 K to 300 K range. Two $Y_5Ba_6Cu_{11}O_y$ samples (HP-1 and HP-2) which were oxygenated by the high-pressure

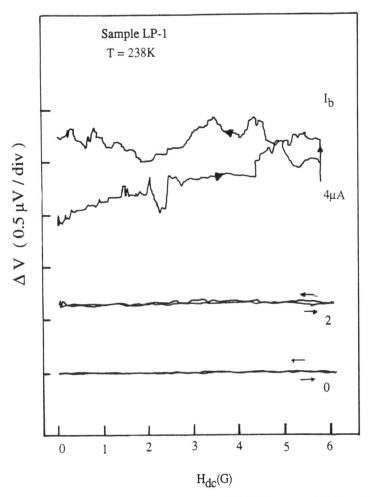

Figure 7. The dc voltage change versus dc magnetic field for several different bias currents at 238 K (T≤T_c) for sample LP-1.

process (ii) had normal-state resistances of 3 mΩ and 4 μΩ, respectively. To obtain the necessary temperature stability, the tail section of the metal dewar was first immersed in a dry ice/acetone bath with the sample cooling to a temperature of 200 K overnight. The sample was then allowed to slowly warm to 290 K over a period of more than 30 hours as the dry ice/acetone bath naturally evaporated. As a result of this approach, the thermal voltage fluctuations were less than the 20 nV instrumental resolution. The I-V characteristics for these samples are shown in Figs. 8 and 9 for temperatures above and below T_c (≈ 265 K). The critical current I_0 for sample HP-1 is shown in Fig. 8 to increase rapidly when the temperature decreases with a maximum I_0 of greater than 4.3 mA (20 nV criterion) at 240 K and below (e.g. at 221 K as shown in Fig. 8). This I_0 value is equivalent to a resistance of less than 4.7 μΩ and represents a resistance change of nearly three orders of magnitude from the 3 mΩ value determined at 283 K from the ohmic I-V behavior. As shown in Fig. 9 for sample HP-2, a zero-voltage current of approximately 10 mA is clearly observable at 254 K in contrast to the ohmic behavior at 305 K. In between these two temperatures, the nonlinearity progressively diminishes as the temperature increases. The nonlinearity observed at 254 K clearly leads to a voltage displacement from the ohmic behavior which is larger than the fluctuation voltage of about 30 nV. The I-V characteristics at 254 K in Fig. 9 indicate why it is difficult, if not nearly impossible, to obtain zero resistance when the temperature is changing continuously as in a resistance vs temperature measurement. The thermal voltage fluctuation can easily be greater than 100 nV if the temperature is not stabilized properly and consequently would require a bias current of greater than 25 mA to measure a resistance change of 4 μΩ at the transition. Such a bias current is greater than the I_0 of 10 mA, and hence only a resistance step may be observable.

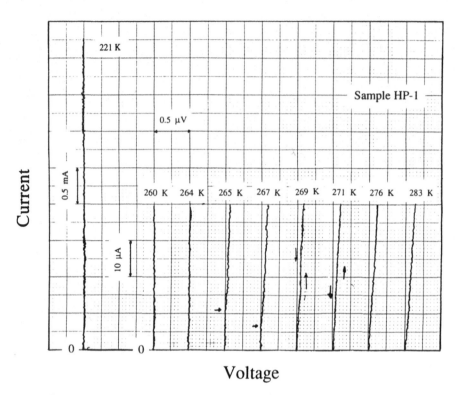

Figure 8. Current-voltage characteristics of sample HP-1 at temperatures below and above the transition temperature of about 265 K. Note that the current scale for the 221 K data is larger than for the other temperatures.

Magnetic measurements on a smaller section of sample LP-1 as well as other samples from the same batch - $Y_5Ba_6Cu_{11}O_y$ oxygenated by the low-pressure, low-temperature process (i) - were performed to determine if diamagnetic contributions (Meissner effect) associated with zero resistance could be detected. The magnetization-versus-temperature curves (see Fig. 10) showed hysteretic behavior and diamagnetic-like deviations from their paramagnetic background in the temperature range of 150 K to 270 K depending upon the applied magnetic field and thermal cycling history. The hysteretic behavior is evidenced in the magnetization data of LP-1 (see Fig. 10(a)) taken after cooling the sample to 150 K in zero magnetic field, then applying a field of 500 Oe, and measuring the magnetization while warming the sample. Below 270 K, this so-called zero-field-cooled (zfc) magnetization data showed a significant difference from the field-cooled (fc) magnetization data which was taken during cooling the sample from 310 K to 150 K in a field of 500 Oe. Similar hysteretic magnetization behavior has also been observed on several $Y_5Ba_6Cu_{11}O_y$ samples oxygenated by the high-pressure process (ii) with diamagnetic-like deviations, albeit on top of a larger paramagnetic background. (See Fig. 10(b).) The strong temperature correlations of these magnetic data to the electrical resistance results provide some collaborative evidence for the onset of superconductivity in these samples at temperatures above 200 K. The lack of an absolute diamagnetic response could be due to the microstructure of these mixed-phase materials. Since the penetration depth of the 90 K superconducting phase is on the order of tenths of microns, it is conceivable that the penetration depth of a 200 K superconductor might

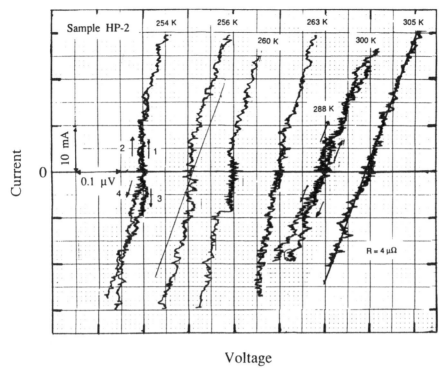

Figure 9. Current-voltage characteristics of sample HP-2 at temperatures below and above the transition temperature of about 265 K. Note that the voltage fluctuations are comparable to the instrumental 20 nV fluctuations. The numbers accompanying the curves at 254 K represent the chronological order that the curves were traced.

be on the order of microns which is comparable to the grain size. Alternatively, the superconducting path may be more filamentary along the grain boundaries or of a reduced dimensionality. Thus the lack of diamagnetism simply implies that the observed phenomena, if related to superconductivity, is not bulk-like.

CONCLUSIONS

Thermally recycleable zero-resistance states with superconducting transition temperatures above 200 K are observed by electrical resistance and I-V characteristics measurements in mixed-phase $Y_5Ba_6Cu_{11}O_y$ samples treated by a low-temperature (<150°C) oxygenation process. The stability of the higher T_c phase(s) is maintained by enclosing the samples in an oxygen atmosphere during electrical and magnetic measurements. Although a full Meissner effect is not obtained in these samples, diamagnetic-like deviations and hysteretic behavior are observed to occur at the same temperatures as the resistive transitions.

ACKNOWLEDGEMENTS

The authors wish to acknowledge Professor Simon Ng for performing the TGA measurements, R.E. Soltis for help with the high pressure materials preparation, R. Ager, S. Shinozaki, J. Hangas, and C.R. Peters for materials characterization, and Professor R.L. Thomas of the WSU-IMR and Dr. M.A. Roberts of Ford for their continuous support of this research over the past two years.

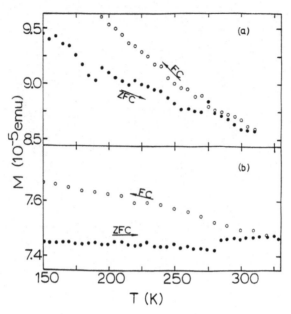

Figure 10. (a) The temperature-dependent magnetization for a small section of the $Y_5Ba_6Cu_{11}O_y$ sample (LP-1). Significant irreversibility in the magnetization is observed below 270 K between zero-field-cooled (zfc) and field-cooled (fc) data for H=500 Oe. (b) The temperature-dependent magnetization for a $Y_5Ba_6Cu_{11}O_y$ sample from the HP process in a field of 300 Oe. It should be noted that these samples are also surrounded by an oxygen atmosphere during the magnetization measurements.

REFERENCES

[1] J.G. Bednorz and K.A. Müller, *Z. Phys. B* **64**, 189 (1986).

[2] M.K. Wu, J.R. Ashburn, C.J. Tong, P.H. Hor, R.L. Wong, L. Gao, Z.J. Huang, Y.Q. Wang and C.W. Chu, *Phys. Rev. Lett.* **58**, 908 (1987); P.H. Hor, L. Gao, R.L. Meng, Z.J. Huang, Y.O. Wang, K. Forester, J. Vassiliow and C.W. Chu, *Phys. Rev. Lett.* **58**, 911 (1987).

[3] R.J. Cava, B. Batlogg, R.B. van Dover, D.W. Murphy, S. Sunshine, T. Siegrist, J.P. Remeika, E.A. Rietman, S. Zahurak and G.P. Espinosa, *Phys. Rev. Lett.* **58**, 1676 (1987).

[4] S.S. Parkin, V.Y. Lee, E.M. Engler, A.I. Nazzal, T.C. Huang, G. Gorman, R. Savoy and R. Beyers, *Phys. Rev. Lett.* **60**, 2539 (1988).

[5] L.C. Bourne, M.L. Cohen, W.N. Creager, M.F. Crommie, A.M. Stacy and A. Zettl, *Phys. Lett. A* **120**, 494 (1987).

[6] C.Y. Huang, L.J. Dries, P.H. Hor, R.L. Meng, C.W. Chu and R.B. Frankel, *Nature* **328**, 403 (1987).

[7] S.R. Ovshinsky, R.T. Young, D.D. Alfred, G. DeMaggio and G.A. Van der Leeden, *Phys. Rev. Lett.* **58**, 2579 (1987).

[8] H. Ihara, N. Terada, M. Jo, M. Hirabayashi, M. Tokumoto, Y. Kimura, T. Matsubara and R. Sugise, *Jpn. J. Appl. Phys.* **26**, L1413 (1987).

[9] J. Narayan, V.N. Shukla, S.J. Lukasiewicz, N. Biunno, R. Singh, A.F. Schreiner and S.J. Pennycook, *Appl. Phys. Lett.* **51**, 940 (1987).

[10] H.D. Jostarndt, M. Galffy, A. Freimuth and D. Wohlleben, *Solid State Commun.* **69**, 911 (1989).

11. J.T. Chen, L.E. Wenger, C.J. McEwan and E.M. Logothetis, *Phys. Rev. Lett.* **58**, 1972 (1987); J.T. Chen, L.E. Wenger, E.M. Logothetis, C.J. McEwan, W.Win, R.E. Soltis and R. Ager, *Chin. J. Phys.* **26**. S93 (1988).
12. A.T. Wijeratne, G.L. Dunifer, J.T. Chen, L.E. Wenger and E.M. Logothetis, *Phys. Rev. B* **37**, 615 (1988).
13. E.M. Jackson, G.J. Shaw, R. Crittenden, Z.Y. Li, A.M. Steward, S.M. Bhagat and R.E. Glover, *Supercond. Sci. & Technology* (to be published).
14. T. Laegreid, K. Fossheim, E. Sandvold and S. Julsrud, *Nature* **330**, 637 (1987).

Study of High T_c Superconductor/Metal-Oxide Composites

M. K. Wu*

Department of Applied Physics
School of Engineering and Applied Science
Columbia University
New York, NY 10027

C. Y. Huang

Lockheed Missiles & Space Co. Inc.
Research & Development Division
Palo Alto, CA 94304

F. Z. Chien

Department of Physics
Tamkang University
Tamsui, Taiwan

INTRODUCTION

An interesting problem concerning the high T_c oxides is the metastability of certain high temperature superconducting phase. For example, superconducting transitions from 40 K to 80 K were reported in the Y-Sr-Cu-O system [1-4]. However, the superconducting phase has not been determined due to the existence of multiphase in this system. Oda et al. [2] proposed a tetragonal phase and Wu et al. [4] suggested an orthorhombic structure for the superconducting compound of stoichiometry $YSr_2Cu_3O_{6+y}$ (YSCO).

Another important issue concerning the application and understanding of high T_c superconductors is the low critical current densities, J_c, in the presence of magnetic fields. For the Cu-O based systems, J_c decreases rapidly with a field applied perpendicular to the copper oxide planes [5]. This implies that highly oriented material is required for most applications. It is also reported that the critical current densities derived from the magnetization measurements of polycrystalline samples are 10^3 - 10^6 times higher than their transport critical current densities [6]. These results suggest that these materials consist of regions of excellent superconductivity separated by extensive networks of weak links. Other experimental results, such as the evidence for intragrain Josephson junctions and the development of a tail in the resistive transition in the presence of even moderate applied fields, indicate that the phenomenology of the high T_c oxides is significantly different from that of the classical superconductors [7]. It also creates serious problem when trying to define the upper critical field and the critical current [8,9]. Recently, several models have been proposed to explain the broadening of the resistive transition under magnetic fields [8,10,11,12]. However, there are results confirming that high J_c's can be achieved (in zero field) in high quality single crystal films and highly textured bulk polycrystals [13].

In order to explore the possibility of improving the critical current density in the bulk oxide superconducting material and the formation of single phase oxide superconductors, we performed experiments based on the formation of superconducting oxides through the dispersion of fine metal oxide. In this paper, we report two major results in these studies: (1) The preparation of single phase superconducting $YSr_2Cu_3O_{6+y}$ compound by the addition of small amount of MoO_3, and (2) the fabrication of strong flux pinning RE-123/AgO composites using a relatively high temperature annealing conditions.

EXPERIMENTAL

High Tc YSCO compounds were prepared by mixing appropriate amounts of metal oxides (nominal composition of $YSr_2Cu_{3-x}Mo_xO_{6+y}$ with x in the range of 0.02 - 0.05), pressed into pellets, heated at 950°C for 12 hours, and quenched to room temperature (RT). The material was then reground, pressed, reheated to 1150°C for 6 h in O_2, and slowly cooled to RT. High purity alumina crucibles were used in the sample preparation.

The superconducting RE-123/AgO composites were prepared according to that reported [14]. Electrical resistivity measurements were made with the conventional 4-probe technique. The DC magnetic moment measurements were made with a Quantum Design MPMS SQUID magnetometer. High field measurements were performed at Francis Bitter National Magnetic Laboratory at MIT. A standard 4-probe using pulse current was used to determine critical current density at zero field. Structural and phase determinations were made by x-ray diffraction. A Cambridge 250 SEM, equipped with a Kevex EDS system was employed for microstructure study.

RESULTS AND DISCUSSION

A. Study of High T_c Y-Sr-Cu-O

The temperature dependence of resistance of YSCO with small amount of Mo is shown in Figure 1. The superconducting onset temperature is about 80 K, which is in agreement with that of the magnetic measurement, as shown in the inset. A linear temperature dependence of R before the onset of superconductivity was observed. Based on the magnetic signal, it was estimated that the superconducting phase is about 60 percent. Without the addition of MoO_3, the 80 K superconducting phase does not form when the samples are prepared at a temperature lower than 1300°C. This indicates that the 80-K phase in the YSCO compound is thermodynamically stable only at higher temperatures. However, with the addition of MoO_3 during sintering resulted in the reduction of the sintering temperature. Detailed x-ray diffraction patterns of the compound are shown in Fig. 2. A tetragonal structure with lattice constants a = 3.818 A and c = 11.555 A was determined using TREOR program [15]. The atomic positions of space group P4/mmm and the isotropic Debye Waller factors listed in table I were determined with program based on the stoichiometry of $YSr_2Cu_3O_{6+y}$, where the values of the Debye Waller factors were confined between 0.3 and 4.0 (the corresponding atomic rms displacement are 0.05 A and 0.25 A). The final R factor is 12.6%. Atoms O(3) and Cu(1) on the basal plane can be considered relatively unstable due to their large Debye Waller factors. Cu(1) were draged along by the unstable O(3). The distance between the CuO_2 planes is determined to be 3.305 A which is smaller than that of the $YBa_2Cu_3O_{6+y}$, 3.388 A.

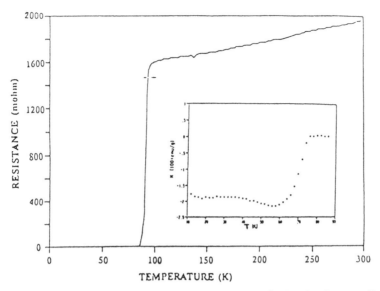

Figure 1. Temperature dependence of resistance of $YSr_2Cu_3O_{6+y}$. Inset is the temperature dependence of the magnetic susceptibility.

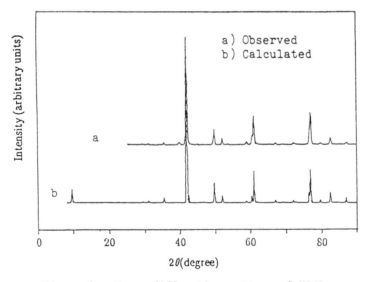

Figure 2. X-ray diffraction pattern of YSCO.

The existence of high temperature superconducting phase ($T_c > 77K$) with tetragonal "123" structure in YSCO is rather unusual. It is well known that in YBCO system high T_c phase exists only in the orthorhombic phase. This result further confirms the existence of an ordered copper oxygen chain is not essential for high temperature superconductivity [16]. The presence of Mo element to the formation of high T_c phase may be related to the enhancement of phase stability through the dispersion of fine metallic oxide as usually observed in the oxide dispersion strengthening alloys [17]. It is interesting to note that similar results have been observed in the Bismate and Thallate [18] with the addition of lead oxide. The results of the enhencement in electrical and magnetic properties in the 123/AgO composites, which is to be discussed later, may also come from the same origin.

TABLE I. Positional parameters and isotropic Debye-Waller factor for the tetragonal structure of $YSr_2Cu_3O_{6+y}$ based on space group P4/mmm and a = 3.818 A. and c = 11.555 A.

Atoms	Wyckoff	x	y	z	B	Occupancy notation
Y	1d	0.5	0.5	0.5	0.3	1
Sr	2h	0.5	0.5	0.197	0.7	1
Cu 1	1a	0	0	0	2.6	1
Cu 2	2g	0	0	0.357	1.3	1
O 1	2f	0	0	0.148	1.1	1
O 2	4i	0	0.5	0.372	4.0	1
O 3	2g	0	0.5	0	4.0	0.6

B. Study of RE-123/AgO Composites

The optimal heat treatment conditions for the formation of strong flux pinning RE-123/AgO composites are shown in Table II. It was found that their sintering temperatures are different for different RE-123 compounds. In general, the temperatures required are higher than those needed to form the corresponding RE-123 compounds, and they do not depend on the weight ratio of RE-123 to AgO. It is interesting to note that the annealing temperature depends on the ionic size of the rare earth element. The empirical trend is that larger rare earth ion requires higher temperature. It is also found that these strong flux pinning RE-123/AgO composites form only in a narrow annealing temperature range.

The sensitivity of the observed pinning effect to the sintering temperature could be due to the change of the interdiffusion rate of silver metal, which is from the decomposition of AgO at temperature above 220°C, with the superconducting particles, particle surfaces and particle grain boundaries. If interdiffusion of Ag is critical, then AgO particle size and RE-123 grain size will be important variables as well as the reaction temperature and time. Our preliminary results indeed show that finer starting material particles give stronger pinning effect.

The M-H hysteresis loops for 3Y123/AgO at various temperature are displayed in Fig. 3. For comparison, we have also obtained the M-H loops for nY123/AgO (n=5 and 7), which are identical to those for n=3. Therefore, it is most likely that the pinning samples is independent of n (< 7). This implies that the residual magnetization in a sample with a smaller value on n, there is more Ag in the sample, and then the residual magnetization in the Y123 part of the sample is larger than that in the sample with larger n. The temperature dependence of the residual magnetization is depicted in Fig. 4, and as expected, the residual magnetization increases with decreasing temperature. As shown in the figure, similar results are also observed in the 3Eu123/AgO and 3Gd123/AgO samples. Typical temperature dependence of the resistance of RE-123/AgO composites is very much similar to that of the corresponding RE-123 compound except that its overall resistance is lower. The superconducting transition tempera-

Table II. RE-123 Superconductors and RE-123/AgO Composites

Sample	Annealing Temp. (C)	T_c	R(R.T.) (µΩ-cm)
Nd-123/AgO	1030	90	423
Nd-123	970	89	520
Sm-123/AgO	1035	91	187
Sm-123	960	93	486
Eu-123/AgO	1025	96	530
Eu-123	950	96	526
Gd-123/AgO	1020	93	1125
Gd-123	950	95	445
Dy-123/AgO	965	95	52
Dy-123	955	96	516
Ho-123/AgO	960	95	39
Ho-123	950	91	413
Er-123/AgO	960	91	165
Er-123	955	92	3205
Y-123/AgO	990	93	252
Y-123	950	94	803

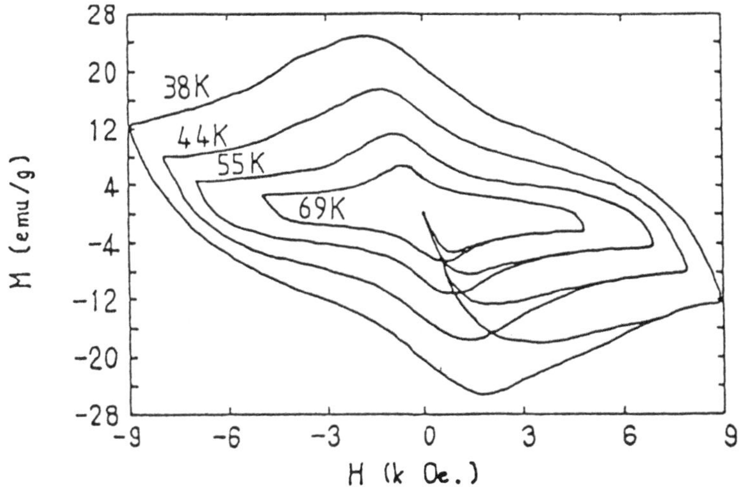

Figure 3. The M-H Hysteresis loops of 3Y123/AgO at various temperature.

ture T_c of the composite remains almost unchanged compared with that of the starting RE-123 compounds. The decrease in the normal state resistivity with the addition of silver oxide suggest a lowering of contact resistances between grains. Near identical T_c's suggest that grain interiors experienced little change from the starting RE-123 compounds. One important advantage derived from these RE-123/AgO composites is the reduction of overall contact resistance.

We have also measured the resistance of the Y-123/AgO, Gd-123/AgO and Eu-123/AgO composites near T_c as a function of magnetic field in a Bitter Magnet. Figure 5 demonstrates the field dependence of the resistance near T_c (91.6K) for 2Y123/AgO. Similar results were also observed in the other RE-123/AgO composites studied. To study in detail the resistive transition, we compare the data with the model proposed by Tinkham [11]. This model treats the resistance at the transition as arising from phase slippage at a complicated network of channels, where floxons are slipping past one another over energy barriers between local minima. Based on the theory of a single heavily damped current-driven Josephson junction by Ambegaokar and Halperlin [19], he found that the resistive transition width, $T_c = T_c(0) - T_c(H)$, scales as $B^{2/3}$. To compare our data with the theory, we deduced the data at $R/R_n = 0, 25, 50$ and 75 % for field up to 20 T. The results are displayed in Fig. 6. All of them show that T_c $H^{2/3}$ within our experimental errors. Similar results are also obtained for other samples. From Tinkham's theory, we estimate that $J_{co}(0)$ 9×10^6 A/cm^2, which is a reasonable value for the 123 superconductor. A particular significant result is the unusually large pinning effect in these composites, as further evidenced by the still zero resistance state of several samples at about 80K and under a magnetic field of 8 T [14].

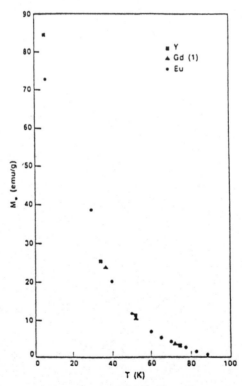

Figure 4. Temperature dependence of residual magnetization for nY123/AgO (n=3,5,7), 3Gd123/AgO and 3Eu123/AgO.

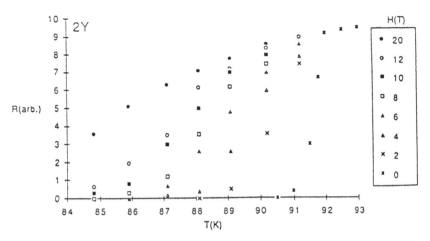

Figure 5. Temperature dependence of R for 2Y123/AgO at various fields.

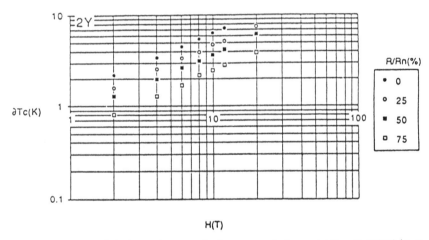

Figure 6. Field dependence of the transition width of 2Y123/AgO.

Figure 7 is the polarized light micrograph for a 3:1 Y-123/AgO. EDS of the electron micrograph of the same sample has identified the presence of Y-123 phase, silver particles, and voids. It shows a microstructure of dispersed silver in a matrix of the Y-123 phase. Intergranular silver was observed in some areas and some voids have been filled up by silver, resulting in the reduction of the normal resistivity. More interesting is the observation of large grain growth in the Ag dispersed composite. As display in figure 7, the grain size of the 3:1 Y-123/AgO composite can be as large as 0.5mm. While the same microscopic picture of a Y-123 pellet reveals average grain sizes of only 20 micron. Similar results are also observed in other composites exhibiting large flux pinning effects. These results suggest that silver could serve as the agent to clean out unwanted nucleation centers, thus allowing the superconducting "123" grains to grow. It is interesting to note that for those samples containing larger grains exhibit large values of residual magnetization and strong pinning. It is also found that silver particles, in the order of micron, are present dispersively inside grains. It is conceivable that these dispersed silver particles may serve as pinning center for the strong flux pinning observed. However, whether the larger grain sizes or the fine dispersion of the silver particles contributes more to the strong pinning has yet to be determined.

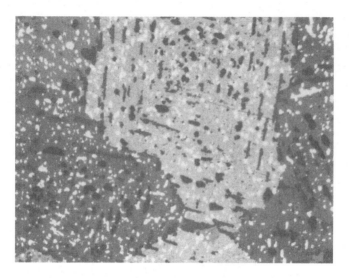

Figure 7. The polarized light micrograph for a 3Y123/AgO.

It has also been shown that the addition of Ag improves the high temperature mechanical properties and reduce the amount of microcracking in higher density sintered specimens [20]. Silver appears to wet the surfaces of the RE-123 quite well, segregating to prior particle boundaries and grain boundaries. A much better workability has also been demonstrated by the inclusion of Ag in RE-123 [20]. It is believed that Ag on boundaries enhances oxygen diffusion. This may be a most necessary feature for bulk material since preliminary data have shown oxygen diffusion to be slow in dense oxide superconductors.

SUMMARY

We have achieved in preparing single phase high T_c Y-Sr-Cu-O compound by the addition of small amount of MoO_3 to the stoichiometry $YSr_2Cu_3O_{6+y}$. The presence of Mo atom reduces the required reaction temperature to form the high T_c phase. On the other hand, the presence of Ag metal in the RE-123 compound helps the grain growth and serves as clean impurities in the grains, results in the strong flux pinning materials. Based on this study, inclusion of "proper" metal oxides in the formation of high T_c oxides will enhance the physical properties of the original compound. More detailed studies are necessary to have better understanding of the underlying mechanism for the formation these high T_c oxides.

ACKNOWLEDGEMENTS

The authors like to acknowledge P. T. Leong, H. Chou, H. H. Tai, T.E. Dann, W.Y. Yu, J.B. Lee, K.Y. Pan and T.L. Kuo for their contributions. The work at Columbia was supported by NASA grants NCC8-6, IBM through its Materials and Materials Processing Grant, and the Grant from New York State Institute on Superconductivity. The work at Lockheed was supported by the Lockheed Independent Research funds and at Tamkang University by R. O. C. National Science Council Grant NSC78-0208-M032-13.

REFERENCES

* Also at Physics Department, National Tsing-Hua University, Hsin-Chu, Taiwan, Republic of China.

[1]. Mei, Y., et al., Novel Mechanisms of High Temperature superconductors. ed. V. Kresin, and S. A. Wolf, p. 1041 (1987).

[2]. Oda, M., et al., Jpa. J. Appl. Phys. 26, L804 (1987).

[3]. Zhang, Q. R., et al., Solid State Commun. 63, 535 (1987).

[4]. Wu, M. K., et al., Phys. Rev. B37, 9765 (1988).

[5]. Dinger, T. R. et al., Phys. Rev. Lett. 58 2687 (1987).

[6]. Larbalestier, D. et al., J. Appl. Phys. 62 3308 (1987).

[7]. Esteve, D. et al., Europhys. Lett. 3 1237 (1987).

[8]. Yeshrun, Y., and Malozemoff, A. P., Phys. Rev. Lett. 60, 2202 (1988).

[9]. Palstra T. T., et al., Phys. Rev. Lett. 61, 1662 (1988).

[10]. Muller K. A., et al., Phys. Rev. Lett. 58, 1143 (1987).

[11]. Tinkham M. K., Phys. Rev. Lett., 61, 1658 (1988).

[12]. Yeh, N. C. and Tsuei, C. C. Phys. Rev. B39, 9708 (1989)

[13]. Chaudhari, P. et al., Phys. Rev. Lett. 58 2684 (1987). Jin, S. et al., Appl. Phys. Lett. 52, 2074 (1988).

[14]. Huang, C. Y., et al., Modern Phys. Lett. B2, 869 (1988).

[15]. Werner, P. E., et al., J. Appl. Crystallogr. 18, 367 (1985).

[16]. Xiao, G., et al., Phys. Rev. Lett. 59, 1967 (1988).

[17]. Borofka, J. C., Tien, J. K., and Kissinger, R. D., "Superalloys, Supercomposites and Superceramics", ed. by J. K. Tien, Academic Press, Inc., 1989, pp. 237.

[18]. Luo, H. L. et al., Superconductivity and Applications, ed. by Y. H. Kao and H. S. Kwok, Plenum Publishing Co. 1990.

[19]. V. Ambegaokar and B. L. Halperin, Phys. Rev. Lett., 22, 1364 (1969).

[20]. J. K. Tien, private communication.

MAGNETIC ALIGNMENT OF Bi AND Tl CUPRATES CONTAINING RARE EARTH ELEMENTS

F. Chen[a,c], R.S. Markiewicz[b,c] and B.C. Giessen[a,c]

Department of Chemistry (a) and Physics (b) and Barnett Institute (c)
Northeastern University, Boston, MA 02115

INTRODUCTION

Magnetodiffractometry is a new technique in which an intense magnetic field is used to align crystalline grains[1] in preparation for X-ray diffraction analysis. The field orientation greatly simplifies the resulting spectra, similar to using single crystals. This technique was instrumental in identifying the first four-Cu layer superconductor, $TlBa_2Ca_3Cu_4O_x$ (Tl-1234), T_c = 122 K.[2] Furthermore, since well-oriented crystallites produce intense reflections, this technique allows the detection of minority phases.

This technique is only successful for crystals with magnetic anisotropy. Hence, in turn, it can be used to provide information about this anisotropy. In rare earth (RE) substituted $YBa_2Cu_3O_{7-\delta}$ (Y-123), it has been used to analyze crystal fields[3]: it is found that most RE moments are oriented parallel to the c-axis, whereas a few (Er, Yb) lie within the a-b plane. Here, we report an extention of this analysis to RE-substituted Bi- and Tl-based superconductors. We find that the RE moments which align in-plane tend to align along a particular axis, but that this axis is different for the RE 123 materials vs the Bi,Tl-families.

MAGNETIC ALIGNMENT

Alignment occurs because of a magnetic anisotropy of the crystals. In the RE-free superconductors, the anisotropy is associated with the two-dimensional CuO_2 plane conductivity, and the grains always align with c-axis parallel to the magnetic field. In RE-substituted materials, the grain orients with the RE moment parallel to the field. The large RE moments allow good alignment in substantially lower fields. Aligned samples have been used in a variety of experiments, studying anisotropy of magnetization[4], critical currents[5], and optical properties[6]. It has recently been shown that field orientation can be used to significantly enhance critical currents[7].

SUPERCONDUCTIVITY OF RE-CONTAINING COMPOUNDS

Finely ground samples of the superconductors were mixed with an epoxy in a volume ratio of about 1:4. Samples were then held in a magnetic field of between 1-15 T at room temperature for ~20 minutes until the epoxy set. Superconducting transition temperatures were determined from SQUID measurements of the temperature-depentent magnetization.

The materials studied were RE-123, with the RE 100 % substituted for the Y, $Bi_2Sr_2Ca_{1-x}RE_xCu_2O_4$ (Bi-2212), and $Pb_{0.25}TlBa_2Ca_{3.5}RE_{0.5}Cu_4O_4$ (starting composition). In the latter compounds, only part of the Ca can be replaced before the superconducting T_c is depressed. For instance, with RE = Er, the Bi compound had T_c = 82 K for $x \leq 0.2$; 40 K for x = 0.3; and was semiconducting for $x \geq 0.4$ (Fig. 1). This is in approximate agreement with earlier work.[8] In contrast, the degree of orientation saturated rapidly as x was increased. Hence, in most of our work on Bi, fixed values x= 0.3 or 0.5 were employed. For x = 0.3, T_c is partially depressed for the larger RE's (Fig. 2).

The degree of alignment was checked by means of X-ray diffraction using CuK_α radiation in the 2θ range of 20-70°. Materials were studied before and after alignment for comparison.

X-ray patterns of the aligned, disc-shaped samples were taken from the face lying perpendicular to the vertical magnetic field. Typical diffractograms of Bi-2212 are presented in Fig. 3, which contrasts an unoriented sample with both a parallel (Ho) and a normal (Er) aligner. As stated, samples orient with their c-axis parallel or perpendicular to the magnetic field direction, depending on which rare earth element is present in the materials. In the case of elements for which c ∥ B, only the (00ℓ) reflections are seen in the X-ray pattern; by contrast, when c ⊥ B, only (hk0) lines appear.

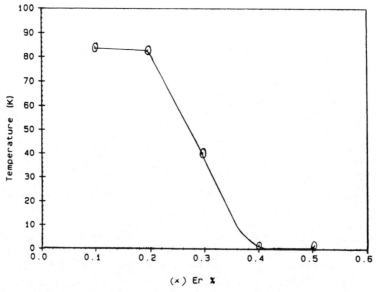

Fig. 1 Decrease of T_c with increasing fraction x of Er relative to Ca in $Bi_2Sr_2Ca_{1-x}Er_xCu_2O_z$.

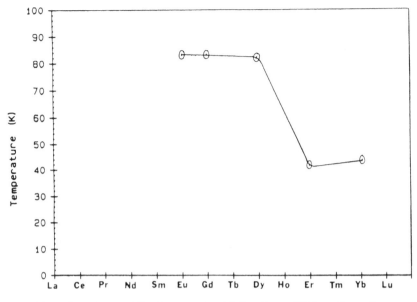

Fig. 2 T_c vs RE element in $Bi_2Sr_2Ca_{0.7}(RE)_{0.3}Cu_{2.2}O_x$.

Table 1. Magnetic alignment behavior of RE-123 and Bi and Tl cuprates containing RE elements, showing orientation of c axis relative to magnetic field.

	1 2 3	Bi	Tl
Ce		∥	
Pr			
Nd	∥	∥	
Sm	∥	∥	
Eu	⊥	⊥	
Gd	⊥	∥	∥
Tb	(∥)	∥	
Dy	∥	∥	∥
Ho	∥	∥	
Er	⊥	⊥	⊥
Tm	(⊥)	⊥	
Yb	⊥	⊥	⊥
Lu		∥	

Note the different behavior of Bi and Tl cuprates with Gd compared to 123.

Table I summarizes the results of our study, separating the various RE into parallel aligners (c ∥ B) and normal aligners (c ⊥ B). Symbols in parentheses were taken from the literature. The degree of orientation may be quantified by an orientation factor $P_{\alpha,\beta}$, which we define as

$$P_{\alpha,\beta} = \left[1 - \frac{(I_\beta/I_\alpha)_{oriented}}{(I_\beta/I_\alpha)_{random}}\right] \times 100\ \%,$$

where α,β refer to particular (h,k,ℓ), (h',k',ℓ') reflections, measured both on an oriented and an unoriented sample, assumed to be random. For

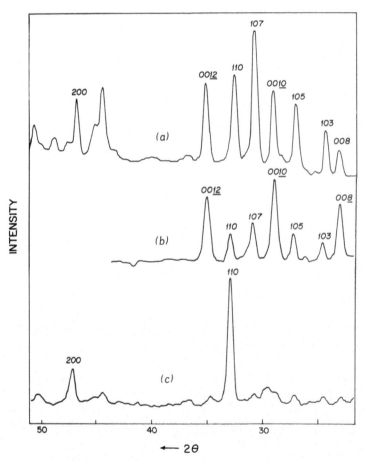

Fig. 3   X-ray diffractograms (Cu-K$_\alpha$ radiation) of Bi-2212 compounds.

(a)   Unoriented $Bi_2Sr_2Ca_{1.5}Er_{0.5}Cu_2O_x$

(b)   Oriented $Bi_2Sr_2Ca_{1.5}Ho_{0.5}Cu_2O_x$ ("parallel" aligner)

(c)   Oriented $Bi_2Sr_2Ca_{1.5}Er_{0.5}Cu_2O_x$ ("normal" aligner) instance, for a normal aligner, α is chosen to be a particular (h,k,0) direction, and β an (00ℓ). For perfect orientation $(I_\beta)_{oriented} = 0$, so $P_{\alpha\beta} = 100$ %. While $P_{\alpha\beta}$ depends weakly on β, by comparing $P_{\alpha\beta}$ for several α's but the same β, we can assess the degree of order in the (a,b)-plane. This is done in Fig. 4 for the normal aligners. The results are quite striking. For the Bi-2212 compounds, there is a pronounced tendency to align along the (110)-direction, whereas for the RE-123's, the preferred axis is the (020). [Note that the difference between (020) and (200) is expected to be small, since our grains are naturally twinned.] In both cases Er shows the least tendency for preferential in-plane orientation. Indeed, in Er-123, there appears to be no preferential orientation within the a-b plane, even though there is a high degree of aligning the c-axis perpendicular to the applied field.

Fig. 5 shows that the degree of orientation also differs among parallel aligners. At the field used (15T) there is only an approximate correlation of P with the RE magnetic moment.

Gd presents a very interesting case: whereas Gd-123 is a normal orienter, Gd-doped Bi and Tl-1212 are both parallel orienters. This may be related to the relatively low degree of orientation found in Gd-doped Bi-1212. It may be that Gd aligns within the a,b plane, but at the relatively low doping levels, $x = 0.3$, the net sample magnetization is still dominated by the tendency of pure ($x = 0$) Bi-2212 to parallel align. Doping dependence studies are planned to test this hypothesis.

We should finally note that this technique is not restricted to conducting compounds. We have found that Yb-211 (the "green phase") is an extremely strong orienter.

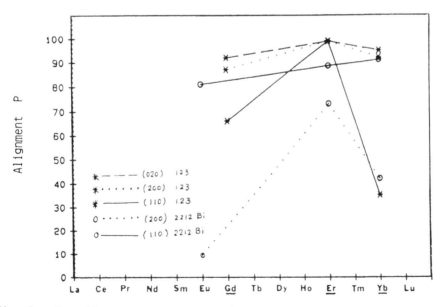

Fig. 4 P = alignment percentage of "normal" RE aligners (c $\perp$ B) for both $Bi_2Sr_2Ca_{0.7}RE_{0.3}Cu_{2.2}O_x$ (=2212) and $(RE)Ba_2Cu_3O_{7-\delta}$ (= 123).

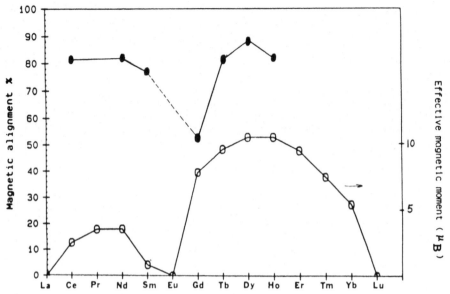

Fig. 5 Alignment percentage P (solid circles) and RE magnetic moment (open circles) for Bi-2212 "parallel" aligners.

ACKNOWLEDGEMENTS

These samples were field-oriented in the superconducting magnet facility of the Francis Bitter National Magnet Laboratory, Cambridge, Mass. 02139 and the magnetization studies were carried out at the Magnet Lab's SQUID facility. The National Magnet Lab is supported at MIT by the NSF. Our research was supported by SDIO under subcontract from Intermagnetics General Corporation.

REFERENCES

1. D.E. Farrell, B.S. Chandrasekhar, M. R. De Guire, M.M. Fang, V.G. Kogan, J.R. Clem, and D.K. Finnemore, Phys. Rev. B36:4025 (1987).
2. P. Haldar, K. Chen, R. Maheswaran, A. Roig-Janicki, N. Jaggi, R. Markiewicz and B.C. Giessen, Science 241: 1198 (1988).
3. J. D. Livingston, H.R. Hart, Jr., and W.P. Wolf, J. Appl. Phys. 64:5806 (1988).
4. R.S. Markiewicz, K. Chen, B. Maheswaran, A.G. Swanson, and J.S. Brooks, J. Phys. Condensed Matt., to be published.
5. R.H. Arendt, A.R. Gaddipati, M.F. Grabauskas, E.L. Hall, H.R. Hart, Jr. K.W. Lay, J.D. Livingston, F.E. Luborsky, and L.L. Schilling, in "High Temperature Superconductivity", ed. by M.B. Brodsky, R.C. Dynes, K. Kitazawa, and H.L. Tuller, Amsterdam, North-Holland (1988), p. 203.
6. F. Lu, C.H. Perry, K. Chen, and R.S. Markiewicz, J. Opt. Soc. Amer. B6:396 (1989).
7. K. Chen, B. Maheswaran, Y.P. Liu, B.C. Giessen, C. Chan, and R.S. Markiewicz, Appl. Phys. Lett. 55:289 (1989).
8. J.-M. Tarascon, P. Barboux, G.W. Hull, R. Ramesh, L.H. Greene, M. Giroud, M.S. Hegde, and W.R. McKinnon, Phys. Rev. B 39:4316 (1989).

The Removal of Organics from Tape Cast $YBa_2Cu_3O_{7-x}$

B.A. Whiteley and J.A.T. Taylor

New York State College of Ceramics
Alfred University
Alfred NY, 14802

ABSTRACT

The decomposition of tape cast $YBa_2Cu_3O_{7-x}$ during the removal of organics was studied using X-ray diffraction (XRD) and thermal gravimetric analyses (TGA). The $YBa_2Cu_3O_{7-x}$ phase decomposes in the presence of carbon dioxide and water produced from the volatilization of organic matter, forming an intermediate carbonate phase. This carbonate phase was found to decompose at higher temperatures upsetting the stoichiometry of the $YBa_2Cu_3O_{7-x}$ phase and producing some barium cuprate phase ($Ba_2Cu_3O_5$).

INTRODUCTION

Tape casting (doctor blading) is a well established technique for fabricating large-area, thin, flat ceramic parts and is most extensively used in fabrication of multilayered electrical components. A tape is produced by dispersing a ceramic powder in a solvent system and adding a binder and plasticizer to give the tape mechanical strength in the green state [1,2]. Thin tapes (typically 1-5 mil thick) are laid on a carrier material and allowed to dry (evaporation of solvents) prior to removal from the carrier material. The dried tape contains a binder and plasticizers making it easy to handle in the green state. These organics can account for 15 to 20 wt% of the green body (depending on the organics being used) and must be totally removed prior to sintering [1,2]. The organics used in the present binder solution are listed in Table 1.

Binders and plasticizers are used extensively in the ceramic industry to impart green strength and flexibility to a body before firing [1,3]. The removal of organics is very dependant on the composition and structure of the organic material, the composition of the local gas environment in and around the material, the rates of diffusion of the decomposition product gases, the furnace gas through the product and the permeability of the material [1,6]. Incomplete pyrolyses of organics can adversely affect subsequent densification, degrade the physical properties and decompose a ceramic material [4].

In the present study the partial decomposition of $YBa_2Cu_3O_{7-x}$ during the volatilization of organics was investigated using thermal gravimetric analyses (TGA)[1] and X-ray diffraction (XRD)[2].

Table 1. Binder-solvent-plasticizer system for tape casting.

Binder	poly(vinyl butyral)
Solvents	toluene-ethanol
Plasticizer	phthalates

EXPERIMENTAL

Reagent grade Y_2O_3, $BaCO_3$ and CuO were combined in the correct stoichiometric ratio to achieve $YBa_2Cu_3O_{7-x}$. The powders were ball milled for 12 hours in deionized water to achieve intimate mixing, dried and calcined at 920°C for 16 hours in flowing air. The calcined powder was then ball milled in hexane or toluene for 72 hours to reduce the particle size.

---

[1]TGA balance, Harrop, Columbus, OH.

[2]Siemens D500, Siemens Energy and Automation, Cherry Hill, NJ.

The milled powder was carefully dried and mixed with the tape casting solution[3] and intimately mixed in a ball mill for a further 6 to 12 hours.

The tape solutions were then doctor bladed on a glass sheet with green thicknesses ranging from 2 to 4 mil. After evaporation of the solvents (drying of the tape) the tapes were peeled from the carrier glass and cut into 1 inch squares. Stacks of 15 to 20 one inch squares were laminated in a die at 55°C and 34.5MPa for 2 minutes.

Thermal gravimetric analyses (TGA) was performed on the laminates to determine an appropriate heating schedule. A setter consisting of 20 laminates was then loaded into the furnace and a pair of samples was removed and quenched in air at predetermined temperatures labeled on the TGA curve in Figure 1.

X-ray analyses was performed on powders of each of the samples quenched at different temperatures using a powder diffractometer with CuK-alpha radiation and a diffracted beam graphite monochrometer.

RESULTS AND DISCUSSION

Thermal gravimetric analyses (TGA) in flowing air of a tape cast and laminated $YBa_2Cu_3O_{7-x}$ material is shown in Figure 1. Samples were quenched at the temperatures indicated on the TGA curve in Figure 1. X-ray plots of the samples quenched at these temperatures are displayed in Figures 2 through 5. The diffraction patterns in Figure 2 show the $YBa_2Cu_3O_{7-x}$ phase remains stable up to 150°C. Figure 1 shows that at approximately 170°C the organics begin to decompose and volatilize according to the general reaction:

$$C_xH_y =====> m\ H_2O + n(C/CO/CO_2)$$

The amount of $C/CO/CO_2$ present depends upon the oxygen partial pressure in the local environment of the sample. From 208°C to 212°C the organics rapidly burn-out of the ceramic matrix, and at 212°C the sample has incurred a weight loss of 9 wt% associated with the rapid decomposition and volatilization of the organics. Figure 6 shows a TGA curve for the binder and plasticizers. A residual ash content of 4 wt% was calculated from this curve. The initial pyrolysis of

---

[3]Binder Solution b73-210, Metoramic Science Inc., Carlsbad, CA.

poly(vinyl butyral)(PVB) is dominated by the elimination of side groups (butyral, hydroxyl)[4]. During early stages of pyrolysis, crosslinking and cyclization of the main polymer chain occurs increasing the temperature required to break down the degradation resistant cyclic-crosslinked structure, leading to a higher residual ash content.

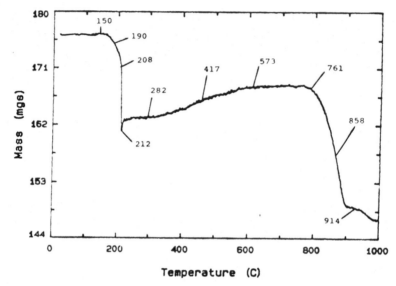

Figure 1. TGA of tape cast and laminated $YBa_2Cu_3O_{7-x}$. The heating rate was 5°C/minute in flowing air (0.5L/min.).

The plasticizers (phthalates) have a very low molecular weight compared to PVB (average molecular weight 30,000). As a consequence of the relatively simple, low molecular weight structure of the phthalates they will burn out of the ceramic matrix relatively cleanly leaving behind a low ash residue.

Previous research has shown that when fine powders of $YBa_2Cu_3O_{7-x}$ are exposed to carbon dioxide in the presence of water the superconducting phase decomposes [7,8]. When the laminated samples went through the binder burn out stage a 9 wt% loss was incurred as the organics volatilized into water

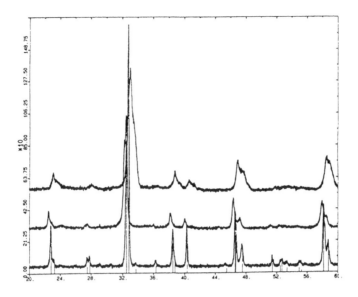

Figure 2. XRD of $YBa_2Cu_3O_{7-x}$ in various forms. a) initial calcined material. b) Taped and laminated $YBa_2Cu_3O_{7-x}$. c) Laminated $YBa_2Cu_3O_{7-x}$ heated to 150°C.

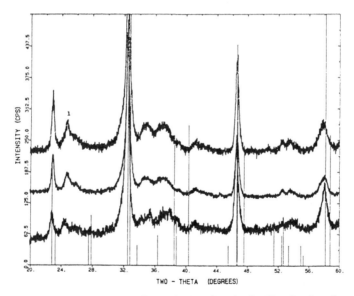

Figure 3. XRD of tape cast and laminated $YBa_2Cu_3O_{7-x}$ heated to a) 190°C b) 208°C and c) 212°C. The peak labelled 1 at 25.5° (2-theta) is the 100% peak for $BaCO_3$.

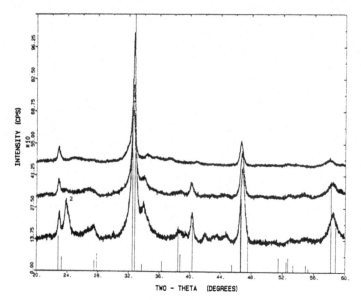

Figure 4. XRD of tape cast and laminated $YBa_2Cu_3O_{7-x}$ heated to a) 282°C b) 417°C and c) 573°C. The intensities of the diffraction patterns a) and b) are approximately half that of c). The peak labelled 2 is the 100% peak for Witherite ($BaCO_3$).

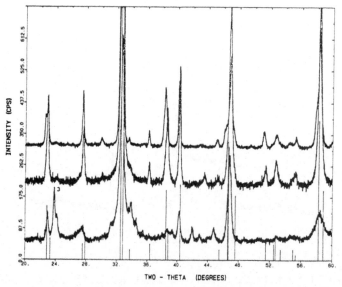

Figure 5. XRD of tape cast and laminated $YBa_2Cu_3O_{7-x}$ heated to a) 761°C b) 858°C and c) 914°C. The peak labelled 3 at 29.2° in diffraction patterns b) and c) is $Ba_2Cu_3O_5$.

and $C/CO/CO_2$. The volatile components reacted with the $YBa_2Cu_3O_{7-x}$ matrix to form appreciable amounts of carbonates as shown in Figure 3. The main peak at 25.5°(2-theta) was identified as barium carbonate ($BaCO_3$).

In the temperature range 282°C to 573°C the samples gain approximately 3 wt% as shown by the positive slope in Figure 1. It is proposed that the residual carbon (approximately 4 wt%) dispersed through-out the $YBA_2Cu_3O_{7-x}$ matrix after organic

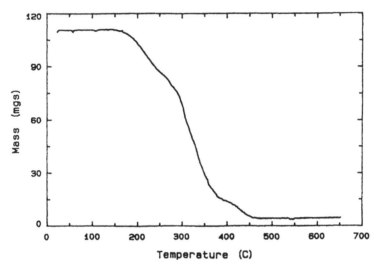

Figure 6. TGA of a stoichiometric mixture of the binder and plasticizer used to tape cast $YBa_2Cu_3O_{7-x}$. The TGA was run at 5°C/min. in flowing air (0.5L/min).

combustion picks up oxygen in this temperature regime forming further carbonate contamination. Or some oxygenation of the $YBa_2Cu_3O_{7-x}$ phase may also be occurring (from 282°C to 573°C) to replace oxygen displaced during the combustion of the organics. At 282°C the barium carbonate produced during combustion of the organics has decomposed (Figure 4a). Between 417°C and 573°C the carbonate peak reappears at 23.9°(2-theta). The barium carbonate peak has shifted from 25.5° to 23.9° (2-theta) corresponding to a change in crystalline form from barium carbonate ($BaCO_3$) to witherite ($BaCO_3$).

At 761°C the YBa$_2$Cu$_3$O$_{7-x}$ phase enters the reactive region (Figure 1) where the carbonates begin to decompose, altering the stoichiometry of the YBa$_2$Cu$_3$O$_{7-x}$ phase. From the diffraction patterns in Figure 5, at 858°C the carbonate peak has all but disappeared and the 100% peak for the Ba$_2$CO$_3$O$_5$ (barium cuprate) phase has appeared. The 100% peak for the barium cuprate is more pronounced in the diffraction pattern at 914°C (Figure 5c).

CONCLUSIONS

A major processing problem arises during the removal of large quantities of organic matter from tape cast YBa$_2$Cu$_3$O$_{7-x}$. Organic matter generally burns to form H$_2$O and C/CO/CO$_2$ depending upon the oxygen partial pressure in the local environment of the sample. Previous research has shown that the YBa$_2$Cu$_3$O$_{7-x}$ structure decomposes in the presence of carbon dioxide and water [7,8]. Diffraction patterns in Figure 3 clearly show the formation of barium carbonate during the combustion of large amounts of organic matter. Between 212°C and 573°C the barium carbonate decomposes and reappears in a different crystalline form. At approximately 780°C witherite decomposed altering the stoichiometry of the YBa$_2$Cu$_3$O$_{7-x}$ and forming a barium cuprate phase. A comparison of Figure 2b (taped cast material) and Figure 5c (sample quenched at 914°C) shows the formation of a barium cuprate phase which is a consequence of organic removal. This current work has confirmed the degradation of YBa$_2$Cu$_3$O$_{7-x}$ in the presence of carbon dioxide and water as being a major processing problem in the removal of large quantities of organic material from tape cast cuprate ceramic superconductors. Further research may show the need to develop a new binder system which decomposes to something other than carbon dioxide and water.

REFERENCES

1 J.S.Reed, "Introduction to the Principles of Ceramic Processing", John Wiley and Sons, New York, 1988.
2 R.Mistler, D.Shanefield, and R.Runk, Tape Casting of Ceramics, in "Ceramic Processing Before Firing". G.Onoda and L.Hench (eds), Wiley-Interscience, New York, 1976.
3 J.P.Pollinger and G.L.Messing, Thermal Analyses of Organic Binders for Ceramic Processing, pp. 359-370 in "Advances in Materials Characterization II" R.L.Snyder et al. (eds), Plenum, New York, 1985.

4 W.K.Shih, M.D.Sacks, G.W,Scheiffele, Y.N.Sun, and J.W.willams, "Pyrolysis of Poly(Vinyl Butyral) Binders: I, Degradation Mechanisms". To be published.

5 G.W.Scheiffele and M.D.Sacks, "Pyrolysis of Poly(Vinyl Butyral) Binders: II, Effects of Processing Variables". To be published.

6 M.J.Cima, J.A.Lewis, and A.D.Devoe, "Binder Distribution in Ceramic Greenware during Thermolysis", J.Am.Ceram.Soc.,72[7] 1192-99, 1989.

7 J.J.Simmins, R.H.McCluskey, G.S.Fishman, and R.L.Snyder, "High Temperature CO_2 Decomposition of the $YBa_2Cu_3O_{7-x}$ Superconductor". To be published.

8 D.R.Clarke, T.M.Shaw, and D.Dimos, "Issues in the Proceedings of Cuprate Ceramic Superconductors", J.Am.Ceram.Soc., 72[7] 1103-13, 1989.

ANISOTROPIC CRITICAL CURRENTS IN

ALIGNED SINTERED COMPACTS OF $YBa_2Cu_3O_{7-\delta}$.

J.E. Tkaczyk, K.W. Lay, and H.R. Hart

GE Corporate Research and Development, Schenectady, N.Y.

ABSTRACT

Measurements of the critical current anisotropy in grain aligned $YBa_2Cu_3O_{7-\delta}$ ceramics are presented. The low zero-field critical current $J_c \sim 10^2 - 10^3$ A/cm^2 and the sensitivity of J_c to weak magnetic fields indicates that the weak-link character of the intergranular coupling has not been improved by alignment of the grain boundaries. The unusual dependence of the critical current on the direction of the applied field is explained in terms of the anisotropic intragranular properties and the anisotropic nature of Josephson weak links. In the favorable field direction, critical currents at 1/2 tesla above 200 A/cm^2 have been achieved. This represents an order of magnitude improvement over non-aligned ceramics.

INTRODUCTION

The characteristic layered crystal structure of the high temperature superconductors strongly influences the superconducting and normal state properties. The resistivity of $YBa_2Cu_3O_{7-\delta}$ single crystals[1] shows an in-plane versus out-of-plane anisotropy of about 75 and the temperature dependence suggests metallic behavior in-plane and semi-insulating behavior between planes. Magnetic hysteresis measurements[2] yield an intragranular critical current anisotropy of 10. Direct transport measurements of the critical current along the c-axis are difficult to make in thin, plate-like single crystals; however, measurements of Martin et al[3] show an enormous anisotropy $\sim 10^3$ for $Bi_2Sr_2CaCu_2O_8$. Due to the tendency of the currents to lie in the plane, the magnetization of the Abrikosov lattice is not parallel to the applied field. Measurements[4] of the resulting intrinsic torque yield an effective mass anisotropy of $\gamma=(m_c/m_{ab})^{1/2}=\xi_{ab}/\xi_c \sim 5$. Measurements of the fluctuation diamagnetism indicate in-plane and out-of-plane coherence lengths of $\xi_{ab} = 13.6$ and $\xi_c = 1.2$ Å (i.e. $m_c/m_{ab} \sim 100$).[5]

As a direct consequence of the anisotropy and small coherence lengths, the intergranular critical currents in a polycrystalline material are expected to be reduced. Measurements in thin films of Dimos et al[6] showed that rotations of two grains about the c-axis reduced the intergranular J_c by as much as a factor of 50. By studying the boundary conditions between two anisotropic superconductors, Kogan[7] has proposed an intrinsic mechanism for reduced critical currents between misaligned grains. Deutscher[8] points out that due to the small coherence length in the c-axis direction, the order parameter and critical current is significantly suppressed at grain boundaries oriented along the ab plane. Ekin et al[9] make the argument that current transfer from the high intragranular critical current, ab-direction of one grain into the low critical current, c-axis direction a neighboring grain is bound to be lossy. The inevitable presence of such boundaries along the current path in an unoriented polycrystalline ceramic may explain the weak link character of the transport critical current.

From such evidence it appears that high critical currents will be improved in grain-aligned ceramics. The fabrication of grain aligned ceramics using the magnetic alignment technique of Farrell *et al*[10] was demonstrated by Arendt *et al*[11]. A detailed study of the sintering and annealing process has recently been presented.[12] Increased critical currents and reduced normal state resistivities were obtained as a result from the alignment of the grains. In the present paper, the same results of this study are summarized and additional transport measurements characterizing the anisotropy of the critical current are presented. Emphasis is placed on obtaining a better understanding of the factors limiting the intergranular critical current, specifically the role of alignment on the weak link character of grain boundaries.

SAMPLE PREPARATION

A detailed explanation of sample fabrication has been given previously,[12] so only an outline is given here. Stoichiometric mixtures of the elemental oxides are calcined in air at 870-920C. The reacted powder is dry milled and suspended in heptane with the use of a deflocculant. The suspension is left for 15 hours in a 4 tesla magnetic field until the heptane has evaporated. The c-axis of the grains tends to align along the field direction. Reference will be made to the c-axis direction and a,b-plane in describing the net current path through the sample. Densification, grain growth and intergranular connectivity were accomplished by sintering. The density of the samples continued to increase with increasing sintering temperature up to 89% at the highest sintering temperature used, 1000C. The evidence suggests that the rapid densification above a sintering temperature of ~930C is due to the presence of low melting point, Ba-Cu-O rich phases in the calcined powders. Accompanying the densification is a drop in the normal state resistivity and increase in the critical current.

In accord with previous work[13] the critical current of non-aligned samples was maximized for sintering temperatures in the range of 930-950C and fell rapidly at higher temperatures. In contrast, for the aligned samples critical current remained high up to the highest sintering temperature used 1000C. The difference was attributed to reduced microcracking in the aligned ceramics for which anisotropic thermal contraction between the c and ab planes is accommodated by changes in the sample dimensions. In non-aligned ceramics, differential contraction of misaligned grains leads to microcracking. The presence of Ba-Cu-O rich phases does not appear to be the most important factor limiting J_c. In fact, off-stoichiometric samples doped to promote melting have the lowest normal state resistivities and highest critical currents at 1/2 tesla yet measured in a ceramic material.

Data on stoichiometric, aligned samples are presented in this report. The sample MAL15E is the same used in reference 12. The density is 79% and room temperature resistivity in the plane is 340 µΩ–cm. The sample MAL34 was sintered under about the same conditions as sample MAL15E (890C for 24 hours).

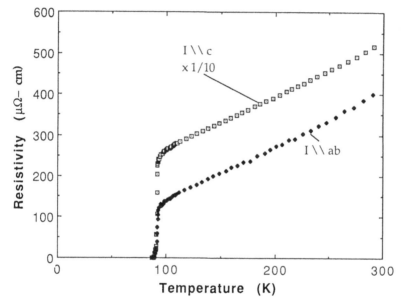

Fig. 1 The resistivity as a function of temperature of sample MAL34. For current applied along the c-axis direction, one tenth of the resistivity is plotted.

However, the density is slightly lower ~77% and the transport properties are not as good. Yet this sample has the advantage that pieces were sectioned such that current can be applied either along the ab-plane or along the c-axis. Contacts were made with Ag epoxy annealed at 900C for 9 hr yielding a contact resistance of the order of 1-10 µΩ–cm^2 with current density up to 100 A/cm^2 through the contact.

EXPERIMENTAL RESULTS

As shown in figure 1, the resistance of aligned sample MAL34 is linear as a function of temperature from room temperature down to the fluctuation region just above T_c. The linear behavior is present for the current applied in either the a,b or c-axis direction, but the magnitude of the resistivity is an order of magnitude larger for the current applied along the c-axis direction.

At the fixed temperatures, 4.2 and 77k, the critical current of sample MAL15E has been measured as a function of magnetic field using a 2μV/cm criterion. The sample is immersed in liquid helium or nitrogen and placed at the center of an iron core electromagnet. In order to avoid flux trapping effects, the sample is cooled in zero field from above T_c. As shown in figure 2, with the current applied in the a,b plane direction, the application of as little as 10mT depresses the critical current by as much as an order of magnitude. The drop is larger by a factor of 2 to 3 if the field is applied parallel to the c-axis direction as compared with the field applied in the a-b plane. For 1/2 tesla applied in the favorable orientation, the critical current of the aligned material ~200 A/cm^2 is an order of magnitude above that of the non-aligned material ~15 A/cm^2.

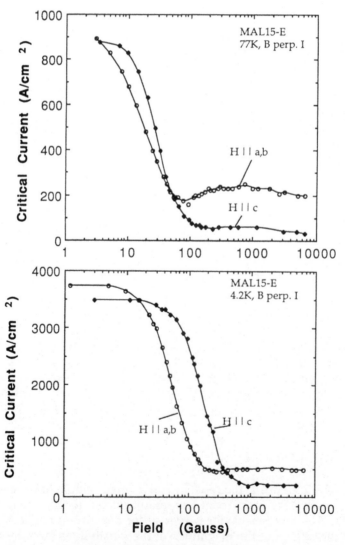

Fig. 2 The critical current of sample MAL15E is plotted as a function of field after zero field cooling.

The dependence of the critical current as a function of the angle between field, current and crystalline axes was measured at 77K with the sample MAL34 immersed in liquid nitrogen. The 0.28T field was applied with a rotatable permanent magnet. This magnet was a prototype NMR imaging magnet having a large bore and high homogeneity. As shown in figure 3, the critical current is severely depressed for fields applied out of the a,b-plane. This result is independent of whether the current is applied in the ab-plane or c-axis direction. Note that with a field of 0.28T applied in the ab-plane, the critical current in the low conductivity, c-axis direction is ~50 A/cm^2 which is three times larger than that in non-aligned ceramics. In zero field, the critical current parallel and perpendicular to the c-axis was ~400 A/cm^2 and 600 A/cm^2 respectively.

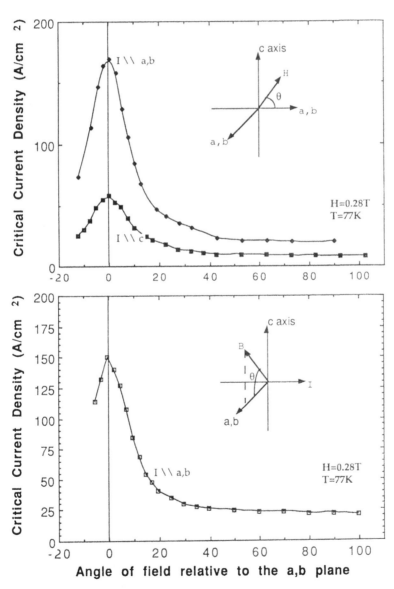

Fig. 3 The critical current of sample MAL34 as a function of the angle between a 0.28T applied field and the ab-plane.

INTERPRETATION

In an attempt to understand the peculiar anisotropic transport properties just described, it is useful to focus on four currents entering into a percolative picture. There are two intra- and two intergranular currents which are considered. In a well sintered material, the normal state resistive contribution of the grain boundaries is negligible, and the normal state resistivity is primarily determined by the <u>intra</u>granular currents in the ab-plane and along the c-axis. In contrast the critical current is dominated by the intergranular coupling. In a grain aligned material, few are the grain boundaries where the ab-plane of one grain contacts the c-axis of another. Therefore, the two relevant <u>inter</u>granular currents are the current across a basal plane grain boundary (i.e. parallel to the c-axis direction) and the current crossing parallel to the ab-plane.

As discussed in the introduction, the evidence from single crystals is that the ab-plane is the high conductivity direction. The resistivity of aligned ceramics is about twice that of single crystals when the current is applied in the ab-plane direction, while the resistivity is about half that of single crystals when the current is applied in the c-axis direction. Furthermore, the temperature dependence of the resistivity in the c-axis direction fails to show the semi-insulating behavior seen in single crystals. Rather, the resistivity decreases with decreasing temperature in accord with the behavior of the resistivity in the ab-plane of single crystals. This suggest that the normal state percolation paths through the material tend to primarily follow the ab-plane whether the net current applied through the sample is along or perpendicular to this plane.

The atomic scale coherence length along the c-axis direction suggests that little (if any?) supercurrent travels across basal plane grain boundaries. Thus, it appears that intergranular supercurrents travel parallel to the ab-plane. The complicated behavior of the critical current in an applied field displayed in figures 2 and 3 can be accounted for within a picture incorporating this anisotropy of the intergranular supercurrents as well as the anisotropic properties of Josephson weak links. As shown in figure 2, with the application of as little as 10mT the critical current can be depressed by an order of magnitude. This effect is also present in unaligned samples and has been attributed in the literature to quenching of the intergranular Josephson current. In the aligned ceramic, the drop is larger by a factor of 2 to 3 if the field is applied parallel to the c-axis direction as compared with the field applied in the a-b plane. Glowacki and Evetts have seen a similar anisotropy in uniaxially compacted powders and interpret their results in terms of the anisotropic character of Josephson weak links.[14] It is characteristic of Josephson weak links that the critical current is suppressed for fields applied perpendicular to the current flow. Thus, a field applied in the c-axis direction quenches supercurrents traveling parallel to the ab-plane. The transport current must then travel some distance along the c-axis direction in order to find a percolating path through the sample. In contrast, with the field in the a-b plane, currents along the field direction are not affected by the field and a percolating path lying entirely in the a,b plane may be found.

In figure 3, one observes that for the net current applied along the c-axis direction the critical current is again maximized for the field parallel to the ab-plane. Within the weak link picture just described, this result suggests that the current criss-crosses from grain to grain in the ab-plane thereby establishing a percolative path which advances in the c-direction intragranularly. In a figuratively sense, the current path resembles the path of a boat making a series of tacks while sailing into the wind.

In figure 2, one observes two additional features differentiating the behavior for H parallel and perpendicular to the c-axis. At 77K an increase in J_c (so called peak effect) is observed as the field is increased between 7.5 to 15 mT in the case that H is applied parallel to the ab-plane. The field required to reduce the critical current to its field independent value, (i.e. decoupling field) is smaller for H parallel to the ab-plane. This effect is most noticeable at 4.2K. The large difference in the demagnetization factor of the large plate-like grains accounts for the field orientation dependence these observations. For H parallel to the ab-plane, the field is parallel to the plates (demagnetization factor ~0) and below H_{c1}~10mT flux is largely excluded from the grains. Most of the flux is excluded from the sample, but the residual flux through the sample tends to concentrate in the grain boundaries thereby suppressing the intergranular critical current. This is presumably the reason for the smaller decoupling field for H parallel to the ab-plane. Above H_{c1}, flux abruptly enters the grains thereby reducing the flux density in the intergranular regions whereupon J_c increases. In order to further test this explanation, note that at temperatures above 77K, H_{c1} of the grains will be closer to the decoupling field resulting in a even more significant peak effect. Experiments testing this prediction are in progress. Negative magnetoresistance behavior has been reported which is explained in a similar manner.[15]

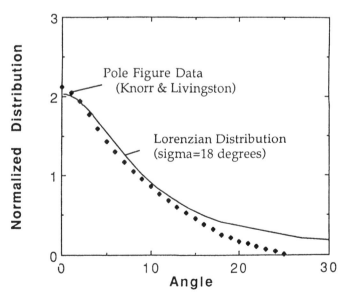

Fig. 4 The angular distribution of the grain c-axes in magnetically aligned $YBa_2Cu_3O_{7-\delta}$ ceramics is approximated by a Lorentzian distribution. The pole figure data was obtained from Knorr and Livingston of reference 15.

In order to make this weak link picture more quantitative consider the following simple averaging procedure which gives remarkably good agreement with the data. The sample is modeled as a array of weak links with a decoupling field of $H_D=10mT$. The critical current of each weak link is assumed to fall exponentially with the component of the applied field perpendicular to the current through the junction. The exponential dependence is chosen since it is non-singular at H=0. In addition a field independent critical current contribution is assumed for each junction k_0 to account for any "strong links" in the ceramic material.

$$j_c(H,\vartheta)= k_o + j_c(0)\, e^{-H(\sin\vartheta)/H_D}$$

Here, $k_o + j_c(0)$ is the critical current of the weak link at zero field and υ is the angle between the weak link current and the applied field. Knorr and Livingston[16] have measured the angular distribution of grains in magnetically aligned ceramics. Their data has been fitted to a Lorentzian distribution $f(\upsilon)$ as shown in figure 4. The c-axis of the grains are oriented primarily along one direction with a full width at half maximum of $\sigma=18$ degrees. Call this direction the sample c-axis and the normal plane the sample ab-plane. Thus, given the angle θ of the applied field with respect to the sample ab-plane, the angle of the field with respect to the ab-plane of each weak link is spread around θ with this Lorentzian distribution. In the simplest model one takes a simple average of the critical current of all the weak links to arrive at the critical current of the sample.

$$J_c(H,\theta)=\int_{-\pi}^{+\pi} d\eta\; j_c(H,\theta+\eta)\, f(\eta) \qquad \text{where} \qquad f(\vartheta)=\frac{(1/\pi\sigma)}{1+(\frac{\vartheta}{\sigma})^2}$$

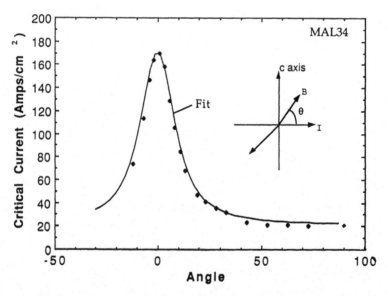

Fig. 5 The fit of a weak link model to one set of data from figure 3.

The resulting fit using the free parameters $k_0=20$ and $j_c(0)=1080$ A/cm^2 is shown in figure 5. Note that the presence of intergranular supercurrents parallel to the c-axis would result in a second peak (not observed) at $\theta=90$ degrees in figure 4.

In conclusion, in spite the low normal state resistivity and the alignment of the grains, the transport critical current remains low and is sensitive to weak magnetic fields. The anisotropy of the critical current in an applied field is consistent with a weak link picture where intergranular currents lie for the most part parallel to the ab-plane. The evidence suggests that some "strong" links exist and make a 20 A/cm^2 contribution to the critical current which is relatively field independent below 1/2 tesla.

ACKNOWLEDGEMENT

Partial support supplied by the Defense Advanced Research Projects Agency contract # N00014-88-C-000681. We would like to thank D.B. Knorr and J.D. Livingston for making available their data on the c-axis alignment in magnetically aligned ceramics.

REFERENCES

[1] Penney, T., S. von Molnar, D. Kaiser, F. Holtzberg, and A. W. Kleinsasser, *Phys. Rev. B*, **38**, 2918 (1988).
[2] Dinger, T.R., T.K. Worthington, W.J. Gallagher, and R.L. Sandstrom, *Phys. Rev. Lett.* **58**, 2687 (1987).
[3] Martin, S., A.T. Fiory, R. M. Fleming, G.P. Espinosa, and A.S. Cooper, *Appl. Phys. Lett.* **54**, 72 (1989).
[4] Farrell, D.E., C.M. Williams, S.A. Wolf, N.P. Bansal, V.G. Kogan, *Phys. Rev. Lett.* **61**, 2805 (1988).
[5] Lee, W.C., R.A. Klemm, and D.C. Johnston, *Phys. Rev. Lett.* **63**, 1012 (1989).
[6] Dimos, D., P. Chaudhari, J. Mannhart, and F.K. LeGoues, *Phys. Rev. Lett.* **61**, 219 (1988);
[7] Kogan, V.G., *Phys. Rev. Lett.* **62**, 3001 (1989).
[8] Deutscher, G., *IBM J. Res. Develop*, **33**, 293 (1989); Deutscher, G., *Physica C*, **153-155**, 15 (1988).
[9] Ekin, J.W., A.I. Braginski, A.J. Panson, D.W. Capone II, N.J. Zaluzec, B. Flandermeyer, O.F. deLima, M.Hong, J. Kwo, and S.H. Liou, *J. Appl. Phys.* **62**, 4821 (1987).
[10] Farrell, D.E., B. S. Chandrasekhar, M.R. DeGuire, M.M. Fang, V.G. Kogan, J.R. Clem, and D.K. Finnemore, *Phys. Rev. B* **36**, 4025 (1987).
[11] Arendt, R.H., A.R. Gaddipati, M.F. Garbauskas, E.L. Hall, H.R. Hart, Jr., K.W. Lay, J.D. Livingston, F.E. Luborsky, and L.L. Schilling, *Mat. Res. Soc. Symp. Proc.* **99**, 203 (1988).
[12] Tkaczyk, J.E., and K.W. Lay, submitted to *J. Material Research*,.
[13] Kikuchi, A., M. Maesuda, M., T. Maeda, M. Ishh, M. Takata, and T. Yamashita, *Jap. J. Appl. Phys.* **27**, 1231 (1988); Shi, D., D.W. Capone II, G.T. Goudey, J.P. Singh, N.J. Zaluzec and K. C. Goretta, *Mat. Lett.* **6**, 217 (1988); Sawano, K., A. Hayashi, T. Ando, T. Inuzuka, and H. Kubo, preprint.
[14] Glowacki, B. A. and J.E. Evetts, *Mat. Res. Soc. Symp. Proc.* **99**, 419 (1988).
[15] Shifang, S., Z. Yong, P. Guoqiang, Y. Daoqi, Z. Han, C. Zuyao, Q. Yitai, K. Weiyan, and Z. Qirui, *EuroPhys. Lett.* **6**, 359 (1988).
[16] Knorr, D.B., and J.D. Livingston, *Supercond. Sci. Technol.* **1**, 302 (1989).

YBa$_2$Cu$_3$O$_{7-\delta}$: ENHANCING j_c BY FIELD ORIENTATION

R.S. Markiewicz[a,d], K. Chen[a,d]†, B. Maheswaran[a,d],
Y.P. Liu[b]*, B.C. Giessen[c,d], and C. Chan[b]

Departments of Physics(a), Electrical Engineering(b), and Chemistry(c)
and Barnett Institute(d),
Northeastern University, Boston, MA 02115

Transport properties of dense, pressed pellets of field-oriented grains of YBa$_2$Cu$_3$O$_{7-\delta}$ have been studied in the superconducting phase (T=77K) as a function of current, annealing time, and magnetic field. These pellets have strikingly different annealing properties from similarly prepared but unoriented samples. The oriented samples show significant enhancements in density and degree of orientation and transport j_c with length of annealing time. The best samples have less than 2% of the x-ray intensity corresponding to misoriented grains, and a density so high that excess low-T oxygen annealing is required for the best j_c (currently, $j_c(77K) > 1500 A/cm^2$). These values are limited by the magnetic field generated by the current itself.

These samples are ideally suited for a variety of studies of material anisotropy, such as transport, magnetization, and optical properties, especially in situations where large single crystals are not available. Preliminary measurements suggest that the Tl- and Bi-based superconductors are even easier to orient.

INTRODUCTION

A fundamental limitation to applications of the new high-T_c superconductors has been the extremely small values found for critical currents, j_c, in polycrystalline samples of these materials, especially in a magnetic field. In YBa$_2$Cu$_3$O$_{7-\delta}$, the small j_c's are known to be associated with weak transmission of current across grain boundaries: much higher j_c's are found in single crystals and epitaxial thin films. In other high-T_c materials, particularly those based on Bi or Tl, there is an intrinsic limitation to j_c, with small values being found even in single crystals. This latter effect may be associated with flux lattice melting or flux creep. In the present paper, we shall be primarily interested in the former problem, grain boundary reduction of j_c.

A number of reasons have been suggested for the large and deleterious effect grain boundaries play in these materials. Fundamentally, the problem arises because the superconducting coherence length is so short that any perturbation at the grain boundary can strongly surpress superconductivity. The precise nature of this perturbation, however, remains controversial, with suggestions ranging from impurities (e.g., carbon), to deviations from stoichiometry, to intrinsic surface phases, to simple mismatch in crystallographic orientation. This last suggestion has received strong experimental support from the work of Dimos et al.[1], who showed that even a misalignment within the conducting (a,b)-plane could reduce j_c

by a factor of 50. Kogan[2] has shown how such an effect might arise. The purpose of our investigations has been to explore one particular aspect of this misalignment-induced reduction of j_c, namely, the role of c-axis orientation on enhancing j_c.

FIELD-ORIENTED PELLETS

Farrell, et al.[3], showed that an intense magnetic field could be used to preferentially orient grains of the high-T_c superconductors with their c-axes parallel to the applied field. This effect has nothing to do with superconductivity (indeed, the orienting is carried out at room temperature), but with the magnetic anisotropy of the grains, with the direction of highest susceptibility aligning along the field. In some rare-earth substituted compounds, this direction actually lies within the (a,b)-plane, due to crystal-field effects[4], but in pure $YBa_2Cu_3O_{7-\delta}$ or the Tl- or Bi-based superconductors, the nearly two-dimensional electronic structure ensures that the preferred direction is the c-axis.

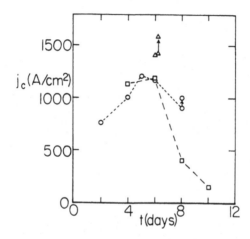

Fig. 1 Critical current j_c (at 77K) vs annealing time at 950°C, for three samples of $YBa_2Cu_3O_{7-\delta}$. Vertical arrows show the effect of additional low-temperature (450-550°C) oxygen anneals.

In the original experiments of Farrell, et al[3], their grains were dispersed in epoxy, which hardened in the field, producing samples with a very high degree of alignment. Arendt, et al.[5], prepared oriented pellets which could be used in transport studies, by allowing a suspension of the superconducting grains to evaporate in the field, and then annealing the settled powder. These pellets retained a high degree of alignment. We have followed a similar procedure[6], but carry out the evaporation inside of a Cu-Be die, so that the resulting pellets can be pressed ($\sim$ 5kBar) prior to annealing. These oriented pellets have been employed in a variety of experiments, usually to measure anisotropic properties, including a.c. magnetization[7] and optical[8] studies. In this paper, we will discuss measurements of critical current in these samples.

CRITICAL CURRENTS

All experiments involved four terminal measurements of voltage vs applied current. Leads were formed by sintering silver paint contacts at 900°C for 1-2 hours in flowing oxygen. The contact resistivity was $\sim 10^{-5} - 10^{-6}$ Ω-cm^2 at 77K. While lower contact resistivities can be obtained by Ag evaporation[9], the present values are too low to substantially affect measured j_c values. The onset of finite voltages is fairly sharp, making our value of j_c insensitive to the criterion used; we take j_c as that value of current for which a $1\mu V$ voltage drop appears across the sample. For comparison, we have prepared a control series of pellets which did not undergo the field-orienting step. After annealing, these pellets had $j_c \sim 150 - 250 A/cm^2$ at 77K, independent of the duration of the annealing. In contrast, the field-oriented samples showed considerably higher j_c values, which had a striking dependence on annealing time at $950^\circ C$, Fig. 1.

While there is some sample-to-sample variation, j_c increases monotonically with annealing time until it has attained values $> 1000 A/cm^2$ after ~ 6 days, but then j_c rapidly decreases. We believe that the anneal is causing a melting of the smallest grains, which then grow onto larger, better oriented grains. There may be an element of stress-relief annealing involved as well. Parallel x-ray studies show that this annealing is accompanied by a significant improvement in the degree of c-axis orientation, with the intensity of reflections associated with misaligned grains decreasing from $\sim 10\%$ that of an unoriented sample to $< 2\%$. There is also a significant increase in the sample density, associated with the filling in of voids as the grains regrow. We believe that this void filling is responsible for the decrease in j_c at long anneal times, since the voids form a convenient pathway for oxygen diffusion into the sample interior, and once they are eliminated, only the much slower diffusion through bulk material remains. A very similar annealing-time dependence of j_c has been observed by Alford, et al[10] on extruded samples of $YBa_2Cu_3O_{7-\delta}$.

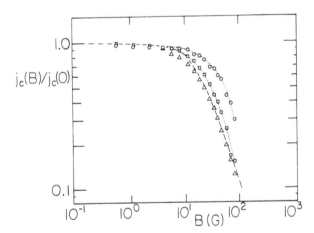

Fig. 2 Scaled critical current, $j_c(B)/j_c(0)$ in field-oriented sample of $YBa_2Cu_3O_{7-\delta}$ as a function of magnetic field, for a variety of geometrical arrangements (j in (a,b)-plane): triangles $= j \perp B \parallel c$; squares $= j \perp B \perp c$; circles $= j \parallel B$. For this sample, $j_c(0) = 1200 A/cm^2$.

As a test of this interpretation, we studied the effect of subsequent low-temperature (450-550°C) oxygen anneals. For most of the data in Fig. 1, there was no subsequent anneal; the samples were merely slow-cooled in flowing oxygen. Well below the peak in j_c, additional anneals have no significant effect on j_c. At the peak, there is a substantial effect; Fig. 1 shows our best result to date: a two hour anneal at 450°C increased j_c from ~ 1400 to $1590 A/cm^2$. Above the peak, annealing also has an effect, but substantially longer anneals are required. Fig. 1 shows the result of a two day anneal at 550°C. It seems likely that in any commercial applications of these new materials, bulk samples of near ideal density will be used, and it will be necessary to overcome this oxidation barrier.

MAGNETIC FIELD DEPENDENCE

Thus, we find that field-aligning can enhance j_c by a factor of ~6. The strong magnetic field dependence of j_c (Fig. 2), however, shows that we have not eliminated the weak links between grains. Peterson and Ekin (PE)[11] have shown that the strong field dependence of j_c could be understood as a suitable average of Josephson effects in a random array of weak-link grain boundaries. For a single link, with interface perpendicular to the field direction, the field dependence is

$$j_c(B) = j_c(0)|sin(\pi B/B_0)|/(\pi B/B_0), \qquad (1)$$

where $B_0 = \Phi_0/dL = 6G/(L/10\mu m)$, Φ_0 is the flux quantum, d is the effective junction thickness, and L is the junction length, comparable to the grain diameter. The numerical estimate of B_0 is from the results of PE. If B is not perpendicular to the interface, only the perpendicular component, $Bsin\theta$ should appear in Eq. 1. For an array of interfaces, Eq. 1 should be suitably averaged over L and θ. For B not too small, PE find

$$j_c(B) = j_c(0)B_1/B, \qquad (2)$$

where $B_1 = 1.35G/(L/10\mu m)$.

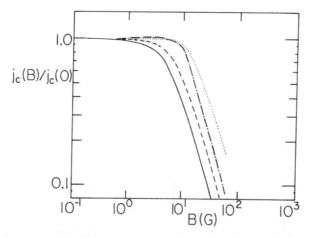

Fig. 3 Scaled critical current, $j_c(B)/j_c(0)$, for non oriented pressed pellets of YBa$_2$Cu$_3$O$_{7-\delta}$ (solid line = $j \perp B$; dashed line = $j \parallel B$; $j_c(0) = 250A/cm^2$) or TlBa$_2$Ca$_3$Cu$_4$O$_x$ (dot-dashed line = $j \perp B$; dotted line = $j \parallel B$; $j_c(0) = 500A/cm^2$).

The data of Fig. 2 approximately follow the 1/B dependence, but with a much larger value of B_1. While this could be due to a smaller value of L (which was not measured), this seems unlikely. Fig. 3 shows that B_1 is lower in the control (non-field-aligned) samples, whereas it seems unlikely that either the aligning or the additional annealing would produce a smaller value of L. A more likely alternative explanation has been suggested by Ekin[12]. He suggested that the critical current is limited by the magnetic field produced by the current itself. To test this, we assume that j_c decreases as $1/B_{net}$, where B_{net} is the total field at the surface of the sample, i.e., the applied field B plus the self field due to the current, $I_c = j_c \pi r^2$, where r is the radius of the sample, approximated as a cylinder. Assuming that, on average, the self-field $B_s = B_{sx} \equiv \mu_0 I_c/2\pi r$ is orthogonal to the applied field, then $B_{net}^2 = B_s^2 + B^2$, and $j_c = j_0/B_{net}$. Writing $\nu = j_c(B)/j_c(0)$, we have $B_s = \nu B_{s0}$ and

$$\nu^2 = \frac{1}{\nu^2 + (\frac{B}{B_{s0}})^2}. \tag{3}$$

Eq. 3 (dashed line in Fig. 2) provides a satisfactory fit to the data, taking B_{s0}=12G to agree with the data at ν=0.5. If B and B_s are not orthogonal, the modified Eq. 3 would require a larger value for B_{s0}; the largest B_{s0}, $\simeq$15G, corresponds to B and B_s being parallel. Surprisingly, this value is *smaller* than the expected value of $B_{sx} \simeq$30G, using the observed values I_c=6A and r$\simeq$0.4mm. Nevertheless, the fact that the expected self field is larger than that needed to explain the j_c reduction is evidence that Ekin's conjecture is essentially correct. Hence, the measured values of $j_c(B=0)$ are limited by the self field, and the improvements we observe with annealing are directly connected with improvements in the intergrain coupling. The true, field-independent value of $j_c(0)$ can be found only by going to much thinner samples, where the self field is less important. However, a rough estimate can be found by combining Eqs. 2 and 3. Assuming that the long anneal has increased the average grain size to $\sim 10\mu m$ (a point which should be checked), then

$$j_c/j_c(0) \simeq B_1/B_{s0}. \tag{4}$$

Since for this sample, the measured $j_c = 1200 A/cm^2$, this gives $j_c(0) \simeq 11,000 A/cm^2$.

Fig. 2 also displays the dependence of j_c on the direction of the applied field; i.e., for $B \parallel j_c$ or $B \perp j_c$, with either $B \parallel$ or $B \perp$ the c-axis. Surprisingly, all three j_c's show approximately the same field dependence, although the Josephson effect should arise when $B \perp j_c$.[11,13] However, it is known[14] that even in the normal state, the weak transmission of the grain boundaries causes the currents to follow an irregular meander path, so that microscopically there will be current flow perpendicular to B regardless of the direction of the macroscopic j_c. Nevertheless, the field dependence is substantially reduced for $j_c \parallel B$.

Preliminary experiments on a four Cu-layer Tl compound, $TlBa_2Ca_3Cu_4O_x$ (T_c=122K), suggest possible improved performance. Only non-oriented pellets were studied, but these showed higher values for both measured zero-field j_c, $\simeq 500 A/cm^2$, and for the Josephson cutoff field, B_1, Fig. 3. Both Tl and Bi compounds show high degrees of field-orientation in epoxy, and we are currently preparing oriented pressed pellets of these materials for j_c studies.

CONCLUSIONS

For practical applications, it is useful to rewrite the results of the above theory as

$$j_c(B_{s0}) = \sqrt{\frac{2j_c(0)B_1}{\mu_0 r}}, \tag{5}$$

$$B_{1/2} = \sqrt{15}\mu_0 j_c(B_{s0})r/4, \tag{6}$$

where $j_c(B_{s0})$ is the critical current measured in zero external field and $B_{1/2}$ is the field at which j_c has fallen to half of this initial value. Equation 5 shows that the appropriate figure-of-merit for j_c measurements is actually $j_c\sqrt{r}$: for two microscopically equivalent samples, the one with the smaller radius will have the larger value of j_c. This explains why the highest critical current values are generally found in films and thin wires. Equation 6 shows that $B_{1/2}$ should scale linearly with the zero-field value of j_c, a fact that is approximately verified by the data of Figs. 2,3.

Based on the results of Dimos, et al.[2], Rhyner and Blatter[15] have shown that even if the c-axis is perfectly aligned (in, e.g., a thin film) random in-plane grain boundary misorientation can reduce the critical current by a factor of 30. Combined with our results for c-axis alignment, this suggests that a fully-aligned bulk pellet would have $j_c(0) \simeq 320,000 A/cm^2$, comparable to values found in good epitaxial films. Thus it seems highly probable that the low measured values of j_c in bulk pellets are predominantly due to intergranular misorientation.

This research has been supported by E.I. DuPont de Nemours & Company, Inc.

REFERENCES

† Present address: Industrial Technology Research Institute, Chutung, Taiwan, ROC

∗ Present address: EIC Labs, Norwood, Mass.

1. D. Dimos, P. Chaudhari, J. Mannhart, and F.K. Le Goues, Phys. Rev. Lett. 61:219 (1988).
2. V.G. Kogan, Phys. Rev. Lett. 62:3001 (1989).
3. D.E. Farrell, B.S. Chandrasekhar, M.R. De Guire, M.M. Fang, V.G. Kogan, J.R. Clem, and D.K. Finnemore, Phys. Rev. B36:4025 (1987).
4. J.D. Livingston, H.R. Hart, Jr., and W.P. Wolf, J. Appl. Phys. 64:5806 (1988).
5. R.H. Arendt, A.R. Gaddipati, M.F. Grabauskas, E.L. Hall, H.R. Hart, Jr., K.W. Lay, J.D. Livingston, F.E. Luborsky, and L.L. Schilling, in "High-Temperature Superconductivity", ed. by M.B. Brodsky, R.C. Dynes, K. Kitazawa, and H.L. Tuller, Amsterdam, North-Holland (1988), p.203.
6. K. Chen, B. Maheswaran, Y.P. Liu, B.C. Giessen, C. Chan, and R.S. Markiewicz, Appl. Phys. Lett. 55:289 (1989).
7. R.S. Markiewicz, K. Chen, B. Maheswaran, A.G. Swanson, and J.S. Brooks, J. Phys. Condensed Matt., to be published.
8. F. Lu, C.H. Perry, K. Chen, and R.S. Markiewicz, J. Opt. Soc. Amer. B6:396 (1989).
9. J.W. Ekin, A.J. Panson, and B.A. Blankenship, Appl. Phys. Lett., 52:331 (1988); Y.P. Liu, K. Warner, C. Chan, K. Chen, and R. Markiewicz, J. Appl. Phys., to be published.
10. N. McN. Alford, W.J. Clegg, M.A. Harmer, J.D. Birchall, K. Kendall, and D.H. Jones, Nature 322:58 (1988).
11. R.L. Peterson and J.W. Ekin, Phys. Rev. B37:9848 (1988).
12. J.W. Ekin, unpublished.
13. Y.B. Kim and M.J. Stephen in "Superconductivity", ed. by R.D. Parks, N.Y., Dekker (1969), p. 1107.
14. R.S. Markiewicz, Sol. St. Commun., 67:1175 (1988).
15. J. Rhyner and G. Blatter, Phys. Rev. B40:829 (1989).

SURFACE PASSIVATION OF Y-Ba-Cu-O OXIDE USING CHEMICAL TREATMENT

Q.X.Jia and W.A.Anderson

State University of New York at Buffalo
Department of Electrical and Computer Engineering
&
Institute on Superconductivity
Bonner Hall, Amherst, NY 14260

ABSTRACT

A novel chemical treatment was used to passivate high temperature superconducting Y-Ba-Cu-O of both bulk oxides and thin films. The water resistance of the Y-Ba-Cu-O was greatly improved after the superconductors were treated with hydrofluoric acid (HF) at room temperature. No obvious etching of the Y-Ba-Cu-O and no degradation of zero resistance temperature were observed after the Y-Ba-Cu-O superconductors were treated with 49% HF or buffer HF commonly used in semiconductor technology for etching silicon dioxide or silicon nitride. The formation of a thin layer of amorphous fluoride on the film surface could be related to the improved water resistance of Y-Ba-Cu-O after HF treatment. It seems that HF destroys the corrosion products formed on the Y-Ba-Cu-O surface due to the reaction of the Y-Ba-Cu-O with water vapor or carbon dioxide in air.

INTRODUCTION

From the point of view of future practical applications, the electrical and physical properties of the superconducting Y-Ba-Cu-O materials must be environmentally stable. However, experimental results have shown that one of the main problems of the high temperature Y-Ba-Cu-O superconducting materials for practical applications is their high sensitivity to environmental ambient, especially with water and carbon dioxide [1-2]. A few techniques have been explored in order to improve the stability or prevent the degradation of the Y-Ba-Cu-O during its exposure to environmental ambient [3-8]. Improved durability of the superconducting Y-Ba-Cu-O materials to moist air or water has been demonstrated by utilizing intrinsic passivation during film formation [3] or extrinsic passivation after film formation [4-8].

In this paper, a detailed analysis of surface passivation of Y-Ba-Cu-O using chemical treatment with HF is presented. The passivation effects were seen both on bulk and thin film Y-Ba-Cu-O after Y-Ba-Cu-O was treated with HF at room temperature. Further investigation of the surface properties of the superconducting Y-Ba-Cu-O included Auger electron spectroscopy (AES) surface scanning and depth profile, X-ray diffraction (XRD), energy dispersive X-ray analysis (EDAX), in-situ measurement of the water resistance of the Y-Ba-Cu-O films, and the resistivity vs temperature (R-T) characteristics.

EXPERIMENTAL DETAILS

The bulk superconducting Y-Ba-Cu-O samples were prepared using conventional processes. Table 1 outlines the fabrication parameters used in the sample preparation. The resulting bulk material was fully superconductive at temperature greater than 85K.

Table 1 Bulk Superconducting Y-Ba-Cu-O Sample Preparation

materials	Y_2O_3, $BaCO_3$, CuO
stoichiometry	Y:Ba:Cu = 1:2:3
calcining	900C, 16hr, in air
sintering	850C, 24hr, in oxygen
sample dimension	10x2x2 mm^3

The Y-Ba-Cu-O thin films were deposited using RF magnetron sputtering. Table 2 gives the typical deposition conditions. No substrate heating or oxygen were utilized during sputtering. High temperature annealing in pure oxygen was used to transfer the as- deposited amorphous films into superconductive ones. The typical annealing conditions included holding at 850°C and 630°C for 60min and 180min, respectively, and naturally cooling to room temperature from 630°C by shutting off the furnace power supply.

HF treatment of superconducting Y-Ba-Cu-O thin films was different form that of bulk Y-Ba-Cu-O oxides. Direct HF soaking was used for bulk Y-Ba-Cu-O oxides and wet HF vapor for Y-Ba-Cu-O thin films. For bulk materials, a nitrogen blow-dry was used after HF immersion. The HF was a 49% solution commercially available or buffer HF commonly used in semiconductor device fabrication for etching silicon dioxide or silicon nitride.

Table2 Conditions for RF Magnetron Sputtering of Y-Ba-Cu-O Thin Film

target	$YBa_2Cu_3O_{7-x}$
target size	1 inch in diameter & 2mm in thickness
substrate	Y-stabilized ZrO_2
RF frequency	13.56MHz
base pressure	5×10^{-6} Torr
RF power	50W
sputtering gas	Ar (7mTorr)
sputtering rate	1.5Å/sec
film thickness	7000-8000Å

To verify the water resistance of the Y-Ba-Cu-O after HF treatment, deionized (DI) water immersion of the Y-Ba-Cu-O was carried out at room temperature, and R-T characteristics were checked before any treatment, after HF treatment, and after water immersion, respectively. In-situ measurement of the film resistance during immersion in DI water was carried out with a two probe method and a Keithley 192 programmable DMM. The surface properties of the Y-Ba-Cu-O after HF treatment were characterized using XRD, AES surface scanning and depth profile, and EDAX.

RESULTS AND DISCUSSION

Y-Ba-Cu-O Oxides

It is well known that H3PO4, HNO3, and HCl etch Y-Ba-Cu-O materials. These chemical solutions have been intentionally used in patterning processes [10]. No obvious etching of Y-Ba-Cu-O oxides in HF (49% HF or buffer HF) was observed even though the sample was immersed in HF for 20hr. It seems that HF only destroys the erosion products on the Y-Ba-Cu-O surface due to exposing to water vapor or carbon dioxide in air. XRD data support this deduction. As can be seen from Figure 1, the X-ray diffraction patterns were essentially identical before (a) and after (b) HF treatment, where the Y-Ba-Cu-O powders in Figure 1 (b) were directly soaked in 49% HF for 20min. However, a relatively higher background in the X-ray

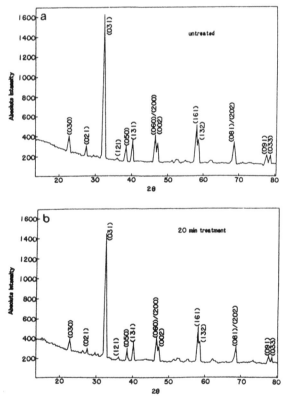

Fig. 1 X-ray diffraction patterns before (a) and after (b) HF treatment, where the Y-Ba-Cu-O powders were immersed in 49% HF at room temperature for 20 min.

patterns was observed with prolonged immersion time, and 60min treatment made the structure of Y-Ba-Cu-O oxides much more complicated. The EDAX analysis showed the HF treated powders to be Ba-rich and Cu- deficient which coincided well with the AES surface scanning data. No Cu and O signals were detected from AES surface survey of the HF immersed Y-Ba-Cu-O oxides. The lack of oxygen on the surface of the sample could be easily understood from the reaction between HF and erosion products on the Y-Ba-Cu-O surface owing to exposing to air [11].

Figure 2 shows the passivation effects from HF treatment of the Y-Ba-Cu-O oxides, where the HF soaking time was 2 min. As can be seen from this figure, a reduced degradation rate of zero resistance temperature of Y-Ba-Cu-O with water immersion was obvious for HF-treated samples compared to untreated ones. It should be mentioned that the top amorphous layer formed due to the reaction between HF and erosion material of Y-Ba-Cu-O should be scraped off in order to form a good contact before the contact pads were attached to the sample surface. The formation of an amorphous fluoride layer on the film surface made the R-T characteristics unstable if the contact pads were directly attached to the treated surface although this layer might be directly related to the improved water resistance of the Y-Ba-Cu-O oxides. The incorporation of fluorine in the surface region of bulk Y-Ba-Cu-O oxides after HF immersion was confirmed by AES surface scanning spectra [11].

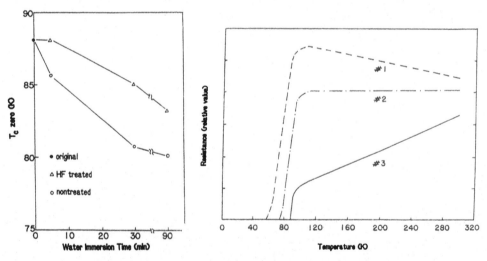

Fig. 2 Comparison of water resistance of Y-Ba-Cu-O before and after HF treatment, where the bulk Y-Ba-Cu-O oxides were immersed in 49% HF at room temperature for 2 min.

Fig. 3 Resistance vs temperature characteristics of Y-Ba-Cu-O thin films with different compositions. The chemical sensitivity of the film toward HF is related to the film composition. Film 1 would etch in HF whereas 2 and 3 would not etch.

Y-Ba-Cu-O Thin Films

No obvious etching of Y-Ba-Cu-O oxides was observed when Y- Ba-Cu-O bulk materials were immersed in 49% HF. However, this became much more complicated in Y-Ba-Cu-O thin films due to the deviation of the film composition from the stoichiometry of $YBa_2Cu_3O_{7-x}$ as in bulk oxides. As illustrated in Figure 3, film #1 with semiconductor-like characteristics above onset temperature was etched by HF, but films #2 and #3 with flat or metallic characteristics above onset temperature, were not etched. It is possible that HF can be used as a preliminary test solution to see if the film composition is right.

To apply this novel passivation technique to thin films, other difficulties were also encountered. The formation of an amorphous layer on the film surface made the electrical measurement much more difficult compared to that of bulk materials. The thickness and uniformity of the surface passivation layer were not able to be easily controlled although a microliter pipette was used to drop HF solution onto the film surface spinning at 3600rpm [8]. The contact problem was solved by forming multi-layer metal Al/Cr/Yb contacts [9] before HF treatment in order to realize stable electrical measurement used in the later characterization of the thin films. High uniformity and controllability were accomplished by treating the films with wet HF vapor. A detailed description of this set-up will be published elsewhere.

Figure 4 shows the AES depth profile on the surface region of the films after wet HF vapor treatment. In order to demonstrate the fluorine signal clearly, only fluorine is included in Figure 4. As can be seen from this figure, fluorine was incorporated in the surface region of the films. It is believed that the formation of this layer is directly related to the improved water resistance.

Figure 5 shows the in-situ measurement of the film resistance during the DI water immersion. Completely different behavior was observed for HF-treated films compared to untreated films. The resistance of the HF-treated films decreased but increased for untreated films with water immersion. R-T measurement showed no degradation of zero resistance temperature at 91K for wet HF vapor treated films immersed in DI water for 30min, but a serious degradation for untreated films. These included the increase of room temperature resistance, a broadening of transition width, the decrease of zero resistance temperature, and the transition from metallic to semiconductive behavior above onset temperature.

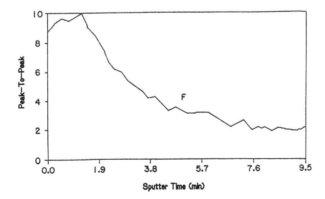

Fig. 4 AES depth profile on the surface region of the Y-Ba-Cu-O thin films after wet HF vapor treatment.

CONCLUSIONS

Surface passivation of both bulk and thin film Y-Ba-Cu-O have been realized using HF treatment of the superconducting Y- Ba-Cu-O. The formation of a fluoride compound on the sample surface might be directly related to the improved water resistance of the Y-Ba-Cu-O after HF treatment. Chemical passivation of Y-Ba-Cu-O using HF treatment may provide other advantages, such as simple processing and compatible processing with semiconductor technology.

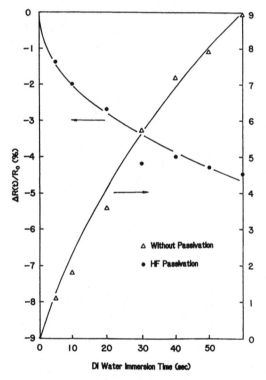

Fig. 5 In-situ measurement of the Y-Ba-Cu-O film resistance during DI water immersion.

ACKNOWLEDGMENT

We gratefully acknowledge the technical assistance from P.Bush and R.Barone in materials analysis studies. This work was supported by the New York Energy Research and Development Authority through the Institute on Superconductivity.

REFERENCES

1. M.F. Yan, R.L. Barns, H.M. O'Bryan, P.K. Gallagher, R.C. Sherwood and S. Jin, Appl. Phys. Lett., 51, 532 (1987).
2. N.P. Bansal and A.L. Sandkuhl, Appl. Phys. Lett., 52, 323 (1988).
3. C.A. Chang, Appl. Phys. Lett., 53, 1113 (1988).
4. D.M. Hill, H.M. Meyer, III, J.H. Weaver and D.L. Nelson, Appl. Phys. Lett., 53, 1657 (1988).
5. Q.X. Jia and W.A. Anderson, J. Appl. Phys., 66, 452 (1989).
6. K. Sato, S. Omae, K. Kojima, T. Hashimato, and H. Koinuma, Jpn. J. Appl. Phys., 27, L2088 (1988).
7. Q.X. Jia and W.A. Anderson, Presented at XVI UNY-VAC Conf., Rochester, NY, June 7-9, 1989.
8. R.P. Vasquez, B.D. Hunt and M.C. Foote, Appl. Phys. Lett., 54, 2373 (1989).
9. Q.X. Jia and W.A. Anderson, J. Phys. D: Appl. Phys., in press.
10. I. Shih and C.X. Qiu, Appl. Phys. Lett., 52, 1523 (1988).
11. Q.X. Jia and W.A. Anderson, J. Mater. Res., in press.

CARBON FIBER REINFORCED TIN-SUPERCONDUCTOR COMPOSITES

C.T. Ho and D.D.L. Chung

Composite Materials Research Laboratory, Furnas Hall
State University of New York
Buffalo, NY 14260, U.S.A.

INTRODUCTION

The high T_c superconductors are brittle, hard to shape, and high in normal-state electrical resistivity. On the other hand, metals are in general ductile, formable, and have low electrical resistivity and high thermal conductivity. In contrast to metals, polymers in general have high electrical resistivity and low thermal conductivity. Therefore, the combination of a superconductor and a metal in the form of a composite[1-4] is more attractive than the combination of a superconductor and a polymer.[5]

Powder metallurgy has been used by a number of workers to fabricate superconductor-metal composites[1-4]. This method involves mixing superconductor powder and metal powder and then sintering[1-3] or mixing metallic Y, Ba, Cu and Ag and oxidizing. The main drawback of this method includes the following:

(a) The metal content is limited to 50 vol.% or below in order to have a continuous superconducting path in the composite. This limits the ductility of the composite.

(b) The choice of metal is limited to metals that are stable at the sintering temperature in oxygen and do not react with the superconductor at the sintering temperature (typically 950°C for $YBa_2Cu_3O_{7-\delta}$).

Yet another method that has been used in forming a superconductor-metal composite involves (i) packing a superconductor powder in a metal tube, (ii) drawing the tube to a smaller diameter, and (iii) sintering[6-8]. The main drawback of this technique is that the choice of metal is limited to metals that are stable at the high sintering temperature required by the superconductor.

In this paper, we have developed a diffusion bonding technique for fabricating superconductor-metal composites. This method solves both of the problems described above for powder metallurgy and metal tube drawing.

Because brittle ceramics are much stronger in compression than in tension, it is necessary to strengthen the composite further, particularly in tension. For this purpose, we have used carbon fibers, because carbon fibers are very strong (especially in tension). They have a nearly zero thermal expansion coefficient (valuable for thermal fatigue resistance), very high thermal conductivity (better than copper), and quite low electrical resistivity and are corrosion resistant. We report here that carbon fiber reinforced tin-superconductor composites provide an effective way of packaging ceramic superconductors (in bulk, film or wire form) so that the package exhibits good mechanical, electrical and thermal properties, for applications in superconducting cables or tapes. The high strength will be particularly valuable for applications under high magnetic field.

FABRICATION

The superconductor was $YBa_2Cu_3O_{7-\delta}$. It was prepared by pressing $YBa_2Cu_3O_{7-\delta}$ powder (W.R. Grace and Co., Davison Chemical Division, super T_c-123, code: S-001) at a pressure of 137 MPa and then sintering at $950^{\circ}C$ for 12 h, followed by annealing at $420^{\circ}C$ in flowing oxygen for 1 h.

The tin used was in the form of foils (Fisher Scientific Co.) containing at least 99.8 wt.% Sn.

The carbon fibers used were continuous, PAN-based (Celion GY-70) and sized, with a tensile modulus of 517 GPa, a tensile strength of 1862 MPa and a tensile ductility of 0.36%. (Desized fibers were found to give similar results as the sized fibers.)

The carbon fiber reinforced tin-superconductor composite was prepared by using a two-step process. The first step involved the preparation of a tin-matrix unidirectional carbon fiber composite (abbreviated MMC) by laying up carbon fibers and tin foils in the form of alternate layers and consolidating by hot pressing (by using a hydraulic press) at $5^{\circ}C$ above the melting temperature of tin and at a pressure in the range from 8,000 to 10,000 psi (from 55 to 69 MPa) for 10-20 s. The minimum pressure for the first step is about 55 MPa. The second step involved hot pressing a layer of superconductor sandwiched by two layers of MMC at $170^{\circ}C$ and 500 psi (4 MPa) for 15 min in order to achieve diffusion bonding. Note that $170^{\circ}C$ is below the melting point of tin ($232^{\circ}C$). The minimum temperature for the second step is $150^{\circ}C$ for composites containing carbon fibers and $100^{\circ}C$ for those containing no carbon fibers. The minimum pressure for the second step is about 500 psi (4 MPa).

CHARACTERIZATION

Fig. 1 shows a scanning electron microscope (SEM) photograph of a portion of a polished section of an MMC-$YBa_2Cu_3O_{7-\delta}$-MMC three-layer composite containing 32.1 vol.% fiber and 50.2 vol.% Sn. The left half of the photograph is the MMC, the right half is the superconductor. No void or crack was observed between the MMC and the superconductor or between the tin and the carbon fibers in the MMC.

Fig. 2 shows the dependence of the electrical resistivity on temperature for the plain superconductor (solid circles), for a composite without fibers but containing 80.1 vol.% Sn (open circles), and for a composite containing 32.1 vol.% fiber and 50.2 vol.% Sn (crosses).

The electrical resistivity was measured with the four-probe

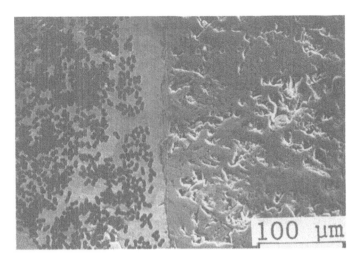

Fig. 1 SEM photograph of a portion of a polished section of an MMC-$YBa_2Cu_3O_{7-\delta}$-MMC three-layer composite containing 32.1 vol.% fiber and 50.2 vol.% Sn. The left half is the MMC; the right half is the superconductor. The dumbbells in the MMC are the tips of carbon fibers.

technique by using a Keithley 181 nanovoltmeter and a Keithley 224 programmable current source, such that the current was around 10^{-3} A. A thermocouple was placed so that it almost touched the sample. In the case of composites containing fibers, the electrical resistivity was measured in the direction of the fibers. (However, the electrical resistivity in the direction perpendicular to the fiber axis was comparable to that in the direction parallel to the fibers.) The critical temperature T_c is essentially the same for the plain superconductor and either composite, though the drop to zero resistivity

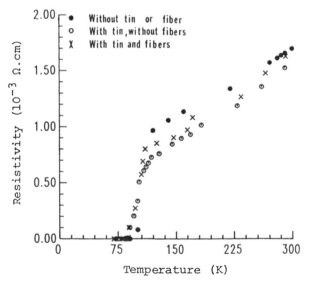

Fig. 2 Dependence of the electrical resistivity on temperature for the plain superconductor (solid circles), for a composite containing 80.1 vol.% Sn and no fibers (open circles), and for a composite containing 32.1 vol.% fibers and 50.2 vol.% Sn (crosses).

is slightly less sharp for either composite than the plain superconductor. Carbon fibers have a lower electrical resistivity than the plain superconductor above T_c, but its value is higher than that of tin. Therefore, the normal-state electrical resistivity of the composite with tin and fibers is lower than that of the plain superconductor and higher than that of the composite with tin and no fiber.

The critical current density J_c was measured at 40 K. Its value was determined by taking the slope of the current-voltage plot at 0.1 μV and extrapolating this slope to zero volt. Table 1 lists the J_c values obtained for the plain superconductor (1.13 mm thick), a composite containing 37 vol.% superconductor (1.82 mm thick) and 63 vol.% tin (3.13 mm thick), and a composite containing 19 vol.% superconductor (1.21 mm thick), 51 vol.% tin and 30 vol.% carbon fibers (4.81 mm thick for the MMC). The J_c values are typical of those of superconductors prepared by the sintering of powders. Table 1 shows that the composite processing does not degrade J_c.

Mechanical testing was performed using a hydraulic Materials Testing System (MTS). The strain in compressive testing was measured by the displacement of the crosshead. The strain in tensile testing was measured by using a strain gage (Measurements Group, Inc., gage type EA-13-120LZ-120, resistance = 120.0 ($\pm$ 0.3%) ohms, gage factor = 1.095 $\pm$ 0.5% at 75°F). The gage length was 34.7 mm for tensile testing and 6 mm for compressive testing. Compressive testing was performed with the force perpendicular to the laminate layers and with the force along the fiber direction. Tensile testing was performed with the force parallel to the fibers in the plane of the laminate. At least three samples were run and the data were averaged for each composite composition in each type of test.

Fig. 3 shows the effects of both fiber and tin contents on the compressive ductility. Tin greatly increased the ductility, but the fibers decreased the ductility.

Fig. 4 shows the effects of both tin and fiber contents on the compressive elastic modulus and compressive strength. The decrease of the modulus with increasing tin content and decreasing fiber content is significant. For the compressive strength, no systematic trend was found, because the strength is very sensitive to small flaws that are bound to be present in the superconductor.

Fig. 5 and 6 show the effects of tin and fiber contents on the tensile test results. The tensile test was performed along the fiber direction. The addition of carbon fibers greatly improved the tensile strength (Fig. 5), but decreased the ductility slightly (Fig. 6). The tensile modulus was increased by increasing the fiber content. Debonding between the MMC and the superconductor and some fiber pull-out were observed from the fracture surface after the tensile test.

Table 1 Critical current densities (J_c)

Material	J_c^* (A/mm^2)
Superconductor	1.47
Superconductor + tin	1.45
Superconductor + MMC	1.44

Each number was the average for three samples, with a data scatter of $\pm$ 2%

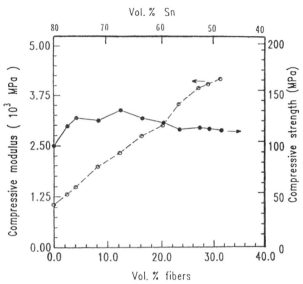

Fig. 3 Effects of tin and fiber contents on the compressive ductility. Each data point is the average value of three samples and the scatter is ± 5%.

A composite containing 31.0 vol.% fibers and 50.2 vol.% tin was subjected to tension up to a tensile stress of 90.6 MPa and a tensile strain of 0.57% (below the stress or strain required for fracture) and then the load was released (allowing the strain to return to essentially zero) and the electrical resistivity was measured as a function of temperature. It remained superconducting, with J_c=1.41 A/mm^2.

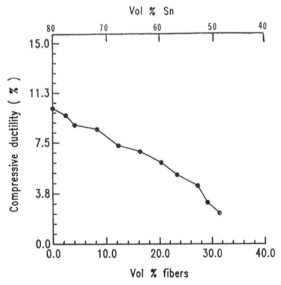

Fig. 4 Effects of tin and fiber contents on the compressive elastic modulus; each data point is the average value of three samples and the scatter is ± 8%. Effects of tin and fiber contents on the compressive strength; each data point is the average value of three samples and the scatter is ± 9%.

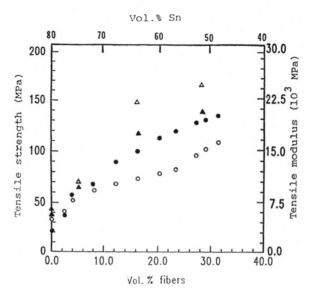

Fig. 5 Effects of tin and fiber contents on the tensile strength at room temperature (●) and 77 K (▲); each data point is the average value of three samples and the scatter is ± 8%. Effects of tin and fiber contents on the tensile modulus at room temperature (○) and 77 K (△); each data point is the average value of three samples and the scatter is ± 7%.

Composites containing from 0 to 30 vol.% fibers were subjected to compression along the fiber direction. Table 2 shows the compressive strength, modulus and ductility thus obtained. The compressive failure was due to delamination at the interface between the superconductor and the tin (or MMC). The presence of the fibers decreased the compressive

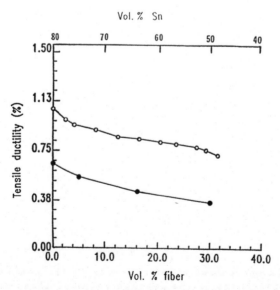

Fig. 6 Effects of tin and fiber contents on the tensile ductility at room temperature (○) and 77 K (●). Each data point is the average value of three samples and the scatter is ± 5%.

Table 2 Mechanical properties of superconducting composites
upon compression along the fiber direction

Carbon fiber content	Sn content	Compressive strength (MPa)	Compressive modulus (GPa)	Compressive ductility
0	54.7%	87.12	2.35	3.63%
0	62.3%	91.12	2.29	3.91%
3.0%	52.1%	71.82	2.46	3.42%
7.0%	54.3%	73.63	2.49	3.32%
30.1%	51.3%	66.23	2.53	3.24%

strength, showing that the fibers weakened the bonding between the superconductor and the tin (or MMC). The presence of the fibers increased the compressive modulus slightly and decreased the compressive ductility slightly.

Comparison of Table 2 and Fig. 4 shows that the compressive strength and modulus values perpendicular to the fiber direction are higher than those along the fiber direction for the same volume percent fibers. Comparison of Table 2 and Fig. 6 shows that the compressive ductility along the fiber direction is lower than that perpendicular to the fiber direction for the same volume percent fibers.

Composites of various fiber contents were subjected to thermal cycling between room temperature and liquid nitrogen temperature (77 K) by immersion in liquid nitrogen for 20 min, followed by room temperature equilibration for at least 30 min, and repeating. After every cycle, each specimen was observed under an optical microscope to look for delamination (slight cracking) between the superconductor and the MMC. Table 3 shows the number of cycles for the start of delamination for each composite composition. The higher was the carbon fiber content, the greater was the number of cycles for the start of delamination.

Because the superconductors are used at liquid nitrogen temperature (77 K) in practice, the mechanical properties of the composites were tested at 77 K also. The tensile testing was performed using the same Materials Testing System at the same strain rate. However, the sample was held with a fixture (OSW-1, Comten Industries), with very severely serated wedges, that was a positive clamp, becoming tighter as force was applied. The fixture and the sample were completed immersed in the liquid nitrogen in a cryostat during the testing. The results are shown in Fig. 5 and 6, where each data point is the average of those of three

Table 3 Thermal fatigue due to cycling between room temperature and liquid nitrogen temperature

Carbon fiber content (vol.%)	Sn content (vol.%)	No. of cycles for delamination to start*
0	43.2	63 ($\pm$ 4%)
3.0	39.4	86 ($\pm$ 3%)
8.3	40.1	102 ($\pm$ 5%)
12.0	43.1	116 ($\pm$ 2%)
15.3	48.3	123 ($\pm$ 5%)
20.1	50.2	127 ($\pm$ 2%)

*Each number is the average of three specimens.

specimens. The modulus was significantly increased while the strength was slightly increased when the temperature was lowered from room temperature to 77 K (Fig. 5). The ductility was decreased when the temperature was lowered from room temperature to 77 K (Fig. 6). In particular, with 29.4 vol.% fibers and 50.3 vol.% Sn, the tensile strength was 140.7 MPa, the modulus was 26.4 GPa and the ductility was 0.40%. This effect of temperature is attributed to the effect of temperature on the mechanical properties of carbon fibers and of tin, both of which have higher moduli at 77 K than room temperature. These results show that the superconducting composites are mechanically sound at 77 K.

DISCUSSION

Comparison of Fig. 4 and 5 shows that, without fibers, the tensile strength (parallel to the fiber direction) was much less than the compressive strength (perpendicular to the fiber layers), but with about 24 vol.% fibers, the tensile strength was approximately equal to the compressive strength. With further increase of the fiber content, the tensile strength exceeded the compressive strength.

The tensile strength of the plain superconductor was roughly 5.2 MPa. The presence of tin (without fibers) increased the value to about 20 MPa (Fig. 5). Further addition of carbon fibers significantly increased the tensile strength, up to 134 MPa for 31 vol.% fibers at room temperature and 141 MPa for 29 vol.% fibers at 77 K. The tensile modulus was 16 GPa at room temperature for 31 vol.% fibers and was 26 GPa at 77 K for 29 vol.% fibers.

Carbon fibers increased the compressive strength and compressive modulus perpendicular to the fiber layers, and also increased the tensile strength, tensile modulus and compressive modulus along the fiber direction, but they decreased the compressive ductility perpendicular to the fiber layers, the compressive strength and ductility along the fiber direction and the tensile ductility. However, because tin was also present and tin is a soft metal, the compressive ductility perpendicular to the fiber layers for the case of 31 vol.% fibers was approximately equal to that for the plain superconductor (without tin or fiber). For carbon fiber contents less than 30 vol.% fibers, the compressive ductility perpendicular to the fiber layers exceeded that of the plain superconductor. In general, the carbon fibers decreased the tensile ductility (Fig. 6) and the compressive ductility along the fiber direction (Table 2) much less than the compressive ductility perpendicular to the fiber layers (Fig. 3).

The superconducting behavior (T_c and J_c) of the composites was maintained after tension almost to the point of tensile fracture.

The fabrication of the composites involved low temperatures. The simplicity of this process makes it possible for an operation to be set up for fabricating continuous superconducting cables which are both shielded and toughened by tin and strengthened by carbon fibers. In contrast to powder metallurgy, the diffusion bonding method allows the metal to be the major phase while still maintaining a continuous superconducting path in the composite. Furthermore, carbon fibers, with their nearly zero thermal expansion coefficient helps match the thermal expansion coefficients of the MMC layer and the superconductor layer. This matching is necessary in order to enhance the durability of the composite to thermal cycling (i.e. thermal fatigue). In addition, carbon fibers are excellent in thermal conductivity (both at ambient and cryogenic temperatures), wear resistant, corrosion resistant, and are low in electrical resistivity.

The process described in this paper can be applied to package multiple layers of ceramic superconductors (separated by tin) in bulk, film or wire form, although only a single layer of bulk superconductor was used here. It can also be applied to package superconductor-metal composites prepared by powder metallurgy and to package superconductors prepared by melt texturing or other techniques. Low temperature metals other than tin (such as tin-lead, indium, etc.) can also be used.

CONCLUSION

Unidirectional and continuous carbon fiber tin-matrix composites were used for the packaging of the high-temperature superconductor $YBa_2Cu_3O_{7-\delta}$ by diffusion bonding at $170^\circ C$ and 500 psi. Tin served as the adhesive and to increase the ductility, the normal-state electrical conductivity and the thermal conductivity. Carbon fibers served to increase the strength and the modulus, both in tension along the fiber direction and in compression perpendicular to the fiber layers, though they decreased the strength in compression along the fiber direction. Carbon fibers also served to increase the thermal conductivity and the thermal fatigue resistance. At 24 vol.% fibers, the tensile strength was approximately equal to the compressive strength perpendicular to the fiber layers. With further increase of the fiber content, the tensile strength exceeded the compressive strength perpendicular to the fiber layers, reaching 134 MPa at 31 vol.% fibers. For fiber contents less than 30 vol.%, the compressive ductility perpendicular to the fiber layers exceeded that of the plain superconductor. At 30 vol.% fibers, the tensile modulus reached 15 GPa at room temperature and 27 GPa at 77 K. The tensile load was essentially sustained by the carbon fibers and the superconducting behavior was maintained after tension almost to the point of tensile fracture. Neither T_c nor J_c was affected by the composite processing.

ACKNOWLEDGEMENT

The authors are grateful to Mr. Shy-Wen Lai of SUNY/Buffalo and Mr. Scott Cooper of Union Carbide Corp. (Linde Division) for technical assistance. This research was supported by an award from the New York State Institute on Superconductivity in conjunction with the New York State Energy Research and Development Authority.

REFERENCES

1. In-Gann Chen, S. Sen and D.M. Stefanescu, Appl. Phys. Lett. 52 (16), 1355 (1988).

2. F.H. Streitz, M.Z. Cieplak, Gang Xiao, A. Gavrin, A. Bakhshai and C.L. Chien, Appl. Phys. Lett. 52, 927 (1988).

3. A. Goyal, P.D. Funkenbusch, G.C.S. Chang and S.J. Burns, Mater. Lett. 6(8-9), 257 (1988).

4. E.A. Early, C.L. Seaman, M.B. Maple and M.T. Simnad, Physica C (Amsterdam), 153-155 (Pt. II), 1161-2 (1988).

5. K. Ravi-Chandar, C. Vipulanandan, N. Dharmarajan and K.P. Reddy, Proc. Intersoc. Energy Convers. Eng. Conf., 23rd (Vol. 2), 525-9 (1988).

6. R.W. McCallum, J.D. Verhoeven, M.A. Noack, E.D. Gibson, F.C. Laabs and D.K. Finnemore, Advanced Ceramic Materials 2 (3B), 388 (1987).

7. S. Jin, R.C. Sherwood, R.B. Van Dover, T.H. Tiefel and D.W. Johnson, Jr., Appl. Phys. Lett. 51, 203 (1987).

8. S. Matsuda, M. Okada, T. Morimoto, T. Matsumoto and K. Aihara, Mat. Res. Soc. Symp. Proc. Vol. 99, 695-698 (1988).

SINGLE CRYSTAL GROWTH OF 123 YBCO AND 2122-BCSCO SUPERCONDUCTORS

K. W. Goeking[1], R. K. Pandey[1], G. R. Gilbert[1,3], S. Nigli[1],
P. J. Squattrito[2] and A. Clearfield[2]

Texas A & M University, College Station, Tx 77843-3128

[1] Electronic Materials Laboratory, Electrical Engineering Department; [2] Chemistry Department; [3] Now with Stewart and Stevenson Services, Inc., Houston, Tx.

INTRODUCTION

Anisotropic properties of electronic materials are exploited in a number of ways to improve the performance of electronic devices. Materials such as magnetics, non-linear dielectrics, superconductors and semiconductors play very important roles in the advancement of electronic technology. The origin of the anisotropy in the physical properties lies in the fundamental crystal structure of materials. Since the discovery of superconductivity in the families of YBaCu-oxides (YBCO), BiCaSrCu-oxides (BCSCO), and TlCaBaCu-oxides, many novel and potential applications have been proposed to utilize the anisotropic nature of these materials exhibited in their basic physical properties, especially in electrical and thermal conductivities and critical current densities. Many efforts have been made to synthesize the high temperature superconductors (HTS) in bulk and thin film forms to attain superior physical, chemical and mechanical properties. For a large number of applications high values of both the critical temperature (T_c) and the critical current density (J_c) are desirable. In this respect, studies done on single crystals are important. Only through the systematic studies of single crystals of the new superconducting oxides the basic understanding of this recently discovered phenomenon can be understood. Technically, on the other hand, many high performance devices can be built using substrates of high quality bulk single crystals. For example, long wavelength infrared detectors (15- 30 μm), millimeter and microwave devices, high performance and dense memory chips, large bandwidth IC devices (5-6 GHz bandwidth; a range far greater than any semiconductor device is capable of performing) etc., can be built for satisfactory performance if large device quality single crystals substrate of HTS materials were readily available. Furthermore, fabrication of highly efficient JOS-FET (Josephson junction field effect transistor) and SAW (surface acoustic wave) device having 10-100 MHz bandwidth can also be built using defect-free substrates of HTS crystals. From these few examples it is evident that single crystals are important both for scientific understanding of the physical mechanism of high temperature superconductivity as well as for the development of new technologies.

CRYSTAL GROWTH

From the vast amount of literature available it is obvious that most of the investigations on HTS materials have been carried out on polycrystalline samples – both in bulk and film forms. In comparison, very limited amount of effort has been directed toward the growth of single crystal. This is because of the chemical complexities of the oxide systems exhibiting high temperature superconductivity. The growth techniques applied to date have been principally of three closely related categories: solid state re-crystallization, flux growth, and self-fluxing. Here we present the results of our research for synthesizing bulk single crystals of Y- and Bi- based superconductors.

123-YBCO ($YBa_2Cu_3O_{6.5+x}$)

The charge consisted of a mixture in which Y, Ba, and Cu were present in the atomic ratio of 1:2:3. Appropriate amounts of high purity grade chemicals were first thoroughly mixed and then calcined at 925°C in air for about 20 to 30 hours. On cooling to room temperature, a solid mass consisting of green and gray portions was obtained. This was reground to obtain fine powder, 40 to 90 μm in particle size, and calcined. After repeating the steps of solid state reaction and calcination several times, a solid mass, uniformly black in color, was obtained. The X-ray diffraction analysis confirmed it to be orthorhombic with lattice constants as reported in literature for single phase 123-YBCO ceramic [1]. The details of processing orthorhombic 123-charge for ceramic pressing, film and single crystals growth have been reported in our previous work [2, 3].

The carefully prepared charge was mixed with $BaCO_3$ and CuO so that the mixture had a nominal composition of $YBa_4Cu_{10}O_{15.5}$. This composition has been reported to produce the best quality crystals [4]. The mixture of $BaCO_3$ and CuO in the molar ratio of 1:3.5 acts as self-fluxing for the growth of YBCO crystals. For a good growth a 100 g mixture of charge and flux consists of: Y_2O_3 = 6.648 g, $BaCO_3$ = 46.51 g, and CuO = 46.84 g. Charge and flux having this mass ratio were packed in a partially split alumina crucible which was supported in a loosely fitting shallow alumina dish. The crystal growth experiments were carried out in a computer controlled resistive furnace in air atmosphere. First the temperature was raised to 1050° C at the rate of 200° Ch^{-1} and held there for the molten mass to soak for about 10 hours. Then the temperature was brought to about 850° C at a cooling rate of 2° Ch^{-1}. Then a faster cooling rate of 100° Ch^{-1} was used to bring the crucible to room temperature. After separating the split crucible from the dish, we observed that the majority of crystals grew outside the crucible and in that region of the dish where the contact angle for the melt owing to its high surface tension had the best chance to support itself. In our previous experiments we had noticed that the 123-YBCO melt has relatively large surface tension and it has therefore, the tendency to creep along the crucible wall and overflow. We have utilized this property of the melt to grow relatively large and well formed crystals by employing split crucible in our experiments. This modification in the standard crystal growth technique, as reported in references [4, 5, 6] has enabled us to grow free standing as well as large single crystals of $YBa_2Cu_3O_{6.5+x}$ superconductor. The fine split in the crucible, 0.5mm to 1mm in width and about 10mm long, apparently allows the super saturated melt to penetrate out of the interior of the crucible where the majority of the materials remains only partially molten. This then facilitates the crystals growth and phase separation between the flux and the crystals. It should be pointed out that harvesting of crystals from the solidified flux has been a major problem when crystals are grown by solid state recrystallization method. It is usually done by mechanical means which invariably damages the crystals and breaks them. The harvesting of the crystals is enormously simplified by using a split crucible. Once the crucible is separated from the dish, individual crystals can simply be picked up one by one.

BCSCO (BiCaSrCu-Oxide)

The single crystal growth of different phases of BCSCO poses a challenge as well as an enormous opportunity. It is now a well accepted fact that three distinct superconducting phases exist in this family. They are: 1. $Bi_2Sr_2Cu_1$-oxide (2021-phase, $T_c \approx$ 10 K; c$\approx$ 24Å); 2. $Bi_2Ca_1Sr_2Cu_2$-oxide (2122-phase, $T_c \approx$ 80 K; c$\approx$ 30Å); and 3. $Bi_2Ca_2Sr_2Cu_3$-oxide (2223-phase; $T_c \approx$ 110 K; c$\approx$ 37Å). This system is extremely sensitive to processing conditions. Therefore, the growth conditions must be carefully optimized for the synthesis of bulk crystals of each individual phase. Some efforts have been made to grow single crystals of BCSCO from eutectic melt of compositions having an excess of CuO [7-10]. This approach has produced tiny crystals which must be separated from the solidified flux by cumbersome mechanical means. Fortunately, KCl has been identified as an effective flux for BCSCO [12-13]. The well known technique of flux growth is the most suitable method for the growth of incongruently melting oxides.

In our initial experiments we used the BCSCO charge of 1112 nominal composition for crystal growth using KCl as the flux. The homogenous charge was prepared by multiple cycles of calcination, grinding, etc., as described previously. Only the temperature of processing was different from 123-YBCO and it was also dictated by the melting point of the specific phase. After the calcination the pellets were cold pressed at the pressing force of about 70 MPa and sintered in air for 10 to 100 hours. These samples showed superconductivity with a mixed phase of $T_c \approx$ 110 K and $T_c \approx$ 80 K. For crystal growth, the pellets were either crushed or kept intact and mixed with 80-98 est. % of KCl and put into a 50cc platinum crucible for crystal growth. The mixtures were heated between 870 and 895° C, held constant at these temperatures for 1-2 hours, and slowly cooled at the rate of 1-10° Ch^{-1} through a 100-200° C temperature. Cooling at the rate of 200° Ch^{-1} was employed thereafter. All crystal growth experiments were done in air, and the crystals were recovered by dissolving the flux in hot water. Small crystals grew by the standard slow cooling technique. However, the size improved substantially by introducing the method of temperature gradient transport (TGTM). In our experience, the TGTM is an excellent method for the growth of large and high quality crystals of oxides of fully or partially solid-solution systems [14]. In this method, the bottom of the crucible is kept at a higher temperature than the top surface of the melt. This introduces a convective current which facilitates the growth of crystals at the top of the crucible.

A series of other fluxes were also tried. They were: NaCl, K_2CO_3, and a mixture of 58% KCl and 42% KF (by mole; melting point $\approx$ 660°C). Only KCl and KCl·KF were found to be effective in synthesizing 2021- and 2122- phases from the charge of composition 1112. The range of crystallization temperatures have been found to be distinctly different for the formation of these two phases.

RESULTS AND DISCUSSION

By the self-fluxing system it has been possible to grow large and easy to harvest 123-YBCO crystals. The split crucible arrangement produces free standing and relatively large crystals which can be harvested without the mechanical means or the usual method of flux dissolution in water or mineral acids. The size of the crystals vary from microscopic in scale to as large as 6mm x 6mm and about 0.1 mm thick. A large number of crystals grew as platelets, Fig. 1. However, a few well faceted bulk crystals could also be crystallized as shown in Fig. 2. They are black in color and highly reflective. As grown crystals are usually of tetragonal structure. The orthorhombic structure is obtained only after prolonged annealing in flowing oxygen at about 950° C. The lattice constants of the two structures have been determined. In orthorhombic phase: a_o=3.831 (3) Å, b_o= 3.889 (4) Å, and c_o=11.703 (8) Å; and in tetragonal phase: a_o=3.861 (3) Å, b_o=3.862 (3) Å, c_o=11.716 (8) Å. These values are in very good agreement with the literature values [1,4]. The superconductivity phase change, in as grown crystals, was determined by the radio-frequency phase impedance technique at 11 MHz. A sharp superconducting transition with the midpoint value of 74 K was obtained. Other characterization of annealed samples have not yet been done.

For the growth of BCSCO crystals, it is important to note that the phase formation depends on the following four factors: 1. initial charge composition, 2. temperature range of crystallization, 3. convective currents, and 4. flux evaporation rates.

The amount of 2122-phase precipitated is directly proportional to the amount of 2223-phase present in the nutrient charge having the nominal composition of 1112. The 2122-phase ($T_c \approx$ 80 K and c$\approx$30Å) crystallizes when the slow cooling in the TGTM experiment is limited between 895 and 875° C. But when the crystallization is attempted between 875 and 855° C the low temperature superconducting 2021- phase ($T_c \approx$ 10K and c=24Å) grows. However, if the entire cooling range of 895 to 855°C is employed both 2021 and 2122 phases form. We have also determined that the 2122-phase crystals can be grown only from the 1112 charge consisting of a large portion of the 2223-phase ($T_c \approx$110K and c$\approx$ 37Å). Compared to the 2122-phase

Figure 1 Platelets of Y-Ba-Cu-Oxide Crystals

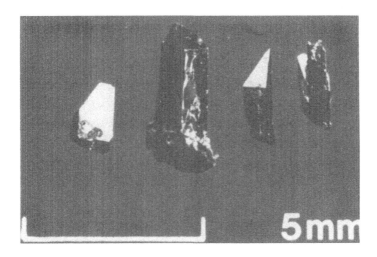

Figure 2 Bulk Crystals of Y-Ba-Cu-Oxide

the 2021-phase appears to be more stable so far as crystal growth is concerned. This phase precipitates irrespective of the fact if the nominal composition of the charge is 1112 or 2122. The details of crystal growth for BCSCO are described in our paper elsewhere [15].

Usually in flux growth technique it is important to achieve the super saturation of the melt which necessitates the proper mass ratio between the charge and the flux. We have determined from a large number of experiments that the charge to flux ratio is not a critical parameter for crystal growth of BCSCO. Selective dissolution of 110 K phase (2223) occurs which crystallizes in 80 K phase (2122). Because of the high vapor pressure of KCl it is important to cool the melt in the crystallization range at a fast rate. Cooling rate of 10° Ch^{-1} gave us the best results. The fast rate of cooling, however, may inhibit the growth and perhaps that may cause the BCSCO crystals to grow as few μm thick plates. This problem can perhaps be solved to some degree by using KCl·KF flux which is far less volatile than the KCl melt. Also, BCSCO is more soluble in KCl·KF than KCl alone, and like KCl it is water soluble. We have used KCl·KF flux in the cooling range of 870 to 850° C at the cooling rate of 2° Ch^{-1} to crystallize 2021-phase. For KCl·KF experiments charges of 1112-composition without 2223-phase were used. The work is in progress to crystallize the 80 K material by optimizing the growth conditions with the help of KCl·KF flux.

BCSCO grows as thin platelets by spontaneous nucleation. Its size varies from a few mm^2 to several cm^2. However, the thickness remains in the order of a few μm. The largest platelets obtained by TGTM using KCl flux measures over 1cm x 1cm. In Fig. 3, we show a few examples of BCSCO crystals grown in our laboratory. The top is shiny black, wavy and it consists of many smaller crystallites of grain size $\approx$ 1mm^2. The bottom surface, on the other hand, is not so well formed as the top surface. Optical microscope examination reveals long filamentary growth habit instead of plate-like morphology. The typical (00l) peaks, characteristic of c-axis oriented BCSCO crystals, are found in x-ray diffraction analysis.

Our samples of BCSCO were tested for superconductivity by high frequency magnetic induction technique. The frequency was 11 MHz and the magnetic field about 50 μT. The crystals as grown of 2122-phase became superconducting at 80 K exhibiting a sharp transition. But for 2021 samples, no superconducting temperature was determined down to 10 K. Further characterization of our samples are in progress and they will be reported at a later time.

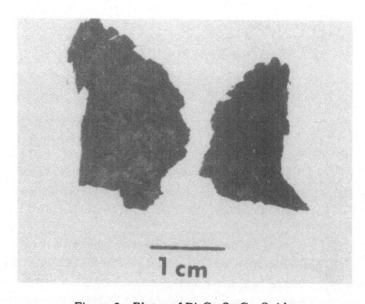

Figure 3 Plates of Bi-Ca-Sr-Cu-Oxide

Efforts were also made to synthesize 110 K phase BCSCO crystals (2223-phase) by using charges prepared with a partial substitution of PbO for Bi_2O_3 [16]. In KCl flux experiments, only 2021-phase could be crystallized and the samples showed no evidence of Pb.

CONCLUSIONS

Single crystals of 2021-phase ($T_c \approx$ 10K, c$\approx$ 24Å) and 2122-phase ($T_c \approx$ 80K, c$\approx$ 30Å) in the BiCaSrCu-oxide family have been grown successfully using KCl flux by the temperature gradient transport method. This technique allows one to grow large crystals by employing the optimized conditions of cooling range, charge composition, etc. This paper also discusses the usefulness of KCl·KF flux for the growth of BCSCO crystals. Pb-doped charges were used for the growth of 110 K phase, but we were unsuccessful in synthesizing this phase as single crystals. It may be possible to synthesize crystals of $Bi_2Ca_2Sr_2Cu_3O_x$ phase with c$\approx$37Å and $T_c \approx$ 110 K if the crystallization range and the suitable starting compositions without Pb can be optimized.

Also, relatively large and free standing crystals of $YBa_2Cu_3O_{6.5+x}$ have been synthesized using BaO·CuO flux by slow cooling the melt in a split crucible configuration. This modification may be utilized to grow larger crystals of YBaCu-oxide. The problem encountered with harvesting crystals is easily solved by using the split crucible configuration.

ACKNOWLEDGMENTS: We acknowledge the support of the following agencies for this research: National Aeronautics and Space Administration (Grant No.: NAWG-1590); Texas A & M University's NASA's Center for Space Power (Grant No.: NAWG-1194); DOE's Battelle Pacific Northwest Laboratories (Contract No.: PNL-030917-A-F1); Texas Engineering Experiment Station (Grant No.: TEES-32125-92870); and Texas A & M University's Board of Regents (Grant No.: AUF-24424-V). We also wish to express our thanks to Dr. W. P. Kirk for magnetic measurements; and to Mr. D. L. Mabius and Mr. C. S. Blanton for technical assistance.

REFERENCES

1. A. Kini, U. Geiser, H-C, I. Kao, K. D. Carlson, H. H. Wang, M. R. Monaghan, and J. M. Williams, Inorg Chem., 26, (1987), 1835.
2. R. K. Pandey, G. R.Gilbert, W. P. Kirk, P. S. Kobiela, A. Clearfield, and P. J. Squattrito, J. Supercond., 1, (1988), 45.
3. R. K. Pandey, M. A. Chopra, and R. G. Garlick, Mater. Lett. (in press).
4. L. F. Schneemeyer, J. J. Waszczak, T. Siegrist, R. B. Van Dover, L.W. Rupp, B. Batlogg, R. J. Cava, and D. W. Murphy, Nature, 328, (1987), 601.
5. J. Z. Liu, G. W. Crabtree, A. Umezawa and Li Zongquan, Phys. Lett. (A), 121, (1987), 305.

6. T. R. Dinger, T. K. Worthington, W. J. Gallagher, and R. L. Sandstrom, Phys. Rev. Lett., 58, (1987), 2687.

7. Y. Hidaka, Y. Enomoto, M. Suzuki, M. Oda, and T. Murakami, J. Crystal Growth, 85, (1987), 581.

8. K. Oka and H. Unoki, Jap. J. Appl. Phys., 26, (1987), L1590.

9. M. Hikita, T. Iwata, Y. Tajima, A. Katsui, J. Crystal Growth, 91, (1988), 282.

10. T. F. Ciszek, J. P. Goral, C. D. Evans, and H. Katayama-Yoshida, J. Crystal Growth, 91, (1988), 312.

11. A. Katsui, Jap. J. Appl. Phys., 27, (1988), L844.

12. A. Katsui and H. Ohtsuka, J. Crystal Growth, 91, (1988), 261.

13. L. F. Schneemeyer, R. B. van Dover, S. H. Glarum, S. A. Sunshine, R. M. Fleming, B. Batlogg, T. Siegrist, J. H. Marshall, J. V. Waszczak, and L. W. Rupp, Nature, 332, (1988), 422.

14. K. W. Goeking, R. K. Pandey, P. J. Squattrito, A. Clearfield, and H. R. Beraton, Ferroelectrics, 92, (1989), 89.

15. K. W. Goeking and R. K.Pandey, Mater. Lett. (in press).

16. M. Takano, J. Takada, K. Oda, H. Kitaguchi, Y. Miura, Y. Ikeda, Jap. J. Appl. Phys., 27, (1988), L1041.

THE EFFECT OF Ag/Ag$_2$O DOPING ON THE LOW TEMPERATURE SINTERING OF SUPERCONDUCTING COMPOSITES

A.Goyal, S.J.Burns and P.D.Funkenbusch

Materials Science Program
Department of Mechanical Engineering
University of Rochester, NY 14627

ABSTRACT

Practical applications of the high-T_c ceramic superconductors will require electrical, mechanical and thermal stabilization of the superconducting phase. Non-noble metal superconducting particulate composites have been formed with various metals, in order to achieve some of the desirable properties. One of the critical parameters in forming superconducting composites with non-noble metals is the processing temperature of the composites, which is dictated primarily by the need to sinter the ceramic particles into a continuous superconducting skeleton. The possilbility to lower the processing temperature of the composites, by the use of Ag/Ag$_2$O is explored. The relative effect of Ag and Ag$_2$O on the low temperature sintering of $Y_1Ba_2Cu_3O_{7-x}$ is discussed. The major difficulty in forming superconducting composites at low temperatures seems to be associated with poor contacts between the individual ceramic particles within the composite.

INTRODUCTION

The recent discovery of high-Tc superconductivity in various quarternery systems[1-4], has caused an unprecedented excitement, based on the potential of these materials for significant technological applications. Many of these applications involve the use of superconductors in bulk form. However it was soon realized that these materials have extremely low critical currents and poor mechanical properties in bulk form. High critical currents and good mechanical properties are necessary to achieve any large scale application.

Commercially useful high temperature superconductors would also require stabilization of the superconducting phase (i.e., adiabatic, enthalpy and cryostabilization) during service, besides the need for enhanced mechanical properties and critical currents. In respect to the considerable enhancement of mechanical properties and the need for stabilization, practical applications of these superconductors would invariably be in composite form. There exists a distinct possibility that one

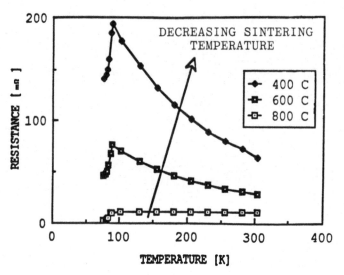

Figure.1. The effect of sintering temperature on pure ceramic samples compacted at 87.5ksi.

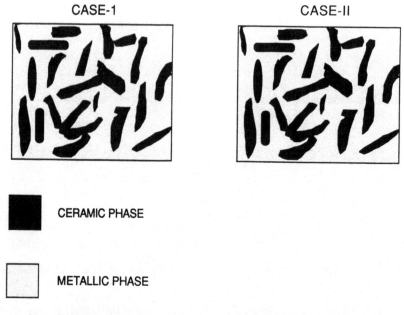

Figure.2a. Schematic microstructure of Type-A superconducting composites.

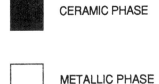

Figure.2b. Schematic microstructure of Type-B superconducting composites.

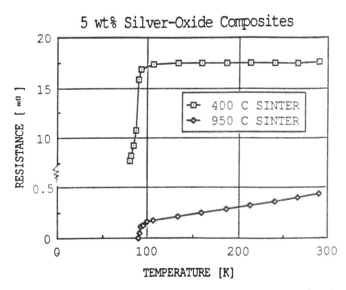

Figure.3. The effect of sintering temperature on the transport behavior of silver-oxide composites, compacted at 16ksi.

could adopt a 'traditional technique' for enhancing the mechanical properties of these ceramics, by incorporating the brittle ceramic phase into a ductile metal matrix. This involves the use of conventional metallurgical techniques i.e., Powder metallurgy and liquid phase sintering. This method of forming particulate superconducting composites would address both the needs for enhancement of mechanical properties and also provide the necessary stabilization of the superconducting
phase. It is also possible that procedures adopted to obtain high transport critical currents, i.e. by aligning or orienting the crystallites, could be employed while fabricating such superconducting composites. Also, with a J_c of $10^3 A/cm^2$, some applications are possible if the mechanical properties of these superconductors could be enhanced, while at the same time achieving stabilization of the superconducting phase.

The properties of these ceramics are not only dependent on the crystal structure and stochiometry of the phases present, but are also very sensitive to all the steps involved in their fabrication. Due to the high reactivity of the ceramic phase, processing of the superconducting composites is still more complex. If a non-noble metallic phase is used in forming a composite, then care has to be taken to keep the reaction between the two phases to a minimum. One of the critical parameters in forming the superconducting composites with non-noble metals is the processing temperature of the composites, which is dictated primarily by the need to sinter the ceramic particles into a continuous superconducting skeleton[5,6]. We have been investigating the possibility to lower the processing temperature for the composites, by the addition of additives i.e., Ag/Ag_2O. A comprehensive report on these experiments is currently being prepared[7]. In this communication we will summarize some of the results and report on preliminary experiments, applying these results to Ni-123 cermet fabrication.

LOW-TEMPERATURE SINTERING CHARACTERISTICS OF THE 123 PHASE

Polycrystalline 123 was prepared according to the procedure described in ref 5. The sintered compact thus obtained was reground into powder form, and sieved to -400 mesh size(~40 um). Magnetization measurements of this powder exhibited the Meissner effect. The resulting powder was pressed into pellets of 13mm diameter at a pressure of either 16ksi(~110MPa) or 87.5ksi(~603MPa). These pellets were then sintered for 0.5-2 hrs in the temperature range 300-800C, under flowing O_2 and subsequently furnace cooled. All these samples showed a visible ''Meissner effect'', tested by cooling the sample to liquid N_2 temperature and bringing it in the vicinity of a strong magnet.

Samples compacted at 16ksi tend to behave in a semiconductor-like fashion (i.e., the resistance increases with decrease in temperature), with no sign of a significant drop in resistivity at the transition temperature. However individual grains within the compact, undergo a superconducting transition upon cooling below the transition temperature, as is evidenced by the visible Meissner effect exhibited by all the samples. The contacts between the individual ceramic particles are poor, and as far as the electrical transport properties are concerned the grains are essentially disconnected or isolated. Hence the compacts thus obtained do not behave in a manner similar to those sintered at high temperatures ~950C, which show a metallic dependence of resistivity in the normal state, and a sharp transition to the superconducting state at ~90K.

Samples compacted at higher compaction pressures i.e., 87.5ksi and sintered in the temperature range 400-600C show a 'semiconductor-like' behavior upto the transition temperature and a significant drop in resistivity at the transition temperature. However the resistance does not go to zero at 77K. Samples sintered at 300C behaved in a manner similar to those compacted at lower compaction pressures(16ksi). However samples sintered at 800C, showed a metallic behavior above the transition temperature and a significant drop in resistivity at the transition temperature. Again, the resistance did not go to zero above 77K. Fig.1 illustrates the effect of sintering temperature on the transport properties of 123 compacted at 87.5ksi. A summary of the transport behavior of polycrystalline 123 in the low temperature sintering regime is given in Table 1.

The above results suggest that the contacts between the ceramic grains are poor and that this factor limits the low temperature ''sinterability''.

NON-NOBLE METAL SUPERCONDUCTING COMPOSITES OR CERMETS

We have demonstrated the sucessful fabrication of superconducting cermets with non-noble metals like, Sn, Cu and Ni, with as high as 60wt% of the metallic phase[5].

Two broad classes of granular superconducting composites or cermets have been identified on the basis of their electrical, magnetic, and microstructural behavior. The first class, designated as Type-A, can have two kinds of microstructure as indicated schematically in Fig.2a. Case-I where a discontinuous ceramic skeleton is embedded within a metallic matrix, and Case-II, where a physically contiguous ceramic network is intermeshed with a metallic matrix. In Case-II, although a physically contiguous ceramic network exists, there is no continuous superconducting skeleton. This class of composites exhibits the Meissner effect, but does not show macroscopic zero resistance above 77K. Composites of 20-60wt% Sn have been fabricated[5,6]. Magnetization measurements, indicating the presence of the Meissner effect in these composites are reported elsewhere[6]. With increasing volume fraction of the metallic phase, the hardness of the composites decreases, approaching the value for that of the pure metallic phase[6].

A schematic microstructure of the second class of composites is ilustrated in Fig.2b. These composites have a continuous superconducting skeleton intermeshed with a metallic skeleton. They show the Meissner effect and also show zero resistivity on a macroscopic scale, above 77K. From sintering experiments on the pure ceramic compacts, it has been shown that in order to achieve a macroscopic zero resistance,

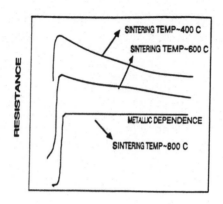

Figure 4. Schematic representation of the transport behavior of Ag-doped composites, compacted at 16ksi.

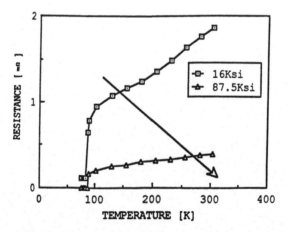

Figure 5. The effect of compaction pressure on the transport behavior of silver-oxide composites, sintered at 800°C.

one has to sinter the composites at high temperatures ~950C. Composites with Cu and Ni as the metallic phases have been successfully fabricated with a Tc onset of ~93K transition widths of 5-7K[5,6]. However the room-temperature resistance of these composites was significantly higher than that of the pure ceramic compacts. This could be due to the reaction between the ceramic and metallic phases, thereby resulting in a tenuous oxide layer around the metallic particles. Oxidation of the metallic phase will also occur due to the highly oxidizing atmosphere used while processing the composites, which is necessary to maintain the stochiometry of the 123 phase. These reactions could be significantly reduced if the processing temperature of the composites can be lowered, thus resulting in superior transport properties in the normal state. Improvement in the contacts between the ceramic grains, by the addition of certain additives(Ag/Ag_2O) can result in lower processing temperatures. The ''benevolent'' nature of silver additions to bulk 123 is now estabilished[8,9]. The use of Ag_2O to enhance the oxidation of the 123 phase has been reported[10,11]. Ag/Ag_2O could also aid in lowering the required sintering temperature of the 123 phase. The addition of silver/silver-oxide on the transport properties of 123 sintered at low temperatures is discussed below.

EFFECT OF Ag/Ag_2O ON THE TRANSPORT PROPERTIES OF 123 IN THE LOW TEMPERATURE SINTERING REGIME

Composites of Ag/Ag_2O with 123 were formed by mixing appropiate fractions of silver/silver-oxide with preformed 123 powders, following a procedure described elsewhere[5,6]. The resulting mixtures were compacted at either 16ksi or 87.5ksi into 13mm discs, and subsequently sintered at varying temperatures. The particle sizes of the Ag and Ag_2O used were -200 and -325 mesh size respectively. The composites were introduced into the furnace at the sintering temperature, isothermally heated for ~2hrs, followed by furnace cooling, all under flowing oxygen. Electrical transport measurements were carried out in a temperature controlled probe with the sample mounted on an isothermal copper block, and using a four point probe technique to eliminate the effect of contact resistances.

For samples compacted at a compaction pressure of 16ksi the results obtained are summarized in a qualitative manner in Table-2. For each composition studied, the sintering temperature, the presence of a visible Meissner effect, the nature of the temperature variation of resistivity, the existence of a significant drop in resistivity at the transition temperature, and the attainment of macroscopic zero resistivity above 77K are indicated in the Table.

All Ag_2O/123 composites sintered in the temperature range 400-800 C show a metallic dependence of resistivity as a function of temperature and a superconducting transition at the appropiate temperature. However the transition is not complete and the resistance does not go to zero above 77K. This is in sharp contrast to the pure ceramic samples compacted at 16ksi

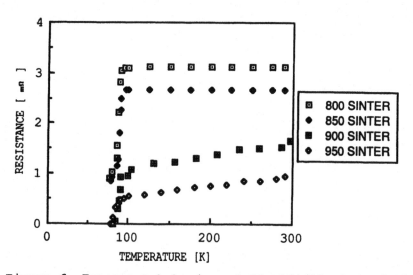

Figure.6. Transport behavior of 10wt%Ni/10wt%Ag$_2$O/123 composites, compacted at 87.5ksi.

and sintered in the same temperature range. It is important to note that as little 5wt% Ag_2O is sufficient to render this change in the transport behavior. Increasing the volume fraction of Ag_2O did not change the observed trend in the transport behavior, however the room-temperature resistance progressively decreased in accordance with the increased volume of the metallic phase. Increasing the sintering temperature to 950C however did result in a complete transition well above 77K. Fig.3 shows the electrical transport behavior of 5wt%Ag_2O samples compacted at 400C and 950C.

Ag/123 composites compacted at 16ksi and sintered in temperature range 400-600C show a 'semiconductor-like' dependence of resistivity with respect to temperature above the transition temperature and a significant drop in resistivity at the transition temperature. The transition is again not complete at 77K. For a sintering temperature of 800C, the temperature dependence of resistivity was metallic. Fig.4 shows the behavior of Ag/123 composites schematically. The difference between the transport behavior of Ag/123 compacts sintered in the temperature range 400-600C and 800C, can be understood in terms of the diffusion coefficient of O_2 in solid silver, which varies drastically as a function of temperature[7]. The difference between the transport behavior of Ag_2O/123 compacts and Ag/123 compacts in the low temperature sintering regime is consistent with the additional oxygen available from the dissociation of Ag_2O at ~400C[7].

DSC and TGA experiments were conducted to find the dissociation temperature of the Ag_2O powder used and to confirm that no undecomposed Ag_2O existed at temperatures higher than the dissociation temperature of silver-oxide, under the conditions of oxygen partial pressure employed to make the composites[7].

Increasing the compaction pressure of uniaxial compression from 16ksi to 87.5ksi drastically changed the resistive behavior of the Ag/Ag_2O-123 composites, as intutively expected, due to better interparticle contact. The results obtained are summarized in a qualitative manner in Table-3. Ag_2O-123 compacts sintered in the temperature range of 400-600C showed the same qualitative behavior as those compacted at lower pressures(Table-2), however the room-temperature resistance is greatly reduced. For a sintering temperature of 800C, macroscopic zero resistance was attained well above liquid nitrogen temperatures. The effect of compaction pressure on the transport properties of the Ag_2O-123 composites is indicated in Fig.5; which depicts the behavior of 20wt% Ag_2O-123 composites sintered at 800C as a function of compaction pressure.

Ag-123 compacts sintered in the temperature range 400-600C and compacted at 87.5ksi show a metallic dependence of resistivity above the transition temperature. This is in contrast to Ag/123 composites compacted at lower pressures, which showed a 'semiconductor-like' dependence of resistivity with respect to temperature above the transition temperature. For a sintering temperature of 800C, macroscopic zero resistivity above liquid nitrogen temperature was attained. Hence the addition of Ag/Ag_2O to the 123 phase does significantly alter the transport behavior of the resulting composite in the low temperature sintering regime. It needs to be mentioned, that in contrast to these results, Chen et al[10,11] reported zero resistivity in both the pure 123 compacts and 123/Ag$_2$O/Al composites sintered at only 600C. Compaction pressures in their work were slightly higher(640MPa), and sintering times slightly longer than those used here. Particle size of the their 123 powder was smaller than 45 um. These distinctions between the processing variables seem too small to account for the observed sintering differences, and their origin remains unclear. In order to determine if time was a significant factor, we sintered a few samples for longer times(~24 hrs), but no significant change was detected in the transport behavior. The sintering temperature appeared to be a far more dominant factor as opposed to the time at the sintering temperature. Unfortunately neither the sintering temperature nor the compaction pressure was varied in their work, and hence a comparison with qualitative trends observed here are not possible.

EFFECT OF Ag/Ag$_2$O ON THE TRANSPORT BEHAVIOR OF Ni-123 SUPERCONDUCTING COMPOSITES

The effect of the addition of additives i.e. Ag/Ag$_2$O, to a non-noble metal superconducting composite as described previously, with Ni as the metallic phase is discussed below. 10wt%Ni/10wt%Ag$_2$O/123 composites were compacted at 87.5 ksi. The particle size of the 123 powder and Ag$_2$O powder was -400 mesh, and that of the Ni powder was -100 mesh. The resulting composites were sintered between 800-950C for 30min under flowing oxygen. The composites were introduced into the furnace at the sintering temperature. The results obtained are illustrated in Fig.6. X-ray diffraction of the composites showed the presence of solid silver, nickel and nickel-oxide, with the 123 diffraction pattern remaining unchanged. Macroscopic zero resistance above liquid nitrogen temperature is attained only for composites sintered at temperatures at or above ~900C. The room-temperature resistivity of these composites is however reduced by an order of magnitude from that observed for composites without the addition of additives and compacted at 16ksi[5,6]. In fact the resistance of the composite sintered at 950C is lower that of the pure ceramic compacts.

SUMMARY

Sintering studies on pure ceramic compacts and Ag/Ag$_2$O-123 composites indicate that the contacts between the individual ceramic particles within the compacts are poor, and dictate the transport properties of polycrystalline 123 in the low temperature sintering regime. Electrical transport in both the normal(above the transition temperature) and the superconducting states(below the transition temperature) improves with increasing sintering temperature, increasing

Table.1. Summary of the transport behavior of polycrstalline 123 compacts as a function of the processing variables.

Compaction Pressure	Sintering Temperature	Meissner Effect	Temp. Dependence of Resistivity	Superconducting Transition	Zero(77K) Resistance
16ksi	400-800C	Yes	Semiconductor-like	No	No
16ksi	950C	Yes	Metallic	Yes	Yes
87.5ksi	300	Yes	Semiconductor-like	No	No
87.5ksi	400-600C	Yes	Semiconductor-like	Yes	No
87.5ksi	800C	Yes	Metallic	Yes	No
87.5ksi	950C	Yes	Metallic	Yes	Yes

Table.2. Summary of the transport behavior of Ag/Ag_2O-123 composites, compacted at 16ksi.

Sample Composition	Sintering Temperature	Meissner Effect	Temp. Dependence of Resistivity	Superconducting Transition	Zero(77K) Resistance
5-20 wt% Ag_2O	400-800C	Yes	Metallic	Yes	No
5-20 wt% Ag_2O	950C	Yes	Metallic	Yes	Yes
10 wt% Ag	400-600C	Yes	Semiconductor-like	Yes	No
10 wt% Ag	800C	Yes	Metallic	Yes	No
10 wt% Ag	~950C	Yes	Metallic	Yes	Yes

Table.3. Summary of the transport behavior of Ag/Ag_2O-123 composites, compacted at 87.5ksi.

Sample Composition	Sintering Temperature	Meissner Effect	Temp. Dependence of Resistivity	Superconducting Transition	Zero(77K) Resistance
20 wt% Ag_2O	300C	Yes	Semiconductor-like	slope change	No
20 wt% Ag_2O	400-600C	Yes	Metallic	Yes	No
20 wt% Ag_2O	800C	Yes	Metallic	Yes	Yes
20 wt% Ag	300C	Yes	Semiconductor-like	slope-change	No
20 wt% Ag	400-800C	Yes	Metallic	Yes	No
20 wt% Ag	800C	Yes	Metallic	Yes	Yes

compaction pressure and the addition of additives. The addition of small quantities of Ag/Ag_2O in forming superconducting composites enhances their electrical transport properties considerably. However the results obtained so far have not permitted lowering of the processing temperature
of our base metal composites below ~900C, in order to achieve a macroscopic zero resistance above 77K.

ACKNOWLEDGEMENT

We would like acknowledge A.Aggarwal, Department of Chemical Engineering, University of Rochester, for his assistance with the TGA experiments. This work was supported by the National Science Foundation Grant #MSM-871893.

REFERENCES

1. J.G. Bednorz and K.A. Muller, Z.Phys. **B64**, 189 (1986).
2. M.K. Wu, J.R. Ashburn, C.J. Torng, P.H. Hor, R.I. Meng,L.Gao, Z.J. Huang, Q. Wang and C.W. Chu, Phys. Rev. Lett., **58**, 908   (1987).
3. Micheal et al., Z. Phys.B, **68**, 421 (1987).
4. Z.Z. Sheng and A.M. Hermann, Nature, **332**,138-139 ,(1988).
5. A. Goyal, P.D. Funkenbusch, G.C.S. Chang and S.J. Burns, Mater. Lett., **6**, 257 (1988).
6. A. Goyal, P.D. Funkenbusch, G.C.S. Chang and S.J. Burns, Superconductivity and its Applications, edited by H.S. Kwok and D.T. Shaw (Elsevier Science, N.Y.,1988), pp.223-229.
7. A. Goyal, S.J. Burns and P.D. Funkenbusch, manuscript in preparation.
8. S. Jin, R. C. Sherwood, R.B. van Dover, T.H. Tiefel, and D.W. Johnson, Jr., Appl. Phys. Lett., **51**, 203-204 (1987).
9. R.W. McCallum, J.D. Verhoeven, M.A. Noack, E.D. Gibson, F.C. Laabs, and D.K. Finnemore, Advanced Ceramic Materials, **2**, 388-400 (1987).
10. In-Gann Chen, S. Sen and D.M. Stefanescu, Appl. Phys. Lett., **52**, 1355-1357 (1988).
11. In-Gann Chen, S. Sen and D.M. Stefanescu, Proceedings of the 4th Annual Regional Meeting, TMS Proceedings and Applications of High-Tc Superconductors:Status and Prospects, Rutgers University May, 1988.

EFFECT OF DOPING ON DEFECT STRUCTURE AND PHASE STABILITY IN Y-BASED

CUPRATES: IN-SITU TEM STUDIES

> K. Rajan and R. Garcia
> L.C. Gupta* and R. Vijayaraghavan*
> Rensselaer Polytechnic Institute
> Troy, N.Y.
> *Tata Institute of Fundamental Research
> Bombay, India.

ABSTRACT

Transmission electron microscopy (TEM) studies of defect structure in $YBa_2Cu_{3-x}M_xO_y$ doped with Mg and Zn (M = Mg, Zn) are reported. The Mg doped material exhibited a large number of non-intersecting ⟨110⟩ twins. In-situ heat pulsing experiments in the TEM showed that only one of the twin variants was unstable to annealing. It was also found that other types of planar defects may replace the disappearance of one of the twin variants. The Zn doped material on the other hand showed remarkable microstructural stability to the in-situ heating experiments. It is suggested that these results may be interpreted in terms of preferential site occupancy of dopants at twin boundaries.

INTRODUCTION

The physical properties of the high temperature superconductors are sensitive to deviations from stoichiometry. Oxygen non-stoichiometry in pervoskites has been extensively studied because of the ability of these materials to reversibly intercalate the oxygen atoms. During such a process the valence of the transition metal element changes in order to balance the electronic charge. It has been shown that valency changes can produce significant changes in the magnetic, electronic and superconducting properties.[1] One of the ways of changing the valency is by metal doping. By studying the effect of doping it is expected that some insight into the possible mechanisms contributing to superconductivity may be gained.

While numerous studies have been carried out on doping effects,[2,3] relatively little has been done on applying transmission electron microscopy (TEM) to the study of metal doped materials. It should be pointed out that most of the TEM studies reported in the literature have dealt with undoped oxides. Numerous types of TEM techniques have been applied, including high resolution microscopy, convergent beam diffraction and in-situ TEM. The latter technique involves making dynamic observations of defect motion during the cooling and heating of the superconductor in the TEM. A number of such studies on undoped

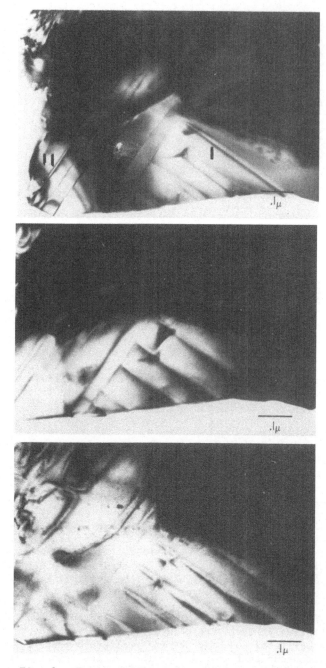

Fig. 1. Heating sequence in Mg doped cuprates

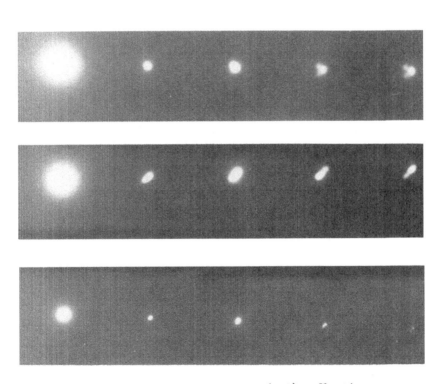

Fig. 2. Effect of heating on (h00) reflections.

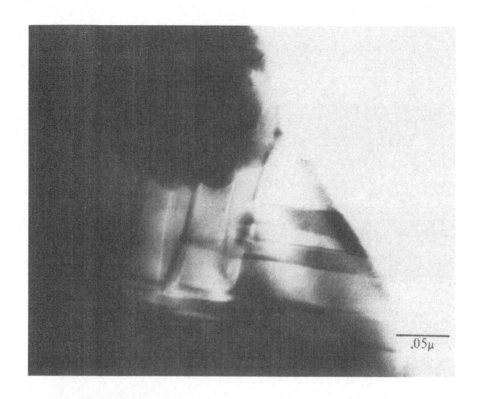

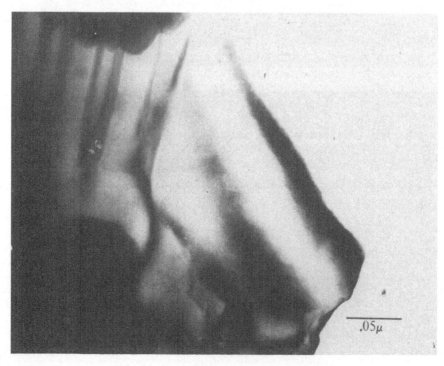

Fig. 3. Formation of domain structures.

Y based cuprates have shown that such experiments can help in understanding the relationship between phase stability and defect structure.[4,7] In this paper we present some of the findings of the first studies of metal doping in YBa$_2$Cu$_3$MO$_{7-K}$ using _in-situ_ TEM techniques.

EXPERIMENTAL DETAILS

For this study, YBa$_2$Cu$_{3-x}$M$_x$O$_y$ was produced by conventional ceramic powder processing techniques. Mg and Zn concentrations of $x = 0.36$ were used. In the electron microscope a local heating can be introduced by focussing and defocussing the electron beam on the specimen. Since the material at room temperature is an insulator, it will heat up under the influence of the electron beam. Upon heating above 750°C, on passes through the orthorhombic-tetragonal transformation, allowing for the disappearance of transformation twins.[4] In this study, we have taken advantage of this temperature rise to observe the stability of transformation twins.

RESULTS

Figure 1 shows a sequence of micrographs of the Mg doped material, after a series of repeated heat pulses. Initially two (110) twin variants are observed. While the two sets meet, they do not cross. Some blunting of the twins (marked I) occurs upon meeting the second set (marked II). The twin interfaces are well defined initially. After heat pulsing only the set II twins have disappeared. However the set I twin interfaces have become diffuse.

Figure 2 shows a series of diffraction patterns corresponding to the same area as shown in Figure 1. The details of the (400) spot are shown in particular. The splitting occurs only in the ⟨110⟩ directions. Initially a two-fold splitting is observed; however, after heating, splitting occurs only in one ⟨110⟩ direction. The enlargement of the (h00) reflections to be consistent with a 90° twin about the (110) plane. Figure 3 also shows the effect of _in-situ_ heating on Mg doped material. As before, only one set of twin variants exhibits instability. However, another interesting feature is the formation of domain-like structures. Preliminary studies that we have conducted indicate these domain structures to be a result of crystallographic shear operations.

Unlike the Mg doped material, the Zn doped crystals were very stable under the influence of the electron beam. As shown in Figure 4, no noticeable change in density or size of twins was observed upon heating. Figure 5 confirms this with a set of diffraction patterns showing no change in spot splitting associated with the heating of the sample.

DISCUSSION

It is well established that twin boundaries exposed to a focussed beam can show a decrease in contrast.[4] Oxygen atoms in twin domains close to the twin boundaries may gain energy through the electron bombardment and diffuse into the twin boundary vacant sites. Thus the contrast at twin boundaries may be expected to smear out.[8] The stability of a twin structure therefore suggests that oxygen diffusion into a twin boundary is inhibited. In our case, this implies that the dopant

Fig. 4. Heating sequence in Zn doped cuprates showing no changes in defect density.

Fig. 5. Effect of heating on (h00) reflections in Zn doped cuprates showing no change in defect structure.

may be residing at normally vacant sites at a twin boundary, inhibiting the diffusion of oxygen into that site. Based on such an argument, the in-situ observations of the Mg doped material would suggest that the Mg segregates to primarily one of the two <110> type twin boundaries. A possible cause for this difference is that for 90° twins, depending upon the atomic coupling, the twin planes need not result in a mirror symmetry plane. An example of this is shown for NiMo in Figure 6 to demonstrate this phenomenon.[9] In the case of the Y-based cuprates, the arrangement of vacant sites may differ between the two twin variants. Consequently, the site occupancy of the Mg atoms between the twin boundaries may also be different. More detailed crystallographic studies are needed to study the validity of this hypothesis. However, it is nonetheless significant that in-situ TEM studies can raise the possibility of preferential site occupancy of dopants to defects.

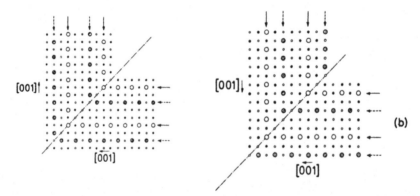

Fig. 6   Structure of perpendicular twins
a) head to tail coupling
b) head to head coupling
Note that in a) the twin plane is a mirror plane for the structure not in b). (By courtesy of Physica Status Solidi)[9].

For the case of Zn doping, no significant change in twin structure could be observed with in-situ heating. No preferential twin stability was observed. This would suggest that the Zn substitution is random. EPR studies of Zn doped Y-based cuprates have been interpreted in terms of random distribution of Zn atoms.[10] Based on the oxygen depleted twin boundary model, Zn atoms may also be inhibiting oxygen diffusion during in-situ heating experiments.

CONCLUSIONS

1. The results of the in-situ heating experiments are in qualitative agreement with vacancy stabilized twin boundary models.

2. In-situ TEM may provide insights into atomic site occupany of dopants associated with defect structure.

REFERENCES

1. J.M. Tarascon, L.H. Greene, B.G. Bagley, W.R. McKinnon, P. Barboux and G.W. Hull, "Chemical Doping and Physical Properties of the New High Temperature Superconducting Pervoskites", in: Poc. Intl. Workshop on Novel Mechanisms of Superconductivity, Berkeley, CA (June 1987) in press.

2. J.D. Jorgensen, D.G. Hinks, H. Shaked, B. Dabrowski, B.W. Veal, A.P. Paulikas, L.J. Nowioki, G.W. Crabtree, W.K. Kwok, A. Umezawa, L.H. Nunez, B.D. Dunlap, C.V. Segre and C.W. Kimball, "Structural and Superconducting Properties of Oxygen-Deficient and Doped $YBa_2Cu_3O_{7-x}$, Physica B, 156 & 157: 877 (1989.

3. J.M. Tarascon, P. Barboux, L.H. Greene, G.W. Hull and B.G. Bagley, "3d-Metal Doping of the High T_c Pervoskite $YBa_2Cu_3O_{7-x}$, in; MRS Proc., 99: 523 (1988).

4. G. Van Tendeloo, J. VanLanduyt and S. Amelinckx, "Electron Microscopy of Static and Dynamic Phenomena", in: Competing Interactions and Microstructures, Statics and Dynamics, R. LeSar, A. Bishop and R. Heffner, eds., Springer-Verlag, Berlin (1988).

5. K. Sasaki, K. Kuroda and H. Saka, "HREM In-situ Observation of a Transformation Interface Between Tetragonal and Orthorhombic Phases in $YBa_2Cu_3O_{7-x}$", Phil. Mag. Letters, 59: 141 (1989).

6. S. Iijima, T. Ichihashi, Y. Kubo and J. Tabuchi, "Twinning of High-T_c $Ba_2YCu_3O_{7-x}$ Oxides", Jap. J. Appl. Phys., 26: L1478 (1987).

7. M. Sugiyama, R. Suyama, T. Inuzuka and H. Kubo, "Phase Transformations in the Superconductor Ba-Y-Cu-O and La-Sr-Cu-O Compounds", Jap. J. Appl. Phys., 26: L1202 (1987).

8. C.J. Jou and J. Washburn, "Formation of Coherent Twins in $YBa_2Cu_3O_{7-x}$ Superconductors", J. Mater. Res., 4: 795 (1989).

9. E. Ruedl, P. Delavignette and S. Amelinckx, "Electron Diffraction and Electron Microscopic Study of Long and Short Range Order in Ni_4 Mo and the Substructure Resulting from Ordering", Phys. Stat. Sol., 28: 305 (1988).

10. A.M. Ponte Goncalves, C.-S. Jee, D. Nichols, J.E. Crow, G.H. Myer, R.E. Solomon and P. Schlottman, "EPR Spectroscopy on Zn-Substituted $YBa_2Cu_3O_7$", in: MRS Proc., 99: 583 (1988).

STABILITY OF YTTRIUM BARIUM CUPRATE IN MOLTEN SALTS

D.B. Knorr and C.H. Raeder*

Materials Engineering Department
Rensselaer Polytechnic Institute
Troy, New York 12180-3590

INTRODUCTION

Low critical current density is an important factor limiting utilization of $YBa_2Cu_3O_{7-x}$ in bulk conductor and power storage applications. The two factors that reduce critical current are the insulating nature of the grain boundaries and the anisotropic current conduction which favors the a-b plane rather than the c-direction in the orthorhombic crystal structure[1,2]. Fabrication of highly oriented polycrystalline bodies that approach single crystal properties would substantially alleviate the later concern. Conventional ceramic processing usually produces randomly oriented grains so special procedures are required to fabricate "grain oriented" ceramics.

Molten salt processing can yield particles with shape anisotropy which are subsequently aligned by techniques such as tape casting. The technique has been successfully applied to a vareity of ferroelectric[3-11] and ferrimagnetic[3,12,13] systems. Usually, constituent oxide powders react in the molten salt solvent to produce the desired product phase with either rod-like or platelet particle morphology depending on the growth anisotropy. Only a very limited number of chloride and sulfate salt systems has been utilized in these studies. They are KCl, NaCl, KCl-NaCl eutectic, Na_2SO_4, Li_2SO_4, and Li_2SO_4-Na_2SO_4 eutectic. The KCl-NaCl eutectic is most widely used for oxide synthesis.

Attempts at molten salt synthesis of $YBa_2Cu_3O_{7-x}$ analogous to the studies in other ceramic systems has not been successful. Decker et al.[14] were unable to synthesize the superconductor in a NaCl-KCl eutectic salt. Furthermore, the $YBa_2Cu_3O_{7-x}$ was found to decompose in NaCl-KCl, $BaCl_2$-$CuCl_2$, $CaCl_2$, and Na_2SO_4-K_2SO_4 salts. Previous work by the present authors[15] more closely examined a variety of chloride molten salt systems. $YBa_2Cu_3O_{7-x}$ (1-2-3 phase) was exposed to molten salts at 900°C for two hours, then examined by X-ray diffraction to identify the phases that were present. Table 1 groups the salt systems and their decomposition products. In Group I salts, the 1-2-3 phase reacts with dichloride ($CaCl_2$, $MgCl_2$, or $CuCl_2$) forming $Cu_2Y_2O_5$, CuO, and $BaCl_2$. High barium activity in the melt promotes Y_2BaCuO_5 as a decomposition product and reaction with CO_2 in the atmosphere to form $BaCO_3$ as seen in Group II

---

*Now Graduate Student, University of Washington, Seattle, WA.

Superconductivity and Applications
Edited by H. S. Kwok *et al.*
Plenum Press, New York, 1990

TABLE 1

DECOMPOSITION OF $YBa_2Cu_3O_{7-x}$ IN CHLORIDE SALTS

Group	Salt Systems	Decomposition Products
I	$CuCl_2$ - 32 mole % KCl	CuO, $Cu_2Y_2O_5$, $BaCl_2$
	$CuCl_2$ - 25 mole % NaCl	
	$CuCl_2$ - 33 mole % $BaCl_2$	
	$CaCl_2$ - 51 mole % NaCl	
	$CaCl_2$ - 33 mole % KCl	
	- 27 mole % NaCl	
	$MgCl_2$ - 37 mole % KCl	
II	$BaCl_2$ - 60 mole % NaCl	$BaCO_3$, Y_2BaCuO_5, CuO or
	$BaCl_2$ - 55 mole % KCl	$CuCl_2$, $BaCl_2$ (trace)
	LiCl - 26 mole % KCl	
	- 10 mole % NaCl	
III	NaCl	Y_2BaCuO_5 (trace), CuO (trace),
	KCl	$BaCl_2$ (trace)
	NaCl - 50 mole % KCl	

TABLE 2

COMPOSITION OF SALT SOLUTIONS

Salt Type	Salt Constituents A	B	Melting Temp. (°C)	Mole % A
Halide	NaCl	NaBr	738	20
	KBr	NaBr	644	53
	KI	NaI	583	45
Nitrate	$Ca(NO_3)_2$	-	501	100
	$Ba(NO_3)_2$	-	604	100
Sulfate	Na_2SO_4	K_2SO_4	845	76
	Li_2SO_4	Na_2SO_4	605	58
	Li_2SO_4	K_2SO_4	695	40
Halide/ Sulfate	KBr	K_2SO_4	655	30
	KCl	K_2SO_4	690	40

salts. Group III salts are most resistant to decomposition, but long times and higher salt loadings encourage some instability as seen by the trace amounts of decomposition products.

The stability of $YBa_2Cu_3O_{7-x}$ is reported for salts in four systems: alkali halides, nitrates, sulfates, and alkali halide/sulfate. Identification of the phases present after molten salt exposure is emphasized. Finally, trends and requirements for 1-2-3 phase stability are explored.

EXPERIMENTAL

$YBa_2Cu_3O_{7-x}$ powders were prepared by the solid state route from reagent grade Y_2O_3, $BaCO_3$, and CuO powders. The powders were calcined in air for 16 hours, ball milled for 8 hours, and annealed in oxygen at 950°C for 24 hours followed by furnace cooling at a rate of 50°C/hour. The powder was analyzed by X-ray diffraction (XRD) using CuK_α radiation to confirm the presence of phase pure orthorhombic 1-2-3.

Single component and binary salt mixtures were prepared in four systems: halide, nitrate, sulfate, and halide/sulfate. The composition are chosen to correspond to the lowest melting temperature in each system as determined from the phase diagram[16]. Systems, compositions, and melting temperatures are listed in Table 2.

A salt portion of five or 7.5 grams and five grams of superconductor were thoroughly mixed with a mortar and pestle, held at 900°C for two hours in air in a Pt crucible, then air cooled. A portion of each salt cake was ground with a mortar and pestle for phase identification. The salt and processed powder were analyzed by an X-ray diffraction scan of 20 to 70° in 2θ. Peaks were matched with compounds in the JCPDS powder diffraction file to identify all constituents and reaction products.

RESULTS AND DISCUSSION

The oxide phases and the salt phases in each molten salt processed superconductor cake are summarized in Table 3 in decreasing order of prominence. In cases where a small quantity ($\lesssim 3\%$) of a phase was found, that compound is followed by (t) for trace. Any appreciable decomposition of the 1-2-3 phase is accompanied by release of barium to the salt. This is reflected in each type of salt. In the halide systems containing a chloride salt such as NaCl-NaBr, barium chloride is observed. $BaCl_2$ is an ubiquitous decomposition product in all chloride systems[15]. In the nitrate salts, barium is incorporated into the $Ca(NO_3)_2$ salt. In sulfate systems, barium sulfate is observed in proportion to the amount of 1-2-3 phase decomposition analogous to the chloride systems. Specific comments will be made for each salt type.

Halide Salts

Using the results in Table 3 and for previous studies of chloride salts[15], several important trends are observed as shown in Table 4. As the cation in the chloride salt becomes larger (move down Column IA in the periodic table), the 1-2-3 phase is increasingly stable judging by the decomposition products. A second comparison involves NaX-KX eutectic salts where X = I, Br, or Cl. As the anion becomes smaller (move up Column VII A on the periodic table), the 1-2-3 phase is more stable. The judgements on stability are based on the type and relative quantity of decomposition products.

TABLE 3

PHASES PRESENT AFTER MOLTEN SALT EXPOSURE AT 900°C

Salt Type	Salt Constituents A	Salt Constituents B	Phases present after exposure* Oxides	Phases present after exposure* Salts
Halide	NaCl	NaBr	$Y_2Ba_2Cu_3O_{6+x}$, CuO, Y_2BaCuO_5	$BaCl_2$ NaBr, NaCl
	KBr	NaBr	$YBa_2Cu_3O_{6+x}$, Y_2BaCuO_5	NaBr, KBr
	KI	NaI	CuO, $Cu_2Y_2O_5$, $YBa_2Cu_3O_{6+x}$, Y_2BaCuO_5	NaI, KI
Nitrate	$Ca(NO_3)_2$	–	$YBa_2Cu_3O_{6+x}$, CuO, $BaCO_3$(t)	$Ba_xCa_{1-x}(NO_3)_2$
	$Ba(NO_3)_2$	–	$Cu_2Y_2O_5$, $Ba_4Y_2O_7$, $Ba_2Y_2O_5$, $BaCO_3$, BaO(t)	$Ba(NO_3)_2$
Sulfate	Na_2SO_4	K_2SO_4	$YBa_2Cu_3O_{6+x}$, $Cu_2Y_2O_5$(t)	$K_2Na_4(SO_4)_3$, K_2SO_4, Na_2SO_4
	Li_2SO_4	Na_2SO_4	$Cu_2Y_2O_3$, CuO, Li_2CuO_2	BaS_2O_7, Na_2SO_4(t), $LiNa(SO_4)$
	Li_2SO_4	K_2SO_4	Y_2BaCuO_5, CuO	BaS_2O_7, K_2SO_4, $KLi(SO_4)$
Halide/ Sulfate	KBr	K_2SO_4	$YBa_2Cu_3O_{6+x}$, CuO(t), Y_2BaCuO_5(t)	KBr, K_2SO_4
	KCl	K_2SO_4	$YBa_2Cu_3O_{6+x}$ CuO, Y_2BaCuO_5	KCl, K_2SO_4

*All phases listed in decreasing order of X-ray signal; (t) refers to trace (<3%) of phase present.

Nitrate Salts

The nitrate salts are quite unstable and do not appear to be useful as solvent salts. Decomposition is observed where the salt cakes are quite porous. Presumably, NO_2 is evolved during the thermal cycle, but vaporization of the salt is another possibility. The X-ray peak heights from the nitrate salts are relatively weak considering the salt to superconductor weight ratio that was tested. In the barium nitrate, salt decomposition promoted 1-2-3 phase decomposition by releasing barium which forms a variety of barium compounds listed in Table 3.

TABLE 4.

COMPARISON OF 1-2-3 PHASE STABILITY IN SEVERAL MOLTEN SALT SYSTEMS

Salt	Resulting Phases*	
	Oxide	Halide
LiCl - 26 mole % KCl-10 mole % NaCl	Y_2BaCuO_5, $YBa_2Cu_3O_{6+x}$, $BaCO_3$, CuO	NaCl/KCl, $BaCl_2$
NaCl	$YBa_2Cu_3O_{6+x}$	NaCl, $BaCl_2$ (t)
KCl	$YBa_2Cu_3O_{6+x}$, Y_2BaCuO_5(t), CuO(t)	KCl
NaI - 45 mole % KI	CuO, $Cu_2Y_2O_5$, $YBa_2Cu_3O_{6+x}$, Y_2BaCuO_5	NaI/KI
NaBr - 53 mole % KBr	$YBa_2Cu_3O_{6+x}$, Y_2BaCuO_5	NaBr/KBr
Na_2SO_4 - 24 mole % K_2SO_4	$YBa_2Cu_3O_{6+x}$, $Cu_2Y_2O_5$(t)	$K_2Na_4(SO_4)_3$, K_2SO_4, Na_2SO_4
NaCl - 50 mole % KCl	$YBa_2Cu_3O_{6+x}$, Y_2BaCuO_5(t)	NaCl/KCl

Sulfate Salts

The sulfates demonstrate reactions analogous to chloride systems[15]. Salts with lithium sulfate promote complete 1-2-3 phase decomposition with loss of the barium to the salt as barium sulfate. Complete 1-2-3 phase decomposition in the presence of LiCl is demonstrated in an analogous chloride system in Table 4. The best sulfate is the sodium potassium eutectic system where the 1-2-3 phase is quite stable with no apparent loss of barium. It compares to the halide eutectic salts as shown in Table 4.

Halide/Sulfate Salts

Although the number of systems is limited, the mixed system of potassium sulfate with either potassium bromide or potassium chloride are moderately stable. The level of stability would rank near KCl alone or KCl-NaCl eutectic[15] in terms of the quantity of decompositoin product. This result opens the possibility of using mixed halide-sulfate combinations to enhance 1-2-3 phase stability.

The requirements for a molten salt system useful for synthesis of $YBa_2Cu_3O_{7-x}$ are quite stringent. The salt should act only as a solvent promoting oxide dissolution, reaction and precipitation, and anisotropic growth of the desired oxide phase[3]. Two undesirable reactions between salt and oxide are: 1) incorporation of salt cations into the product oxide, and 2) dissolution and reaction of an oxide cations with the salt. This later mechanism accounts for barium loss into many of the salts. A general equation that represents this process is:

$$(BaO)_{1-2-3 \text{ phase}} + AX = AO + BaX \qquad (1)$$

where AX is a salt with A cation and X anion. Barium stability requires that this reaction not be thermodynamically favorable.

Two mechanisms for the formation of the desired phase are possible. The desired reactant phase precipitates in the molten salt if its solubility is less than the solubility of the constituents[17]. The lower product solubility indicates a higher thermodynamic stability while the solubility difference between constituents and reactants drives the reaction. The formation of Bi_2WO_6, $Bi_4Ti_3O_{12}$, and $PbNb_2O_6$ are known to occur by this mechanism[3]. If solubilities of the constituents are substantially different, the reaction proceeds by reactant phase formation on particles of the low solubility constituent, e.g. $BaTiO_3$ and ferrites[3].

Synthesis of $YBa_2Cu_3O_{7-x}$ in molten salts is substantially more complicated than synthesis of electroceramics or magnetic ceramics. The 1-2-3 phase has three constituents rather than two in the later systems. Furthermore, constituent solubilities must be in an approximately 1-2-3 molar ratio otherwise reaction products other than the 1-2-3 phase will form. If one constituent is much less soluble than the remaining two, 1-2-3 phase might form on the low solubility constituent, but reaction of the other two soluble constituents alone could occur, negating the formation of 1-2-3 phase.

Two requirements dominate the growth of particles with shape anisotropy. From a thermodynamic standpoint, 1-2-3 phase stability is critical otherwise decomposition occurs. Solubility of the 1-2-3 phase in the salt should be low. This implies that constituent and other decomposition phases should have low solubility otherwise a decomposition reaction is driven. Particle growth occurs by an Ostwald ripening mechanism. Growth rate should be very anisotropic and much higher in the a-b plane than in the c-direction to promote growth of platelet particles. This appears to be the case for $YBa_2Cu_3O_{6+x}$.

CONCLUSIONS

1. Stability of $YBa_2Cu_3O_{6+x}$ is favored in salt system with small anion and large cation sizes.

2. Nitrate salts are subject to decomposition rendering then unacceptable for synthesis or coarsening of 1-2-3 phase powders.

3. Sulfate systems behave similar to chloride salt systems, where an acceptable level of stability is found in the Na_2SO_4 - K_2SO_4 eutectic salt.

4. Data on solubility of 1-2-3 phase, of constituent phases, and of other decomposition products should be obtained in a variety of salt systems. Only then can molten salt formulations that synthesize and coarsen 1-2-3 phase powders be optimized.

ACKNOWLEDGEMENT

This research is supported by an award from the New York State Institute on Superconductivity in conjunction with the New York State Energy Research and Development Authority.

REFERENCES

1. T.R. Dinger, T.K. Worthington, W.G. Gallagher, and R.L. Sandstrom, Direct Observation of Electronic Anisotropy in Single Crystal $Y_1Ba_2Cu_3O_{7-x}$, Phys. Rev. Lett., 58:2687 (1987).

2. Y. Enomoto, T. Murakami, M. Suzuki, and K. Moriwaki, Largely Anisotropic Superconducting Critical Current in Epitaxially Grown $Ba_2YCu_3O_{7-x}$ Thin Film, Jpn. J. Appl. Phys., 26:L1248 (1987).

3. T. Kimura and T. Yamaguchi, Morphology Control of Electronic Ceramic Powders by Molten Salt Synthesis, in "Advances in Ceramics", Vol. 21, G.L. Messing, K.S. Mazkiyasni, J.W. McCauley, and R.A. Haber, eds., American Ceramic Society, Columbus, Ohio (1987).

4. Y. Hayashi, T. Kimura, and T. Yamaguchi, Preparation of Rod-Shaped $BaTiO_3$ Powder, J. Mater. Sci., 21:757 (1986).

5. T. Kimura, H. Chazono, and T. Yamaguchi, Effects of Processing Parameters on the Density and Grain Orientation in $Bi_4Ti_3O_{12}$ Ceramics with Preferred Orientation, in "Advances in Ceramics", Vol. 19, J.B. Blum and W.R. Cannon, eds., American Ceramic Society, Columbus, Ohio (1986).

6. T. Kimura, M. Machida, T. Yamaguchi, and R.E. Newnham, Products of Reaction Between PbO and Nb_2O_5 in Molten KCl or NaCl, J. Amer. Ceram. Soc., 66:C195 (1983).

7. K. Nagata, Y. Uchida, K. Okazaki, and H.-D. Nam, Electrical Properties of Grain-Oriented Ba and La Modified Lead Niobate Ceramics from Molten Salt Synthesized Powders, Jpn. J. Appl. Phys., 24 [Suppl. 24-3]:100 (1985).

8. K. Nagata and K. Okazaki, One-Directional Grain-Oriented Lead Metaniobate Ceramics, Jpn. J. Appl. Phys., 24:812 (1985).

9. S.H. Lin, S.L. Swartz, W.A. Schulze, and J.V. Biggers, Fabrication of Grain-Oriented $PbBi_2Nb_2O_9$, J. Amer. Ceram. Soc., 66:881 (1983).

10. T. Kimura and T. Yamaguchi, Morphology of Bi_2WO_6 Powders Obtained in the Presence of Fused Salts, J. Mater. Sci., 17:1863 (1982).

11. T. Kimura, M.H. Holmes, and R.E. Newnham, Fabrication of Grain-Oriented Bi_2WO_6 Ceramics, J. Amer. Ceram. Soc., 65:223 (1982).

12. T. Kimura, T. Takahashi, and T. Yamaguchi, Preparation and Characteristics of Ni-ferrite Powder Obtained in the Presence of Molten Salts, J. Mater. Sci., 15:1491 (1980).

13. Y. Hayashi, T. Kimura, and T. Yamaguchi, Preparation of Acicular NiZn-ferrite Powders, J. Mater. Sci., 21:2876 (1986).

14. C.T. Decker, V.K. Seth, and W.A. Schulze, Feasibility of Synthesis of Anisotropic Morphology $Ba_2YCu_3O_{7-x}$ Powder by Molten Salt Technique, in "Advanced Superconductors II", M.F. Yan, ed., American Ceramic Society, Westerville, Ohio (1988).

15. C.H. Raeder and D.B. Knorr, Stability of $YBa_2Cu_3O_{7-x}$ in Molten Chloride Salts, submitted to J. Amer. Ceram. Soc.

16. E.M. Levin, C.R. Robbins, H.F. McMurdie, in "Phase Diagrams for Ceramists, 1969 Supplement", M.K. Reser, ed., The American Ceramic Society, Columbus, Ohio (1969).

17. R.H. Arendt, J.H. Rosolowski, and J.W. Szymaszek, Lead Zirconate Titanate Ceramics from Molten Salt Solvent Synthesized Powders, Mater. Res. Bull., 14:703 (1979).

DIFFERENTIAL THERMAL ANALYSIS OF THE $Y_1Ba_2Cu_3O_x$ COMPLEX

G.A. Kitzmann, R.J. Tofte, W.E. Woo & L.E. Rowe

Engineering Physics Research and Testing Laboratories
Physics Department
State University College
New Paltz, New York 12561

Since the investigation of the superconductivity of the Ba-Y-Cu-O system was initiated the nature of the phase relationships of the superconducting compound and the steps in the reaction mechanism that lead to this product have been of great interest. Some studies have attempted to determine aspects of the phase behavior at elevated temperature [1,2]. Other experimental programs have used thermal analysis methods to study the changes in enthalpy, phase, and composition that take place when the material is synthesized [3], or subjected to thermal cycling such as annealing [4,5].

This paper reports the results of thermal studies of this material during its synthesis from the yttrium oxide and copper oxide with barium carbonate or barium oxide. Differential thermal analysis (DTA) was used to follow the course of the reaction and in some cases the thermal stability of the product. The Meissner test was used to find superconductivity in the product. Implications of the DTA results were in some cases scaled up by producing material in muffle and flow ovens using air and oxygen environments.

EXPERIMENTAL

Stoichiometric mixtures of the 1-2-3- barium yttrium cuprate were prepared by grinding appropriate amounts of CuO, Y_2O_3, and either BaO or $BaCO_3$. The yttrium oxide/copper oxide and barium carbonate were obtained from Cerac with at least three nines purity and the barium oxide was catalog number B-58 from Fisher. Weighed mixtures of the salts were ground together in agate ware using acetone to produce a slurry. The acetone subsequently evaporated at 100C. Samples prepared from barium oxide were processed in a glove box using a nitrogen atmosphere. A Tracor DTA Model 202 was used at program rates of 20C per minute with air at rates of up to 0.1SCFH. Samples of up to 10 mg were used in platinum cups with up to 5 mg MgO as a reference. Larger scale mixes were processed using a Lindberg Furnace Model 51232 or a Lindberg flow oven Model 54352A. Magnetic behavior was tested using a Samarium cobalt magnet at liquid nitrogen temperature.

RESULTS

Data representative of a series of seven experiments in which the superconductor was formed in the DTA cup while the reagent mixture made from barium oxide was heated to 1050C are shown in Figure 1. If this data is compared with that of Bellosi et. al. [3], several points of comparison will be noted. In the first place, the present work uses BaO as a starting material. Bellosi, starting from $BaCO_3$, first identified peak production in the stoichiometric mixture at the carbonate transformation/decomposition temperature, 820C. In this work no significant activity was observed as expected.

Another important thermal event observed is a sharp peak near 400C followed by some broad response. This and a series of small peaks after 800C may be associated with reversible peroxide formation that has been reported. Precise identification of these events and the sharp lower temperature activity possibly related to gas adsorption/desorption are of kinetic interest but that task was not pursued in this investigation. Figure 2 shows that the thermal activity up to 925C is apparently produced by the processes associated with BaO.

Figure 3 represents our work using $BaCO_3$ as a reagent. The BaO activity is absent and we found as did Bellosi et. al. [3] that the first event was the carbonate transformation at 820C followed by strong activity after 950C which they interpreted as arising from CuO - $BaCO_3$ interaction. Their temperature limit with oxygen for the reaction was 960C. This work in air was taken to 1050C. With the carbonate a new sharp peak appeared near the reported peritectic point of the superconductor [5,6] and a difference in pattern occurred when BaO and $BaCO_3$ were used for the reaction.

Having gone to 1050C with the reagents we tried to recover the product from the Pt cup. In some cases it could be separated as a small thin disc. A porous residue adhered that has been reported to be indicative of a surface platinum reaction [5]. Assuming that the recovered bulk product was not affected by this surface reaction, a magnet was used to test for the Meissner effect. The results of these tests when product could be recovered, showed a positive Meissner effect. It seemed that when reagents were heated to 1050C with an air flow of 0.075 SCFH and cooled back to ambient temperature superconducting product was produced. When these products were subsequently annealed in flowing oxygen or oxygen flow introduced during cooling in the DTA, the Meissner effect was enhanced. To determine if this rapid production of superconductivity could be extrapolated to larger scale, reaction mixtures that could be pressed into two gram pellets were prepared and oven heated to 1050C for periods that ranged to several hours instead of approximately a day. The results are shown in Table 1.

Table 1. Meissner Test Data at 77 K for 2cm Processed Bulk Pellets Prepared from $BaCO_3$, Y_2O_3, CuO in Stoichiometric Mixture Muffle Heated in Ambient Air.

Sample Identification	Temp	Time	Meissner Effect
1	1050	2 Hr	Negative
2	1050	4 Hr	Positive
3	1050	4 Hr	Positive
4	1050	4 Hr	Positive

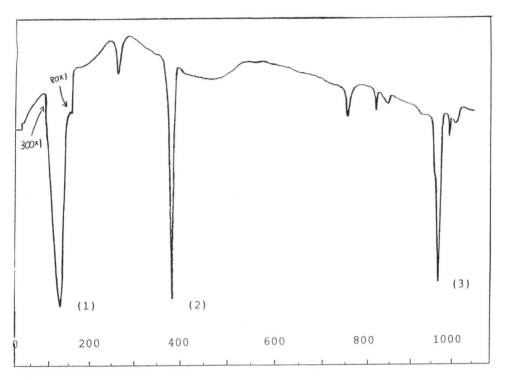

Fig. 1. DTA graph of BaO-Y$_2$O$_3$-CuO.
The labeled major peaks (1) and (2) are associated with BaO activity as can be seen by comparing them to the labeled peaks (1) and (2) in figure 2 below. The third labeled peak (3) is associated with reagent interaction for superconductor formation.

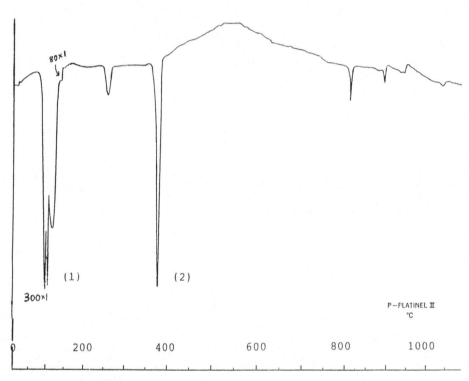

Fig. 2. DTA graph of BaO.
The labeled major peaks (1) and (2) as associated with BaO activity. Compare these peaks to the peaks in figure 1 above. Note that peak (1) is 3.75 times larger than (2) as a result of a scale change in the DTA. All peaks are endothermic.

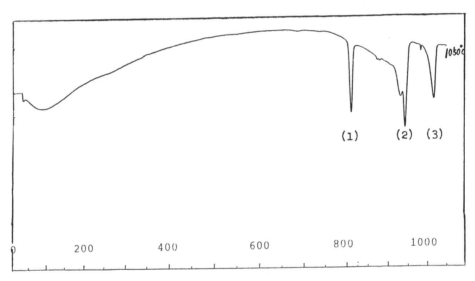

Fig. 3. DTA graph of $BaCO_3$-Y_2O_3-CuO.
The first peak (1) is associated with a crystal transformation. The second peak (2) which is a doublet is associated with reagent interactions and the third (3) peak is associated with product peritectic melt. Compare the second peak (2) from this figure to the third peak (3) in figure 1.

CONCLUSIONS

This work has confirmed the utility of thermal analysis methods in showing the existence of distinct thermal events in the mechanism of kinetically complex reactions that form the high temperature superconductor as well as in the study of the phase relationships of the isolated product. It has its limitations in that these events are designated but not necessarily specified. Recent work being published [6] shows that for this reaction steps that take place at temperatures up to 1500C involve several peritectic reactions which are very important in simplifying the mechanism and increasing the yield of the bulk product. For this laboratory and others this indicates the value of the new high temperature range thermal analysis instruments in continued study of these materials.

Another interesting fact, demonstrated by this work and confirmed by the high temperature studies, is that going to temperatures 100 or more above what are quoted as the normal production temperatures, even flashing to the high temperatures, makes some superconductor quickly even before annealing is carried out [6].

While the purpose of the study was to examine the process of formation, some experiment using pure crystals of Y-Ba-Cu-O super-conductor were performed to examine the process of recrystalization of product.

It was found that recrystalization patterns varied with heat content above peritectic temperature. This has led to additional work which will be published elsewhere.

REFERENCES

1. H. Steinfink, J.S. Swinnea, Z.T. Sui, H.M. Hso, and J.B. Goodenough, JACS, 109 #11 (May 1987) P 334.
2. Lay, K.W. and Renlund, G.M, submitted to Journal of the American Ceramic Society.
3. Bellosi, A.; DePortu, G.; Babini, G.N.; Matacotta, F.C.; Olzi, E. and Masini, F., Br. Ceram Proc, 40 P 205-12, 1988.
4. Shelby, J.E.; Bhargava, A.; Simmins, J.; Corah, N.; McCluskey, P.; Sheckler, C. and Synder, R., Materials Letter 5, Volume 5, Number 11, 12 (1987) PP 420-424.
5. Cook, L.; Chiang, C.; Wong-NG, W. and Blendell, J., Advanced Ceramic Materials, Vol. 2, No. 3B, Special Issue, 1987 PP 656-661.
6. Murakami, M.; Morita, M.; Koyama, N.; Doi, K. and Miyamoto, K. to be published.

EXPERIMENTAL RESEARCH OF THE CONSTRUCTION AND

STRUCTURAL RELAXATION OF Bi-Sr-Ca-Cu-O SUPERCONDUCTIVITY SYSTEM

Jian Enyong, *Qu Yuanfang, Yin Zhiying, [+]Yang Zhian
*Wan Yubin, [+]Liou Wenxi

Department of Physics, Tianjin University, Tianjin, China
(*Department of Material Science, [+]Analytic Center)

Since the high-temperature superconductive ceramics came into being, many kinds of superconductors have been discovered in succession. And for the same prescription, because of the different techniques, the transition temperatures of the superconductors gained are also different. Sometimes there are also reports about high-temperature superconductors over 200k. This shows that the superconductive ceramics have many superconductive phases. Expecially, the coexistence of superconductive phases of the Bi-Sr-Ca-Cu-O systems has been already recognized[1].

In order to obtain stable high-temperature phase and determine the main factors affecting Tc, we have modified the prescription and technique of making superconductive materials of Bi-Sr-Ca-Cu-O system, and gained superconductive monocrystal of this system. Then we have measured the properties of resistance-temperature and analyzed the structure of the monocrystal obtained.

1 Sample-Making

Grade AR. of Bi_2O_3, $SrCO_3$, CaO and CuO has been used as the material in making monocrystal. After being baked, the weight percentage ratio of the ingradients is as follows: $Bi_2O_3:SrCO_3:CaO:CuO=61.82:20.97:3.08:14.12$. Then it is pressed and precombined (or not precombined). The constituents were heated and melted at a temperature of 900-1100°C in an aluminum oxide crucible or a platinum crucibel. It is essential to have a proper control over the cooling rates and making crystals appear. If the cooling rates are controlled properly, we can obtain certain amount of monocrystals in the form of thin layers. The normal size of the monocrystals is $10 \times 2 \times 0.1 mm^3$. The extral part in the original prescription is used as the composition of flux and is removed onto the surface of the meltings or the side of the crucible. The normal of the crystal's large plane is at right angle with the gradient of the temperature. The direction of longest side is in agreement of the gradient of the temperatuer and grows in the forms of thin layers. The crystal is soft and fragile, and easy to be broken. No striking effect of atmosphere has been observed on the crystal growth and the crystal formed.

2. Structural Relaxation

The measurement of the properties of resistance - temperature shows that there exists structural relaxation in the monocrystal samples formed according to this prescription. Let us take the B7 sample obtained in 1988 as an example. The measuring position is the large crystal plane. The measurement which was engaged in our laboratory, shows obviously a sharp drop (about 180k) of the resistance of the sample, which was newly prepared and treated in oxygen.(Fig.1) After a standing of 60 days, the sample was measured again, the results of the remeasurement are as follows: Tc^{on}=85.6k, T_C^o=69.0k, ΔT_C=16.6k, It stood for another 20 days and the property of resistance-temperature becames: T_C^{on}=80.4k, T_C^o=67.3k. ΔT=13.1k. The change of the starting transition temperature is shown in Fig.2.

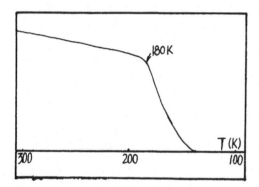 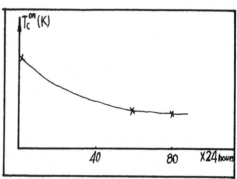

Fig.1 R-T diagram of a fresh sample   Fig.2 T_C^{on} — t curve of BSCCO monocrystal

The properties of the resistance-temperature were measured in the Institute of Physics of the Chinese Academy of Sciences. The results are shown in Fig.3 and Fig.4.

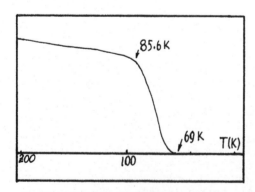 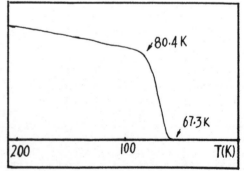

Fig.3 R-T curve of BSCCO crystal   Fig.4 R-T curve of BSCCO crystal
      (standing of 60 days)   (standing for another 20 days)

Fig.2 shows that this sample's starting transition temperature decreases with the increase of the standing time, and gradually tends to a more stable and lower value (about 80k), but the difference between T_C^{on} and T_C^o diminishes and the diminishing rate of the resistance speeds up. Such a relation proves the presence of structural relaxation, and the structural relaxation indicates the instability of high-temperatuer superconductivity phases.

3 Constrcution Analysis

Whe have made construction analysis of the monocrystals with x-ray difraction and transmission electron microscope. Because that monocrystal is soft and fragile, it is very difficult to strip large piece of the samples available for analysis, so the we have still adopted powder samples. It is unevidable to have some foreign substance in such situation.

Fig.5 and Fig.6 show that the x-ray diffraction spectrums of the samples newly made and standing for a long time. The measurement has been carried out by the Measuring and Calculating Center of Nankai University. The crystal system, spectrum index and lattice parameter are measured with the wiser unknown compound crystal system, index and lattice parameter measurement programs of Stockholm University of Sweden.

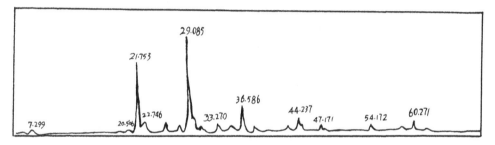

Fig.5 x-ray diffraction pattern of BSCCO crystal

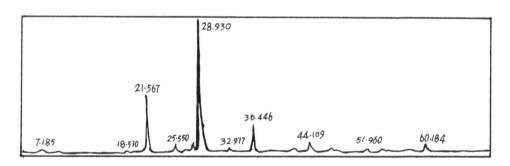

Fig.6 x-ray diffraction pattern of BSCCO crystal

In the course of measuring, the crystal system and crystal cell parameter of this sample was first determined by a computer program for unknown crystal system, then based on the resulte, the parameter was more occurately calculated. The measurement result shows that the newly-made sample is nearly a single-phase. Its crystal cell parameter is $a_1=4.45\text{Å}$, $b_1=5.69\text{Å}$, $C_1=24.64\text{Å}$. $\alpha=\beta=\gamma=90°$, which is categorized as orthorhombic system. The volume of the crystal cell is $V_1=625.10\text{Å}^3$. The long-laid up samples have more mixphases. Without considering the unevidable affectrs of foreign matter that has been brought in there are two kinds of crystal phases belonging to orthorhoimbic system composition, i.e, crystal phase (Phase I) above-mentioned and another one (Phase II). The ratio of phase I and II is about 2:1. The crystal cell parameter of phase II is $a_2=4.56\text{Å}$, $b_2=6.47\text{Å}$, $C_2=29.18\text{Å}$, $V_2=862.20\text{Å}^3$. The diffraction data of x-ray of phase I are given in Table I.

The electron diffraction pattern of the sample is shown in Fig.7. After a standing of 30 days, the diffraction pattern has fewer spots than Fig.7 when the same sample was measured.

Fig.7 EDP of BSCCO crystal

4 Discussion

From the result of the analysis we can see that the Bi-Sr-Ca-Cu-O system monocrystal superconductors we made are orthorhombic system, but the crystal cell parameter is different from that ever reported[2]. Morevover, there are two kinds of different crystal phases coexisting. Maybe the properties of resistance-temperature change is related to the coexisting of these two kinds of crystal phase. Futhen research on relationship between them is being continued.

References

(1) Lichunhong, Cryogenics and Superconductivity Feb. 1988, Vol.17, No.I, P53
(2) Nomiki, Jpn. J. Appl. phys. 27(1988) L567

Appendix

Table I   The data of the diffraction of BSCCO crystal

CYCLE RESULTS
```
  .030253   .018371   .000975   .000000   .000000   .000000
  .029910   .018298   .000976   .000000   .000000   .000000
```
NUMBER OF SINGLE INDEXED LINES = 10
TOTAL NUMBER OF LINES = 18
A = 4.453696   .015016 A   ALFA = 90.000000   .000000 DEG
B = 5.694233   .003669 A   BETA = 90.000000   .000000 DEG
C =24.648880   .014578 A   GAMMA= 90.000000   .000000 DEG
UNIT CELL VOLUME = 625.10 A^3

H	K	L	SST-OBS	SST-CAL	DELTA	2TH-OBS	2TH-CALC	D-OBS	FREE PARAM.
0	0	2	.003900	.00390	-.000006	7.161	7.166	12.3340	.00
0	0	4	.015600	.015624	-.000024	14.350	14.361	6.1670	.00
				.026064		18.581		4.7710	.00
0	0	6	.035105	.035154	-.000049	21.598	21.613	4.1110	.00
1	1	0		.048208			25.367		
1	1	1	.048962	.049184	-.000223	25.568	25.627	3.4810	.00
0	0	8	.062378	.062496	-.000117	28.926	28.954	3.0840	.00
1	0	6	.064878	.065054	-.000186	29.513	29.556	3.0240	.00
0	1	8	.080487	.080793	-.000306	32.963	33.027	2.7150	.00
0	1	9	.097482	.097393	.000089	36.386	36.369	2.4670	.00
0	2	5		.097693			36.410		
0	0	10		.097649			36.419		
1	1	8	.110704	.110704	.000000	38.868	38.868	2.3150	.00
1	1	9		.127304			41.807		
1	2	5		.127513			41.843		
1	0	10	.127871	.127560	.000311	41.905	41.851	2.1540	.00
2	0	3		.128430			42.000		
0	0	12	.140352	.140615	-.000263	44.004	44.047	2.0560	.00
0	1	12	.159111	.158913	.000198	47.017	46.986	1.9310	.00
0	2	11	.191096	.191346	-.000250	51.844	51.880	1.7620	.00
0	0	14		.191393			51.887		
				.206258		54.021		1.6960	
0	3	8	.227186	.227173	.000012	56.932	56.931	1.6160	.00
				.233207		57.752		1.5950	.00
1	0	15	.249514	.249621	-.000107	59.936	59.950	1.5420	.00
0	0	16		.249982			59.998		

NUMBER OF OBS. LINES = 18
NUMBER OF CALC. LINES = 23
M(18) = 6 AV.EPS. = .0001428
F 18 = 5.(.028249. 137)
 M   CF. J. APPL. CRYST. 1(1968)108
 F   CF. J. APPL. CRYST. 12(1979)60
     3 LINES ARE UNINDEXED

ANODIC OXIDATION OF METALLIC SUPERCONDUCTING PRECURSOR

Joseph F. Chiang

Department of Chemistry
State University of New York
College at Oneonta
Oneonta, New York 13820

Since the discovery of the high temperature, T_c ceramic superconductor of cuprate, different techniques have been used to prepare superconducting oxides with good ductibility and strength. Recently, Yurek et al.[1-4] developed method of high temperature oxidation of a metallic alloy (a metallic precursor) that contains the metallic constituents of the ceramic and a noble metal to produce flexibility and strength to the final product. After oxidation, $BaCuO_2$, Yb_2BaCuO_5, $YbBa_2Cu_3O_x$, $Yb_2Ba_4Cu_8O_x$ and $Yb_2Ba_4Cu_7O_x$ have have been observed by X-ray diffraction and high resolution transmission microscopy[5,6]. The proportions of the different phases in the oxidized samples depend critically on the processing parameters (annealing and cooling temperatures and gas compositions).

In order to avoid the difficult high temperature stage in the production of the superconducting material, we have used anodic oxidation method at room temperature. The advantage of this method is that a thin superconducting film on the surface of a metallic precursor can be produced at room temperature. Eventually a low cost product can be obtained.

This work deals with anodic oxidation of metallic precursor ribbons containing $YbBa_2Cu_3$ and 50% silver in a suitable electrolytic solvent and supporting electrolyte at room temperature. The metallic precursor (about 5x2x0.03 mm) is inserted in a cell containing oxygen saturated electrolyte. A saturated calomel electrode (SCE) is used as reference electrode and a platinum plate as counter electrode.

Our major purpose of activity is to search for a suitable electrolytic solvent and supporting electrolyte for the anodic oxidation of 123 metallic precursor in order to produce electrochemically a superconducting oxide film. Water-methanol mixture, pure methanol, water, acetonitrile have been used as electrolytic solvents. Supporting electrolytes have been studied. Passivation was avoided for such investigation.

Figure 1

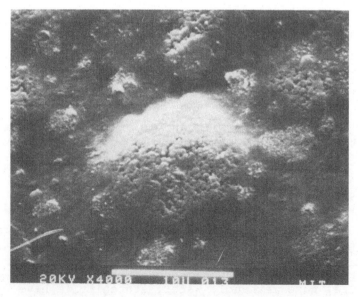

Figure 2

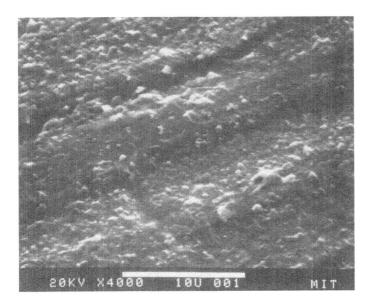

Figure 3

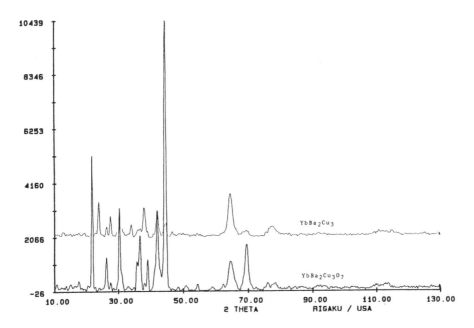

Figure 4

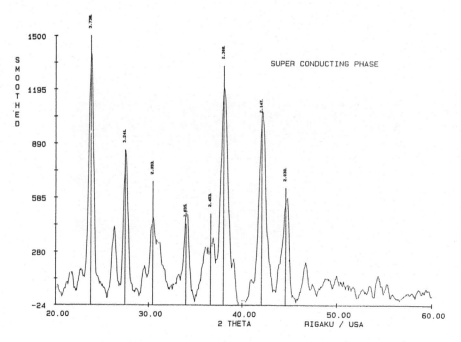

Figure 5

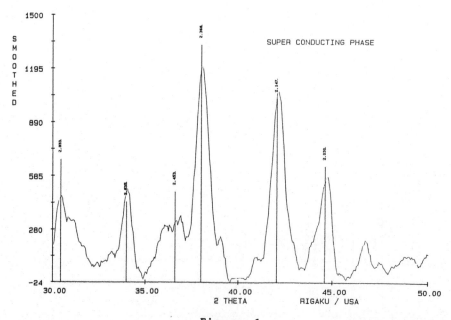

Figure 6

Potentiostat experiments have been performed for the above-mentioned electrolytic solvents. The 123 superconducting thin film was studied by Scanning Electron Microscopy and X-ray diffraction methods. The diffraction technique is probably the most useful method. The intensity vs 2 theta for the known phase of superconductor, $YbBa_2Cu_3O_7$ have been plotted. This can be used as a reference (standard). By comparison of the intensity of the oxidized thin film to that of the standard intensity, a superconducting phase can be identified. Anodic oxidation of metallic precursor of $YbBa_2Cu_3$ at potentials (vs SCE) of 250 mV, 375 mV and 500 mV has been performed. A thin film superconductor at different phases has been developed at room temperature.

Preliminary results were pbtained by using pure methanol as electrolytic solvent and 1.00 M lithium perchlorate as supporting electrolyte. The solution was saturated with air which supplies oxygen, i.e., air was bubbled into the solution for 10 to 15 minutes. Oxidation was carried out by the setup described below:

The cathode is made of platinum. Saturated calomel electrode (SCE) is used as a reference electrode. Platinum foil is used as anode. the precursor is attached to the Pt foil by silver paint. Oxidation takes place about 10 minutes. The final products, a thin film superconductor is obtained and has been studied by SEM and X-ray diffraction.

Scanning electron micrographs of the oxidized products of metallic precursor, $YbBa_2Cu_3$ at 250 mV, 375 mV and 500 mV are shown in Figure 1, 2, and 3 respectively. X-ray diffraction intensity vs 2 theta for the precursor and the oxidized product is shown in Figure 4. The superconducting phase of $YbBa_2Cu_3O_7$ is shown in Figure 5 (intensity vs 2 theta: 20 - 60) and Figure 6 (intensity vs 2 theta:30 - 50).

Acknowledgement: The author wishes to thank New York State Institute on Superconductivity for the financial support of this research and Professor Ron Latanission of the Department of of Materials Science and Engineering, MIT for the initial work of this research.

References

1. G. J. Yurek, J. B. Vander Sande, W. X. Wang and D. A. Rudman, J. Electrochem. Soc. 134, 2635(1987).
2. G. J. Yurek, J. B. Vander Sande, W. X. Wang, D. A. Rudman, Y. Zhang and M. M. Matthiesen, Metallurgical Transactions A. 18A, 1813(1987).
3. G. J. Yurek, J. B. Vander Sande, D. A. Rudman and Y. M. Chiang, J. Metals, 1, 16(1988).
4. G. J. Yurek and J. B. Vander Sande, U. S. Patent Application, Serial No. 031407(1987).
5. T. Kogure, R. Kontra and J. B. Vander Sande, Accepted for publication by Acta Metall. (1988).
6. T. Kogure, R. Konta, G. J. Yurek and J. B. Vander Sande, Private Communication, August, 1988.

A New Phase in the Na/Cu/O System Prepared by High Pressure Synthesis

Xing Liu and Robert C. Liebermann

Dept. of Earth and Space Sciences
SUNY, Stony Brook, NY 11794

L Mihaly and Philip B. Allen

Dept. of Physics
SUNY, Stony Brook, NY 11794-3800

Abstract

A new compound $Cu_r Na_s O_t$ has been formed with composition given roughly by r:s:t equal to 5:2:7 and nominal Cu valence roughly 2.4+. The unit cell is close to orthorhombic, with a slight monoclinic distortion. The material was fabricated by a solid state reaction between CuO and $NaCuO_2$ at pressure exceeding 40Kbar and temperature exceeding 800°C. The product is multi-phased, containing CuO and the new material. The electrical properties of the multi-phased product are those of a narrow gap semiconductor. No superconductivity was detected down to 4.2K.

Introduction

With the exception of $(Ba,K)BiO_3$, all known superconductors [1] with $T_c > 25K$ share the following properties, in order of increasing specificity (1) The materials are crystalline and contain both copper and oxygen. (2) The nominal copper valence is near but not equal 2.0+. (The prototype material is CuO. Materials with copper valence exactly 2.0+ are usually antiferromagnetic insulators, as is CuO. Superconductors usually have nominal valence somewhat greater than 2.0+.) (3) Each copper atom has only oxygen nearest neighbors with coordination 4, 5, or 6. (4) Copper atoms form flat planes with bridging oxygens separating neighboring copper atoms. (5) The planes of copper atoms are not closely coupled by any atoms bearing mobile electrons or holes.

Since the mechanism for high T_c superconductivity is still controversial, it is interesting to ask whether all of these properties are absolutely required, or whether materials can be found which violate one or more of these regularities and are still superconducting. If metallic compounds could be found [2] obeying all regularities except, e.g., number (5), the presence or absence of superconductivity in such a compound would indicate whether or not the two dimensional aspect was critical to superconductivity. These considerations have motivated us to look for compounds closely related to CuO which satisfy some of the regularities above. An obvious idea is to try to dope CuO with a monovalent metal such as Na. The alloy $Cu_{1-x}Na_xO$ would have nominal Cu valence $2+x/(1-x)$ (using 1+ and 2- as the Na and O valences.). Alternately, one could seek new intermetallic compounds of the type $Cu_rNa_sO_t$ with nominal Cu valence $(2t-s)/r$ equal to ~2.1 to 2.4+. If such materials occurred in a structure close to tenorite (CuO) then all regularities except (5) and perhaps (4) would be obeyed.

One way to imagine fabricating such a material would be to react m parts of CuO with n parts of $NaCuO_2$. The resulting compound (if single phase with no oxygen escaping) would have valence $2+n/(m+n)$. Because of the high temperature needed to react these materials, $NaCuO_2$ will decompose, so high pressure is a necessary tool. There is an added advantage of high pressure synthesis: one can explore possible compounds which are thermodynamically unstable at $P=0$, but stable at high P and metastable at $P=0$. Our pressure apparatus does not encourage in situ experiments at high P, so we are not able to explore materials stable at high P unless they remain metastable when pressure is released. This paper reports the fabrication and preliminary characterization of a new compound by this route.

Experimental

A powder mixture of $4CuO + NaCuO_2$ was prepared, which would yield a nominal Cu valence of 2.2+ on complete reaction to a single compound. The starting material $NaCuO_2$ was obtained by heating stoichiometric amounts of CuO and $1/2\ Na_2O_2$ at 450°C under oxygen flow for one week; this material is extremely hygroscopic and unstable in air. Therefore, in order not to introduce uncertainty in the stoichiometry during the subsequent mixing of $NaCuO_2$ with CuO, five parts of CuO and half Na_2O_2 were mixed first and then subjected to 450°C in the oxygen atmosphere for one week. The x-ray diffraction pattern of the resulting product showed it to be a pure mixture of CuO and $NaCuO_2$.

This mixture of $4CuO + NaCuO_2$ was then subjected to high pressure and high temperature in a girdle-anvil (P < 60Kbar) and an uniaxial split-sphere apparatus (USSA-2000, P < 100Kbar). A schematic diagram of the girdle-anvil apparatus is shown in Fig. 1 (From Ingrin and Liebermann, ref. 3). For all the high pressure and high temperature experiments, the sample was placed inside a sealed platinum capsule. No reactions were observed between sample and the platinum after the experiment. The sample temperature is measured directly by a Pt/Pt-13%Rh thermocouple placed next to the capsule, pressure was estimated from pressure calibration at high temperature as explained by Gwanmesia [4].

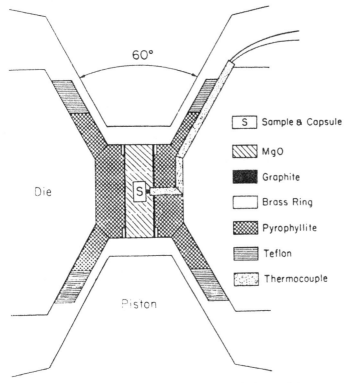

Fig. 1   Girdle-anvil apparatus (schematic).

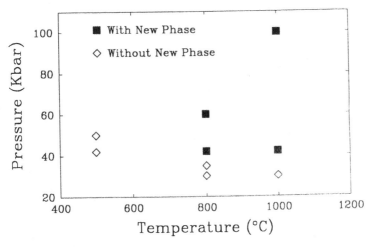

Fig. 2   Synthesis conditions for new phase.

The NaCuO$_2$ starting material vaporizes under 1 atm O$_2$ at 500°C, but is stable up to 1000°C at high pressure. After an experiment at 800°C and 42 Kbar, the x-ray diffraction pattern of the recovered sample indicated the presence of the new phase co-existing with CuO. Various temperatures and pressures were then explored experimentally and Fig. 2 shows the phase diagram for such synthesis experiments. In the cases denoted by open diamonds, the product materials continued to exhibit the same x-ray pattern as that of the starting materials and no new reflections were observed. However, in cases denoted by filled squares, the product materials failed to show any NaCuO$_2$ reflections. Instead, nine new x-ray reflections were seen between 10° and 70° in 2Θ (CuKα x-ray) as indicated in Fig. 3. X-ray patterns of product materials synthesized at the different P-T conditions as denoted by filled squares show exactly the same characteristics in both reflection intensities and positions. We estimate that about 50% of the sample is new phase and 50% is CuO. This suggests a possible composition ~ Cu$_5$Na$_2$O$_7$ with nominal Cu valence of 2.4+.

Table I. Observed and Calculated d-spacings

h	k	l	d(obs.)	d(calc.)
0	0	1	5.754	5.752
1	0	1	4.155	4.156
1	1	-1	3.877	3.873
1	3	-1	2.480	2.483
2	3	0	2.175	2.173
1	4	0	2.146	2.146
4	0	0	1.547	1.548
2	4	2	1.538	1.538
4	1	0	1.527	1.526

The nine x-ray diffraction lines from the new phase were indexed according to a monoclinic unit cell with a = 6.194(2), b = 9.149(6), c = 5.754(8), α = 90°, β = 91.62°, γ = 90°. This is distorted by only ~ 1.6° from an orthorhombic cell. The observed and calculated d-spacings are given in table I.

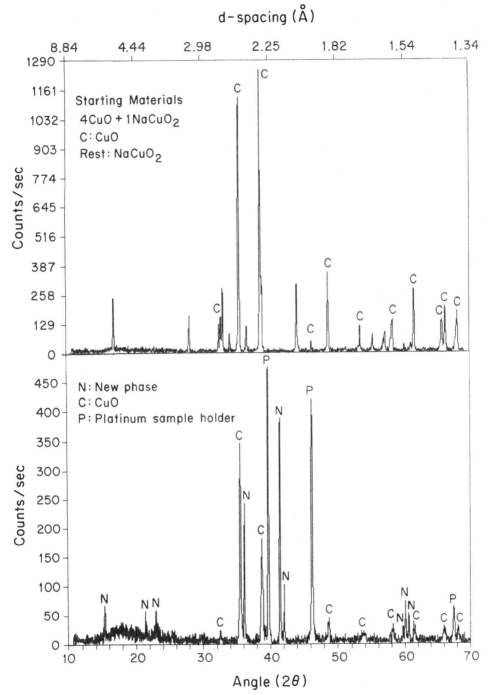

Fig. 3　X-ray diffraction patterns of starting (upper diagram) and recovered materials (bottom diagram).

This new phase is hygroscopic (as is $NaCuO_2$). Our first sample liquefied on exposure to air for 2 days and therefore later samples have been kept under dry conditions. This new phase is also unstable at relatively high temperature (> 200°C). An attempt to analyze the new phase with TEM ended unsuccessfully, due to amorphization of the sample under the electron beam.

Resistivity and susceptibility measurements were also performed at 300K, 77K and 4.2K on the recovered material of this new phase mixed with CuO. No measurable signal was seen in susceptibility. The sensitivity of the apparatus would have been sufficient to detect superconductivity at 4.2K if 1% of the sample had been superconducting. The resistivity increased by $\sim 2 \times 10^3$ as temperature was decreased from 300K to 77K. This is consistent with the behavior of a homogeneous narrow gap semiconductor. Of course, the sample is inhomogeneous with the CuO component having a wider gap.

Similar experiments have also been tried with $KCuO_2$ and CuO. So far no reaction has been achieved.

Conclusions

Syntheses at high P and T offers a route to a potentially rich new family of metastable oxide phases, as illustrated by the new compound described here. No synthesis route for this compound at P=0 is known, and quite likely the material is not thermodynamically stable at P=0. It will be interesting to explore further the crystal chemistry and physical properties of this material, and to continue a search for potentially metallic and potentially superconducting oxides.

Acknowledgments

This work was supported by the New York State Institute on Superconductivity and the Stony Brook High Pressure Laboratory.

References

1. The solid-state chemistry of these materials is discussed in many places, e.g., B. Raveau and C. Michel, in Novel Superconductivity, edited by S. A. Wolf and V. Z. Kresin (Plenum, New York, 1987).

2. J. B. Torrance, Y. Tokura, A. Nazzal, and S. S. P. Parkin, Phys. Rev. Lett. 60, 542 (1988).

3. J. Ingrin and R. C. Liebermann, Phys. Earth Plane. Interiors 54, 387 (1989).

4. G. Gwanmesia, Pressure calibrations in a girdle-anvil and DIA-type high pressure apparati at room temperature and high temperature. M.S. Thesis, State University of New York at Stony Brook, NY 11794 (1987).

EFFECT OF MAGNETIC FIELDS ON SILVER CONTACTS TO YBCO AND YBCO:AG COMPOSITE SUPERCONDUCTORS

Y. Tzeng and M. Belser

Department of Electrical Engineering
Auburn University, Alabama 36849

ABSTRACT

The effect of magnetic fields on the electrical properties of silver-HTSC contacts is investigated. The addition of Ag_2O to YBCO greatly reduces the contact resistance as well as increases the critical current of the contact. The minimum recovery time for magnetically induced contact resistance is about six seconds. For magnetic field greater than 7 Gauss the contact resistance decays to a finite value greater than the value before the contact exposed to the magnetic field and stays there for a long time, as long as the sample is kept below the superconductivity critical temperature.

INTRODUCTION

Much research effort has concentrated on the improvement of the electrical properties of high temperature superconductors. Addition of noble metals such as silver [1] and liquid phase processing [2] are two methods, among many others, employed to enhance the electrical transport properties. All practical applications of these high temperature superconductors require that external leads be connected. Talvacchio has reviewed the requirements and properties of metal-superconductor interfaces for bulk and thin film materials [12].

Microelectronic applications require low resistance, high Jc contacts. Rugged high current carrying contacts are needed for power applications. Several methods for fabricating high performance contacts to superconductors have been previously reported [3-9]. The methods include molten silver processing, vacuum evaporation or sputtering, direct painting of the contact materials, direct wire bonding...etc. One practical application that requires the use of very low resistance contacts is the magnetically controlled opening switch [10]. This device is used to switch a high current to a low resistance load such as an electromagnetic launcher.

Little has been reported concerning the effect of magnetic fields on the electrical properties of the metal-YBCO interface. Wieck reports the use of the two-point four-wire method for characterizing Au-YBCO contacts under the influence of magnetic fields [13]. No explanation was given in [13] whether the resistance of the bulk superconductor under a magnetic field was subtracted from the measured two-point resistance or not. In this work we report the use of a three-point, four-wire technique, which excludes the resistance of the bulk superconductor, for determining I-V curves, contact resistance, and recovery time of the contact resistance as a function of applied magnetic fields.

EXPERIMENTAL PROCEDURES

Silver contacts to YBCO and YBCO:10wt%Ag_2O are fabricated by placing silver disks (2mm diameter) onto superconductor samples and heating them to 962°C [8]. At this temperature the silver disks melt and diffuse into the HTSC forming very good electrical contacts. It is then allowed to cool slowly (1°C/min) to 450°C and this temperature is held for two hours. The sample is then cooled to 130°C slowly before being removed from the furnace. The slow cooling allows further diffusion of silver through the non-superconducting surface layer and into the underlying superconducting material. The restoration of the oxygen stoichometry and the prevention of micro-cracks are also achieved during this process. The YBCO is prepared in the same manner as described in Ref. [9] except the powder is pressed into bars of 5mm x 2mm x 30mm in size.

A three-point, four-wire configuration is used to characterize the metal-superconductor interface. Using this method, current is passed between the center contact and another neighboring contact and the voltage is measured using the center contact and the remaining one. The I-V curves for the metal-HTSC contacts as a function of applied magnetic field are obtained by applying a magnetic field up to 10G perpendicular to the contact area. The recovery time of the contact resistance is measured using a Keithly 195A digital multimeter to record the voltage across the contact as a function of time after the magnetic field is removed.

Results and Discussion

Silver contacts made by melting silver disks onto YBCO and YBCO:Ag_2O have specific contact resistance, i.e., the product of the contact resistance and the contact area, on the order of 10-8 ohm-cm^2 or lower at 77K. This value corresponds to the measurement limit of the micro-voltmeter used in the experiment. It has been shown that the addition of Ag_2O to YBCO greatly reduces the resistance of the bulk material [1-2]. The contact resistance also decreases due to an increase in the effective contact area.

Fig. 1 shows the contact resistance as a function of the applied magnetic field. When a current less than the critical current at zero magnetic field is applied, the magnetic field causes a contact resistance which increases with increasing field strength. The silver oxide doped samples have lower contact resistance than the undoped samples. The magnetic field causes a quenching of the superconductivity in the area under the contact pad. The critical current of the contact under the influence of a magnetic field also decreases with increasing field.

The current-voltage curves as a function of magnetic field is shown in Fig. 2 and 3. These curves are similar to those seen for different temperatures. The magnetic field has the effect of reducing the critical current as increasing the temperature does. For applied fields greater than about 7G, the further increase in magnetic field has less an effect on the contact resistance.

It was observed that contact resistance increases with increasing magnetic field. When the field is removed, the high conductivity of the contact does not recover immediately. Approximately six seconds are needed for the magnetic field induced contact resistance to decay for fields up to 7G (Fig. 4-5). For all applied fields there is a rapid initial decay of the contact resistance after the magnetic field is removed. For fields greater than 7G a significant amount of magnetic flux is trapped in the contact area as long as the sample temperature is lower than the superconductivity critical temperature. The contact resistance decreases rapidly to some finite value greater than the original ones and remains there. The residual resistance increases with increasing applied field. This is shown in Fig. 6-8.

Peters et al showed that when a YBCO/Ag_2O composite is exposed to a magnetic field, magnetic flux is pinned and remains trapped as long as the superconductor is below the transition temperature [11]. At the silver to YBCO contact, a YBCO:Ag composite interface is formed which will enhance the flux trapping at the interface and cause the observed slow decay of the contact resistance.

An undoped sample with four contact pads is subjected to magnetic fields up to 10G. The four-point, four-wire configuration is used to determine if the slow resistance decay detected in the 3-point, four-wire measurement is due to flux trapped in the contact pad or in the bulk superconductor. A current less than the bulk critical current at zero magnetic field is applied and the magnetic field is turned on. A non-zero bulk resistance was detected as expected. When the magnetic field was removed the resistance decayed rapidly. This shows that the three-point, four-wire method does detect the resistance decay due to flux trapped in the contact area.

CONCLUSIONS

The effect of magnetic field on metal-HTSC contacts has been reported. It takes several seconds for contact resistance to recover after being exposed to a low magnetic field. The contact resistance can only decay to a finite value after being exposed to a magnetic field of higher strength. The addition of Ag_2O to YBCO enhances the electrical properties by (1) increasing the critical current density, (2) decreasing the specific contact resistance and (3) decreasing the sensitivity of the contact resistance to magnetic fields.

For applications that require extremely low contact resistance, the contacts must be carefully shielded from both static and transient magnetic fields.

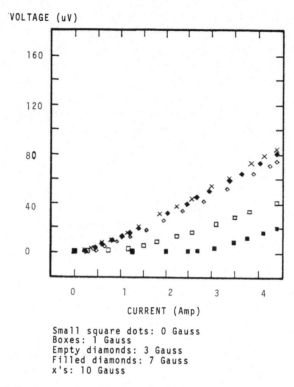

Figure 1. I-V curves for AG-YBCO:10 WT% ago contacts vs B field.

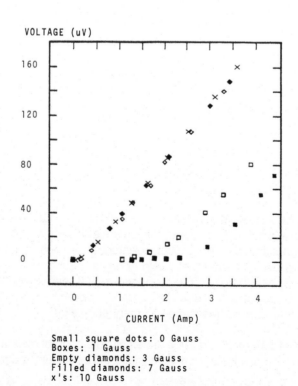

Figure 2. I-V curves for AG-YBCO contacts vs B field.

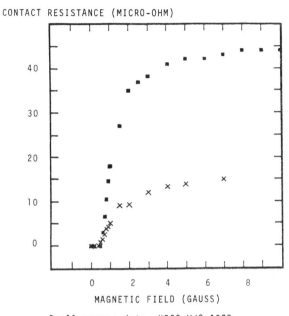

Small square dots: YBCO W/O AG2O
x's: YBCO W/ 10 WT% AG2O

Figure 3. AG-YBCO contact resistance vs. B field.

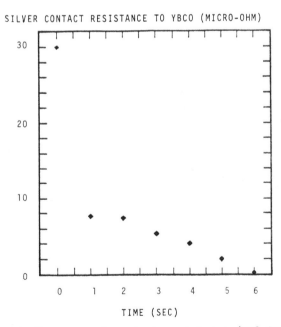

Figure 4. Recovery of contact resistance (B=3 Gauss)

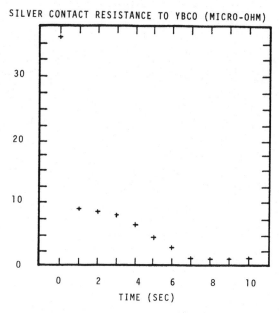

Figure 5. Recovery of contact resistance (B=7 Gauss)

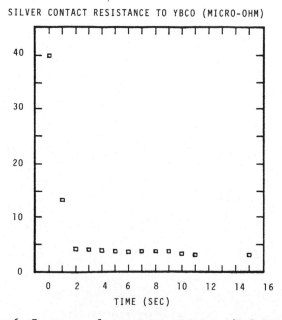

Figure 6. Recovery of contact resistance (B=8.5 Gauss)

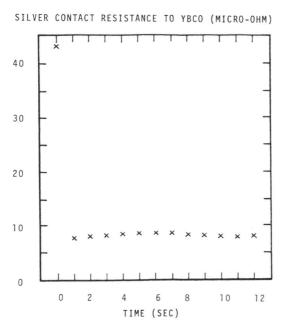

Figure 7. Recovery of contact resistance (B=10 Gauss)

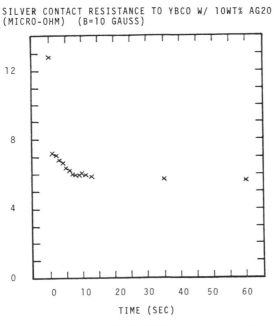

Figure 8. Recovery of AG-YBCO contact resistance.

REFERENCES

1. B. Dwir, M. Affronte, and D. Pavuna, Applied Physics Letters 55, 399 (1989).

2. K. Salama, V. Selvamanickam, L. Gao, and K. Sun, Applied Physics Letters 54, 2352 (1989).

3. Y. Tzeng and M. Belser, in Superconductivity and Its Applications, Elsevier Science Publishing Co., Inc. H.S. Kwok and D.T. Shaw (eds), p. 174 (1988).

4. S. Jin, M.E. Davis, T.H. Tiefel, R.B. vanDover, R.C. Sherwood, H.M. O'Brian, G.W. Kammlott, and R.A. Fastnacht, Applied Physics Letters 54, 2605 (1989).

5. Yoshihiko Suzuki, Tadoaki Kusaka, Takahiro Aoyama, Tsutom Yotsuya, and Soichi Ogawa, Applied Physics Letters 54, 666 (1989).

6. R. Caton and R. Selim, Applied Physics Letters, 52, 1014 (1988).

7. A.D. Wieck, Applied Physics Letters, 52, 1017 (1988).

8. Y. Tzeng, Journal of Electrochem. Soc. 135, 1309 (1988).

9. Y. Tzeng, A. Holt, and R. Ely, Applied Physics Letters, 52, 155 (1988).

10. Y. Tzeng, C. Cutshaw, T. Roppel, C.Wu, C.W. Tanger, M. Belser, R. Williams, and L. Czekala, Applied Physics Letters, 54, 949 (1989).

11. P.N. Peters, R.C. Sisk, E.W. Urban, C.Y. Huang, M.K. Wu, Applied Physics Letters, 52, 2066 (1988).

12. J. Talvacchio, IEEE Transactions on Components, Hybrids, and Manufacturing Technology, 12, 21 (1989).

13. A.D. Wieck, Applied Physics Letters, 53, 1216 (1988).

FRACTURE BEHAVIOR OF SUPERCONDUCTING FILMS PREPARED BY

PLASMA-ASSISTED LASER DEPOSITION

D.D.L. Chung

Department of Mechanical and Aerospace Engineering
State University of New York
Buffalo, NY 14260

INTRODUCTION

High T_c superconductors in the form of films in general have critical current densities (J_c) higher than those of high T_c superconductors in bulk form. In particular, $YBa_2Cu_3O_{7-\delta}$ films prepared by plasma-assisted laser deposition at a substrate temperature of 500°C or above (without post-deposition annealing) exhibited J_c values as high as 5×10^{-6} A/cm^2 [1-5]. This paper is focused on the fracture behavior of such films on metal and ceramic substrates, as this behavior is relevant to the mechanical integrity of these film/substrate combinations.

EXPERIMENTAL

$YBa_2Cu_3O_{7-\delta}$ films deposited on stainless steel, MgO (100) and ZrO_2 (100) were used. The film thickness was 1 μm for the film on stainless steel and was 0.5 μm for the films on MgO and ZrO_2.

The structure of the film on stainless steel was studied by fractography. In this method, the film on stainless steel was subjected to tensile fracture (force parallel to the film plane) and then the fracture surface and its vicinity were examined by scanning electron microscopy (SEM). Figure 1 shows SEM photographs of (a) the film surface in the vicinity of the fracture surface, and (b) the fracture surface, which is perpendicular to the plane of the film. These photographs reveal that the film fractured in the form of strips that were three-dimensionally aligned with respect to the substrate. This fracture behavior indicates the strong crystallographic texture in the film.

The scratch resistance of the films on stainless steel, MgO and ZrO_2 were characterized by the diamond scratch test, which involves the use of a diamond tip under a controlled load (10 - 1000g) to scratch a surface. A Teledyne Taber scratch tester was used.

For the film on steel, the scratch did not go through the film for loads up to 100 g. At a load of 250 g or above, the film on steel was completely removed at the scratch. However, even at the lowest load of 20 g, the film cracked and partially flaked off in the vicinity of the scratch. For the films on MgO and ZrO_2, the scratch did not go through the film even at the highest load of 1000 g, at which cracking in the vicinity of the scratch was still not observed. Figure 2 shows SEM

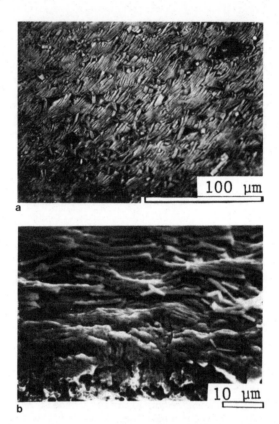

Fig. 1 SEM photographs of a $YBa_2Cu_3O_{7-\delta}$ film on stainless steel after tensile fracture. (a) top view of film, (b) side view of film.

photographs of the films on steel and MgO after scratching at a load of 500 g. Therefore, the scratch resistance of the films on MgO and ZrO_2 are superior to that of the film on steel. This is related to the better adhesion between the film and single crystal oxide substrates than between the film and a polycrystalline metal substrate.

CONCLUSION

The fracture behavior of a $YBa_2Cu_3O_{7-\delta}$ film on stainless steel was intergranular in nature and fractography showed that the grains were elongated and three-dimensionally aligned. This strong texture is consistent with the high J_c value of the film.

The scratch resistance of the film on stainless steel was poor, whereas that of films on MgO and ZrO_2 were very good.

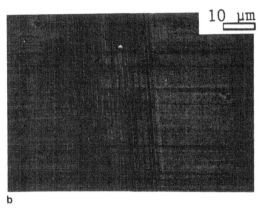

Fig. 2　SEM photographs of $YBa_2Cu_3O_{7-\delta}$ film on (a) stainless steel and (b) MgO (100), after scratching at 500 g. The scratch corresponds to the broad line running nearly vertically in the plane of each photograph.

ACKNOWLEDGEMENT

The author is grateful to D.T. Shaw of New York State Institute on Superconductivity for providing the samples, and to R.C. Chan and J.A. Schmidt of State University of New York at Buffalo for technical assistance. This research was supported by an award from the New York State Institute on Superconductivity in conjunction with the New York State Energy Research and Development Authority.

REFERENCES

1. Q.Y. Ying, D.T. Shaw and H.S. Kwok, Appl. Phys. Lett. 53(18), 1762 (1988).

2. S. Witanachchi, S. Patel, H.S. Kwok and D.T. Shaw, Appl. Phys. Lett. 54(6), 578 (1989).

3. J.P. Zheng, Q.Y. Ying, S. Witanachchi, Z.Q. Huang, D.T. Shaw and H.S. Kwok, Appl. Phys. Lett. 54(10), 954 (1989).

4. S. Witanachchi, S. Patel, D.T. Shaw and H.S. Kwok, Appl. Phys. Lett. 55(3), 295 (1989).

5. Q.Y. Ying, H.S. Kim, D.T. Shaw and H.S. Kwok, Appl. Phys. Lett. 55(10), 1041 (1989).

Multilayer Flexible Oxide Superconducting Tapes

S. Witanachchi, D.T. Shaw, H.S. Kwok, E. Narumi, Y.Z. Zhu, and S. Patel

New York State Institute on Superconductivity
State University of New York at Buffalo
Buffalo, New York 14260

Abstract

Superconducting Y-Ba-Cu-O films have been grown on mirror-finished flexible stainless steel substrates. Plasma assisted laser deposition (PLD) technique enabled us to grow these films at a reduced substrate temperature of 550°C. Superconducting properties of these as-deposited films depend largely on the surface condition of the substrate. In-situ oxidation of the substrate, Ar ion milling, and buffer layer growth, prior to Y-Ba-Cu-O deposition, produced films with critical temperatures ranging from 78-83K. Critical currents of $3 \times 10^3 - 10^4$ A/cm^2 at 40K have been obtained. Scanning electron microscopy and Auger spectroscopy have been utilized to study the surface morphology and the interfacial diffusion of the films.

Introduction

Most of the recent developments in high T_c superconductors are in the area of thin films for electronic and microwave device applications. Films as thin as .1 μm with critical currents in excess of million amps/cm^2 have been grown on single crystal substrates by various vacuum evaporation techniques. So far, the best results have been obtained using the laser ablation technique[1][2]. This extremely simple and versatile technique enables one to carry out reactive evaporation in an oxygen environment, which would lead to the formation of in-situ superconducting films with smooth surfaces.

We have developed an excimer laser evaporation technique[3] that incorporates an oxygen DC plasma in the evaporation region to produce as deposited superconducting films with good superconducting properties.

This plasma assisted laser evaporation (PLD) technique enabled us to fabricate films at substrate temperatures around 500°C on single crystal substrates[3][4][5]. Critical currents of the order of 10^5 A/cm^2 at 80K were obtained for films on SrTiO$_3$ substrates[3]. Even though, large number of research groups have reported the superconducting properties of the films deposited on single crystal substrates, not much work has been done in film growth on metallic substrates. However, in the fabrication of superconducting tapes and ribbons for practical applications, flexible metallic foils, such as stainless steel, are required as substrates. To obtain cryogenic stability in the conductor against flux jumping, a composite structure such as, metal substrate/superconducting film/Cu film, may have to be developed, eventually. In such a structure, if the superconductor is driven normal, the major part of the current will be carried by the copper film until the superconductivity is re-established.

The first step towards the fabrication of this composite structure is the growth of superconducting films on metallic substrates. There are few major concerns in obtaining a good quality superconducting film on a substrate such as stainless steel. (A) At high substrate temperatures interfacial diffusion degrades the film. (B) Even mirror finished substrates have a large amount of surface defects, which lead to poor crystallinity of the films. (C) The polycrystalline nature of the substrate gives rise to poor crystal orientation. Since c-axis normal orientation is desired for high critical current densities, random orientation of the film would lead to low critical currents.

Utilizing the PLD process, we have been able to deposit good film on stainless steel at 550°C substrate temperatures. Due to the increased adatom mobility of the depositing material on the substrate, by the high energy collisions in an oxygen plasma, proper crystal structure of the superconductor is formed at a lower temperature than the 650°C substrate temperature required for non-activated reactive evaporation[2].

Experimental

PLD process has been discussed in previous publications[3]. A schematic of the laser evaporation chamber is shown in Figure 1. This turbo-pumped vacuum system can attain a base pressure of about 5×10^{-6} torr. The evaporation source is a pellet of $Y_1Ba_2Cu_3O_{7-x}$ pressed into about 90% density. An ArF excimer laser (λ = 193 nm, pulse width = 15 nsec.) focused on the rotating pellet produces a molecular beam of the material, which forms a stoichiometric film on a heated substrate placed above. Evaporation is carried out in about 20 mTorr oxygen back pressure. By applying a +300 V DC voltage to the middle ring electrode, a weak oxygen plasma is formed in the evaporation zone.

Y-Ba-Cu-O films were deposited on 8 mill thick stainless steel foil with a mirror finish surface. The substrates were ultrasonically cleaned in trichloroethylene, acetone, methanol and deionized water, in that order, and then mounted on a heating block, which could be heated resistively. In a typical run, a deposition rate of about 2 A/Sec. was maintained at 2-3 J/cm^2 laser fluence at the target.

We have studied the dependence of the superconducting properties of the films on substrate surface condition. Substrate surface conditions were altered by oxidation, or Ar ion milling, or Ag buffer layer growth, prior to the Y-Ba-Cu-O film deposition.

Oxidation: The substrate was heated to 550°C and oxidized at 10-20 m torr oxygen pressure for about 20 minutes. The purpose of the surface oxidation was to form an oxide buffer layer that would eliminate surface deformities and produce a smooth surface.

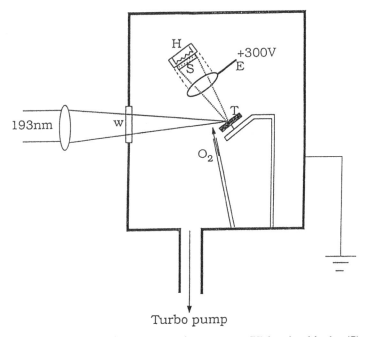

Figure 1. Schematic of the laser evaporation system: (H) heating block; (S) substrate; (E) ring electrode; (T) rotating $Y_1Ba_2Cu_3O_{7-x}$ target' (W) MgF_2 laser window.

In the surface oxidation of stainless steel, the formation of crystalline Cr_2O_3, Fe_3O_4 or $FeO.Cr_2O_3$ in relatively thin layers (~100Å) has been observed[6].

Ion Milling: The substrate was heated to 550°C and ion-milled, in-situ, for 10 minutes with 100-150 eV Ar ions by using a Kauffman ion beam (Ion Tech Model MPS-3000).

Ag Buffer Layers: The substrate was oxidized at 550°C, as the previous sample, and an Ag layer was deposited. Silver deposition was done by laser evaporating a pressed silver pellet in vacuum.

Subsequent to the above surface treatments, Y-Ba-Cu-O films were deposited. After the film growth, all the samples were cooled to room temperature in about 200 torr of oxygen.

Results and Discussion

The resistive transitions and the critical currents of the films were measured by the four point probe technique. Figure 2 shows the resistance-temperature characteristics of three films made under different surface conditions. Films grown on the substrates without any surface treatment showed non metallic behavior with critical temperatures below 65K. Oxidation of the substrate produced films that show metallic behavior with critical temperatures close to 77K. Highest critical temperature of 83K was obtained for films that were ion milled, and the films with a silver buffer layer.

An interesting observation made during this process is the gradual change of the resistance-temperature characteristics of the films from non-metallic to metallic behavior as the duration of surface pre-oxidation is increased from zero to half an hour (Fig. 3). This may be due to the improved surface adhesion and smoothness of the substrate, obtained as the oxide layer is grown. SEM micrographs in Fig. 4. reveal the variations produced in the film and substrate surfaces under different surface treatments.

A large difference in critical currents of the films was observed between the oxidized and ion milled substrates. At 40K, films on oxidized substrates and ion milled substrates produced critical current of 3×10^3 and 10^4 A/cm^2, respectively. Critical currents of the silver buffered films at 40K was 4×10^3 A/cm^2. Dependence of J_c on temperature for these films is shown in Fig. 5.

Auger depth profiles of all the films show low diffusion of Fe at the interface. However, formation of an oxide layer is indicated by a hump in the oxygen profile (Fig. 6) at the interface.

A high percentage of c-axis oriented film growth is revealed by the (OOL) x-ray diffraction peaks (Fig. 7). The presence of other peaks and the broad (OOL) diffraction peaks indicate a degree of misalignment. Large broadening of the low angle peaks may have been due to the slight curvature of the surface.

We have also studied the effect of the oxygen plasma on the superconducting properties of the deposited films by growing Y-Ba-Cu-O films on stainless steel with and without the influence of the plasma, under similar conditions. As our experimental results indicate (Fig. 8), the oxygen plasma is responsible for the increase in the critical temperature of the film and the transformation of the non-metallic to metallic behavior.

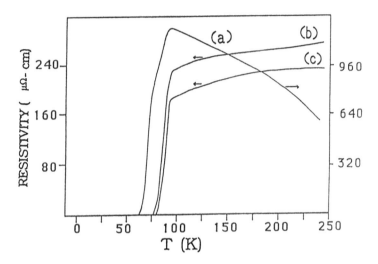

Figure 2. Resistivity vs. temperature for YBaCuO films on stainless steel with (a) no surface treatment, (b) in-situ oxidation, (c) Ag buffer layer.

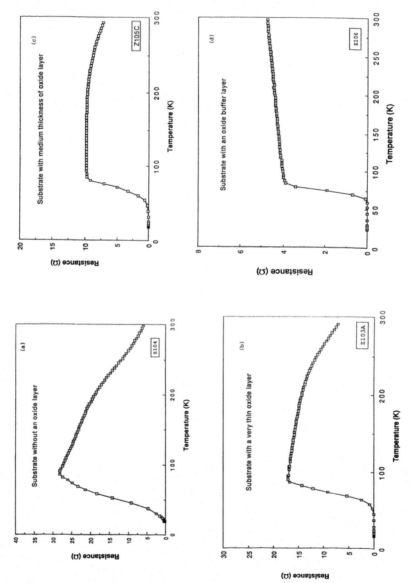

Figure 3. Effect of surface oxidation on the superconducting transition.

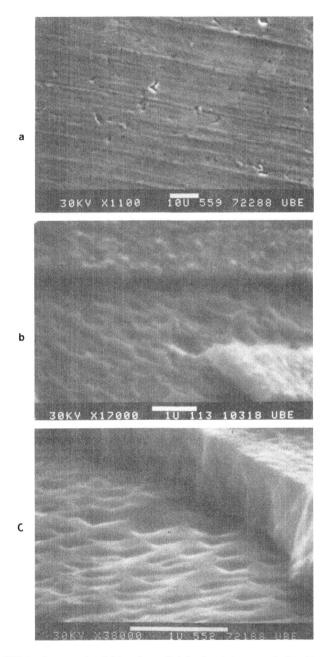

Figure 4. SEM micrograph of (a) mirror finished stainless steel, (b) film on untreated stainless steel, (c) film on oxidized stainless steel.

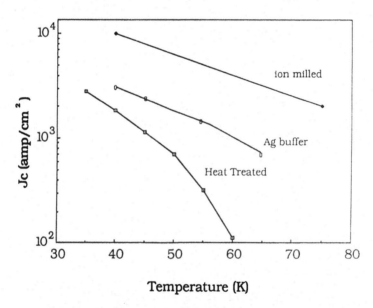

Figure 5. Jc vs. temperature for films on (a) ion milled, (b) Ag buffered, (c) oxidized, stainless steel substrates.

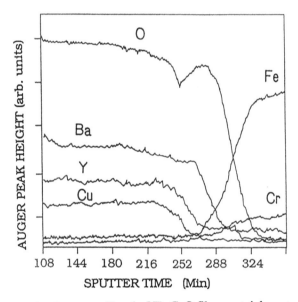

Figure 6. Auger profile of a YBaCuO film on stainless steel.

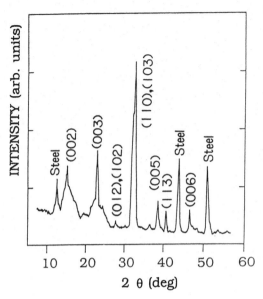

Figure 7. X-ray diffraction pattern of a YBaCuO film on stainless steel.

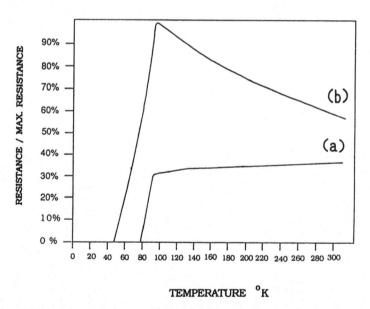

Figure 8. YBaCuO thin films on stainless steel, (a) at 550°C in an oxygen plasma, (b) at 570°C without an oxygen plasma.

In summary, we have grown as-deposited superconducting Y-Ba-Cu-O films on stainless steel with mirror finished surfaces. In-situ oxidized and buffer layered substrates produced better superconducting properties, compared to the untreated substrates. Ion milling of the substrate with low energy Ar ions produced similar critical temperatures, but J_c was improved by about an order of magnitude. Smooth surface texture and improved adhesion obtained by heat treating and ion milling may be responsible for the improved superconducting properties. We have also demonstrated the necessity of the oxygen plasma in producing good quality superconducting films on stainless steel.

Acknowledgement

We wish to thank P. Bush and R. Barone for their technical assistance. This research is supported by an award from the New York State Institute on Superconductivity in conjunction with the New York Energy Research and Development Authority.

References

(1) G. Foren, A. Gupta, E. A. Giess, A. Segmuller, and R. B. Laibowitz, App. Phys. Lett., 54, 1021 (1989).
(2) A. Inam, M. S. Hegde, X. D. Wu, T. Venkatesan, P. England, P. E. Miceli, E. W. Chase, C. C. Chang, J. M. Tarascon, and J. B. Wachtman, App. Phys. Lett., 53, 908 (1988).
(3) S. Witanachchi, H. S. Kwok, X. W. Wang, and D. T. Shaw, App. Phys. Lett., 53, 234 (1988).
(4) S. Witanachchi, S. Patel, D. T. Shaw, and H. S. Kwok, App. Phys. Lett., 54, 578 (1989).
(5) S. Witanachchi, S. Patel, D. T. Shaw, and H. S. Kwok, App. Phys. Lett., 55, 295 (1989).
(6) H. H. McCullough, M. G. Fontana and, F. H. Beck, Tran. Amer. Soc. Metals, 43, 404 (1951).

JOSEPHSON AC EFFECT IN Tl-BASED HIGH-T_c SUPERCONDUCTING BRIDGE

C. Wang, H.L. Chang, M.L. Chu, J.Y. Juang‡, T.M. Uen‡
and Y.S. Gou‡

Institute of Electro-optical Engineering
‡Institute of Electrophysics
National Chiao Tung Univ., Hsinchu, Taiwan, R. O. C.

ABSTRACT

The Josephson ac effect manifested by both the microwave induced constant voltage steps and a dc voltage shift has been observed in Tl-based high-T_c superconducting bridges. The induced dc voltage (V_{dc}) is caused by the non-reflecting vortex motion. From the temperature dependence of the V_{dc}, we inferred three different structures of vortices in this thin film; vortex glass, fluid and unbinding vortex-gas. The nature of the last state is believed to be closely related to the occurrence of Kosterlitz-Thouless transition in a quasi-two dimensional system.

The phenomenon of rf-induced dc voltage has been extensively studied in the subject of Josephson ac effect of superconductors, either in the conventional material[1,2] or in the high-T_c ones.[3,4] D. N. Langenberg et al.[1] in 1966 firstly observed the induced dc voltage in both high and low resistance junctions and speculated the induced voltage V_{dc} related to the microwave frequency ν by the Josephson-type voltage frequency equation $V_{dc}=nh\nu/2e$, where n is an integer. In the work of Chen, Todd and Kim,[2] the induced voltage were observed as an evidence of an inverse ac Josephson effect and were sensitive to the radiation and fluctuation, observable in a large dc magnetic field and of a systematic power dependence. While in the high-T_c materials, this induced voltage effect, first reported by J. T. Chen et al.[3], was also considered as a result of an inverse process of the ac Josephson effect and was used as evident that superconductivity might exist at temperature above 200 K. O. G. Symko et al.[4] currently observed this induced dc voltage effect in $YBa_2Cu_3O_7$ bulk sample and suggested this effect may be due to motion of fluxons. Actually, although both Chen

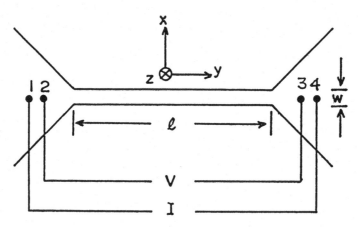

Fig.1   Schematic representation of the geometrical arrangement of the bridge. The points 1,4 and 2,3 represent the current and voltage leads respectively.

et al. and Symko et al. speculated that this phenomenon is closely attributed to Josephson ac effect, the mechanism of this induced voltage observed in the Josephson tunnel junction or weak link of the high-T_c superconductor has not been thoroughly understood yet. To clarify this effect, the direct observation of the induced voltage in a Josephson device of the high-T_c materials is significant in this work. In this paper, we report our observation on a long wide Tl-based high-T_c superconducting bridge radiated by an X-band microwave and a possible mechanism related to non-reflecting vortex motion to account for the induced V_{dc} is suggested. From the temperature dependence of V_{dc} we inferred that there are three vortex states: vortex glass, fluid and unbinding gas states.

The Tl-Ca-Ba-Cu-O superconducting thin film (~ 1 μm thick) used to fabricate the bridge was made by dc sputtering followed by post-annealing at ~ 895 °C for 3 minutes. A typical photolithography process was used to etch the film into bridges of about 10 μm wide and 125 μm long. Details of sputtering conditions and photolithography process were published elsewhere.[5,6] Figure 1 shows the typical appearance and geometrical arrangement of the bridge. Electrical contacts to the films were made using conductive silver paint. The

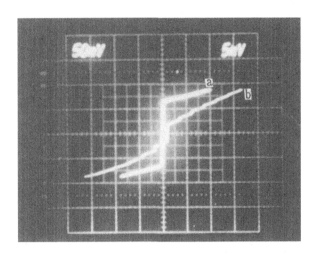

Fig.2 The current-voltage characteristics (IVC) of the bridge at 12.5 K. The horizontal scale is 50 μV/div and the vertical scale is 50 μA/div. Curve (a) is the IVC measured without microwave radiation. Curve (b) is the IVC measured under 12.359 GHz microwave radiation.

resulted bridge has a zero-resistance temperature around 72 K with the criterion of 1 μV. Scanning electron microscopy and X-ray diffraction studies show that although the thin film is c-axis oriented, it exhibits a highly disordered permutation of the grains. Measurements were made in an Air Products 10 K closed-cycle refrigerator system and temperature were controlled to within ±0.5 K. The microwave frequencies between 8 - 12.4 GHz were generated by a gunn diode which can be either operated in the constant-frequency and constant-power mode or in the continuous power sweeping mode. The maximum radiation power is 10 mW. The microwave was guided into the system via a coaxial cable terminated with an antenna. The antenna (~ 1.5 cm long) was extended into the closed-cycle system and placed above the sample along the long direction of the bridge. To measure the induced voltages, the voltage leads on the bridge were connected to a high-impedance differential amplifier followed by an oscilloscope, a digital multimeter or an X-Y recorder. For the current-voltage characteristics (IVC) measurement, a standard four-probe method was employed. While

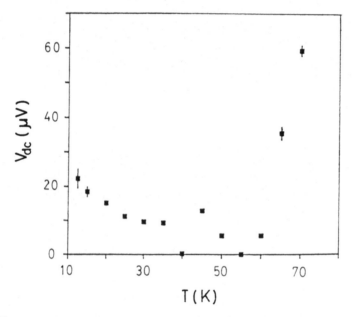

Fig.3　Temperaature dependence of the induced dc voltage (V_{dc}) at 15 dB frequency fixed at 12 GHz.

the vibration and the electromagnetic radiation of the compressor and the expander of our closed-cycle refrigerator are unavoidable, the phenomena observed remain unchanged at low temperatures whether the power of the closed-cycle refrigerator is on or not.

Figure 2 is the IVC's of the sample at 12.5 K. Curve (a) is the IVC without microwave radiation. When microwave radiated on the bridge, we can observed current steps at constant voltages in the normal resistance state as shown in curve (b). The voltage difference between the steps is about 26 μV, which agrees well with the calculated value of $h\nu/2e$. We also observed a clear shift of dc voltage in the curve (b). This really means that a dc voltage in the bridge was induced by the microwave radiation. Notice that as there was no dc current flowing through the bridge, this induced dc voltage still existed. Figure 3 shows the temperature dependence of the induced voltage with a constant power level at 15 dB. Note that there are two nodes at temperatures 40 and 55 K where the induced voltages drop significantly for all of the power level ranging from -2 to 15 dB. Above 35 K, the induced voltage gave an oscillatory conduct with respect to temperature. While below this temperature the V_{dc} increases slowly and tended to a certain value as the temperature decreases.

The induced V_{dc}, we suggest, comes from the motion of vortices in our bridge. The reasons are expected as follows; the vortices were created from one of the edges of the bridge and then penetrated into the bridge under the influence of an ac field current flowing through the specimen when the microwave radiated on the sample. The vortices were acted by Lorentz force and were pushed in a direction normal to the current (i.e. the x-direction shown in Fig.1) and tended to establish a continuous flow of additional vortices across the conductor (dynamic mixed state). The dc voltage drop would be nonzero when the vortices move across the bridge without reflection.

Owing to the above argument, we can estimate the propagation velocity of the vortices. The velocity of the vortices can be approximated as $v = wV_{dc}/n_1\Phi_0 \sim 2 \times 10^5/n_1$ m/s, where w is the width of our bridge, V_{dc} is the induced voltage we observed and n_1 is the number of vortices passing the bridge per unit time. Naturally we find that the collect motion of vortices in this case possesses the ultra sound velocity.

Based on observation of the different behaviors exhibited in the regimes, which are characterized and divided at temperature 35 K in the Fig.3, it thus provides a distinct evidence for the occurrence of the phase transition around this temperature, accordingly referred this temperature as freezing temperature T_f in the Tl-based system. In other words, we speculate that the vortex-glass state would appear below T_f in this system. As the temperature increases from 35 K to 40 K, the vortex-glass was melting and formed a vortex fluid, and the temperature 40 K may correspond to the melting temperature T_m of the vortex-glass state. The behavior between 40 and 55 K may imply another vortex-fluid state and 55 K is another T_m of the Tl-based system. For the region above 55 K, the induced-voltage linearly increasing very fast, we accordingly suggest, may correspond to a state dominated by the unbinding vortices due to the thermal activation: a typical characteristic of the K-T transition. It is interesting to note that the K-T temperature around 55 K in this system is much lower (even in reduced scale) than that of the YBCO system reported by Yeh and Tsuei[7] indicating a more 2-D like nature in our system. However, the understanding of the detail mechanism of the induced voltage shall invoke further studies on this topic.

In summary, we directly observed microwave induced dc voltage in a Tl-based superconducting long wide bridge which can be regarded as a kind of Josephson device. This induced voltage increased continuously as the microwave power increased. The temperature dependent induced voltage showed two nodes at 40 and 55 K. Above 35 K, the induced voltage gave an oscillatory behavior with respect to temperature, while below 35 K this induced voltage increased slowly and tended to a certain value as temperature decrease. The induced V_{dc} can be interpreted as due to the vortices motion without reflection in the bridge. From the temperature dependence of V_{dc}, it is suggested that there are three vortex states existent in these Tl-based high-T_c superconductors. The K-T transition occurred at a much lower reduced temperature as compared to that of YBCO system strongly implies a more 2D-like nature of these Tl-based thin films.

REFERENCE

1. D. N. Langenberg, D. J. Scalapino, B. N. Taylor and R. E. Eck, Microwave Induced D. C. Voltages Across Josephson Junctions, Phys. Lett. 20:563 (1966).
2. J. T. Chen, R. J. Todd and Y. V. Kim, Investigation of Microwave-Induced dc Voltages across Unbiased Josephson Tunnel Junctions, Phys. Rev. B5:1843 (1972).
3. J. T. Chen, L. E. Venger, C. J. McEwan and E. M. Logothetis, Observation of the Inverse ac Josephson Effect at 240 K, Phys. Rev. Lett. 58:1972 (1987).
4. O. G. Symko, D. J. Zheng, R. Durng, S. Duchorme and P. C. Taylor, Dissipative Flow of Josephson and Abrikosov Fluxons in High-T_c Superconductors, Phys. Lett. A134:72 (1988).

5. H. L. Chang, C. Wang, M. L. Chu, T. M. Uen and Y. S. Gou, Preparation of $Tl_2Ca_2Ba_2Cu_3O_x$ Superconducting Thin Films by DC Sputtering, Jpn. J. Appl. Phys. 28:L631 (1989).
6. H. L. Chang, M.L. Chu, C. Wang, Y.S. Gou and T.M. Uen, Josephson Effects in Tl-Ca-Ba-Cu-O Superconducting Thin Film Bridges, unpublished.
7. N.-C. Yeh and C. C. Tsuei, Quasi Two-Dimensional Phase Fluctuations in Bulk Superconducting $YBa_2Cu_3O_7$ Single Crystals, Phys. Rev. B39:9708 (1989).

SPUTTERED HIGH-T_c SUPERCONDUCTING FILMS AS FAST OPTICALLY TRIGGERED SWITCHES

P. H. Ballentine, A. M. Kadin, and W. R. Donaldson

Laboratory for Laser Energetics and
Department of Electrical Engineering
University of Rochester
Rochester, NY

ABSTRACT

Thin films of superconducting $Y_1Ba_2Cu_3O_7$ have been used as the active element in an optically triggered fast opening switch. Both granular and highly c-axis oriented films, ranging in thickness from 2000 Å to 7000 Å, were subject to 150 psec pulses from a Nd:YAG laser and exhibited switching from the superconducting to the normal state in times ranging from 1 nsec to 10 nsec, followed by a much slower decay over 100 nsec to 1 μsec. The magnitude of the voltage response and the slow decay can be explained in terms of a simple bolometric effect, while the rapid switching of even the thicker films is indicative of nonequilibrium hot electron transport from the absorbing surface layer to the remaining film thickness. At higher temperatures the response time is slower than 10 nsec, indicating a more diffusive behavior to the electron heat transport. These results give some indication of the electron-phonon collision time and help set limits to the maximum film thickness useable for laser triggered fast opening switches.

INTRODUCTION

One area of potential application for high T_c superconductors (HTSC) is in optoelectronics, where the unique combination of high electrical conductivity and optical absorption opens the possibility for a new class of optically triggered switches. Potential applications range from small scale integrated optoelectronic devices to high current fast opening switches for pulsed power applications. Additional motivation of this research is to investigate the nature of the photoresponse in HTSC materials. In particular, it is useful to know whether the response is due to simple heating, to nonequilibrium pair breaking, or to some other mechanism.

Several groups have investigated the photoresponse of HTSC, with some reporting observation of a nonequilibrium component[1-3] while others[4,5], including our own[6], report only a thermal response. The exact nature of the response no doubt depends on the operating conditions (temperature, bias current, optical fluence, etc.) and possibly the quality of the film as well. In particular, it has been suggested that the grain boundaries in granular films provide sites for a nonequilibrium effect, whereas high quality epitaxial films will show only a bolometric response. In this study we use both granular and epitaxial films for laser activated switching and find that, in both cases, the magnitude of the response is consistent with a bolometric model but that, at least in some cases, the speed of the response reflects a nonequilibrium heating of the electrons.

The use of laser pulses to trigger fast *closing* switches is well developed.[7] High speed, high voltage switches, constructed out of photoconductive material such as bulk silicon, have been used to switch 10 kV in 100 psec.[8] For some applications, however, a fast *opening* switch capable of handling large currents is required, and it has been proposed that a superconducting thin film be used as the active element in this case.[9] Such a superconducting opening switch could be used in conjunction with a superconducting current storage loop for a compact Terawatt pulsed power generator. The desirable properties for a fast opening switch are high current carrying capability and low resistance in the closed state, high resistance in the open state, efficient use of optical energy, and fast switching times. HTSC thin films appear to be excellent candidates because of their high critical current densities, high normal state resistance, and strong optical absorption. We demonstrate here that under certain circumstances, the switching speeds of HTSC can be quite fast as well.

EXPERIMENTAL

The YBCO films were prepared in a CVC-601 sputtering system by rf magnetron sputtering from a single 8-inch diameter stoichiometric oxide target as described in an earlier publication.[10] Two types of films were tested. Films deposited at substrate temperatures below 400 °C were amorphous as deposited and had to be annealed at 850 °C in oxygen in order to crystallize the superconducting phase. These films were granular, with T_c ranging from 60 to 83 K and low critical current densities, on the order of 1000 A/cm^2 at low T. In comparison, films deposited at 600 to 700 °C were superconducting without further annealing, having critical temperatures ranging from 55 to 75 K and higher critical currents, up to 10^5 A/cm^2. These films exhibited almost perfect c-axis orientation with no evidence of secondary phases present. Although the T_c in these as-deposited films was somewhat depressed, the superconducting transition remained fairly sharp. Furthermore these films exhibited an extended c-axis lattice parameter as measured by x-ray diffraction. The depressed T_c and extended c-axis, which have been observed by other groups sputtering YBCO from a single target[11], are no doubt related and remain a subject of research.

After deposition the films were patterned by laser ablation with a pulsed Nd:YAG laser[12] into an "H" structure with a central bridge region 200 μm wide and 2 mm long. The samples were mounted in an evacuated optical access dewar on a temperature controlled stage with a dc current applied to one end of the H and the load connected to the other. Connections from the planar switch to the connecting coaxial cables were made with SMA stripline launchers, and both the current (input) and voltage (output) lines were matched with 50 Ω loads. This was done to minimize signal dispersion and avoid reflections. The bridge region was exposed to 150 psec pulses from a Nd:YAG laser operating at 1.06 μm. The beam was focused down to a spot of approximately 400 μm wide by 4 mm long, so that the entire central bridge region was covered by the pulse. The energy per pulse could be adjusted up to 100 μJ and the repetition rate varied from 50 Hz to 1 KHz. The voltage at the load was measured with a digitizing oscilloscope and a fast analog scope providing temporal resolution down to 1 nsec. Further details of the experimental set-up are given in ref. 6.

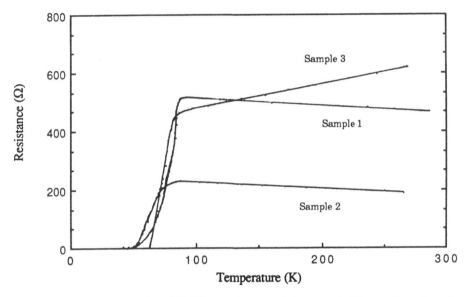

Fig. 1. R vs T curves for three YBCO samples used in laser switching studies. Sample 1: 7000 Å granular film on zirconia, T_c = 60 K, I_c = 1.5 mA (J_c = 1000 A/cm^2) @ 12 K, sample 2: 5000 Å epitaxial film on MgO, T_c = 50 K, I_c = 5 mA (J_c = 5000 A/cm^2) @ 40 K, sample 3: 2000 Å epitaxial film on MgO, T_c = 55 K, I_c = 12 mA (J_c = 10^5 A/cm^2) @ 45 K.

Figure 1 shows the dc R vs T characteristics for three samples used for laser switching studies. Sample 1 is a 7000 Å granular film on a zirconia substrate with a T_c of 60 K and an I_c of 1.5 mA (J_c = 1000 A/cm^2) at 12 K. Sample 2 is a 5000 Å epitaxial film on MgO with a T_c of 50 K and a I_c of 5 mA at 40 K (J_c = 5000 A/cm^2). Sample 3 is a 1750 Å epitaxial film on MgO with a T_c of 55 K and a I_c of 12 mA at 45 K (J_c = 10^5 A/cm^2). Most of the epitaxial films tested showed metallic behavior above T_c and sharp resistive transitions. Although sample 2 showed semiconducting behavior above T_c and a broadened transition, SEM photographs and XRD patterns indicated a highly c-axis oriented single phase film with a smooth surface.

Figures 2-4 show the transient response to 150 psec laser pulses for three switches tested. Figure 2 shows the response of sample 1 for the conditions of T = 17 K, I_{dc} = 2 mA, and a laser fluence of 1.6 mJ/cm^2 per pulse. In this case, the switch was biased slightly above the critical current so the initial voltage is not zero. The signal has a rise time of around 1 nsec followed by a slow decay with a characteristic time of around 1 μsec. This slow decay is believed to have been limited by poor thermal contact between the substrate and the sample holder. Figure 3 shows the response for sample 2 for the conditions of T = 28 K with I_{dc} = 1 mA and a laser fluence of 0.93 mJ/cm^2 per pulse. In this case there is an initial fast rise in 1 nsec, followed by a slower rise of about 10 nsec and a decay time of about 40 nsec. Figure 4 shows the response for sample 3 for T = 28 K, I = 1 mA, and laser fluence = 0.48 mJ/cm^2. Here again there is a 1 nsec rise followed by a 20 nsec decay. We discuss below the significance of the magnitude and time dependence of these signals and present an explanation of the observed behavior.

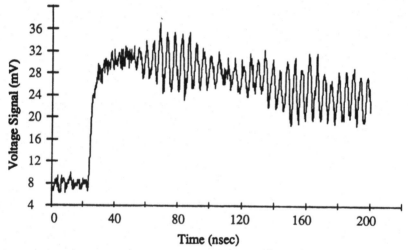

Fig. 2. Response of sample 1 to a pulse of 1.6 mJ/cm^2 for T = 18 K and I = 2 mA.

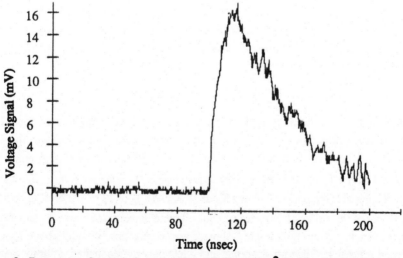

Fig. 3. Response of sample 2 to a pulse of 0.93 mJ/cm^2 for T = 28 K and I = 1 mA.

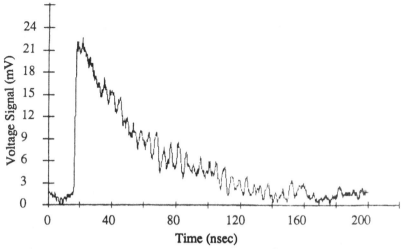

Fig. 4. Response of Sample 3 to a pulse of 0.48 mJ/cm² for T = 35 K and I = 1 mA.

ANALYSIS

In Fig. 5 we plot the magnitude of the peak response as a function of laser fluence for fixed temperature and bias current along with the results expected for a simple heating model. In this model we assume all of the laser fluence is deposited into the film which is then heated to a uniform temperature before there is time for cooling via the substrate. The resistance is then obtained from the dc R vs. T curves and the signal voltage is calculated for this resistance in parallel with the two 50 Ω transmission lines. There is a general agreement between our data and this simple bolometric model (with no adjustable parameters) for both granular and epitaxial films. The voltage saturates at about the fluence required to heat the film to its fully normal state. The measured saturation voltages lie a little below the expected values because part of the legs of the "H" pattern are in the beam, causing some series voltage drop which reduces the measured signal. There should also be a threshold fluence required to raise the temperature to T_c (R = 0). In the threshold region the data lie above the theory because our R vs. T curves were taken with lower currents than were used for the switching studies and the resistance of the films is sensitive to current close to T_c.

Although the magnitude of the response is in agreement with a simple bolometric model, the rise time of the signal is too fast to be explained in terms of heat diffusion, at least for the thicker films. The optical absorption depth for these films was measured using a spectrophotometer to be about 120 nm. Thus the optical energy is initially deposited in the top 2000 Å or so. The time required for the energy to diffuse to the remainder of the film is on the order of order h^2/D were h is the depth and D is the thermal diffusivity. For YBCO at 20 K, D is about 0.1 cm²/sec[13], thus the diffusion time for a 5000 Å film is on the order of 25 nsec, much slower than the observed switching times. This is demonstrated more clearly in Fig. 6 which presents the results of a one dimensional finite element calculation of the thermal diffusion process in a 2 μm thick YBCO film at T=20 K subject to a 150 psec pulse of 2 mJ/cm².

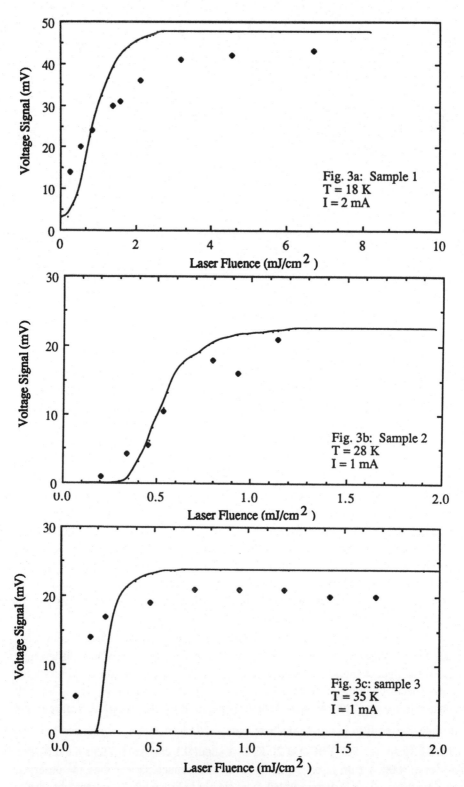

Fig. 5. Measured (•) and calculated (—) voltage response as a function of laser fluence. The calculated response is based on a simple bolometric model.

If only the top few thousand Å of the film is heated in the first 10 nsec, then how are the observed switching times to be explained? Current flowing in the heated region will produce a voltage, but this current should redistribute itself to the lower, unheated portion of the film shunting out the voltage signal. The characteristic time for this redistribution can be calculated from Maxwell's equations and is given by h^2/D_{em} where $D_{em} = \rho/\mu_0 \approx 10^5$ cm^2/sec is the electromagnetic diffusivity. This gives a time of about 1 fsec. Hence one would expect a fsec pulse followed by a slow rise due to thermal diffusion of the heat. In our case we should not even see the fast pulse because of the limited bandwith of measurement setup. For the response shown in Fig. 2, the switch was current biased slightly above the critical current. Here the fast switching may be

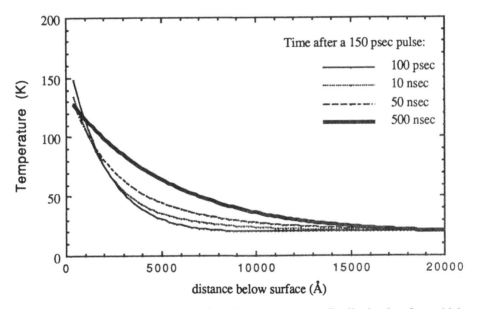

Fig. 6. Results of a numerical simulation of the temperature distribution in a 2 μm thick YBCO film at 20 K exposed to a 150 psec pulse of 2 mJ/cm^2. Values of temperature dependent diffusivity were calculated form data in Ref. 13.

explained on the basis of a rapid redistribution of current from the heated surface to lower portions of the film. This would lead to higher current density and higher voltage drop even without heat redistribution. However, even for films biased well below I_c, as in the case of Fig. 3, the signal shows a rapid rise to the value expected by assuming a uniform film temperature. Clearly the observed switching transients cannot be explained by simple thermal diffusion alone.

The explanation lies in the fact that these fast optical pulses heat the electrons to hundreds of degrees above the lattice temperature. These hot electrons are able to travel throughout the thickness of the film, either ballistically or diffusively, before loosing their energy through interactions with the lattice[14]. Hot electron cooling of laser heated surfaces is well established for normal metal thin films[15] and has recently been observed in single crystal La$_2$CuO$_4$.[16] For cases where hot electron transport is active, the signal rise time would depend not on the equilibrium thermal diffusivity of the film but on the electron-phonon relaxation time τ_{e-ph} and the velocity of the electrons. Recent estimates of τ_{e-ph} in YBCO[17] are 100 psec at 1.5 K and much less than 10 psec at 77 K. Using a Fermi velocity of 10^7 cm/sec[18] and τ_{e-ph} = 1 psec, the distance hot electrons may travel is on the order of 0.1 μm. Since the electron-phonon relaxation time decreases with increasing temperature, one would expect hot electron cooling to be more effective at low temperatures. In fact the voltage signal for sample 2 in Fig. 3, which was at a higher temperature than sample 1 (28 K vs. 17 K), shows signs of a slower component after the initial fast rise. At 28 K, the hot electrons may only travel partially through the 5000 Å film before thermalizing with the lattice. For a given temperature, there is expected to be a limit to the maximum film thickness that can be heated by hot electrons. In addition, by measuring the transient response for a number of films of different thickness it should be possible to estimate the τ_{e-ph}.

CONCLUSION

We have used superconducting YBCO thin films for a laser activated opening switch and have observed rise times on the order of 1 to 10 nsec for films up to 7000 Å thick. The magnitude of the response and the slow decay appear to indicate a bolometric behavior for both epitaxial and granular films. The fast rise time of the signal, however, indicates that hot electrons are responsible for heat transfer. At higher temperatures (greater than 30 K) the rise becomes slower, indicating that the electron-phonon relaxation time is less than the time required for the electrons to traverse the film thickness. These results indicate it should be possible to estimate τ_{e-ph} by testing films with a range in thickness. In addition, this places limits on the maximum film thickness useable when nsec switching is required.

ACKNOWLEDGEMENTS

This work was supported in part by NSF Grant DMR-8913524 and by the Laser Fusion Feasibility Project at the Laboratory for Laser Energetics. The thin films were deposited at CVC Products, Inc., Rochester, NY with help from R. Rath and J. Allen. X-ray diffraction patterns and ICP analysis of film composition was provided by Xerox Corp. and Kodak Corp. We also wish to acknowledge the help of R. Sobolewski and L. Kingsley for help with the optical measurements.

REFERENCES

1. J.C. Culbertson, U. Strom, S.A. Wolf, P. Skeath, E.J. West, and W.K. Burns, Phys Rev. **B 39**, 12362 (1989).

2. D.P. Osterman, R. Drake, R. Patt, E.K. Track, M. Radparvar, and S.M. Faris, IEEE Trans. Magn. **MAG-25**, 1323 (1989).

3. H.S. Kwok, J.P. Zheng, and Q.Y. Ying, Appl. Phys. Lett. **54**, 2473 (1989).

4. M.G. Forrester, M. Gottlieb, J.R. Gavaler, and A.I. Braginski, Appl. Phys. Lett. **53** 1332 (1989).

5. W.S. Brocklesby, *et al.*, Appl. Phys. Lett. **54**, 1175 (1989).

6. W.R. Donaldson, A.M. Kadin, P.H. Ballentine, and R. Sowbelewski, Appl. Phys. Lett. **54**, 2470 (1989).

7. C.H. Lee, editor, *Picosecond Optoelectronic Devices* (Academic, New York, 1984).

8. W.R. Donaldson, in *Picosecond Electronics and Optoelectronics II*, edited by F.J. Leonberger, C.H. Lee, F. Capasso, and H. Morkoc (Springer-Verlag, New York, 1987).

9. T.L. Francavilla, D.L. Peebles, H.H. Neslon, J.H. Claassen, S.A. Wolf, and D.U. Gubser, IEEE Trans. Magn. **MAG-25**, 1397 (1987).

10. A. M. Kadin, P.H. Ballentine, J. Argana, and R.C. Rath, IEEE Trans. Magn. **MAG-25**, 2437 (1989).

11. C.B. Eom, J.Z. Sun, K. Yamamoto, A.F. Marshall, K.E. Luther, T.H. Geballe, and S.S. Laderman, Appl. Phys. Lett. **55**, 595 (1989).

12. P. H. Ballentine, A. M. Kadin, M.A. Fisher, and D.S. Mallory, IEEE Trans. Magn. **MAG-25**, 950 (1989).

13. H.E. Fischer, S.K. Watson, and D.G. Cahill, Comments on Condensed Matter Physics, **14**, 65 (1988).

14. S.D. Brorson, J.G. Fujimoto, and E.P. Ippen, Phys. Rev. Lett. **59**, 1962 (1987).

15. A.M. Kadin, W.R. Donaldson, P.H. Ballentine, and R. Sobolewski, Proc. of the Int. Conf. on the Materials and Mechanism of High Temperature Superconductivity, Stanford, CA, July 23-28, 1989.

16. G.L. Doll, *et al.*, Appl. Phys. Lett. **55**, 402 (1989).

17. E.M. Gershenzon, M.E. Gershenzon, G.N. Gol'tsman, A.D. Seminov, and A.V. Sergeev, in Extended Abstracts of 1989 International Superconductivity Electronics Conference, Tokyo, Japan, June 12-13, 1989.

18. V.Z. Kresin and S.A. Wolf, Solid State Comm. **63**, 1141 (1987).

RADIATION DETECTION MECHANISMS IN HIGH TEMPERATURE SUPERCONDUCTORS

Young Jeong and Kenneth Rose

Center for Integrated Electronics
Rensselaer Polytechnic Institute
Troy, NY 12181

ABSTRACT

All radiation detection modes in superconductors and superconducting structures can be characterized by the dependence of resistance on temperature, current, or voltage. This allows their performance to be estimated and compared by measuring film resistance as a function of temperature and current. We present measurements of the current dependence of film resistance for high temperature oxide superconductors (HTOS) which indicate a high (> 10KV/W) responsivity for radiation detection. We will relate this high responsivity to resistance mechanisms in these films.

INTRODUCTION

Superconducting films and structures exhibit several modes of photodetection. These include a bolometric mode which depends on changes in film temperature and a rectification mode which depends on the nonlinear dependence of resistance on current or voltage. These modes and the resulting detector performance were reviewed at the last NYSIS Conference[1] and, more recently, at the Thirteenth International Conference on Infrared and Millimeter Waves[2].

Summarizing these reviews, the performance of bolometric detectors can be estimated quite accurately from the dependence of film resistance on temperature. A sharp drop in resistance over a narrow temperature range is required for a high sensitivity detector. Speed and sensitivity depend on heat capacity and thermal conductance of the film to a temperature bath. A very high speed bolometer mode is possible in which light pulses cause a nonequilibrium excitation of quasiparticles (and phonons) in the superconducting films. The response is fast because the cooling rate is limited by the time for nonequilibrium phonons to escape from the film. Both effects have been observed in HTOS films[3]. As in low temperature superconductors, nonequilibrium excitation has been found to be orders of magnitude less sensitive than conventional bolometer action.

The rectification mode can be understood by examining the behavior of SIS junctions and weak links. In an ideal SIS (Superconductor/Insulator/Superconductor) junction no current flows until the applied voltage exceeds the energy gap, $eV \geq 2\Delta$. At higher voltages quasiparticle tunneling occurs and the IV characteristic approaches the normal state resistance. Tucker and Feldman[4] have derived a general formula for current responsivity, $r_i = i_{out}/P_{in}$, from quantum tunneling theory. In the classical limit this reduces to

$$r_i = \frac{1}{2}\frac{d^2I/dV^2}{dI/dv} \qquad (1)$$

while the quantum limit is

$$r_i = e/hf. \qquad (2)$$

The quantum limit corresponds to one additional electron tunneling across the barrier for each absorbed photon, making SIS detectors the most sensitive detectors of electromagnetic radiation. The key point is that nonlinearity of the IV characteristic, as described by (1), provides a good guide to detector sensitivity.

The weak link is a dual of the SIS detector. No voltage appears across the weak link until a critical current I_c is exceeded. At higher currents the VI characteristic approaches the normal state resistance. For weak links one generally considers the voltage responsivity

$$r_v = \frac{1}{2}\frac{d^2V/dI^2}{dV/dI}. \qquad (3)$$

As for the SIS detector, the classical responsivity is determined by the nonlinearity of the VI characteristic. An appropriate quantum limit should reduce to the reciprocal of a minimum current. A reasonable estimate of this current is to divide the voltage quantum set when single photon absorption breaks an electron pair, hf/2e, by the microbridge resistance R. Thus, a quantum upper limit for voltage responsivity would be

$$r_v \approx \frac{2e}{hf}R. \qquad (4)$$

In the resistively shunted junction (RSJ) model of a weak link, R encompasses the flow of quasiparticle tunneling currents, resistive shorts, and flux flow resistance.

Granular films can act directly as radiation detectors without being patterned into particular structures. Early work on granular tin films showed a high responsivity detection mode which could be related to the nonlinearity of the VI characteristic[5]. Subsequent work showed that such films could behave as coherent arrays of weak links[6]. Examining the nonlinear VI characteristics of HTOS films should allow us to estimate their potential sensitivity as radiation detectors.

R(I,T) CHARACTERISTICS

Fig. 1 shows R(I,T) measurements on an unpatterned $Y Ba_2 Cu_3 O_{7-x}$ film deposited on an MgO substrate by sequential electron beam evaporation. The as deposited film was 0.6 μm thick and the substrate temperature was at ambient temperature. Subsequently, this film (90220) was sintered at 900°C for a half hour and annealed in flowing oxygen at 450°C for seven hours. The R(I,T) behavior of this HTOS film differs significantly from that of granular tin films we have studied previously[5,6]. In particular, at higher temperatures, the horizontal lines indicate the presence of a residual resistance caused by normal, ohmic regions. Only as the temperature drops to lower values, around 40K for this film, do we see the sort of R(I) characteristics previously observed with granular tin films. A more detailed comparison is made in [2]. Other films we have prepared show similar characteristics; however, the transition to previously observed behavior may occur at higher temperatures, e.g. 60K.

The gradual decrease of resistance with temperature implies low responsivity in the bolometer mode. For film 90220 at $I_b = 75$ μA and T = 84K, $\gamma = \partial R/\partial T \approx 0.2$ Ω/K. If K = 1mW/K this would respond to a responsivity. $r_v \approx I_b/K = 15$ mV/W. Higher currents would increase r_v. However, as we will show, such a film can have a high responsivity in the rectification mode.

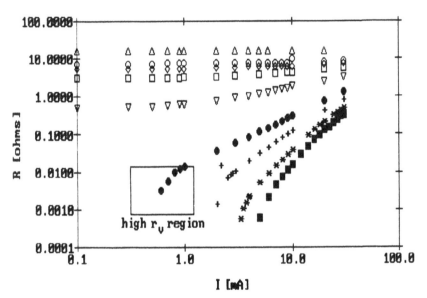

Fig. 1 Resistance dependence on current and temperature. Log-log plot for unpatterned $YBa_2Cu_3O_{7-x}$ film 90220. ($\triangle$: 81.7K, $\bigcirc$: 67.4K, $\diamond$: 64.2K, $\square$: 60.4K, $\triangledown$: 51.4K, $\bullet$: 41.6K, +: 38.2K, *: 33.7K, $\blacksquare$: 31K.)

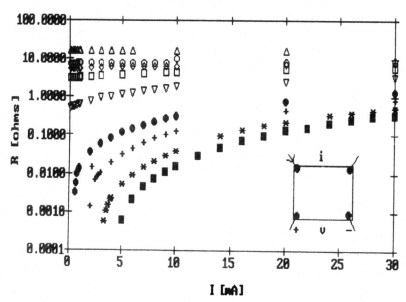

Fig. 2 Resistance dependence on current and temperature. Semilog plot for unpatterned $YBa_2Cu_3O_{7-x}$ film 90220. ($\triangle$: 81.7K, $\bigcirc$: 67.4K, $\diamondsuit$: 64.2K, $\square$: 60.4K, $\triangledown$: 51.4K, ●: 41.6K, +: 38.2K, *: 33.7K, ■: 31K.)

Although shallow, near horizontal slopes in Fig. 1 indicate a low rectification or current mode responsivity, steep slopes do not guarantee a high responsivity. (3) can be rewritten as

$$r_v = \frac{1}{2}\frac{dR/dI}{R}. \qquad (5)$$

(R in (5) should be the ac resistance dV/dI rather than the dc resistance V/I plotted in Fig. 1. However, power series expansion of the characteristic shows that use of the dc resistance will give a conservative estimate of r_v.) One can show that [2]

$$r_v = (log\text{–}log\ slope)/2I \qquad (6)$$

so that high slopes at low currents are required for high responsivity as indicated by the box in Fig. 1.

Plotting log R as a function of I on semilog paper gives a direct indication of r_v[2].

$$r_v \approx (semilog\ slope). \qquad (7)$$

The data of Fig. 1 has been replotted on semilog paper in Fig. 2 which clearly shows the importance of low bias currents for high responsivity.

Film 90220 is a square of $(5mm)^2$ area with current and voltage contacts on the corners as indicated in Fig. 1. Higher responsivities are achieved with a strip structure as shown in Fig. 3. Film 90418 was prepared in the same fashion as Film 90220. It is 0.68 μm thick and was deposited through a shadow mask to produce a pattern 1 mm wide with 3 mm between voltage terminals.

The current mode voltage responsivity can be calculated from the data of Fig. 3 for Film 90418 using (5). The results are shown in Fig. 4. Even if we discount the extreme values which approach 100 kV/W it seems clear that $r_v > 10$ kV/W should be achievable around 50K in this film! Further work is required to verify these conclusions by direct measurement of responsivity.

RESISTANCE MECHANISMS

To optimize the performance of current mode radiation detectors in these films requires a deeper understanding of the resistance mechanisms leading to these nonlinear characteristics. There are several potential sources of resistance in these films. In addition to inclusions of normal material, they include fluctuation effects, proximity effects, and current or magnetic field induced transitions to the normal state. Magnetic field effects in our measurement should be minor, limited to the effects of the earth's magnetic field (≈ 0.5 gauss). The critical current, however, is more likely determined by fluxon pinning than electron depairing.

Fluctuation effects are of interest because they allow estimation of the critical temperature and coherence length for a material. Fig. 5 shows a reasonable fit to the 3D fluctuation model[7] for Film 90220 with $T_c = 83$K and $\xi_s = 1.3$ nm. Since the temperatures in Fig. 1 are all below 83K, we conclude that the residual resistance we observe is not caused by fluctuation effects. Note that a 2D model is a poor fit for a film of this thickness.

We can explain the decrease of residual resistance with temperature in terms of the proximity effect. We assume the presence of normal regions of length L within the superconducting film. These may be grain or tetragonal twin boundaries. The probability amplitude for finding a Cooper pair at a distance x from the S/N boundary has the asymptotic form

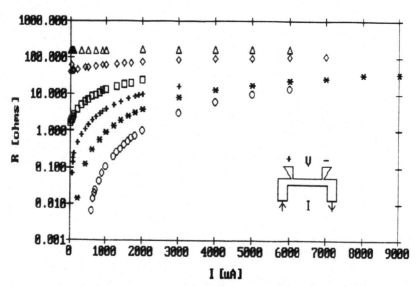

Fig. 3 Resistance dependence on current and temperature. Semilog plot for $YBa_2Cu_3O_{7-x}$ film in a strip structure(90418). ($\triangle$: 85K, $\diamond$: 71K, $\square$: 57K, +: 50K, *: 42K, $\bigcirc$: 32K)

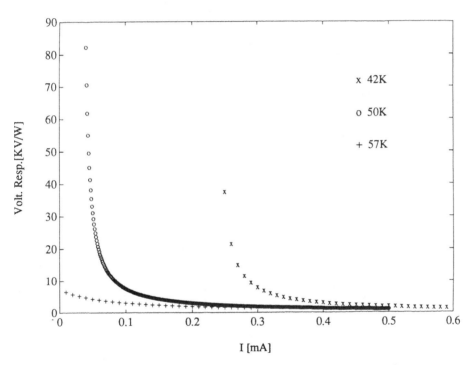

Fig. 4   Voltage responsivity calculated for $YBa_2Cu_3O_{7-x}$ strip structure 90418.

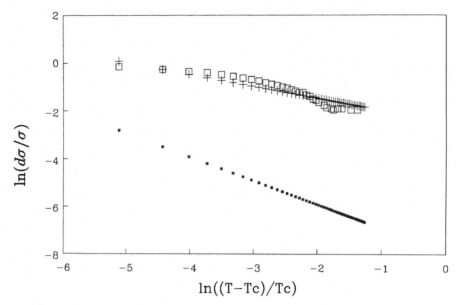

Fig. 5 Fit of resistance to fluctuation model for unpatterned film 90220 at temperatures above 83K. (⊙: 2D model for thickness(d) of .6 μm, +: 3D model for $\xi = 1.3nm$, □: experimental data of 90220.)

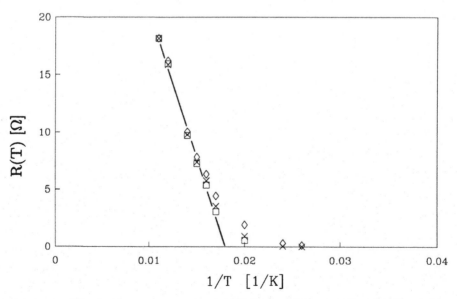

Fig. 6 Fit of resistance to proximity effect model for unpatterned film 90220 at temperatures between 50 and 82K. (□: 0.5mA, ×: 3mA, ◇: 10mA.)

$$F \approx F_o exp(-x/\xi_N) \qquad (8)$$

where $\xi_N = \hbar\, v_N/2\pi k_B\, T^8$. v_N is the Fermi velocity of the normal material and we assume that scattering is negligible, $\ell_N \gg \xi_N$. Roughly, superconducting behavior should extend a distance ξ_N into the normal region on either side. As the temperature drops ξ_N will increase, reducing the length of the normal region. Thus, we expect R(T) to have the form

$$R = R_o(1 - 2\xi_n/L) = R_o(1 - C/T). \qquad (9)$$

A plot of R(T) as a function of 1/T at various current levels is shown in Fig. 6 for Film 90220. Eq. (9) provides a good fit to the observed behavior. Extrapolating to R=0 gives C=0.0176 K^{-1}/ The appropriate value of v_N to use when evaluating ξ_N is not clear but should be between the values of 10^7 cm/s for the $Y\,Ba_2\,Cu_3\,O_{7-x}$ superconductor and 10^8 cm/s for a normal metal. This would correspond to values of L between 43 and 430 nm. Similar measurements on a strip structure such as Film 90418 show greater curvature as R approaches zero. Since the curvature increases with increasing current we attribute it to critical current effects.

CONCLUSIONS

We have prepared HTOS films which exhibit residual resistance below their fluctuation–determined transition temperature. These films have a low responsivity in the bolometer detection mode but high responsivity (> 10 kV/W) as current mode detectors. Fabrication of these films as strip structures improves their responsivity. The observed residual resistance can be accounted for by proximity effects.

REFERENCES

1. K. Rose, "The potential of high temperature superconductors as radiation detectors", in: "Superconductivity and Its Applications", H.S. Kwok and D.T. Shaw, eds., p. 201, Elseviever, NY (1988).

2. J.-Y. Jeong and K. Rose, to be published in the International Journal of Infrared and Millimeter Waves.

3. W.S. Brocklesby et al., "Electrical response superconducting $Y\,Ba_2\,Cu_3\,O_{7-\delta}$ to light", Appl. Phys. Lett., 54: 1175 (1989).

4. J.R. Tucker and M.J. Feldman, "Quantum detection at millimeter wavelengths", Rev. Mod. Phys., 57:1055 (1985).

5. C.L. Bertin and K. Rose, "Enhanced-mode radiation detection by superconducting films", J. Appl. Phys., 42:631 (1971).

6. W.J. Ayer and K. Rose, "Radiation detection by coherent Josephson phenomena in agglomerated tin films", IEEE Trans. Magn., MAG-11:61 (1975).

7. P.P. Freitas et al., "Thermodynamic fluctuations in the superconductor $Y_1\,Ba_2\,Cu_3\,O_{9-\delta}$: evidence for three-dimensional superconductivity", Phys. Rev. B, 36:833 (1987).

8. G. Deutscher and P.G. DeGennes, "Proximity effects", in: "Superconductivity", R.D. Parks, ed., p. 1005, Marcel Dekker, NY (1969).

RF SURFACE RESISTANCE OF A MAGNETICALLY ALIGNED

SINTERED SAMPLE OF Y $Ba_2Cu_3O_7$

H. Padamsee[†], J. Kirchgessner[†], D. Moffat[†],
D. Rubin[†], Q. S. Shu[†], H. R. Hart[#], and A. R. Gaddipati[#^]

[†] Laboratory of Nuclear Studies, Cornell University,
  Ithaca, N. Y.
[#] GE Corporate Research and Development,
  Schenectady, N. Y.
[^] Now at Knolls Atomic Power Laboratory,
  Schenectady, N. Y.

INTRODUCTION

The present study is described in detail in a paper by Padamsee, et al.,[1] which will appear shortly in the Journal of Applied Physics. This report is thus an extended abstract which includes references which have appeared since the submission of the original manuscript.

High-temperature superconductors offer several potential advantages over low-temperature superconductors for microwave superconducting cavities for particle accelerators. However, initial results on sintered polycrystalline samples of Y-Ba-Cu-O have shown very poor DC and RF properties.[2] Much better properties have been observed in single crystals[3,4] and oriented thin films.[5-7] Grain-aligned sintered samples of Y-Ba-Cu-O have shown superior DC superconductive properties[8-10] and might be expected to show superior RF properties as well. Thus, in this work, measurements were made of the 6 GHz surface resistance as a function of temperature and RF magnetic field for a Y-Ba-Cu-O sample in which substantial grain alignment had been achieved by magnetic alignment before sintering. Since the properties of Y-Ba-Cu-O are known to be very anisotropic, the RF measurements were made with the RF currents flowing in two directions, either parallel to or perpendicular to the c-axis.

In existing superconducting accelerator cavities, niobium is the material of choice,[11] yielding an accelerating field of 5-10 MeV/m; in test cavities fields approaching 25 MeV/m have been observed. The theoretical maximum possible accelerating field, set by the superheating critical magnetic field, is expected to be 50 MeV/m for niobium. For Y-Ba-Cu-O, with its higher superheating critical magnetic field, the maximum accelerating field should approach 200 MeV/m. One should recognize that the achievement of a significant fraction of the theoretical maximum field in niobium cavities has involved a major developmental effort in the suppression of problems such as field emission. One needs very high purity niobium with surfaces having a high degree of perfection.

Conventional superconducting accelerators operate at liquid helium temperatures with the concomitant refrigeration inefficiencies and costs. Operation at higher temperatures with high-temperature superconductors would allow substantial savings in refrigeration costs, both capital and operational, even with a RF surface resistance somewhat higher than that of niobium.

SAMPLE PREPARATION AND CHARACTERIZATION

The preparation of the sample was based on the discovery by Farrell and his coworkers[12] that single-crystal grains of Y-Ba-Cu-O suspended in epoxy could be c-axis aligned at room temperature in a magnetic field of 9.4 Tesla. The present sintered samples were prepared by suspending the particles in heptane and aligning them in a 4 T magnetic field at room temperature. The dry cake resulting from the evaporation of the heptane was then sintered to yield a solid, conducting sample.

The preparation and characterization of the sample are described by Arendt, et al.,[8] and Padamsee, et al.[1] The anisotropy ratio of the room-temperature resistivity is 14:1, typically 450 µOhm-cm in the nominal ab-direction and 6300 µOhm-cm in the nominal c-direction. Likewise, anisotropies of 15:1 in the critical current were observed at 77 K and 0.28 T for current flow (B⊥I) in the ab-plane and in the c-direction. The alignment was not complete in two senses: there is no alignment of the a- and b-axes in the nominal ab-plane, and the c-axis misalignment yields a rocking curve angle of about 14° FWHM.

RF MEASUREMENTS

Measurements of RF surface resistance, Rs, were made in a 6 GHz niobium cavity operated in the TE011 mode ($Q \sim 10^8$), as described by Rubin, et al.[3] The sample (1.0 cm x 0.7 cm x 0.38 cm) was thermally isolated from the cavity so that its temperature could be varied while the temperature of the niobium cavity was held at 2 K. The earth's magnetic field was nulled to below 10 mOe. Rs was measured for two orientations of the sample, one with the c-axis perpendicular to the RF current plane and one with the c-axis parallel to the RF current plane. For low RF fields (<0.1 Oe), Rs was measured as a function of temperature from 4 K to 100 K. In addition, at a reasonably low temperature (<20 K), Rs was measured as a function of RF magnetic field strength from 0.01 Oe to 35 Oe.

The measurements at low RF magnetic field (<0.1 Oe) indicate that at 4 K the anisotropy of the low-field 6 GHz surface resistance is 28:1, with the lower resistance for RF current flow (0.25 mΩ) in the nominal ab-plane. At 77 K the surface resistance is 3 mΩ in the better orientation, a significant improvement over randomly oriented sintered material (10-20 mΩ), but not as good as for single crystals or high-quality thin films (<0.5 mΩ).[3-7] The drop in Rs is much more gradual with decreasing temperature below Tc than for thin films and high-quality single crystals.

At low temperatures (<20 K) the surface resistance increases with RF magnetic field strength above a threshold level. For current flow in the ab-plane the resistance increases from 0.25 mΩ for RF fields below 0.1 Oe, to 0.6 mΩ at 1 Oe, 4 mΩ at 10 Oe, and 10 mΩ at 35 Oe. Somewhat surprisingly the absolute increase is essentially the same for the two orientations of the sample, approximately 5 mΩ between 0.01 and 10 Oe. Thus the anisotropy decreases as the RF field is increased. The increase in surface resistance with RF field strength observed in the grain-aligned sintered sample becomes significant at a much lower field (~1-2 Oe) than for high-quality thin films (~10-30 Oe)[6,7] or single crystals (~90 Oe)[3]. Note that the RF magnetic fields in existing niobium accelerator cavities are in the range 200-400 Oe.

DISCUSSION

These results indicate that, while the low-temperature low-field RF behavior is significantly improved by the orientation of the grains, the behavior at higher fields or higher temperatures is much inferior to that of high-quality crystals and the best laser-ablated epitaxial thin films. These deficiencies can be attributed to the poor current-carrying ability of the weak links associated with the boundaries between the grains. Note that the alignment of the grains is imperfect on two counts: the c-axes are only roughly aligned and the a- and b- axes are randomly oriented in the ab-plane. Thus, even for chemically clean boundaries, weak links may govern.[13]

Several models for the RF surface resistance in oxide superconductors have recently been presented, for example by Hylton, et al.,[14] and Portis, et al.[15] In the model of Hylton, et al., the RF surface resistance is the sum of two contributions, the intrinsic resistance of the grains, which decreases to zero exponentially with temperature, and the losses caused by the weak links between the grains. At low temperatures the contributions of the weak-links, modeled as resistively-shunted Josephson junctions, should yield a temperature-independent residual resistance which depends inversely on the critical-current density. Such a prediction is in qualitative agreement with the low-field results for the grain-aligned sample, for the residual surface resistance is very anisotropic and is much lower in the direction showing the higher critical-current density. The increase in surface resistance with RF field strength is essentially isotropic and is thus not easily explained, for the critical-current density, the coherence length, and the normal-state resistivity are all extremely anisotropic.

A recent preprint by Hein, et al.,[16] describes RF surface resistance measurements at 21.5 GHz on 10-20 μm thick textured or grain-aligned films of Y-Ba-Cu-O electrophoretically deposited on silver in an 8 T magnetic field. As in the present experiment the surface resistance was found to be lower than for similar, but unaligned samples. The measured surface resistances are 18 mΩ at 77 K and less than 3 mΩ at 4.2 K. Scaling these 21.5 GHz results to 6 GHz using a 1.7 power law in frequency[6] yields values very close to those found in the present study: 2 mΩ at 77 K, slightly lower than the 3 mΩ reported above, and less than 0.34 mΩ at 4.2 K to be compared with 0.25 mΩ for the present study. No measurements were reported of the effects of stronger RF fields on the surface resistance.

The results of the RF measurements on the grain-aligned, sintered sample of Y-Ba-Cu-O, when compared to those of good epitaxial thin films and single crystals, indicate that both extremely good grain alignment and clean grain boundaries will be necessary for good RF behavior.

ACKNOWLEDGMENTS

The RF portion of this study was supported by the National Science Foundation with supplementary support from the US-Japan Collaboration. One of the authors (H.R.H.) wishes to thank the School of Applied and Engineering Physics of Cornell University for its hospitality during a portion of this work. We wish to thank the following for valuable assistance: T. W. Noh, J. D. Livingston, M. J. Curran, A. J. Barbuto, J. Sears, J. Potts, and K. Green.

REFERENCES

1. H. Padamsee, J. Kirchgessner, D. Moffat, D. Rubin, Q. S. Shu, H. R. Hart, and A. R. Gaddipati, J. Appl. Phys. (in press).

2. H. Padamsee, 1988 Linac Conference, Williamsburg, Va., October, 1988 (in press).

3. D. L. Rubin, K. Green, J. Gruschus, J. Kirchgessner, D. Moffat, H. Padamsee, J. Sears, Q. S. Shu, L. F. Schneemeyer, and J. V. Waszczczak, Phys. Rev. B $\underline{38}$, 6538 (1988).

4. Dong-Ho Wu, W. L. Kennedy, C. Zahopoulos, and S. Sridhar, Appl. Phys. Lett. $\underline{55}$, 696 (1989).

5. N. Klein, G. Müller, H. Piel, B. Roas, L. Schultz, U. Klein, and M. Peiniger, Appl. Phys. Lett. $\underline{54}$, 757 (1989).

6. A. Inam, X. D. Wu, L. Nazar, M. S. Hegde, T. Vankatesan, R. Simon, K. Daly, H. Padamsee, J. Kirchgesser, D. Moffat, D. Rubin, Q. S. Shu, E. Belohoubek, L. Drabeck, G. Mihaly, G. Gruner, R. Hammond, F. Gamble (submitted).

7. H. Padamsee, J. Kirchgessner, D. Moffat, D. Rubin, Q. S. Shu, L. Nazar, M. S. Hegde, T. Vankatesan, A. Inam, and X. D. Wu (to be published).

8. R. H. Arendt, A. R. Gaddipati, M. F. Garbauskas, E. L. Hall, H. R. Hart, Jr., K. W. Lay, J. D. Livingston, F. E. Luborsky, and L. L. Schilling, in High-Temperature Superconductors, Mater. Res. Soc. Symp. Proc., Vol. 99, edited by M. B. Brodsky, R. C. Dynes, K. Kitazawa, and H. L. Tuller (Materials Research Society, Pittsburgh, 1988) p. 203.

9. K. Chen, B. Maheswaran, Y. P. Liu, B. C. Giessen, C. Chan, and R. S. Markiewicz, Appl. Phys. Lett. $\underline{55}$, 289 (1989).

10. J. E. Tkaczyk, K. W. Lay, F. E. Luborsky, L. L. Schilling, A. R. Gaddipati, and H. R. Hart, Proceedings of the Third Annual Conference on Superconductivity and Applications, H. S. Kwok, editor., Plenum Press, N. Y., 1989; J. W. Ekin, H. R. Hart, and A. R. Gaddipati (to be published)

11. H. Padamsee, J. Superconductivity $\underline{1}$, 377 (1988).

12. D. E. Farrell, B. S. Chandrasekhar, M. R. DeGuire, M. M. Fang, V. G. Kogan, J. R. Clem, and D. K. Finnemore, Phys. Rev. B $\underline{36}$, 4025 (1987).

13. D. Dimos, P. Chaudhari, J. Mannhart, and F. K. LeGoues, Phys. Rev. Lett. $\underline{61}$, 219 (1988).

14. T.L. Hylton, A. Kapitulnik, M. R. Beasley, J. P. Carini, L. Drabeck, and G. Gruner, Appl. Phys. Lett. $\underline{53}$, 1343 (1988).

15. A. M. Portis, D. W. Cooke, and H. Piel, Physica C (in press).

16. M. Hein, G. Müller, H. Piel, L. Ponto, M. Becks, U. Klein, and M. Peiniger (submitted)

FABRICATION OF A TECHNOLOGICALLY USEFUL CONDUCTOR USING CERAMIC SUPERCONDUCTOR*

D.W. Hazelton, R.D. Blaugher, M.S. Walker
Intermagnetics General Corporation
Guilderland, New York 12084

ABSTRACT

Several critical design, fabrication and operational issues must be resolved in order for the new high T_c materials to be transitioned into useful conductor forms. This paper is an initial progress report of a NYSIS sponsored assessment of design requirements for a high T_c technologically useful conductor for various energy related applications. Initial results indicate that a tape conductor configuration provides sufficient mechanical support and thermal stability to function at higher operating temperatures than the conventional low temperature superconductors.

INTRODUCTION

The new ceramic superconductors with critical temperatures above liquid N_2 (77K) enable one to envision operation of superconducting magnets at temperatures substantially above the current 4.2K liquid helium cooled superconducting magnets utilizing conventional superconductors such as NbTi and Nb_3Sn. This paper is an initial report of a NYSIS sponsored assessment of the requirements to transition these newer materials into a technologically useful conductor form.

Historically, and for good reasons, brittle superconductors were first made in a tape with multifilamentary development lagging at least a decade behind. Following this trend, it is likely that the first realization of a practical ceramic based superconductor will be in tape form. The need for

---

* This work was supported in part by the New York Energy Research and Development Authority through the New York State Institute on Superconductivity.

multifilament oxide conductors might not be as critical as in the lower T_c materials. Studies by several researchers[1,2,3] suggest a reduced need for multifilament oxide conductors for stability and losses and indicate that tape conductors are well suited to the higher operating temperatures. For these reasons, the focus of our evaluation will be concentrated on a tape configuration.

Traditionally, (Low T_c) superconductivity has been readily accepted for applications such as magnetic resonance imaging (MRI), high energy accelerators, magnetic confinement fusion, defense related magnetic energy storage (SMES), MHD, high field research magnets, NMR spectrometers and various electronics applications. However, superconductivity has been slow to gain acceptance for conventional power related technology such as electrical transmission lines, generators, commercial magnetic energy storage, motors, propulsion and magnetic separation. While the projected advantages of superconductors offer substantial performance and efficiency advantages, the liquid helium requirement presents concerns about the long term reliability of superconducting systems.

The high T_c conductors provide incremental improvement in efficiency and operational advantages over the low T_c superconductors for most of the conventional technologies. The high T_c materials should, however, simplify manufacturing and improve reliability of refrigerators for continuous long term operation. New applications could conceivably result from the ability to operate at higher temperature.

PROGRAM APPROACH

The approach used in the evaluation of the requirements for transitioning the ceramic superconductors into a useful conductor form starts with a generic consideration of the following parameters:

1. Mechanical requirements
2. J_c vs. strain properties
3. Stability issues
4. J_c vs. operating temperature
5. Refrigeration and cryostat requirements

Initial work in addressing these issues is described in this paper. The generic analysis with respect to the following specific requirements is to be followed by a case by case examination of applications. Where applicable information and analyses already exist in the literature, they will be utilized.

1. Stability requirements
2. Mechanical requirements
3. J_c vs. strain considerations
4. Long term conductor stability
5. AC loss considerations
6. Manufacturability of commercially usable lengths
7. Property enhancement
8. Magnet protection
9. Cost
10. Applicability of tape conductor

Cases to be examined include:

1. Large bore (traditionally cryostable) windings
   a. Generic circular coils
   b. Magnetic confinement fusion
   c. Superconducting leads and busses
   d. Superconducting magnetic energy storage
2. Transmission lines
3. Very high field, small bore, dynamically stable magnets
4. Conductors for more complex winding shapes
   a. Accelerator and high energy magnets
   b. MHD magnets
   c. Generators
5. Conductors for other AC or swept field applications

This paper presents initial results of the scoping analysis for some of the specific requirements for the first of the applications listed, 1a. Generic, large bore circular coils.

GENERIC CONSIDERATION OF STABILITY, J_c VS. OPERATING TEMPERATURE, REFRIGERATION AND CRYOSTAT REQUIREMENTS

There are several tradeoffs to be considered when determining the operating temperature of the superconducting system. Operating above 20K provides long term reliable refrigeration technology with possible reduction of the need for thermal shields. Operating at as low a temperature as possible allows the highest usable flux-flow free J_c with the least sensitivity to temperature changes. However, the temperature of operation should be high enough to take advantage of the increased heat capacity of solid winding components for stability (primary) and, for some applications, for reduced refrigeration costs.

For most magnet systems, a conduction and helium gas cooled tape winding operating at 25K seems to meet all the above requirements. As an example, consider a magnet with a 12 Tesla central solenoid field parallel to the a-b plane of an oriented conductor with c-axis perpendicular to the conductor direction. Such a magnet would have a 6 Tesla perpendicular end turn component (parallel to the c-axis) which would limit J_c in the conductor. As shown in Figure 1 from Rowell[4], at Hc (perpendicular) equal to 6 Tesla, T_c is equal to 70K, a high enough temperature to accommodate a transient temperature rise to 35K with $J_{op} \ll J_c$. Allowing a temperature rise from 25K to 35K, preliminary analysis shows that a tape winding interleaved with support is highly adiabatically stable using only the heat capacities of the solid components. This 10K temperature rise from 25K to 35K can absorb roughly 40 times the estimated local energy from a worst case credible event disturbance, 1800 times the energy of a 1K rise from 4.2K. The minimum propagating zone at the 25K operating temperature is possibly as high as 1000 times that at 4.2K. Such windings may be so stable that a quench may have to be actively propagated in order to properly protect the system.

GENERIC CONSIDERATION OF MECHANICAL AND J_c VS. STRAIN REQUIREMENTS FOR Y123

In order to look at the stresses and strains developed in a composite conductor, a mechanical model has been developed for a composite consisting of $YBa_2Cu_3O_{7-\delta}$ (Y123) layers, MgO buffer layers and a Ni-200 substrate. A schematic of the composite conductor is shown in Figure 2.

Consider the composite to be fully bonded consisting of a number of layers, each designated by the subscript i. For each layer i:

- t_i = layer thickness
- A_i = layer cross-sectional area = $t_i * w$
- $(dL/Lo)_i$ = thermal expansion (normalized to Lo)
- E_i = modulus of elasticity
- ε_i = strain
- σ_i = stress

At the various bonding temperatures (i.e., buffer to substrate, Y123 to buffer/substrate), assume all of the layers are of the same length and width. As the composite's temperature is varied from the bonding temperatures and the composite is kept flat (aided by its symmetry), the same strain, ε_c, must occur in each layer to prevent cracking. This composite strain, ε_c, is given by:

$$\varepsilon_c = \varepsilon_{i\text{-thermal}} + \varepsilon_{i\text{-mechanical}}$$

The thermal strain, $\varepsilon_{i\text{-thermal}}$, is determined by taking into account the following for each component:

1. The normalized thermal expansion, $(dL/Lo)_i$, as a function of temperature.
2. The presence of any phase changes and resultant expansions/contractions of the material due to the phase change (i.e., Y123 [tetragonal] <> Y123 [orthorhombic]).
3. The effects of orientation in anisotropic materials. For example, Y123 has a nominal coefficient of thermal expansion of 10 ppm/K along the a-axis, 12 ppm/K along the b-axis, and 40 ppm/K along the c-axis. This model assumes the c-axis is perpendicular to the buffer since this configuration is needed to maximize usuable current densities. The a- and b-axis orientations are either random or directed.

The mechanical strain, $\varepsilon_{i\text{-mechanical}}$, is a function of σ_i, under the constraint that $\Sigma A_i \sigma_i = F$, where F is the applied force on the conductor. The following factors must be taken into account when evaluating the mechanical strain:

1. The σ-ε relationship for a given material vs. temperature.
2. Any phase changes within the components
3. The anisotropic nature of some of the materials

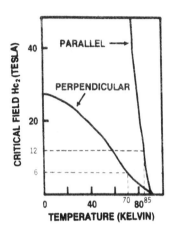

Figure 1. Anisotropy of critical field vs. temperature for a 1:2:3 single crystal from data from Osaka University and NTT as reported by Rowell[4], with dashed lines added by IGC. A 6 Tesla field normal to the tape surface will limit T_c to 70K.

Figure 2. Symmetrical Y123/MgO/Ni/MgO/Y123 composite tape conductor used in the model.

The forces on the conductor originate at several sources:

1. Fabrication stress
2. Winding stress
3. Cooldown stress
4. Magnetic stresses during operation

Another strain component, $\varepsilon_{i\text{-bending}}$, enters into consideration whenever the conductor is wound over a radius. For a radius of curvature r_o:

$$\varepsilon_{i\text{-bending}} = x_i/r_o \text{ - where } x_i \text{ is the distance from the neutral (zero strain) axis}$$

Preliminary computer analysis using the above model has confirmed that the tape configuration is favorable in both the thermal and mechanical environments. Cooling of the composite results in a compressive stress being generated in the 123 material for a c-axis perpendicular condition. In addition, the thicker the 123 layer relative to the other layers, the lower the overall level of stress (and strain) generated in the 123 layer. Future work includes the following improvements to the model; improved material properties vs. temperature values, examination of shear stresses generated at the component interfaces, evaluation of other superconductor, buffer and substrate systems as well as inclusion of other structural components.

SPECIFIC CASE 1a: LARGE BORE CIRCULAR COILS

For the first trial design, a 12 Tesla solenoid with a nominal 1 meter bore size was selected. The operating current within the conductor element has been set at 10000 Amps with a current density J_c (free of flux flow) > 250 A/mm^2 at 12 Tesla, 35K (to allow for a 10K temperature rise from a T_{op} of 25K). A conductor scheme is given in Figure 3. This design utilizes a stainless steel interleaved superconductor tape pancake winding with grading by width and parallel layers. The resultant system is adiabatically stable based upon the heat capacities of the solid components only. Each element consists of layered 123/buffer/substrate/buffer/123 tapes with intermediary OFHC Cu stabilizer soldered into a half hard Cu structural case. These winding elements are then cowound into conductor layers interleaved with stainless steel and insulation. The layers of thin superconductor tape may be prepared using either vapor deposited films or processed thick coatings. Future work includes analysis of the dynamic stability, and coupled hysteresis and eddy current charging losses.

CONCLUSIONS

Preliminary examination of some of the critical design issues of using the ceramic superconductors lead to the initial conclusion that high T_c materials in a tape form may able to be used in several applications. Mechanically, a composite conductor approach is necessary to put the conductor into compression during cooldown in order for the tensile stresses induced during operation to be tolerated. Proper selection of the component materials, fractions and application will lead to this condition. While the operating temperature of 25K considered for some applications is below the "magical" 77K LN$_2$

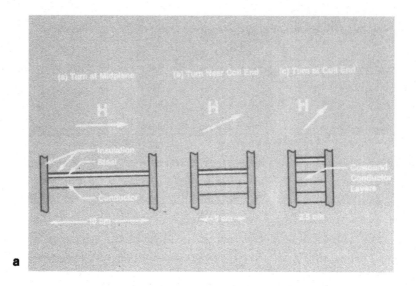

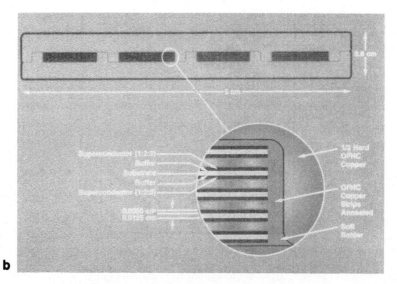

Figure 3. Initial design concept for conductor to be used in large bore circular coils.
  a. Graded turn configurations at various coil locations
  b. Detail of stabilized conductor element

operating point, it does provide enough temperature margin below T_c for high J_c, high stability and can readily be achieved using He gas and simplified refrigeration and cryostat technology. For space applications, ambient temperature may prove to be adequate for operation of such a system. Active protection will probably be needed and this will be needed in the case of a quench. The selected current density of 250 A/mm^2 at 12 Tesla, 35K (to allow for disturbing temperature excursions) is aggressive, but within grasp, especially if one looks to the use of thin or thick film technologies for laying down oriented superconductor films onto a tape substrate.

REFERENCES

(1) E.W. Collings, "Design Considerations for High Tc Ceramic Superconductors," Proceedings of Materials Research Society Meeting on Advanced Materials, Tokyo, Japan, May 30 - June 3, 1988, Published in Cryogenics.

(2) H.L. Laquer, F.J. Edeskuty, W.V. Hassenzahl and S.L. Wipf, "Stability Projections for High Temperature Superconductors," IEEE Trans. Mag. 25, 1989, pp. 1516-1519.

(3) Y. Iwasa, "Design and Operational Issues for 77K Superconducting Magnets," IEEE Trans. Mag. 24, 1988, pp. 1186-1189.

(4) J.M. Rowell, "The Status, Recent Progress and Promise of Superconducting Materials for Practical Applications," IEEE Trans. Mag. 25, 1989, pp. 759-768.

(5) J. Ekin, private communication.

(6) Togano et al, "Fabrication and Processing of Flexible Tapes of Bi(Pb) Compounds," Paper BX-02, Presented at the ICMC Meeting at UCLA, July, 1989.

RAPIDLY SOLIDIFIED SUPERCONDUCTING FILM ON METALLIC WIRE

S. A. Akbar, M. L. Chretien and J. Huang

Department of Materials Science and Engineering
The Ohio State University
Columbus, OH 43210

Abstract

For technological applications of high-T_c superconductors, it will be necessary to overcome the inherent problem of brittleness, to develop materials with high current carrying capacity, and to devise ways of joining superconductors with other materials. This article attempts to address these aspects by focusing on superconducting films coated on metallic wires by solidification from melt. The composite systems are expected to produce flexible wires with desirable properties. Results of some preliminary investigations are reported.

I. Introduction

The discovery of superconductivity in ceramic oxides prompted intensive worldwide research. Almost two and a half years after this discovery, however, these materials remain a long way from technological applications. The inherent problem of brittleness limits the flexibility of fabricating these materials in desired shapes for device applications. Moreover, it will be necessary to develop these materials with high current carrying capacity, and to devise ways for joining them with other materials.

Superconducting thin films with high current density (J_c) have been produced on oxide substrates such as $SrTiO_3$ [1] and MgO [2]. There is, however, a need for superconducting films produced on semiconductors, metals and other ceramics so that they can be integrated into microelectronic devices. Superconducting films coated on metallic wires by directly solidifying from the melt may lead to the development of manufacturing processes for economically producing flexible wires of ceramic superconductors.

II. Past Developments

Number of methods have been reported in the literature [3-15] to fabricate wires of these materials. Since these materials are brittle, they cannot be drawn out to form wires as copper and other ductile metals. Fabricating composites of metal and ceramic superconductors has, thus, gained extensive interest [3-6,11-15]. Best known among these is the so-called "powder in tube" technique. In this method, preform composites of normal metal tube (mostly Ag alone or Cu with Ni/Au diffusion barrier) are filled with heat treated and pulverized superconducting powders and then cold worked to the desired outer diameter (~ 1 mm). The superconducting powder core deforms continuosly in conformation with composite wire geometry during wire drawing by particle/particle sliding. As the ceramic core is cladded by ductile metal, the drawn wire can be wound into coil shape. When the desired final shape is obtained, it is heat treated for sintering of powders to obtain the desired compound structure (orthorhombic for $YBa_2Cu_3O_{7-y}$) and oxygen content. The ceramic core in this type of wire, can therefore be shielded from the thermal, mechanical and environmental stresses. The metal can also serve as a parallel electrical conduction path to carry current in case of local loss of superconductivity.

However, due to a low compaction density, the resulting material is very porous and consequently the current carrying capacity is low. The difficulty of oxygen supply to the core of the wire may also affect the critical current density [11]. The tensile stress arising from the sintering of the ceramic core introduces many large cracks [3] between the core and the metal sheath, and such discontinuous structure lowers the critical current density. The maximum value for J_c in wires produced by this technique has been reported [12,13] to be of the order of 10^3 amp/cm^2 which is about three orders of magnitude lower than in thin films, and too low to be practically useful.

In a second method, ceramic powders in stoichiometric proportions are mixed with an organic binder, and then the mixture is slowly extruded into a thin wire shape through a die [7-10,16]. At this stage, the extruded wire can be easily bent or deformed. But it must then be fired to burn off the binder, and heat treated to sinter the powders into a current carrying filament. After the heat treatment, the wire regains its brittleness, and also, because of the presence of the grain boundaries and the relatively low density of the sintered material, the value of J_c is very low (10^2 - 10^3 amp/cm^2).

Since newly discovered superconductors have anisotropic crystal structures [17,18], a hot forging method has been considered to achieve preferred grain orientation. Bulk samples have been produced by this method in which the c-axis (the poor current carrying direction) is found to be normal to the forging direction [19,20]. With regard to wire fabrication, the ceramic superconducting phase can be mixed with a metallic phase to get the ductility, and then such a composite can be forged into a wire through a die. However, since the wire shape product has not yet been produced, the grain orientation effect (c-axis to be normal to the wire axis) remains uncertain. Furthermore, the current density is another concern for this method as a large amount of metallic phase could block the superconducting current path which lowers the current density.

Several other techniques such as melt drawing, melt spinning and preform-wire melting have been tested, however, the value of J_c was below 10^3 amp/cm^2 in every case [21]. Although superconducting fibers with J_c about 6×10^5 amp/cm^2 have been fabricated by laser-heated pedestal growth method [22], the intrinsic brittleness of these fibers needs to be overcome before commercial application is possible.

III. Current Approach

A new approach of fabricating wires of ceramic superconductors by coating metallic wires with superconducting films is underway. The project focuses on the Bi-Ca-Sr-Cu-O systems, and the reasons for this choice are: (1) better stability, (2) reproducible zero resistivity above 100 K [23] in the 2223 phase with nominal composition $Bi_2Sr_2Ca_2Cu_3O_y$ and (3) relatively low melting temperature (~ 1050 °C).

In order to form a layer of superconducting phase on a metallic wire by solidifying from a melt, the right composition and temperature have to be chosen. Published data on the phase diagrams of ceramic superconductors are very limited; some on the Y-Ba-Cu-O systems [24-28] and none so far on the Bi-Ca-Sr-Cu-O systems. In the absence of such information, the films were prepared simply by so-called dipping coating method.

Melts of the materials were prepared with exact stoichiometric compositions (2223 phase) in which the metallic wires were immersed. The substrate wires were passed through the melt fast enough to avoid precipitation of any nonsuperconducting phases. For the preliminary studies, Pt wires were used. Other metals such as Ag and Au (which do not react with the melt) can be used as substrate materials.

The production of bulk superconductors (with $T_c > 100$ K) in the Bi-Ca-Sr-Cu-O systems by rapid solidification from melt has been reported in the literature [29-31]. These works showed that the solidified samples were amorphous, and needed subsequent

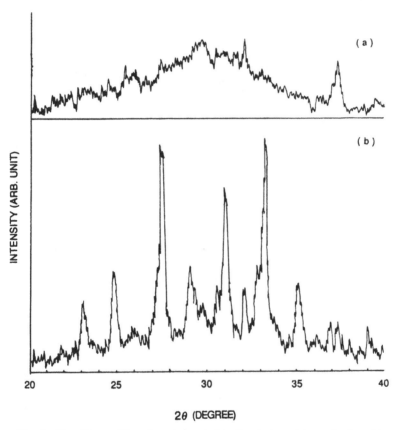

Fig.1 The X-ray diffraction data of the film before (a), and after (b) heat treatment.

heat treatment for the superconducting phase to crystallize. The X-ray diffraction data shown in Fig.1a clearly indicates that the films produced are in the amorphous state. These films were then heat treated (~ 860 °C) to convert to the superconducting phase. Comparison of the diffraction data before (Fig.1a) and after (Fig.1b) the heat treatment indicates that the transition from the amorphous state to a crystalline phase has occured. Superconductivity in the heat treated coating has also been confirmed by the resistance vs. temperature measurement as shown in Fig.2.

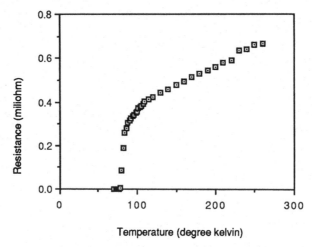

Fig.2 Resistance vs. temperature of a heat-treated coating

The metallic wire can be etched or faceted to help hold the coating. A high degree of homogeneity in the film could be achieved, if the composition is maintained constant by adding (to the melt) the same amount of the compound which is consumed during solidification. The thickness of the coated layer can be controlled by the moving rate of the metal substrate. An experimental setup for continuous wire coating can be designed as shown schematically in Fig.3.

In forming a coating by solidification from a melt, one is concerned with the wettability of the melt with the substrate material. The wettability of the superconductor with Pt was found to be very good in our studies. Figure 4a shows the photograph of a typical coating formed on a Pt wire, Fig.4b is a metallographic picture of the cross-section of the composite wire, and Fig.4c is an SEM picture of the cross-section highlighting the Pt/film interface.

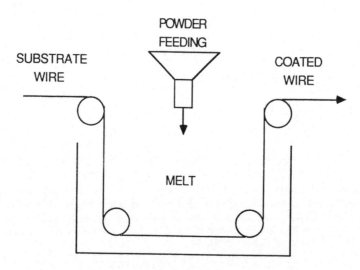

Fig.3 Schematic showing a continuous wire coating setup

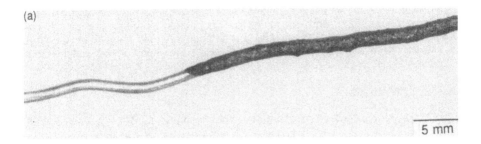
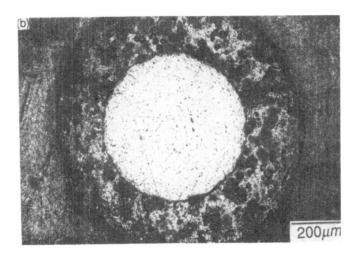

Fig.4 a. Photograph of a superconducting film on a Pt wire.
b. Optical micrograph of the cross-section of the Pt/film composite.
c. SEM micrograph of the cross-section showing the Pt/film interface.

In addition to measuring the temperature dependence of the electrical resistivity of the fabricated films, the critical current densities need to be estimated. The right way to describe the critical current density would be to divide the maximum supercurrent by the cross-sectional area of the film plus the substrate. This allows comparison with commercially available wires, and a rough estimate of the number would be useful. The microstructure of the films would be studied by X-ray diffraction (XRD) and scanning electron microscope (SEM), and correlated to the growth conditions such as melt composition, solidification temperature, substrate temperature and time of immersion of metallic wire in the melt. Also, the mechanical properties need to be examined by measuring the residual stress and adhesive properties of the films to metallic substrate.

Anticipated Properties

1. Density: Since the coatings are produced by precipitation from melt, large porosity and defects such as cracks can be avoided, and hence, a highly dense product can be expected.

2. Mechanical Stability: Due to the difference in thermal expansion coefficients between ceramic superconductor and metal substrate, the coating layer is expected to be under a compressive force. This force is estimated to be of the order of 5000 psi, and it should be possible to bend the composite wire into desirable shapes. It should also be possible to regain such compressive force by annealing the bent wire, because when it is cooled down, a similar compressive force is generated.

3. Current Carrying Capacity: As opposed to bulk superconductors, thin coated films are expected to have better aligned microstructure (c axes of the precipitated crystallites or grains are normal to the substrate). This alignment, along with the improved density, is expected to contribute to a higher critical current density.

4. Other Properties: For this composite wire, if the film inadvertently loses its superconductivity and becomes resistive, the metallic substrate would act as a conduit for the current until the superconductivity is recovered as in the case of commercially used Nb-based superconductor which is connected to Cu. The metallic substrate may also act as a thermal-stabilizing material for the oxide superconducting layer to minimize local heating and prevent catastrophic loss of superconductivity.

References

1. Y. Enomoto, T. Marakami, M. Suzuki and K. Moriwaki, Jpn. J. Appl. Phys., 26, L1248 (1987).

2. R. E. Somekh et al., Nature, 326, 857 (1987).

3. Y. Yamada, N. Fukushima, S. Nakyama, H. Yoshino and S. Murase, Jpn. J. Appl. Phys., 26, L865 (1987).

4. O. Kohno, Y. Ikeno, N. Sadakata, S. Aoki, M. Sugimoto and N. Nakagawa, Jpn. J. Appl. Phys., 26, L1527 (1987).

5. S. Jin, R. S. sherwood, T. H. Tiefel, R. B. van Dover and D. W. Johnson, Jr., Appl. Phys. Lett., 51, 203 (1987).

6. R. W. McCallum et al., Adv. Ceram. Mater., 2, 388 (1987).

7. Y. Tanaka, K. Yamada and T. Sano, Japan. J. Appl. Phys., 27, L799 (1987).

8. T. Goto and M. Kada, Japan. J. Appl. Phys., 26, L1327 (1987).

9. M. B. Brodsky, MRS Meeting, Anaheim, CA (1987).

10. S. Jin, T. H. Tiefel, R. C. Sherwood, G. W. Kammlot and S. M. Zahurak, Appl. Phys. Lett., 51, 943 (1987).

11. N. Alford, W. J. Clegg, M. A. Harmer, J. D. Birchall, K. Kendall and D. H. Jones, Nature, 332, 58 (1988).

12. N. Sadakata, Y. Ikeno, M. Nakagawa, K. Gotoh and O. Kohno, MRS Proc., 99, 293 (1988).

13. M. Kawashima, M. Nagata, Y. Hosoda, S. Takano, N. Shibuta, H. Mukai and T. Hikata, Applied Superconductivity Conference, San Francisco, CA (1988).

14. D. Shi and K. C. Goretta, Materials Lett., (1989).

15. K. Heine, J. Tenbrink and M. Thoner, preprint (1989).

16. N. McN Alford, J. D. Birchall, T. W. Button, W. J. Clegg and K. Kendall, Annual Meeting of the Am. Ceram. Soc., Indianapolis (1989).

17. T. R. Dinger, T. K. Worthington, W. J. Gallagher and R. L. Sandstrom, Phys. Rev. Lett., 58, 2687 (1987).

18. Y. Kioke, T. Nakanomyo and T. Fukase, Japan. J. Appl. Phys., 27, L841, (1988).

19. T. Takena, H. Nada, A. Yoneda and K. Sakata, Jpn. J. Appl. Phys., 27, L1209 (1988).

20. J. W. Ekin, Adv. Ceram. Mater., 2, 586 (1987).

21. S. Jin et al., Appl. Phys. Lett., 51, 943 (1987).

22. R. S. Feigelson, D. Gazit, D. K. Fork and T. H. Geballe, Science, 240, 1624 (1988).

23. S. A. Akbar, M. J. Botelho, M. S Wong and M. Alauddin, "A Systematic Study of Superconductivity in Bi-Pb-Sb-Sr-Ca-Cu-O System," submitted to Physica C.

24. K. Oka, K. Nakane, M. Ito and M. Saito, Jpn. J. Appl. Phys., 27, L1065 (1988).

25. R. S. Roth, K. L. Davis and J. R. Dennis, Adv. Ceram. Mater., 2, 303 (1987).

26. S. Takekawa and N. Iyi, Jpn J. Appl. Phys., 26, L851 (1987).

27. D. L. Kaiser, F. Holtzberg, M. F. Chisholm and T. K. Worthington, J. Cryst. Growth, 85, 593 (1987).

28. H. J. Scheel, Physica C, 153-155, 44 (1988).

29. D. Shi, M. Blank, M. Patel, D. G. Hinks and A. W. Mitchell, Physica C, 156, 822 (1988).

30. D. G. Hinks, L. Soderholm, D. W. Capone II, B. Dabrowski, A. W. Mitchell and D. Shi, Appl. Phys. Lett., 53, 423 (1988).

31. D. Shi, M. Tang, K. Vandervoort and H. Claus, Phys. Rev. B, in print.

SUPERCONDUCTING PROJECTILE ACCELERATOR

X.W. Wang, J.D. Royston

Alfred University
Alfred, NY

ABSTRACT

We report results on the design and testing of superconducting projectile accelerators. The barrel of the projectile accelerator is mainly made of $Y_1 Ba_2 Cu_3 O_{7-x}$ superconductors (T_c = 93K), which is mounted on a copper seat cooled with liquid nitrogen (77K). Before acceleration forces are applied, a magnetised projectile is suspended inside of the superconducting barrel, without touching the walls of the barrel. Electromagnetic coils with the pulse power supply are used to accelerate the projectile successively. Due to the contactless motion of the projectile travelling along the barrel, the kinetic energy of the projectile is close to the magnetic energy provided by the power supply. That is, the efficiency of energy-conversion from the magnetic energy to the kinetic energy is about ninety percent or higher. The entire firing process is controlled by a microprocessor. The detailed system will be described. The experimental results will be presented. The applications will also be discussed.

I. INTRODUCTION

The discoveries of ceramic superconductors [1-4] have stimulated a wide range of researches on new materials.[5] Some of the known ceramic superconductors with the critical temperature, Tc, higher than 77K can be summarized as $R Ba_2 Cu_3 O_{7-x}$ (R = rare earth earth element such as Y, Gd, Ho, Er, etc.), and $(AO)_m M_2 Ca_{n-1} Cu_n O_{2n+2}$, (A = Tl, or Bi, or BiPb; m = 1,2 for Tl, m = 1 for Bi; M = Ba, or Sr; n = 1,2...).[6] Initial studies on the ceramic materials showed that such superconductors are useful in different applications.[7]

The ceramic superconductors can be classified as type II

superconductors which are characterized by first and second values of critical field, H_{c1} and H_{c2}.[8] Some preliminary results indicate that the first critical field value, H_{c1}, of ceramic superconductors is about 10 Gauss to 100 Gauss; while the second critical field value, H_{c2}, of the ceramic superconductors is about 1 Tesla to 100 Tesla.[9]

When a magnet is placed on a superconductor, the magnet will be levitated above the superconductor via Meissner effect.[10] Hellman, et al. have given a theoretical treatment on the levitation of a magnet over a flat type II superconductor.[11] The stability of the magnet levitated above the superconductor is provided by flux pinning.[12] Such equilibrium situation is further discussed by R. Williams, et al.[13] Synchronous rotation of a floating magnet has been studied by J.C. Macfarlane, et al.[14] We have worked on the planar motion of magnetized objects levitated on superconductors since 1988.[15] Several contactless mass transfer systems have been designed and constructed.[16] In this paper, we shall be concentrating on a superconducting projectile accelerator. The system design will be introduced in Section II, and the experimental results will be given in Section III.

II. SYSTEM DESIGN

A. EXISTING SYSTEMS

Projectile accelerators have been in existance for a long period of time. The common projectile accelerator is the firing mechanism of gun powders. However, even after hundreds of years of development the gun and its numerous relatives have encountered many inefficiencies and limitations. Another type of mass accelerator is called "railgun", whose root is in the "eletromagnetic cannon", invented by K. Birkelnad.[17] In conventional rail guns, projectiles with or without carrier are accelerated along rail tracks by means of the Lorentz force.[18] Other versions of rail guns may have superconducting electromagnetic coils,[19] or rails consisting of both superconductor and normal conductor.[20] However, the contact between a projectile carrier and rail track is inevitable in conventional rail guns.[21] Large amounts of energy are dissipated due to such rail contact resistances, (about seventy percent or so). Thus the efficiency of conversion from electromagnetic energy to kinetic energy is rather low. In order to avoid contact problems, we designed a contactless system.

B. CONTACTLESS SYSTEM

Our system includes three parts as shown in Figs. 1 and 2. The first part is the cylindrical superconducting chambers (14 of Fig. 1, 40 of Fig. 2) which are used to levitate and stabilize magnetized projectiles (38 of Fig. 2). These superconducting chambers are surrounded by a copper cooling chamber (22) filled with liquid nitrogen (12). The second part is a series of electromagnetic coils (24, 26) which are used to generate magnetic fields by passing pulsed current sequentially through the coils. Under the magnetic fields, the projectile will be accelerated along the barrel without touching the walls of the barrel. The third part is the current controller, consisting of a microprocessor (32), a power supply (30) and electronic circuits, [power transistor (34), lead (28), connection (36), and 100 Ohm resistor].

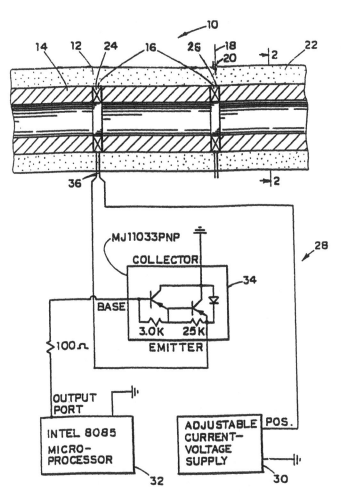

Fig. 1. Block diagram of the system.

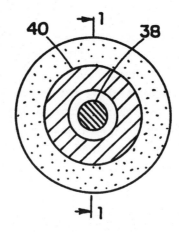

Fig. 2. Cross sectional view of both superconducting chamber and cooling chamber in Fig. 1.

C. SUPERCONDUCTING BARREL

When a magnet is placed on a flat superconductor, the levitation height d, (distance from the center of the sphere to the top surface of the superconductor) is given by Hellman's formular.[11] Our experimental results are close to the formula. For example, for a $Y_1 Ba_2 Cu_3 O_{7-x}$ superconductor pellet (thickness = 3mm) the levitation height of a magnet is about 0.7 cm, when the mass of the magnet is 0.5 gram, and its magnetic moment is 8000 Gauss. When the $Y_1 Ba_2 Cu_3 O_{7-x}$ material is doped with silver, the levitation height is increased to 1.6 cm.[22]

Cylindrical superconducting chambers were formed from $Y_1 Ba_2 Cu_3 O_{7-x}$ with dimensions specified in Fig. 3. Solid state reaction method was used to produce the shapes. The procedure includes mixing, grinding/calcination (two times), grinding, pressing into cylinder (about 20,000 psi), sintering and annealing with oxygen. In order to avoid the deformation of cylinders, sintering temperature is around 920 C (24 hours), which is lower than that of previous experiments.[23] The x-ray diffraction patterns of samples matched with the pattern of pure $Y_1 Ba_2 Cu_3 O_{7-x}$ phase. The superconducting transition temperature was around 93 Kelvin as measured by the four-point method.[24]

When a magnet is put in such a hollow cylinder, the magnet will be levitated near the center, see Fig. 2. Besides the gravity force tending to pull the magnet downwards, the repulsive Meissner forces are applied to the magnet from almost all (radial) directions. Due to the symmetrical arrangement of the chamber, the magnet will not touch the walls of the chamber.

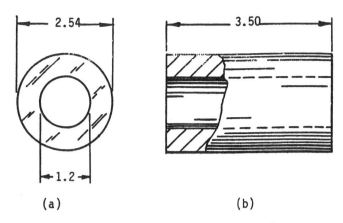

Fig. 3. Dimensions of a superconductor cylinder (in centimeters). (a) Cross sectional view; (b) Front view of the cylinder. It is partially broken away to show the interior of the device.

D. ELECTROMAGNETIC COILS

To accelerate magnetized objects, electromagnetic coils are used to produce desired magnetic fields. The width of a coil is about 3mm with the outside and inside diameters matching with that of the cylinder in Fig. 3. This width should be large enough to allow the proper amount of coil windings. However, it should not be too wide in order to eliminate a non-levitating void. Coils are constructed with AWG No. 26 gauge "magnet" wires (135 turns).

E. MICROPROCESSOR AND ELECTRONICS

An Intel 8085 microprocessor (32 in Fig. 1) was used to control currents passings through coils. Since the microprocessor has eight bit output ports, the minimum number of coils is eight. A software was written to send out control signals to output ports. The control signals will allow a coil to receive certain current from power supply (30 in Fig. 1) via power transistor (34) during a specified time period.

III. EXPERIMENTAL RESULTS

A test setup is illustrated in Fig. 4. In this configuration, only one coil (24) was used. The projectile (38) was a rare earth cobalt magnet with a magnetic moment of 8200 Gauss, mass of 0.54 grams, and dimension of 0.89cm (length) by 0.48cm (width) by 0.34cm (height). The separation (56) between the coil (14) and the start position of the magnet (38) was 2.8cm; the separation (54) between the coil and the exit of the barrel was 13.5cm; and the separation (58) between the exit and target (57) was 63.2cm. The maximum of distance (60) is 14.1cm, which is the height from the dotted line to the bottom of the target.

The superconducting barrel (40) was cooled below its critical temperature by liquid nitrogen in copper cooling chamber (22). The cooling chamber was then fixed on brick support (52).

After a pulsed current (3.2 Amperes) was supplied to the coil, the magnet started acceleration motion towards the coil. Before the magnet reached the coil, the current was then turned off by the controller. The magnet then moved along the barrel. As soon as it left the exit of the barrel, its trajectory could be described by a projectile motion.[25] The projectile hit the target (57) at location y (60) measured from the dotted line. We repeated the measurement nine times, and obtained the average velocity of 4.7 meters/second as the exit velocity. The energy conversion coefficient between the kinetic energy of the projectile and the electromagnetic energy provided by the power supply is more than ninety percent. The dissipated energy was mainly due to the magnetic flux pinning.[12] Air friction was negligible in the measurement.

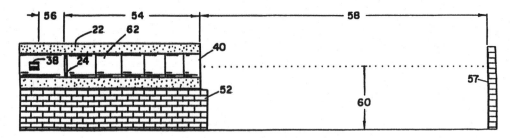

Fig. 4. Velocity measurement setup.

IV. DISCUSSION

Our experiments have shown that superconductors can be used as the barrels, and magnetized objects can be accelerated efficiently. We have also conducted other experiments and found that such designs can be applied to gun firing, material handling, and conveyor systems, etc. with some modifications.

ACKNOWLEDGEMENT

This work is partially supported by CACT.

REFERENCES

1. J.G. Bednorz, and K.A. Mueller, Z. Phys. B 64, 189(1986).

2. M.K. Wu, J.R. Ashburn, C.J. Torng, P.H. Hor, R.L. Meng, L. Gao, Z.J. Huang, Y.Q. Wang, and C.W. Chu, Phys. Rev. Lett. 58, 908(1987).

3. H. Meada, et al., Jpn. J. Appl. Phys. <u>27</u>, L. 209(1988); C.W. Chu, J. Bechtold, L. Gao, P.H. Hor, Z.J. Wang, and Y.Y. Xue, Phys. Rev. lett. <u>60</u>, 941(1988).

4. Z.Z. Sheng, A.M. Hermann, A. EL Ali, C. Almasan, J. Estrada, T. Datta, and R.J. Matson, Phy. Rev. Lett. <u>60</u>, 937(1988).

5. See, for example, "Ceramic Superconductors", ed. W.J. Smothers, et al., a special supplementary issue of Advanced Ceramic Materials <u>2</u>, 273-733(1987); "Ceramic Superconductors II", ed M.F. Yan, (American Ceramic Soc., Inc., Westerville, Ohio, 1988).

6. A.W. Sleight, Science <u>242</u>, 1519(1988).

7. See, for example, a special issue on superconductors and applications, IEEE transactions on Magnetics <u>25</u>, No. 2, March 1989.

8. M. Tinkham, "Introduction to Superconductivity", (Krieger, Florida, 1980).

9. D.W. Murphy, D.W. Johnson, Jr., S. Jin, and R.E. Howard, Science <u>241</u>, 922(1988); A. Bezinge, J.L. Jorda, A. Junod, and J. Mueller, Solid State Comm. <u>64</u>, 79(1987); F. Seidler, P. Boehm, H. Geus, W. Braunisch, E. Braun, W. Schnelle, I. Felner, and Y. Wolfus, preprint.

10. The levitation experiments have been carried out routinely all over the world. For an experimental design, see, for example, D. Prochnow, "Superconductivity: Experimenting in a New Technology", (TAB, Pennsylvania, 1989).

11. F. Hellman, E.M. Gyorgy, D.W. Johnson, Jr., H.M. O'Bryan, and R.C. Sherwood, J. Appl. Phys. <u>63</u>, 447(1987).

12. L.C. Davis, E.M. Logothetis, and R.E. Soltis, J. Appl. Phys. <u>64</u>, 4212(1988).

13. R. Williams, and J.R. Matey, Appl. Phys. Lett. <u>52</u>, 751(1988).

14. J.C. Macfarlane, K.M. Mueller, and R. Driver, IEEE Mag. <u>25</u>, 2515(1989).

15. Some preliminary results have been reported. See, for example, X.W. Wang, in "Proceedings of 1989 ASEE Annual Conference", ed. L.P. Grayson, and J.M. Biedenbach, pp 947-9, (ASEE, Washington, DC, 1989).

16. X.W. Wang, and J.D. Royston, patent pending.

17. A. Egeland, IEEE Trans. Plasma Science 17, 73(1989).

18. J.L. Upshaw, IEEE Mag 22, 1779(1986); S. Usuba, Y. Kakudate, K. Aoki, M. Yoshida, K. Tanaka, and S. Fujiwara, IEEE 22, 1785(1986).

19. C.G. Homan, C.E. Cummings, C.M. Fowler, and M.L. Hodgdon, in "Megagauss Technology and Pulsed Power Applications", ed. C.M. Fowler, R.S. Caird, D.J. Erickson, (Plenum, New York, 1987).

20. L. Jasper, Jr., "Electromagnetic Launcher with Cryogenic Cooled Superconducting Rails", US Patent No. 4,813,332, Mar. 21, 1989.

21. C. Persad, C.J. Lund, and Z. Eliezer, IEEE Mag 25, 433(1989).

22. Silver has been used to increase the critical current of $Y_1Ba_2Cu_3O_{7-x}$ superconductors in other institutions.

23. X.W. Wang, H.S. Kwok, L. Shi, J.P. Zheng, P. Mattocks, and D.T. Shaw, J. Mater Res. 3, 1297(1988).

24. M.N. Pitsakis, and X.W. Wang, Rev. Sci. Instrum. 60, 135(1989).

25. F.W. Sears, M.W. Zemansky, and H.D. Young, "University Physics", 6th ed. (Addison, Massachusetts, 1983).

MILLIMETER WAVE TRANSMISSION STUDIES OF $YBa_2Cu_3O_{7-\delta}$ THIN FILMS IN THE 26.5 TO 40.0 GHz FREQUENCY RANGE.

F.A. Miranda and W.L. Gordon
Department of Physics, Case Western Reserve University
Cleveland, Ohio 44106

K.B. Bhasin, V.O. Heinen and J.D. Warner
National Aeronautics and Space Administration
Lewis Research Center, Cleveland, Ohio 44135

G.J. Valco
Department of Electrical Engineering
The Ohio State University
Columbus, Ohio 43210

ABSTRACT

Millimeter wave transmission measurements through $YBa_2Cu_3O_{7-\delta}$ thin films on MgO, ZrO_2 and $LaAlO_3$ substrates, are reported. The films (0.2 to 1.0 μm) were deposited by sequential evaporation and laser ablation techniques. Transition temperatures T_c, ranging from 89.7 K for the laser ablated film on $LaAlO_3$ to approximately 72 K for the sequentially evaporated film on MgO, were obtained. The values of the real and imaginary parts of the complex conductivity, σ_1 and σ_2, are obtained from the power transmitted through the film, assuming a two fluid model. The magnetic penetration depth is evaluated from the values of σ_2. These results will be discussed together with the frequency dependence of the normalized transmission amplitude, P/P_c, below and above T_c.

INTRODUCTION

Millimeter wave measurements of the new high T_c superconductors are of fundamental importance due to the potential applicability of these oxides in the fabrication of devices operational in these frequency ranges.[1] Through these measurements, information on the nature of superconductivity in these new superconductors can be obtained from the temperature dependence of parameters such as the surface resistance,[2-6] and the complex conductivity.[7-9] Another important question is the applicability of millimeter wave measurements for the characterization of superconducting thin films. While dc resistance versus temperature measurements give no further information once the zero resistance state is achieved, millimeter wave transmission and absorption measurements provide a sensitive, contactless technique, which yield important information about the microstructure of superconducting films[10] and their behavior at temperatures below the critical temperature (T_c). Millimeter and microwave absorption studies in low and high T_c superconductors have been performed using resonant cavities.[10-16] Usually, those studies applying millimeter or microwave transmission analysis, have reported results at just one particular frequency.[8,9]

In this work we have measured the power transmitted through $YBa_2Cu_3O_{7-\delta}$ thin films at frequencies within the frequency range from 26.5 to 40.0 GHz and at temperatures from 20 to 300 K. From these measurements and assuming a two fluid model, we have obtained values of the normal and complex conductivities above and below T_c respectively. The zero temperature magnetic penetration depth has been obtained using the value of the imaginary part of the complex conductivity, σ_2.

ANALYSIS

We have applied the two fluid model due to its simplicity and because in the past it has given good results for the microwave properties of metallic type II superconductors in cases for $\hbar\omega \ll E_{gap}$.[17] Since the energy gap for $YBa_2Cu_3O_{7-\delta}$ superconductors corresponds to frequencies in the terahertz range, we expect the model to be applicable in the frequency range studied. In this phenomenological model, the complex conductivity is defined as

$$\sigma = \sigma_1 - i\sigma_2 \qquad (1)$$

with

$$\sigma_1 = \sigma_c t^4 \quad \text{and} \quad \sigma_2 = \sigma_c(1 - t^4)/\omega\tau . \qquad (2)$$

Here, σ_c is the normal conductivity at $T = T_c$, $\omega = 2\pi f$ is the angular frequency, t is the reduced temperature T/T_c, and τ is the mean carrier scattering time. Thus, to determine either σ_1 or σ_2 we need to know the transition temperature T_c and the value of σ_c. Furthermore, the value of τ must be known beforehand if σ_2 is to be obtained from Eq. (2).

In this study, the value of T_c was determined from the standard four-point probe versus temperature measurements. To determine the normal and complex conductivities, we used the method applied by Glover and Tinkham.[18] In this method, the transmission of a normally incident plane wave through a film of thickness d ($\ll$ wavelength or skin depth) deposited on a substrate of thickness ℓ and index of refraction n, is measured. Following the notation of Glover and Tinkham[18] the power transmission is given by

$$T = \frac{8n^2}{A + B \cos 2k\ell + C \sin 2k\ell} \qquad (3)$$

where

$$A = n^4 + 6n^2 + 1 + 2(3n^2 + 1)g + (n^2 + 1)(b^2 + g^2)$$

$$B = 2(n^2 - 1)g - (n^2 - 1)^2 + (n^2 - 1)(b^2 + g^2)$$

$$C = 2(n^2 - 1)nb$$

$$k = n\omega/c$$

and

$$y = g - ib = YZc = (G - iB)Z_c = (\sigma_1 - i\sigma_2)dZ_c$$

is the dimensionless complex admittance per square of the film in units of the characteristic admittance, Z_c^{-1}, of the wave guide ($Z_c = Z_0/\sqrt{1 - (f_c/f)^2}$, $Z_0 = 377 \, \Omega$, mks; $Z_0 = 4\pi/c$, cgs; f_c = cutoff frequency of the TE mode wave guide and f is the operational frequency).

In the normal state, Eq. (3) becomes

$$T_N = \frac{8n^2}{\sigma_N^2 d^2 Z_c^2 Q + \sigma_N dZ_c R + P} \tag{4}$$

where

σ_N = normal conductivity

$Q = (n^2 + 1) + (n^2 - 1)\cos 2k\ell$

$R = 2(3n^2 + 1) + 2(n^2 - 1)\cos 2k\ell$

$P = n^4 + 6n^2 + 1 - (n^2 - 1)^2 \cos 2k\ell.$

The normal state conductivity of the film can be expressed conveniently in terms of the power transmission as

$$\sigma_N = \frac{-RT_N \pm \sqrt{R^2 T_N^2 - 4QT_N(PT_N - 8n^2)}}{2QT_N dZ_c} \tag{5}$$

where only the expression with the + sign has physical relevance. It is convenient to use the ratio T_S/T_N in the analysis of the superconducting state, where T_S refers to the transmission in the superconducting state given by Eq. (3). Thus,

$$\frac{T_S}{T_N} = \frac{\sigma_N^2 d^2 Z_c^2 Q + \sigma_N dZ_c R + P}{A + B \cos 2k\ell + C \sin 2k\ell} \tag{6}$$

Solving (6) for the imaginary part, σ_2, of the conductivity, and using the value of σ_N at $T = T_c$ we have

$$\sigma_2/\sigma_c = -\beta/2 \frac{1}{\sigma_c dZ_c} + \left\{ \frac{1}{(\sigma_c dZ_c)^2}\left[(\beta/2)^2 - \gamma\right] - \frac{\alpha \sigma_1}{\sigma_c^2 dZ_c} - \left(\frac{\sigma_1}{\sigma_c}\right)^2 \right.$$
$$\left. + (T_c/T_S)\left[1 + \frac{\alpha}{\sigma_c dZ_c} + \frac{\gamma}{(\sigma_c dZ_c)^2}\right] \right\}^{1/2} \tag{7}$$

where σ_c and T_c are the conductivity and the transmissivity at $T = T_c$, and

$$\alpha = \frac{1}{D}[6n^2 + 2 + 2(n^2 - 1)\cos 2k\ell]$$

$$\beta = \frac{1}{D}[-2n(n^2 - 1)\sin 2k\ell]$$

$$\gamma = \frac{1}{D}[n^4 + 6n^2 + 1 - (n^2 - 1)\cos 2k\ell]$$

$$D = n^2 + 1 + (n^2 - 1)\cos 2k\ell .$$

Thus, from the relation for σ_1 in Eq. (2), and Eq. (7), the real and imaginary parts of the complex conductivity can be determined.

The magnetic penetration depth, λ, can be obtained from the London expression

$$\lambda = \left(\frac{1}{\mu_o \omega \sigma_2}\right)^{1/2} \quad (8)$$

which can be written in terms of the superfluid density N_S, as

$$\lambda = \left(\frac{m}{\mu_o N_S e^2}\right)^{1/2} \quad (9)$$

where m is the effective mass of the charge carriers. From the two fluid model

$$\frac{N_S}{N} = 1 - t^4 \quad (10)$$

where $N = N_n + N_s$ is the total number of carriers per unit volume, we have

$$\lambda = \left[\frac{m}{\mu_o N e^2}\right]^{1/2} (1 - t^4)^{-1/2} = \lambda_o (1 - t^4)^{-1/2} \quad (11)$$

From this expression the zero-temperature penetration depth, λ_o, can be obtained. Because Eq. (9) applies to homogeneous superconductors, the values of λ_o obtained in this method are larger than those that would be obtained for homogeneous films.

Our measurements were made on thin films (0.2 to 1.0 μm thickness) of $YBa_2Cu_3O_{7-\delta}$ on $LaAlO_3$, MgO and ZrO_2 substrates. The substrates were generally between 0.025 and 0.100 cm thick. The deposition techniques used for the preparation of the films used in this study are described in Refs. 19 and 20. For the laser ablated films, X-ray diffraction data showed that the films were c-axis oriented on $LaAlO_3$ and partially c-axis oriented for those on MgO and ZrO_2. They had T_c's ranging from 89.7 K for the film on $LaAlO_3$ to 79 and 78 K for those deposited on MgO and ZrO_2 respectively. The film deposited by sequential evaporation on MgO had a T_c of approximately 72 K.

The power transmission measurements were made using a Hewlett-Packard model HP-8510 automatic network analyzer connected to a modified closed cycle refrigerator by Ka-band (26.5 to 40.0 GHz) waveguides. Inside the vacuum chamber of the cryosystem, the sample was clamped between two waveguide flanges which were in direct contact with the cold head of the refrigerator. The power transmitted through the sample was obtained by measuring the scattering parameters as described in Ref. 21. The temperature gradient of the waveguide flanges between the top and bottom of the sample, was estimated to be 2.5 K or less at 90 K. The system was properly calibrated with short, open, load and through calibration standards before each measurement cycle was started.

RESULTS

Figures 1 and 2 show the temperature dependence of the normalized power transmitted through $YBa_2Cu_3O_{7-\delta}$ thin films deposited by laser ablation on $LaAlO_3$ and MgO respectively. The data are normalized with respect to the transmitted power at the critical temperature T_c. The measurements of the power transmitted through the films were started at room temperature and then carried out during sample cooling. In Fig. 1, it can be observed that the rapid decrease in transmitted power occurs at T_c. This is typical of films with a high degree of homogeneity, where all the regions of the film undergo the superconducting transition simultaneously. This is not the case for the film considered in Fig. 2, for which the transmitted power starts to decrease rapidly at temperatures just below an onset temperature (~90 K) approximately 11 K above its transition temperature of 79 K. This behavior may be associated with the presence of inhomogeneities, resulting in a distribution of transition temperatures. For temperatures below T_c both films are characterized by a smooth decrease of the power transmitted through them.

The behavior shown in Figs. 1 and 2 for the power transmitted through the film-substrate combination, as a function of decreasing temperature, was also observed for the laser ablated film on ZrO_2 and for

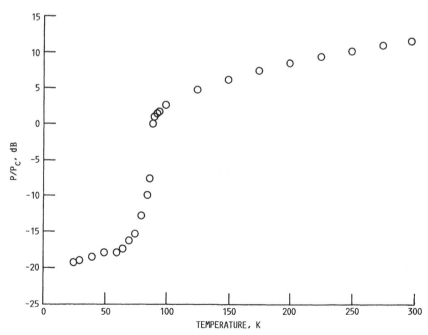

FIGURE 1. - NORMALIZED TRANSMITTED POWER VERSUS TEMPERATURE FOR A LASER ABLATED $YBa_2Cu_3O_{7-\delta}$ THIN FILM (0.7 MICRONS) ON $LaAlO_3$ AT 37.0 GHz.

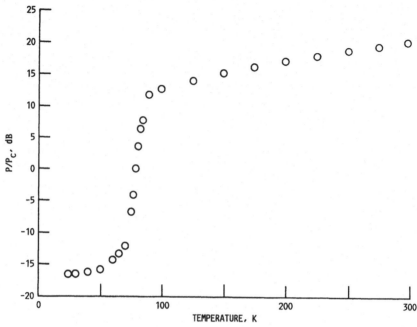

FIGURE 2. - NORMALIZED TRANSMITTED POWER VERSUS TEMPERATURE FOR A LASER ABLATED $YBa_2Cu_3O_{7-\delta}$ THIN FILM (0.2 MICRONS) ON MgO AT 28.5 GHz.

the sequentially evaporated film on MgO. For the latter film the transmission data suggest a lower film quality when compared to the film deposited on MgO by laser ablation. The films on ZrO_2 and sequentially evaporated on MgO also show a wide transition region. This temperature behavior was verified to be frequency independent for the frequencies employed in this study, and our analysis suggest that it is related to the degree of homogeneity and quality of the films.

Figures 3 to 10 and Table I, show the results for the conductivity above and below T_c, and at different frequencies, for the various films considered in this study. Figures 3 and 4 show the real and imaginary parts of the conductivity, σ_r and σ_2 respectively, corresponding to the $YBa_2Cu_3O_{7-\delta}$ film deposited on $LaAlO_3$ by laser ablation. The value for the normal conductivity at room temperature, 2.0×10^5 S/m, compares reasonably well with reported values of the dc conductivity in this type of film.[22,23] The cusp in σ_r at the transition temperature can be observed clearly in Fig. 3 and again indicates the high level of homogeneity and quality of this film. The imaginary part of the conductivity increases as a function of decreasing temperature, as can be seen in Fig. 4. Values of 5.17×10^6 S/m and 6.80×10^6 S/m are obtained at 70 and 40 K respectively. Using Eq. (8) we find $\lambda = 0.81$ μm at 70 K and $\lambda = 0.70$ μm at 40 K. From the value of λ at 40 K we found $\lambda_o = 0.69$ μm.

Figures 5 to 10 show the real and imaginary parts of the complex conductivity for the laser ablated films on MgO and ZrO_2, and for the sequentially evaporated film on MgO. Note that the normal to the superconducting transition region has been clearly identified in Figs. 5, 7 and 9. In the absence of a physical model which can account for the

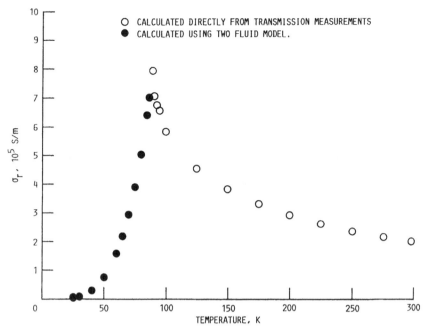

FIGURE 3. - REAL PART OF THE CONDUCTIVITY, σ_r, VERSUS TEMPERATURE FOR A LASER ABLATED $YBa_2Cu_3O_{7-\delta}$ THIN FILM (0.7 MICRONS) ON $LaAlO_3$ AT 37.0 GHz. $\sigma_r = \sigma_N$ FOR $T > T_c$ AND $\sigma_r = \sigma_1$ FOR $T < T_c$.

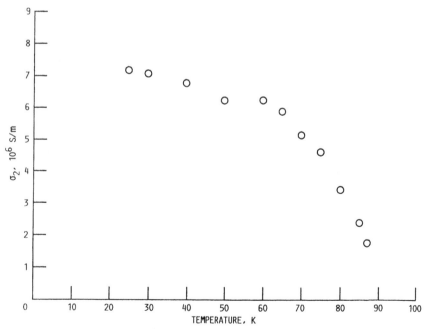

FIGURE 4. - IMAGINARY PART OF THE CONDUCTIVITY, σ_2, VERSUS TEMPERATURE FOR A LASER ABLATED $YBa_2Cu_3O_{7-\delta}$ THIN FILM (0.7 MICRONS) ON $LaAlO_3$ AT 37.0 GHz.

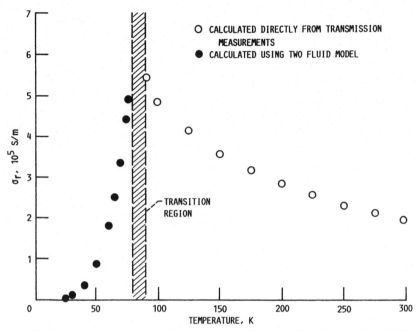

FIGURE 5. – REAL PART OF CONDUCTIVITY, σ_r, VERSUS TEMPERATURE FOR A LASER ABLATED $YBa_2Cu_3O_{7-\delta}$ THIN FILM (0.2 MICRONS) ON MgO AT 28.5 GHz. $\sigma_r = \sigma_N$ FOR $T > T_c$ AND $\sigma_r = \sigma_1$ FOR $T < T_c$.

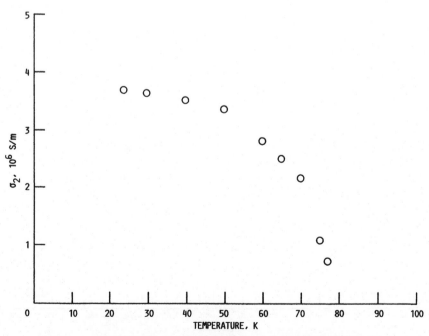

FIGURE 6. – IMAGINARY PART OF THE CONDUCTIVITY, σ_2, VERSUS TEMPERATURE FOR A LASER ABLATED $YBa_2Cu_3O_{7-\delta}$ THIN FILM (0.2 MICRONS) ON MgO AT 28.5 GHz.

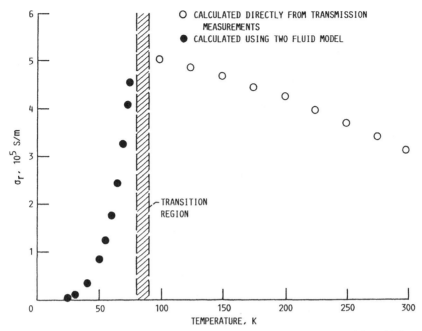

FIGURE 7. - REAL PART OF THE CONDUCTIVITY, σ_r, VERSUS TEMPERATURE FOR A LASER ABLATED $YBa_2Cu_3O_{7-\delta}$ THIN FILM (0.75 MICRONS) ON ZrO_2 AT 37.0 GHz. $\sigma_r = \sigma_N$ FOR $T > T_c$ AND $\sigma_r = \sigma_1$ FOR $T < T_c$.

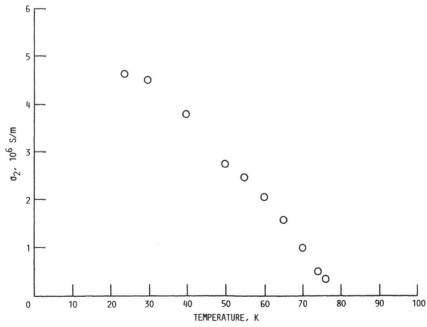

FIGURE 8. - IMAGINARY PART OF THE CONDUCTIVITY, σ_2, VERSUS TEMPERATURE FOR A LASER ABLATED $YBa_2Cu_3O_{7-\delta}$ THIN FILM (0.75 μm) ON ZrO_2 AT 37.0 GHz.

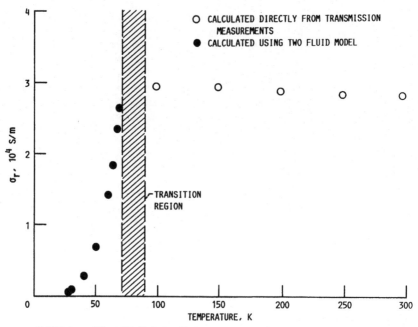

FIGURE 9. – REAL PART OF THE CONDUCTIVITY, σ_r, VERSUS TEMPERATURE FOR A SEQUENTIALLY EVAPORATED $YBa_2Cu_3O_{7-\delta}$ THIN FILM (1.0 MICRON) ON MgO AT 33.0 GHz. $\sigma_r = \sigma_N$ FOR $T > T_c$ AND $\sigma_r = \sigma_1$ FOR $T < T_c$.

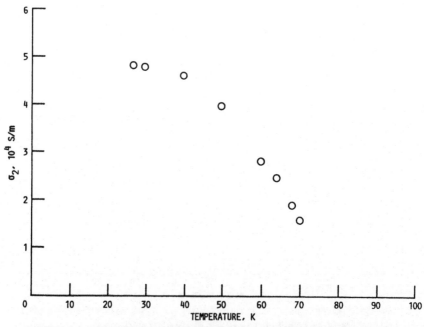

FIGURE 10. – IMAGINARY PART OF THE CONDUCTIVITY, σ_2, VERSUS TEMPERATURE FOR A SEQUENTIALLY EVAPORATED $YBa_2Cu_3O_{7-\delta}$ THIN FILM (1.0 MICRON) ON MgO AT 33.0 GHz.

TABLE I. – MILLIMETER WAVE CONDUCTIVITIES (σ_1, σ_2) AND ZERO TEMPERATURE PENETRATION DEPTH (λ_0) AT 35.0 GHz FOR YBa$_2$Cu$_3$O$_{7-\delta}$ THIN FILMS DEPOSITED ON DIFFERENT SUBSTRATES BY LASER ABLATION (LA) AND SEQUENTIAL EVAPORATION (SE)

Parameter	Substrates			
	MgO		LaAlO$_3$	ZrO$_2$
	SE	LA	LA	LA
σ_1 (70K)	3.0x10^4 S/m	3.9x10^5 S/m	3.3x10^5 S/m	1.7x10^5 S/m
σ_2 (70K)	1.9x10^4 S/m	1.1x10^6 S/m	6.4x10^6 S/m	1.1x10^6 S/m
σ_1 (40K)	3.1x10^3 S/m	4.1x10^4 S/m	3.5x10^4 S/m	1.9x10^4 S/m
σ_2 (40K)	7.1x10^4 S/m	4.0x10^6 S/m	7.7x10^6 S/m	3.6x10^6 S/m
λ_0	6.8 µm	0.91 µm	0.67 µm	0.96 µm

distribution of normal and superconducting material in the transition region, we can not accurately determine the normal conductivity down to the transition temperature T_c. Therefore, we have considered the critical conductivity to be the conductivity at or just above the onset temperature. Since the two fluid model approximation is based upon the assumption that the normal to the superconducting state transition is a sharp one, as for the film on LaAlO$_3$, the values of σ_1 obtained using $\sigma_c = \sigma_{onset}$ in Eq. (2) will be less than those expected for a sharp transition. The magnitude of this difference will depend upon the width ΔT of the transition region and the overall film quality. To estimate the size of the discrepancy between using σ_c at T_{onset} and σ_c at T_c, one can extrapolate σ_r above T_{onset} to T_c. When this is done, the σ_c obtained is 12 percent larger for the laser ablated film on MgO, 3.3 percent for the laser ablated film on ZrO$_2$ and 1.7 percent larger for the sequentially evaporated film on MgO. In the better films the discrepancy between σ_{onset} and the extrapolated value of σ_r at T_c, is larger due to the larger slope of σ_r for temperatures above the onset temperature as can be seen in Figs. 5, 7, and 9. This discrepancy becomes smaller as T_{onset} nears T_c, as for the film on LaAlO$_3$.

Figures 6, 8 and 10 show the imaginary part of the complex conductivity for the laser ablated films on MgO and ZrO$_2$, and for the sequentially evaporated film on MgO. Using Eq. (8) we obtain values for λ of 1.1, 0.95, and 9.1 µm, at 40 K, for the laser ablated films on MgO and ZrO$_2$ and for the sequentially evaporated film on MgO respectively. Additional values for the conductivities and for λ_0 at 35.0 GHz are given in Table I. The value for λ_0 obtained for the laser ablated film on LaAlO$_3$, compares favorably with that reported by Kobrin, et al.[24] ($\lambda_0 \sim 0.48$ µm, at 60.0 GHz) for ion-beam sputtered YBa$_2$Cu$_3$O$_{7-\delta}$ films on LaAlO$_3$.

CONCLUSIONS

Millimeter wave power transmission studies have been performed on $YBa_2Cu_3O_{7-\delta}$ thin films at frequencies within the frequency range from 26.5 to 40.0 GHz and at temperatures from 20 to 300 K. The normal, σ_N, and complex, $\sigma_1 - i\sigma_2$, conductivities have been determined for laser ablated films on $LaAlO_3$, MgO and ZrO_2. The conductivities of films on MgO grown by laser ablation and sequential evaporation have been compared. From the results obtained in this study, it is apparent that at least for films deposited on MgO, films deposited by laser ablation appear to have a higher quality than those deposited by the sequential evaporation technique. We have also shown that millimeter wave transmission and conductivity measurements can be used as a test of thin film quality. It was observed that for a film with a narrow transition region, the two fluid model should be more applicable than for those films with a wide transition region. Finally, values for the zero-temperature magnetic penetration depth have been determined from the obtained values of σ_2.

ACKNOWLEDGMENT

The authors are pleased to acknowledge helpful suggestions by Dr. S. Sridhar and Dr. J. Halbritter. Our thanks to Dr. S. Alterovitz, Dr. M. Stan and Dr. T. Eck for helpful discussions.

REFERENCES

1. Hartwig, W.; and Passow, C.: RF Superconducting Devices -- Theory, Design, Performance, and Applications. Applied Superconductivity, vol. 2, V.L. Newhouse, ed., Academic Press, New York, 1975, pp. 541-639.

2. Martens, J.S.; Beyer, J.B.; and Ginley, D.S.: Microwave Surface Resistance of $YBa_2Cu_3O_{6.9}$ Superconducting Films. Appl. Phys. Lett., vol. 52, no. 21, 23 May 1988, pp. 1822-1824.

3. Carini, J.P., et al.: Millimeter-Wave Surface Resistance Measurements in Highly Oriented $YBa_2Cu_3O_{7-\delta}$ Thin Films. Phys. Rev. B, vol. 37, no. 16, 1 June 1988, pp. 9726-9729.

4. Newman, H.S., et al.: Microwave Surface Resistance of Bulk Tl-Ba-Ca-Cu-O Superconductors. Appl. Phys. Lett., vol. 54, no. 4, 23 Jan. 1989, pp. 389-390.

5. Klein, N., et al.: Millimeter-Wave Surface Resistance of Epitaxially Grown $YBa_2Cu_3O_{6-x}$ Thin Films. Appl. Phys. Lett., vol. 54, no. 8, 20 Feb. 1989, pp. 757-759.

6. Sridhar, S.; Shiffman, C.A.; and Handed, H.: Electrodynamic Response of $Y_1Ba_2Cu_3O_y$ and $La_{1.85}Sr_{0.15}CuO_{u-s}$ in the Superconducting State. Phys. Rev. B, vol. 36, no. 4, 1 Aug. 1987, pp. 2301-2304.

7. Cohen, L., et al.: Surface Impedance Measurements of Superconducting $YBa_2Cu_3O_{6+x}$. J. Phys. F: Met. Phys., vol. 17, 1987, pp. L179-L183.

8. Ho, W., et al.: Millimeter-Wave Complex-Conductivity Measurements of Bi-Ca-Sr-Cu-O Superconducting Thin Films. Phys. Rev. B, vol. 38, no. 10, 1 Oct. 1988, pp. 7029-7032.

9. Nichols, C.S., et al.: Microwave Transmission Through Films of $YBa_2Cu_3O_{7-\delta}$. To be published in Phys. Rev. B.

10. Tyagi, S., et al.: Low-Field AC Susceptibility and Microwave Absorption in YBaCuO and BiCaSrCuO Superconductors. Physica C, vol. 156, 1988, pp. 73-78.

11. Maxwell, E.; Marcus, P.M.; and Slater, J.C.: Surface Impedance of Normal and Superconductors at 24,000 Megacycles per Second. Phys. Rev. vol. 76, no. 9, 1 Nov. 1949, pp. 1332-1347.

12. Pippard, A.B.: The Surface Impedance of Superconductors and Normal Metals at High Frequencies. Proc. R. Soc. A, vol. 203, no. 1072, 7 Sept. 1950, pp. 98-118.

13. Gittleman, J.I.; and Bozowski, S.: Transition of Type-I Superconducting Thin Films in a Perpendicular Magnetic Field: A Microwave Study. Phys. Rev., vol. 161, no. 2, 10 Sept., 1967, pp. 398-403.

14. Durny, R., et al.: Microwave Absorption in the Superconducting and Normal Phases of Y-Ba-Cu-O. Phys. Rev. B, vol. 36, no. 4, 1 Aug. 1987, pp. 2361-2363.

15. Tyagi, S., et al.: Frequency Dependence of Magnetic Hysteresis in the Field-Induced Microwave Absorption in High-T_c Superconductors at $T \ll Tc$. To be published in Phys. Lett. A.

16. Jackson, E.M., et al.: Study of Microwave Power Absorption in Yttrium-Barium-Copper Based High Temperature Superconductors and Allied Compounds. To be published in Supercond. Sci. Technol.

17. Gittleman, J.I.; and Rosemblum, B.: Microwave Properties of Superconductors. IEEE Proc., vol. 52, no. 10, Oct. 1964, pp. 1138-1147.

18. Glover III, R.E.; and Tinkham, M.: Conductivity of Superconducting Films for Photon Energies Between 0.3 and 40 KTc. Phys. Rev., vol. 108, no. 2, 15 Oct. 1957, pp. 243-256.

19. J.D. Warner, J.E. Meola and K.A. Jenkins: "Study of Deposition of YBa2Cu3O7-x on Cubic Zirconia," NASA TM-102350 (1989).

20. G.J. Valco, N.J. Rohrer, J.D. Warner and K.B. Bhasin: "Sequentially Evaporated Thin Y-Ba-Cu-O Superconducting Films on Microwave Substrates" NASA TM-102068 (1989).

21. Miranda, F.A., et al.: Measurements of Complex Permittivity of Microwave Substrates in the 20 to 300 K Temperature Range From 26.5 to 40.0 GHz. NASA TM-102123, 1989.

22. Gurvitch, M.; and Fiory, A.T.: Resistivity of $La_{1.825}Sr_{0.175}CuO_4$ and $YBa_2Cu_3O_7$ to 1100K: Absence of Saturation and Its Implications. Phys. Rev. Lett., vol. 59, no. 12, 21 Sept. 1987, pp. 1337-1340.

23. Collins, R.T., et al.: Comparative Study of Superconducting Energy Gaps in Oriented Films and Polycrystalline Bulk Samples of Y-Ba-Cu-O. Phys. Rev. Lett., vol. 59, no. 6, 10 Aug. 1987, pp. 704-707.

24. Kobrin, P.H., et al.: Millimeter-Wave Complex Conductivities of Some TlBaCaCuO and $YBa_2Cu_3O_{7-f}$ Films, Presented at the M^2s-HTSC Conference, Stanford, CA, July 24-28, 1989. To be published in Physica C.

FABRICATION OF $YBa_2Cu_3O_{7-x}$ SUPERCONDUCTING WIRES BY SPRAY DEPOSITION

J. G. Wang and R. T. Yang*

Department of Chemical Engineering
State University of New York at Buffalo
Buffalo, NY 14260

ABSTRACT

$YBa_2Cu_3O_{7-x}$ superconducting wires have been fabricated by spray deposition/pyrolysis of metal nitrate solutions using commercial carbon fibers as the substrate, followed by heating in air and annealing in O_2. The T_c of the wires have been found to be 82 K and the J_c to be 2150 A/cm^2 in zero field at 77 K. The heating/pyrolysis process was further studied by TGA and SEM. SEM pictures showed that the superconducting phase formed by slow heating/annealing was well-aligned and well-connected along the axial direction of the wires.

Y-Ba-Cu-O superconducting wires have been fabricated by a number of methods[1-10]. At present, one of the major impediments to practical application of the wires is the low critical current density (J_c). The values of J_c of the wires prepared by the known methods[1-10] range from 0.4 A/cm^2 to 1000 A/cm^2 at 77 K in zero field. The J_c values of 10^4 - 10^5 A/cm^2 in a field of several tests is required for the wires to be useful for major applications[3]. It is known that the poor values of J_c are not intrinsic, but are the result of the random orientation of the grains and the low density. The molten-oxide processing technique[5-6] could result in wires with higher density than the samples prepared by sintering, but the grain alignment is not significantly improved by the latter technique.

In this paper, we report a new process for the fabrication of Y-Ba-Cu-O superconducting wire by spray deposition of nitrate solutions on carbon fiber substrate,[11] and the possible mechanism involved in this process.

Spray deposition techniques have been developed only for the preparation of films of Y-Ba-Cu-O and Bi-Sr-Ca-Cu-O.[12-17] Various substrates have been used for fabricating these films. However, interactions between the substrate and the superconducting phase occur at the high

---
*Address all correspondence to R. T. Yang.

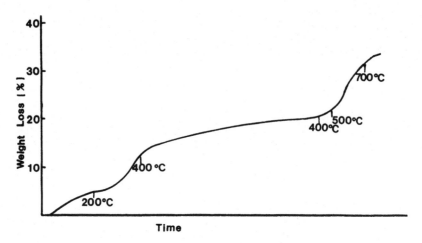

Fig. 1. TGA curve for carbon fibers coated with nitrates, heated in air at 5°C/min from 25 to 400°C, held at 400°C for 8 hrs., and heated at 2°C/min from 400 to 750°C.

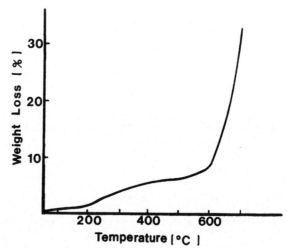

Fig. 2. TGA curve for uncoated carbon fibers, heated in air with the same temperature history as in Fig. 1 except not held at 400°C.

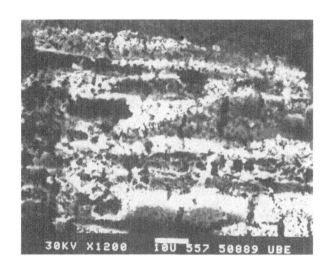

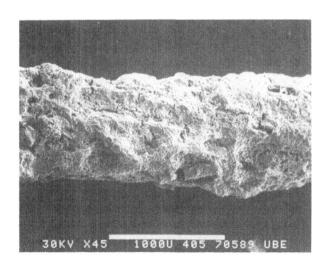

Fig. 3. SEM of wires obtained by heating in air at 5°C/min from 500 to 950°C.

annealing temperatures and these interactions result in poor-quality superconductors[18]. The interactions can be minimized by using carbon as the substrate. The commercial carbon fiber Thornel P55 was used in this work.

The solution for spray deposition was prepared by dissolving Y_2O_3 (99.99%), CuO (99.99%), $Ba(NO_3)_2$ (99.9%) in a dilute nitrate acid in the ratio of Y : Ba : Cu = 1 : 2 : 3. The solution was repeatedly sprayed on a strand of carbon fibers which was kept at 200-300°C, until a desired amount of nitrates was deposited. The sprayed fibers were then suspended in a quartz reactor for heating and annealing.

A thermogravimetric analysis (TGA) of the sprayed fiber was conducted in order to understand the processes occurring during the heating step. The TGA was Cahn System 113. The TGA curve of the sprayed fiber is shown in Fig. 1. The same heating procedure was used in the fiber fabrication process and the TGA measurement, except in fiber fabrication a final annealing step (in O_2) at 950°C was required. The heating atmosphere was air and the heating rate was as follows: 5°C/min from 25°C to 400°C, holding at 400°C for 8 hrs., then heating at 2°C/min from 400°C to 750°C. The TGA curve for the uncoated carbon fibers is shown in Fig. 2, using the same heating history except without holding at 400°C. The data in Fig. 2 show that significant carbon burnoff occurred only at temperatures higher than 600°C -- and this was without any protective coating. In the spray/ deposition process, the carbon fibers were coated repeatedly with the nitrates at 200 - 300°C. This treatment formed a thick (several times the carbon fiber diameter) and dense coating. In the ensuing heat-treatment step, complex processes took place: decomposition of the nitrates below 500°C,[15] and carbon burnoff. The mixed oxides are undoubtedly good catalysts for carbon oxidation.[20] However, they also formed a protective layer limiting the oxygen flux. The net result is shown in Fig. 1. The weight loss below 200°C was due to moisture which was gained during the time period before the TGA analysis when the coated sample was exposed to the ambient air. The weight loss below 500°C was due to the decomposition of nitrates.[15] Carbon burnoff took place at higher temperatures.

The spray deposited fibers were then heated to 950°C and held at that temperature for 5 hrs. (in air). The subsequent cooling procedure was: cooling at 10°C/min to 700-750°C and held for 10 hrs., cooling at 0.5°C/min to 400°C and held for 10 hrs., followed by cooling to room temperature.

The heating rate from 400 to 950°C was important in the process. Figures 3 and 4 are the SEM pictures of wires obtained by using different heating rates. Heating at 5°C/min produced wires with cracks and crevices (Fig. 3) whereas heating at 2°C/min produced dense, uniform fibers (Fig. 4). (The same cooling procedure was used in both cases.)

Figure 5 shows the temperature dependence of resistance for the wires prepared in the present work. The superconducting transition temperature (onset) was 85 K and the zero resistance temperature was 82 K. The J_c value was 2150 A/cm^2, which was measured by the dc I-V curve technique at 77 K in zero field. This value of J_c is about 4 times that for the sintered bulk $YBa_2Cu_3O_{7-x}$.[22] The reason for the high J_c value was apparently the good alignment of the crystals, as shown in Fig. 6. XRD results showed[11] that the wires were orthorhombic superconducting phase and that the carbon fibers were absent.

From the high J_c values, it is believed that the current carrying a-b basal planes of the orthorhombic structure were parallel to the fiber axis. This alignment may be attributed to the epitaxy between graphite and the

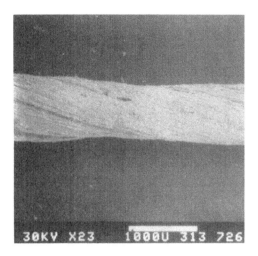

Fig. 4. SEM of wire obtained by heating in air at 2°C/min from 500 to 950°c.

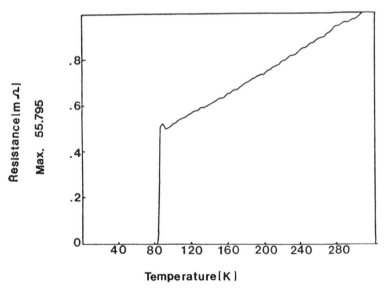

Fig. 5. Resistance vs. temperature for the wire.

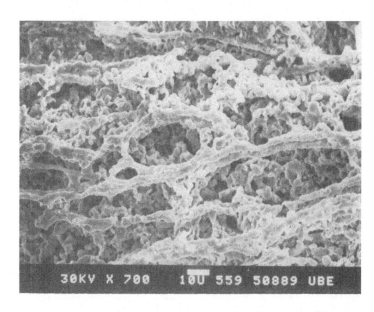

Fig. 6. SEM of superconducting wire showing crystal alignment along the wire axis.

superconductor. It is known that a significant amount of the surface on the carbon fiber is exposed prismatic edge planes of graphite, and that the d-spacing in commercial carbon fibers is larger than that of the ideal graphite (3.354Å), being larger at lower heat-treatment temperatures during the manufacturing of the carbon fibers.[21] These edge-plane surfaces would provide epitaxy for the a-b planes of the superconductor (a = 3.8185 Å and b = 3.8856 Å) during its early formation stage (during heating at 400°C to 950°C). Although epitaxial growth of the 1-2-3 superconducting phase usually takes place at approximately 650°C as reported in the literature, it could occur at temperatures as low as 400 - 450°C on a number of substrates (ZrO_2, MgO and $SrTiO_3$).[19] However, in our process the ensuing high temperature annealing step is necessary. This indicates that a superconductor precursor was formed at the carbon-coating interface during the heat treatment at 400°C and higher temperatures before carbon burnoff. This precursor served as a seed for further crystal growth during the sintering step.

Further studies on both fibers and films grown by spray deposition/ pyrolysis using various carbon substrates as well as repeated spray pyrolysis to further increase J_c are in progress in our laboratory.

This work was partially supported by the New York State Institute of Superconductivity and the National Science Foundation (to RTY). We also wish to thank Y.Z. Zhu and J.P. Zeng for T_c and J_c measurements, and P.J. Bush for SEM and XRD analyses.

REFERENCES

1. F. Uchikawa and J.D. MacKenzie, J. Mater. Res., 4, 787 (1989).
2. H. Konish, T. Takamura, H. Kaga, and K. Katsuse, Jpn. J. Apppl. Phys., 28, L241 (1989).
3. T.H. Tiefel, S. Jin, R.C. Sherwood, R.B. van Dover, R.A. Fastnacht, M.E. Davis, D.W. Johnson, Jr., and W.W. Rhodes J. Appl. Phys., 64, 5896 (1988).
4. T. Goto, Jpn. J. Appl. Phys., 27, L680 (1988).
5. M. Miljak, E. Babic, A. Hamzic, G. Bratina, and Z. Marohnic, Supercon. Sci. Tech., 1, 141 (1988).
6. S. Jin, T.H. Tiefel, R.C. Sherwood, G.W. Kammlott and S.M. Zahurak, Appl. Phys. Lett., 51, 943 (1987).
7. M.A. Lusk, J.A. Lund, A.C.D. Charklader, M. Bubank, A.A. Fife, S. Lee, B. Taylor, and J. Viba, Supercon. Sci. Tech., 1, 137 (1988).
8. Y. Yamada, N. Fukushima, S. Nakayama, H. Yoshino, and S. Murase, Jpn. J. Appl. Phys., 26, L856 (1987).
9. T. Goto and M. Kada, Jpn. J. Appl. Phys., 26, L1527 (1987).
10. T. Goto and I. Horiba, Jpn. J. Appl. Phys., 26, L1970 (1987).
11. J.G. Wang and R.T. Yang, J. Appl. Phys., in press.
12. R.L. Henry and H. Lessof, J. of Crys. Growth, 85, 615 (1987).
13. A.K. Saxena, S.P.S. Arya, B. Das, A.K. Singh, R.S. Tiwari and O.N. Srivastava, Solid State Commun., 66, 1063 (1988).
14. M. Kawai, T. Kawai, H. Mashuhira, M. Takahashi, Jpn. J. Appl. Phys., 26, L1740 (1987).
15. A. Gupta, G. Koren, E.A. Giess, N.R. Moore, E.J.M. O'Sullivan, and E.I. Cooper, App. Phys. Letter, 52, 163 (1988).
16. H.M. Hsu, I. Yee, J. Deluca, C. Hilbert, R.F. Miraky, and L.N. Smith, Appl. Phys. Letter, 54, 975 (1989).
17. E.I. Cooper, E.A. Giess, and A. Gupta, Mater. Lett., 7, 5 (1988).
18. P.R. Broussard and S.A. Wolf, J. of Crys. Growth, 91, 340 (1988).
19. S. Witanachchi, H.S. Kwok, X.W. Wang and D.T. Shaw, Appl. Phys. Letters, 53, 234 (1988).
20. R.T. Yang, in "Chemistry and Physics of Carbon," Ed. P.A. Thrower,

Vol. 19, pp. 163-211, Dekker, New York (1984).
21. D. Ghosh, R. Gangwar, and D.D.L. Chung, Carbon, 22, 325 (1984).
22. S. Jin, T.H. Tiefel, R.C. Sherwood, M.E. Davis, R.B. van Dover, G.W. Kammlott, R.A. Fastnacht, and H.D. Keith, Appl. Phys. Lett., 52, 2074 (1988).

MICROWAVE LOSS MEASUREMENTS ON YBCO FILMS USING A STRIPLINE RESONATOR

A.M. Kadin, D.S. Mallory, and P.H. Ballentine

Dept. of Electrical Engineering, University of Rochester
Rochester, NY 14627

M. Rottersman and R.C. Rath

CVC Products, Inc.
Rochester, NY 14603

ABSTRACT

Thin films of $YBa_2Cu_3O_7$ were deposited by rf magnetron sputtering onto heated substrates, and were superconducting without a subsequent anneal. Several of these films were placed as ground planes in a stripline resonator in a temperature-controlled cryostat. The frequency dependence of the reflection coefficient was measured up to 3 GHz using a network analyzer. From the quality factor Q of the resonances, the microwave surface resistance R_s of the superconducting films was determined over a range of temperatures. For a YBCO film on (100) $LaGaO_3$, R_s (2 GHz) was less than 3 mΩ for T<40 K.

INTRODUCTION

The techniques for depositing high-quality films of YBCO have reached the stage that a range of device applications are becoming feasible.[1] The simplest of these will be patterned out of a single thin film on a substrate, for such applications as electronic interconnects for high-speed digital circuits, and passive microwave devices. These two classes of applications have a common set of requirements, since fast digital pulses have frequency components in the microwave range. These requirements include high-quality, smooth films, with high critical current densities and low microwave losses. In addition, the films must be deposited on substrates that are compatible with high-frequency propagation, should be uniform over areas several inches across, and should ideally be able to operate at liquid nitrogen temperature (77 K).

A superconductor has true zero resistance only for dc signals, since there are always some "normal electrons" present that can produce dissipation at non-zero frequencies. However, low-temperature superconductors such as Pb and Nb exhibit extremely low microwave surface resistances for $T<T_c$, several orders of magnitude below those for normal metals such as Cu or Al. Early measurements of microwave surface resistance on YBCO yielded values that were rather high, even far below T_c.[2] Since then, several measurements on very high-quality single crystals and epitaxial films have yielded values much closer to ideal superconducting performance.[3,4] Evidently, microwave measurements provide a more stringent test of film quality than do dc resistance measurements.

With this in mind we have prepared films using an approach that is likely to result in good microwave performance, and we have developed a test system to evaluate this performance. We have chosen a stripline resonator,[5] of a geometry that will permit us to study this performance over a wide range of frequencies. We present here the measurements and analysis on our first two YBCO samples chosen for this study. Although the microwave performance of these films falls short of our goals, these preliminary results suggest some directions for further improvement.

THIN FILM DEPOSITION

We have used rf magnetron sputtering from a single stoichiometric YBCO target, onto a substrate heated during deposition to $\approx 675°C$. This approach is designed to yield smooth, crystallographically oriented films that are superconducting as deposited, without the need for a further high-temperature anneal. As in our previous work,[6] we have made use of a large target 8 inches in diameter, so that negative ion bombardment is localized to a ring directly above the target, about 6 inches in diameter (see Fig. 1). This leaves the region above the center of the target essentially unaffected by this bombardment, and we are able to obtain stoichiometric 1-2-3 films from a stoichiometric target, over a wide range of sputtering powers and substrate temperatures. For the films of this study, our target consisted of loose powder, although we have also used a solid sintered target with similar results.

Sputtering is carried out in 10 mTorr of Ar with 10% oxygen, using 600 W of rf power (13 MHz). The substrate is located about 3 cm above the target, yielding a deposition rate of about 70Å/minute, for a total of about 100 minutes for a $0.7\mu m$ film. For this study, substrates used include (100) MgO, (100) $LaGaO_3$, and (100) $La_{0.9}Gd_{0.1}GaO_3$. These substrates are known to promote epitaxial growth of YBCO films, and in addition should have reasonable values of dielectric constant and relatively low dielectric loss.[7] In contrast, $SrTiO_3$ has a huge dielectric constant which depends strongly on temperature and frequency, as well as large dielectric losses, and is completely unsuitable as a substrate for high-frequency applications.

The substrate was heated from above using a quartz lamp, which could produce temperatures up to $\approx 700°C$. Temperature was measured using a thermocouple sandwiched between two substrates, located right next to the substrate to be used for the film. Immediately after the deposition is complete, the oxygen pressure in the chamber is raised to 1 torr, and the substrate is permitted to cool slowly (≈ 30 minutes) in order to incorporate sufficient oxygen to convert fully to the orthorhombic superconducting phase.

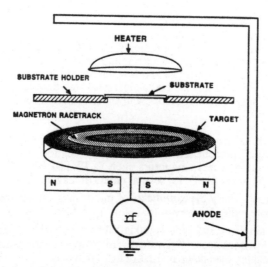

Fig. 1. Configuration of rf magnetron sputtering system used to deposit YBCO films.

MICROSTRUCTURE AND DC SUPERCONDUCTING PROPERTIES

The films deposited in the manner indicated above were black, smooth, and generally featureless under microscopic examination, at least for substrate temperatures of ≈ 675 °C or less. The composition, as measured using an energy-dispersive x-ray microprobe in a scanning electron microscope, was essentially stoichiometric and uniform on both the macroscopic and microscopic scales. This correlated well with the x-ray diffraction measurements, which indicated single phase highly oriented c-axis growth of $YBa_2Cu_3O_{\approx 7}$ perpendicular to the substrate.[8] This was true for samples deposited at temperatures of 600-700 °C, even for those that did not exhibit very good superconducting performance.

The samples were also characterized by their dc superconducting properties. Four-terminal measurements were carried out using evaporated Ag contacts. Resistance measurements indicated in all cases a depressed onset of superconductivity to $\leq 80K$. Some samples showed a sharp transition below that, with T=0 as high as 78 K. Two examples are shown below in Figs. 6 and 7. Others showed a more broadened transition, becoming fully superconducting as low as ≈ 50 K. The broadening in these films was not due to excessive test currents, since most of these films showed rather large critical currents (approaching 10^5 A/cm^2 or more) at low temperatures. Critical current measurements was made using a patterned film to narrow the conducting path to ≈ 100 μm. Films as thin as 0.2 μm showed similar behavior, suggesting that substrate reaction was not a limiting factor. In addition, some films were subjected to a post-anneal in one-atmosphere of oxygen at 500°C, without improvement of T_c. We suspect that atomic-level defects other than oxygen may be responsible for the depression in superconductivity,[9] but this is a matter for continuing investigation.

MICROWAVE RESONATOR MEASUREMENTS

A schematic representation of the microwave resonator measurement is given in Fig. 2, with a more detailed picture of the resonator itself in Fig. 3. The resonator is located in an evacuated chamber, typically with a small amount of He "exchange gas", which could be immersed either in liquid nitrogen or liquid helium. In either case, a resistive heater was available to heat the resonator above the temperature of the bath. Together with a Si-diode thermometer, this allowed the temperature to be feedback-controlled over a wide temperature range. A frequency-swept microwave signal was emitted from a microwave network analyzer (Hewlett-Packard 8753A, 0.3-3 GHz), and propagated down to the resonator via a 50-Ω rigid coaxial line. The reflected wave was propagated back along the same line, and the vector ratio of reflected to the incident wave, the reflection coefficient S_{11}, was determined by the network analyzer. It was important to minimize partial reflections from connectors on the input line, since these reflections may cause resonances that with the observation of the specific resonances of interest.

The resonator consisted of an open-ended stripline,[5] with a straight center strip about 2 mm wide by 22 mm long, sandwiched symmetrically between two groundplanes (12 mm side) spaced 3 mm apart. The strip typically consisted of Al tape, 50μm thick. The input/output line was connected to an SMA stripline launcher, which was in good electrical contact with a short length of Al strip at the edge of the substrate. Weak coupling between the resonator and the external circuit was achieved by means of a $\approx 2mm$ gap in the line (see Fig. 3). The ground planes were superconducting YBCO films, and the substrates were included within the cavity. Aternatively, other ground planes (such as Al tape, the Cu surface of the resonator, or superconducting Pb foil) could be used for calibration purposes.

Fig 4 shows the measured frequency dependence of the magnitude of the reflection coefficient $|S_{11}|$ for a resonator with YBCO ground planes on MgO substrates at T=6 K, in the region near a resonance. This is a printout of the screen display of the network analyzer. The important features about this resonance are its location, its width, and the depth of the dip.

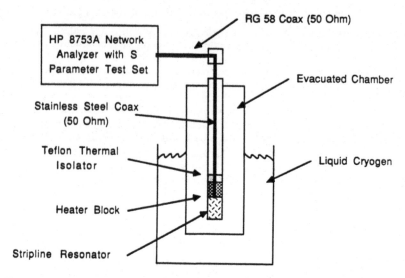

Fig. 2. Experimental setup for measurements of microwave surface resistance.

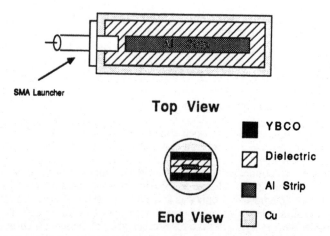

Fig. 3. Schematic view of the stripline resonator used for the determination of R_s.

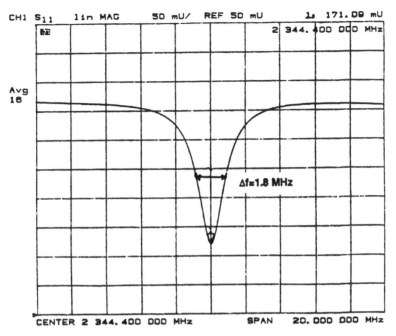

Fig. 4. Amplitude of the reflection coefficient $|S_{11}|$ vs. frequency (screen image from HP-8753A Network Analyzer) near the half-wave resonance for a stripline resonator with Al center strip and YBCO ground planes (corresponding to Fig. 6 at T=6 K). The center frequency is at 2.344 GHz, with a width at half-maximum Δf = 1.78 MHz.

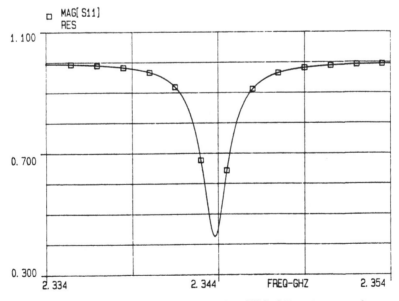

Fig. 5. Simulation of measurement in Fig. 4, using EESof Touchstone microwave CAD package[10] on IBM-AT. For open-ended stripline 22 mm long with 1.5 mm dielectric (ε_r=8) on each side of 2-mm-wide center strip, with conductor resistivity 0.2μΩ-cm and C=0.04 pF coupling to the 50-Ω input line.

DATA REDUCTION AND ANALYSIS

In analyzing resonance curves such as that in Fig. 4, we have adopted two complementary approaches. In one, we have made use of a microwave CAD package, EESof Touchstone,[10] by adjusting parameters to provide a detailed quantitative fit to the resonance curve. The result of this, based on the curve in Fig. 4, is shown in Fig. 5. This corresponds to the theoretical reflection coefficient from a stripline with a center strip 2 mm wide, 50 µm thick, and 22 mm long, between two ground planes 3 mm apart, separated by a (lossless) dielectric with $\varepsilon_r=8$, near its half-wavelength resonance. The external coupling capacitance is 0.04 pF, and the conductor resistivity $\rho=0.2\mu$m-cm. The dielectric constant (which is needed to match the center frequency $f_0=2.344$ GHz) seems a bit smaller than the value $\varepsilon_r \approx 10$ expected for MgO, but air gaps in the physical line may be expected to lower the effective value somewhat. The program permits only a single conductor material obeying the usual relation for the sheet resistance $R_s = (\pi\rho f\mu_0)^{1/2}$, so that this value of ρ corresponds to an average value $R_s \approx 4$mΩ. If we know the the R_s value for the center strip, then we can estimate that for the ground planes (see discussion below).

Alternatively, we can also approach the problem more analytically. A key aspect to any resonator is its dimensionless quality factor Q, which can be defined as the ratio of the energy stored to the energy dissipated per radian. An equivalent definition is $Q=f_0/\Delta f$, where Δf is the frequency width of the resonance, the frequency difference between the points where the resonator is at half its maximum power. Our measured quantity S_{11} corresponds not to the power in the resonator, but to the amplitude reflected. Strictly speaking, we should take the half-maximum of the factor $S_{bg}^2 - S_{11}^2$, where S_{bg} is the background value of S_{11} (corresponding to loss on the input line). However, this factor is the product of $(S_{bg}-S_{11})*(S_{bg}+S_{11})$, and the latter factor is essentially constant for small dips, so that it is generally sufficient to consider Δf at half the depth of the dip on the curve of S_{11} vs. f. In Fig. 4, $Q \approx 1300$.

It is frequently convenient to discuss Q in terms of its reciprocal Q^{-1} ($\approx 7.7 \times 10^{-4}$), since independent contributions to the loss in a resonator are additive components of Q^{-1}:

$$Q^{-1} = Q_L^{-1} + Q_d^{-1} + Q_c^{-1}.$$

Here Q_L^{-1} is the coupling loss due to the loading of the resonator (which is proportional to the capacitance C of the gap in the present case), $Q_d^{-1}=\tan\delta$ is the dielectric loss tangent, and $Q_c^{-1} = \alpha\lambda/\pi$ is the conductor loss (α is the attentuation constant on the line and λ the wavelength).[11] If we can estimate the coupling and dielectric losses, then we can determine the conductor loss from the total Q. Since α is proportional to R_s for a single conductor (or a linear combination of values for separate center strip and ground planes), we can also determine R_s if we know this proportionality constant.

For MgO and LaGaO$_3$ substrates, we estimate that the dielectric loss $\tan\delta \ll 10^{-4}$, so we have neglected it.[11] The coupling loss can be estimated from the simulation by extrapolating to the limit of low conductor loss, or alternatively by comparing Q^{-1} for the half-wavelength resonance on a stripline of length ℓ and that for the full-wavelength resonance on a stripline of length 2ℓ (the center frequencies should be the same). Since

$Q_L^{-1} \sim 1/\ell$ while Q_c^{-1} is constant, the difference between the two measurements of Q^{-1} yields Q_L^{-1} for the longer line. For the resonance in Fig. 4, this gives $Q_L^{-1} \approx 1.6 \times 10^{-4}$, yielding a conductor loss $Q_c^{-1} \approx 6 \times 10^{-4}$. This comparison also indicates that the loading of the resonator by the external circuit is significant but not overwhelming.

The conductor loss can in turn be expressed as a sum of the losses due to the individual conducting planes:

$$Q_c^{-1} = Q_c^{-1}(\text{center strip}) + Q_c^{-1}(\text{ground planes})$$
$$= K_1 * R_s(Al) + K_2 * R_s(YBCO),$$

where $K_1 = 7.5 \times 10^{-5}$ and $K_2 = 4.5 \times 10^{-5}$ are constants characteristic of the geometry of the stripline. These two proportionality constants are different because while the currents in the center strip are restricted to a width w=2mm, those in the ground plane are spread over a larger width w'≈w+2h, where h is the height of the dielectric between the strip and the ground. The larger effective width causes a smaller contribution to the attenuation and hence the value of Q_c^{-1}. These values of K_1 and K_2 were determined using the relation[5] $K_2 = (2/\mu_0\omega)\partial \ln Z_0/\partial h$, together with values of the characteristic impedance Z_0 and the attenuation α of the stripline calculated by the commercial program EESof LineCalc.[10]

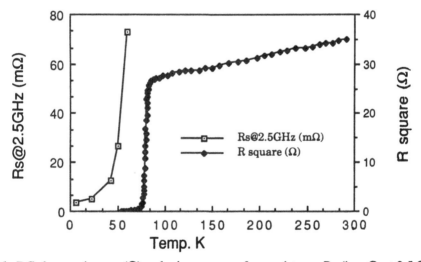

Fig. 6. DC sheet resistance (Ω) and microwave surface resistance R_s (in mΩ, at 2.5 GHz) for YBCO sputtered films (≈0.4μm thick) on (100) MgO substrates. The dc resistance goes to zero at T_c=68 K.

Based on this analysis, and on measurements of an all-Al line, we have determined that $R_s(Al) = 6$ mΩ at 2.5 GHz for low temperatures, and that therefore from the results of Fig. 4, $R_s(YBCO) \approx 4$ mΩ at 6 K. A similar analysis for a wider range of temperatures is shown in Fig. 6 for this film on MgO, and in Fig. 7 for a similar film on LaGaO$_3$, with the dc R(T) also shown on the same plots.

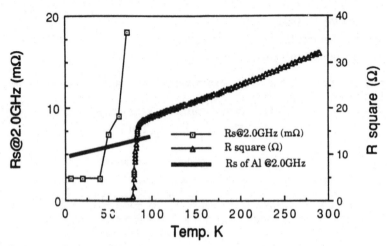

Fig. 7. DC sheet resistance (Ω) and microwave surface resistance R_s (in mΩ, at 2.0 GHz) for YBCO sputtered films $\approx$0.4μm thick on (001) LaGaO$_3$ and La$_{0.9}$Gd$_{0.1}$GaO$_3$ substrates. The dc resistance goes to zero at T_c=78 K. For comparison, R_s measured for Al at the same frequency is also shown.

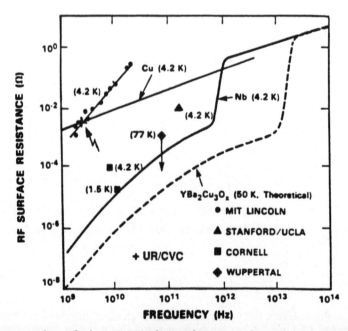

Fig. 8. Literature values of microwave surface resistance vs. frequency for YBCO and other materials (adapted from ref. 11). The UR/CVC result $R_s\approx$4 mΩ at 2.5 GHz (for low temperatures) is shown (labeled with a +) for comparison.

DISCUSSION AND CONCLUSIONS

The temperature dependence of R_s in Figs. 6 and 7 shows a fairly sharp drop just below T_c, followed by a more gradual decline at lower temperatures. In addition, the sample on $LaGaO_3$ (Fig. 7), which shows the sharper dc transition, also shows the sharper rf transition, becoming independent of T below about 40 K. However, the residual value at low T is about the same, particularly if the lower frequency in Fig. 7 (reflecting the higher dielectric constant in $LaGaO_3$) is taken into account.

This residual value of R_s (about 4 mΩ at 2.5 GHz) is comparable to some earlier measurements on YBCO films,[5] but is much greater than some more recent measurements on single crystals and laser-deposited epitaxial YBCO films.[3,4] This is illustrated in Fig. 8, adapted from ref. 11, which shows the frequency dependence of R_s for a number of measurements on YBCO, together with the theoretical dependence for Cu, Nb, and YBCO. Our residual values of R_s are comparable to a good normal conductor such as Cu (cooled to low T) only at relatively low frequencies. Although we have not yet evaluated the frequency dependence of R_s in our samples, superconductors typically exhibit an f^2 dependence, in contrast to normal conductors which go as $f^{1/2}$. Therefore, for potential applications which require wave propagation at 10 GHz or higher, our samples are likely to be more lossy than Cu or Al.

Although we do not yet have a clear understanding of the origin of the excess R_s for our films, we plan to carry out several tests to evaluate alternative possibilities. We will deposit thicker films, to see if perhaps 0.4μm is too thin for films of this quality. In addition, we will invert the films and measure R_s with the substrate outside the effective resonant cavity, to determine whether the excess loss may be localized at the film-substrate interface or in the dielectric itself. We will also look for a nonlinear dependence of the loss on the microwave power, to determine if the microwave magnetic field (or induced currents) in the resonator may be too large. This is probably unlikely, since for 1 mW input power and a Q≈1000, we estimate that the maximum amplitude of the microwave magnetic field is H≈0.1 Oe, a rather small value. Finally, there is the possibility that the excess rf loss is related to the same defects that are depressing our T_c.

In conclusion, we have fabricated smooth, uniform, highly-oriented YBCO films on microwave-compatible substrates, by sputtering at ≈ 675 °C withough a high-temperature anneal. However, these films exhibit a significantly depressed T_c, that may reflect atomic-level defects frozen in by the deposition process. We have also used reflection from a stripline resonator to obtain quantitative measurements of the microwave surface resistance of these films. While the technique has been quite successful, the values of R_s have been somewhat disappointing. We anticipate that as the dc properties are improved, the rf losses will likewise improve to values that are closer to ideal. If that turns out to be the case, then our sputtering method may offer advantages in scaling up to the larger areas required for some microwave and electronic applications.

ACKNOWLEDGEMENTS

The research at the University of Rochester was supported in part by the New York State Institute for Superconductivity under contract #88F001. Work at CVC Products, Inc. was supported in part by NASA contract NAS7-1045. One of us (PHB) has received fellowship support from the UR Laboratory for Laser Energetics. We would like to thank M. Ece and R.W. Vook of Syracuse University for x-ray diffraction scans. Finally, we also wish to thank M.F. Bocko, M.J. Wengler, and M.A. Fisher for valuable discussions on microwave measurements.

REFERENCES

1. A.I. Braginski, "Material Constraints on Electronic Applications of Oxide Superconductors", Physica C153-155, 1598 (1988).
2. L. Cohen, I.R. Gray, A. Porch, and J.R. Waldram, "Surface Impedance Measurements of Superconducting $YBa_2Cu_3O_{6+x}$", J. Phys. F: Met. Phys. 17, L179 (1987).
3. D.L. Rubin, et al., "Observation of a Narrow Superconducting Transition at 6 GHz in crystals of $YBa_2Cu_3O_7$", Phys. Rev. B38, 6538 (1988).
4. N. Klein, et al., "Millimeter Wave Surface Resistance of Epitaxially Grown $YBa_2Cu_3O_{7-x}$ Thin Films", Appl. Phys. Lett. 54, 757 (1989).
5. M.S. DiIorio, A.C. Anderson, and B.-Y. Tsaur, "RF Surface Resistance of Y-Ba-Cu-O Thin Films", Phys. Rev. B38, 7019 (1988).
6. A.M. Kadin, P.H. Ballentine, J. Argana, and R.C. Rath, "High Temperature Superconducting Films by RF Magnetron Sputtering", IEEE Transac. Magnet. 25, 2437 (1989).
7. R.L. Sandstrom, et al., "Lanthanum Gallate Substrates for Epitaxial High-Temperature Superconducting Thin Films", Appl. Phys. Lett. 53, 1874 (1988).
8. M. Ece, private communication, and manuscript in preparation.
9. C.B. Eom, et al., "In-situ Grown $YBa_2Cu_3O_{7-d}$ Thin Films from Single-Target Magnetron Sputtering", Appl. Phys. Lett. 55, 595 (1989).
10. Touchstone version 1.7 (1989), and LineCalc version 1.1 (1986), EESof, Inc., Westlake Village, CA.
11. D.E. Oates, A.C. Anderson, and B.S. Shih, "Superconducting Stripline Resonators and High-T_c Materials", p. 627 in Proc. 1989 IEEE MTT-S International Microwave Symposium, ed. by K.J. Russell, IEEE, New York, 1989.

RESPONSE OF HIGH TEMPERATURE SUPERCONDUCTORS TO A STEP IN MAGNETIC FIELD

Charles P. Bean

Physics Department
Rensselaer Polytechnic Institute
Troy, N.Y.

INTRODUCTION

The "critical state" model that has been used to characterize the response of strongly Type II superconductors assumes that a critical current is caused to flow locally by any electric field induced in the superconductor[1]. This simple postulate is sufficient to describe the losses in commercial superconductors[2] and the details of their responses to alternating magnetic fields[3]. High temperature superconductors near the critical temperature show a more complicated behavior. There is significant thermal activation of the movement of flux[4]. In this paper we use a simple extension of the critical state model that includes a flux flow resistivity. It is used to calculate the response to a step in magnetic field.

EXTENDED CRITICAL STATE MODEL

We start with Maxwell's equations expressed in practical units.

$$\text{div } \underline{D} = \rho \qquad \text{curl } \underline{E} = -10^{-8} \delta \underline{B}/\delta t$$
$$\text{div } \underline{B} = 0 \qquad \text{curl } \underline{H} = 4\pi \underline{J}/10 + \delta D/\delta t \qquad (1)$$

Since we deal with conductors, we may assume that $\rho=0$ and that the displacement current is negligible compared to the currents carried by the motion of charge. The equations then become

$$\text{div } \underline{D} = 0 \qquad \text{curl } \underline{E} = -10^{-8} \delta \underline{B}/\delta t$$
$$\text{div } \underline{B} = 0 \qquad \text{curl } \underline{H} = 4\pi \underline{J}/10. \qquad (2)$$

In a strong type II superconductor, i.e. $H_{c2}/H_{c1} >> 1$, the local B and H may be considered to be identical. In addition, if D is collinear with E, we have

$$\text{div } \underline{E} = 0 \qquad \text{curl } \underline{E} = -10^{-8} \delta \underline{B}/\delta t$$
$$\text{div } \underline{B} = 0 \qquad \text{curl } \underline{B} = 4\pi \underline{J}/10 \qquad (3)$$

These equations must be supplemented by a constitutive relationship between J and E. I use, in this paper, the approximation introduced by Kim, Hempstead and Strnad[5] of a flux flow resistivity, ρ_f, defined by

$$E = \rho_f(J-J_c) \qquad E>0. \qquad (4)$$

If E is negative, the negative sign is replaced by a positive sign. (The simple critical state model assumes that ρ_f is infinite so that $J=J_c$ for any electric field and, of course, is parallel to that field. The relationship of (4) is sketched below.

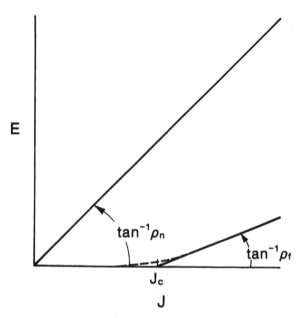

Figure 1. Current density as a function of field for a normal conductor and for an inhomogeneous type II superconductor. The slopes of the curves are the normal state resistivity, ρ_n, and the flux flow resistivity, ρ_f. The critical current, J_c, of the superconductor is indicated by a tick on the horizontal axis.

ELECTRODYNAMIC EQUATIONS IN DIMENSIONLESS UNITS FOR A PLANAR GEOMETRY EXPOSED TO A PULSED FIELD

For the purposes of calculation and of clarity it is useful to consider a specific geometry and to put (3) and (4) into dimensionless units. I consider a semi-infinite slab of superconductor whose inward surface normal points in the positive x direction. The magnetic field is applied in the z direction and currents are induced along the y axis. Equations (3) and (4) become

$$\delta B_z/\delta x = -4\pi J_y/10$$

$$\delta E_y/\delta x = -10^{-8} \delta B_z/\delta t \tag{5}$$

$$E_y = \rho_f(J_y - J_c)$$

To simplify these equations, I introduce the natural units for magnetic field, current density, distance and time. Specifically, the magnetic field is measured in terms of the magnitude of the magnetic field ($B_z(0)$, applied to the sample at t=0. The current density is measured in terms of J_c and distance is measured in terms of the full penetration distance1 for the flux. The electric field is measured in terms of the product of the flow resistivity and critical current density. Lastly, time is measured in terms of a diffusion relaxation time[6] for the motion of flux in a medium of resistivity ρ_f over a distance comparable to the full penetration distance. Denoting reduced units by a prime superscript and with suppression of subscripts,

$$B' = B(x)/B(o)$$

$$J' = J(x)/J_c$$

$$x' = x/x_{max}, \quad x_{max} = 10B(o)/4\pi J_c \tag{6}$$

$$E' = E/\rho_f J_c$$

$$t' = t/\tau, \quad \tau = 4\pi \times 10^{-9} x_{max}^2/\rho_f = 10^{-7} B(o)^2/4\pi J_c^2 \rho_f$$

With these normalizations, (5) and (6) become

$$\frac{\delta B'}{\delta x'} = -J'$$

$$\frac{\delta E'}{\delta x'} = -\frac{\delta B'}{\delta t'} \tag{7}$$

$$E' = J'-1, \quad E>0$$

$$E' = J'+1, \quad E<0$$

For some purposes, it is useful to note that E' may be eliminated among Equations (6) to obtain

$$\frac{\delta^2 B'^2}{\delta x'^2} = \frac{\delta B'}{\delta t'} \tag{8}$$

This is the canonical form of the diffusion equation that may be integrated analytically or numerically subject to appropriate boundary conditions.

CALCULATION OF TRANSIENT RESPONSE TO A STEP IN MAGNETIC FIELD

The profile of B'(x',t') that may be used to calculate all aspects of the response of a slab subjected to a step in magnetic field is found by integration of (7) subject to the boundary conditions:

$$B'(x',0) = 0$$

$$B'(0,t') = 1 \quad (9)$$

$$\left.\frac{\delta B'}{\delta x'}\right|_{B'=0} = -1$$

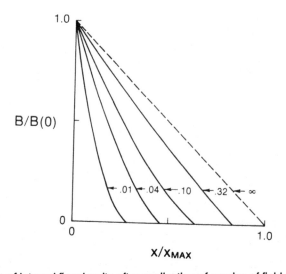

Figure 2. Profile of internal flux density after application of a pulse of field at t=0. The units are

The first boundary condition expresses the fact that the initial flux density is uniform throughout the sample while the second expresses the assumption that the step of field is constant in time. The third expresses the assumption that a critical current, J_c, is caused to flow by the infinitesimal electric field at the moving boundary of the pulse.

The integration is performed directly in a standard fashion to yield the results below.

normalized. Flux density is measured in terms of the applied field, B(0). Distance is measured in terms of the full penetration distance, 10 B(0)/4πJ_c. Time is measured in terms of $10^{-7} B(0)^2/4\pi J_c^2 \rho_f$.

The surface electric field E'(0,t') is found by combination of the first and third of the equations given in (7). Results are presented below.

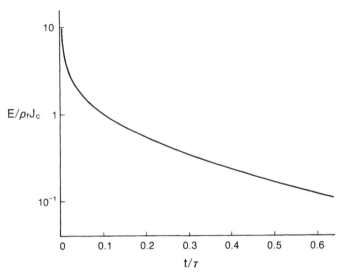

Figure 3. Surface electric field as a function of time after application of a pulse magnetic field at t=0. The electric field is measured in terms of $\rho_f J_c$ while time is measured in terms of $10^{-7} B(o)^2/4\pi J_c^2 \rho_f$.

DISCUSSION

The dynamics of flux penetration is seen to proceed as a front whose velocity decreases rapidly with time and approaches the critical state model as a limit. The surface electric field, observable with a pick up coil, is seen to have a general magnitude of $\rho_f J_c$ and of exponential time constant measured in terms of $\tau = 10^{-7} B(0)^2/4\pi J_c^2 \rho_f$. Around t=τ the field can be approximated by a simple negative exponential with a time constant of 0.37τ and a pre-exponential factor equal to 0.61 $\rho_f J_c$. For high T_c superconductors, ρ_f is not well known and J_c is structure dependent. To obtain order of magnitude guesses for these quantities we may note that normal state resistivities of ~100μΩ-cm and $H_{c2}(0)$~2×10⁶G may be inferred[7]. If the concepts of scaling introduced by

Kim et al are valid whereby $\rho_f = \rho_n B/H_{c2}$, the flow resistivity might be as low as $1\mu\Omega$-cm. If this were true and if one assumed a critical current density of $10^4 A/cm^2$ and step in field of 10^2 Gauss, the time constant would be on the order of a microsecond and the surface electric fields on the order of 10 millivolts per centimeter.

This treatment concentrates on the flux flow regime and not that of flux creep as measured by Yeshuraun et al[4]. By use of a suitable relationship between E and J that included thermal activation, one could derive the dynamics of flux penetration or retention appropriate to that case.

ACKNOWLEDGEMENTS

This work was initiated in a contract with the Air Force Materials Laboratory (AFML-TR-65-431) and continued with partial support by the New York State Institute on Superconductivity. The extensions reported here were aided by conversations with Timothy Gallagher of RPI. Dominick Fanelli and Taj Mahabub helped effectively with the calculations.

REFERENCES

1. C.P. Bean, Phys. Rev. Lett. 9, 309 (1962).
2. E.B. Forsyth, Science 242, 391 (1988).
3. C.P. Bean, Rev. Mod. Phys. 36, 31 (1964).
4. Y. Yeshurun, A.P. Malozemoff, T.K. Worthington, R.M. Yandrofski, L. Krusin-Elbaum, F.H. Holtzberg, T.R. Dinger and G.V. Chandrashekhar, Cryogenics 29S, 258 (1989).
5. Y.B. Kim, C.F. Hempstead and A.R. Strnad, Phys. Rev. 129, 528 (1963).
6. C.P. Bean, R.W. DeBlois and L.B. Nesbitt, J. Appl. Phys. 30, 1976 (1959).
7. T.P. Orlando, K.A. Devlin, S. Foner, E.J. McNiff Jr., J.M. Tarascon, L.H. Greene, W.R. McKinnon and G.W. Hall, Phys. Rev. Lett. 36, 2394 (1987).

DEVELOPMENT OF A COMPOSITE TAPE CONDUCTOR OF Y-Ba-Cu-O*

R.D. Blaugher, D.W. Hazelton and J.A. Rice

Intermagnetics General Corporation

Guilderland, New York 12084

ABSTRACT

Samples of a "prototype" composite tape conductor have been successfully prepared consisting of a thick layer of Y Ba$_2$Cu$_3$O$_7$ (123) on a Ni substrate. Thick continuous layers of 123 were applied to a Ni substrate by multiple applications using an aerosol technique. Sintered composite samples were prepared at 910°C and equilibrated in oxygen at 450°C. These samples demonstrated promising mechanical properties with good adhesion to the substrate, tolerance to thermal cycling and ability to withstand nominal bending deformation without any detrimental influence, i.e., cracking of the superconductor. The superconducting properties showed 123 orthorhombic formation with T_c onset of approximately 92K. Processing approaches to improve the current density are currently under investigation, including the use of a buffer layer to inhibit Ni diffusion and melt-texturing as a means for increasing bulk transport currents. Preliminary melt processing results show evidence for texturing and current density near 10^3 A/cm^2. Similar values for the critical current were also obtained for vapor deposited "polycrystalline" 123 films on MgO/Ni substrates.

---

*Supported in part by SDIO Contract SDIO 84-88-C-0049.

INTRODUCTION

Superconductivity has been successfully applied to large scale magnets for fusion, power generation, magnetic energy storage, magnetic resonance imaging and high energy physics[1]. These applications, without exception, utilized conventional low temperature superconductors Nb-Ti or Nb$_3$Sn, which can achieve a current density of approximately 10^5 A/cm^2 at 5T and a 4.2K operating temperature. Wires or tapes fabricated from the High Tc superconducting oxides (HTS) must achieve current densities comparable to those of conventional superconductors before similar large scale applications can be seriously considered.

In addition, the HTS wires or tapes must have the mechanical ability to withstand fabrication and winding stresses, thermal contraction during cooldown to cryogenic operating temperature, and the Lorentz forces produced by high magnetic fields. For these reasons, it appears impractical to consider the development of a HTS wire or tape in any configuration other than a composite, which incorporates a substrate (or matrix) for mechanical reinforcement of the ceramic superconductor. These requirements can be satisfied by a composite tape, which we have selected as the format for development of a HTS based conductor. This paper will present our processing methods and results to date for an approach which will hopefully produce a high current density, mechanically rugged HTS composite tape. The present approach is concentrated on the Y-Ba-Cu-O (123) compounds but could be extended to the Tl or Bi based materials.

HTS WIRES AND TAPES

Since the discovery of the superconducting high temperature ceramics there has been a large worldwide effort directed at the materials understanding and processing necessary to achieve a HTS conductor. Much progress has been made but, to date, no one has demonstrated a technologically useful HTS wire or tape. Recent work by Sumitomo reports the highest current density data published to date for an actual wire or tape configuration; silver sheathed tape samples of Bi-Pb-Sr-Ca-Cu-O showed a current density of 1.26×10^4 A/cm^2 at 77K in zero field. The current density, however, rapidly degraded in applied magnetic field showing approximately 10^3 A/cm^2 at a low field of 0.1T[2]. The best data for Y-Ba-Cu-O (123) material in a similarly fabricated Ag sheathed tape is 4×10^3 A/cm^2 at 77K and zero field. This tape displayed a rapid decrease in J_c with applied field[3].

The best wire or tape thus shows over two orders of magnitude lower critical current and severe degradation in a magnetic field as compared to high quality thin films. In bulk material the formation of "weak links," which are attributed to microstructural, crystalline anisotropy and mechanical causes such as micro-cracking, apparently limit the ability to maintain a high transport current.

Recent work on the so-called "melt processed" material presents encouraging data with significant improvement in current density for bulk samples, showing greater than $10^4 A/cm^2$ at zero field with field degradation dependence comparable to single crystals showing approximately 10^3 A/cm^2 at $1.0T^{(4)}$. These results show a field dependence comparable to published data on single crystal bulk material but are lower in current density by close to two orders of magnitude.

COMPOSITE TAPE APPROACH

The most immediate weak link problem that must be solved for eventual realization of a useful conductor is the tendency for the ceramic materials to show micro-cracking. As shown in Table I, the 123 compound exhibits a marked crystalline anisotropy in its thermal expansion properties. This would produce high stress concentrations at the high angle grain boundaries, which would lead to micro-cracking.

In addition, the wide temperature difference between the high temperature processing and the cryogenic operating temperature could produce sufficient stresses to fracture the superconductor if the substrate does not match the superconductor in thermal expansion properties.

Published data on Ag sheathed 123 wires have shown that the high temperature sintering heat treatment produces sufficient uniaxial elongation in the sheath due to thermal expansion mismatch to severely crack the superconductor core, thus destroying the ability to transport current[5]. The processing approach for the present tape should promote strong a-b crystalline development planar to the substrate with c-axis normal to the plane. Matching the oxides expansion properties in the a-b direction to those of the substrate should offer a reasonable approach for a composite configuration.

Nickel or nickel base alloys appear to offer a potentially good substrate material. Nickel, in addition, has a fairly high melting point of 1455°C and good oxidation resistance which is needed for the high temperature melt processing.

Numerous methods have been described for preparing 123 in bulk thick coatings or thin films which can be adapted to the present tape requirements. A 10-50 μm layer of stoichiometric 123 must be applied to long lengths of substrate to obtain a continuous uniform thickness. Precursor powder methods or vapor deposition techniques appear to be well suited for tape application. These methods, followed by melt processing, are presently being explored as a technique for densification and refinement of the microstructure to reduce weak links and improve the current density.

TABLE 1. PHYSICAL PROPERTIES OF SUBSTRATE AND BUFFER MATERIALS COMPARED TO 123

Material	Crystal Structure	Lattice Constant (A°)	Thermal Expansion Coefficient α ppm/(°C)	Melting Temperature (°C)
YBaCuO (123)		a=3.859	12	1015
	perovskite	b=3.920	10	
		c=11.843	40	
Cu			17.7	1083
Ni			15.3	1455
MgO		a=4.213	13.5	2800
NaCl	rocksalt			
α-Al$_2$O$_3$	hexagonal	a=4.76	8.8	2045
		c=12.95		
SrTiO$_3$	perovskite	a=3.905	7.9	
Au			14.2	1063
Ag			19.0	961
ZrO$_2$ (YSZ)	fluorite	a=5.140	10.0	

Several powder precursor methods were explored by us to identify a technique suitable for tape fabrication. We initially examined a diffusion method reported by Tachikawa[6], which offers the ability to apply mixed oxides which are then reacted to form a 123 layer. This approach was found to be undesirable because of the formation of an insulating surface layer of 211 which forms over the 123 compound, compromising electrical contact and limiting further attempts to improve the superconductor properties. A successful approach which allowed the application of a uniform layer of 123 to a substrate is described in the following.

SINTERED COATINGS ON Ni

Pre-reacted powders of 123 with particle sizes between 0.5 - 3 μm were introduced under controlled atmospheric conditions to anhydrous ethyl alcohol. The alcohol suspended powder solution was then converted to an aerosol using an inert gas carrier sprayed onto the substrate and subsequently heated to drive off the alcohol. The resultant coatings were then sintered in flowing oxygen at 900-925°C for 2-8 hours and cooled to 450°C for oxygen equilibration. The superconducting transitions were characterized using both DC resistivity and AC susceptibility. Current densities were measured as a function of temperature using a standard four probe technique at 1 μV/cm sensitivity. Current contacts were applied with silver paste followed by a diffusion reaction at elevated temperature.

Characterization of the superconducting orthorhombic phase development and impurity phases was done using an X-ray diffractometer. Scanning Electron Microscopy (SEM) and optical metallography were used for microstructural evaluation. Energy dispersive analysis (EDX) was used for detailed analysis of the microstructural composition. The superconducting properties of the sintered coatings on Ni showed broad resistive transitions of approximately 10 degrees wide with an onset temperature near 92K. The 123 adhesion on the Ni was remarkably good and showed no evidence of cracking or flaking even after being bent around a 2cm radius. The current density was typical of sintered material showing only 10 A/cm^2 at 70K. Attempts to improve the current density by carefully controlling the atmosphere to minimize carbonate formation resulted in improvement in current density by factors of 2-4. Detailed SEM analysis of the 123 surface (Figure 1a) showed high porosity and evidence of Ni diffusion. The degradation of 123 due to Ni diffusion has been studied by others, with the 123 able to tolerate fairly high levels of Ni before experiencing significant depression of T_c. These studies showed that 123 would gradually degrade from 92K to approximately 70K which was correlated with a 10% substitution for the copper[7].

The x-ray diffractometer measurements of the sintered coatings on Ni showed only 123 orthorhombic development with a weak Ni peak. The 123 coating thickness was approximately 10 um which may not be sufficiently thick to prevent reflection from the substrate.

MELT PROCESSING OF SINTERED COATINGS

The 123 on Ni coatings were then subjected to melt processing heat treatment above the peritectic from 1050-1200°C. As the temperature was increased to 1200°C, even for short times of approximately 1 minute, violent reaction of the 123 with the Ni was observed. The samples treated at 1200°C were so badly attacked that no detailed analysis was

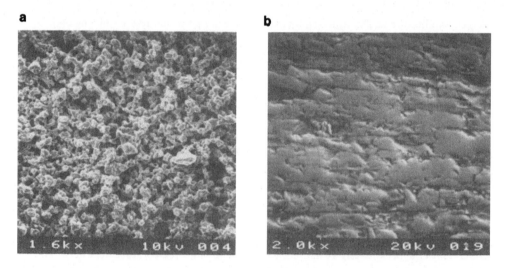

Figure 1. SEM evaluation of a) sintered, b) melt processed samples of 123

possible. Samples reacted at 1100°C showed only modest reaction compared to 1200°C with no evidence for 211 or orthorhombic structure. The x-ray diffractometer showed high Ni was present throughout the sample and could not completely resolve all of the phases. These initial melt processing attempts on 123 coated Ni thus indicate that a non-reactive buffer layer must be incorporated with the Ni substrate to facilitate high temperature melt processing. Table I and published data on 123 thin film-substrate interaction indicate that MgO should be a good candidate for a buffer layer on Ni.

MgO BUFFER LAYER

An MgO buffer layer at thicknesses ranging from 0.5 μm to 3 μm was applied by RF sputtering or E-beam evaporation to the Ni substrates. Heat treatment of the MgO/Ni up to 1200°C showed no evidence of any reaction. Good adhesion was observed providing either the Ni was roughened and/or back-sputtered. Clean, non-roughened, non-back sputtered surfaces showed poor adhesion.

Aerosol coatings of 123 were applied to the MgO/Ni substrates and subjected to sintering at 910°C followed by melt processing heat treatment up to 1200°C. The T_c and onset temperature for the sintered 123 on MgO/Ni was essentially the same as the previous samples on Ni with only a slight depression in the onset temperature. X-ray measurements showed orthorhombic 123 with a weak MgO and Ni peak. Samples of 123 on MgO/Ni which were melt processed at approximately 1200°C for short times of approximately 1 minute showed significant reduction in reaction between the 123 and Ni. Some reaction was seen around the MgO/Ni edges which was attributed to Ba-Cu-O liquid phase migration to the Ni. At temperatures above the peritectic the 123 decomposes to 211 + Ba-Cu-O liquid phase. The liquid phase migrates over the MgO surface and will attack any exposed Ni not covered by MgO. This attack for prolonged heat treatments was quite severe and would eat through and erode the Ni under the MgO coating.

X-ray measurements of these samples showed mostly 211 plus Ni oxides and Y_2O_3 with no evidence for 123 orthorhombic phase. It is quite clear from these results that the Ba-Cu-O liquid migration is limiting the 123 development. The sample geometry must be optimized to insure that the Ba-Cu-O liquid be contained and remain in close proximity to the 211 phase. A thick precursor coating applied to both sides of a completely MgO buffered Ni substrate would provide an ideal configuration. It was thus decided to evaluate a completely buffered approach by using sapphire substrates. The desired information on the melt processing reaction conditions could be obtained by simulating this buffered configuration.

MELT PROCESSING ON SAPHIRE

Thick layers of 123 were melt processed on sapphire substrates. Although sapphire is not as good a substrate as MgO under most reaction conditions, highly dense inexpensive substrates of Al_2O_2 can be readily obtained. The degradation in T_c due to Al diffusion although greater than Mg is fairly well understood and impurity limits are known.

Encouraging data on sapphire would establish reaction conditions and could be easily followed by similar experiments on MgO. Melt processed results on sapphire, moreover, would also present some indication as to the level of Ni degradation on the superconducting properties. Thick layers of 123 were thus melt processed on sapphire between 1050-1200°C.

Samples were rapidly heated to temperatures above the peritectic and held at temperature in flowing O_2 for 1-10 minutes. Samples were then either quenched to RT by removal from the hot zone or furnace cooled to intermediate temperatures above the peritectic (1015°C). The samples quenched to RT were subsequently up quenched above the peritectic and slow cooled through the peritectic to maximize the 123 development.

Figure 2 presents resistance and susceptibility measurements for typical melt processed 123 on sapphire samples. The superconducting transition is fairly sharp, $\Delta T_c \approx 3K$, showing an onset at 93K and zero resistance around 90K. The sintered samples in comparison were much broader in T_c and showed zero resistance at approximately 75K. SEM studies of the melt processed sample shown in Figure 1(b) indicate high densification and evidence for texturing with the development of elongated grains. $BaCuO_2$ and CuO was also distributed throughout the 123 matrix with no evidence for the 211 phase. Critical current measurements were performed on the melt processed sample using the four probe transport method. Electrical contacts to the sample using Ag paste followed by a diffusion reaction did not produce good results. The contacts showed poor bonding and could be easily removed. Numerous attempts were made to improve the contacts by sectioning and surface abrasion with no marked improvement. The high sample density and surface oxides appear to limit the ability to make good contacts. It is quite clear that an in-situ method using back sputtering and Au or Ag evaporation is needed to implement good contacts. Contact heating and poor definition of the sample cross-section thus compromised the transport measurement. Critical current densities at zero field and 70K for the melt processed material showed approximately 10^3 A/cm^2 which is a conservative estimate. The active superconducting region was difficult to quantify, hence it is suspected that a more uniform stoichiometric cross section would show much higher critical current density than reported. It is important to note that the published data or melt processed results are highly optimized in that the reported data was obtained on "selected" samples.

VAPOR DEPOSITION ON MgO/Ni

In order to explore the practicality of a lower temperature method for processing dense coatings of 123 on a substrate, samples were prepared by E-beam vapor deposited 123 on

MgO/Ni substrates. Cu, Ba and Y were sequentially E-beam evaporated onto the MgO/Ni substrates and heat treated in flowing O_2 at 900°C to form the 123 compound. Following O_2 equilibration at 450°C Ag, contacts were attached to the film and resistivity/critical current measurements performed. Samples showed a T_c onset at approximately 92K with a 2K transition width. Current density measurements were compromised by heating at

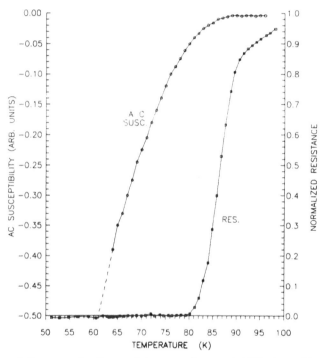

Figure 2. Superconducting resistance and susceptability transitions for typical samples of 123 melt processed on sapphire.

the contacts but showed at least 10^3 A/cm^2 at 70K in zero field. Decreasing the temperature did not change the onset of resistance indicating that the contact heating was severe and actually normalizing the sample, i.e., no temperature dependance in the critical current could be observed. X-ray measurements of the E-beam film showed complete orthorhombic 123 development. Microstructural evaluation basically showed a polycrystalline development of the 123 with a trace of the 211 phase.

SUMMARY AND CONCLUSIONS

Good progress has been made toward the development of a 123 composite tape using a MgO buffered nickel substrate. Melt processing and vapor deposition has been explored as a means for improving current density.

Future work will be directed at improvements in the buffer layer to hopefully completely isolate the 123 from the nickel substrate. In addition, the melt processing heat treatment will be extended to higher temperatures (1300-1500°C) following the approach suggested by Murakami[8]. He has suggested that improved development of 123 and enhanced pinning can be obtained by producing a fine dispersion of Y_2O_3 followed by a remelt above the peritectic and then a slow cool. We will continue to explore low temperature vapor deposition methods as an alternative to the melt processing approach.

ACKNOWLEDGEMENTS

We would like to thank Bill Giessen and Bob Markiewicz at Northeastern University for their advice and x-ray characterization. We would also like to acknowledge Ken Rose and Young Jeong at RPI for assistance in the E-beam vapor deposition.

References

(1) R.D. Blaugher, Proc. of the Tokai University Symposium on Superconductivity, Tokai, Japan, November, 1988, World Scientific, p. 183.

(2) H. Hitosuyanagi, K. Sato, S. Tokano, M. Nagata, to be published in Proc. of Magnet Technology, Tskuba, 1989.

(3) M. Nagato, K. Ohmata, H. Mukai, T. Hikata, Y. Hosoda, N. Shibuta, K. Sato, H. Hitosuyanagi, M. Kawashima, Materials for Cryogenic Technology Symposium, May 15-16, 1989, Japan.

(4) S. Jin, T.H. Trefel, R.C. Sherwood, M.E. Davis, R.B. VanDover, G.W. Kammloff, R.A. Fastnacht, H.D. Keith, Appl. Phys. Lett., 52, p. 2075, 1988.

(5) O. Kohno, Y. Ikeno, N. Sadakota, K. Goto, J. Jour of Appl. Phys., 27, L77, 1988.

(6) K. Tachikawa, M. Sagimoto, N. Sadokata, O. Kohno, Mat. Res. Soc. Symp. Proc., 99, p. 727, 1988.

(7) N. Ali, X. Zhang, P. Hill, S. Labroo, J. Less Comm. Metals, 149, p. 427, 1989.

(8) M. Murakami, M. Morita, K. Miyamoto, S. Matsuda, to be published in Proc. Osaka Univ. Int. Symp. on New Developments in Applied Superconductivity, Osaka, 1988.

FABRICATION OF HIGH-T_c SUPERCONDUCTING COATINGS

BY ELECTROSTATIC FLUIDIZED-BED DEPOSITION

David W. Kraft and Kenneth S. Gottschalck

University of Bridgeport
Bridgeport, CT 06601

and

Bedrich Hajek

SL Electrostatic Technology, Inc.
New Haven, CT 06511

ABSTRACT

We report the fabrication of high-T_c superconducting coatings by a novel technology, electrostatic fluidized-bed deposition. Fluidized powders of the $YBa_2Cu_3O_{7-x}$ superconductor were charged and deposited electrostatically on substrates of copper, alumina and stainless steel, and subjected to various sintering treatments. Substrate geometries included rectangular and cylindrical shapes. The superconducting transition was observed by both electrical and magnetic means; an AC four-probe technique was used to measure electrical resistance, and inductance measurements were made on copper coils wound on the cylindrical specimens. The effects of powder particle size, substrate preparation, sintering time and sintering temperature were studied. The advantages of this deposition process over thermal spray processes are described, and means to improve densification and critical-current density are discussed. An application to magnetic shielding is presented.

INTRODUCTION

The discovery of superconducting behavior in ceramic copper oxides [1] at easily accessible transition temperatures has stimulated a worldwide research and development effort, and the possible applications have been widely publicized [2]. There remain, however, major problems concerning the fabrication of these materials into the thick coating form needed for large-scale applications [3]. Such applications of surface coatings may include magnetic shielding and electronic devices.

Thermal spray techniques [4-10], i.e. plasma and flame spraying, have produced coatings ranging in thickness from 10 μm to over one mm. Advantages of these processes include high deposition rates, composition control, and the ability to coat complicated shapes and large areas. Major limitations of thermal spray processes are their line-of-sight character which, both, inhibits their ability to coat sharp corners and the edges of holes, and does not allow for removal of coating material from regions where no coating is desired.

The present work reports the fabrication of superconducting coatings by electrostatic fluidized-bed deposition, a proprietary technology [11] of SL Electrostatic Technology, Inc. This process possesses the advantages of the thermal spray processes without their limitations. Also, the process is better suited for continuous industrial processing and has the potential for higher process speeds. Our objective was to determine whether coatings of superconducting materials can be applied by electrostatic fluidized-bed deposition. A description of the process follows.

ELECTROSTATIC FLUIDIZED-BED DEPOSITION

In electrostatic fluidized-bed deposition, powder particles are areated in a fluidizing chamber and are electrically charged by contact with ionized air forced through a porous plate in the base of the chamber. As the powder particles become charged, they repel each other and, carried by the flowing air, rise above the chamber to form a charged cloud in a region where the ionizing electric field is applied. A grounded object placed in this cloud, or conveyed through it, attracts the particles and thereby acquires a powder coating. The coated object is then subjected to heat treatment that affixes the coating to the surface. The deposition process is illustrated in Figure 1.

The process possesses several advantages over thermal spray processes. Since the charged particles are attracted to regions of greatest electric field intensity, they seek the less exposed sites; as a result, coating thickness variations of as little as 2.5 μm can be achieved. The same mechanism also permits the coating of sharp corners, surface flaws and the

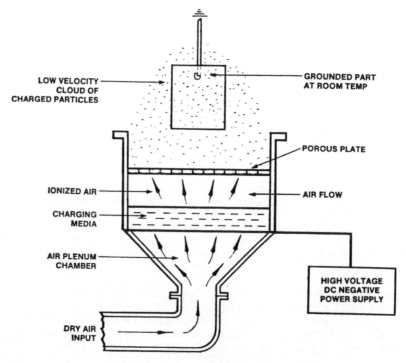

Fig.1. Electrostatic fluidized-bed deposition.

edges of holes. Particles can be wiped, vacuumed or blown from regions where no coating is desired [12].

The surface to be coated can be of any shape or size, and need not be a conductor, for insulators such as paper and fabric have also been coated [13]. Coating materials have included acrylic, epoxy, polyester, urethane, nylon, polyethylene, polypropylene, fluorocarbons, and nonplastic powders such as mica, talc, and glass frit [14]. In principle, any powder that can be charged and fluidized can serve as a coating. Coatings made with the materials listed above contained particle sizes between 2 and 100 µm, with mean values typically between 25 and 50 µm. The limits result from the tendency of fine particles to agglomerate and from the difficulty of charging and lifting the larger particles. Coating thicknesses have ranged from 20 µm to 0.5 mm. The lower limit is determined by the average particle size, while the upper limit depends on the voltage applied to the charging medium, the length of time that the surface is exposed to the charge cloud, and the dielectric properties of the powder. The thermal treatments are typically performed in an oxygen environment at the temperatures required for sintering of the powders.

EXPERIMENTAL

Coating Fabrication

An SL Electrostatic Technology Model C30 Powder Applicator was modified to fabricate superconducting coatings. The powder bed was redesigned and downsized to accommodate smaller volumes of the powder material and to allow for better control of the specimen. A system to conserve and recover undeposited powder material was installed, and a protective chamber was built around the device; the chamber also served to hold coating conditions constant when studying the effects of varying other process parameters.

The coating material selected for this work was $YBa_2Cu_3O_{7-x}$ (YBCO), with a transition temperature above 90 K [15]. Finished superconducting YBCO powders of >99% orthorhombic phase purity were obtained from CPS Superconductor Corp. and from SSC, Inc. Powder particle size distributions as determined X-ray sedigraphs furnished by the suppliers are given below.

Table I. Superconducting Powder Particle Diameters.

Supplier	Mean (µm)	Range (µm)
CPS	1.7	0.8 - 4
SSC	6	1 - 38
SSC	12	2 - 38
SSC	47	30 - 75

The mean value is the median of the cumulative distribution by weight, and the lower limit of the range refers to the lowest decile. The upper limit of the range specifies the largest particles that were present. We will refer to powders by their average particle diameter.

Substrate materials were selected from among those reported for flame or plasma sprayed deposition: copper [6,7,9], stainless steel [8,9], alumina [6,7,8,10], and alumina coated with a thin layer of silver [7]. Most substrates were in the form of 1" x 1" coupons, although some irregular shapes were also used. Two copper substrates were 1.5"-long hollow cylinders of 0.250" and 0.382" O.D. Prior to powder deposition, the substrate surfaces were cleaned by vapor degreasing in a chlorinated hydrocarbon and, in some cases, sandblasted prior to cleaning to increase surface roughness. All combinations of treatments, i.e., with or without sandblasting and with or without a silver print coating, were tried for each substrate material.

The powder particles were negatively charged in a potential difference of 60 kilovolts. Substrates at ground potential were held in the fluidized charged cloud for approximately 1.5 seconds, i.e., for as long as electrostatic attraction of the powder particles to the substrate was observed.

Sintering was performed on an alumina boat in a Lab-Line Carbolite combustion tube furnace capable of temperatures up to 1200°C, with a set point accuracy of 0.5%. The properties of YBCO powders have been found to depend on the details of the sintering and cooling procedures [16]. The specimens were sintered at 950°C for intervals of one or six hours and then furnace-cooled to 550°C and held there for one hour, after which they were furnace-cooled to room temperature. All procedures were performed in an atmosphere of flowing oxygen [16].

Measurements

Electrical resistance was measured by the standard four-probe method. The coatings were contacted by spring-loaded pins arranged in a line at 0.1" spacings, parallel to and equidistant from the sample edges. The pin array bed was affixed to a single-axis optical mount with a micrometer adjustment to ensure reproducible contact pressures.

An AC measurement method was selected to minimize contact-resistance and thermal effects. A constant 10 nA current through the two outer probes was furnished by the internal oscillator of a PARC Model 4510 two-phase lock-in amplifier connected in series with a 0.1% 100 kΩ precision resistor. The resulting voltage response across the two inner probes was amplified by a PARC Model 181 low-noise preamplifier driving the A-channel of the lock-in amplifier. Both the magnitude of the response signal and its phase shift relative to the oscillator signal were observed, from which the in-phase component was recorded. Since it is only the resistive change of the specimen that is of interest, the quadrature component, which is related to the reactance of the system, can be ignored. The measurement frequency was 101 Hz; the prime number feature of this frequency ensured discrimination against line-frequency harmonics and subharmonics.

During measurement, the specimen was in thermal contact with a copper block, but electrically isolated from it by a mica sheet. Thermal contact was enhanced by coating both sides of the mica sheet with a silicone-base thermal conductive paste. Measurements were made at temperatures between 300 K and 77 K, upon both specimen cooling and heating. Type E, chromel-constantan thermocouples were placed at two locations on the coating surface and monitored by a Keithley Model 199 digital multimeter scanner. The silicone paste was again used to improve the thermal contact. Spatial temperature variations in the specimen, as measured by the thermocouples, did not exceed one K. The thermocouples were calibrated against a silicon diode resistance thermometer accurate to within 0.5 K.

Specimens were cooled by immersion of the copper block in a bath of liquid nitrogen, and the temperature was controlled by slowly raising the level of the bath. Heating rates depended on the thermal mass of the copper block and on the rate of vaporization of the bath liquid. Although the heating rate was therefore not controllable, the observed rates were slower than the cooling rates and usually provided the best experimental runs.

An inductor was constructed from a cylindrical specimen by surrounding it with a coil of about 300 turns of 40-AWG magnet wire. The coil was half as long as the 38 mm specimen, and its diameter was just large enough to slide on the specimen whose outside diameter, including the 1.05 mm-thick coating, was 11.8 mm. An iron bolt was inserted through the structure. Inductance measurements were made with a Hewlett-Packard Model 4276A LCZ meter and, separately, by monitoring the voltage across the inductor in an R-L series circuit. The LCZ meter provided absolute values of inductance for frequencies between 100 Hz and 20 kHz, at fixed temperatures of 300 K and 77 K. The R-L circuit technique provided relative values of inductance at a fixed frequency for variable temperatures between 77 K and 300 K.

RESULTS AND DISCUSSION

Coating Fabrication

The best fluidization was achieved for the powders of largest particle size and the poorest fluidization was obtained for the two powders with the smallest particles. The finer particles tended to agglomerate and to form geysers. Cloud formation was also found to depend on powder particle size, but here the two powders with the finest particles were best suited.

None of the YBCO powders accepted or retained electric charge as well as the standard, nonsuperconducting powder coating materials used in this process. A likely reason for the lower charge acceptance is that the YBCO powders have lower dielectric properties. Information furnished by the powder suppliers revealed that the powder particles were of irregular shape and had jagged edges; as a result, charges accepted by a particle might tend to concentrate at the edges where they can also easily leak away. It was observed, however, that the YBCO powders with the largest particles were the best suited, probably because of larger particle surface areas. The reduced ability to accept and retain charge also resulted in depositions that were not as good and not as dense as those for the powders usually used in this process. Judged in terms of coating evenness and smoothness, the 12 µm and 47 µm powders provided the most satisfactory results. The powders with the largest particles also produced the sturdiest and most durable coatings. Coatings made from the powder with the finest particles were uneven and prone to fracture.

The best adhesion of the coating to the substrate was obtained for the copper and for silver-coated alumina substrates, while the poorest adhesion resulted for stainless steel substrates. No major differences in adhesion were observed among the various powders. Coating thicknesses ranged from 0.10 to 1.05 mm; the thickest coatings were obtained on copper substrates. In additon to coating the exterior surface of the cylinders, the process also succeeded in coating much of their interior surface.

Measurements

Figure 2 shows the temperature dependence of the resistance for three specimens made with the 12 µm powder deposited on copper coupons. The coating for curve (a) was sintered for 6 hours and the others for one hour. The resistance scale is in arbitrary units and has been scaled differently for the three cases for visual clarity. The data were taken upon specimen heating and represent the coatings with the highest resistance ratios, i.e. the resistance at 300 K divided by the "noise resistance" at 77 K. The values are 9800, 4100 and 3400. Since the resistance just above the superconducting transition region is about the same as that at room temperature, these values also represent the magnitude of the resistance drop at the transition.

Although the transitions occur at the temperatures expected for YBCO, the structure in the curves suggest the presence of additional phases which may have formed during sintering. The crossover to ohmic response was typically not complete until temperatures in the 100 K range were reached.

This behavior is neither characteristic of YBCO, nor can it be fully attributed to errors in the measurement protocol. It is felt that this behavior may be attributed to several sources: impurities in the specimen resulting from deleterious reactions with the substrate and other elements in the oven environment, and electrical contact complications arising from the mechanical properties of the coating.

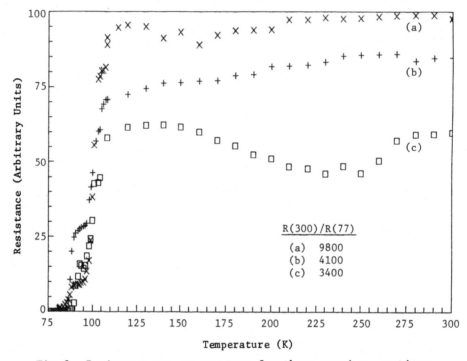

Fig. 2. Resistance vs. temperature for three specimen coatings.

Figure 3 shows the temperature dependence of the resistance and inductance for the inductor specimen, in which the inductance transition serves to confirm the resistance transition. The measurement frequency for each curve was 101 Hz. Note that the inductance curve completes its transition to the superconducting state at a higher temperature and does not contain the structure exhibited by the resistance curve. Since the inductance measurement is a volume effect, it is reasonable to infer that some of the phenomena responsible for the structure in the resistance transition are effects that take place at or near the specimen surface.

The inductance curve also demonstrates the ability of the coating to serve as a magnetic shield. The data, which were taken upon specimen heating, show that the magnetic field of the current-carrying coil penetrated the inductor core only when the specimen was in the normal state. The magnetic shield ability is again demonstrated in Figure 4 where the frequency dependence of the inductance is plotted for the inductor and for reference structures at both 300 K and 77 K. The reference structures

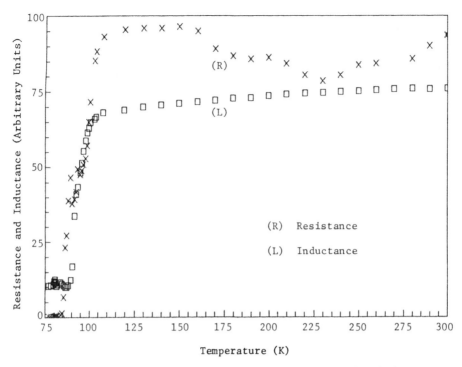

Fig. 3. Resistance and inductance vs. temperature for inductor.

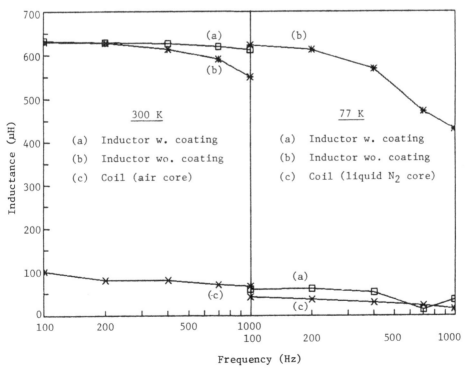

Fig. 4. Inductance vs. frequency for inductor and reference structures.

791

consist of the inductor structure sans superconducting coating, and of the coil winding containing either air or liquid nitrogen as the core. Below T_c, the specimen inductance falls to a level barely higher than that of the coil winding in liquid nitrogen and is less than the inductance of the coil winding in air.

CONCLUSIONS

We have demonstrated the feasibility of fabricating superconducting coatings by electrostatic fluidized-bed deposition. The structures coated by this process are not limited to planar surfaces and coating thicknesses in excess of one mm have been attained.

Both visual observation of the coatings and preliminary measurements of critical current densities reveal that the coatings do not yet possess the densification achieved with materials usually used by this process, or for bulk superconducting materials fabricated by other means. Although a successful application to magnetic shielding has been demonstrated, steps to improve densification must be taken before other useful superconducting structures can be fabricated. First steps should include the optimization of process variables related to the coating and substrate materials and to the post-deposition thermal treatment. There are additional approaches that merit consideration. Highly spherical powder particles, such as those fabricated by the plasma rapid solidification technology developed by Cheney [17], should improve densification for reasons both of geometry and improved charge retention. The inclusion of sinter-forging techniques, which have been shown to increase densification to better than 95% of theoretical density [18], should be studied. Magnetic fields have proved successful for grain alignment [19]; if the powders are single grains, depositions could be attempted in the presence of a magnetic field.

ACKNOWLEDGMENTS

We are indebted to R.F. Cheney for advice on materials and sintering procedures, to C. Messoligitis for software development, to A.F. Kraft for specimen preparation, and to D. Gillette and G. Gerling for encouragement. Portions of the work were supported by the Connecticut Department of Higher Education under Grant 89-301 and by the Strategic Defense Initiative Organization under Contract DASG60-88-C-0091.

REFERENCES

1. For review articles see, for example:
   a. E.M. Engler, Chemtech, Sept. 1987, p.542.
   b. D.R. Clarke, Adv. Ceram. Mater. 2, 3B, 272 (1987).
   c. T.H. Geballe and J.K. Hulm, Science 239, 367 (1988).
2. See, for example, reference 1c and J. Clarke, Nature 333, 29 (1988).
3. See, for example, F. Moon, J.R. Hull and G.F. Berry, Mechanical Engineering 110, 60 (1988).
4. J.P. Kirkland, R.A. Neiser, H. Herman, W.T. Elam, S. Sampath, E.F. Skelton, D. Ganset and H.G. Wang, Adv. Ceram. Mater. 2, 401 (1987).
5. W.T. Elam, J.P. Kirkland, R.A. Neiser and E.F. Skelton, Adv. Ceram. Mater. 2, 411 (1987).
6. J.J. Cuomo, C.R. Guarnieri, S.A. Shivashankar, R.A. Roy, D.S. Yee and R. Rosenberg, Adv. Ceram. Mater. 2, 422 (1987).
7. R. Guarnieri, R.A. Roy, S.A. Shivashankar, J.J. Cuomo, D.S. Yee and R. Rosenberg, AIP Conference Proceedings No. 165, edited by G. Lucovsky (American Institute of Physics, New York, NY, 1988), p.204.

8. G.N. Heintze, R. McPherson, D. Tolino and C. Andrikidis, preprint.
9. K. Tachikawa, I. Watanabe, S. Kosuge, M. Kabasawa, T. Suzuki, Y. Matsuda and Y. Shinbo, Appl. Phys. Lett. 52, 1011, (1988).
10. A. Asthana, P.D. Han, L.M. Falter, D.A. Payne, G.C. Hilton and D.J. Van Harlingen, Appl. Phys. Lett. 53, 799 (1988).
11. U.S. Patent 3,916,826: Electrostatic Coating Apparatus.
12. D.J. Gillette, Insulation/Circuits, June, 1982.
13. F.L. Cook, Final Report, DOE Grant No. DE-FG05-84CE40702, Dec. 1986.
14. W.C. Goodridge, Modern Plastics Encyclopedia, 1971-1972.
15. See, for example, reference 1.
16. See, for example, D.W. Murphy et al, Science 241, 922 (1988).
17. R.F. Cheney, MRS Symposia Proceedings 30, 163 (1983).
18. Q. Robinson et al, Adv. Ceram. Mater. 2 (3B), 380 (1987).
19. D.E. Farrell et al, Phys. Rev. B 36, 4025 (1987).

SUPERCONDUCTING POWER TRANSMISSION LINES COST ASSESSMENT

Dr. James Wegrzyn and Richard Thomas

Brookhaven National Laboratory
Department of Applied Science
Upton, NY 11973

Ms. Miriam Kroon

Kroon Unlimited
275 Woodacres Rd.
East Patchogue, NY 11772

Dr. Tom Lee

Manhattan College
Manhattan College Parkway
Riverdale, NY 10471

INTRODUCTION

A computer model of the component costs of superconducting power transmission lines (SPTL) has been prepared by Brookhaven National Laboratory (BNL) for the Empire State Electric Energy Research Corporation (ESEERCO). Technical design requirements from the work of Brookhaven and Los Alamos National Laboratories have been reviewed for low T_c cables. Similar performance values have been assumed as the input parameters for costing the high T_c transmission line. Special emphasis has been placed on the interplay between the design requirements of a superconducting cable and their cost. Thus, as new research findings become available, these improvements can effectively be incorporated into the working cost model. The study has benefit to the utility sector by bringing the recent developments in high T_c superconducting materials into a clearer focus, while also assisting the research scientist and developer through identifying the cost/design criteria that controls the implementation of a superconductive power transmission line.

BACKGROUND

The difficulties experienced in obtaining overhead transmission line rights-of-way, coupled with the more recent environmental concerns with electromagnetic field effects, are expected to produce an

increasing demand for underground transmission systems. The cost of
meeting the anticipated demand for new transmission systems will be
prohibitive unless current technology can be improved in an efficient,
reliable, economical and environmentally acceptable manner. The cost
algorithm (in the model) has been developed to qualify and quantify the
cost of a SPTL, which takes full advantage of the higher operating
temperatures of the new class of superconductors.

Past studies[1,2] have shown that superconducting power transmission
lines can be cost effective in high capacity situations. However, it is
now quite clear that the need to deliver large amounts of electrical
power to urban loads centers has diminished along with the lesser
prospects of remote 10 GVA generation parks. With projected demand for
new transmission lines at 3 GVA or less,[3] the question raised is whether
superconducting power transmission lines have a future in the nations
electrical energy transmission infrastructure. It may be presumptious
to think that a single study could answer such a complex question. This
paper, however, makes an attempt to arrive at answers by formulating a
cost algorithm that uses the method of annualized revenue requirement
for a "postulated" 90°K SPTL which takes advantage of the allowable
higher operating temperatures. The model assumes that the research will
eventually develop a high temperature superconducting cable which has
electrical and physical attributes at least as good as the currently
available low T_c cable. Even if the required advancements are not
developed for several decades, it is still not too soon to pursue this
technology, since many years of R&D are necessary to meet the stringent
reliability requirements of the electric utilities.

The objective then of this work is to evaluate SPTL as an
economically and technically viable option for the electric utility
industry by gauging the cost and necessary performance criteria of key
components of the transmission line. To make these judgments without
such specific inputs as power levels, site location and contingency
requirements, a generic cost algorithm was developed that uses best
available information with reasonable (or at least clearly stated)
assumptions to project costs in terms of annualized revenue
requirements. The work selects key material properties, component
cable design and thermal insulation systems that, when taken in total,
represent a realistic system. The technical design of the cable was
taken from past BNL work on SPTL,[4,5,6] while the thermal insulation
requirements for liquid nitrogen cooled transmission lines were
specified by using previous studies on cryogenic cable systems.[7]

THERMAL MANAGEMENT

Before cost estimates of a SPTL can be assessed, the configuration
of the system and the dimensions of the components must be known. From
past studies, a design using flexible cable which can be field-installed
in a rigid thermally insulated pipe has emerged as the configuration of
choice. Other features which were adopted for this study are:

o   the electrical and mechanical properties of the high
    temperature superconductors improve to the performance
    level of the niobium-tin superconductors;

o   the critical temperature of the candidate supercon-
    ductors are greater than 90°K;

o the coaxial cable has full current capability in both the inner and outer conductor;

o the cable is cooled by liquid nitrogen flowing first through the bore of the cable, and then returning through the interstitial region between the cables and the enclosure;

o the module length between refrigeration sites is taken as 25 miles;

o the dielectric insulation consists of lapped polymeric tape and is permeated by the coolant;

o the thermal insulation is provided by a rigid evacuated multi-layer enclosure;

o the circuitry is 3-phase, single circuit with no redundancy.

In order to specify redundancy for a power transmission it is necessary to have detailed knowledge of the overall system and the mandated contingency requirements. As in the PECO study,[1] the final ranking of various alternative transmission systems was strongly affected by its contingency policy. The "no redundancy" statement merely represents the simplest system. More realistic systems can be an outgrowth from this study.

The initial task is to determine the radial dimensions of the cable system for given voltage and power levels. The bore radius is determined by the requirement that for ac systems that transmissions occur near the matched impedance condition of

$$\overline{Z}_o = \sqrt{Z_L Z_C}/(V_{L-N}/I) = 1 \qquad [1]$$

where the per unit impedance $\sqrt{Z_L Z_C}$ has been normalized to its base impedance (V/I). Since the inductive reactance Z_L of a coaxial cable is given as

$$Z_L = \frac{\omega \mu o}{2\pi} \ln\left[\frac{R2}{R1}\right] \qquad [2]$$

and since the capacitive reactance of a coaxial cable is given as

$$Z_C = \frac{1}{2\pi \omega \epsilon_r \epsilon_o \ln(R2/R1)} \qquad [3]$$

and since the electrical field (E_s) on the outer surface of the inner conductor is given by

$$E_s = \frac{V_{L-N}}{R1 \ln\left[\frac{R1}{R2}\right]} \qquad [4]$$

and finally, since the surface current density (Jl_s) of the inner conductor is

$$J1_s = \frac{I}{2\pi R1} \quad [5]$$

then when combining Equations [2-5] with the impedance equation yields the interesting result that

$$\frac{J1_s}{E_s}\sqrt{\frac{\mu_o}{\epsilon_o \epsilon_r}} = 1$$

or

$$\frac{E_s}{J1_s} = 254 \text{ ohms} \quad [6]$$

for $\epsilon_r = 2.2$ for polymeric insulation.

For 3-phase systems $P = \sqrt{3}$ IV. The bore radius for a power level P and the line-to-line voltage V can now be written as

$$R1 = \frac{I}{2\pi J1_s} = \frac{P}{2\pi\sqrt{3} \, J1_s \, V} = \frac{254P}{2\pi\sqrt{3} \, V \, E_s} \quad [7]$$

With the bore radius being fixed by Equation [7], the remaining dimensions are then determined as follows:

$$R2 = 2(R1) \, \text{Exp}\left[\frac{V_{L-N}}{(R1)E_s}\right] = 2(R1) \, \text{Exp}\left[\frac{V}{\sqrt{3}(R1)E_s}\right] \quad [8]$$

R3 = R2 + Thickness of Cable Wrap [9]

R4 = R3 x Jamming Ratio [10]

R5 = R4 + Thickness of Inner Enclosure Wall [11]

$$R6 = (R5) \times \text{Exp}[2\pi k_{eff}\Delta T/Q_F] \quad [12]$$

where k_{eff} is the effective thermal conductivity of the insulated enclosure in $\left[\frac{W}{m^\circ K}\right]$; ΔT is the temperature difference across the enclosure [°K] and QF is the heat in-take of the enclosure $\left[\frac{W}{M}\right]$

R7 = R6 + Thickness of Outer Enclosure Wall [13]

R8 = R7 + clearance to trench wall [14]

Equations [7-14] fix the physical dimensions of the cable assembly from which both thermal and cost analysis have been made.

Since the radii of the cable assembly can be expressed in terms of power, a thermal analysis can now be performed to see whether the energy balance is maintained for this particular configuration. The cable assembly consists of the coaxial cable (Figure 1), and the enclosure which contains the three coaxial cables (Figure 2). The three types of heat loss mechanisms associated with superconducting AC transmission lines are: current dependent losses QR1 and QR2; the voltage dependent loss QD; and the heat in-leak loss QF (Figure 1).

The current dependent losses are defined in terms of the AC-loss property of the cable. A value[2] of 1.25 [W/M^2] for a surface current density of 50,000 [A/M] is used and is consistent with Low T_c cables. Note that the current measured losses on high T_c materials are several orders of magnitude higher than the 1.25 [W/M^2] value, and more research is needed before an the High T_c cable can be expected to achieve this value. Expressions for the current dependent losses QR1 and QR2 can be written as:

$$QR1 = 1.25 \left[\frac{J1_s}{50,000} \right] 2\pi R1 \qquad [15]$$

and

$$QR2 = 1.25 \left[\frac{J2_s}{50,000} \right] 2\pi R2 \qquad [16]$$

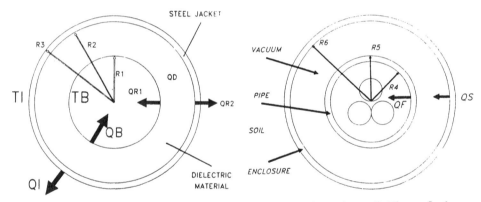

Figure 1. Thermal Flux of the Cable.

Figure 2. Thermal Flux of the Enclosure.

The model assumes that the heat generated by the inner conductor is removed by the coolant in the bore, while the heat generated by the outer conductor is removed by the return coolant in the interstitial region. Voltage dependent losses are determined by the Poission heat conduction equation, because the heat is first being generated within the dielectric layer, and then flows by conduction to either R1 or R2. The Poission heat conduction equation is

$$k_D \nabla^2 T(r) = - \frac{QD}{\text{Area}} \qquad [17]$$

where QD is in the units of watts per meter and area is the cross sectional area of the dielectric.

It can be shown that the voltage dependent losses QD can be written as

$$QD = \left[\frac{V}{\sqrt{3}}\right]^2 \frac{2\pi\omega \tan(\delta_D) \epsilon_r \epsilon_o}{\ln(R2/R1)} \quad [18]$$

where $\tan(\delta_D)$ is the dielectric loss tangent and (ω) is the angular frequency [radians/sec].

The allowable heat in-take is determined from the energy balance equation by setting

$$QF = \frac{\omega Cp(T_{out}-T_{in})}{\text{module length}} - (QD + QR1 + QR2) \quad [19]$$

where the mass flow (ω) of coolant in [kg/sec] is given by

$$\omega = \frac{2\rho(\Delta P)}{\left[\left[\frac{1}{A_B^2}\right]\left[\frac{L}{D_B}\right]f_B + \left[\frac{1}{A_I^2}\right]\left[\frac{L}{D_I}\right]f_I\right]^{\frac{1}{2}}} \quad [20]$$

where Cp = specific heat of Liquid Nitrogen

ΔP = Pressure Drop $f_{I,B}$ = Friction Factors
ρP = Nitrogen Density $D_{I,B}$ = Pipe Diameters
$A_B = \pi R1^2$ $A_I = \pi(R4^2-3R2^3)$
L = Module Length

Gathering terms for the heat generated per unit length to the bore (QB) and to the interstitial (QI) regions and the heat conducted from the interstitial to the bore region yields

$$QI = \frac{2\pi k_d (T_B-T_I)}{\ln\left[\frac{R2}{R1}\right]} + QD\left[\frac{1}{2 \ln\left[\frac{R2}{R1}\right]} - \frac{(R2)^2}{(R2)^2-(R1)^2}\right] + QR2 \quad [21]$$

$$QB = \frac{2\pi k_d (T_B-T_I)}{\ln\left[\frac{R2}{R1}\right]} + QD\left[\frac{1}{2 \ln\left[\frac{R2}{R1}\right]} - \frac{(R1)^2}{(R2)^2-(R1)^2}\right] + QR1 \quad [22]$$

where the factors following the term QD came from the solution of the Poission Equation [17].

The requirement of energy balance has been reduced to solving two simultaneous first order ordinary differential equations.

$$\omega C_p \frac{dT_B}{dz} = 3QB \qquad [23]$$

$$\omega C_p \frac{dT_I}{dz} = -(QF + 3QI) \qquad [24]$$

The minus sign in the second equation is a result of the return coolant flowing in the negative (z) direction. The two unknowns to be solved are the temperatures of the coolant in the bore (T_B) and the temperature (T_I) in the interstitial cavity, both of which vary along the length of the cable. Mathematically, the thermal management of the SPTL constitutes a boundary-value problem where the inlet and outlet temperatures have been preselected. In the model, equations [23] and [24] are solved by iterating a fourth-order Runge Kutta method along the length of the cable until the temperatures at the 25 mile mark agree to within one degree Kelvin.

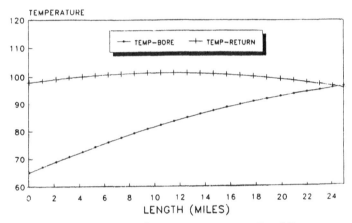

Figure 3. SPTL/90°K Temperature Profiles.

Figure 3 shows the solutions to the temperature profile over the length of 25 miles for 800 MW at a voltage of 138 kV, our base case. Acceptable solutions are defined as those which keep the maximum temperature below the 105°K boiling point of N_2 at 10 atm and have heat in-leak values above the minimum value of .3 watts/meter.

This simplified thermal analysis does not address the problem of instabilities caused by fluctuations in localized heating. The more stringent analysis would consider the proper use of the copper stabilizers, and would further lower the allowable operation temperature. However, the results from this analysis are adequate for costing purposes, since it forces energy conservation and relates the physical size of the system to maximum allowable temperature (whether 105°K or lower). This analysis underlines the fact that the use of a 90°K superconductor requires distances of much less than 25 miles between refrigeration sites when using liquid N_2 as the coolant.

SPTL COSTING ALGORITHM

This section discusses the development of a costing model for a SPTL. Since the evaluation of cost for a cable system is a utility-specific and circuit-specific problem, a goal of this work is to provide the reader with a "simple" method for comparing costs. Costs in this context are defined as "the present worth of annual revenue requirements". This method is well suited for predicting costs for the utility sector since it indicates the potential impact of a project on future consumer's electrical rates.

The revenue requirement is the total revenue needed to compensate the utility for all expenditures associated with a project. This is the sum of the carrying charges plus expenses, where the carrying charge is the amount of revenue needed to support an investment and is equal to the sum of the return on debt, return of equity, federal and state taxes, insurance costs and depreciation. Over the life-time of a project the annual carrying charge declines as a result of depreciating the asset. Because this is a stream of payments over time, the effects of inflation and discounting is factored in as well. Levelizing equalizes the revenue requirement for each year over the life-time of the project. The carrying charge rates for different inflation rates were calculated using a modified version of the TAG program ECONCC,[8] adjusted so that the life-time of the cable was specified to be 40 years. The TAG program only allowed for a 30 year maximum book life.

Care should be taken when comparing the results of this study to other cost studies which may have used another type of economic analysis. The strength of this work is the ability to compare alternative cable systems using the same methodology. The complete study, as commissioned by ESEERCO (Empire State Electric Energy Research Corporation), compares the alternative underground cable technologies of AC-SPTL/90°K, DC-SPTL/90°K, AC-SPTL/4.2°K and conventional underground cable. However, this paper is limited to discussions on AC-SPTL/90°K. The interested reader is referred to the program's final report.[9]

The total cost of the cable system is the sum of operating cost (cost of losses) and capital cost. Technical specification for the determinants of cable costs were gathered from best available sources and adjusted by the GNP growth factor to 1989 dollars. These determinants are treated in the model as cost functions which relate component costs to the design parameters such as the power, voltage, physical size, heat in-leak, etc. Therefore, any design change is reflected directly as a cost variation. Figure 4 is a chart of the algorithm for calculating the annualized revenue requirement. Only major cost groupings were considered to retain the generic nature of the study. The cost of series/shunt compensation is negligible for SPTL, but has been included in the study so that comparison with conventional underground cable systems is possible.

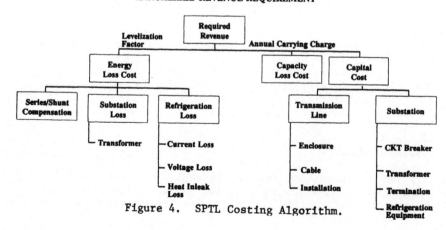

Figure 4. SPTL Costing Algorithm.

The costing algorithm is best reviewed by discussing the key statements in the working program. The required revenue is given by the statement

REQ_REV = LEV_FAC * ELOSS$ + YR_CARRYING_CHARGE * CAP_LOSS$ +
    YR_CARRYING_CHARGE * CAPITAL$

where as stated earlier, the levelizing factor and the yearly carrying charge were determined using EPRI's Technical Assessment Guide. The cost of Losses (ELOSS$) is the product of energy loss times fuel cost, or

ELOSS$ = ELOSS * FUEL$

where

ELOSS = REFR_YR_LOSS + YR_SUB_ENGY_LOSS + YR_SER_COMP_LOSS

In addition to the fuel cost of the lost energy, there are costs associated with the additional system capacity required to offset electrical energy losses. There is a wide range of opinion as to how these costs should be assessed, or if they should be considered at all. A reasonable approach seems to evaluate capacity loss at a value of 80% of creating new generating facilities. The capitalized losses used in the base model is $1200/kw.

Capital costs are the sum of the various equipment and installation costs, where the individual costs functions are given by:

1. ENCLOSURE COST -
   ENCL$ = 2*(R3 * 1070. + 100. * (HT_INTAKE*1000.)**(-0.3))

2. CABLE COST -
   CABLE$ = (CAB_MAT$ + DIE_MAT$) * CAB_LENGTH

3. TRENCH COST -
   TRENCH$ = TR$_PER_VOL * TR_VOL

4. CIRCUIT BREAKER COST:
   TOT_CKT_BRK$ = CKT_BRK$ * NUM_CKT_BRK

5. TRANSFORMER COST:[11]
   TRANSFORM$ = (3.7E6 + 1.7E-3 * POWER) * 1.2

6. TERMINATION COST:
   TOT_TERM$ = TERMINATION$ * NUM_TERMINATION

7. REFRIGERATION EQUIPMENT COST:
   RE_EQP$ = 15000.*(TOT_REFR_LOSS *HT_LOSS_CHA*0.001)**(0.7)

The pie chart of Figure 5 shows a breakdown of the total annualized revenue requirement costs for a 90°K SPTL. The detached section "Energy Cost" represents the cost of losses which for this example consists mainly of the transformer losses, which is large when compared to refrigeration losses. Assumptions in the cost functions are: the installation cost of a rigid, underground pipe was taken to be the same as the installation cost of laying conventional underground cable. For

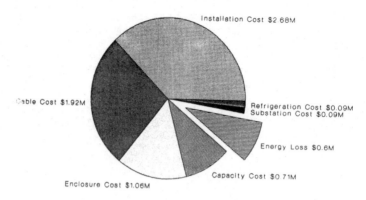

Figure 5. AC-SPTL/90°K Component Cost for the Base Case.

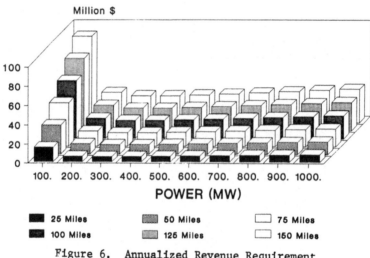

Figure 6. Annualized Revenue Requirement for AC-SPTL/90°K.

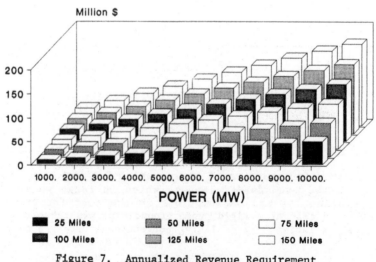

Figure 7. Annualized Revenue Requirement for AC-SPTL/90°K.

the northeast region this cost is roughly $16 per cubic foot of trench volume. The trench volume is the product of Length times Leeway times Depth, where the pipe is buried to a depth of four feet. This cost includes such items as excavation and backfill, cable pulling cost, field installation, monitoring and maintenance. The enclosure cost is a function of the physical size (R3) of the cable and heat in-take (QF). The cost function for the enclosure was taken from a Los Alamos[12] study which has

$$ENCL\$ = \left[1070*2*R3 + \frac{100}{(1000*QF)^{0.3}}\right].$$

This empirical relationship agrees reasonably well with two other cost estimates,[7,13] and has the feature that the more effective thermal enclosures (smaller QF) have higher capital costs.

The cable cost for the high T_c superconductor has been estimated using the cost of low temperature superconductor wire. This cost of $12/M for 3/8" diameter multi-filament wire (which includes copper stabilization) produces a cost function for the conductor of

CAB MAT$ = CAB$_PER_M*2*PI*(R1+R2)*1.3*105.

The above equation has a 30% fabrication charge, while also showing that it takes 105 turns of 3/8" diameter wide wire to make a meter length. The total cable cost is the sum of cable and dielectric materials costs.

The costing algorithm was exercised to generate revenue requirements for a 138 kV SPTL as a function of power levels and lengths. The cable's length was divided into 25 miles modules and the power was varied from 100 to 10,000 MW. Figure 5 shows the results for the power levels of 100 to 1000 MW, and has two interesting features. The increase in cost for the 100 MW line is a direct result of equation [8], which causes an exponential increase in R2 (dielectric thickness) when $R1<(V/\sqrt{3}E_s)$. The relatively large value of R2 for the 100 MW line forces the remaining dimensions to also increase substantially, and thereby raising overall costs.

The flat cost profiles for the 200 to 2000 MW levels also are a result of the radial dimensions of the cable assembly. In these cases the radial sizes of the lines are essentially constant. This constant size produces the observed flat cost profiles. Figure 6 shows the annualized costs increasing with power levels. The rate of increase is less than one, so that on a dollars per kilowatt argument, the larger the power delivery system the more cost effective is the system.

DISCUSSIONS

A "generic" cost model has been developed for superconducting power transmission which associate the electrical, material and physical properties of the cable with the annualized revenue requirement for financing and operating the system. A hypothetical 90°K system enjoys a 10% savings over a similar 4.2°K, 800 MW and 138 kV SPTL. However, the 90°K superconductor has restricted applications as a conductor of choice (a 110°K or higher is preferable), because of thermal management problems when cooling with 77°K liquid nitrogen. The cable assembly used for this study was originally designed for large power delivery (10,000 MW). Consequently, when downsizing power levels to less than 2,000 MW, costs were found to be relatively independent of the power level. The reason for this flat cost response in power systems between 200 to 2000 MW is that the physical size of the cable assembly remain essentially constant for these cases.

Maintenance and right-of-way costs were not considered with this study and any comparison of underground cable to overhead cable transmission must consider the higher right-of-way charges of overhead lines.

Finally, the cost algorithm developed is a relatively simple and direct method of computing the cost of underground cable systems. Uncertainties in the component's cost functions does not allow the projection of absolute costs with a high degree of confidence. However, the model is sufficiently responsive to evaluate alternative electrical transmission scenarios.

ACKNOWLEDGMENT

Special recognition is given to Dr. Reza Ghafurian (Consolidated Edisor Co.) and Shalom Zelingher (New York Power Authority) for the valued guidance they contributed to this program in their capacity as committee member and committee chairperson, respectively.

REFERENCES

1. Philadelphia Electric Co., Evaluation of the Economical and Technological Viability of Various Underground Transmission Systems for Long Feeds to Urban load Areas, U.S. Dept. of Commerce, HCP/T-2055/1 (1977).
2. Thomas, R. A. and E. G. Forsyth, in "Advances in Applied Superconductivity: A Preliminary Evaluation of Goals and Impacts" Chap. 6., Ed., A. M. Wolsky, et al., ANL/CNSV-69 (1988).
3. Bechtel Group, Superconducting Power Transmission Assessment, Empire State Electric Energy Research Corporation, Research Report EP 85-36 (1986).
4. Forsyth, E. B., Energy Loss Mechanism of Superconductors Used in Alternating Current Power Transmission Systems, Science, Vol. 242, (1988).
5. Forsyth, E. B., Superconducting Cables, "Encyclopedia of Physal Science and Technology,: Vol. 13 (1987).
6. Forsyth, E. B. and R. A. Thomas, Performance Summary of the Brookhaven Superconducting Power Transmission System, Cryogenic, Vol. 26 (1986).
7. General Electric Co., Resistive Cryogenic Cable, Phase III, for Electric Power Research Institute, EPRI EL-503 (1977).
8. Technical Assessment Group, TAG™ Technical Assessment Guide, EPRI, p. 4463-SR, Vols. [1-4] (1988).
9. Wegrzyn, J., Superconductivity Applications Assessment, prepared for Empire State Electric Energy Research Corp., Final Report (1989).

10. Walldorf, S. P. and Eich, E. D., "Evaluation of the Cost of Losses for Underground Transmission Cable Systems," *IEEE Transactions on Power Apparatus and Systems*, Vol. PAS-102, No. 10 (1983).
11. Bonneville Power Administration, Superconductivity Applications on an Electric Power System: A System Engineering Study (1988).
12. Laquer, H. L., Electrical Cryogenic and Systems Design of a DC Superconducting Power Transmission Line, *IEEE Transactions on Magnetics*, MAG-13, Vol. 1 (1977).
13. Schauer, F., "Assessment of Potential Advantages of High T_c-Superconductors for Technical Application of Superconductivity, Kernforschungszentrum Karlsruhe, KFK-4308 (1986).

CONTRIBUTORS

Adelmann, P., 273
Akbar, S.A., 719
Alexander, M., 273
Allen, P.B., 647
Alp, E.E., 313
Alterovitz, S.A., 139
Anderson, W.A., 573
Balachandran, U., 265
Ballentine, P.H., 685, 757
Bean, C.P., 767
Belser, M., 653
Bharathi, A., 335
Bhasin, K.B., 735
Blaugher, R.D., 709, 773
Boyce, J.B., 303
Brass, A., 507
Bridges, F., 303
Brooks, K.C., 185
Burns, S.J., 599
Bush, P., 285
Caracciolo, R., 127
Chaki, T.K., 399
Chan, C., 567
Chang, C.-A., 175
Chang, H.L., 405, 677
Chen, F., 541
Chen, J.T., 517
Chen, K., 567
Chern, C., 127
Cheung, N.H., 47
Chi, C.C., 241
Chiang, J.F., 641
Chien, F.Z., 531
Chretien, M.L., 719
Chu, C.W., 335
Chu, M.L., 405, 677
Chung, D.D.L., 581, 661

Claeson, T., 303
Clearfield, A., 591
Cohen, R.E., 463
Coppens, P., 285
Cormack, A.N., 425
Crabtree, G.W., 379
Dabrowski, B., 379
Danyluk, S., 265
Darab, J.G., 441
Decker, D.L., 379
DeMarco, M., 419
Donaldson, W.R., 685
Dwivedi, A., 425
Enyong, J., 635
Farrell, C.E., 175
Feng, A., 185
Fink, J., 273
Funkenbusch, P.D., 599
Gaddipati, A.R., 705
Gallois, B., 127
Gao, Y., 285
Garcia, R., 441, 611
Garhart, J., 185
Geohegan, D.B., 153
George, T.F., 485
Ghosh, A.K., 27
Giessen, B.C., 371, 541, 567
Gilbert, G.R., 591
Goeking, K.W., 591
Goodman, G.L., 313
Gopalakrishnan, S., 411
Gordon, W.L., 735
Goretta, K.C., 265
Gottschalck, K.S., 785
Gou, Y.S., 405, 677
Goyal, A., 599
Graafsma, H., 285

Green, S.M., 389
Greer, J.A., 117
Guan, W.-Y., 249
Gupta, A., 241, 611
Gurvitch, M., 193
Hajek, B., 785
Halbritter, J., 351
Hao, L., 335
Hart, H.R., 557, 705
Hazelton, D.W., 709, 773
Heinen, V.O., 735
Hinks, D.G., 379
Ho, C.T., 581
Holma, M., 185
Hor, P.H., 335
Huang, C.Y., 531
Huang, J., 719
Huang, Z.J., 335
Hyun, O.B., 313
Ignjatovic, D., 139
Jean, Y.C., 335
Jenkins, K.A., 163
Jeong, Y., 695
Jia, Q.X., 573
Jing, T.W., 323
Jorgensen, J.D., 379
Juang, J.Y., 677
Kadin, A.M., 685, 757
Kallin, C., 507
Kaloyeros, A.E., 185
Kao, Y.H., 281, 363
Kawai, T., 61
Kear, B., 127
Kim, H.S., 47
Kirchgessner, J., 705
Kitzmann, G.A., 629
Knorr, D.B., 621
Koren, G., 241
Kraft, D.W., 785
Krakauer, H., 463
Krol, A., 281
Kroon, M., 795
Kwok, H.S., 47, 665
Laibowitz, R.B., 45
Lay, K.W., 557
Lee, H.R., 485
Lee, P., 285
Lee, S.-G., 241
Lee, T., 795

Licata, T.J., 175
Liebermann, R.C., 647
Lin, C.S., 281
Lin, D.L., 475
Liu, S.H., 455
Liu, X., 647
Liu, Y.P., 567
Logothetis, E.M., 517
Luo, H.L., 389
Ma, Q.Y., 175
MacCrone, R.K., 441
Maheswaran, B., 567
Mallory, D.S., 757
Manzi, A.E., 389
Markiewicz, R.S., 371, 541, 567
Mashburn, D.N., 153
Mattocks, P.G., 419
McGuire, M.J., 265
Mei, Y., 389
Meola, J.E., 163
Meng, R.L., 335
Mihaly, L., 647
Ming, Z.H., 281
Mini, S.M., 313
Miranda, F.A., 735
Moffat, D., 705
Moodenbaugh, A.R., 27
Nakai, S., 273
Narayan, J., 71
Narumi, E., 109, 665
Naughton, M., 419
Nigli, S., 591
Noh, D.W., 127
Norris, P., 127
Nücker, N., 273
Ong, N.P., 323
Padamsee, H., 705
Pandey, R.K., 591
Papaconstantopoulos, D.A., 463
Patel, S., 99, 109, 665
Pei, S., 379
Phillips, J.C., 41
Pickett, W.E., 463
Poeppel, R.B., 265
Qian, L.-X., 517
Raeder, C.H., 621
Rajan, K., 441, 611
Ramanathan, M., 313
Rath, R.C., 757

Rice, J.A., 773
Richards, D.R., 379
Romberg, H., 273
Rose, K., 695
Rossi, D.V., 175
Rottersman, M., 757
Rowe, L.E., 629
Royston, J.D., 727
Rubin, D., 705
Schluter, M., 1
Schmidt, M.T., 175
Schulze, W.A., 411
Segmüller, A., 241
Shah, A., 99
Shaw, D.T., 47, 99, 109, 281, 665
Sher, C.J., 281
Shi, D., 265
Shu, Q.S., 705
Singh, D., 463
Singh, R.K., 71
Smith, G.C., 281
Song, L.W., 281, 363
Squattrito, P.J., 591
Sridhar, S., 207
Stan, M.A., 139
Suenaga, M., 27
Sundar, C.S., 335
Tabata, H., 61
Tarascon, J.M., 323
Taylor, J.A.T., 547
Thomas, R., 795
Ting, C.S., 495
Tkaczyk, J.E., 557
Tofte, R.J., 629
Tompa, G.S., 127
Trbovich, G., 419
Tseng, S.C., 399
Tzeng, Y., 653
Uen, T.M., 405, 677
Uher, C., 217
Valco, G.J., 735

Vandervoort, K.G., 379
Vanfleet, H.B., 379
Van Hook, J., 169
Vijayaraghavan, R., 611
Walker, M.S., 709
Wang, C., 405, 677
Wang, E., 323
Wang, J.G., 749
Wang, L.-Q., 517
Wang, T.C., 405
Wang, X.W., 419, 727
Wang, Z.Z., 323
Warner, J.D., 163, 735
Wegrzyn, J., 795
Welch, D.O., 27
Weng, Z.Y., 495
Wenger, L.E., 517
Wenxi, L., 635
Whiteley, B.A., 547
Williams, W.S., 185
Witanachchi, S., 665
Woo, W.E., 629
Wu, M.K., 531
Wu, W., 507
Xu, Y., 27
Xu, Y.-H., 249
Yang, E.S., 175
Yang, R.T., 749
Ye, J., 285
Ying, Q.Y., 47
Youngdahl, C.A., 265
Yuanfang, Q., 635
Yubin, W., 635
Zawadski, P., 127
Zeibig, K., 249
Zheng, H., 475
Zheng, J.P., 47
Zhian, Y., 635
Zhiying, Y., 635
Zhu, Y., 27
Zhu, Y.Z., 281, 665

INDEX

Absorption spectroscopy (*see* X-ray absorption spectroscopy)
Accelerator, projectile, 727-732
Accelerator cavities, 705-707
AC effects
   structural defects, 441
   Tl-based bridges, 677-682
AC losses, 258, 259, 261, 262
Acoustic response, YBCO, 241-247
Aligned materials (*see* Magnetic ordering)
Andreev reflection, 325-330
Anisotropy
   critical currents in aligned sintered YBCO, 557-565
   thermal transport mechanisms, 220, 221, 233
Annealing
   and interface problem, 323
   film-substrate interdiffusion, 169-174
   YBCO
      enhancement of critical current, 567-572
      LT-MOCVD, 186-188
      substrate and process effects, 175-182
Anodic oxidation, precursor, 641-646
Anomalies
   scattering of Bi distribution in 2212 materials, 286-291
   tunneling in NCCO, 331-332
Antiferromagnetic state, 485-493
   correlation effects, 470-472
   oxygen defects, phase separation, and superconductivity, 379-386
Applications, 653
   accelerator cavities, 705-707
   fast optically triggered switches, 685-692
   microwave devices
      electrodynamic properties, 213, 214
      millimeter wave transmission studies, 735-746
   power transmission line cost assessment, 795-807
   projectile accelerator, 727-732
   radiation detection mechanisms, 695-703
   single crystal growth of YBCO and BSCCO, 591-597
   weak links and planar defects, 351-361

$Ba_2Tl_2CuO_6$, 5, 7
$Ba_6K_4BiO_3$, 343
Ball milling, 441-452
Band structure, 273-280 (*see also* Electronic structure)
   copper oxides, 4, 5
   correlation effects, 464-466
Bardeen, Cooper, and Schreiffer (BCS) theory, 475
   electrodynamics of YBCO, 208-209, 212, 214
   quasiparticle tunneling, 200-203
   theory of strongly fluctuating superconductivity, 455-460
Bardeen, Rickhyazen and Tewordt (BRT) theory, 219-220, 233
B-H curves, 442, 443
$Bi_2Ca_2Sr_2Cu_2O_8$, 425-439
$Bi_2Ca_2Sr_2Cu_3O_{10}$, 425-439
$Bi_2Sr_2CuO_6$
   electronic properties, 20
   single crystal growth, 591-592, 593-594, 595, 596
   superstructure energetics, 425-439
Bi-series (*see also* BSCCO)
   Bi-Ca substitutions, BSCCO, 393-395
   Bi-O interactions, 429-430
   BPSCCO, grain boundary diffusion, 399-403
   lattice energy calculations, 428-429
   and layered cuprates, 463
   rare earth element substitutions, magnetodiffractometry, 541-546
   superstructure energetics, 425-439
Bloch-Gruneisen transport theory, 470
Bolometric detection
   fast optically triggered switches, 692
   photodetection, 695-703
Born-Mayer central force potential, 427-428
Born model, 427-429
Boundary layer thickness, width of, 36
BPSCCO (*see* Pb-BSCCO)
Breathing mode in $Ba_{1-x}K_xBiO_3$ compounds, 495-503
Bridges
   oxygen-stabilized zero-resistance states in YBCO, 517-528

Bridges (continued)
  Tl-based, 677-682
Brittleness, 581-582
BSCCO (*see also* Bi-series)
  composition, 389-396
  construction and structural relaxation, 635-638, 639
  critical current anisotropy, 557
  doping, 399
  fabrication, 109-115
  heat conduction mechanism, 235-236
  positron annihilation studies, 340-341
  progress in modeling, 434
  site-selective substitution in tailored thin films, 61-69
  structure
    complications, 285-300
    Cu-O plane buckling in, 431-439
    electronic, 273, 276, 277
    microstructure and electrical properties, 265-270
    vortex glass, 406
    X-ray adsorption techniques, 281-284
  superstructure energetics, 425-439
  thermal transport mechanisms, 234
Bulk materials
  heat conduction mechanisms (*see* Heat conduction mechanisms)
  surface passivation, 573-578
Bulk superconductivity, magnetization studies, 249-265

$Ca_2Ba_2Tl_2Cu_3O_{10}$, 5
$Ca_2Sr_2Tl_2Cu_3O_{10}$, 5
$CaBi_2Sr_2Cu_2O_8$, 6
Carbonate phase, decomposition of tape cast YBCO, 547-554
Carbon fibers
  wires, spray deposition, 749-755
  tin composites, 584, 585, 586, 587, 588
Carriers
  local pairing and antiferromagnetism, 485-493
  thermal transport mechanisms, 226, 227, 231
Cavities, accelerator, 705-707
Cell dimensions, BSCCO crystals, 297
Channeling, YBCO epitaxial growth studies, 91, 92, 93
Charge
  copper compounds, 315, 319
  dopant, classification by, 376
  X-ray absorption spectroscopy
    doping effects, 316-318, 318-320
    La_2CuO_4 systems, 319-320
    principles, 314-316
    YBCO systems, 318-319
Charge density wave (CDW) state, $Ba_{1-x}K_xBiO_3$, 498-503
Charge transfer gap, scaling, 12, 13
Chemical substitution (*see* Substitution)
Chemical vapor deposition, 127-136, 185-190

Chloride salts, YBCO decomposition, 622
Circular coils, large bore, 715-716
Classification, structural, 371-377
  functional groups, 373-376
  types of doping, 371-373, 376
Cluster sequence mapping, 14, 16, 18-21
Coatings, sintered
  composite tape fabrication, 777-778
  fluidized-bed deposition, 785-792
Cobalt doping, YBCO, 308-310, 311
Coherence length
  and flux pinning, 29
  substitution and, 31
Coherent potential approximation, 466
Coils, 715-716, 727-732
Composites
  sintering, doping effects, 599-610
  YBCO
    carbon fiber reinforced, 581-589
    magnetic field effects, 653-659
Composition (*see also* Substitution)
  BSCCO
    Bi-Ca, subsitutions between, 393-395
    oxygen, 395-396
    Pb replacement of Bi, 391
    Sr/Ca ratio, 391-393
  oxygen content (*see* Oxygen stoichiometry)
Compressive forces, carbon fiber reinforced tin composites, 584, 585, 586, 587, 588
Conductivity, optical, 1
Constrained Density Functional (CDF) approach, 8-9
Cooper pairs
  Andreev reflection spectrum, 325-330
  Josephson tunneling, 197
  quasiparticle tunneling, 198
  thermal transport mechanisms, 220
  tunneling, weak links, 357-358
  YBCO, 194
Coordination numbers, copper compounds, 319
Copper-oxygen-copper angles, and copper displacements, 296
Correlation effects
  antiferromagnetic insulating phase, 470-472
  disordered phases, 466-467
  size of, 463-472
  spectroscopy, 461-466
  structural energies and phonon frequencies, 467-469
  transport properties, 470
  weak links, 357-358
Coulomb interactions, local pairing and, 485-473
Counter-ion doping, 373-376
Cracks
  path systems, 353
  YBCO film fracture behavior, 661-663
Creep, weak links and, 356
Critical current density
  carbon fiber reinforced tin composites, 584
  flux line free energy and, 32

Critical current density (continued)
   magnetic hysteresis and, 255-258, 259
   YBCO
      anisotropy in aligned sintered YBCO, 557-565
      field orientation and, 567-572
      flux pinning and, 28-32
      molten salt processing, 621-627
      polycrystalline, 139-150
      pure vs. alloyed, 27-28
      silver doped, 363-369
Critical field
   design issues, 713
   electrodynamics of YBCO, 211
   and flux pinning, 29
Critical state model, response to step in magnetic field, 767-772
Critical temperature, squeezed polaron model, 478-479, 480, 481
Crystal structure (*see* Structure)
Curie-Weiss susceptibility, 262, 264
Current-voltage characteristics (*see also* Critical current density; Electrical transport)

Decomposition, tape cast YBCO, 547-554
Defect-assisted percolation model, 41-43
Defect-enhanced electron-phonon interactions, 42
Defects
   EPR spectra, 441-452
   doping effects, 611-618
   oxygen, and phase separation, 379-386
   planar, 351-361
   positron annihilation studies, 345-346
   and thermal transport mechanisms, 229
   thin films, Tl-Ca-Ba-Cu-O, 405-410
   tunneling, Josephson, 196
   and X-ray absorption, 284
Density functional method, 2-4, 5, 6, 463-472
Density of states
   vs. local density aproximation, 273
   quasiparticle tunneling, 200
   weak links, 361
Deposition (*see* Fabrication)
Design, 709-717 (*see also* Applications; Fabrication)
   films on wires, 719-724
   tape, composite, 773-782
Dielectric function, YBCO, 12, 13
Differential thermal analysis, YBCO, 629-634
Diffusion, film-substrate, 169-174
Diffusion bonding, 581-589
Dimensionality, and metallic conductivity, 42
Disorder
   correlation effects, 466-467
   thin films, Tl-Ca-Ba-Cu-O, 405-410
Doctor blading, 547-554
Domain structure formation, YBCO, 614
Doping
   $Ba_{1-x}K_xBiO_3$, softening of breathing mode, 498-503

Doping (continued)
   BPSCCO, and grain boundary diffusion, 399-403
   BSCCO
      atomic modulation amplitudes, 295
      temperature dependence of modulation, 300
   and charge on copper in $Nd_{2-x}Ce_xCuO_4$, 316-318
   EELS spectrum, 464, 465
   electronic structure, 273-280
   energy spectra of phases, 16, 17
   low-temperature in-situ processing, 91, 94, 97
   magnetic field dependence of critical current density, 363-369
   modulation wave vector, 294
   Mossbauer effect studies, 419-423
   $Nd_{2-x}Ce_xCuO_4$, 316-318
   positron annihilation studies, 338-339, 343, 344
   and sintering of composites, 599-610
   and thermal transport mechanisms, LSCO, 229-230
   types of, 371-373
   X-ray absorption, local structure and distortions, 303-311
Ductility, carbon fiber reinforced tin composites, 581, 582, 584, 585, 586, 587, 588

Effective charge
   copper compounds, 319
   X-ray absorption spectroscopy, 313-321
Electrical transport (*see also specific materials*)
   carbon fiber reinforced tin composites, 582-584
   films on wires, 724
   magnetic field dependence of critical current density, 363-369
   oxygen-stabilized zero-resistance states in YBCO, 517-528
   structure and in bulk HTS, 265-270
   Tl-based bridges, 677-682
   Tl-Ca-Ba-Cu-O thin film, 406-409, 410
   weak links and planar defects, 351-361
   YBCO films, polycrystalline, 139-150
   YBCO-silver interfaces, magnetic field effects, 653-659
   YBCO, 531-538
   weak links and planar defects, 351-361
Electrodynamics
   cavities and fields, 213, 214
   intrinsic vs. granular response, 213-214
   lower critical field, 211
   microwave measurements, 207-214
   penetration depth, 209-211
   surface resistance, 211-212
Electron beam evaporation, YBCO deposition, 175-182
Electron energy loss spectroscopy (EELS), correlation effects, 464-466
Electronic structure, 273-280
   Andreev reflection, 325-330
   composite tape, 776

Electronic structure (continued)
  disordered phases, 466-467
  positron annihilation studies, 335-347
  tunneling in NCCO, 330-332
  X-ray absorption techniques, 281-284
Electronic structure of copper oxide materials
  density functional description, 2-4, 5, 6, 7
  mapping procedures, 4, 7-21
    finite cluster results, 14, 16, 18-21
    generalized Hubbard Hamiltonian, 4, 7-13, 14, 15, 17
Electron-phonon interactions
  defect-enhanced, 42
  fast optically triggered switches, 685-692
  squeezed polaron model, 475-483
Electrostatic fluidized-bed deposition, 785-792
EMCORE System 5000, 127-136
Energy density, pulsed laser, 84-87
Epitaxial growth
  methods of study, 91, 92, 93
  temperature limits, 96
Er-containing superconductors, magnetization studies, 249, 253, 255-265
Etching, quasiparticle tunneling, 198-199
Eu-containing superconductors, magnetization studies, 249-265
Extended impurity band states, 41
Extended X-ray absorption fine structure, 290

Fabrication (see also Laser ablation)
  anodic oxidation of precursor, 641-646
  ball milling, microstructural defects, 441-452
  BSCCO, 635-638, 639
  chemical vapor deposition (MOCVD), 127-136, 185-190
  $Cu_xNa_8O_t$, high pressure synthesis, 641-646
  design requirements, 709-717
    large bore circular cores, 715-716
    mechanical stresses, 712-715
    operating conditions, 712
  films on wires, 719-724
  fluidized-bed deposition, 785-792
  fracture behavior of films, 661-663
  laser ablation
    in-situ processing, 87-94
    neutral and ion transport during laser ablation, 153-161
    technique, 72-73, 74
    theoretical model, 73, 75-81
  laser deposition
    BSCCO films, 109-115
    physics of, 47-59
    YBCO films on stainless steel substrates, 665-666
  molten salt processing, 411-418, 621-627
  single crystal growth of YBCO and BSCCO, 591-597
  spray deposition, 749-755
  sputtered films, microwave surface resistance measurement, 757-765

Fabrication (continued)
  SQUIDS, 45-46
  surface passivation, 573-578
  tailored thin films, site-selective substitution, 61-69
  tapes
    composite, 773-782
    removal of organics from, 547-554
    stainless steel substrates, 665-675
  wires, spray deposition, 749-755
Fatigue, thermal, 587
Fermi surface, positron annihilation studies, 344-345
Fibers (see Carbon fibers)
Filamentous flow, T_c and, 139-150
Films
  acoustic response, 241-247
  electrodynamic properties, 214
  fast optically triggered switches, 685-692
  fracture behavior, 661-663
  in-situ diagnostics, 56-59
  microwave surface resistance measurement, 757-765
  millimeter wave transmission studies, 735-746
  physics of, 54-56
  radiation detection systems, 695-703
  surface passivation, 574, 576-578
  Tl-Ca-Ba-Cu-O, dissipative structures in, 405-410
  weak links, 359-361
  on wires, 719-724
  X-ray absorption techniques, 281-284
Finite cluster mapping, 14, 16, 18-21
Finite systems, semion pairing, 507-515
Fluctuation dynamics, 455-460
Fluidized-bed deposition, 785-792
Flux creep, 32-35
Flux flow resistivity, response to step in magnetic field, 767-772
Flux pinning
  microstructures and, 35-38
  RE-123/AgO composites, 534, 535
  temperature dependence, 33
  thermodynamic properties and, 38
  YBCO, 28-32
Flux quantization, 510
Functional groups, structural classification, 373-376

Gaps
  electrodynamic properties, 214
  quasiparticle tunneling, 200-202
Gd-containing superconductors, magnetization studies, 249, 253, 255-265, 545
Generalized Hubbard Hamiltonian, 4, 7-13, 14, 15, 17
Ginzburg-Landau description, flux pinning, 29-30
Ginzburg theory, 455-460
Grain boundaries
  BPSCCO, 399-403

Grain boundaries (continued)
    insulators, 353-355
    magnetic field dependence of critical current density, 363-369
    tunneling, 193
Grain growth in RE-123, silver and, 534-538
Grain orientation, YBCO, 411-418
    critical current anisotropy, 557-565
    molten salt processing, 621-627
    rf surface resistance, 705-707
Ground state neutrals, during laser ablation, 153-161
Growth anisotropy, molten salt processing, 621-627

Halides, YBCO molten salt processing, 622, 623, 624, 625, 626-627
Heat
    and defect structure and phase stability, 611-618
    differential thermal analysis of YBCO complex, 629-634
    power transmission line cost assessment, 796-801
Heat conduction mechanisms, 217-237
    experimental technique, 218-219
    theoretical aspects, 219-221
    thermal conductivity, 221-236
        BSCCO, 235-236
        LSCO, 229-235
        YBCO, 221-229
High pressure synthesis, $Cu_rNa_sO_t$, 641-646
Holes
    copper oxide materials, 4, 8
    doped
        EELS spectrum, 464, 465
        t-t'-J model, 17, 18
    localization onto distinct Cu (II) sites, 449
Hole superconductors, 486
Holstein-Primakoff transformation, 459
Hopping parameters, 485-493
Hubbard Hamiltonian, generalized, 4, 7-13, 14, 15, 17, 18
Hydrodynamic model, thin film deposition, 94-96
Hydrofluoric acid treatment, YBCO, 573-578

Impurity band states, 41
In-situ deposition, 47-59, 109-115
In-situ oxidation, 665-675
In-situ TEM, 611-618
Insulators
    mapping, 14, 15
    metal-insulator transition in Si:P, 41
Interatomic distances, copper compounds, 319, 320
Interatomic potential model, Bi-series compounds, 429-430
Intercalation layer doping, 371-376
Interface behavior
    film-substrate, 169-174

Interface behavior (continued)
    magnetic field effects, 653-659
    tunneling, 193-194
    YBCO tapes, 665-675
Interface resistance
    Andreev reflection, 332-335
    tunneling, 330-332
Ion channeling, 91, 92, 93
Ionization potentials, transition metals, 2, 3
Ion transport during laser ablation, YBCO, 153-161
Iron, 310, 311
Irradiation, and YBCO thermal transport mechanisms, 229
Isotope effect, squeezed polaron model, 482-483

Josephson effect, 193-198
    silver doping and, 363-369
    theory of strongly fluctuating superconductivity, 455-460
    Tl-based bridges, 677-682
Josephson junction field effect transistor, 591
Junctions, tunneling, 193, 193-194, 195

Knudsen layer, 159
Kosterlitz-Thouless transition, 405-410
    Tl-based system, 681-682
    Tl-Ca-Ba-Cu-O thin film, 409

$La_{3-x}Ba_{3+x}Cu_6O_{14-delta}O_{4-delta}$, 42
$(La_{1-x}Sr_x)_2CuO_4$ (see LSCO)
La_2CuO_4
    electron and hole doped, t-t'-J model, 17
    insulating, Raman scattering data, 14, 15, 16
    interatomic potential model, 430
    Local Density Functional band structure, 3, 4
    three-band Hubbard model, 12
$La_2CuO_4Nd.Nd_2CuO_4$ systems, 319-320
$La_2MO_{4+delta}$, 379-386
$LaAlO_3$, 735-746
Lanthanide rare-earth ions (see Magnetization studies; YBCO)
Large bore circular core design requirements, 715-716
Laser ablation
    atomic oxygen source, 109-115
    deposition characteristics, 81
    deposition dyamics, 82-84
    films, 735-746
    in-situ processing, 87-94
    neutral and ion transport during, 153-161
        ion probe measurements, 159-161
        optical absorption technique, 155-157
        transport measurements, 157-159
    pulse energy density effects, 84-85
    site-selective substitution in tailored thin films, 61-69
    spatial compositional dynamics, 85-87
    SQUIDS, 45-46
    theory, 73, 75-81

Laser ablation (continued)
    zirconia, deposition on, 163-167
Laser deposition
    BSCCO films, 109-115
    fracture behavior of films, 661-663
    physics of, 47-59
        free expansion zone, 51-54
        in-situ diagnostics, 56-59
        laser-target interaction zone, 49-51
        schematic of process, 47-49
        thin-film formation, 54-56
Lattice constant, composite tape, 776
Lattice energy calculations, Bi-series, 428-429
Lattice instabilities, quantum percolation theory, 41-43
Layer number, modulation dependence in BSCCO materials, 292-294
LEDE process, 48-49
Linearized Augmented Plane Wave (LAPW) calculation, 465, 467-469
Linear muffin-tin orbital (LMTO) technique, 9
Local density approximation (LDA)-type density functional calculations, 2-4, 5, 6, 7, 273
    antiferromagnetic insulating phase, 470-471
    correlation effects, 463-472
Localization, electron, 42
Localized impurity band states, 41
Localized states, weak links, 358
Local spin density approximation, self-interaction correction to, 471-472
Low-pressure metal-organic chemical vapor deposition (LP-MOCVD), 127-136
Low-temperature metal-organic chemical vapor deposition (LT-MOCVD), 185-190
LSCO
    doping, 373-376
    heat conduction mechanism, 229-235
    positron annihilation studies, 340, 344-345

Madelung potential, 9
Madelung shifts, 469
Magnetic flux pinning, 28-32
Magnetic ordering
    thermal transport mechanisms, 232-233
        LSCO, 229-230
        YBCO, 227
    YBCO
        critical current anisotropy, 557-565
        critical current enhancement, 567-572
        sintered samples, rf surface resistance, 705-707
Magnetic susceptibility
    BSCCO systems, composition and, 389-396
    oxygen-stabilized zero-resistance states in YBCO, 517-528
Magnetization studies, 249-265
    RBCO
        ac losses, 258, 259, 261, 262
        coexistence of superconductivity and paramagnetism, 258, 262, 263, 264

Magnetization studies (continued)
    RBCO (continued)
        magnetic hysteresis and critical current densities, 255-258, 259
        types of superconductivity, 251-255
    RE-123 and composites, 534-538
    response to step in, 767-772
    semion pairing in finite systems, 507-515
    YBCO, silver-doped
        and critical current density, 363-369
        YBCO-silver contacts and alloys, 653-659
Magnetodiffractometry of Bi and Tl cuprates containing rare earth elements, 541-546
Magnetron sputtering
    microwave surface resistance measurement, 757-765
    YBCO thin films, 579-580
Mapping
    finite cluster results, 14, 16, 18-21
    generalized Hubbard Hamiltonian, 4, 7-13, 14, 15, 17
Mean field approximation, strongly fluctuating superconductivity, 455-460
Mechanical properties
    carbon fiber reinforced tin composites, 581-589
    design issues, 709-717
    films on wires, 719-724
    tape, composite, 773-782
    weak links and planar defects, 351-361
Meissner effect, 609
Melt processing, composite tape fabrication, 778-780, 781
Metal-insulator transition in Si:P, 41
Metallic conductivity, dimensionality and, 42
Metalorganic chemical vapor deposition, low-pressure, 127-136
Metalorganic chemical vapor deposition, low-temperature, 185-190
MF-Hubbard model, 9-10
Mg doping, YBCO defect structure and phase stability, 611-618
MgO buffer layer, 776, 779
MgO substrates
    interdiffusion, 169-174
    structure, X-ray absorption techniques, 281-284
    tapes, vapor deposition on, 781
    YBCO films
        fracture behavior, 661-663
        millimeter wave transmission studies, 735-746
        rapid thermal annealing, 179, 180, 181
Microbridges, 355, 356, 517-528
Microcracks, 193, 196
Microstructure (see also Structure)
    and electrical properties of bulk HTS, 265-270
    Type-A composites, 600
    X-ray absorption techniques, 281-284
    YBCO
        doping effects on defect structure and phase stability, 611-618

Microstructure (continued)
  YBCO (continued)
    EPR monitoring, 441-452
Microwave devices
  cavities and filters, 213, 214
  films, millimeter wave transmission studies, 735-746
Microwave radiation
  electrodynamics of YBCO surface resistance, 211-212
  stripline resonator measurements, 757-765
  Tl-based system, current-voltage characteristics, 680-681
Millimeter wave transmission studies, YBCO films, 735-746
Milling time, and EPR spectra, 443, 444, 445
Modulated solids, 291-296
Molecular beam epitaxy, 56
Molten salts
  grain orientation, 411-418
  stability of yttrium barium cuprate, 621-627
Molten silver processing, 653
Mossbauer effect, 419-423
Multilayer oxide tapes, 665-675

$NaCuO_2$, 319
NCCO
  Andreev reflection in single crystals, 330-332
  copper charge, 316-318
  electronic structure, 278
  positron annihilation studies, 343, 344
  tunneling spectrum in, 330-332
Nd_2CuO_4, 4, 5
Neutral and ion transport during laser ablation, 153-161
Neutron diffraction
  BSCCO single crystals, 294
  oxygen stoichiometry, 382-385
Neutron irradiation, and thermal transport mechanisms, 229
Neutron scattering, LDA calculations of structural energies and phonon frequencies vs., 467-469
Nickel doping, 310, 311
Ni composites, 776, 773-782
  physical properties, 776
  silver doping, 603, 608-610
  sintered coatings on, 777
  vapor deposition on, 781
Nitrate salts
  YBCO molten salt processing, 622, 624
  wire spray deposition, 749-755
NMR, 1
  energy spectrum clusters, 18
  temperature dependence of relaxation rates, 21
Non-local change in potential, 469
Normal state resistivity, 1
N-type superconductors, charge carrier location, 486

Off-diagonal long range order, semion pairing in finite systems, 510, 512
One-band Hubbard model, 17
One-dimensional band, YBCO, 4
One-electron parameters, LDA bandstructure, 8
One-electron potential, 469
Operations (*see also* Applications; Mechanical properties)
  design issues, 709-717
  weak links and planar defects, 351-361
Optical absorption spectroscopy, neutral and ion transport during laser ablation, 153-161
Optical conductivity, 1
Optically triggered switches, 685-692
Organics, removal from tape cast YBCO, 547-554
Oriented powders
  $La_2CuO_4.Nd_2CuO_4$ systems, 319-320
  weak links, 353-355
  YBCO, X-ray absorption spectroscopy, 318-319
Oxidation
  precursor, 641-646
  stainless steel substrate, 665-675
Oxygen annealing, YBCO, 567-572
Oxygen-stabilized zero-resistance states, YBCO, 517-528
Oxygen stoichiometry
  and acoustic response, 245, 246, 247
  and Andreev reflection, 330
  BSCCO, 294, 296, 395-396
  doping, types of, 371-373
  phase separation and superconductivity, 379-386
    neutron powder diffraction, 382-385
    theromgravimetric analysis, 381-382
    transport properties, 385-386
  doping, types of, 371-373
  positron annihilation spectroscopy, 345
  thermal transport mechanisms, 221-222
  YBCO
    local structure and distortions, X-ray absorption studies, 303-311
    Mossbauer effect studies, 419-423
    thin film fabrication technique, 109-115
  and X-ray absorption, 284

Pairing
  local, 485-493
  semion, in finite systems, 507-515
Pair potentials, Bi-series, 430
Paramagnetism
  coexistence with superconductivity, 258, 262, 263, 264
  in $ErBa_2Cu_6O_{7-y}$, 249, 250
Particle accelerators, 705-707
Passivation, YBCO, 573-578
Pb, tunneling in NCCO, 330-332
$Pb_2CuSr_2O_x$, 375
Pb-BSCCO
  grain boundary diffusion, 399-403

Pb-BSCCO (continued)
    microstructure and electrical properties, 265-270
    powder x-ray diffraction, 391, 392
Peak splitting, BSCCO, 297, 298, 299
Percolation model, 41-43
Perturbation theory, fourth-order, 488
Phase behavior
    differential thermal analysis of YBCO, 629-634
    doping effects in YBCO, 611-618
    modulation amplitudes in BSCCO, 294
    oxygen defects and, 379-386
    theory of strongly fluctuating superconductivity, 455-460
Phase locking model, 140
Phase transitions
    BSCCO materials, 287-300
    Tl-based system, 681-682
Phonons (*see also* Electron-phonon interactions)
    $Ba_{1-x}K_xBiO_3$, 498-503
    correlation effects, 467-469
    thermal transport mechanisms, 220, 224, 229
Photodetection, radiation detectors, 695-703
Pinning, flux, 28-32
    RE-123AgO composites, 534, 535
    weak links and, 356
Plasma
    laser-target interaction zone, 49-51
    neutral and ion transport during laser ablation, 153-161
    pulsed laser deposition, 75, 76, 77, 78-79
Plasma-assisted laser deposition
    fracture behavior of films, 661-663
    YBCO films, 99-106
Polarized X-ray absorption spectroscopy, 318-320
Polarons, strong-coupling narrow-band systems, 475-476
Polycrystalline materials
    critical current anisotropy, 558
    thermal transport mechanisms, 218, 235
    weak links, 353-355, 359-361
    YBCO
        electrodynamics of YBCO, 139-150, 212
        fracture behavior, 661-663
        rf surface resistance, 705-707
Positron annihilation studies, 335-347
    Fermi surface measurements, 344-345
    new materials, 340-344
    structural and defect properties, 345-346
    temperature, 336-3540
Potential models, Bi-series compounds, 427-430
Powder metallurgy, 601
    BSCCO, neutron diffraction data, 294
    $La_2CuO_4.Nd_2CuO_4$ systems, 319-320
    YBCO, X-ray absorption spectroscopy, 318-319
Power transmission line cost assessment
    costing algorithm, 801-805
    thermal management, 796-801
PrBCO, tunneling, 197

Precursors
    anodic oxidation, 641-646
    doping effects, 611-618
Processing (*see also* Annealing)
    bulk materials, microstructure and electrical properties, 265-270
    composite tape fabrication, 778, 779-780, 781
    YBCO
        molten salt, 621-627
        surface passivation, 573-578
Projectile accelerator, 727-732
    barrel, 730
    electromagnetic coil, 731
    electronics, 731
    system design, 728-729
Properties
    acoustic response, 241-247
    charge determination on copper, 313-320
    electrical, 265-270
    local pairing and antiferromagnetism, 485-493
    magnetization, 249-265, 653-659
        magnetic field dependence of critical current density, 363-369
        response to step in magnetic field, 767-772
    Mossbauer effect studies, 419-423
    positron annihilation studies, 335-347
    thermal conductivity, 217-237
    weak links and planar defects, 351-361
Pseudo-spin formalism, strongly fluctuating superconductivity, 456-460
Pulsed laser ablation (*see* Laser ablation)

Quantum chemistry, copper oxide materials, 2-4, 5, 6
Quantum Interference Devices, 45-46
Quantum percolation theory, 41-43
Quantum statistics, semion pairing in finite systems, 507-515
Quasiparticles
    Andreev reflection spectrum, 325-330
    band structure, angle-resolved photoemission spectroscopy, 465-466
    tunneling, 198-203, 406-410
Quasi-two-dimensional system, Tl-Ca-Ba-Cu-O thin film, 409

Radiation detection mechanisms, 695-703
    resistance mechanisms, 699, 702-703
    R(II,T) characteristics, 696-699, 700, 701
Radiofrequency fields
    electrodynamic response of YBCO, 207-214
    magnetically aligned sintered YBCO, 705-707
    stripline resonator measurements, 757-765
Radiofrequency sputtering, 579-580
    and Andreev reflection, 330
    microwave surface resistance measurement, 757-765
    YBCO films, 99-106
Radiofrequency transmission, millimeter wave transmission studies, 735-746

Raman scattering, 1
   cluster sequence mapping, 14, 15, 16
   LDA calculations of structural energies and phonon frequencies vs., 467-469
   YBCO, 18, 19
Rapid thermal annealing, 175-182
Rare-earth compounds (*see* Magnetization studies; YBCO)
Rare earth elements
   Bi and Tl cuprates, magnetodiffractometry, 541-546
   subsititutions
      charge determinations, 319-320
      silver oxide composites, 534-538
Reflective high energy electron diffraction (RHEED), 56
Resistivity
   $Bi_2Sr_2CuO_6$, temperature and, 20
   in-situ measurement, 57
   normal state, 1
Resonant tunneling, 358
Ruddelsden-Popper compounds, 426, 431
Rutherford backscattering spectroscopy, 91, 119

Sandwich heterostructures, tunneling, 194
Sapphire substrates
   composite tape, 776, 779-780, 781
   interdiffusion between films and, 169-174
   rapid thermal annealing, 179, 180, 181
Satellite reflections, lead doped BSCCO, 300
Selective atom diffraction, 285-301
Self-interaction correction to local spin density approximate, 471-472
Semion pairing in finite sytems, 507-515
Shape anisotropy, molten salt processing, 621-627
Silicon substrate, 179, 180, 181
Silver doping (*see also* Doping)
   critical current density, magnetic field dependence of, 363-369
   magnetic field effects, 363-369, 653-659
   and sintering of composites, 599-610
Silver substrate, composite tape, 776
Single carriers, 485-493
Single crystal diffraction techniques, BSCCO characterization, 285-301
Single crystals, 713
   BSCCO
      construction and structural relaxation, 635-638, 639
      thermal transport mechanisms, 234
   electrodynamic properties (*see* Electrodynamics)
   plasma-assisted laser deposition of YBCO film onto, 665-666
   thermal transport mechanisms
      BSCCO, 234
      LSCO, 232
      YBCO, 226, 227
   tunneling, YBCO and NCCO, 330-332
Single-phase oxide superconductors, 531-538

Single-wavelength X-ray diffraction, 286
Sintered materials, 581
   composite tape, 777-779
   doping and, 599-610
   fluidized-bed deposition, 785-792
   grain boundary diffusion in, 399-403
   magnetization studies, 249-265
   processing, microstructure and electrical properties, 265-270
   thermal transport mechanisms
      BSCCO, 234
      YBCO, 221-229
   YBCO
      critical current anisotropy, 557-565
      resistive transition in, 144-149, 150
      rf surface resistance, 705-707
      thermal transport mechanisms, 221-229
      weak links, 361
Site occupancies, BSCCO, 291
Solid state reaction, $Cu_rNa_sO_t$, 641-646
Specific conductance, $Bi_2Sr_2CuO_6$Pb tunneling junction, 20
Specific heat, 43
Spectroscopy (*see also specific methods*)
   correlation effects, 461-466
   materials evaluation (*see specific fabrication methods*)
Spherical topology, semion pairing in finite systems, 507-515
Spinel phase, film-substrate interdiffusion and, 169, 173, 174
Splitting, BSCCO peak profiles, 297, 298, 299
Spray deposition, YBCO wires, 749-755
Sputtered films, 579-580
   Andreev reflection, 330
   fast optically triggered switches, 685-692
   microwave surface resistance measurement, 757-765
   substrate interdiffusion, 169-174
Squeezed polaron model, 475-483
SQUIDS, 45-46
$Sr_2Bi_2CuO_6$, structure, 5
Sr-Ca ratio, BSCCO, 391-393
$SrTiO_3$, 163
   composite tape, physical properties, 776
   LP-MOCVD, 134-135
   plasma-assisted laser deposition, 665-666
Stability
   BSCCO systems, 389
   molten salts of yttrium barium cuprate, 621-627
Stainless steel substrates
   fracture behavior, 661-663
   laser-assisted deposition, 665-675
Stresses, design issues, 709-717
Stripline resonator, 757-765
Strong-coupling narrow-band systems, 475-483
Strongly fluctuating superconductivity, theory of, 455-460
Structural energies, correlation effects, 467-469

Structural modulation, BSCCO, 291-296
Structural relaxation, BSCCO, 635-638, 639
Structure, 5
   classification
      functional groups, 373-376
      types of doping, 371-373
   complications, 285-300
      chemical disorder, 258-291
      phase transitions, 287-300
      structural modulation, 291-296
   composite tape, 776
   defects
      doping effects, 611-618
      planar, 351-361
   and electrical properties of bulk HTS, 265-270
   electronic, 273-280 (*see also* Electronic structure of copper oxide materials)
   and flux pinning, 35-38
   positron annihilation studies, 335-347
   site-selective substitution in tailored thin films, 61-69
   thin films, vortex-glass, 405-410
   YBCO, 35-38
      in-situ processed, 87-94
      local structure and distortions in pure and doped materials, 303-311
Structure, of Bi-series, 291-296, 635-638, 639
   buckling of CuO planes in BSCCO compounds, 431-434
   computational techniques
      lattice energy calculations, 428-429
      potential models, 427-428
   energetics of superstructures, 425-439
   model development, 429-430
   model structures, 433, 435-438
   Ruddleston-Popper-type compounds, 431
   two-body potential model, 432
Substitutions
   BSCCO, 389-396
      anomalous scattering study, 286-288
      modulation dependence on, 292-294
   classification, functional groups, 373-376
   and flux pinning, 30
   and interatomic distances, 320
   rare earth element
      magnetization studies (*see* Magnetization studies)
      magnetodiffractometry of Bi and Tl cuprates, 541-546
   site-selective, 61-69
   and thermal transport mechanisms, 228
   twin boundary layer thickness, 36
   YBCO, 91, 94, 228
Substrates (*see also* Fabrication)
   films on
      fracture behavior, 661-663
      interdiffusion between films and, 169-174
      millimeter wave transmission studies, 735-746
      temperature, and film growth, 118

Substrates (continued)
   mechanical strains, 712, 713, 714
   tape, composite, 667-675, 773-782
   tunneling, 193-194
   wires, carbon and, 749, 752
   YBCO effects
      rapid thermal annealing, 175-182
      zirconia, deposition on, 163-167
   YBCO films, fracture behavior, 661-663
   YBCO tapes, 665-675
Sulfates, YBCO molten salt processing, 622, 626-627
Superconducting Quantum Interference Devices (SQUIDS), 45-46
Supercurrent, 557-565
Superexchange coupling, and EPR silence, 448
Superexchange energy, 485-493
Superfluid correlations, semion pairing in finite systems, 507-515
Superstructures, Bi-series of superconductors, 425-439
Surface acoustic wave device, 241-247, 591
Surface conditions, YBCO
   passivation, 573-578
   tapes on stainless steel substrates, 665-675
Surface resistance
   electrodynamics of YBCO, 211-212
   magnetically aligned sintered YBCO, 705-707
   stripline resonator measurements, 757-765
   magnetically aligned sintered YBCO, 705-707
Switches, fast optically triggered, 685-692
Symmetry, hole doping and, 16, 17, 18, 21

Tapes
   design issues, 709-717
   development of, 773-782
      MgO buffer layer, 776, 779
      properties, 775-777
      sapphire, melt processing on, 779-780, 781
      sintered coatings, melt processing of, 778
      sintered coatings on Ni, 777-778
      vapor deposition on MgO/Ni, 781
      wires and tapes, 774-775
   processing, microstructure and electrical properties, 265-270
   YBCO, 665-675
      removal of organics from, 547-554
TBCCO
   electronic structure, 276-277
   positron annihilation studies, 342
TEM, in-situ, 611-618 (*see also specific fabrication processes*)
Temperature
   and $Bi_2Sr_2CuO_6$ resistivity, 20
   and film growth, 96, 118
   and film-substrate interdiffusion, 169-174
   magnetic susceptibility vs., 30
   positron annihilation studies, 336-340
Thermal conductivity (*see* Heat conduction mechanisms)

Thermal fatigue, 587
Thermal management, power transmission line cost assessment, 796-801
Thermal studies
 differential thermal analysis of YBCO, 629-634
 thermogravimetric analysis (*see* Thermogravimetric analysis)
Thermodynamic limit, scaling to, 12, 13
Thermodynamic properties, and flux pinning, 29, 38
Thermogravimetric analysis
 decomposition of tape cast YBCO, 547-554
 oxygen stoichiometry, 381-382
 spray deposited wires, 752
Thickness, film, 119, 685
Three-band Hubbard model, 10, 11, 12, 15, 17, 18
Three body potential, 430, 432
Time-of-flight absorption spectroscopy
 laser plume, 53, 54
 neutral and ion transport during laser ablation, 158, 159
Tin composites, 581-589, 603
Tin doping, Mossbauer effect studies, 419-423
Tl cuprates, rare earth element substitutions, 541-546
Topology, semion pairing in finite systems, 507-515
T-phase superconductors (*see* NCCO)
Transient optical absorption spectroscopy, 153-161
Transition, superconducting
 composites, 609
 positron annihilation studies, 336
Transition metal ionization potentials, 2, 3
Transition temperature
 $Ba_{1-x}K_xBiO_3$, doping and, 501, 502
 RE-containing superconductors, 542, 543
 Tl-Ca-Ba-Cu-O, 406
Transmission line cost assessment, 795-807
Transport properties
 correlation effects, 470
 critical current anisotropy in sintered aligned YBCO, 557-565
 critical current enhancement by field orientation, 567-572
 oxygen stoichiometry and, 385-386
 silver doped composites, 602, 604, 605-610
T structure, La_2CuO_4, 4
t-t'-J model, 16, 17, 21, 22
Tunneling, 43, 46
 Josephson, 193-198
 quasiparticle, 198-203
 thermal transport mechanisms, 224, 236
 weak links, 357-358
 YBCO and NCCO single crystals, 330-332
Tweed-like microstructure, 447
Twinning, 441, 619
Two-dimensional angular correlation (2D-ACAR) positron annihilation spectroscopy, 344-345

Two-dimensional Hubbard Hamiltonian, 485-493
Two-fluid model
 electrodynamics of YBCO, 209
 millimeter wave transmission studies, 736-739
Type-A SC, 600
Type-B SC, 602

Ultrasound, acoustic response of YBCO, 241-247
Unit cell size, Bi-series, 426

Vapor deposition, composite tape fabrication, 781 (*see also* Sputtering)
Vapor phase, neutral and ion transport during laser ablation, 153-161
Vortex glass transition, thin films, 405-410

Weak coupling theory (*see* Bardeen, Cooper, and Schreiffer (BCS) theory)
Weak Link Filament Array model, 144-149, 150
Weak links, 351-361
 critical current anisotropy in aligned sintered YBCO, 557-565
 electronic make up of, 357-358
 magnetization studies, 249-265
 silver doping and, 363-369
Wires, 774-775
 films on, fabrication of, 719-724
 spray deposition, 749-755

X-ray absorption spectroscopy
 charge on copper, 313-320
 electron localization, 486
 laser plume, 53, 54
 local structure and distortions in pure and doped materials, 303-311
 neutral and ion transport during laser ablation, 153-161
 polarized, 318-320
X-ray diffraction
 decomposition of tape cast YBCO, 547-554
 rare earth element susbstituted Bi and Tl cuprates, magnetodiffractometry for, 541-546
 single wavelength, 286
 YBCO, oxygenated, 519
 YSCO, 533
X-ray photoemission spectroscopy, angle-resolved (AR-XPS), 359-361, 465-466
YBCO
 annealing, substrate and process effects, 175-182
 anodic oxidation of precursor, 641-646
 applications
  design issues, 709-717
  fast optically triggered switches, 685-692
  projectile accelerator, 727-732
  radiation detectors, 695-703
 bulk processing, microstructure and electrical properties, 265-270
 classification, 373, 374-376

823

YBCO (continued)
  composite tape, 773-782
  doping
    bulk, 399
    defect structure and phase stability, 611-618
  electrical transport measurements
    anisotropic critical currents, 557-565
    critical current enhancement by field orientation, 567-572
    polycrystalline films, 139-150
  fabrication, 73, 75-81
    fluidized-bed deposition, 785-792
    free-expansion zone, 51-54
    in-situ growth, 109-115
    in-situ processing, 87-94
    laser-target interaction zone, 49-51
    low-pressure metalorganic chemical vapor deposition, 127-136
    LT-MOCVD, 185-190
    molten salt flux, grain orientation with, 411-418
    laser ablation, neutral and ion transport during, 153-161
    pulsed laser techniques, 72-73, 74
    removal of organics from, 547-554
    RF plasma deposition, 99-106
    single crystal growth, 591-593, 595
    SQUIDS, 45-46
    zirconia, deposition on, 163-167
  flux creep, 32-35
  Hubbard parameters, 12, 13
  magnetically aligned sintered samples, 705-707
  magnetization studies, 249-265
    critical current enhancement by field orientation, 567-572
    magnetic field dependence of critical current density, 363-369
    magnetic susceptibility vs. temperature, 30
  molten salts, stability, 621-627
  oxygen-stabilized zero-resistance states, 517-528
  positron annihilation spectroscopy, 335-347
    Fermi surface measurements, 344-345
    structural and defect properties, 345
  properties
    Andreev reflection, 330-332
    acoustic response, 241-247

YBCO (continued)
  properties (continued)
    design issues, 709-717
    differential thermal analysis, 629-634
    electrodynamics, 207-214
    heat conduction mechanism, 229-235
    millimeter wave transmission studies, 735-746
    rf surface resistance, 705-707, 757-765
  quantum percolation theory, 41-43
  sintering of composites, silver doping and, 599-610
  structure, 6, 273, 278
    EPR spectra, 441-452
    finite cluster mapping, 18, 19
    and flux pinning, 35-38
    one-dimensional band, 4
    twin boundary layer thickness, 36
    weak links and planar defects, 351-361
    vortex glass structure, 406
    X-ray absorption techniques, 281-284
  surface passivation, 573-578
  tape, composite, 773-782
  tunneling, 194
    Josephson, 193-198
    quasiparticle, 198-203
    in single crystals, 330-332
  wires, spray deposition, 749-755
  X-ray absorption spectroscopy
    charge on copper, 318-319
    local structure and distortions in pure and doped materials, 303-311
YBCSO, Mossbauer effect studies, 419-423
YSCO, properties, 531-538

Zirconium
  films on
    fracture behavior, 661-663
    millimeter wave transmission studies, 735-746
  structure, X-ray absorption techniques, 281-284
  YBCO materials
    deposition on, 163-164
    fracture behavior, 661-663
Zn doping
  defect structure and phase stability, 611-618
  YBCO and BSCCO, 399-403

Printed by Printforce, the Netherlands